STATISTICS FOR BUSINESS AND ECONOMICS

Paul Newbold

University of Illinois, Urbana-Champaign

Prentice-Hall, Inc., Englewood Cliffs, New Jersey 07632

Library of Congress Cataloging in Publication Data

Newbold, Paul.
    Statistics for business and economics.

    Includes index.
    1. Commercial statistics.   2. Economics—Statistical
methods.   3. Statistics.   I. Title.
HF1017.N48   1983   519.5   83-22992
ISBN   0-13-845140-0

7 6 2840601

Editorial/production supervision by Margaret Rizzi
Interior design by Suzanne Behnke
Cover photograph by Paul Silverman
Cover design by Suzanne Behnke
Manufacturing buyer: Ed O'Dougherty

Printed in the United States of America

10 9 8 7 6 5 4 3 2 1

ISBN 0-13-845140-0

Prentice-Hall International, Inc., *London*
Prentice-Hall of Australia Pty. Limited, *Sydney*
Editora Prentice-Hall do Brasil, Ltda., *Rio de Janeiro*
Prentice-Hall Canada Inc., *Toronto*
Prentice-Hall of India Private Limited, *New Delhi*
Prentice-Hall of Japan, Inc., *Tokyo*
Prentice-Hall of Southeast Asia Pte. Ltd., *Singapore*
Whitehall Books Limited, *Wellington, New Zealand*

To My Parents

contents

3 PROBABILITY   82

4 DISCRETE RANDOM VARIABLES AND PROBABILITY DISTRIBUTIONS   134

5 CONTINUOUS RANDOM VARIABLES AND PROBABILITY DISTRIBUTIONS   185

6 SAMPLING AND SAMPLING DISTRIBUTIONS   228

7 POINT ESTIMATION   261

8 INTERVAL ESTIMATION   276

9 HYPOTHESIS TESTING   331

10 SOME NONPARAMETRIC TESTS   394

11 GOODNESS OF FIT TESTS AND CONTINGENCY TABLES   415

12 LINEAR CORRELATION AND REGRESSION   440

13 MULTIPLE REGRESSION   494

14 ADDITIONAL TOPICS IN REGRESSION ANALYSIS   549

15 ANALYSIS OF VARIANCE   609

16 INDEX NUMBERS   657

17 TIME SERIES ANALYSIS AND FORECASTING   688

18 SURVEY SAMPLING METHODS   745

19 STATISTICAL DECISION THEORY   795

APPENDIX: TABLES   849

ANSWERS TO SELECTED EVEN-NUMBERED EXERCISES   870

INDEX   888

preface

Statistics courses are offered in virtually all college business programs, feared by students in most, and rarely enjoyed. At least at the outset, students tend to suspect that such courses will be both difficult and boring. I would like to reassure the minority of students who actually read prefaces by asserting that the subject matter of business statistics is both easily grasped and invariably entertaining. However, honesty prohibits my doing so.

Experience suggests that students may have problems in immediately grasping some of the material in this book. Many new procedures for the analysis of business data will be met and, while the temptation is to try to memorize the ingredients of each ("cookbook" style), this can eventually lead to confusion. To avoid such problems, I have tried to set out carefully the rationale for each procedure, keeping in mind that the vast majority of readers of this book will be consumers rather than producers of statistical analyses. To facilitate understanding of the material, I have included a large number of numerical examples. Nevertheless, in the final analysis, statistics is a subject in which "learning by doing" is crucially important. This being the case, the text also contains a great many exercises. It is hoped that, by working through several exercises on each topic, the student's understanding of the subject matter will be sharpened.

In writing this text, I have been conscious of the importance of persuading students to invest the effort needed to absorb the methods of statistical analysis. My strategy has been to incorporate many realistic examples and exercises taken from the business literature, including, for example, the fields of accounting, economics, finance, marketing, and industrial organization. It is hoped that the accumulation of this illustrative material will convince the reader of the relevance of statistics to the modern business environment, and therefore heighten his or her interest in the subject.

This text is suitable for a one- or two-semester introductory course aimed at business or economics majors. I have included more than enough material for a two-semester course, and many instructors will not wish to cover every chapter in detail, particularly if a large amount of time is to be devoted to project work. All, or parts, of Chapters 10, 11, and 14–19 can be omitted without loss of continuity. Also, Chapter 19 can be covered at any stage after Chapter 4, and Chapter 18 at any time after Chapter 8. One of the many possibilities for a one-semester course is to cover Chapters 1–9, 12, and perhaps 13. In such a course, instructors may wish to omit Sections 4.4, 4.6, 4.7, 7.3, 9.9, and 12.2.

This book developed from courses I have taught to students of business and economics in a number of universities, most recently the University of Illinois at Urbana-Champaign. I am very grateful to my colleagues, to graduate teaching assistants, and, perhaps most of all, to students for helpful discussions, suggestions, and criticisms. All, or parts, of earlier drafts were read by a number of reviewers, including Professor Craig F. Ansley, University of Chicago; Professor Paul D. Berger, Boston University; Professor Frank P. Buffa, Texas A&M University; Professor Virginia Knight, Western New England College; Professor Zenon S. Malinowski, University of Connecticut; Professor Glenn W. Milligan, Ohio State University; Professor Charles F. Mott, Bowling Green State University; Professor Chester Piascik, Bryant College; Professor Leonard Presby, William Paterson College; Professor Susan L. Reiland, North Carolina State University; Professor Terry G. Seaks, University of North Carolina; Professor Ronald E. Shiffler, University of Louisville; and Professor Craig W. Slinkman, University of Texas. Their comments were invaluable and greatly enhanced the presentation in many places. I must also acknowledge the help and constructive prodding of two Prentice-Hall editors, Doug Thompson and Margaret Rizzi.

The final manuscript and its earlier drafts were expertly typed by Dixie Trinkle. I must thank Alice Newbold for preparing the graphs, and also for her patience and understanding throughout this long project.

Finally, to any student who has read this far through the preface, I wish you success in your studies of business statistics. In particular, I hope that I will be able to convince you of the importance and relevance of statistics in many branches of business.

Paul Newbold

applications index

Throughout this text, statistical methods are illustrated through real and realistic business examples. Many examples and exercises are drawn from the published literature in different areas of business. These applications are indexed below.

what is statistics?

CHAPTER ONE

Since the majority of readers tend to skip the introductory chapter of a text, it is tempting to answer the question of the title with, "Statistics is what statisticians do," and to proceed immediately to the next chapter. Unfortunately, my editor argued forcibly that rather more might be expected. In the end, I decided to compromise with convention and to keep these introductory remarks as brief as possible.

Compulsory statistics courses have received a bad press. Indeed, on the typical college campus, enrollment in such a course ranks somewhere in popularity between laws establishing a minimum drinking age and mandatory draft registration. In part, this reputation is deserved. The concepts involved are not always easy to grasp at first, and it is certainly necessary to work hard to keep up with any worthwhile course in the subject. However, one complaint heard occasionally is definitely unjustified. **Statistics is not irrelevant.** The remainder of this chapter is devoted to an expansion of this point, particularly with respect to business and economic problems.

Statistics *is* what statisticians do. What is remarkable is the range of activities in which statisticians are involved. These activities impinge on virtually every aspect of daily business and economic life. We will group these activities under six broad headings.

1.1   MAKING SENSE OF NUMERICAL INFORMATION

Any manager operating in the business environment requires as much information as possible about the characteristics of that environment. In the modern era, much of the available information is **quantitative.** For example, it may be necessary to assimilate movements in interest rates, stock market prices, money supply, or unemployment.

Market research surveys are carried out to determine the strength of product demand. An auditor is concerned about the number and size of errors found in accounts receivable. A personnel manager may be able to use aptitude test scores, in addition to subjective judgment of candidates for employment. The list is virtually endless.

The common features of these examples are that the information to be absorbed is *numerical,* and that the sheer amount of that information renders it, in its raw form, virtually impossible to comprehend fully. The statistician's role involves the extraction and synthesis of the important features of a large body of numerical information. One objective is to try to make sense of numerical data by summarizing it in such a way that a readily understood picture emerges while little of importance is lost.

Many issues are involved in a comprehensive analysis and synthesis of numerical data. The most appropriate method will depend on the nature of the numerical information and how it is to be used. In some circumstances it will be desirable to employ some of the heavy artillery of formal techniques to be discussed in the later chapters of this book. On other occasions, a relatively straightforward numerical or graphical summary may be sufficient and should, in any event, provide a good basis for a deeper analysis. In Chapter 2, we will consider some useful techniques for summarizing numerical information.

1.2 DEALING WITH UNCERTAINTY

A second answer to the question posed at the beginning of this chapter is, "Statistics is the science of uncertainty." In statistics, we do not deal with questions of what *is,* but of what *could be,* what *might be,* or what *probably is.* Consider the following statements:

"The price of IBM stock will be higher in six months than it is now."

"If the federal budget deficit is as high as predicted, interest rates will remain high for the rest of the year."

"If a bid of this level is submitted, it will be lower than competitors' bids and the contract will be secured."

"The best opportunities for improvement in market share for this product lie in an advertising campaign aimed at the 18–25 year age group."

Each of these statements contains language suggesting a spurious amount of certainty. At the time the assertions were made, it would have been impossible to be *sure* of their truth. Although an analyst may believe that anticipated developments over the next few months are such that the price of IBM stock is likely to rise over the period, he or she will not be certain of this. Thus, from a purely semantic point of view, the above statements should be modified, as indicated by the following examples:

"The price of IBM stock *is likely to be* higher in six months than it is now."

"If a bid of this level is submitted, *it is probable that* it will be lower than competitors' bids and the contract will be secured."

However, our concern about uncertainty is not merely semantical. All we have done so far is to replace unwarrantedly precise statements by unnecessarily vague statements. After all, what is meant by "is likely to be" or "it is probable that"? Perhaps the two modified statements could be interpreted as assertions that the events of interest are more likely than not to occur. But, *how much* more likely? The English language is rich in words that describe uncertainty, and, indeed, some of these suggest a gradation from the impossible to the certain. Nevertheless, language alone is inadequate to provide a satisfactory description of the degree of uncertainty attached to the occurrence of a particular event. Rather, we need a more formal structure for this purpose.

In the majority of this book, we will be discussing procedures for attacking problems where the conclusion will necessarily be couched in the formal language of uncertainty. As a prelude, that language—**probability**—is introduced in Chapter 3.

1.3 SAMPLING

Before bringing a new product to market, a manufacturer wants to arrive at some assessment of the likely level of demand, and a market research survey may be undertaken. The manufacturer is, in fact, interested in the **population** of all potential buyers. However, it is prohibitively expensive, if not impossible, for a typical market research survey to contact every member of that population. Rather, a small subset— or **sample**—of population members will be contacted, and any conclusions about the population will be based on information obtained from the sample.

The technique of sampling large populations is commonly used in business. For example, decisions about whether a production process is operating correctly are based on the quality of a sample from its output. Again, an audit of accounts receivable will generally be based on a sample of all accounts.

When we have information on a sample from a population, it is generally straightforward to summarize the numerical sample data. However, taking a sample is merely a means to an end. The objective is not to make statements about the sample, but rather to draw conclusions about the wider population. Thus, an important problem for the statistician involves the extent to which it is possible to generalize about a population, based on results obtained from a sample.

Of course, if a sample is taken from a population, we will not be able to learn *precisely* the population characteristics. For example, suppose that a sample of accounts receivable is examined, and it is found that 8.2% of these are in error. It does not follow that *exactly* 8.2% of all the accounts receivable in the population are in error. We will have learned something about this population percentage, but we will not know its exact value. Some uncertainty will remain. Hence, in making inferences about a population based on the results of a sample, any conclusions will naturally involve the language of uncertainty, as discussed in the previous section.

We begin our exploration of procedures for the analysis of sample data in Chapter 6, postponing to Chapter 18 a comprehensive discussion of methods for selecting samples.

1.4 ANALYZING RELATIONSHIPS

Does the rate of growth of the money supply influence the inflation rate?

If General Motors increases the price of subcompact cars by 5%, what will be the effect on the sales of these cars?

Are companies whose dividends are a high percentage of total cash flow viewed as high or low risk?

Are utilities more profitable in areas where they have local monopoly power than where they are subject to competition?

Does minimum wage legislation affect the level of unemployment?

Each of these questions is concerned with the possibility and nature of a relationship between two or more variables of interest. For example, how might we begin to answer the question about the effect on the demand for automobiles of a 5% increase in prices? Simple economic theory tells us that, all other things being equal, an increase in price will lead to a decrease in demand. However, such theory is purely qualitative. It does not tell us *by how much* demand will fall. Subject matter theory is extremely valuable in suggesting the influential factors for such quantities of interest as product demand. In order to proceed further, we must collect numerical quantitative information in order to assess how demand has responded to price changes in the past. We would then base our assessment on the premise that what happened in the past is likely to be repeated after the proposed current price increase.

In the automobile example, the objective is to use numerical information to learn something about the relationship between the variables of interest. Procedures for analyzing relationships are discussed in Chapters 11–14.

1.5 FORECASTING

The desire to be able to foretell the future is a very human characteristic. However, the need for reliable predictions in business goes far beyond curiosity. Investment decisions must be made well ahead of the time at which a new product can be brought to market, and forecasts of likely market conditions some years into the future would obviously be desirable. For established products, short-term sales forecasts are important in the setting of inventory levels and production schedules. Predictions of future interest rates are important to a company deciding whether to issue new debt. In formulating a coherent economic policy, the government requires forecasts of the likely outcomes for variables such as gross domestic product, unemployment, and inflation under various policy options.

Essentially, forecasts of future values are obtained through the discovery of regularities in past behavior. Thus, data are collected on the past behavior of the variable to be predicted, and on the behavior of other related variables. The analysis of this information may then suggest likely future trends.

Some of the methods of business forecasting are introduced in Chapter 17.

1.6 DECISION-MAKING IN AN UNCERTAIN ENVIRONMENT

In any business, decisions are made regularly in an environment where the decision-maker cannot be certain of the future behavior of those factors that will eventually affect the outcomes following from the various options under consideration.

In submitting a bid for a contract, a manufacturer will not be completely certain of the total future cost involved in fulfilling it. Moreover, he will not know the levels of the bids to be submitted by his competitors. In spite of this uncertainty, a decision as to where to pitch the bid must be made. An investor, deciding how to balance her portfolio among stocks, bonds, and money market instruments must make this decision when future market movements are unknown. She may take some view on probable future developments, but will not be able to predict the future with perfect accuracy.

These examples demonstrate that, in order to think about possible options when business decisions are to be made, it is inevitable that techniques for dealing with uncertainty will be relevant. Some useful procedures will be outlined in Chapter 19.

summarizing numerical information

CHAPTER TWO

2.1 POPULATIONS AND SAMPLES

It is commonplace to come across the assertion that in today's world we are assailed on all sides by a veritable barrage of numbers. It is impossible to read a newspaper or listen to a news report without having to digest the impact of such statements as "the Dow-Jones average fell 6 points today," "the Consumer Price Index rose by 0.8% last month," or "the latest survey indicates that the President's approval rating now stands at 40%." Now, issues such as the state of the stock market, the rate of price inflation, and the electorate's opinion of the performance of the President are likely to be of concern to many of us. It is becoming the case that, in order to obtain an intelligent appreciation of current developments in such fields, one must absorb and interpret substantial amounts of numerical information. Certainly, the amount of such information that is collected has grown at a phenomenal rate over the past few years. Government has contributed to this development, both through its own collection efforts and through requirements on corporations to release information. The private sector, too, has played its part. The well-publicized Gallup surveys of voters' attitudes and Nielsen ratings of the week's television shows are merely the tip of a vast iceberg of market research studies. The annual reports of large corporations are rendered somewhat indigestible by the sheer mass of numerical information contained on the many facets of the organization. To a considerable extent, the electronic computer is responsible for this trend. Developments in computing have made relatively straightforward the storage, retrieval, and analysis of information in quantities that would have been completely overwhelming a few years ago. In consequence, it is a fact of life that we are being forced more and more to think about the numerical aspects of any issue of interest. To many people, this development is a source of confusion, to many others a source of irritation. To statisticians it is a source of income.

Somebody has to make sense of all the numerical information. Certainly we are all free to interpret the numbers as we choose, just as we are all free to carry out our own electrical repairs. Nevertheless, the task of making sense of all the numerical information is by no means trivially easy. Faced with a mass of data on, say, the annual incomes of every family in a large city, what can be done? Concentration on the individual incomes will produce nothing more rewarding than a headache. The trick is to extract the essence from the data in as straightforward and simple a form as possible. The statistician's objective is to summarize succinctly, bringing out the important characteristics of the numbers in such a way that a clear and accurate picture emerges. One wants to reduce the mass of information as far as possible, while guarding against the possibility of obscuring important features through too extreme a reduction. As we shall see in later chapters, there is much science involved in the analysis of numerical data, but there is considerably more art.

In this chapter, our aim is to survey some of the methods employed in the summarization of numerical information. Some of the techniques involve the production of numerical summary measures, while others are graphical in nature. All have attractions, and can, in particular applications, have drawbacks. Unfortunately, there is no single "right way" to analyze data. Rather, the appropriate line of attack is typically problem-specific, depending on the characteristics of the data and the purposes of the analysis.

Before beginning our discussion on the summarization of numerical data, we pause to distinguish between two types of data sets. In a study of household incomes in a small town of 1,000 households, one might conceivably obtain the income of every household. The data would then constitute the complete set, or **population,** of household incomes of this town. Assuming that the information gained was accurate, we would then have discovered everything that was to be learned about this population. However, as we noted in the previous chapter, it is very often prohibitively expensive to obtain the complete set of data from a population. Far more often, an investigator will be able to gather only a subset, or **sample,** of the population values. Thus, we may collect a sample of, say, fifty household incomes from this town. However, if only a sample has been obtained, the analysis does not end at this point, for the investigator's objective is to learn about the population. In subsequent chapters of this book, we will discuss many techniques for making inferences about a population based on sample information. In this chapter, we will deal only with procedures for the summarization of numerical information, whether it originates from a sample or from the whole population. The techniques are very similar in either case, though we will find it convenient to make some notational distinctions on occasion.

Definitions

The **population** is the complete set of numerical information on a particular quantity in which an investigator is interested.

A **sample** is an observed subset of the population values.

In Sections 2.2 and 2.3 we will consider numerical summary measures of data sets, while some graphical procedures will be presented in Sections 2.4–2.6. In many practical applications, the use of both numerical and graphical approaches is likely to be the best strategy.

2.2 NUMERICAL SUMMARY: MEASURES OF CENTRAL TENDENCY

Table 2.1 shows the annual salaries of the seven shop-floor supervisory staff employed by a small corporation.

A casual glance at the numbers in the table reveals one simple fact. They are not all the same; that is to say, there is a spread, or **distribution,** of salaries. This is most easily illustrated by plotting the data along a line, as in Figure 2.1, where each point represents a single observation.

In trying to summarize the numbers in Table 2.1, we may begin by looking for the center of their distribution. Summary measures with such an objective as their goal are called **measures of central tendency.** Several measures of this kind are used in business problems. In this section we will discuss three of them.

(i) THE MEAN

One measure of central location that springs readily to mind is the average. In statistics, an average is referred to as a **mean.**

Definition

The **mean** of a set of numerical observations is their sum divided by the number of observations.

Thus, the mean annual salary for the shop-floor supervisory staff of Table 2.1 is

$$\text{Mean} = \frac{24{,}500 + 20{,}700 + 22{,}900 + 26{,}000 + 24{,}100 + 23{,}800 + 22{,}500}{7}$$

$$= \$23{,}500$$

FIGURE 2.1 Annual salaries of shop-floor supervisory staff: Data of Table 2.1

TABLE 2.1 Annual salaries of shop-floor supervisory staff

$24,500,	$20,700,	$22,900,	$26,000,	$24,100,	$23,800,	$22,500

The mean, or average, salary for these staff is $23,500.

Now, the computations involved in finding a mean are not difficult to describe verbally. However, later on we will encounter arithmetic manipulations that are more conveniently described algebraically. Let us see how to produce a simple algebraic formula to describe the operation of finding the mean for any arbitrary set of data. First, if population data are involved, we let N denote the number of observations, so that in our example,

$$N = 7$$

Next, we let x_1 denote the first observation, x_2 the second, and so on, so that x_N denotes the last observation. Thus, for the data of Table 2.1,

$$x_1 = 24,500; \quad x_2 = 20,700; \quad x_3 = 22,900; \quad x_4 = 26,000;$$

$$x_5 = 24,100; \quad x_6 = 23,800; \quad x_7 = 22,500$$

With sample data, the notation is slightly different, as n is used to denote the number of observations. The notation is summarized in the accompanying box.

Notation

Population N observations, labelled $x_1, x_2, \ldots, x_N$
Sample n observations, labelled $x_1, x_2, \ldots, x_n$

The mean is found by adding up all the observations and dividing the result by the number of observations. Therefore, we can write an algebraic expression for the population mean as

---

$$\text{Population mean} = \frac{x_1 + x_2 + \cdots + x_N}{N} \qquad (2.2.1)$$

Now, the numerator of the right-hand side of equation 2.2.1 is, of course, the *sum* of the numbers $x_1, x_2, \ldots, x_N$. In subsequent sections of this book, we will want to refer regularly to the act of **summation**. It is therefore convenient to introduce a notation to represent this act. For this purpose, the symbol Σ (Greek capital "sigma") is employed, and rather than write

$$x_1 + x_2 + \cdots + x_N$$

we will write

$$\sum_{i=1}^{N} x_i$$

This is read as: "Take the numbers x_i, and beginning at $i = 1$ (that is, with x_1), form their sum, adding in turn the numbers x_2, x_3, and so on, terminating the process at $i = N$ (that is, when x_N has been added in)."

Summation notation can be put to more general use. Suppose our interest is not primarily in the numbers x_i, but in some function (such as the logarithm or the square) of the individual numbers. In general, we will use $g(x_i)$ to denote a function of the quantity x_i. Then, the sum of these functions for our N numbers is

$$g(x_1) + g(x_2) + \cdots + g(x_N)$$

Using summation notation, this is more compactly written as

$$\sum_{i=1}^{N} g(x_i)$$

Summation Notation

Let $x_1, x_2, \ldots, x_N$ be a set of numbers. Then:

(i)
$$\sum_{i=1}^{N} x_i = x_1 + x_2 + \cdots + x_N$$

(ii) If $g(x_i)$ is any function of x_i, then

$$\sum_{i=1}^{N} g(x_i) = g(x_1) + g(x_2) + \cdots + g(x_N)$$

To illustrate the second formula in the box, we can write the sum of the squares of the numbers $x_1, x_2, \ldots, x_N$ as

$$\sum_{i=1}^{N} x_i^2 = x_1^2 + x_2^2 + \cdots + x_N^2$$

Expressions for the mean can now be written using summation notation. We use the symbol μ (Greek "mu") to denote a population mean, and $\bar{x}$ (read "x bar") to denote a sample mean.

Algebraic Expressions for Means

(i) Let $x_1, x_2, \ldots, x_N$ be the N population members. Then, the **population mean** is

---

$$\mu = \frac{\sum\limits_{i=1}^{N} x_i}{N}$$

(ii) Let $x_1, x_2, \ldots, x_n$ be the n sample members. Then, the **sample mean** is

$$\bar{x} = \frac{\sum\limits_{i=1}^{n} x_i}{n}$$

Example 2.1

A sample of eight United States corporations showed the following percentage increases in earnings per share in the current year compared with the previous year:

13.6%; 25.5%; 43.6%; −19.8%; −13.8%; 12.0%; 36.3%; 14.3%

Find the sample mean percentage increase in earnings per share.
The sample contains $n = 8$ observations, so the mean is

$$\bar{x} = \frac{\sum\limits_{i=1}^{n} x_i}{n}$$

$$= \frac{13.6 + 25.5 + 43.6 + (-19.8) + (-13.8) + 12.0 + 36.3 + 14.3}{8}$$

$$= 13.9625\%$$

Thus, the mean percentage increase in earnings per share for this sample is 13.9625%.

Example 2.2

Over a 7-year period the annual percentage returns on common stocks were

4.0%; 14.3%; 19.0%; −14.7%; −26.5%; 37.2%; 23.8%

Find the population mean percentage return over this period.
Regarding the seven observations as the population of interest, we have $N = 7$, so that their mean is

$$\mu = \frac{\sum\limits_{i=1}^{N} x_i}{N}$$

$$= \frac{4.0 + 14.3 + 19.0 + (-14.7) + (-26.5) + 37.2 + 23.8}{7}$$

$$= 8.1571\%$$

The mean percentage return from investment in common stocks over this 7-year period was 8.1571%.

(ii) THE MEDIAN

The mean provides an intuitively plausible and easily interpreted measure of central tendency, and, indeed, is calculated more often than any other. However, an alternative measure may be preferable for some purposes.

Consider again the seven salaries of Table 2.1. Arranging these in ascending order, we have

20,700;   22,500;   22,900;   23,800;   24,100;   24,500;   26,000

Now, when the data are arranged in this fashion, the middle value is $23,800. Three of the staff have higher salaries and three have lower. This middle value is called the **median** of these observations. In this particular instance, the number of observations is odd and their median is easily located. When there is an even number of values, however, there is not a single middle one. In such situations, the median is conventionally taken as the average of the middle pair when the observations are arranged in ascending order.

Definition

The **median** of a set of observations is the middle one if the number of observations is odd, and the average of the middle pair if their number is even, when these observations are arranged in increasing order. Thus, if there are N observations arranged in increasing order, the median is the $[(N + 1)/2]$th observation when N is odd, and the average of the $(N/2)$th and $[(N + 2)/2]$th observations when N is even.

For the data of Table 2.1, we have seen that the mean is $23,500 and the median is $23,800. These two values are very close, and, as can be seen in Figure 2.2, it makes very little difference which is used in acquiring a feel for the center of these observations. However, such is not invariably the case.

My journey from home to the University of Illinois takes me by three banks, each of which has a device that displays the current temperature. On one particular day, the temperature recorded at the first bank was 19°F, while that at the second was 31°F. This was rather perplexing, but if at that stage I had been forced to estimate the true temperature, perhaps the best that could have been done was to take the average of these two numbers. According to the third bank on my route, the temperature was

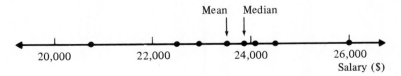

FIGURE 2.2 Annual salaries of shop-floor supervisory staff, with mean and median salaries shown

20°F. Given the three readings of 19, 31, and 20, the option of estimating the true temperature by their mean (23 ⅓°F) remained open. However, this seemed rather unattractive. Given two other readings so close together, it was natural to suspect that equipment registering 31°F was malfunctioning. Thus, the median of the three observations (20°F) provided a more plausible estimate of the true temperature. Since I was due to teach a statistics class on this day, and the clamor of my students for real examples could not justifiably be resisted much longer, I detoured to pass a fourth bank. Here the temperature was given as 20°F. At this stage, then, the available data consisted of four observations:

$$19; \quad 20; \quad 20; \quad 31$$

The mean of these values is 22.5°F and the median is 20°F, as shown in Figure 2.3. In this case, the mean and median do differ somewhat, and most of us would probably prefer the median. The observation of 31°F differs so much from the other three that it inevitably arouses suspicion. Thus, in finding a measure of central tendency, one's inclination is to give relatively little weight to this observation.

The above simple example serves to illustrate a general point. The mean can be quite markedly affected by extreme observations, whereas the median is not susceptible to such strong influence. For example, the median temperature would remain unchanged if the highest of the four recordings had been 21°F or 41°F, rather than 31°F.

In circumstances where it is deemed inappropriate to give much weight to extreme observations, the median is often preferred to the mean as a measure of central tendency. Distributions of the incomes of households in a city, state, or country tend to contain a relatively small proportion of very high values. As a result, the mean of such distributions is typically quite a bit higher than the median, and the median would be preferred as the measure of central tendency. It is easily interpreted as that level of income or wealth exceeded by half the households in the population. On the other hand, the mean, which is pushed up by the very wealthy, gives rather too rosy a picture of the economic well-being of the typical household in the community.

In spite of its advantage in discounting extreme observations, the median is used less frequently than the mean. The reason for this is that the theoretical development of inferential procedures based on the mean, and measures related to it, is considerably more straightforward than the development of procedures based on the median. Accordingly, most statisticians work with the mean and related measures, incorporating into their analyses special techniques to deal with those situations in which it is suspected that extreme outlying observations could exert undue influence.

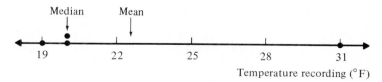

FIGURE 2.3 Four temperature recordings, with mean and median recordings shown

(iii) THE MODE

We briefly note another measure of central location, which is defined in the box.

Definition

The **mode** of a set of observations is the value that occurs most frequently.

The concept of a mode is relevant in cases of multiple occurrences of observation values, as the following example illustrates.

Example 2.3

A manufacturer of portable radios obtained a sample of fifty radios from a week's output. The radios were thoroughly checked, and the number of defects were recorded as follows.

NUMBER OF DEFECTS	0	1	2	3
NUMBER OF RADIOS	12	15	17	6

Find the modal number of defects for this sample.

Since two defects occur more than any other number, the mode of this sample is 2.

The mode is used less than either the mean or the median in business applications. Perhaps its most obvious use is by manufacturers who produce goods, such as clothing, in various sizes. The modal size of items sold is then the one in heaviest demand.

2.3 NUMERICAL SUMMARY: MEASURES OF DISPERSION

In Table 2.2 we show the salaries of seven shop-floor workers in a second corporation.

Now, the data of Table 2.2 have precisely the same mean ($23,500) as those of Table 2.1. Moreover, both data sets have the same median ($23,800). Thus, if we restrict ourselves to measures of central location, we have no basis for distinguishing between the salary distributions in the two corporations. Yet these distributions do differ in an important way, as is apparent from Figure 2.4. Clearly, the second set of data is more *dispersed* than the first.

A measure of central location is almost never, by itself, sufficient to provide an adequate summary of the characteristics of a set of data. We will usually require, in

TABLE 2.2 Annual salaries of shop-floor supervisory staff in a second corporation

$24,900,	$17,500,	$21,600,	$29,700,	$25,300,	$23,800,	$21,700

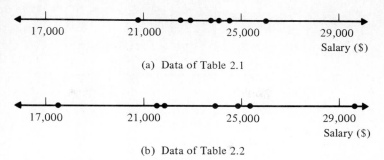

(a) Data of Table 2.1

(b) Data of Table 2.2

FIGURE 2.4 Annual salaries of shop-floor supervisory staff

addition, a measure of the amount of **dispersion** in the data. In this section we consider several such measures.

(i) *THE VARIANCE AND THE STANDARD DEVIATION*

Let $x_1, x_2, \ldots, x_N$ represent a population of numerical values, with mean μ. Since our interest is in the dispersion of these values, it is natural to look at their discrepancies from the mean, that is, the differences

$$x_1 - \mu, x_2 - \mu, \ldots, x_N - \mu$$

Since some population members will be higher than the mean and some lower, some of these differences will be positive and some negative. Indeed, they "balance out" in the sense that their sum is zero.[1] However, in assessing spread, the *sign* of the discrepancy between an observation and the mean is of no interest. We want to treat a negative discrepancy in exactly the same fashion as a positive discrepancy of the same amount. For example, a salary that is $1,000 below the mean should be treated in the same way as a salary that exceeds the mean by $1,000. One way to achieve this is to look not at the discrepancies themselves, but at their *squares*:

$$(x_1 - \mu)^2, (x_2 - \mu)^2, \ldots, (x_N - \mu)^2$$

The average of these squared discrepancies provides a measure of dispersion called the **variance** of the observations. Using summation notation, we may express the variance, denoted by the symbol σ^2 (σ is Greek lowercase "sigma"), as

$$\sigma^2 = \frac{\sum_{i=1}^{N} (x_i - \mu)^2}{N}$$

---

[1] This follows, since

$$\sum_{i=1}^{N} (x_i - \mu) = (x_1 - \mu) + (x_2 - \mu) + \cdots + (x_N - \mu)$$

$$= (x_1 + x_2 + \cdots + x_N) - N\mu = N\mu - N\mu = 0$$

TABLE 2.3 Calculations for the variance of the data of Table 2.1

x_i	$x_i - \mu$ $= x_i - 23{,}500$	$(x_i - \mu)^2$
24,500	1,000	1,000,000
20,700	−2,800	7,840,000
22,900	−600	360,000
26,000	2,500	6,250,000
24,100	600	360,000
23,800	300	90,000
22,500	−1,000	1,000,000
Sums	0	16,900,000

The calculations for the variance of the salaries of the staff of Table 2.1 are set out in Table 2.3. We have used the fact that the population mean is $23,500.

From Table 2.3, we see that the sum of squared discrepancies of the observations about their mean is

$$\sum_{i=1}^{N} (x_i - \mu)^2 = 16{,}900{,}000$$

Therefore, the variance of this population is

$$\sigma^2 = \frac{\sum_{i=1}^{N} (x_i - \mu)^2}{N} = \frac{16{,}900{,}000}{7} = 2{,}414{,}286$$

In exactly the same way, we can find the variance of the salaries of Table 2.2, which again have mean $23,500. The calculations are set out in Table 2.4.

The sum of squared discrepancies of the observations about their mean is

$$\sum_{i=1}^{N} (x_i - \mu)^2 = 86{,}580{,}000$$

TABLE 2.4 Calculations for the variance of the data of Table 2.2

x_i	$x_i - \mu$ $= x_i - 23{,}500$	$(x_i - \mu)^2$
24,900	1,400	1,960,000
17,500	−6,000	36,000,000
21,600	−1,900	3,610,000
29,700	6,200	38,440,000
25,300	1,800	3,240,000
23,800	300	90,000
21,700	−1,800	3,240,000
Sums	0	86,580,000

Hence, the variance of this population is

$$\sigma^2 = \frac{\sum\limits_{i=1}^{N} (x_i - \mu)^2}{N} = \frac{86,580,000}{7} = 12,368,571$$

We see, then, that the variance of the salaries in the second corporation is higher than that of those in the first corporation. This confirms the visual impression, gained from Figure 2.4, of greater dispersion in the second population.

For purposes of computational convenience, the following alternative but equivalent[2] formula is sometimes used for calculating the population variance:

$$\sigma^2 = \frac{\sum\limits_{i=1}^{N} x_i^2}{N} - \mu^2$$

This formulation is easily remembered as *the mean of the squares, less the square of the mean.*

Now, the variance can be used to compare the dispersions of two or more population distributions. However, since the discrepancies from the mean are squared in computing a variance, it is rather difficult to interpret the variance of a single population. A simple way of returning to the original units of measurement is to take the square root of the variance. The resulting quantity, denoted σ, is called the **standard deviation.** For the shop-floor supervisory staff of the first corporation, the standard deviation of salaries is

$$\sigma = \sqrt{\sigma^2} = \sqrt{2,414,286} = \$1,554$$

Similarly, for the staff of the second corporation, the salaries have standard deviation

$$\sigma = \sqrt{\sigma^2} = \sqrt{12,368,571} = \$3,517$$

Definitions

Let $x_1, x_2, \ldots, x_N$ denote the N members of a population with mean μ. The **population variance,** σ^2, is the average of the squared discrepancies of these values from their mean. Thus,

$\longrightarrow$

---

[2] The equivalence is shown as follows:

$$\sum_{i=1}^{N} (x_i - \mu)^2 = \sum_{i=1}^{N} (x_i^2 - 2\mu x_i + \mu^2)$$

$$= \sum_{i=1}^{N} x_i^2 - 2\mu \sum_{i=1}^{N} x_i + \sum_{i=1}^{N} \mu^2$$

$$= \sum_{i=1}^{N} x_i^2 - 2N\mu^2 + N\mu^2 = \sum_{i=1}^{N} x_i^2 - N\mu^2$$

$$\sigma^2 = \frac{\sum\limits_{i=1}^{N} (x_i - \mu)^2}{N}$$

$$= \frac{\sum\limits_{i=1}^{N} x_i^2}{N} - \mu^2$$

The **population standard deviation,** σ, is the (positive) square root of the variance.

Example
2.4
In Example 2.2, we analyzed the annual percentage returns on common stocks over a 7-year period. Over the same period, the annual percentage returns on U.S. Treasury bills were

6.5%; 4.4%; 3.8%; 6.9%; 8.0%; 5.8%; 5.1%

Compare these two population distributions in terms of means and standard deviations.

For the common stocks, we found in Example 2.2 that the mean percentage return over this 7-year period was

$$\mu = 8.1571\%$$

The sum of squares of these returns is

$$\sum_{i=1}^{7} x_i^2 = (4.0)^2 + (14.3)^2 + (19.0)^2 + (-14.7)^2 + (-26.5)^2 + (37.2)^2 + (23.8)^2$$

$$= 3,450.11$$

Hence, the variance is

$$\sigma^2 = \frac{\sum\limits_{i=1}^{N} x_i^2}{N} - \mu^2 = \frac{3,450.11}{7} - (8.1571)^2 = 426.3346$$

and the standard deviation is

$$\sigma = \sqrt{\sigma^2} = \sqrt{426.3346} = 20.6479\%$$

The mean percentage return on Treasury bills in this period was

$$\mu = \frac{\sum\limits_{i=1}^{N} x_i}{N} = \frac{6.5 + 4.4 + 3.8 + 6.9 + 8.0 + 5.8 + 5.1}{7}$$

$$= 5.7857\%$$

The variance of these percentage returns may be found by first computing the sum of squares

$$\sum_{i=1}^{7} x_i^2 = (6.5)^2 + (4.4)^2 + (3.8)^2 + (6.9)^2 + (8.0)^2 + (5.8)^2 + (5.1)^2$$

$$= 247.31$$

The variance is then

$$\sigma^2 = \frac{\sum_{i=1}^{N} x_i^2}{N} - \mu^2 = \frac{247.31}{7} - (5.7857)^2 = 1.8557$$

Therefore, the standard deviation is

$$\sigma = \sqrt{\sigma^2} = \sqrt{1.8557} = 1.3622\%$$

We summarize the data on percentage returns on these two types of investment over this 7-year period in the following table.

	COMMON STOCKS	U.S. TREASURY BILLS
MEAN	8.1571%	5.7857%
STANDARD DEVIATION	20.6479%	1.3622%

It can be seen that, while investment in common stocks yielded the higher average rate of return, the returns on Treasury bills were considerably less variable.

INTERPRETATION OF THE POPULATION STANDARD DEVIATION

So far, we have seen how the variance or the standard deviation can be used to compare the dispersions of two populations. It is also possible to interpret the standard deviation for a single population. Specifically, this quantity can be used to estimate the percentage of population members that lie within a specified distance of the mean. Two rules are commonly used for forming such estimates. The first is true for *any* population.

Tchebychev's Rule

For *any* population with mean μ and standard deviation σ, *at least* $100(1 - 1/m^2)\%$ of the population members lie within m standard deviations around the mean, for any number $m > 1$.

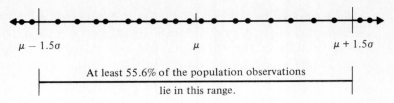

$\mu - 1.5\sigma$ μ $\mu + 1.5\sigma$

At least 55.6% of the population observations
lie in this range.

FIGURE 2.5 Illustration of Tchebychev's rule

To see how Tchebychev's rule works in practice, we construct the following table:

m	1.5	2	2.5	3
$100(1 - 1/m^2)\%$	55.6%	75%	84%	88.9%

Hence, according to Tchebychev's rule, at least 55.6% of the population values lie within 1.5 standard deviations around the mean and so on. Tchebychev's rule is shown diagramatically in Figure 2.5. There we show a set of population values, with the range from $(\mu - 1.5\sigma)$ to $(\mu + 1.5\sigma)$ indicated. Tchebychev's rule guarantees that *at least* 55.6% of the population members lie within this range. (In fact, for this particular population, the actual percentage of observations within this range is much higher than 55.6%).

To provide a numerical illustration, consider the population of Table 2.1, which has mean $23,500 and standard deviation $1,554. It follows from Tchebychev's rule that, for this population, at least 55.6% of the salaries must fall within $(1.5)(1,554) =$ $2,331 around the mean—that is, within the range $21,169 to $25,831. Similarly, at least 75% of the salaries in this population must fall within $3,108 around the mean—that is, within the range $20,392 to $26,608.

The advantage of Tchebychev's rule is that its applicability extends to *any* population. However, it is within this guarantee that its major drawback lies. For many populations, the percentage of values falling in any specified range is much higher than the minimum assured by Tchebychev's rule. For one "standard" distribution,[3] which describes the shape of many large populations in the real world, it is possible to state a "rule of thumb" that often provides reliable estimates.

Rule of Thumb

For many large populations, approximately 68% of the population members lie within one standard deviation of the mean, and approximately 95% lie within two standard deviations of the mean.

---

[3] The "normal distribution" will be introduced in Section 5.5. It has been found to describe well the distributions of many actual populations.

Suppose we have a large population of salaries with mean $23,500 and standard deviation $1,554. The rule of thumb would then estimate that roughly 68% of the salaries are between $21,946 and $25,054, and that approximately 95% fall within the range $20,392 to $26,608.

SAMPLE VARIANCE AND STANDARD DEVIATION

We can define also the variance and standard deviation for a sample of n observations, $x_1, x_2, \ldots ; x_n$. Again, we will denote the sample mean by $\bar{x}$. The sample variance is based on the squared discrepancies of the sample values from their mean, that is

$$(x_1 - \bar{x})^2, (x_2 - \bar{x})^2, \ldots, (x_n - \bar{x})^2$$

However, in computing the sample variance, we do not average these squared discrepancies. Instead, their sum is divided by one less than the number of observations. Thus, the **sample variance**, s^2, is defined as

$$s^2 = \frac{\sum_{i=1}^{n} (x_i - \bar{x})^2}{n - 1} \tag{2.3.1}$$

As we have defined it, the sample variance has desirable properties as an estimate of the corresponding population variance. One way to explain division by $(n - 1)$ rather than n in equation 2.3.1 is that we have to use, in that formula, the sample mean rather than the population mean as the measure of central location. If the population mean, μ, were known, a natural quantity to calculate when looking at the dispersion in a sample would be the average of the squared discrepancies

$$(x_1 - \mu)^2, (x_2 - \mu)^2, \ldots, (x_n - \mu)^2$$

However, since μ will not be known in practice, it must be replaced by a suitable proxy—the sample mean, $\bar{x}$. Essentially, it is as compensation for using the sample mean as proxy for the population mean in equation 2.3.1 that the divisor is $(n - 1)$, rather than n. We will return to this point in considerably more detail in Chapters 6 and 7.

As in the case of a population, the **standard deviation** for a sample is the square root of the variance.

Definitions

Let $x_1, x_2, \ldots, x_n$ denote the n members of a sample, whose mean is $\bar{x}$. The **sample variance**, s^2, is defined as

$$s^2 = \frac{\sum_{i=1}^{n} (x_i - \bar{x})^2}{n - 1}$$

An equivalent formula for computation is

$$s^2 = \frac{\sum\limits_{i=1}^{n} x_i^2 - n\bar{x}^2}{n - 1}$$

The **sample standard deviation**, s, is the (positive) square root of the variance.

Example 2.5

Find the sample standard deviation of percentage increase in earnings for the eight corporations of Example 2.1.

In Example 2.1, we found

$$n = 8; \quad \bar{x} = 13.9625\%$$

The sum of squares of the sample values is

$$\sum_{i=1}^{n} x_i^2 = (13.6)^2 + (25.5)^2 + (43.6)^2 + (-19.8)^2$$
$$+ (-13.8)^2 + (12.0)^2 + (36.3)^2 + (14.3)^2$$
$$= 4,984.83$$

The sample variance is, therefore,

$$s^2 = \frac{\sum\limits_{i=1}^{n} x_i^2 - n\bar{x}^2}{n - 1} = \frac{4,984.83 - (8)(13.9625)^2}{7} = 489.3170$$

Therefore, the sample standard deviation is

$$s = \sqrt{s^2} = \sqrt{489.3170} = 22.1205\%$$

The variance and standard deviation are the most commonly used numerical measures of dispersion in a set of data. The reason for their popularity lies in the fact that making inferences about a population, based on a sample, is most conveniently carried out using these measures. However, there are occasions on which alternative measures of spread might be preferable. In the remainder of this section, we briefly discuss three alternatives.

(ii) THE MEAN ABSOLUTE DEVIATION

Consider a population of N members, $x_1, x_2, \ldots, x_N$, with population mean μ. In assessing dispersion, we noted earlier that the deviations from the mean

$$x_1 - \mu, x_2 - \mu, \ldots, x_N - \mu$$

might be used to provide relevant information, provided a negative deviation is treated the same way as a positive deviation of the same amount. In forming the variance, this requirement is met by squaring the discrepancies. Another solution is provided by looking at their absolute values. (The **absolute value** of a positive number is the number itself, while that of a negative number is obtained by multiplying the number by -1.) The **mean absolute deviation** is then the average of the absolute deviations.

Definition

Let $x_1, x_2, \ldots, x_N$ denote the N members of a population whose mean is μ. Their **mean absolute deviation** is the average of the absolute discrepancies from their mean; that is,

$$\text{MAD} = \frac{\sum\limits_{i=1}^{N} |x_i - \mu|}{N}$$

The **sample mean absolute deviation** is defined analogously as the average of the absolute deviations of the sample observations from their mean.

Example 2.6

Find the mean absolute deviation of the annual salaries of shop-floor supervisory staff, given in Table 2.1.

The mean salary for these staff members is

$$\mu = \$23,500$$

The calculations for the mean absolute deviation are set out in the table.

x_i	$x_i - \mu$ $= x_i - 23,500$	$|x_i - \mu|$
24,500	1,000	1,000
20,700	−2,800	2,800
22,900	−600	600
26,000	2,500	2,500
24,100	600	600
23,800	300	300
22,500	−1,000	1,000
Sums	0	8,800

From this table, we see that

$$\sum_{i=1}^{N} |x_i - \mu| = 8,800$$

Thus, the mean absolute deviation is

$$\text{MAD} = \frac{\sum\limits_{i=1}^{N} |x_i - \mu|}{N} = \frac{8,800}{7} = \$1,257$$

Hence, the average absolute discrepancy of these salaries from their mean is $1,257.

The mean absolute deviation has two advantages over the standard deviation as a descriptive measure of the amount of dispersion in a set of data. First, it is conceptually easier to interpret. It is far simpler to form a mental picture of "the average absolute deviation from the mean" than of "the square root of the average squared deviation from the mean." Second, because the individual deviations are squared in the calculation of the variance and the standard deviation, these two measures are more seriously influenced by odd extremely large or extremely small observations than is the mean absolute deviation. In spite of these points, the mean absolute deviation is employed relatively infrequently in practice because complications can arise from its use in making inferences about a population, based only on sample observations.

(iii) THE RANGE

Perhaps the simplest, and most obvious, measure of the dispersion in a set of numerical observations is the difference between the largest and the smallest values. This is known as the **range.**

Definition

   The **range** of a set of data is the difference between the largest and smallest observations.

Example 2.7

Find the range of the annual salaries of shop-floor supervisory staff, given in Table 2.1.

From the table, we see that the highest salary is $26,000 and the lowest is $20,700. The range is, therefore,

$$\text{Range} = 26{,}000 - 20{,}700 = \$5{,}300$$

The range is certainly easy to interpret, and in some applications may itself be of interest. However, because it takes into account only the largest and smallest observations, it is susceptible to considerable distortion if there is an odd extreme value. Moreover, its value is likely to be influenced by the number of observations. In general, we would expect that the range of a large sample would be higher than that of a small sample taken from the same population.

To illustrate the drawback of looking at only the largest and smallest values, suppose that a corporation has seven employees, one with salary $26,000, one with salary $20,700, and each of the other five with salary $23,800. We would all find it easy to agree that there is less dispersion in these salaries than in those of Table 2.1, yet in each case the range is $5,300.

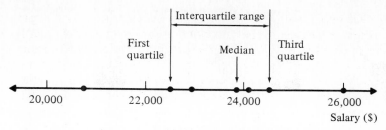

FIGURE 2.6 The quartiles for the salary data of Table 2.1

(iv) THE INTERQUARTILE RANGE

We have indicated that the range may be an unsatisfactory measure of dispersion because it is too much influenced by a single very high or very low observation. One way around this difficulty is to arrange the observations in ascending order, discard a few of the highest and lowest, and find the range of those remaining. A particular possibility of this sort involves dividing the data into four groups of equal size. In Figure 2.6, we have done this with the seven salaries of Table 2.1. When these seven observations are ranked in ascending order, the second observation is referred to as the **first quartile,** the fourth as the **second quartile** (which is the same as the median), and the sixth as the **third quartile.** In general, quartiles are defined in such a way that the same number of observations (in this case, one) occur before the first quartile, between the first and second quartiles, between the second and third quartiles, and after the third quartile. For these data, then, the first quartile is $22,500, the second quartile is $23,800, and the third quartile is $24,500.

Now, the difference between the third and first quartiles provides a measure of dispersion that is particularly attractive when the median is employed as the measure of central tendency. This difference is called the **interquartile range** of the data. In our example,

$$\text{Interquartile range} = 24,500 - 22,500 = \$2,000$$

The general procedure for finding this measure of dispersion is described in the accompanying box.

Quartiles and Interquartile Range

Suppose that N observations are arranged in ascending order. Then, the **first quartile** is the $[(N + 1)/4]$th observation and the **third quartile** is the $[3(N + 1)/4]$th observation. The **second quartile** (the median) is the $[(N + 1)/2]$th observation.

When $(N + 1)$ is not an integer multiple of 4, the quartiles are found by interpolation. For example, suppose we have $N = 12$ observations, so that $(N + 1) = 13$. Then $(N + 1)/4 = 3\frac{1}{4}$, and the first quartile is taken to be that number which is one-quarter of the way from the third observation to the fourth. Similarly,

---

$3(N + 1)/4 = 9\frac{3}{4}$, so that the third quartile is taken as that number which is three-quarters of the way from the ninth observation to the tenth.

The difference between the third and first quartiles provides a measure of dispersion called the **interquartile range.**

Example 2.8

Find the interquartile range of the eight sample percentage increases in earnings per share for the corporations of Example 2.1.

Arranged in ascending order, these observations are:

$$-19.8; \quad -13.8; \quad 12.0; \quad 13.6; \quad 14.3; \quad 25.5; \quad 36.3; \quad 43.6$$

For these data, the median is the average of the fourth and fifth observations—that is, 13.95%.

Since there are $n = 8$ observations, we have $(n + 1)/4 = 2\frac{1}{4}$. Hence, the first quartile is one-quarter of the way from the second observation (-13.8) to the third (12.0). Therefore,

$$\text{First quartile} = -13.8 + \frac{1}{4}[12.0 - (-13.8)] = -7.35\%$$

Similarly, since $3(n + 1)/4 = 6\frac{3}{4}$, the third quartile is three-quarters of the way from the sixth observation (25.5) to the seventh (36.3). Thus, we have

$$\text{Third quartile} = 25.5 + \frac{3}{4}(36.3 - 25.5) = 33.60\%$$

Finally, the interquartile range is the difference between the third and first quartiles:

$$\text{Interquartile range} = 33.60 - (-7.35) = 40.95\%$$

The interquartile range provides a measure of dispersion that is very little influenced by an occasional extreme observation. On theoretical grounds, however, it is not very convenient as a basis for making inferences about a population from sample data. For this reason, statisticians usually employ the mean and standard deviation as summary numerical measures of central location and dispersion, respectively.

EXERCISES

1. A department store manager is interested in the number of complaints received by the customer service department about the quality of electrical products sold by the store. Records over a 10-week period yield the data shown in the table.

WEEK	1	2	3	4	5
NUMBER OF COMPLAINTS	12	15	8	16	8

WEEK	6	7	8	9	10
NUMBER OF COMPLAINTS	4	21	11	3	12

(a) Find the mean number of weekly complaints for this population.

(b) Find the median number of weekly complaints for this population.

2. The dean of a business college is interested in the enrollments in the college's upper-division undergraduate classes. A sample of eight classes yielded the following enrollments:

$$35; \quad 21; \quad 14; \quad 46; \quad 31; \quad 27; \quad 20; \quad 40$$

(a) For this sample, find the mean number of students enrolled.

(b) For this sample, find the median number of students enrolled

3. A company owns twelve commercially zoned parcels of land. The assessment rates (in percentages) assigned to these in 1982 were:

$$20; \quad 22; \quad 27; \quad 36; \quad 22; \quad 29; \quad 22; \quad 23; \quad 22; \quad 28; \quad 36; \quad 31$$

For this population:

(a) Find the mean percentage assessment rate.

(b) Find the median of these percentage rates.

(c) Find the mode of these percentage assessment rates.

4. The defensive coach of a college football team takes a sample of eleven plays run on first down by his team's opponents. The numbers of yards gained in these plays were:

$$2; \quad 3; \quad 0; \quad 27; \quad -5; \quad 10; \quad 5; \quad 2; \quad 2; \quad 3; \quad 4$$

(a) Find the sample mean number of yards gained per play.

(b) Find the median for this sample.

(c) Find the mode for this sample.

5. Over a 10-year period, the annual inflation-adjusted percentage total returns on long-term corporate bonds were:

$$-2.35\%; \quad -3.08\%; \quad -7.79\%; \quad -2.06\%; \quad -13.45\%;$$
$$12.25\%; \quad 7.42\%; \quad 3.73\%; \quad -7.06\%; \quad -13.73\%$$

For this population:

(a) Find the mean percentage inflation-adjusted annual return.

(b) Find the median of these rates of return.

6. Over a 5-year period, uranium production (in short tons) in the United States was as shown in the table.

YEAR	1976	1977	1978	1979	1980
PRODUCTION	17,600	11,800	15,800	15,100	18,000

(a) For this population, find the mean annual uranium production.

(b) For this population, find the median annual production level.

7. Develop a realistic example in which the most appropriate measure of central tendency is:

(a) The mean.

(b) The median.

(c) The mode.

8. The accompanying table gives the 1980 percentage net profit margins of the ten major firms in the medical services industry in the United States.

FIRM	PERCENTAGE NET PROFIT MARGIN
1. American Medical International	6.0
2. Beverly Enterprises	2.9
3. Cenco, Inc.	9.4
4. Community Psychiatric Centers	14.2
5. Hospital Corp. of America	6.6
6. Humana	5.8
7. Lifemark Corp.	5.7
8. National Medical Care, Inc.	10.2
9. National Medical Enterprises, Inc.	4.8
10. Shared Medical Systems Corp.	12.5

(a) Find the mean percentage net profit margin.

(b) Find the median net profit margin.

(c) Suppose you wanted an overall measure of industry net profit margin. Since the market shares of these firms are not identical, the simple mean calculated in part (a) is not necessarily an appropriate measure. In fact, the market shares in 1980 were as shown in the table.

FIRM	PERCENTAGE MARKET SHARE	FIRM	PERCENTAGE MARKET SHARE
1	11.9	6	24.8
2	6.5	7	3.9
3	2.4	8	5.4
4	1.4	9	13.5
5	27.3	10	2.4

Use this information to find a measure of industry net profit margin in 1980.

9. Refer to the data of Exercise 1 on weekly complaints received by a store's customer service department over a 10-week period.

(a) Find the population variance and standard deviation.

(b) Find the mean absolute deviation.

(c) Find the range.

(d) Find the interquartile range.

10. Refer to the sample of enrollment sizes in business courses, given in Exercise 2.

(a) Find the sample variance and standard deviation.

(b) Find the mean absolute deviation.

(c) Find the range.

(d) Find the interquartile range.

11. Refer to the percentage assessments data of Exercise 3.

(a) Find the population variance and standard deviation.

(b) Find the mean absolute deviation.

(c) Find the range.

(d) Find the interquartile range.

12. The data of Exercise 4 show yards gained in a sample of eleven first down running plays against a college football team.
    (a) Find the sample variance and standard deviation for yards gained.
    (b) Find the mean absolute deviation.
    (c) Find the interquartile range.

13. Refer to the data of Exercise 5 on annual inflation-adjusted returns on long-term corporate bonds.
    (a) Find the population variance and standard deviation of percentage returns.
    (b) Find the interquartile range.

14. Refer to the data of Exercise 6 on uranium production over a 5-year period.
    (a) Find the population variance.
    (b) Find the range.

15. Refer to the data of Exercise 8 on percentage net profit margins of the ten major firms in the medical services industry.
    (a) Find the population standard deviation of these net profit margins.
    (b) Find the interquartile range.

16. A sample of twelve supermarket customers found that their total costs of food purchased that day were (in dollars):

    22.50; 38.40; 53.50; 18.20; 12.80; 41.70;
    52.90; 63.00; 17.90; 28.40; 31.30; 37.20

    (a) Find the sample mean.
    (b) Find the sample median.
    (c) Find the sample variance and standard deviation.
    (d) Find the interquartile range.

17. Consider the following four populations:

    $$\text{(i)} \quad 1, \quad 2, \quad 3, \quad 4, \quad 5, \quad 6$$

    $$\text{(ii)} \quad 1, \quad 1, \quad 1, \quad 6, \quad 6, \quad 6$$

    $$\text{(iii)} \quad 1, \quad 3, \quad 3, \quad 4, \quad 4, \quad 6$$

    $$\text{(iv)} \quad -4, \quad -1, \quad 2, \quad 5, \quad 8, \quad 11$$

    All of these populations have the same mean. *Without doing the calculations,* arrange the populations according to the magnitude of their variances, from smallest to largest. Then check your intuition by calculating the four population variances.

18. An auditor finds that the values of a corporation's accounts receivable have mean $185 and standard deviation $41.
    (a) Find a range in which it can be guaranteed that 84% of these values lie.
    (b) Find a range in which it can be guaranteed that 50% of these values lie.

19. A credit card company finds that the mean amount of its overdue accounts is $240.50, and the standard deviation is $41.60.
    (a) Find a range in which it can be guaranteed that 75% of the amounts of these delinquent accounts lie.
    (b) Find a range in which it can be guaranteed that 60% of these amounts lie.

20. The accompanying table shows eleven consecutive biweekly percentage changes in the prices of stock in the Safeco Growth Fund and in General Motors in the period from October 1980 to March 1981. Compare the population standard deviations of these two sets of percentage price changes.

SAFECO GROWTH FUND	GENERAL MOTORS	SAFECO GROWTH FUND	GENERAL MOTORS
5.56	−1.20	−1.69	−0.82
−10.07	−3.17	−4.42	−0.83
8.24	−2.02	2.63	7.52
0.12	−12.60	2.94	3.37
−5.70	1.18	1.03	4.01
4.35	6.10		

2.4 GROUPED DATA AND HISTOGRAMS

When a data set of interest contains only a few observations, the presentation of numerical measures of central location and dispersion, together with a plot such as Figure 2.1, typically provides an adequate summary. The purpose of including the plot is to give a visual impression of the distribution of the observations. However, most data sets met in practice contain many observations and it is generally desirable to obtain a better picture of the distribution of such data.

To illustrate the methods to be examined in this section, Table 2.5 shows inflation-adjusted annual percentage returns on common stocks over a period of 30 years. To summarize these data, we could certainly calculate a mean and variance. However, our objective here is to provide additional visual feel for the information they contain.

The task of interpreting the data of Table 2.5 might be made somewhat easier by reducing the amount of information that must be absorbed, possibly by *grouping* the observations. We could subdivide the range of the data and count the number of returns in each subinterval. This has been done in Table 2.6, which shows the number of values between −39.95 and −19.95, the number between −19.95 and 0.05, and so on. This tabulation allows us rather easily to see simple facts that cannot be absorbed so readily from Table 2.5. For example, there are twice as many positive real returns as negative ones, 60% of all returns are positive but less than 40.05%, and so on.

The subintervals into which the data are broken down are called **classes,** and the numbers of observations in each class are called **frequencies.** For any particular class, the **cumulative frequency** is the total number of observations in that and previous classes. For example, from Table 2.6 we see that 21 of the 30 returns are less than 20.05%.

TABLE 2.5 Inflation-adjusted annual percentage returns on common stocks over a 30-year period

−3.2	17.4	−13.4	−9.9	20.4	15.1
2.7	−1.6	41.0	20.8	6.1	−21.8
20.9	53.4	10.3	15.1	−13.8	−34.8
24.6	31.1	−1.0	10.3	−1.5	28.3
17.2	3.6	26.0	−13.0	10.6	18.2

TABLE 2.6 A subdivision of the inflation-adjusted returns of Table 2.5 into classes

INFLATION-ADJUSTED RETURNS (CLASSES)			NUMBER OF YEARS (FREQUENCIES)	CUMULATIVE FREQUENCIES
−39.95%	to	−19.95%	2	2
−19.95%	to	0.05%	8	10
0.05%	to	20.05%	11	21
20.05%	to	40.05%	7	28
40.05%	to	60.05%	2	30

Tabular information can also be represented pictorially, using a diagram called a **histogram.** The histogram corresponding to Table 2.6 is shown in Figure 2.7. The class boundaries are marked along a horizontal scale. On top of each class interval is drawn a rectangle, whose area is proportional to the frequency in that class. Since the class intervals are all of the same width (20%) in this example, the heights of the rectangles also are proportional to the frequencies. The histogram allows us to form a quick and reliable visual impression of the proportions of observations falling in particular ranges.

Now, Table 2.6 quotes the actual number of observations in each class, while our instinct is to think in terms of the *proportion* of observations in each class. It seems desirable, then, that these proportions, or **relative frequencies,** should be shown. In addition, we often want to consider the proportion which are either in that or one of the earlier classes. These proportions are called **cumulative relative frequencies.** These two modifications are easily incorporated, as can be seen in Table 2.7.

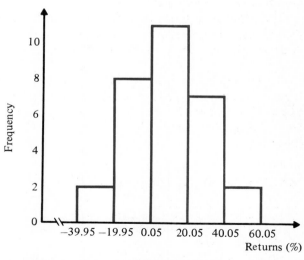

FIGURE 2.7 Histogram for inflation-adjusted returns on common stocks, using the classification of Table 2.6

TABLE 2.7 Classification of inflation-adjusted returns, showing relative frequencies and cumulative relative frequencies

CLASSES		FREQUENCIES(f_i)	RELATIVE FREQUENCIES (f_i/N)	CUMULATIVE RELATIVE FREQUENCIES
−39.95% to −19.95%		2	$2/30$	$2/30$
−19.95% to 0.05%		8	$8/30$	$10/30$
0.05% to 20.05%		11	$11/30$	$21/30$
20.05% to 40.05%		7	$7/30$	$28/30$
40.05% to 60.05%		2	$2/30$	1
	Sums	$N = 30$	1	

The relative frequencies are obtained by dividing the frequencies by the total number of observations. In drawing a histogram with equal class widths, we can use relative frequency rather than frequency along the vertical axis, the picture otherwise being unchanged, as shown in Figure 2.8. We see, for example, that $7/30$ of all the returns are between 20.05% and 40.05%.

The cumulative relative frequencies are the cumulated sums of the relative frequencies. Thus, for the first class, the cumulative relative frequency is the same as the relative frequency. For subsequent classes, the cumulative relative frequency is obtained by adding the relative frequency for the class to the cumulative relative frequency of the previous class. The interpretation of these quantities is straightforward, and often valuable. For example, $21/30$ of all yields are in the class

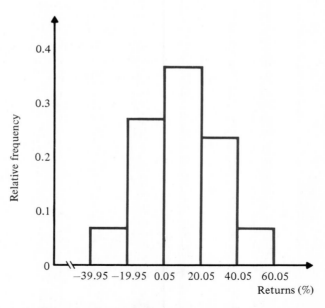

FIGURE 2.8 Histogram for inflation-adjusted returns on common stocks, scaled in terms of relative frequencies

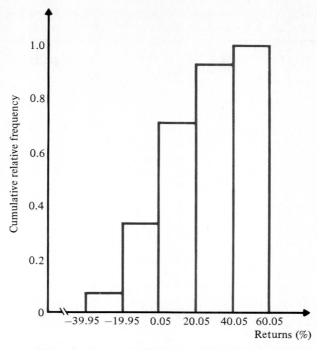

FIGURE 2.9 Cumulative relative frequencies for inflation-adjusted returns on common stocks

0.05%–20.05%, or in one of the previous classes; more succinctly, $^{21}/_{30}$ of all yields are less than 20.05%. The information contained in the cumulative relative frequencies can also be presented pictorially, as in Figure 2.9. Unlike the histogram, the areas of the rectangles drawn over the class intervals here are proportional to the *cumulative* relative frequencies.

The accompanying box summarizes the terminology and notation introduced thus far.

Definitions and Notation

Suppose that a set of N numerical observations is subdivided into K classes. Then:

(i) The numbers of observations falling into each of these classes are called **frequencies,** and are denoted $f_1, f_2, \ldots, f_K$. Since N is the total number of observations, we must have

$$\sum_{i=1}^{K} f_i = N$$

(ii) The proportions of observations falling into each of the classes are called **relative frequencies.** Hence, the relative frequency for the ith class is f_i/N.

(iii) The proportion of all the observations that are less than the upper boundary of the ith class is called the **cumulative relative frequency** of that class. This proportion is given by $(f_1/N + f_2/N + \cdots + f_i/N) = (f_1 + f_2 + \cdots + f_i)/N$.

In any practical application, the objective of constructing a histogram is to bring out the interesting and important features of the data, and the most important question, almost inevitably, is how much detail to include. Presentation of too little detail can mask important characteristics, while, at the other extreme, these can be lost by a mass of detail. The best guide is common sense, although it is possible to set out a few general guidelines:

(a) The range of possible observations should be subdivided into *nonoverlapping* classes, so that any particular observation must fall into one, but only one, of these classes. This may be accomplished by specifying the class boundaries in units on a finer grid than the data. For example, the percentage returns on common stocks in Table 2.5 are given to the nearest tenth of a percent. Thus, using values such as 20.05%, 40.05%, and so on, as class boundaries in Table 2.6 and Figure 2.7 ensures that each observation will fall strictly within one of the classes, rather than on a boundary.

(b) In general, because of the resulting ease of interpretation, it is preferable to have all class intervals of equal width. Thus, in our example, the range of possible values was broken down into five intervals, each of width 20%. On occasion, however, this principle must be abandoned. If a data set is such that many observations fall into a relatively narrow part of the range, while others are widely dispersed, it may be desirable to have narrow classes where the bulk of the observations lie, and broader ones elsewhere. If this is done, it is important to remember that it is the *areas,* rather than the heights, of the rectangles of the histogram that must be proportional to the frequencies. In Example 2.9, we will illustrate the construction of a histogram when class widths are not all the same.

(c) It is important to ensure that the midpoints of the class intervals are representative of the values of the class members. For example, many items in stores are priced at $19.99, $29.99, and so on. If we divide the range of prices into intervals from $10 to $20, $20 to $30, etc., it is likely that a preponderance of prices in each class will be near its upper boundary. A better solution would be to have classes from $15.50 to $25.50, $25.50 to $35.50, and so on. One reason for having class midpoints be representative of their members' values is that the histogram will present a more reliable visual picture. Also, as we will see in the next section, measures of central tendency and dispersion are often calculated from grouped data. These calculations rely on an assumption that the class midpoints are representative.

(d) Often, the most difficult decision is the number of classes to include. If too few classes are employed, the resulting coarse classification can obscure important aspects of the data. On the other hand, if there are too many classes, a choppy and uneven picture, which is difficult to interpret, can result. It is generally felt that at least five, but no more than twenty, classes should be used. To some extent, the more observations, the more classes it is reasonable to include. The subdivision of, say, twenty observations into ten to fifteen classes will inevitably lead to many empty or near-empty classes. On the other hand, this is likely to be less of a problem if we have 200 observations. Even with this factor taken into consideration, the choice will often not be clear-cut. It is a good idea, in such circumstances, to try one or two possibilities and see which of the resulting histograms appears to present the truest and clearest picture.

Example
2.9

The accompanying table shows nonaudit fees as a proportion of total auditor remuneration for 692 Australian companies that were charged nonaudit fees.[4]

$\dfrac{\text{NONAUDIT FEE}}{\text{TOTAL AUDITOR REMUNERATION}}$	NUMBER OF COMPANIES
0.00–0.05	84
0.05–0.10	113
0.10–0.15	112
0.15–0.20	85
0.20–0.25	77
0.25–0.30	58
0.30–0.40	75
0.40–0.50	48
0.50–1.00	40

Find the relative frequencies and cumulative relative frequencies, and draw the histogram.

The fact that the class intervals are not of equal width does not affect the calculations for the relative frequencies and cumulative relative frequencies, which are shown in the next table.

CLASSES	f_i	RELATIVE FREQUENCIES	CUMULATIVE RELATIVE FREQUENCIES
0.00–0.05	84	$84/692$	$84/692$
0.05–0.10	113	$113/692$	$197/692$
0.10–0.15	112	$112/692$	$309/692$
0.15–0.20	85	$85/692$	$394/692$
0.20–0.25	77	$77/692$	$471/692$
0.25–0.30	58	$58/692$	$529/692$
0.30–0.40	75	$75/692$	$604/692$
0.40–0.50	48	$48/692$	$652/692$
0.50–1.00	40	$40/692$	1
Sums	692	1	

The quantities in this table are interpreted in the usual way. Thus, a proportion $113/692$, or 16.3%, of all these companies had nonaudit fee as a proportion of total auditor remuneration in the range 0.05 to 0.10. Again, a proportion $309/692$, or 44.7%, had nonaudit fee as a proportion less than 0.15 of total auditor remuneration.

In constructing the histogram, it is crucial to keep in mind that it is the *areas* of the rectangles drawn over the class intervals that must be proportional to their fre-

---

[4] The data are given by J. R. Francis and B. M. Pollard, "An investigation of nonaudit fees in Australia," *Abacus, 15* (1979), 136–44, published by Sydney University Press.

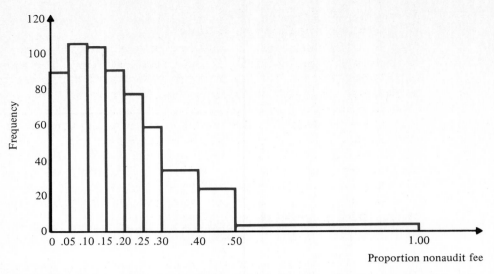

FIGURE 2.10 Histogram for nonaudit fee as proportion of total auditor remuneration, using data of Example 2.9

quencies. Since each of the first six classes has width 0.05, we can draw rectangles of heights 84, 113, 112, 85, 77, and 58 over these class intervals. The next two classes are of width 0.10—that is, twice the width of the first six classes. Thus, in order for their areas to be proportional to the frequencies, the rectangles drawn over these class intervals should have heights that are one-half of the corresponding frequencies—that is, 37.5 and 24. Finally, the last class has width 0.50, ten times the width of each of the first six classes. It follows that the height of the rectangle drawn over this last class interval should be one-tenth of the class frequency—that is, 4. The completed histogram is shown in Figure 2.10.

2.5 NUMERICAL SUMMARY OF GROUPED DATA

A histogram provides a very convenient visual summary of a large set of numerical observations. However, an investigator will frequently want, in addition to this picture, some numerical summary measures of central tendency and dispersion. When the original data are available, this can be accomplished using the procedures discussed in Sections 2.2 and 2.3. Given modern computing resources, this typically provides only a modest computational burden, even for very large data sets. However, it sometimes happens that only grouped data, rather than the raw values, are available. In that case it will not be possible to determine precisely the values of such quantities as the mean and variance. It is then desirable to have methods for estimating these measures from the recorded group frequencies. Such estimates are also useful if a quick approximation is wanted, even when all the data are available. In this section, we will discuss procedures for finding numerical summary measures based only on grouped data.

MEAN AND VARIANCE FOR DATA
WITH MULTIPLE-OBSERVATION VALUES

Suppose that the data are such that only a few different observation values, which may occur repeatedly, are possible. For example, a publisher receives from a printer a copy of a 500-page textbook. The page proofs are carefully read, and the number of errors on each page is recorded, producing the data shown in Table 2.8.

Suppose that we want to find the mean and variance of the number of errors per page for this population. The data presented here are simply a special case, in which there happens to be multiple-observation values, of the general data sets considered in Sections 2.2 and 2.3. Thus, no new principles are involved in computing these numerical summary measures. We simply have 102 observations taking the value 0, 138 taking the value 1, 140 taking the value 2, and so on.

The mean number of errors per page is just the total number of errors divided by the number of pages. The total number of errors is

$$(102)(0) + (138)(1) + (140)(2) + (79)(3) + (33)(4) + (8)(5) = 827$$

Thus, the mean number of errors is

$$\mu = \frac{827}{500} = 1.654$$

Hence, the mean for this population is 1.654 errors per page.

The variance of the number of errors per page is the average of the squared discrepancies of all the observations from their mean. Now, we have 102 discrepancies of $(0 - 1.654)$, 138 discrepancies of $(1 - 1.654)$, and so on. Therefore, the sum of all squared discrepancies is

$$(102)(0 - 1.654)^2 + (138)(1 - 1.654)^2 + (140)(2 - 1.654)^2 + (79)(3 - 1.654)^2$$
$$+ (33)(4 - 1.654)^2 + (8)(5 - 1.654)^2 = 769.1420$$

Thus, the population variance is

$$\sigma^2 = \frac{769.1420}{500} = 1.5383$$

Hence, the standard deviation is

$$\sigma = \sqrt{\sigma^2} = \sqrt{1.5383} = 1.240$$

The data have a standard deviation of 1.240 errors per page.

TABLE 2.8 Number of errors found in a textbook of 500 pages

NUMBER OF ERRORS	NUMBER OF PAGES
0	102
1	138
2	140
3	79
4	33
5	8
Sum	500

We can now consider the general case, in which there are K possible observation values, $m_1, m_2, \ldots, m_K$, and the number of occurrences are, respectively, $f_1, f_2, \ldots, f_K$. The formulas for both the population and sample cases are presented in the box.

Mean and Variance for Multiple-Observation Values

Suppose that a data set contains observation values $m_1, m_2, \ldots, m_K$, occurring with frequencies $f_1, f_2, \ldots, f_K$, respectively.

(i) For a *population* of N observations, so that

$$N = \sum_{i=1}^{K} f_i$$

the **mean** is

$$\mu = \frac{\sum_{i=1}^{K} f_i m_i}{N}$$

and the **variance** is

$$\sigma^2 = \frac{\sum_{i=1}^{K} f_i(m_i - \mu)^2}{N} = \frac{\sum_{i=1}^{K} f_i m_i^2}{N} - \mu^2$$

(ii) For a *sample* of n observations, so that

$$n = \sum_{i=1}^{K} f_i$$

the **mean** is

$$\bar{x} = \frac{\sum_{i=1}^{K} f_i m_i}{n}$$

and the **variance** is

$$s^2 = \frac{\sum_{i=1}^{K} f_i(m_i - \bar{x})^2}{n - 1} = \frac{\sum_{i=1}^{K} f_i m_i^2 - n\bar{x}^2}{n - 1}$$

The arithmetic is most conveniently set out in tabular form. For the data on errors in a textbook, this is done in Table 2.9. Note that we will compute the variance using the alternative, computationally efficient, formula.

Reading directly from the table, we have

$$\sum_{i=1}^{K} f_i = N = 500; \qquad \sum_{i=1}^{K} f_i m_i = 827; \qquad \sum_{i=1}^{K} f_i m_i^2 = 2{,}137$$

TABLE 2.9 Calculations for the mean and variance of data on errors found in textbook

m_i	f_i	$f_i m_i$	$f_i m_i^2$
0	102	0	0
1	138	138	138
2	140	280	560
3	79	237	711
4	33	132	528
5	8	40	200
Sums	500	827	2,137

Thus, the population mean is

$$\mu = \frac{\sum_{i=1}^{K} f_i m_i}{N} = \frac{827}{500} = 1.654$$

as before. For the variance, we find

$$\sigma^2 = \frac{\sum_{i=1}^{K} f_i m_i^2}{N} - \mu^2 = \frac{2,137}{500} - (1.654)^2 = 1.5383$$

confirming our previous calculations.

MEAN AND VARIANCE FOR GROUPED DATA

Suppose that an investigator has available only data grouped into classes, such as the information in Table 2.6 on inflation-adjusted returns of common stocks. Given only this information, we want to obtain estimates of the mean and variance. We will not, of course, be able to find the precise values of these summary measures if the raw data are unavailable. For example, from Table 2.6 we know that seven of the returns are between 20.05% and 40.05%. However, we do not know where in this range these seven returns lie. In order to make further progress, some approximation is needed. Since the exact location in a particular class of all its members is unknown, one obvious possibility is to proceed as if they were all located at the midpoint of the class interval. Thus, we would take each of the seven returns between 20.05% and 40.05% to have the value 30.05%. If this is done, we are in the position of having multiple-observation values, and can proceed to calculate their mean and variance exactly as described above. When employed in this way, the class midpoints are often referred to as **class marks**.

Approximate Mean and Variance for Grouped Data

Suppose that we have data grouped into K classes, with frequencies $f_1, f_2, \ldots, f_K$. If the midpoints of these classes are $m_1, m_2, \ldots, m_K$, the mean and variance of the

---

grouped data are estimated by using the formulas for multiple-observation values given previously.

The calculations for estimating the mean and variance of the inflation-adjusted returns on common stocks, when grouped as in Table 2.6, are set out in Table 2.10. Once the class midpoints have been located, the computations proceed precisely as those for multiple-observation values.

From the table, we find

$$\sum_{i=1}^{K} f_i = N = 30; \qquad \sum_{i=1}^{K} f_i m_i = 281.5; \qquad \sum_{i=1}^{K} f_i m_i^2 = 15,028.075$$

The population mean is then estimated by

$$\mu = \frac{\sum_{i=1}^{K} f_i m_i}{N} = \frac{281.5}{30} = 9.383$$

Thus, the mean return over this 30-year period is estimated as 9.383%.

Next, we estimate the population variance by

$$\sigma^2 = \frac{\sum_{i=1}^{K} f_i m_i^2}{N} - \mu^2 = \frac{15,028.075}{30} - (9.383)^2 = 412.8952$$

The population standard deviation is obtained by taking the square root:

$$\sigma = \sqrt{\sigma^2} = \sqrt{412.8952} = 20.320$$

Therefore, the standard deviation of these returns is estimated as 20.320%.

A word of caution is in order. If one or another of the extreme classes is much wider than the others, it is particularly important that the midpoint of that class be

TABLE 2.10 Calculations for mean and variance of grouped inflation-adjusted returns on common stocks

CLASSES			MIDPOINTS	FREQUENCIES		
			m_i	f_i	$f_i m_i$	$f_i m_i^2$
−39.95%	to	−19.95%	−29.95	2	−59.90	1,794.0050
−19.95%	to	0.05%	−9.95	8	−79.60	792.0200
0.05%	to	20.05%	10.05	11	110.55	1,111.0275
20.05%	to	40.05%	30.05	7	210.35	6,321.0175
40.05%	to	60.05%	50.05	2	100.10	5,010.0050
		Sums		30	281.50	15,028.0750

representative of its members' values, in order to obtain reasonably good estimates of the mean and variance.

Example 2.10

A sample of 20 batches of a chemical was tested for concentration of impurities. The results obtaned were:

PERCENTAGE IMPURITIES	0–2	2–4	4–6	6–8	8–10
BATCHES	2	3	6	5	4

Find the sample mean and standard deviation of these percentage impurity levels.

The computations are set out in the table below.

CLASSES	m_i	f_i	$f_i m_i$	$f_i m_i^2$
0–2	1	2	2	2
2–4	3	3	9	27
4–6	5	6	30	150
6–8	7	5	35	245
8–10	9	4	36	324
Sums		20	112	748

From this table, then,

$$\sum_{i=1}^{K} f_i = n = 20; \quad \sum_{i=1}^{K} f_i m_i = 112; \quad \sum_{i=1}^{K} f_i m_i^2 = 748$$

The sample mean is estimated by

$$\bar{x} = \frac{\sum_{i=1}^{K} f_i m_i}{n} = \frac{112}{20} = 5.6$$

Since these are sample data, the variance is estimated by

$$s^2 = \frac{\sum_{i=1}^{K} f_i m_i^2 - n\bar{x}^2}{n - 1}$$

$$= \frac{748 - (20)(5.6)^2}{19} = 6.3579$$

Hence the sample standard deviation is estimated as

$$s = \sqrt{s^2} = \sqrt{6.3579} = 2.52$$

FIGURE 2.11 Illustration of the assumed spacings of observations within classes in the estimation of the median and quartiles for grouped data

Therefore, for this sample, the mean impurity concentration is estimated to be 5.6%, with the sample standard deviation 2.52%.

MEDIAN AND INTERQUARTILE RANGE FOR GROUPED DATA

In Section 2, we defined the median as the middle value when the observations are arranged in ascending order. Now, if we have available only data grouped into classes, so that the original observation values are unknown, we will not be able to arrange them in ascending order. Still, we can do part of that job. Referring to the returns on common stocks, in Table 2.7, we know that the two smallest returns are in the class −39.95% to −19.95%, the eight next smallest are in the class −19.95% to 0.05%, the eleven next smallest in the class 0.05% to 20.05%, and so on.

Now, since there are 30 observations in all for this data set, the median is the average of the 15th and 16th values when these observations are arranged in ascending order. How can we estimate what, for example, this 15th observation is? In fact, we know quite a bit about it. It is in the class 0.05% to 20.05%, and is, moreover, the 5th value of the 11 observations in that class.

We see, then, that the estimation of the median, and similarly that of the quartiles, boils down to estimating, for example, the location of the fifth observation when the eleven observations in a particular class are arranged in ascending order. We do not have enough information to answer such questions with certainty, but can achieve a reasonable approximation by assuming that, within a class, the observations are equi-spaced. To see how this is done in practice, consider the simple example illustrated in Figure 2.11, where each of three consecutive classes, 0–10, 10–20, 20–30, contains five observations. The width of each class interval is 10 units, and dividing this by five suggests that, if the observations are to be equi-spaced, they should be separated by two units. This can be achieved, as illustrated in the figure, by placing the first observation one unit after the lower boundary of the class, and ending with the last observation one unit before the upper boundary. Thus, for the middle class, the five observations are placed at 11, 13, 15, 17 and 19.

This simple illustration suggests the general rule stated below.

Estimating the Position of an Observation in a Class

Suppose that a class, with lower boundary L and upper boundary U contains f observations. If these observations were to be arranged in ascending order, the jth is estimated by

---

$$L + (j - \tfrac{1}{2})\frac{(U - L)}{f}$$

for $j = 1, 2, \ldots, f$.

We will illustrate this rather forbidding looking formula by estimating the median and quartiles of the inflation-adjusted returns on common stocks. As a first step, we re-tabulate the data in Table 2.11. Here we show the cumulative frequencies, that is, the total number of observations up to, and including, those of the corresponding class.

Since there are 30 observations, the median is, as we have already noted, the average of the 15th and 16th when they are arranged in ascending order. To begin, we will estimate the location of the 15th observation. Clearly, from the cumulative frequencies, this is the fifth observation in the class 0.05 to 20.05, which contains 11 observations in total. Thus, in our notation,

$$j = 5; \quad f = 11; \quad L = 0.05; \quad U = 20.05$$

The location of the fifth of these observations is illustrated in Figure 2.12. The class width

$$U - L = 20.05 - 0.05 = 20$$

is divided by the number of observations, $f = 11$, so that

$$\frac{U - L}{f} = \frac{20}{11}$$

To locate the fifth observation, we count half a step for the first, and one step for each of the next four, so that we have

$$(j - \tfrac{1}{2}) = (5 - \tfrac{1}{2}) = 4\tfrac{1}{2}$$

of these steps to add to the lower limit, $L = 0.05$. We therefore have

$$L + (j - \tfrac{1}{2})\frac{(U - L)}{f} = 0.05 + (4\tfrac{1}{2})\frac{(20)}{11} = 8.23$$

Thus we estimate this observation to be 8.23%.

TABLE 2.11 Classification of inflation-adjusted returns, showing cumulative frequencies

CLASSES	FREQUENCIES	CUMULATIVE FREQUENCIES
−39.95% to −19.95%	2	2
−19.95% to 0.05%	8	10
0.05% to 20.05%	11	21
20.05% to 40.05%	7	28
40.05% to 60.05%	2	30
Sums	30	

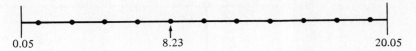

0.05   8.23   20.05

FIGURE 2.12 Estimation of the location of the fifth of the eleven observations in the class 0.05–20.05

Similarly, the 16th observation is the sixth one in this class, so that we now have $j = 6$, with everything else as before. Our estimate is therefore

$$L + (j - \tfrac{1}{2})\frac{(U - L)}{f} = 0.05 + (5\tfrac{1}{2})\frac{(20)}{11} = 10.05$$

The median, then, is the average of the 15th and 16th observations, so that

$$\text{Median} = \frac{8.23 + 10.05}{2} = 9.14$$

Thus, our estimate of the median inflation-adjusted return is 9.14%.

In Section 2.3, we defined the first and third quartiles. Since there are $N = 30$ observations, we have

$$\frac{N + 1}{4} = \frac{31}{4} = 7\tfrac{3}{4}$$

Hence, the first quartile is three-quarters of the way from the 7th observation to the 8th. From Table 2.11, we see that the 7th observation is the fifth value in the class -19.95% to 0.05%. In our notation, then,

$$j = 5; \quad f = 8; \quad L = -19.95; \quad U = 0.05$$

The 7th observation is then estimated by

$$L + (j - \tfrac{1}{2})\frac{(U - L)}{f} = -19.95 + (4\tfrac{1}{2})\frac{(20)}{8} = -8.7$$

Similarly, the 8th observation is the sixth value in the same class, so that now, with $j = 6$, we have

$$L + (j - \tfrac{1}{2})\frac{(U - L)}{f} = -19.95 + (5\tfrac{1}{2})\frac{(20)}{8} = -6.2$$

Since the first quartile is three-quarters of the way from the 7th observation to the 8th, we have

$$\text{First quartile} = -8.7 + \tfrac{3}{4}[-6.2 - (-8.7)] = -6.825$$

To locate the third quartile, we have

$$\frac{3(N + 1)}{4} = \frac{93}{4} = 23\tfrac{1}{4}$$

Therefore, when the observations are arranged in ascending order, the third quartile is one-quarter of the way from the 23rd to the 24th.

Looking at Table 2.11, we see that the 23rd observation is the second value in the class 20.05% to 40.05%, which contains 7 observations. We have, then,

$$j = 2; \quad f = 7; \quad L = 20.05; \quad U = 40.05$$

Thus, the 23rd observation is estimated by

$$L + (j - \tfrac{1}{2})\frac{(U - L)}{f} = 20.05 + (1\tfrac{1}{2})\frac{(20)}{7} = 24.336$$

Similarly, the 24th observation is the third value in this same class, so that, with $j = 3$, we estimate it by

$$L + (j - \tfrac{1}{2})\frac{(U - L)}{f} = 20.05 + (2\tfrac{1}{2})\frac{(20)}{7} = 27.193$$

Hence, since the third quartile is one-quarter of the way from the 23rd observation to the 24th, we have

$$\text{Third quartile} = 24.336 + \tfrac{1}{4}[27.193 - 24.336] = 25.050$$

Finally, then, the inter-quartile range is the difference between the third and first quartiles, so that

$$\text{Interquartile range} = 25.050 - (-6.825) = 31.875$$

Thus, if the interquartile range is to be used as a measure of dispersion, we estimate it by 31.875%.

MODAL CLASS

When the raw data are available, we saw in Section 2.2 that an occasionally used measure of central tendency is the mode, which is defined as the most frequently occurring value. A similar concept can be defined when the data are grouped.

Definition

For grouped data, the **modal class** is that class with the highest frequency.

For the inflation-adjusted returns on common stock, grouped as in Table 2.6, the modal class is 0.05% to 20.05%; that is, more of the returns are in this class than in any other.

SKEWNESS

In much applied work, only measures of central tendency and dispersion—typically the mean and standard deviation—are calculated. Thus, a very large set of data is often reduced to just two numbers. For many real data sets, such extreme parsimony is justifiable and little more would be learned through the calculation of further summary measures. However, it can certainly happen that valuable insight into the form of the population distribution is lost when the data are reduced to just a measure of central tendency and one of dispersion. We will now illustrate one possibility of this sort.

Consider the three histograms of Figure 2.13. The histogram in part (a) of the figure depicts a situation in which the data are distributed **symmetrically** about their central value. Extremely large observations are no more likely than extremely small ones. By contrast, the histogram in part (b) of the figure has a very long tail to the right, with a far more abrupt cutoff to the left. Such distributions, which are said to be **skewed to the right,** have the characteristic that their mean exceeds their median. Distributions of population income or wealth generally have this shape. A large proportion of the population have relatively modest incomes, but the incomes of, say, the highest 10% of all earners extend over a considerable range. The histogram in part (c) of Figure 2.13 depicts the opposite situation. Here, the distribution is **skewed to the left,** so that the lowest observations extend over a wide range, but the highest do not.

The property of skewness can be of considerable interest in the characterization of a distribution of observations. However, the mean and standard deviation contain no information about the skewness of a distribution. The most straightforward way to detect skewness is through inspection of the histogram. Indeed, we have already seen one example of a skewed distribution. It is quite clear from Figure 2.10 that the distribution of nonaudit fees as a proportion of total auditor remuneration is quite severely skewed to the right. Indeed, for the bulk of the companies, this proportion is relatively modest (less than 20% for 394 of the 692 companies). Yet, for an appreciable number of companies the proportion is considerably higher.

FIGURE 2.13 (a) Symmetric distribution

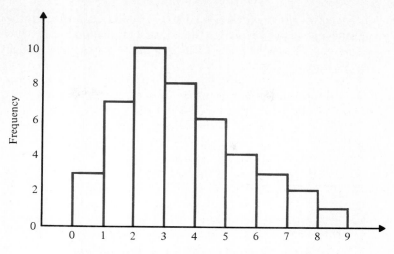

FIGURE 2.13 (b) Distribution skewed to the right

Although it is possible to find a numerical summary measure to indicate the direction and extent of the skewness of a set of observations, the details need not detain us here. Our objective, rather, has been to indicate the value of drawing a histogram, and the potential for the histogram to reveal interesting and important information that would be missed if only a mean and standard deviation were calculated. Viewed in this light, the histogram is an important pictorial tool for summarizing data. In the

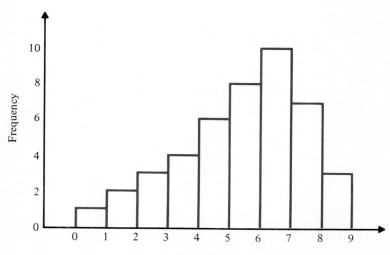

FIGURE 2.13 (c) Distribution skewed to the left

following section, we will discuss other graphical methods that can be of value in understanding and presenting numerical information.

EXERCISES

21. The accompanying table shows the percentage of business tax to total tax in the fifty states and the District of Columbia for 1977.

STATE	PERCENTAGE BUSINESS TAX	STATE	PERCENTAGE BUSINESS TAX
Alaska	63.3	New Jersey	29.8
Louisiana	52.2	Indiana	29.6
Wyoming	51.4	Pennsylvania	29.4
West Virginia	45.4	Utah	28.9
Texas	42.6	Kentucky	28.8
New Mexico	39.7	Michigan	28.5
Montana	38.1	Virginia	28.4
Oklahoma	35.1	Oregon	28.3
Washington	35.1	Arkansas	27.6
Tennessee	34.9	Idaho	27.6
Delaware	34.2	North Carolina	26.7
Connecticut	33.4	North Dakota	26.7
Kansas	33.0	Rhode Island	26.7
Ohio	32.9	Vermont	26.6
Nevada	32.7	Minnesota	26.2
Alabama	32.6	South Carolina	26.1
Florida	32.0	Georgia	25.7
Illinois	32.0	Hawaii	25.3
District of Columbia	31.9	Maine	24.6
Arizona	31.8	Maryland	24.1
Mississippi	31.7	Massachusetts	23.5
Colorado	31.1	South Dakota	22.8
California	31.0	Wisconsin	22.5
Missouri	30.9	Iowa	21.5
New York	30.4	Nebraska	20.8
New Hampshire	30.1		

Construct an appropriate histogram to summarize the data.

22. The accompanying table shows the percentage unemployment rate in the fifty states in 1980.

STATE	PERCENTAGE UNEMPLOYED	STATE	PERCENTAGE UNEMPLOYED
Alabama	7.1	Montana	5.1
Alaska	9.2	Nebraska	3.2
Arizona	5.1	Nevada	5.1
Arkansas	6.2	New Hampshire	3.1

STATE	PERCENTAGE UNEMPLOYED	STATE	PERCENTAGE UNEMPLOYED
California	6.2	New Jersey	6.9
Colorado	4.8	New Mexico	6.6
Connecticut	5.1	New York	7.1
Delaware	8.0	North Carolina	4.8
Florida	6.0	North Dakota	3.7
Georgia	5.1	Ohio	5.9
Hawaii	6.3	Oklahoma	3.4
Idaho	5.7	Oregon	6.8
Illinois	5.5	Pennsylvania	6.9
Indiana	6.4	Rhode Island	6.6
Iowa	4.1	South Carolina	5.0
Kansas	3.4	South Dakota	3.5
Kentucky	5.6	Tennessee	5.8
Louisiana	6.7	Texas	4.2
Maine	7.2	Utah	4.3
Maryland	5.9	Vermont	5.1
Massachusetts	5.5	Virginia	4.7
Michigan	7.8	Washington	6.8
Minnesota	4.2	West Virginia	6.7
Mississippi	5.8	Wisconsin	4.5
Missouri	4.5	Wyoming	2.8

Construct an appropriate histogram to summarize the data.

23. The following table shows the percentage unionization in the fifty states in 1980.

STATE	PERCENTAGE UNIONIZED	STATE	PERCENTAGE UNIONIZED
Alabama	19.2	Montana	24.1
Alaska	26.2	Nebraska	15.3
Arizona	13.8	Nevada	22.9
Arkansas	15.0	New Hampshire	13.3
California	23.7	New Jersey	23.0
Colorado	15.2	New Mexico	12.1
Connecticut	21.9	New York	39.2
Delaware	21.7	North Carolina	6.5
Florida	11.7	North Dakota	14.7
Georgia	13.6	Ohio	29.5
Hawaii	32.1	Oklahoma	13.5
Idaho	14.3	Oregon	23.1
Illinois	31.5	Pennsylvania	34.2
Indiana	29.3	Rhode Island	27.1
Iowa	19.2	South Carolina	6.7
Kansas	12.8	South Dakota	10.3
Kentucky	22.4	Tennessee	17.7
Louisiana	16.0	Texas	11.0
Maine	18.3	Utah	13.0
Maryland	21.0	Vermont	17.5
Massachusetts	24.4	Virginia	12.7

STATE	PERCENTAGE UNIONIZED	STATE	PERCENTAGE UNIONIZED
Michigan	34.6	Washington	33.1
Minnesota	24.4	West Virginia	36.8
Mississippi	12.4	Wisconsin	27.8
Missouri	30.0	Wyoming	14.9

Construct an appropriate histogram to summarize the data.

24. The accompanying table shows the life expectancy (in years) in thirty-five low-income countries in 1979.

COUNTRY	LIFE EXPECTANCY	COUNTRY	LIFE EXPECTANCY
Laos	42	Mozambique	47
Bhutan	44	Sierra Leone	47
Bangladesh	49	China	64
Chad	41	Haiti	53
Ethiopia	40	Pakistan	52
Nepal	44	Tanzania	52
Somalia	44	Zaire	47
Mali	43	Niger	43
Burma	54	Guinea	44
Afghanistan	41	Central African Republic	44
Vietnam	63	Madagascar	47
Burundi	42	Uganda	54
Upper Volta	43	Mauritania	43
India	52	Lesotho	51
Malawi	47	Togo	47
Rwanda	47	Indonesia	53
Sri Lanka	66	Sudan	47
Benin	47		

Construct an appropriate histogram to summarize this information.

25. The accompanying table shows appropriations (in dollars) for higher education per thousand dollars of personal income for the fifty states in 1979–80.

STATE	APPROPRIATIONS FOR HIGHER EDUCATION	STATE	APPROPRIATIONS FOR HIGHER EDUCATION
Alabama	16.02	Montana	11.42
Alaska	16.42	Nebraska	12.72
Arizona	13.41	Nevada	9.13
Arkansas	13.00	New Hampshire	4.65
California	14.14	New Jersey	6.23
Colorado	11.41	New Mexico	15.78
Connecticut	7.68	New York	10.57
Delaware	10.71	North Carolina	15.82
Florida	9.37	North Dakota	16.18
Georgia	11.30	Ohio	7.93

STATE	APPROPRIATIONS FOR HIGHER EDUCATION	STATE	APPROPRIATIONS FOR HIGHER EDUCATION
Hawaii	15.95	Oklahoma	11.13
Idaho	13.58	Oregon	12.62
Illinois	8.76	Pennsylvania	8.12
Indiana	9.93	Rhode Island	10.23
Iowa	13.10	South Carolina	16.31
Kansas	12.91	South Dakota	10.54
Kentucky	12.98	Tennessee	11.15
Louisiana	12.39	Texas	13.08
Maine	8.34	Utah	16.93
Maryland	9.34	Vermont	8.46
Massachusetts	6.88	Virginia	11.24
Michigan	10.37	Washington	14.59
Minnesota	14.53	West Virginia	12.88
Mississippi	17.59	Wisconsin	13.30
Missouri	8.81	Wyoming	14.12

Summarize the data in a histogram.

26. One question on a particular examination was worth a maximum of 5 points. The scores of the twenty-five students on this question are summarized in the table.

SCORE	0	1	2	3	4	5
NUMBER OF STUDENTS	5	3	2	4	4	7

(a) Find the mean score.
(b) Find the median of these scores.
(c) What is the modal score?
(d) Find the variance and standard deviation of these scores.

27. A company manufactures packages of candy. A sample of forty of the packages were examined, and the number of pieces of candy in each package was counted, yielding the results shown in the table.

NUMBER OF PIECES	18	19	20	21	22
NUMBER OF PACKAGES	5	8	15	10	2

(a) What is the mean number of pieces per package for this sample?
(b) Find the sample median number of pieces.
(c) Find the modal number of pieces for this sample.
(d) Find the sample variance and standard deviation.

28. Refer to the data of Exercise 21 on the percentages of business taxes to total taxes.
    (a) Estimate the population mean, based on the data grouping used in constructing the histogram.
    (b) Estimate the population standard deviation, based on the data grouping used in constructing the histogram.

(c) Estimate the population median.

(d) Estimate the population interquartile range.

(e) Now calculate the mean and standard deviation directly from the original data, and compare with your answers to parts (a) and (b).

29. Refer to the data of Exercise 22 on the percentage unemployment rates in the fifty states.

(a) Estimate the mean of these percentage unemployment rates, based on the data grouping used in constructing the histogram.

(b) Estimate the population standard deviation for these rates, based on the data grouping used in constructing the histogram.

(c) Estimate the median unemployment rate.

(d) Estimate the interquartile range of these rates.

(e) Now calculate the mean and standard deviation directly from the fifty observations, and compare with your answers to parts (a) and (b).

30. Refer to the data of Exercise 23 on the percentage unionization in the fifty states.

(a) Estimate the mean of these unionization rates, based on the data grouping used in constructing the histogram.

(b) Estimate the population standard deviation of these rates, based on the data grouping used in constructing the histogram.

(c) Estimate the median unionization percentage.

(d) Estimate the interquartile range of these percentages.

(e) Now calculate the mean and standard deviation directly from the fifty observations, and compare with your previous answers.

31. Refer to the data of Exercise 24 on life expectancy in thirty-five low-income countries.

(a) Estimate the mean of these life expectancies, based on the data grouping used in constructing the histogram.

(b) Estimate their population standard deviation, based on the data grouping used in constructing the histogram.

(c) Estimate the median life expectancy.

(d) Estimate the interquartile range of these life expectancies.

(e) Now calculate the population mean and standard deviation directly from the raw observations, and compare with your previous answers.

32. A village council is considering a new tax on dogs, by weight, to help support the dog shelter. A census of the 120 dogs in the village produced the results shown in the table.

WEIGHT (IN POUNDS)	0–20	20–40	40–60	60–80
NUMBER OF DOGS	13	32	51	24

(a) Draw the histogram of these weights.

(b) Find the relative frequencies.

(c) Find the cumulative relative frequencies, and draw the corresponding histogram.

(d) Estimate the population mean weight.

(e) Estimate the population variance and standard deviation for the weights of these dogs.

(f) Estimate the median weight.

(g) Estimate the interquartile range of these weights.

(h) What is the modal weight class?

33. A sample of twenty-five batches of a chemical were tested for impurities. The accompanying table shows the percentage impurity levels for this sample.

PERCENTAGE IMPURITIES	0–2	2–4	4–6	6–8	8–10
NUMBER OF BATCHES	3	3	8	7	4

(a) Draw the histogram.
(b) Find the relative frequencies for this sample.
(c) Find the cumulative relative frequencies, and draw the corresponding histogram.
(d) Estimate the sample mean percentage impurity level.
(e) Estimate the sample variance and standard deviation.

34. The American Popcorn Institute is interested in the ability of the consumer public to cook popcorn successfully. They selected a sample of twenty people and observed their efforts to pop kernels of corn. For each person, the percentage of unpopped kernels was recorded. The results are given in the table.

PERCENTAGE KERNELS UNPOPPED	0–2	2–4	4–6	6–8	8–10
NUMBER OF PEOPLE	6	4	7	2	1

(a) Draw the histogram.
(b) Find the sample relative frequencies.
(c) Find and interpret the sample cumulative relative frequencies.
(d) Estimate the sample mean percentage of unpopped kernels.
(e) Estimate the sample variance and standard deviation.
(f) Estimate the sample median percentage of unpopped kernels.
(g) Estimate the sample interquartile range.
(h) What is the modal class for this sample?

35. During a winter flu epidemic, waiting times at a student health center were longer than usual. The accompanying table summarizes the distribution of waiting times for a sample of twenty students who visited the health center in this period.

WAITING TIME (IN HOURS)	0–1	1–2	2–3	3–4
NUMBER OF STUDENTS	5	9	5	1

(a) Draw the histogram.
(b) Find the sample relative frequencies.
(c) Find and interpret the sample cumulative relative frequencies.
(d) Estimate the sample mean waiting time.
(e) Estimate the sample variance and standard deviation of waiting times.
(f) Estimate the sample median waiting time.
(g) Estimate the sample interquartile range.
(h) What is the modal class for this sample?

36. The athletic director of a major university is interested in the amount of money donated to the Athletic Department by season ticket holders of the preferred seats for basketball games. The table shows donations by a sample of thirty of these ticket holders.

DONATION (IN DOLLARS)	NUMBER OF TICKET HOLDERS
0–400	3
400–800	5
800–1,200	12
1,200–1,600	6
1,600–2,000	4

(a) Draw the histogram.
(b) Find the sample relative frequencies.
(c) Find and interpret the sample cumulative relative frequencies.
(d) Estimate the sample mean donation.
(e) Estimate the sample standard deviation for these donations.
(f) Estimate the sample median donation.
(g) Estimate the sample interquartile range.
(h) What is the modal class for this sample?

37. For a sample of fifty new full-size cars, miles per gallon figures were obtained and summarized in the accompanying table.

MILES PER GALLON	14–16	16–18	18–20	20–22	22–24
NUMBER OF CARS	4	6	13	20	7

(a) Draw the histogram.
(b) Find the sample relative frequencies.
(c) Find and interpret the sample cumulative relative frequencies.
(d) Estimate the sample mean miles per gallon.
(e) Estimate the sample standard deviation of miles per gallon.
(f) Estimate the sample median miles per gallon.
(g) Estimate the sample interquartile range.
(h) What is the modal class for this sample?

38. For households in a large town, the tabled information on incomes is available.

HOUSEHOLD INCOME (IN DOLLARS)	RELATIVE FREQUENCY
10,000–15,000	.22
15,000–20,000	.18
20,000–25,000	.14
25,000–30,000	.12
30,000–40,000	.14
40,000–50,000	.12
50,000–60,000	.08

(a) Draw the histogram.

(b) Estimate the population mean household income.

(c) Estimate the population standard deviation of household incomes.

(d) Estimate the population median income.

(e) Compare your estimates in (b) and (d), and comment on their difference.

2.6 SOME OTHER GRAPHICAL METHODS

The presentation of numerical information is often most conveniently and attractively achieved through the use of graphs and charts. These have the advantage that features of data are more easily absorbed visually than through the contemplation of a numerical tabulation. In this section, we will briefly outline a few of the graphical techniques in common use.

(i) BAR CHARTS

Bar charts provide a convenient way of comparing numerical quantities that are distributed either spatially or through time. To illustrate, we will consider some of the information contained in the 1980 annual report of International Telephone and Telegraph Corporation (ITT). Table 2.12 shows the number of employees of the corporation's five business groups in 1980. An alternative, visual, presentation of this information is provided by the bar chart of Figure 2.14. The chart is constructed so that the height of the rectangle drawn for each business group is proportional to its number of employees. Charts of this kind are easily and quickly absorbed, and are very frequently employed in the reporting of business information.

The same technique can be used to illustrate the progression of a variable through time. Table 2.13 gives the dividends per common share declared by ITT over an 8-year period. This same information is pictured in the bar chart of Figure 2.15. It is easier to form a quick and reliable assessment of dividend growth through inspection of the bar chart than through examination of the numbers themselves.

An interesting and useful extension to the simple bar chart arises when components of the individual categories are also of interest. For example, Table 2.14 gives research and development expenditures for ITT over a 5-year period. Expenditures by ITT and total expenditure by ITT plus customers are shown separately. This information is portrayed in a bar chart by breaking up the total expenditure bars so that expenditures by ITT and customers are separately indicated by differences in shading, as shown in Figure 2.16. It is easy to use such charts to make comparisons of both

TABLE 2.12 Number of employees (in thousands) in the business groups of ITT in 1980

BUSINESS GROUP	NUMBER OF EMPLOYEES (IN THOUSANDS)
Telecommunications and Electronics	163
Engineered Products	93
Consumer Products and Services	53
Natural Resources	11
Insurance and Finance	28

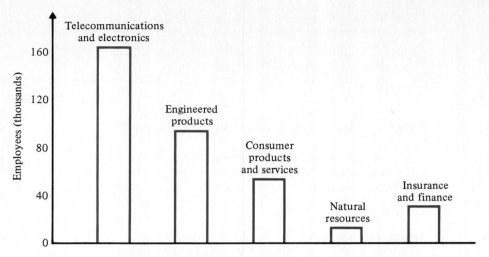

FIGURE 2.14 Bar chart for number of employees in ITT business groups in 1980

TABLE 2.13 Dividends per common share (in dollars) declared by ITT, 1973–80

YEAR	DIVIDENDS PER SHARE	YEAR	DIVIDENDS PER SHARE
1973	1.32	1977	1.82
1974	1.46	1978	2.05
1975	1.54	1979	2.25
1976	1.64	1980	2.45

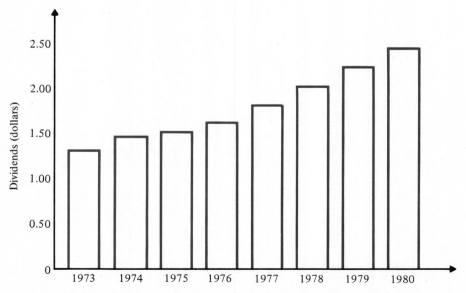

FIGURE 2.15 Bar chart for dividends per common share paid by ITT, 1973–80

TABLE 2.14 ITT research and development expenditures (in millions of dollars), 1976–80

YEAR	ITT	ITT PLUS CUSTOMERS
1976	247	526
1977	282	610
1978	371	799
1979	436	959
1980	505	1,116

the total and the individual components. This kind of chart is sometimes called a **component bar chart.**

(ii) TIME PLOTS

As an alternative to a bar chart, the progression of a numerical quantity through time can also be illustrated by graphing its value against time. Measuring time along the horizontal axis and the numerical quantity of interest along the vertical axis yields a point on the graph for each observation. Joining points adjacent in time by straight lines then produces a time plot, which provides an easily read visual impression of the historical record.

To illustrate, Table 2.15 gives the number of employees of ITT at year-end for the period 1971–80. The corresponding time plot is shown in Figure 2.17.

From Figure 2.17 one readily sees employment rising to a peak in 1973, and thereafter steadily declining, this pattern being broken only by a modest rise in 1977–78. On the other hand, any more detailed inference than this from such a plot

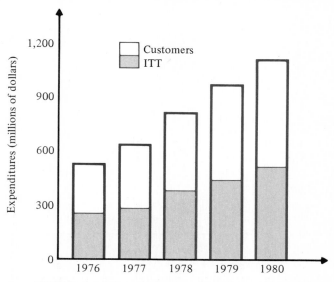

FIGURE 2.16 Component bar chart for research and development expenditures by ITT and customers, 1976–80

TABLE 2.15 Number of employees (in thousands) of ITT at year-end, 1971–80

YEAR	NUMBER OF EMPLOYEES (IN THOUSANDS)	YEAR	NUMBER OF EMPLOYEES (IN THOUSANDS)
1971	398	1976	375
1972	428	1977	375
1973	438	1978	379
1974	409	1979	368
1975	376	1980	348

is risky. It may be tempting, from the visual examination of a time plot, to extrapolate apparent "trends" into the future. Unfortunately, experience suggests that this is a very unreliable strategy for predicting the future. The difficulty lies in the fact that, when viewed on a time plot, chance variability can appear to present the picture of a systematic pattern.

If one is interested in using a historical record to obtain forecasts, a time plot alone is inadequate. Rather, as will be discussed in Chapter 17, a more thorough analysis of the data is necessary.

(iii) PIE CHARTS

Pie charts are used to depict the division of a whole into its constituent parts. To illustrate, Table 2.16 shows the 1980 expenditures of the Halifax Building Society,[5] broken down into five segments.

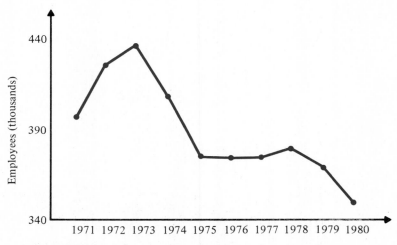

FIGURE 2.17 Number of employees of ITT at year-end, 1971–80

---

[5] British building societies serve roughly the same role as savings and loan associations in the United States.

TABLE 2.16 Expenditures (in millions of pounds) of Halifax Building Society, 1980

ITEM	EXPENDITURE
Investors' interest	979
Income tax on interest	280
Management and other expenses	111
Corporation tax	26
Surplus added to reserves	53
Total	1,449

This breakdown is illustrated in the pie chart of Figure 2.18. The circle (or "pie") represents total expenditures, and the segments (or "pieces of the pie") cut from its center depict shares of that total. The pie is constructed so that the area of each segment is proportional to the corresponding expenditure. Thus, for example, "Management and other expenses" take up a proportion $^{111}/_{1,449} = 0.077$ of total expenditure. Thus, the area of the corresponding piece of the pie is a proportion 0.077 of the total area of the pie. Pie charts are easily drawn. Beginning at the center of the

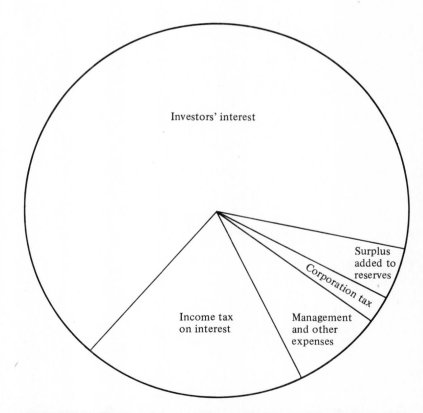

FIGURE 2.18 Pie chart for expenditures of Halifax Building Society, 1980

circle, the total of 360° is subdivided to determine the angle cut off at the center by each segment. For "Management and other expenses," this angle is the proportion 0.077 of 360°—that is, 27.7°.

(iv) SCATTER PLOTS

A common problem in statistical inference is the assessment of the relationship, if any, between a pair of numerical variables. For example, Table 2.17 shows the expenditure per child on education and the income per adult for sixty-four countries. We would expect educational expenditures to be highest in the richest countries, and, indeed, a glance at the values in the table is sufficient to verify that, in general, this is so. To obtain a fuller picture of the relationship between these two quantities, we have graphed the data on the scatter plot of Figure 2.19. Due to the large number of observations and the very wide range of values, the data have been broken into three

TABLE 2.17 Educational expenditures per child and income per adult (in U.S. dollars, 1970)

COUNTRY	EDUCATIONAL EXPENDITURE PER CHILD	INCOME PER ADULT	COUNTRY	EDUCATIONAL EXPENDITURE PER CHILD	INCOME PER ADULT
Argentina	71	1,514	Jordan	23	528
Australia	341	3,239	Kenya	14	241
Austria	288	1,934	Korea	20	350
Bolivia	12	276	Liberia	11	317
Brazil	19	473	Libya	185	2,696
Bulgaria	179	1,117	Malawi	7	143
Burma	6	133	Mexico	32	1,074
Canada	760	3,786	Nepal	1	133
Ceylon	19	322	Netherlands	476	2,411
Chile	59	836	Nicaragua	20	731
Colombia	10	547	Norway	510	2,880
Costa Rica	61	981	Pakistan	3	200
Cuba	68	406	Panama	74	1,179
Czechoslovakia	251	1,803	Paraguay	11	444
Dominica	18	528	Peru	29	600
Ecuador	20	462	Philippines	14	396
El Salvador	18	527	Poland	168	1,306
Finland	480	2,676	Portugal	25	718
France	344	3,280	Sudan	11	208
Ghana	16	345	Switzerland	493	3,506
Greece	74	1,120	Syria	24	491
Guatemala	17	648	Thailand	13	281
Honduras	18	491	Trinidad/Tobago	82	1,508
Hungary	236	1,392	Turkey	24	603
India	7	190	Uganda	10	203
Iran	19	648	United States	989	5,808
Iraq	39	596	Uruguay	80	778
Ireland	172	1,609	USSR	291	1,667
Israel	257	2,343	Venezuela	96	1,887
Italy	251	1,842	Yugoslavia	112	806
Jamaica	54	1,019	Zaire	8	138
Japan	244	1,882	Zambia	35	537

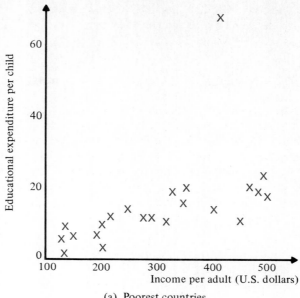

(a) Poorest countries

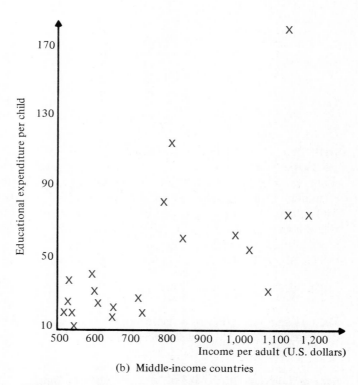

(b) Middle-income countries

FIGURE 2.19 Scatter plot of educational expenditure per child

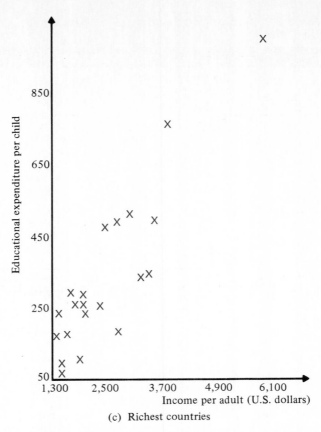

FIGURE 2.19 Scatter plot of educational expenditure per child (cont.)

parts. Part (a) of the figure represents the twenty-two poorest countries (in terms of income per adult), while part (c) of the figure shows the plot for the twenty-one richest countries. The data for the twenty-one middle-income countries are shown in part (b). Examination of this graph reveals the low levels of educational expenditure in the poorest countries, and shows the extent to which educational expenditures climb for richer countries. Such a pictorial representation can serve to provide a general overview of the relationship between a pair of variables. Statistical techniques to be discussed in Chapters 12–14 allow a more detailed analysis of this type of data.

(v) BOX AND WHISKER PLOTS

The box and whisker plot[6] is useful for the graphical summary of a batch of data. To illustrate, Table 2.18 shows educational expenditures per child for three groups of

---

[6] This and other graphical techniques are discussed in J. W. Tukey, *Exploratory Data Analysis,* Addison-Wesley (1977). See also S. Leinhardt and S. S. Wasserman, "Teaching regression: An exploratory approach," *The American Statistician, 33* (1979), 196–203.

TABLE 2.18 Educational expenditures per child for three groups of countries (in U.S. dollars, 1970)

POOREST COUNTRIES			
COUNTRY	EDUCATIONAL EXPENDITURE PER CHILD	COUNTRY	EDUCATIONAL EXPENDITURE PER CHILD
Bolivia	12	Liberia	11
Brazil	19	Malawi	7
Burma	6	Nepal	1
Ceylon	19	Pakistan	3
Cuba	68	Paraguay	11
Ecuador	20	Philippines	14
Ghana	16	Sudan	11
Honduras	18	Syria	24
India	7	Thailand	13
Kenya	14	Uganda	10
Korea	20	Zaire	8

MIDDLE-INCOME COUNTRIES			
COUNTRY	EDUCATIONAL EXPENDITURE PER CHILD	COUNTRY	EDUCATIONAL EXPENDITURE PER CHILD
Bulgaria	179	Jordan	23
Chile	59	Mexico	32
Colombia	10	Nicaragua	20
Costa Rica	61	Panama	74
Dominica	18	Peru	29
El Salvador	18	Portugal	25
Greece	74	Turkey	24
Guatemala	17	Uruguay	80
Iran	19	Yugoslavia	112
Iraq	39	Zambia	35
Jamaica	54		

RICHEST COUNTRIES			
COUNTRY	EDUCATIONAL EXPENDITURE PER CHILD	COUNTRY	EDUCATIONAL EXPENDITURE PER CHILD
Argentina	71	Japan	244
Australia	341	Libya	185
Austria	288	Netherlands	476
Canada	760	Norway	510
Czechoslovakia	251	Poland	168
Finland	480	Switzerland	493
France	344	Trinidad/Tobago	82
Hungary	236	United States	989
Ireland	172	USSR	291
Israel	257	Venezuela	96
Italy	251		

countries—the twenty-two poorest (as measured by income per adult), the twenty-one richest, and the twenty-one middle-income countries of the countries of Table 2.17.

Consider first the poorest group of countries. For this group, the median expenditure is 12.5, while the first and third quartiles are 7.75 and 19, respectively. An unusual feature of these data is the relatively high value (68) for Cuba. With the exception of this observation, all expenditures range from 1 to 24. This information is summarized in the plot of Figure 2.20. The scale shows educational expenditures. The rectangle (or "box") is drawn so that its lower and upper boundaries correspond to the first and third quartiles, while a line at the value of the median is drawn in the interior of the box. The unusual value for Cuba is shown separately, while lines run from the edges of the box to dashed lines (or "whiskers") drawn at the levels of the largest and smallest of the remaining observations.

For the middle-income countries, the median educational expenditure per child is 32, while the first and third quartiles are 19.5 and 67.5. Two countries—Bulgaria at 179 and Yugoslavia at 112—have expenditures considerably higher than the others, which range from 10 to 80. These facts are summarized in the box and whisker plot of Figure 2.21, from which one can quickly read the median and quartiles, the outlying observations, and the range of the remaining data.

Finally, for the richest countries, the median educational expenditure per child is 257, the first and third quartiles being 178.5 and 478. Three countries in this group—Argentina at 71, Trinidad/Tobago at 82, and Venezuela at 96—have very low educational expenditures, while the United States at 989 and Canada at 760 are very high. Expenditures in the remaining countries range from 168 to 510. This information is depicted in Figure 2.22. It clearly emerges from that picture that expenditures are considerably more disperse in the upper ranges of this distribution; that is, a few of

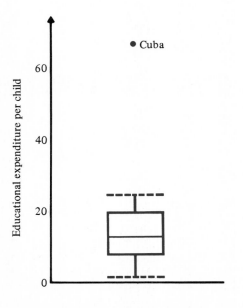

FIGURE 2.20 Box and whisker plot for educational expenditures per child in poorest countries

Summarizing Numerical Information

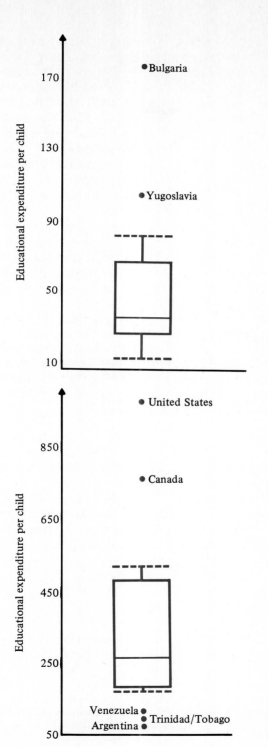

FIGURE 2.21 Box and whisker plot for educational expenditures per child in middle-income countries

FIGURE 2.22 Box and whisker plot for educational expenditures per child in richest countries

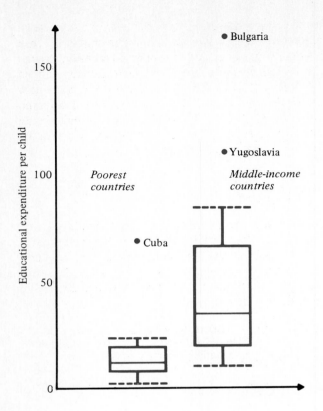

FIGURE 2.23 Box and whisker plots for educational expenditures per child in poorest and middle-income countries

the countries have expenditures considerably higher than the median for this group as a whole.

Another use of box and whisker plots is to compare two or more batches of data. For example, if we want to compare educational expenditures per child in the poorest and the middle-income countries, the two plots can be drawn side-by-side to the same scale, as is done in Figure 2.23. This allows us to see, at a glance, the extent to which educational expenditures are lower in the poorest countries.

2.7 LYING WITH STATISTICS[7]

We have now considered several procedures that might be appropriate for summarizing and presenting numerical information. Used sensibly and carefully, these can be

---

[7] The title of this section is inspired by D. Huff and I. Geis, *How to Lie with Statistics,* Norton (1954). This delightful little book is essential reading for anyone with a serious interest in the presentation of statistical information.

excellent tools for extracting the essential information from what would otherwise be an indigestible mass of numbers. Unfortunately, however, it is not invariably the case that an attempt at data summarization is carried out either sensibly or carefully. In such circumstances one can easily be misled by the manner in which the summary is presented. The statistician's art involves drawing from data as clear and accurate a picture as possible. Improper use of the various techniques can produce a distorted picture, yielding a false impression. It is possible to "lie with statistics" without being deliberately dishonest. In this section we present six suggestions for how to do so, the intent being not to encourage their use, but to caution against their dangers.

(i) EMOTIVE AND LOADED STATEMENTS

Numbers, in and of themselves, contain no value judgments. Data simply provide factual material, which could, of course, be useful on one side or another of a particular argument. However, it is possible, through simple verbal tricks of presentation, to color the numbers in a suggestive manner. Suppose that a census of blue-collar production workers in a particular plant reveals their mean annual income to be $25,000. This fact can be presented as:

> "The mean annual income for these employees is $25,000."

A union representative might report the same information as:

> "The mean annual income for these employees is only $25,000."

The injection of the word "only" into this sentence clarifies nothing. Rather, it carries the suggestion—certainly unwarranted in the absence of further information—that the average income is unduly low.

A more colorful, and more extreme, example is the following statement.[8]

> "If all the nation's federal bureaucrats were laid end to end, they would reach from New York City to beyond Houston, Texas."

The purpose of such a statement is not to present numerical information, but to convey the impression that the number of federal bureaucrats is very large. The reader is encouraged to deduce that there are "too many" such government employees. Here the underlying numerical information is obscured by the spurious appeal to the irrelevant fact that, in the unlikely event that these people were to be laid end to end, a considerable distance would be covered. Only by working backwards can we extract the fact that there are 1.4 million federal white-collar employees.

---

[8] This is based on a total of 1.4 million federal white-collar employees. Assuming an average height of 5 feet 9 inches gives a total distance of 1,525 miles.

(ii) INADEQUATE NUMERICAL SUMMARIES

The reduction of a vast amount of data to one or two summary measures, intended to carry as much information as possible, is often necessary. Without such summary statistics, interpretation would frequently be impossible. Still, the process can be taken too far. If too little summary information is provided, a false impression can be created.

An example[9] of this type of distortion concerns the population density in Washoe County, Nevada, which has an average 13.5 people per square mile. In fact, 80% of the inhabitants of this county live in Reno and Sparks, where there are, respectively, 4,362 and 6,155 people per square mile. The remainder of Washoe County—99.8% of its area—has 2.66 people per square mile. The average population density for this county then tells us very little. The great majority of its inhabitants live in parts that are substantially more densely populated, while the overwhelming majority of the area is much more sparsely populated than the average. Yet, presented with just the mean number of people per square mile, we might jump to the erroneous conclusion of uniformity of population density in this county.

(iii) CHOICE OF SCALE FOR TIME PLOTS

Figure 2.17, which for convenience is reproduced as Figure 2.24(a), shows the number of employees of ITT over a 10-year period. The general downward drift in these numbers after 1973 is quite apparent. Precisely the same information is plotted

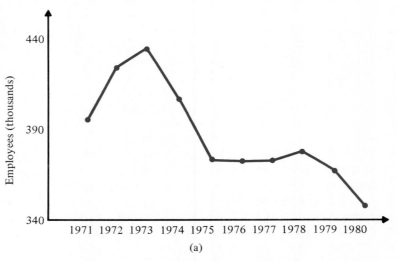

(a)

FIGURE 2.24 Time plots of numbers of ITT employees, 1971–80, drawn to two different scales

---

[9] This illustration is given by J. W. Tukey, "Methodology and the statistician's responsibility for both accuracy and relevance," *Journal of the American Statistical Association, 74* (1979), 786–93.

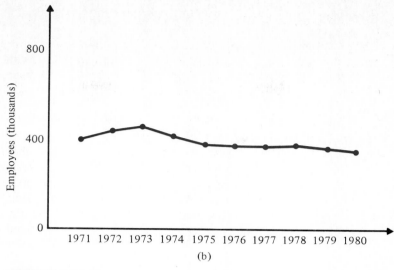

FIGURE 2.24 Time plots of numbers of ITT employees, 1971–80, drawn to two different scales (cont.)

in Figure 2.24(b), but now a far coarser scale is used on the vertical axis. In consequence, the resulting picture is quite flat, giving the impression of virtually no change in the numbers employed over the whole period.

By selecting a particular scale of measurement, one can, in a time plot, create an impression either of relative stability or of substantial fluctuations over time. There is no "correct" choice of scale for any particular plot. Rather, the conclusion from examples such as this is that looking at the shape of the plot alone is inadequate for obtaining a clear picture of the data. It is also necessary to keep in mind the scale on which the measurements are made.

(iv) IMPROPER GRAPHICAL SIZE COMPARISONS

From Table 2.14, it can be seen that expenditures on research and development by ITT in 1980 were a little more than twice the 1976 level. One simple possibility would be to illustrate this by drawing a one-unit square to depict the 1976 figure, and a two-unit square to represent the 1980 value, as is done in Figure 2.25(a). But, a casual look at this diagram does not suggest a doubling of expenditures over the period. It suggests considerably more, because we associate size with area. Hence, since the second square has area four times that of the first, the impression created is of a quadrupling of research and development expenditures.

It is easy enough to put matters right. The area of the square for 1980 should be twice the area of that for 1976. Therefore, as in Figure 2.25(b), the sides of the 1980 square should be of length $\sqrt{2}$ units. This latter figure now creates the correct impression of the relative sizes of research and development expenditures in these two years.

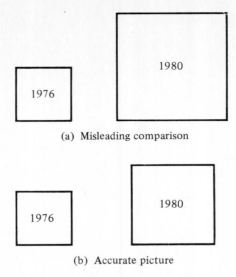

(a) Misleading comparison

(b) Accurate picture

FIGURE 2.25 ITT research and development expenditures in two years

This same point is important to keep in mind when constructing histograms in situations where the classes are not all of equal width. In Example 2.9 we looked at some data on non-audit fees as a proportion of total auditor remuneration. Correctly drawn, as in Figure 2.10, the *areas* of the rectangles over the class intervals should be proportional to the frequencies. For convenience, Figure 2.10 is reproduced as Figure 2.26(a). Now, suppose instead, as in Figure 2.26(b), the histogram is (incorrectly) drawn so that the *heights* of the rectangles over the class intervals are proportional to the frequencies. The change is quite dramatic. Visual inspection of this latter figure gives the mistaken impression of a very large proportion of observations in the highest classes.

(v) COINCIDENCES THAT REALLY ARE JUST THAT

It often happens that we are told of a conjunction of events, and are invited to conclude that this conjunction is "too much of a coincidence" to be coincidental. My own favorite is the "Super Bowl Stock Market Predictor Theory." According to this "theory," if a member of the old American Football League wins the Super Bowl, the outlook for the stock market is bearish, while if an original National Football League member wins, the outlook is bullish. This simple rule correctly predicted the overall direction of the stock market in thirteen of the years following the first fourteen Super Bowls.

One should react with skepticism to this theory. The notion that the outcome of a football game could influence the direction of the stock market is implausible, and it remains implausible even given the evidence of its apparent success as a predictor.

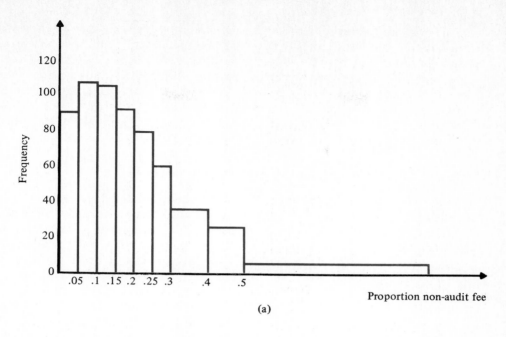

(a)

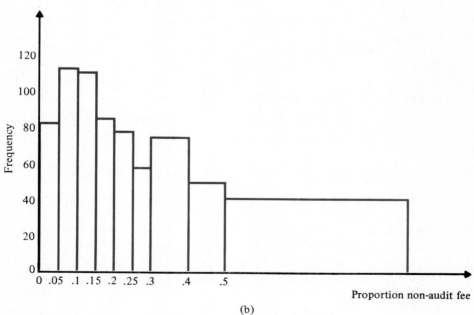

(b)

FIGURE 2.26 Histograms for non-audit fee as proportion of total auditor remuneration in Australia (a) drawn correctly, with areas proportional to frequencies; (b) drawn incorrectly, with heights proportional to frequencies

This evidence can be put into better perspective when one considers just how many people—amateurs and professionals—look for indicators of future trends in the stock market. How many possible correlates had been looked at before this particular one was discovered? Probably hundreds, possibly thousands! An appropriate attitude might be to argue that, among the very many that have been tried, this one seems to be a fairly close correlate of past behavior. However, a look at how well this *single* predictor performs over the next few years would allow a fair assessment as to whether the record of the first fourteen years really was coincidental!

(vi) GENERALIZING FROM VERY SMALL SAMPLES

"Two out of three beer drinkers prefer Blast to Schlush." How should one react to this statement? If the opinions of 3,000 beer drinkers had been solicited, and a preference was expressed for Blast by about 2,000 of them, this would provide substantial evidence in favor of the contention that Blast is preferred by a majority of drinkers. On the other hand, if only three drinkers had been questioned, and it turned out that two of them preferred Blast, we would be much less impressed. Rather, we would believe it quite likely that for three other drinkers, the position could be reversed, or that all would prefer Schlush.

On the basis of a very small sample, it is extremely dangerous to generalize about a population. Such samples may contain insufficient evidence to allow a conclusion with a reasonable level of certainty. As we will see in subsequent chapters, generalization from a sample to conclusions about a population becomes more precise as the amount of sample evidence increases. Very small samples may tell us little, if anything, about the wider population.

EXERCISES

39. The accompanying table shows 1979 interest rates on corporate bonds, according to their Moody classification. Graph this information on a bar chart.

CLASSIFICATION	Aaa	Aa	A	Baa
INTEREST RATE	9.63	9.94	10.20	10.69

40. The accompanying table shows the number of business failures in the United States in 1978, by industry grouping. Show these data on a bar chart.

INDUSTRY	NUMBER OF FAILED FIRMS
Construction	1,204
Manufacturing and Mining	1,013
Wholesale Trade	740
Retail Trade	2,889
Commercial and Services	773

41. The table given here shows farm population as a percentage of total population in seven different years. Draw a bar chart to summarize these data.

YEAR	1967	1969	1971	1973	1975	1977	1979
PERCENTAGE FARM POPULATION	5.5	5.1	4.6	4.5	4.2	3.6	3.4

42. Draw a bar chart to summarize the accompanying data on average household size in the United States in seven different years.

YEAR	1967	1969	1971	1973	1975	1977	1979
HOUSEHOLD SIZE	3.28	3.19	3.14	3.01	2.94	2.86	2.78

43. The accompanying table shows the number of M.B.A. degrees awarded in the United States in six different years. Draw a bar chart to depict the data.

YEAR	1963	1966	1969	1972	1975	1978
NUMBER OF M.B.A. DEGREES	5,787	12,988	19,398	30,511	36,450	48,484

44. The table shows the percentage returns on average common stockholders' equity for Humana, Inc., over a period of 5 years. Draw a bar chart to summarize the data.

YEAR	1977	1978	1979	1980	1981
RETURN	14.9	19.1	26.7	29.5	33.3

45. The accompanying table shows income (in millions of dollars) for the five divisions of ITT over a 5-year period. Use a bar chart to show the progression of total ITT income through this period. On the same chart, show the course of the breakdown of income among the five divisions.

	1976	1977	1978	1979	1980
Telecommunications and Electronics	225	239	277	219	338
Engineered Products	130	167	180	206	203
Consumer Products and Services	18	18	45	12	102
Natural Resources	50	43	28	105	94
Insurance and Finance	192	239	300	349	354

46. The percentage unemployment rate in the United States from 1967 to 1980 is shown in the table. Draw a time plot of the data, and verbally interpret the resulting picture.

YEAR	1967	1968	1969	1970	1971	1972	1973
PERCENTAGE UNEMPLOYMENT	3.8	3.6	3.5	4.9	5.9	5.6	4.9

YEAR	1974	1975	1976	1977	1978	1979	1980
PERCENTAGE UNEMPLOYMENT	5.6	8.7	7.7	7.0	6.0	5.9	7.2

47. The accompanying table gives interest rates on 4–6 months commercial paper from 1967 through 1979. Draw the time plot, and provide a verbal interpretation.

YEAR	1967	1968	1969	1970	1971	1972	1973
INTEREST RATE	5.10	5.90	7.83	7.72	5.11	4.69	8.15

YEAR	1974	1975	1976	1977	1978	1979
INTEREST RATE	9.87	6.33	5.35	5.60	7.99	10.91

48. The table shown here gives yields on public utilities bonds for a 14-month period in 1980–81. Draw the time plot, and give a verbal interpretation.

1980	June	July	Aug.	Sept.	Oct.	Nov.	Dec.
YIELD	11.87	12.12	12.82	13.29	13.53	14.07	14.48

1981	Jan.	Feb.	March	April	May	June	July
YIELD	14.22	14.84	14.86	15.32	15.84	15.27	15.87

49. The accompanying table shows social insurance and related payments in the United States as a percentage of personal income, for the period 1970–80. Draw and interpret the time plot for these data.

YEAR	1970	1971	1972	1973	1974	1975
PERCENTAGE INCOME	8.0	8.9	8.9	9.3	9.9	11.3

YEAR	1976	1977	1978	1979	1980
PERCENTAGE INCOME	11.4	11.3	11.0	11.0	11.7

50. The table shows the debt of developing countries, as a percentage of their gross national product, for the years 1970 through 1979. Draw, and verbally interpret, the time plot.

YEAR	1970	1971	1972	1973	1974	1975	1976	1977	1978	1979
DEBT	12.3	13.1	13.5	13.1	12.6	13.9	15.5	17.0	18.3	17.8

51. It has been found that 32.3% of all plate steel imports to the Midwest come from Belgium and Luxembourg, 22.9% from Canada, 16.5% from Norway, 12.9% from Romania, 8.0% from Netherlands, and 7.4% from West Germany. Show these figures on a pie chart.

52. The table shows the 1980 market shares for beer sales in the United States. Draw the pie chart for these data.

COMPANY	MARKET SHARE
Anheuser-Busch	28.0%
Miller	21.0%
Pabst	8.6%
Schlitz	8.5%
Coors	7.8%
G. Heileman	7.5%
Others	18.6%

53. Draw a box and whisker plot for the data of Exercise 21 on the percentage of business tax to total tax in the fifty states and the District of Columbia.

54. Draw a box and whisker plot for the data of Exercise 22 on the percentage unemployment rate in 1980 in the fifty states.

55. **(a)** Draw a box and whisker plot for the data of Exercise 23 on the percentage unionization in 1980 in the fifty states.

   (b) Ignoring Alaska and Hawaii, divide the country into five geographic regions. Draw on the same scale separate box and whisker plots for unionization in each region.

56. Draw a box and whisker plot for the data of Exercise 24 on life expectancy at birth in thirty-five low-income countries in 1979.

57. The accompanying table shows the average annual inflation rate, over the period 1970–79, in thirty-one low-income countries. Summarize the data in a box and whisker plot.

COUNTRY	INFLATION RATE	COUNTRY	INFLATION RATE
Bangladesh	15.8	Sierra Leone	11.3
Chad	7.9	Haiti	10.9
Ethiopia	4.3	Pakistan	13.9
Nepal	8.7	Tanzania	13.0
Somalia	11.3	Zaire	31.4
Mali	9.7	Niger	10.8
Burma	12.1	Guinea	4.4
Afghanistan	4.4	Central African Republic	9.1
Burundi	11.2	Madagascar	10.1
Upper Volta	9.8	Uganda	28.3
India	7.8	Mauritania	10.1
Malawi	9.1	Lesotho	11.6
Rwanda	14.6	Togo	10.3
Sri Lanka	12.3	Indonesia	20.1
Benin	9.2	Sudan	6.8
Mozambique	11.0		

58. The table shows total state government revenues per thousand dollars of personal income (in dollars) for each state in 1979–80. Draw a box and whisker plot to summarize the data.

STATE	REVENUES	STATE	REVENUES
Alabama	138.48	Montana	156.57
Alaska	661.13	Nebraska	103.86
Arizona	124.35	Nevada	112.25
Arkansas	138.85	New Hampshire	90.75
California	129.83	New Jersey	100.02
Colorado	110.42	New Mexico	211.45
Connecticut	98.65	New York	137.25
Delaware	164.44	North Carolina	129.71
Florida	96.46	North Dakota	167.74
Georgia	117.39	Ohio	88.01
Hawaii	195.89	Oklahoma	127.41
Idaho	133.88	Oregon	136.33
Illinois	100.38	Pennsylvania	112.33
Indiana	93.41	Rhode Island	150.09
Iowa	116.19	South Carolina	140.44
Kansas	98.83	South Dakota	136.90
Kentucky	143.62	Tennessee	111.06
Louisiana	157.30	Texas	101.43
Maine	152.65	Utah	158.34
Maryland	124.87	Vermont	174.41
Massachusetts	131.55	Virginia	112.81
Michigan	118.71	Washington	128.64
Minnesota	145.95	West Virginia	157.32
Mississippi	165.42	Wisconsin	139.75
Missouri	91.40	Wyoming	179.50

59. The accompanying table shows both higher education expenditures and public welfare expenditures as a percentage of total expenditures for each state in 1979–80. Write an essay on these data. Use any numerical or graphical summary measures that you think are appropriate for the extraction of information contained in the data.

STATE	HIGHER EDUCATION	PUBLIC WELFARE	STATE	HIGHER EDUCATION	PUBLIC WELFARE
Alabama	15.4	13.5	Montana	9.7	9.3
Alaska	8.5	5.6	Nebraska	17.0	12.4
Arizona	16.7	5.6	Nevada	10.3	6.4
Arkansas	12.0	14.8	New Hampshire	13.4	14.9
California	9.2	11.0	New Jersey	6.6	10.1
Colorado	20.0	7.5	New Mexico	14.8	9.3
Connecticut	8.7	18.1	New York	5.1	6.4
Delaware	17.3	11.1	North Carolina	13.6	11.0
Florida	8.8	9.5	North Dakota	16.2	8.8
Georgia	12.0	14.2	Ohio	11.8	11.8
Hawaii	12.7	13.8	Oklahoma	16.1	16.8

STATE	HIGHER EDUCATION	PUBLIC WELFARE	STATE	HIGHER EDUCATION	PUBLIC WELFARE
Idaho	12.7	10.8	Oregon	11.8	13.1
Illinois	8.3	19.8	Pennsylvania	5.5	19.2
Indiana	17.2	8.8	Rhode Island	9.8	20.8
Iowa	13.9	14.4	South Carolina	15.1	10.7
Kansas	14.7	16.2	South Dakota	15.5	13.7
Kentucky	12.0	13.9	Tennessee	14.6	14.7
Louisiana	11.3	13.1	Texas	17.0	12.7
Maine	9.1	19.5	Utah	19.6	10.3
Maryland	8.8	13.8	Vermont	15.7	14.1
Massachusetts	5.9	23.9	Virginia	15.4	10.3
Michigan	11.3	19.8	Washington	15.1	13.7
Minnesota	12.7	10.3	West Virginia	9.7	9.3
Mississippi	11.3	14.3	Wisconsin	13.2	12.4
Missouri	12.2	16.2	Wyoming	11.4	5.3

60. Turn to the current day's issue of the *Wall Street Journal*. Discuss the procedures that are used therein for the graphical summarization of numerical information. Do the methods used present clear pictures? Can you suggest any alternative or additional graphical methods that might have been used?

61. Obtain the annual report of a major United States corporation. Describe the graphical techniques for data presentation used in the report, and suggest any improvements that might be made.

62. Collect data on any business or economic phenomenon of interest to you. Provide a graphical summary that gives a clear and accurate picture of these data. Can you now produce a *misleading* graphical summary?

REVIEW EXERCISES

63. Explain what can be learned about a population from each of the following measures.
    (a) The mean
    (b) The median
    (c) The standard deviation
    (d) The interquartile range

64. If the standard deviation of a population is 0, what can you say about the members of that population?

65. (a) Two populations each contain two members. The means of the two populations are the same, as are their standard deviations. Are the numerical values of the members of the first population necessarily the same as those of the second?
    (b) Two populations each contain three members. The means of these two populations are the same, as are their standard deviations. Are the numerical values of the members of the first population necessarily the same as those of the second?

66. Draw two histograms to represent two populations with a common mean, but with the standard deviation of the first population larger than that of the second.

67. A supermarket chain has ten stores in a city. The following values are the numbers of cases of generic baked beans sold in these stores in a particular week:

| 27 | 23 | 18 | 12 | 16 | 31 | 22 | 11 | 9 | 36 |

For this population:
(a) Find the mean.
(b) Find the median.
(c) Find the variance.
(d) Find the standard deviation.
(e) Find the interquartile range.

68. A sample of nine gas stations in a large city showed the following prices (in cents) for a gallon of regular gas:

| 136 | 139 | 142 | 133 | 131 | 134 | 137 | 132 | 136 |

For this sample:
(a) Find the mean.
(b) Find the median.
(c) Find the variance.
(d) Find the standard deviation.
(e) Find the interquartile range.

69. The accompanying table* shows the accumulated years of service before voluntary resignation of 355 managerial, professional, and technical employees of a large oil company.

YEARS OF SERVICE	NUMBER OF EMPLOYEES	YEARS OF SERVICE	NUMBER OF EMPLOYEES
0–1	4	8–9	11
1–2	41	9–10	7
2–3	67	10–11	14
3–4	82	11–12	6
4–5	28	12–13	14
5–6	43	13–14	5
6–7	14	14–15	2
7–8	17		

*Data taken from G. F. Dreher, "The role of performance in the turnover process," *Academy of Management Journal*, 25 (1982), 137–47.

For this sample:
(a) Draw the histogram.
(b) Estimate the mean.
(c) Estimate the variance.
(d) Estimate the standard deviation.
(e) Estimate the median.
(f) Estimate the interquartile range.

70. Let $x_1, x_2, \ldots, x_N$ denote the N observations in a population with mean μ. Let K be any number. Show that

$$\sum_{i=1}^{N} (x_i - K)^2 = \sum_{i=1}^{N} (x_i - \mu)^2 + N(K - \mu)^2$$

Hence, deduce that the value of K for which $\sum_{i=1}^{N} (x_i - K)^2$ is smallest is $K = \mu$.

71. An investment adviser selects fifty stocks to recommend to clients. The table shows the percentage increases in their prices over the following year.

PERCENTAGE INCREASE	2–4	4–6	6–8	8–10	10–12	12–14
NUMBER OF STOCKS	2	7	12	17	9	3

For this sample:
(a) Draw the histogram.
(b) Estimate the mean.
(c) Estimate the variance.
(d) Estimate the standard deviation.
(e) Estimate the median.
(f) Estimate the interquartile range.

72. The accompanying table shows the yields of a sample of fifty money market funds on June 11, 1982.

13.61	13.42	13.28	13.45	13.42
13.42	13.37	12.95	14.18	13.44
13.92	12.88	13.15	13.31	13.24
13.47	13.24	13.63	13.48	13.26
13.10	13.70	13.64	13.39	13.30
13.41	13.31	13.15	13.57	13.58
13.51	13.59	13.34	13.26	13.37
13.04	13.34	13.57	13.59	13.60
13.31	14.06	13.34	13.73	13.54
13.23	13.33	13.09	13.40	13.36

(a) Represent these data with a histogram.
(b) Draw a box and whisker plot for these data.

73. The following values are the average maturities (in days) of a sample of thirty money market funds on June 11, 1982.

31	23	11	32	26	29
27	32	21	20	35	13
20	27	34	27	16	24
35	18	23	31	31	12
41	28	43	22	28	32

(a) Draw a histogram for these data.

(b) Draw a box and whisker plot for these data.

74. The table given here shows the median number of days of gross revenue in accounts receivable for the three months ending December 31 for hospitals with over 400 beds. Draw a bar chart to represent this information.

YEAR	NUMBER OF DAYS	YEAR	NUMBER OF DAYS
1974	63.6	1978	62.7
1975	59.4	1979	63.2
1976	61.1	1980	59.5
1977	61.9	1981	59.0

75. The accompanying table shows earnings per share (in dollars) of Schlumberger Limited over a 6-year period. Draw a bar chart for the data.

YEAR	EARNINGS	YEAR	EARNINGS
1976	1.01	1979	2.30
1977	1.39	1980	3.47
1978	1.75	1981	4.37

76. The table shows the value of funds, in billions of dollars, raised on international capital markets. For each year, the total is broken down into medium-term Euro-credits and external bond offerings. Draw a component bar chart for these data, and comment on the resulting picture.

YEAR	TOTAL	MEDIUM-TERM EURO-CREDITS	EXTERNAL BOND OFFERINGS
1974	40.7	28.5	12.2
1975	43.4	20.6	22.8
1976	62.2	27.9	34.3
1977	69.9	33.8	36.1
1978	110.0	74.2	35.8
1979	116.5	79.1	37.4
1980	117.9	79.9	38.0
1981	185.4	137.5	47.9

77. The table shows the total assets (in billions of dollars at market value) and asset mix of private noninsured pension funds over an 8-year period. Draw a component bar chart to represent these data, and discuss the picture that emerges.

YEAR	TOTAL	STOCKS	CORPORATE AND OTHER BONDS	CASH, GOVERNMENT SECURITIES, AND OTHER ASSETS
1973	132.2	83.6	27.7	20.9
1974	111.7	63.3	30.8	17.6
1975	145.6	88.6	34.5	22.5
1976	173.9	109.7	37.9	26.3
1977	181.6	101.9	42.8	36.9
1978	201.5	107.9	48.6	45.0
1979	225.2	123.8	51.3	50.1
1980	297.2	175.8	60.0	61.4

78. The accompanying table shows the value of the U.S. dollar in Canadian dollars over a 12-year period. Draw a time plot of these data, and discuss the resulting graph.

YEAR	VALUE	YEAR	VALUE	YEAR	VALUE
1970	1.044	1974	0.978	1978	1.140
1971	1.010	1975	1.017	1979	1.172
1972	0.991	1976	0.986	1980	1.169
1973	1.000	1977	1.064	1981	1.199

79. The table shown here gives the unemployment rate as a percentage of the labor force in Canada over a period of 16 years. Draw a time plot and discuss the resulting picture.

YEAR	UNEMPLOYMENT RATE	YEAR	UNEMPLOYMENT RATE
1966	3.4	1974	5.3
1967	3.8	1975	6.9
1968	4.5	1976	7.1
1969	4.4	1977	8.1
1970	5.7	1978	8.4
1971	6.2	1979	7.5
1972	6.2	1980	7.5
1973	5.5	1981	7.6

80. Over the first 4 months of 1981, 73.6% of the Canadian automobile market went to cars made in North America, 20.9% to cars made in Japan, and 5.5% to automobiles manufactured in Europe. The corresponding percentages for the same period in 1982 were 68.4%, 25.6%, and 6.0%. Draw a pie chart for each of these years to depict this information.

probability

CHAPTER THREE

3.1 **INTRODUCTION**

In Chapter 1 we stressed the importance of the problem of making inferences about a population, based on observations drawn from a sample. The sample is taken in order to gain knowledge of the population, but will typically not produce certain knowledge. For example, a new product may be test-marketed in a limited number of retail outlets to get an assessment of consumer reaction. The results are used to form a judgment of the likely demand if the product were to be marketed nationally. Of course, based on this sample information, it is impossible to *know exactly* the reaction of the whole population; any measure of that reaction will inevitably involve *uncertainty*.

Although it is not possible, on the basis of a sample, to derive certain knowledge about a population, it may be possible to make precise statements about the nature of our uncertainty. Such statements are couched in the language of *probability,* which is therefore a concept of fundamental importance in statistical inference. It is also a notion frequently met in everyday life. For example, the "probability of precipitation" is an important element in daily weather forecasts; investment decisions are based on the investor's assessment of probable future returns; a baseball fan will use information such as past records and starting pitchers to form a judgment of the probability of a team's winning a particular game.

In this chapter a formal structure is developed for making probability statements and some basic results are derived. For purposes of exposition it is often simplest to illustrate the concepts by reference to games of chance, although their applicability is far broader and will be demonstrated in subsequent chapters.

In order to make statements about an uncertain environment, we need to develop a language. One can think of probability as the language in which we discuss uncertainty. Before we can communicate with one another in this language, we need to acquire a common vocabulary. Moreover, as in any other language, "rules of grammar" are needed so that clear statements can be made with our vocabulary. It will be necessary, therefore, to introduce a good deal of new terminology and to become acquainted with the manipulation of these terms in the production of probability statements.

3.2 RANDOM EXPERIMENT, OUTCOMES, EVENTS

Suppose that a process which could lead to two or more different outcomes is to be observed and there is uncertainty beforehand as to which outcome will occur. Some examples are the following:

1. A coin is thrown.
2. A die is rolled.
3. A voter is asked which candidate he or she prefers.
4. An item from a set of accounts is examined by an auditor.
5. The daily change in the price of gold is observed.
6. A batch of a chemical produced by a particular process is tested to determine whether it contains more than an allowable percentage of impurity.

Each of these examples involves a **random experiment.**

Definition

    A **random experiment** is a process leading to at least two possible outcomes with uncertainty as to which will occur.

In each of the first three experiments listed, it is possible to specify what outcomes might arise. If a coin is thrown the result will be either "head" or "tail." If a die is rolled the result will be one of the numbers 1, 2, 3, 4, 5, or 6. A voter might indicate a preference for the Republican candidate or the Democratic candidate or no preference. In each case the different possible outcomes, called **basic outcomes,** have been listed. The set of all these outcomes exhausts the possibilities, and is called the **sample space** of the random experiment.

Definition

    The possible outcomes of a random experiment are called the **basic outcomes** and the set of all basic outcomes is called the **sample space.**

Notice that basic outcomes are defined in such a way that no two can occur simultaneously; moreover, the random experiment must necessarily lead to the occurrence of one of the basic outcomes. The symbol S will be used to denote the sample space.

Example 3.1

A die is rolled. The basic outcomes are the numbers 1, 2, 3, 4, 5, 6. Thus, the sample space is

$$S = [1, 2, 3, 4, 5, 6]$$

Here we see that there are six basic outcomes. No two can occur together, and one of them must occur.

Example 3.2

An investor follows the stock market, and is particularly interested in the Dow-Jones industrial index. Consider the following two outcomes:

"At the close of trading today the Dow-Jones index is higher than at yesterday's close."

"At the close of trading today the Dow-Jones index is not higher than at yesterday's close."

One or the other of these outcomes must occur, but they cannot occur simultaneously. Therefore, these two outcomes together constitute a sample space.

Frequently, interest is not in the basic outcomes themselves but in some subset of all the outcomes in the sample space. For example, if a die is rolled an event that might be of interest is whether the resulting number is even—a result that will occur if one of the basic outcomes 2, 4, or 6 arises. Such sets of basic outcomes are called **events.**

Definition

An **event** is a set of basic outcomes from the sample space, and is said to *occur* if the random experiment gives rise to one of its constituent basic outcomes.

In many applications we are concerned simultaneously with two or more events. For example, if a die is thrown two events that might be considered are "The number resulting is even" and "The number resulting is at least 4." One possibility is that all of the events of interest might occur; this will be the case if the basic outcome of the random experiment belongs to all these events. The set of basic outcomes belonging to every event in a group of events is called the **intersection** of these events.

Definition

Let A and B be two events in the sample space S. Their **intersection,** denoted

---

$A \cap B$, is the set of all basic outcomes in S that belong to both A and B. Hence, the intersection $A \cap B$ occurs if and only if both A and B occur.

More generally, given K events E_1, E_2, . . . , E_K, their intersection, $E_1 \cap E_2 \cap \cdots \cap E_K$ is the set of all basic outcomes that belong to every E_i ($i = 1$, 2, . . . , K).

A useful pictorial mechanism for thinking about intersections and other set relations is the *Venn diagram*. Figure 3.1 shows diagrams for pairs of sets A and B. In part (a) of the figure, the rectangle S represents the sample space, while two closed figures denote the two events A and B. So, for example, a basic outcome belonging to A will be inside the corresponding figure. The shaded area where the figures intersect is $A \cap B$. Clearly, a basic outcome will be in $A \cap B$ if and only if it is in both A and B. Thus, in rolling a die, the outcomes 4 and 6 both belong to the two events "Even number results" and "Number at least 4 results."

It is possible that events A and B have no common basic outcomes, in which case the figures will not intersect, as in part (b) of Figure 3.1. Such events are said to be **mutually exclusive.** For example, if a set of accounts are audited, the events, "Less than 5% contain material errors" and "More than 10% contain material errors" are mutually exclusive.

Definition

If the events A and B have no common basic outcomes they are called **mutually exclusive** and their intersection $A \cap B$ is said to be the **empty set.** It follows then that $A \cap B$ cannot occur.

More generally, the K events $E_1, E_2, \ldots , E_K$ are said to be mutually exclusive if every pair of them is a pair of mutually exclusive events—that is, if $E_i \cap E_j$ is the empty set for all $i \neq j$.

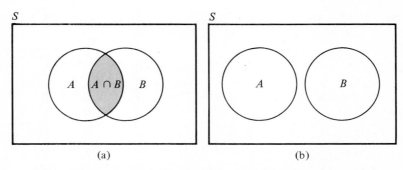

(a)   (b)

FIGURE 3.1 Venn diagrams for the intersection of events A and B:
(a) $A \cap B$ is the shaded area; (b) A and B are mutually exclusive

When considering jointly several events, another possibility of interest is that at least one of them will occur. This will happen if the basic outcome of the random experiment belongs to at least one of the events. The set of basic outcomes belonging to at least one of the events is called their **union.** For example, in the die throw experiment, the outcomes 2, 4, 5, and 6 all belong to at least one of the events "Even number results" or "Number at least 4 results."

Definition

Let A and B be two events in the sample space S. Their **union,** denoted $A \cup B$, is the set of all basic outcomes in S that belong to at least one of these two events. Hence, the union $A \cup B$ occurs if and only if either A or B (or both) occurs.

More generally, given K events $E_1, E_2, \ldots, E_K$, their union $E_1 \cup E_2 \cup \cdots \cup E_K$ is the set of all basic outcomes belonging to at least one of these K events.

The union of a pair of events is illustrated in the Venn diagram in Figure 3.2, from which it is clear that a basic outcome will be in $A \cup B$ if and only if it is in either A or B (or both).

A case of special interest concerns a collection of several events whose union is the whole sample space S. Since every basic outcome is always contained in S, it follows that every outcome of the random experiment will be in at least one of this collection of events. These events are then said to be **collectively exhaustive.** For example, if a die is thrown, the events "The result is at least 3" and "The result is at most 5" are together collectively exhaustive—at least one of these two events must occur.

Definition

Let $E_1, E_2, \ldots, E_K$ be K events in the sample space S. If $E_1 \cup E_2 \cup \cdots \cup E_K = S$, then these K events are said to be **collectively exhaustive.**

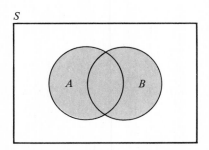

FIGURE 3.2 Venn diagram for the union of events A and B; $A \cup B$ is the shaded area

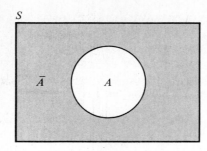

S

FIGURE 3.3 Venn diagram for the complement of event A; $\bar{A}$ is the shaded area

Using the terminology just introduced, it follows that the set of all basic outcomes contained in a sample space are both mutually exclusive and collectively exhaustive. We have already noted that these outcomes are such that one must occur, but no more than one can simultaneously occur.

Next, let A be an event, and suppose our interest is that A not occur. This will happen if the basic outcome of the random experiment lies in S (as it must), but *not* in A. The set of basic outcomes belonging to the sample space but not to a particular event is called the **complement** of that event, and is denoted $\bar{A}$. Clearly the events A and $\bar{A}$ are mutually exclusive (no basic outcome can belong to both) and collectively exhaustive (every basic outcome must belong to one or the other). The complement of the event A is illustrated on the Venn diagram in Figure 3.3.

Definition

Let A be an event in the sample space S. The set of basic outcomes of a random experiment belonging to S but not to A is called the **complement** of A, and denoted $\bar{A}$.

We have now met three important concepts—the intersection, the union, and the complement. All will prove useful in our subsequent discussions of probability. The following examples should serve to make these ideas more concrete.

Example 3.3

A die is rolled. Let A be the event "Number resulting is even" and B be the event "Number resulting is at least 4." Then

$$A = [2, 4, 6] \qquad \text{and} \qquad B = [4, 5, 6]$$

The complements of these events are respectively

$$\bar{A} = [1, 3, 5] \qquad \text{and} \qquad \bar{B} = [1, 2, 3]$$

The intersection of A and B is the event "Number resulting is both even and at least 4," so that

$$A \cap B = [4, 6]$$

The union of A and B is the event "Number resulting is either even or at least 4, or both" and so

$$A \cup B = [2, 4, 5, 6]$$

Note also that the events A and $\bar{A}$ are mutually exclusive, since their intersection is the empty set, and collectively exhaustive, since their union is the sample space S, that is,

$$A \cup \bar{A} = [1, 2, 3, 4, 5, 6] = S$$

The same statements also apply for the events B and $\bar{B}$.

Example 3.4

Consider the observation of the Dow-Jones industrial average over two consecutive days. We will designate the four basic outcomes as follows:

O_1: Dow-Jones average rises on both days.
O_2: Dow-Jones average rises on the first day, but does not rise on the second day.
O_3: Dow-Jones average does not rise on the first day, but rises on the second day.
O_4: Dow-Jones average does not rise on either day.

Clearly, one of these outcomes must occur but not more than one can occur at the same time. We can therefore write the sample space as $S = [O_1, O_2, O_3, O_4]$.

Now, let us consider the two events

A: Dow-Jones average rises on the first day.
B: Dow-Jones average rises on the second day.

We see that the event A occurs if either basic outcome O_1 or O_2 occurs, so that we can write $A = [O_1, O_2]$. Similarly, we have $B = [O_1, O_3]$.

The intersection of A and B is the event "Dow-Jones average rises on the first day and rises on the second day." This is the set of all basic outcomes belonging to both A and B, so that $A \cap B = [O_1]$.

The union of A and B is the event "Dow-Jones average rises on at least one of the two days." This is the set of all basic outcomes belonging to either A or B, or both. It follows that $A \cup B = [O_1, O_2, O_3]$.

Finally, the complement of A is the event "Dow-Jones average does not rise on the first day." This is the set of all basic outcomes in the sample space S that do not belong to A. Hence, $\bar{A} = [O_3, O_4]$.

It is also possible to examine other unions or intersections involving event complements. To illustrate, the intersection of the events $\bar{A}$ and B is shown in Figure 3.4. This intersection contains all those outcomes that are in both $\bar{A}$ (that is, not in A) and B. In Example 3.3 the intersection of these two events is $\bar{A} \cap B = [5]$, as the only outcome that is both "Not even" and "At least 4" is 5.

We now introduce three results involving unions and intersections of sets.

(i) Let A and B be two events. Then the events $A \cap B$ and $\bar{A} \cap B$ are mutually exclusive and their union is B, as illustrated in the Venn diagram in Figure 3.5. Clearly,

$$(A \cap B) \cup (\bar{A} \cap B) = B$$

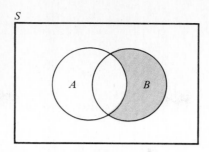

FIGURE 3.4 Venn diagram for the intersection of $\bar{A}$ and B; $\bar{A} \cap B$ is the shaded area

(ii) Let A and B be two events. Then the events A and $\bar{A} \cap B$ are mutually exclusive and their union is $A \cup B$. Again this result can best be seen from inspection of the Venn diagram, shown in Figure 3.6. It is clear from this figure that

$$A \cup (\bar{A} \cap B) = A \cup B$$

(iii) Let $E_1, E_2, \ldots , E_K$ be K mutually exclusive and collectively exhaustive events, and let A be some other event. Then the K events $E_1 \cap A, E_2 \cap A, \ldots , E_K \cap A$ are mutually exclusive and their union is A.

To see the truth of the third statement, consider the Venn diagram in Figure 3.7. The large rectangle denoting the whole sample space is subdivided into smaller rectangles, depicting the K mutually exclusive and collectively exhaustive events E_1, $E_2, \ldots , E_K$. The event A is represented by the closed figure. We see that the events comprised of the intersection of A and each of the E_i are indeed mutually exclusive, and that their union is simply the event A. We can therefore write

$$(E_1 \cap A) \cup (E_2 \cap A) \cup \cdots \cup (E_K \cap A) = A$$

Example 3.5
The truth of results (i) and (ii) is established for the die rolling experiment of Example 3.3. As before, let

$$A = [2, 4, 6]; \qquad B = [4, 5, 6]; \qquad \bar{A} = [1, 3, 5]$$

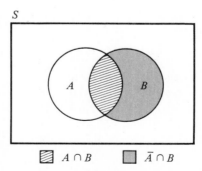

$\boxtimes \; A \cap B$ $\blacksquare \; \bar{A} \cap B$

FIGURE 3.5 Venn diagram for $A \cap B$ and $\bar{A} \cap B$

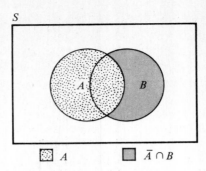

$$A \qquad \bar{A} \cap B$$

FIGURE 3.6 Venn diagram for A and $\bar{A} \cap B$

Then,

$$A \cap B = [4, 6] \qquad \text{and} \qquad \bar{A} \cap B = [5]$$

Then $A \cap B$ and $\bar{A} \cap B$ are mutually exclusive and their union is

$$B = [4, 5, 6]$$

Also, A and $\bar{A} \cap B$ are mutually exclusive and their union is

$$A \cup B = [2, 4, 5, 6]$$

Example 3.6

The truth of result (iii) is established for another die rolling example. Define

$$A = [2, 4, 6]; \qquad E_1 = [1, 2]; \qquad E_2 = [3, 4]; \qquad E_3 = [5, 6]$$

so that E_1, E_2, E_3 are mutually exclusive and collectively exhaustive. Then

$$E_1 \cap A = [2]; \qquad E_2 \cap A = [4]; \qquad E_3 \cap A = [6]$$

Clearly these three events are mutually exclusive and their union is

$$(E_1 \cap A) \cup (E_2 \cap A) \cup (E_3 \cap A) = [2, 4, 6] = A$$

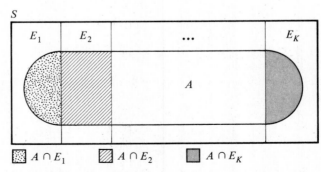

$$A \cap E_1 \qquad A \cap E_2 \qquad A \cap E_K$$

FIGURE 3.7 Venn diagram for $A \cap E_1, A \cap E_2, \ldots, A \cap E_K$

Example 3.7

A problem often faced in market research is that some of the questions we would like to ask are so sensitive that many subjects will either refuse to reply or will give a dishonest answer. One way of attacking this problem is through the method of *randomized response*.[1] This technique involves pairing the sensitive question with a nonsensitive question. For instance, we might have the following pair:

(a) Have you purposely shoplifted in the last 12 months?
(b) Have you made a purchase from a catalog in the last 12 months?

Subjects are asked to flip a coin and then to answer question (a) if the result is "head" and (b) otherwise. Since the investigator cannot *know* which question is answered, it is hoped that honest responses will be obtained in this way. The nonsensitive question is one for which the investigator already has information about the population under study. Thus, in our example, the investigator knows what proportion of the population made a purchase from a catalog in the last 12 months. (Later in this chapter we will see how the responses can be analyzed to produce the information required.)

Now, we define the following events:

A: Subject answers "Yes."
E_1:   Subject answers sensitive question.
E_2:   Subject answers nonsensitive question.

Clearly, the events E_1 and E_2 are mutually exclusive and collectively exhaustive. Thus, the conditions of result (iii) are satisfied and it follows that the events

$A \cap E_1$:   Subject both responds "yes" and has answered the sensitive question

and

$A \cap E_2$:   Subject both responds "yes" and has answered the nonsensitive question

are mutually exclusive. Furthermore, their union must be the event A; that is,

$$A = (A \cap E_1) \cup (A \cap E_2)$$

3.3 WHAT IS PROBABILITY?

Suppose a random experiment is to be carried out and we are interested in the chance of a particular event occurring. The concept of probability is intended to provide a numerical measure for the likelihood of an event's occurrence. Probability is measured on a scale from 0 to 1. At the extremes of this range, a probability of 0 implies that

---

[1] See, for example, M. D. Geurts, "Using a randomized response research design to eliminate nonresponse biases in business research," *Journal of Academy of Marketing Science*, 8 (1980), 83–90.

the event is impossible (it is certain not to occur), while a probability of 1 implies that the event is certain to occur. For uncertain events, we want to attach a probability between 0 and 1 such that the more likely the event is to occur, the higher the probability. In practice, such ideas are frequently met. It is known that rain is more likely in certain meteorological conditions than others. An experienced manager may judge that one product is more likely to achieve substantial market penetration than another.

To take a very simple example, suppose a coin is thrown. The statement "The probability that a head results is ½" may be viewed through two distinct ideas—*relative frequency* and *subjective probability*.

RELATIVE FREQUENCY

Suppose that a random experiment can be replicated in such a way that, after each trial, it is possible to return to the initial state and repeat the experiment so that the resulting outcome is unaffected by previous outcomes. For example, a coin or die can be thrown repeatedly in this way.

If some number N of experiments are conducted and the event A occurs in N_A of them (N_A clearly depending on N) then we have

$$\text{Proportion of occurrences of } A \text{ in } N \text{ trials} = \frac{N_A}{N}$$

Now, if N is very large, we would not expect much variation in the proportion N_A/N as N increases; that is, the proportion of occurrences of A will remain approximately constant. This notion underlies the **relative frequency** concept of probability.

Definition

   Let N_A be the number of occurrences of event A in N repeated trials. Then, under the **relative frequency** concept of probability, the probability that A occurs is the limit of the ratio N_A/N as the number of trials N becomes infinitely large.

Under this definition, if we say "The probability of a head resulting from a single throw of a coin is ½," we mean that if the coin is thrown repeatedly, the proportion of heads resulting will get very close to ½ as the number of trials gets very large.

The relative frequency notion provides a convenient framework for thinking about probability, but it does involve conceptual difficulties. These are illustrated, in increasing order of magnitude, in the following examples:

(a) Do we really have to throw a coin a very large number of times before concluding that the probability of a head resulting is ½? Certainly, this could be done, but it would be a very tedious business indeed if all probability assessments had to be made in this fashion. One way around the difficulty is to regard repeated experimentation as a purely conceptual notion, without requiring that the experiments actually be carried out. For example, it might be reasonable to conclude that the coin appears to be perfectly fair, so that the two outcomes, "Head" and "Tail," are equally likely. Hence, one might reasonably infer that, if the coin were thrown repeatedly, the proportion of heads would approach ½. (Notice that, in deciding that "Head" and "Tail" are equally likely, we have introduced an element of subjectivity.) More generally, suppose a random experiment can lead to n mutually exclusive and collectively exhaustive possible basic outcomes, where each basic outcome is equally likely. If the event A occurs in n_A of these outcomes, it could be inferred that the probability of A occurring is n_A/n. For example, if a fair die is rolled there are six basic outcomes, each equally likely. The event "An even number results" involves three of these outcomes, so that its probability is $\frac{3}{6} = \frac{1}{2}$. Unfortunately, it is not always the case that the sample space of an experiment is made up of equally likely basic outcomes.

(b) A meteorologist announces that the probability of rain today is 0.7. Now, once today has ended, it is impossible to go back to its beginning and start again from the same initial conditions. However, in assessing a probability for rain, the meteorologist could argue that meteorological conditions essentially the same as those prevailing at the time the forecast was made had been experienced many times in the past and rain had resulted on 70% of those occasions. Thus, a long-run frequency interpretation of the meteorologist's probability statement could still be made. Similarly, we might reasonably assert that the probability that the Dow-Jones index will close higher than yesterday is ½, because historically the market has risen 50% of all trading days.

(c) The 1984 Kentucky Derby is to be run. A bettor is interested in a particular horse and concludes that the probability that it will win the race is 0.4. In contrast to the previous example, essentially similar races involving essentially similar horses will not have been run frequently in the past, and certainly the bettor's probability judgment will not have been based on such a notion. It seems more reasonable to suggest that, taking various relevant factors (such as past performance of the horse) into account, the bettor has formed a personal subjective judgment as to the chance of the horse winning, and this is reflected in the probability statement. In a similar vein, corporate executives must often face decisions as to whether to make potentially lucrative investments in countries that have unstable political climates. Either formally or informally, it is necessary to enter into the decision-making process some assessment of the likelihood of a revolution, resulting in the nationalization of the corporation's assets and the receipt of unfavorable compensation terms for such a take-over. Such an assessment must surely be subjective.

SUBJECTIVE PROBABILITY

An alternative view, which does not depend on the notion of repeatable experiments, regards probability as a personal subjective concept, expressing an individual's degree of belief about the chance that an event will occur. One way to understand this idea is in terms of *fair bets*.

For example, if I assert that the probability of a head resulting from the throw of a coin is ½, what I have in mind is that the coin appears to be perfectly fair and that the throw is just as likely to produce a head as a tail. In assessing this subjective probability, I am not necessarily thinking in terms of repeated experimentation, but am concerned with only a single throw of the coin. My subjective probability assessment implies that I would view as fair a bet in which I had to pay $1 if the result was tail, and would receive $1 if the result was head. If I were to receive more than $1 if the throw yielded a head, I would regard the bet as in my favor. Similarly, if I believe that the probability of a horse winning a particular race is 0.4, I am asserting the personal view that there is a 40–60 chance of its winning. Given this belief I would regard as fair a bet in which I lost $2 if the horse did not win and gained $3 if it did.

It should be emphasized that subjective probabilities are personal; there is no requirement that different individuals considering the same event should arrive at the same probabilities. In the coin throwing example, most people will conclude that the appropriate probability for a head is ½. However, an individual with more information about the coin in question might believe otherwise. In the example of the horse race it is likely that two bettors will reach different subjective probabilities. They may not, for example, have the same information, and even if they do they might not interpret it in the same way. It is certainly clear that individual investors do not all hold the same views on the likely future behavior of the stock market! Their subjective probabilities might be thought of as depending on the knowledge they have, and the way they interpret it.

3.4 PROBABILITY AND ITS POSTULATES

It is necessary to develop a framework in which probabilities can be assessed or manipulated. In order to do this, we first set down three rules (or postulates) that probabilities will be required to obey, and show that these requirements are "reasonable."

PROBABILITY POSTULATES

Let S denote the sample space of a random experiment, O_i the basic outcomes, and A an event. Then, using the notation $P(A)$ for "Probability event A occurs," we have the requirements stated in the box.

---

1. If A is any event in the sample space S,

$$0 \leq P(A) \leq 1$$

2. Let A be an event in S, and let O_i denote the basic outcomes. Then

$$P(A) = \sum_A P(O_i)$$

where the notation implies that the summation extends over all the basic outcomes in A.

3. $$P(S) = 1$$

The first postulate simply requires that a probability lie between 0 and 1. The second postulate can be motivated in terms of relative frequencies. Suppose a random experiment is repeated N times. Let N_i be the number of times the basic outcome O_i occurs and N_A the number of times event A occurs. Then, since the basic outcomes are mutually exclusive, N_A is just the sum of the N_i for all the basic outcomes in A; that is,

$$N_A = \sum_A N_i$$

and, on dividing by the number of trials N, we obtain

$$\frac{N_A}{N} = \sum_A \frac{N_i}{N}$$

But, under the relative frequency concept of probability, N_A/N tends to $P(A)$ and each N_i/N tends to $P(O_i)$ as N becomes infinitely large. Thus, the second postulate can be seen as a logical requirement when probability is viewed in this way. The third postulate can be paraphrased as "When a random experiment is to be carried out, something has to happen." Replacing A by the sample space S in the second postulate gives

$$P(S) = \sum_S P(O_i)$$

where the summation extends over all the basic outcomes in the sample space. But since $P(S) = 1$ by the third postulate, it follows that

$$\sum_S P(O_i) = 1 \qquad (3.4.1)$$

That is, the sum of the probabilities for all the basic outcomes in the sample space is 1.

CONSEQUENCES OF THE POSTULATES

We now list and illustrate some immediate consequences of the three postulates.

(i) If the sample space S consists of n equally likely basic outcomes $O_1, O_2, \ldots, O_n$, then each of these has probability $1/n$; that is,

$$P(O_i) = \frac{1}{n} \qquad (i = 1, 2, \ldots, n)$$

The first consequence follows from equation 3.4.1: If $P(O_i)$ is the same for each basic outcome and $\sum_{i=1}^{n} P(O_i) = 1$, then $P(O_i) = 1/n$ for each outcome. For example, if a fair die is rolled, then the probability for each of the six basic outcomes is $1/6$.

(ii) If the sample space S consists of n equally likely basic outcomes and the event A consists of n_A of these outcomes, then

$$P(A) = \frac{n_A}{n}$$

This follows from consequence (i) and the second postulate. Every basic outcome has probability $1/n$ and, by postulate 2, $P(A)$ is just the sum of the probabilities (each $1/n$) of the n_A basic outcomes in A. For example, if a fair die is rolled and A is the event "An even number results," there are $n = 6$ basic outcomes and $n_A = 3$ of these are in A. Hence, $P(A) = 3/6 = 1/2$. Notice that this result agrees with our intuitive reasoning in the previous section.

(iii) Let A and B be mutually exclusive events. Then the probability of their union is the sum of their individual probabilities; that is,

$$P(A \cup B) = P(A) + P(B)$$

More generally, if $E_1, E_2, \ldots, E_K$ are mutually exclusive events,

$$P(E_1 \cup E_2 \cup \cdots \cup E_K) = P(E_1) + P(E_2) + \cdots + P(E_K)$$

This result is a consequence of the second postulate. The probability of the union of A and B is

$$P(A \cup B) = \sum_{A \cup B} P(O_i) \qquad (3.4.2)$$

where the summation extends over all the basic outcomes in $A \cup B$. But, since A and B are mutually exclusive, no basic outcome can belong to both, so that the right-hand side of equation 3.4.2 can be broken down into the sum of two parts:

$$\sum_{A \cup B} P(O_i) = \sum_A P(O_i) + \sum_B P(O_i)$$

The right-hand side of this equation is just $P(A) + P(B)$, by postulate 2. The more general result follows through similar reasoning.

(iv) If $E_1, E_2, \ldots, E_K$ are collectively exhaustive events, the probability of their union is

$$P(E_1 \cup E_2 \cup \cdots \cup E_K) = 1$$

Since the events are collectively exhaustive, their union is the whole sample space S and the result follows from the third postulate.

Example 3.8

Suppose one card is drawn from a full deck of fifty-two cards. Let A be the event "The card is a heart" and B the event "The card is a diamond."

The sample space consists of 52 equally likely basic outcomes, of which 13 are contained in event A. Hence, $P(A) = 13/52 = 1/4$. Similarly, $P(B) = 1/4$. Now, A and

B are mutually exclusive events (the card cannot be both a heart and a diamond) and their union is the event "The card drawn is red." Hence,

$$P(A \cup B) = P(A) + P(B) = \frac{1}{4} + \frac{1}{4} = \frac{1}{2}$$

This result could have been deduced directly from the fact that half the cards in the deck are red.

Example 3.9

A charitable organization sells 1,000 lottery tickets. There are 10 major prizes and 100 minor prizes, all of which must be won. The process for choosing winners is such that, at the outset, each ticket has an equal chance of winning a major prize, and each has an equal chance of winning a minor prize. No ticket can win more than one prize. What is the probability of winning a major prize with a single ticket? What is the probability of winning a minor prize? What is the probability of winning *some* prize?

This problem can be viewed in exactly the same way as that of the previous example. Of the 1,000 tickets, 10 will win major prizes, 100 will win minor prizes, and 890 will win no prize. Our single ticket can be regarded as one selected from the 1,000. Let A be the event "Selected ticket wins major prize." Since there are 1,000 equally likely outcomes, 10 of which correspond to event A, we have

$$P(A) = \frac{10}{1,000} = 0.01$$

Similarly, for event B, "Selected ticket wins minor prize," it follows that

$$P(B) = \frac{100}{1,000} = 0.10$$

Now, the event "Ticket wins some prize" is simply the union of the events A and B. Moreover, since only one prize per ticket is permitted, these events are mutually exclusive. It follows that the probability required is

$$P(A \cup B) = P(A) + P(B) = 0.01 + 0.10 = 0.11$$

Example 3.10

In Example 3.4, we considered the course of the Dow-Jones average over two days, and defined the four basic outcomes

O_1: Average rises on both days.
O_2: Average rises on first day, but does not rise on second day.
O_3: Average does not rise on first day, but does rise on second day.
O_4: Average does not rise on either day.

It is reasonable to assert that these four basic outcomes are equally likely. In that case, what is the probability that the market will rise on at least one of the two days?

The event of interest, "Market rises on at least one of the two days," contains three of the four basic outcomes—O_1, O_2, O_3. Since the basic outcomes are all equally likely, it follows that the probability of this event is ¾.

1. A builder's probability assessment of the number of days in which a project will be completed is shown in the table. Let A be the event "The project will take at most 2 days to complete" and B the event "The project will take either 3 or 4 days to complete."

NUMBER OF DAYS	1	2	3	4	5
PROBABILITY	0.10	0.25	0.40	0.20	0.05

   (a) Find the complement of event A.
   (b) Find the union of the events A and B. Are these events collectively exhaustive?
   (c) Are the events A and B mutually exclusive?
   (d) Find the individual probabilities of the events A and B.

2. An investment fund manager is considering investment in a portfolio of energy stocks. The manager's assessment of probabilities for this portfolio over the next year are summarized in the accompanying table. Let A be the event "The portfolio will increase in value" and B the event "The portfolio will decrease in value."

EVENT	PROBABILITY
Increase in value by more than 20%	0.15
Increase in value by 10%–20%	0.35
Increase in value by 0%–10%	0.30
Decrease in value by 0%–10%	0.15
Decrease in value by 10%–20%	0.05

   (a) Are the events A and B mutually exclusive?
   (b) Are the events A and B collectively exhaustive?
   (c) Find the individual probabilities of the events A and B.
   (d) Find the probability that the portfolio will increase in value by at least 10%.

3. A contractor plans to submit a bid for a new project. Before doing so, he forms a probabilistic assessment as to what the lowest bid of his competitors is likely to be. These assessments are given in the table. Let A be the event "Competitors' lowest bid is at least $150,000" and B the event "Competitors' lowest bid is at most $175,000."

COMPETITORS' LOWEST BID	PROBABILITY
Over $200,000	0
$175,000–$200,000	0.15
$150,000–$175,000	0.45
$125,000–$150,000	0.30
$100,000–$125,000	0.10
Under $100,000	0

(a) Are the events A and B mutually exclusive?

(b) Are the events A and B collectively exhaustive?

(c) Find the individual probabilities for events A and B.

(d) What is the complement of the event B? Find its probability.

4. A company receives shipments of machine tools. Past experience indicates that 3% of the tools have serious defects that render them worthless. An additional 9% have minor defects that can easily be repaired, while the remainder have no defects.

(a) If a tool is selected at random from the shipment, what is the probability that it has some defect?

(b) What is the probability that it has no defect?

5. In the early stages of the development of the Hibernia oil site in the Atlantic Ocean, the Petroleum Directorate of Newfoundland estimated the probability that economically recoverable reserves exceeded 2 billion barrels to be 0.1. The probability for reserves in excess of 1 billion barrels was estimated to be 0.5, while the probability that reserves exceeded 500,000 barrels was given as 0.9.

(a) Find the probability that reserves are below 500,000 barrels.

(b) Find the probability that reserves are between 500,000 and 1 billion barrels.

(c) Find the probability that reserves are between 1 and 2 billion barrels.

6. The accompanying table shows probabilities for the number of breakdowns of a machine in a week.

NUMBER OF BREAKDOWNS	0	1	2	3	4 or more
PROBABILITY	0.21	0.37	0.20	0.12	0.10

(a) What is the probability that there will be fewer than two breakdowns in a particular week?

(b) What is the probability that there will be more than two breakdowns in a particular week?

3.5 PERMUTATIONS AND COMBINATIONS

A practical difficulty that sometimes arises in computing the probability of an event is counting the numbers of outcomes in the sample space and the event of interest. For some problems the use of *permutations* or *combinations* can be helpful.

We begin with the problem of ordering. Suppose that we have some number x of objects that are to be placed in order. Each object may be used only once. How many different sequences are possible? We can view this problem as a requirement to place one of the objects in each of x boxes arranged in a row, as illustrated in Figure 3.8. Beginning with the first box, there are x different ways to fill it. Once an object is put in that box, there are $(x - 1)$ objects remaining, and so $(x - 1)$ ways to fill the second box. That is, for each of the x ways to place an object in the first box, there are $(x - 1)$ possible ways to fill the second box, so that the first two boxes can be filled in a total of $x(x - 1)$ ways. Given that the first two boxes are filled, there are now $(x - 2)$ ways of filling the third box, so that the first three boxes can be filled in a total

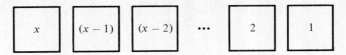

FIGURE 3.8 The orderings of x objects

of $x(x-1)(x-2)$ ways. Finally, when we arrive at the last box there is only one object left to put in it. Hence, the total number of possible orderings is $x(x-1)(x-2) \cdots 2 \cdot 1$, which for notational convenience is written $x!$ (read "x factorial").

The number of possible orderings of x objects is
$$x(x-1)(x-2) \cdots 2 \cdot 1 = x!$$

Example 3.11

The three letters A, B, C can be arranged in $3! = 6$ different orders. These are:

ABC ACB BAC BCA CAB CBA

This example is illustrated in the *tree diagram* of Figure 3.9. We begin at the intersection on the left-hand side of the figure by choosing one of the three letters to fill the first position. Following each of the emerging branches, we then have two

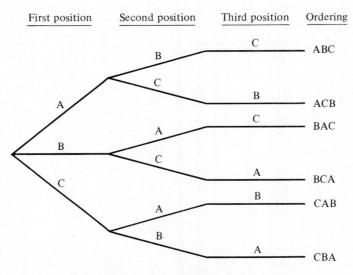

FIGURE 3.9 Tree diagram for Example 3.11

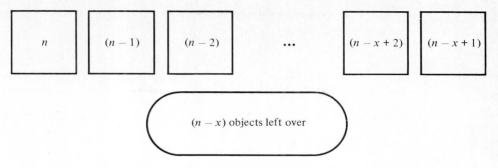

FIGURE 3.10 The permutations of x objects chosen from n

possibilities for filling the second position. For example, if the letter A is in the first position, either B or C must be placed in the second position. Finally, once the first two positions have been filled, there is just one letter available to put in the final position. On the right-hand side of the figure we show the six possible orderings achieved in this way.

Example 3.12
A consumer is asked to rank, in order of preference, the taste of five brands of beer. If the consumer is in fact indifferent among these brands, what is the probability that a specific ordering will be selected by chance?

There are $5! = 120$ different possible orderings. Thus, the probability of selecting any particular one, if each is equally likely to be picked, is $\frac{1}{120}$.

Suppose now that we have a number n of objects with which the x ordered boxes could be filled (with $n > x$). Each object may be used only once. The number of possible orderings is called the number of **permutations** of x objects chosen from n, and is denoted by the symbol $_nP_x$. Now, we can argue precisely as before, except that there will be n ways to fill the first box, $(n - 1)$ ways to fill the second box, and so on, until we come to the final box. At this point there will be $(n - x + 1)$ objects left, each of which could be placed in that box, as illustrated in Figure 3.10. Thus, the total number of permutations is

$$_nP_x = n(n - 1)(n - 2) \cdots (n - x + 1)$$

A more convenient expression is obtained by multiplying and dividing by $(n - x)(n - x - 1) \cdots 2 \cdot 1 = (n - x)!$ giving

$$_nP_x = \frac{n(n - 1)(n - 2) \cdots (n - x + 1)(n - x)(n - x - 1) \cdots 2 \cdot 1}{(n - x)(n - x - 1) \cdots 2 \cdot 1} = \frac{n!}{(n - x)!}$$

Definition

The number of **permutations**, $_nP_x$, of x objects chosen from n is the number of possible arrangements when x objects are to be selected from a total of n and arranged in order. This number is

$$_nP_x = \frac{n!}{(n-x)!}$$

Example 3.13

Suppose two letters are to be selected from A, B, C, D, E, and arranged in order. The number of permutations, with $n = 5$ and $x = 2$, is $n!/(n-x)! = 5!/3! = 20$. These are

AB	AC	AD	AE	BC
BA	CA	DA	EA	CB
BD	BE	CD	CE	DE
DB	EB	DC	EC	ED

We illustrate this example in the tree diagram of Figure 3.11. At the left-hand intersection, there are five possible letters to put in the first position. Now, following any of the emerging branches, we have four letters from which to fill the second position. For instance, if A is in the first position, then B, C, D, or E can be placed in the second position. Thus, we see on the right-hand side of the figure that there are twenty possible permutations.

Example 3.14

Refer to Example 3.12, where five beers are ranked in order of preference. What is the probability that an individual who is truly indifferent will select a specific ordering for the first two places?

The number of permutations is $_5P_2 = 5!/(5-2)! = 5!/3! = 20$. Thus, the probability of selecting any one of these, if each is equally likely, is $\frac{1}{20}$.

Finally, suppose we are interested in the number of different ways that x objects can be selected from n (where no object may be chosen more than once), but are not concerned about the order. In Example 3.13, there are ten possibilities for selecting two objects from a group of five. Notice that the entries in the second and fourth rows are just rearrangements of those directly above them, and thus are ignored. The number of possible selections is called the number of **combinations,** and is denoted $_nC_x$, where x objects are to be chosen from n. If this choice is to be made, note first that the number of possible permutations is $_nP_x$. However, many of these will be rearrangements of the same x objects, and so are irrelevant. In fact, since x objects can be ordered in $x!$ ways, we are concerned with only a proportion $1/x!$ of the permutations; that is, the number of combinations is

$$_nC_x = \frac{_nP_x}{x!} = \frac{n!}{x!\,(n-x)!}$$

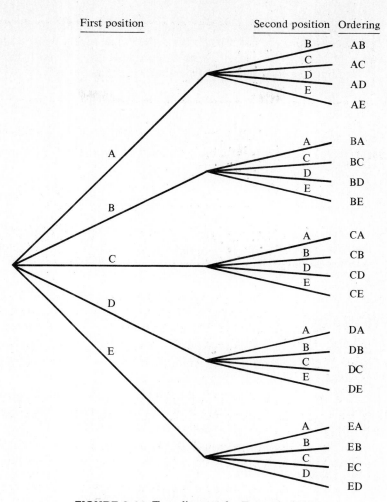

FIGURE 3.11 Tree diagram for Example 3.13

Definition

The number of **combinations**, $_nC_x$, of x objects chosen from n is the number of possible selections that can be made. This number is

$$_nC_x = \frac{n!}{x!\,(n-x)!}$$

Example
3.15
A personnel officer has available eight candidates to fill four positions. Five candidates are men, and three women. If, in fact, every combination of candidates is equally likely to be chosen, what is the probability that no women will be hired?

First, the total number of possible combinations of candidates is

$$_8C_4 = \frac{8!}{4! \ 4!} = 70$$

Now, in order for no women to be hired, it follows that the four successful candidates must come from the available five men. The number of such combinations is

$$_5C_4 = \frac{5!}{4! \ 1!} = 5$$

Therefore, if at the outset each of the 70 possible combinations was equally likely to be chosen, the probability that one of the 5 all-male combinations would be selected is $5/70 = 1/14$.

3.6 PROBABILITY RULES

It is often the case that our interest centers on some compound of events, such as the union or intersection. In this section we will develop rules for computing probabilities for events of this kind.

First, let A be an event in the sample space S. We have noted already that A and its complement $\overline{A}$ are mutually exclusive and collectively exhaustive. Hence, by result (iii) of Section 3.4, since A and $\overline{A}$ are mutually exclusive,

$$P(A \cup \overline{A}) = P(A) + P(\overline{A})$$

and by result (iv) of that section, since A and $\overline{A}$ are collectively exhaustive,

$$P(A \cup \overline{A}) = 1$$

Putting these equations together yields

$$P(A) + P(\overline{A}) = 1 \qquad \text{or} \qquad P(\overline{A}) = 1 - P(A)$$

so that the probability that an event does *not* occur is 1 minus the probability that it *does* occur. For example, when a die is rolled the probability of getting a 1 is $1/6$, so the probability that the result is not 1 is $1 - 1/6 = 5/6$.

Let A be an event and $\overline{A}$ its complement. Then
$$P(\overline{A}) = 1 - P(A)$$

Example 3.16

In Example 3.15, there were five men and three women available for four positions. Assuming again that every combination is equally likely to be chosen, what is the probability that at least one woman will be selected?

Let A be the event "No women are selected." The event "At least one woman is selected" is then $\bar{A}$, the complement of A. In Example 3.15 we found $P(A) = \frac{1}{14}$. Therefore, the required probability is

$$P(\bar{A}) = 1 - P(A) = 1 - \frac{1}{14} = \frac{13}{14}$$

Now let A and B be two events. In Section 3.4 we showed that if A and B are mutually exclusive, the probability of their union is the sum of their individual probabilities. We now want to find the probability of the union when the events are not mutually exclusive. Two results derived in Section 3.2 will be used. First, the events $(A \cap B)$ and $(\bar{A} \cap B)$ are mutually exclusive and their union is B. Hence,

$$P(B) = P(A \cap B) + P(\bar{A} \cap B) \qquad (3.6.1)$$

Also, the events A and $(\bar{A} \cap B)$ are mutually exclusive and their union is $A \cup B$, so that

$$P(A \cup B) = P(A) + P(\bar{A} \cap B) \qquad (3.6.2)$$

Eliminating $P(\bar{A} \cap B)$ from equations 3.6.1 and 3.6.2 then gives

$$P(A \cup B) = P(A) + P(B) - P(A \cap B)$$

This is called the **addition rule of probabilities.** Notice that the rule implies that the probability of a union is *not* the sum of the individual probabilities, unless the events are mutually exclusive.

Addition Rule of Probabilities

Let A and B be two events. Then the probability of their union is
$$P(A \cup B) = P(A) + P(B) - P(A \cap B)$$

Example 3.17

A hamburger chain found that 75% of all customers use mustard, 80% use ketchup, and 65% use both, when ordering a hamburger. What is the probability that a particular customer will use at least one of these?

Let A be the event "Customer uses mustard" and B the event "Customer uses ketchup." From the statement of the example, we then have

$$P(A) = 0.75; \qquad P(B) = 0.80; \qquad P(A \cap B) = 0.65$$

The probability required is

$$P(A \cup B) = P(A) + P(B) - P(A \cap B)$$
$$= 0.75 + 0.80 - 0.65 = 0.90$$

Suppose that we are interested in a pair of events A and B, and are given the extra piece of information that event B has occurred. A question of interest then is, what is the probability that A occurs, *given that B has occurred?* This type of problem can be approached through the notion of **conditional probability.** The basic idea is that the chance of any event occurring is likely to depend on whether or not other events occur. For example, a manufacturer planning to introduce a new brand may test-market the product in a few selected stores. This manufacturer is likely to be much more confident about the brand's success in the wider market if it is well accepted in the test market than if it is not. The firm's assessment of the probability of high sales will therefore be conditioned by the test market outcome.

Again, if I knew that interest rates were going to fall over the next year, I would be far more bullish about the stock market than if I believed that interest rates would rise. Once more, my probabilistic assessment of the likely course of stock prices is conditioned by what I know, or believe, about interest rates. We must therefore be concerned about the probability of occurrence of a particular event, given the occurrence of another.

Definition

Let A and B be two events. The **conditional probability** of event A, given event B, denoted $P(A \mid B)$, is defined as

$$P(A \mid B) = \frac{P(A \cap B)}{P(B)}$$

provided that $P(B) > 0$. Similarly, the conditional probability of B given A is defined as

$$P(B \mid A) = \frac{P(A \cap B)}{P(A)}$$

provided that $P(A) > 0$.

This definition can be motivated in terms of relative frequencies. Suppose that a random experiment is repeated N times, with N_B occurrences of event B and $N_{A \cap B}$ occurrences of A and B together. Then the proportion of times that A occurs, *when B has occurred,* is $N_{A \cap B}/N_B$ and one can think of the conditional probability of A given

B as the limit of this proportion as the number of replications of the experiment becomes infinitely large. But

$$\frac{N_{A \cap B}}{N_B} = \frac{N_{A \cap B}/N}{N_B/N}$$

and, as N becomes large, the numerator and denominator of the right-hand side of this expression approach $P(A \cap B)$ and $P(B)$, respectively. The definition of conditional probability is thus compatible with the relative frequency concept.

Example 3.18

Refer to Example 3.17. If 75% of the chain's customers use mustard, 80% use ketchup, and 65% use both, what are the probabilities that a ketchup-user uses mustard and that a mustard-user uses ketchup?

Let A be the event "Customer uses mustard" and B the event "Customer uses ketchup," so that $P(A) = 0.75$, $P(B) = 0.80$, and $P(A \cap B) = 0.65$. The probability that a ketchup-user uses mustard is the conditional probability of event A, given event B; that is,

$$P(A|B) = \frac{P(A \cap B)}{P(B)} = \frac{0.65}{0.80} = 0.8125$$

In the same fashion, the probability that a mustard-user uses ketchup is

$$P(B|A) = \frac{P(A \cap B)}{P(A)} = \frac{0.65}{0.75} = 0.8667$$

An immediate consequence of the definition of conditional probability is the **multiplication rule of probabilities,** which expresses the probability of an intersection in terms of probabilities for individual events and conditional probabilities.

Multiplication Rule of Probabilities

Let A and B be two events. Then the probability of their intersection is
$$P(A \cap B) = P(A|B)P(B)$$
Also

$$P(A \cap B) = P(B|A)P(A)$$

The following example illustrates an interesting application of the multiplication rule of probabilities and ties together some of the ideas introduced in previous sections of this chapter.

Example 3.19

In Example 3.7 we briefly met the *randomized response* approach for the solicitation of honest answers to sensitive questions in surveys. In a survey of this kind carried out in Hawaii by Geurts, each respondent was faced with the following two questions:

(a) Is the last digit of your Social Security number odd?

(b) Do you believe there should be fewer Caucasian professors in the College of Business at the University of Hawaii?

Respondents were asked first to flip a coin, and then to answer question (a) if the result was "head" and (b) otherwise. Thirty-seven percent of all respondents gave the answer "yes." What is the probability that a respondent who was answering the sensitive question (b) replied "yes"?

Let us define the following events:

A: Respondent answers "yes".
E_1: Respondent answers question (a).
E_2: Respondent answers question (b).

We now summarize the given information. First, since 37% of all respondents answered "yes," it follows that $P(A) = 0.37$. Further, since the question answered is determined by the flip of a coin, the probability that either question will be answered is 0.5—that is, $P(E_1) = 0.5$ and $P(E_2) = 0.5$.

We also know something about the answers to question (a). Since half of all Social Security numbers have an odd last digit, it must be that the probability of a "yes" answer, *given that question (a) has been answered,* is 0.5—that is, $P(A|E_1) = 0.5$.

However, what we require is $P(A|E_2)$, the conditional probability of a "yes" response, given that question (b) was answered. In order to obtain this probability we make use of two results from previous sections. First, since the events E_1 and E_2 are mutually exclusive and collectively exhaustive, we know that the two intersections $E_1 \cap A$ and $E_2 \cap A$ are mutually exclusive, and that their union is A. It therefore follows that the sum of the probabilities of these two intersections is the probability of A, so that

$$P(A) = P(E_1 \cap A) + P(E_2 \cap A)$$

Next, we use the multiplication rule, which in this context implies

$$P(E_1 \cap A) = P(A|E_1)P(E_1) \quad \text{and} \quad P(E_2 \cap A) = P(A|E_2)P(E_2)$$

We therefore have

$$P(A) = P(A|E_1)P(E_1) + P(A|E_2)P(E_2)$$

Substituting known values in this equation gives

$$0.37 = (0.5)(0.5) + P(A|E_2)(0.5)$$

We can now solve for the required conditional probability:

$$P(A|E_2) = \frac{(0.37 - 0.25)}{0.5} = 0.24$$

Hence, we estimate that 24% of the surveyed population believes there should be fewer Caucasian professors in the College of Business at the University of Hawaii.

Notice in general that the probability of an intersection of two events is *not* equal to the product of the individual event probabilities. However, a special case of considerable practical importance arises when this is in fact so. The events are then said to be **statistically independent.**

Definition

Let A and B be two events. These events are said to be **statistically independent** if and only if

$$P(A \cap B) = P(A)P(B)$$

It follows from the multiplication rule that equivalent conditions are

(i) $P(A|B) = P(A)$ if $P(B) > 0$

(ii) $P(B|A) = P(B)$ if $P(A) > 0$

More generally, the events $E_1, E_2, \ldots, E_K$ are statistically independent if and only if

$$P(E_1 \cap E_2 \cap \cdots \cap E_K) = P(E_1)P(E_2) \cdots P(E_K)$$

The logical basis for this definition of statistical independence is best seen in terms of conditional probabilities, and is perhaps most appealing from a subjective view of probability. Suppose I believe that the probability that event A will occur is $P(A)$. I am now given the extra piece of information that event B has occurred. If this does not change my view about the likelihood of occurrence of A, then my conditional probability assessment $P(A|B)$ will be the same as $P(A)$. I will have concluded that knowledge of the occurrence of B is of no use in determining whether or not A will occur; that is, A is no more or less likely to occur when B does than otherwise. Thus, this definition of statistical independence agrees with a common sense notion of "independence." In what follows, where the sense is clear, we will drop the word *statistical* and refer to events as being *independent*.

It is often required to check, using the definition, whether or not a pair of events are independent. The method is illustrated in the next example.

Example 3.20   A card is drawn from a complete deck of fifty-two cards. Let A be the event "The card is a heart" and B the event "The card is an ace." Since the deck contains 13 hearts and 4 aces, it follows that $P(A) = \frac{1}{4}$ and $P(B) = \frac{1}{13}$. The intersection of these two events is the event "The card is both a heart and an ace," which can occur only if the card drawn is the ace of hearts, so that $P(A \cap B) = \frac{1}{52}$. In this case, then,

$$P(A \cap B) = P(A)P(B)$$

and the two events are independent.

It is important to distinguish between the terms *mutually exclusive* and *independent*. A pair of events are mutually exclusive if they cannot jointly occur—that is, the probability of their intersection is 0. Independent events are characterized by the fact that the probability of their intersection is the product of their individual probabilities. Thus, in Example 3.20, the events "The card is a heart" and "The card is an ace" are independent, but they are not mutually exclusive, as the ace of hearts could be drawn.

In some circumstances independence can be deduced, or at least reasonably inferred, from the nature of the random experiment, and the probabilities for intersections then calculated as the product of individual probabilities. This is particularly useful in the case of repeated trials, and is an assumption that was tacitly made in discussing the relative frequency concept of probability in Section 3.3.

Example 3.21

A fair coin is thrown three times. Find the probability that a head results every time.

Let E_i ($i = 1, 2, 3$) be the events "A head results on the ith throw." Then, since the coin is fair, $P(E_i) = \frac{1}{2}$. It is reasonable to assume that these three events are independent—that the result on one throw will not depend on that of another. Then, since the event "A head results every time" is just the intersection of E_1, E_2, and E_3, we have

$$P(E_1 \cap E_2 \cap E_3) = P(E_1)P(E_2)P(E_3) = \left(\frac{1}{2}\right)^3 = \frac{1}{8}$$

Example 3.22

Assume that the probability that the Dow-Jones average will rise over any given trading day is 0.5, and that the course of the stock market in any day is independent of what has happened on previous days. What is the probability that the Dow-Jones average will rise over every one of four consecutive trading days? What is the probability that the average will fall or remain constant on at least one of these days?

Let E_i ($i = 1, 2, 3, 4$) be the events "Dow-Jones average rises on the ith day." Then, by our assumptions, $P(E_i) = \frac{1}{2}$ and these events are independent. Hence, the probability that the market will rise on all four days is

$$P(E_1 \cap E_2 \cap E_3 \cap E_4) = P(E_1)P(E_2)P(E_3)P(E_4) = \left(\frac{1}{2}\right)^4 = \frac{1}{16}$$

Further, the event "Dow-Jones average will fall or remain constant on at least one of these days" is simply the complement of the event whose probability we have just found. Its probability is therefore

$$1 - P(E_1 \cap E_2 \cap E_3 \cap E_4) = 1 - \frac{1}{16} = \frac{15}{16}$$

We conclude this section with two examples to illustrate a useful technique—finding the probability of the complement of an event as a first step in deducing the probability for that event.

Example 3.23

How many people must there be in a group so that there is a 50% chance that at least one pair of them has the same birthday? To make the problem manageable, we assign all those born on February 29 to March 1 and assume that all 365 possible birthdays occur equally often in the population at large. (These simplifications have only very small effects on the numerical results.)

Let M be the number of people in the group and A the event "At least one pair has a common birthday." Now, to find the probability of A directly would be very tedious, since we would have to take into account the possibility of more than one pair of matching birthdays. It is more straightforward to find the probability of the complement $\overline{A}$—that is, the event "All M people have different birthdays."

First, since there are 365 possible birthdays for each individual, and each can be associated with every possible birthday of other individuals, the total number of equally likely distinct arrangements for M people is 365^M. Next, we ask how many of these outcomes are contained in the event $\overline{A}$—that is, how many involve the M individuals all having different birthdays. This is precisely the same as asking how many ways M birthdays can be selected from 365 possible birthdays and arranged in order. This is just the number of permutations, $_{365}P_M$. Hence, the probability that all M birthdays will be different is

$$P(\overline{A}) = \frac{_{365}P_M}{365^M} = \frac{365!}{365^M(365 - M)!}$$

$$= \frac{(364)(363) \cdots (366 - M)}{365^{M-1}} \qquad \text{(for } M \geq 2\text{)}$$

Having evaluated the probability of the complement, we obtain the required probability as follows:

$$P(A) = 1 - P(\overline{A}) = 1 - \frac{(364)(363) \cdots (366 - M)}{365^{M-1}}$$

Some values of this probability for specific numbers of people M are shown in the table.

M	10	20	22	23	30	40	60
$P(A)$	.117	.411	.476	.507	.706	.891	.994

Therefore, if there are at least twenty-three people in the group, it is more likely than not that there will be at least one pair of matching birthdays. This probability rises fairly sharply as the group size increases, until, with sixty people in the group, the probability of finding a match is very high indeed.

At first sight, this result may be counterintuitive. After all, the probability that any given pair of individuals will have the same birthday is very small ($\frac{1}{365}$). However, for a group of twenty or more people, the number of possible matches becomes quite large, so that although the event "At least one match" is the union of unlikely events, there are so many of them that, when considered together, they yield a high probability for the union.

Example
3.24

In the spring of 1980, United Airlines introduced a promotion in which customers, or potential customers, were given vouchers. A proportion $\frac{1}{325}$ of these were worth a free round-trip ticket anywhere United flies. How many vouchers would an individual need to collect in order to have a 50% chance of winning at least one free trip?

The event A of interest is "At least one free trip is won from M vouchers." Again, it is easiest to find first the probability of the complement $\bar{A}$—that is, the event "No free trips are won with M vouchers."

The probability of a win with a single voucher is $1/325$, so the probability that this voucher will not win is $324/325$. If the individual has M vouchers, the event that none of these wins is just the intersection of the events "No win" for each of the M vouchers. Moreover, these events are independent, so that

$$P(\bar{A}) = \left(\frac{324}{325}\right)^M$$

Hence, the probability of at least one win is

$$P(A) = 1 - P(\bar{A}) = 1 - \left(\frac{324}{325}\right)^M$$

In order for $P(A)$ to be at least 0.5, the individual needs at least $M = 225$ vouchers.

Again this result might appear counterintuitive. At first sight one might guess that if the probability of a win for a single voucher was $1/325$, then 163 vouchers would be enough to ensure a 50% chance of a win. However, arguing along these lines one would be implicitly assuming that the probability of a union was the sum of the individual probabilities, neglecting to subtract for double counting in the intersections (which in this case would involve more than one win from the M vouchers).

EXERCISES

7. A company knows that a rival is about to bring out a competing product. It believes that this rival has three possible packaging plans (superior, normal, cheap) in mind and that all are equally likely. Also, there are three equally likely possible marketing strategies (intense media advertising, price discounts, and use of a coupon to reduce the price of future purchases). What is the probability that the rival will employ superior packaging in conjunction with an intense media advertising campaign? Assume that packaging plans and marketing strategies are determined independently.

8. A small city has four snow removal crews, to be assigned to four areas.
   (a) How many different arrangements are possible?
   (b) If assignments are made at random, what is the probability that a specific crew will be assigned to a given area?

9. In a taste test, a consumer is given six different brands of beer and asked to rank them in order of preference.
   (a) How many different rankings are possible?
   (b) If the choice is made randomly, what is the probability that the consumer will select some specific ordering?

10. A company has fifty sales representatives. It decides that the most successful representative during the previous year will be awarded a January vacation in Hawaii, while the second most successful will win a vacation in Las Vegas. The other representatives will be required to attend a conference on modern sales methods in Buffalo. How many outcomes are possible?

11. A securities analyst claims that, given a specific list of five common stocks, it is possible to predict, in the correct order, the three that will perform best during the coming year. What is the probability of making the correct selection by chance?

12. From a list of ten candidates, a personnel officer has to fill one position in the marketing department and one in the customer relations department of the company. How many different choices are possible?

13. You are one of nine candidates being considered by a corporation that has three vacancies. If the eventual choices will be made randomly, what is the probability that you will be successful?

14. A committee consists of four men and two women. A subcommittee of four members is to be chosen randomly. What is the probability that there will be no women on the subcommittee?

15. A supermarket chain has ten stores in a city. Three of these stores are to be used for a new promotional campaign. How many different combinations are possible?

16. A work crew for a building project is to be made up of two craftsmen and four laborers selected from a total of five craftsmen and six laborers available.
    (a) How many different combinations are possible?
    (b) The brother of one of the craftsmen is a laborer. If the crew is selected at random, what is the probability that both brothers will be selected?
    (c) What is the probability that neither brother will be selected?

17. A broker is recommending investment in six bonds and five stocks. A client wants to have three bonds and three stocks in her portfolio.
    (a) How many different selections of portfolios are possible from the broker's recommendations?
    (b) Unknown to the broker and the client, two of the bonds on the broker's list will turn out to be "losers." If the client's portfolio is chosen randomly from the list, what is the probability that it will contain at least one "loser"?

18. In a particular city, 40% of all households have a color television set and 20% of all households have a microwave oven. If 15% of all households have both a color television set and a microwave, what is the probability that a randomly chosen household will have at least one of these appliances?

19. A survey of shoppers at a convenience grocery store showed that 70% purchase milk and 40% purchase bread. If 25% of all the shoppers purchase both milk and bread, what is the probability that a randomly selected shopper will purchase at least one of these commodities?

20. Components from an assembly line are checked by two inspectors. The first inspector correctly classifies 80% of all the defective components, while the second inspector detects 75% of all the defectives. If both inspectors correctly identify 60% of the defective components, what is the probability that a defective item will be missed by both inspectors?

21. Candidates for employment in a large corporation must pass through two initial screening procedures—a written aptitude test and an oral interview. Fifty percent of the candidates are unsuccessful on the written test, 40% are unsuccessful in the interview, and 25% are unsuccessful in both. The corporation gives further consideration only to candidates who are successful in both of these procedures. What is the probability that a randomly chosen candidate will receive further consideration for employment?

22. A mail-order firm considers three possible errors in filling an order:

    A: The wrong item is sent.
    B: The item is lost in transit.
    C: The item is damaged in transit.

Assume that the events A, B, and C are independent of one another with $P(A) = .02$, $P(B) = .01$, and $P(C) = .05$. What is the probability that none of these errors will occur for a randomly selected order?

23. Every day, a salesman meets three contacts for potential sales. From past experience, he knows that the probability that he will be successful in making a sale is 0.4, and that the three outcomes are independent of one another. What is the probability that he will fail to make any sales during the day?

24. Refer to Exercise 19.

   (a) What is the probability that a customer who has purchased bread will also purchase milk?

   (b) What is the probability that a customer who has purchased milk will also purchase bread?

   (c) Are the events "Purchase milk" and "Purchase bread" independent?

25. Refer to Exercise 21.

   (a) What is the probability that a candidate who is unsuccessful in the written test will be successful in the interview?

   (b) What is the probability that a candidate who is successful in the interview will also be successful on the written test?

   (c) Are successes in the two screening procedures independent of one another?

26. It is known that 20% of the items produced on an assembly line are defective. An inspector examines items coming from the assembly line, and a review of her record indicates that she accepts only 10% of all defective items. What proportion of all the items produced will be both defective and accepted by the inspector?

27. (a) Develop an example showing two events that are mutually exclusive, but not statistically independent.

   (b) Develop an example showing two events that are statistically independent, but not mutually exclusive.

28. According to recent figures from the National Center for Educational Statistics, 47% of all bachelor's degrees are obtained by women and 27% of bachelor's degrees in business are obtained by women. Further, 17.5% of all bachelor's degrees are in business.

   (a) What is the probability that a randomly selected new bachelor's degree holder will be a woman with a degree in business?

   (b) What is the probability that a randomly selected woman bachelor's degree holder will have a degree in business?

29. A convention begins with an evening series of lectures, attended by 60% of the delegates. The following morning 80% of the delegates attend lectures. Seventy percent of those attending this session had attended the previous evening.

   (a) What is the probability that a randomly selected delegate attended both lecture sessions?

   (b) What is the probability that a delegate who attended the evening session also attended the following morning?

   (c) What is the probability that a randomly selected delegate attended at least one of the two sessions?

   (d) Are attendances at the two sessions statistically independent?

30. A bank classifies borrowers as high risk or low risk. Only 20% of its loans are made to those in the high risk category. Of all its loans, 5% are in default, and 40% of those in default are to high risk borrowers. What is the probability that a high risk borrower will default?

31. An insurance salesman reviews past records and finds that 30% of all calls have led to sales. Forty percent of these sales have been to previous policy holders in the company.

The salesman estimates that the probability of making a sale to a previous policy holder is 0.6. What is the probability that a person on whom the salesman calls is a previous policy holder?

32. A conference began at noon with two parallel sessions. The session on portfolio management was attended by 40% of the delegates while the session on chartism was attended by 50%. The evening session consisted of a talk titled, "Is the random walk dead?" This was attended by 80% of the delegates.

    (a) If attendance at the sessions on portfolio management and chartism are mutually exclusive, what is the probability that a randomly chosen delegate attended at least one of these sessions?

    (b) If attendance at the portfolio management and evening sessions are statistically independent, what is the probability that a randomly chosen delegate attended at least one of these sessions?

    (c) Of those attending the chartism session, 75% also attended the evening session. What is the probability that a randomly chosen delegate attended at least one of these two sessions?

33. An editor may use all, some, or none, of three possible strategies to enhance the sales of a book. The three strategies are

    A: An expensive prepublication promotion
    B: An expensive cover design
    C: A bonus for sales representatives who meet predetermined sales levels

    In the past, these three strategies have been applied simultaneously to only 2% of the company's books. Twenty percent of the books have had expensive cover designs, and of these 80% have had expensive prepublication promotion. A rival editor learns that a new book is to have both expensive prepublication promotion and cover design, and now wants to know how likely it is that a bonus scheme for sales representatives will be introduced. Compute the probability of interest to the rival editor.

3.7 BIVARIATE PROBABILITIES

In this section we introduce a class of problems that can be handled using the material developed previously. However, they are sufficiently important to treat them separately.

The general setup is as follows. A random experiment is to be conducted and interest centers on two distinct sets of events that could occur. We label these A_1, A_2, . . . , A_h and B_1, B_2, . . . , B_k. The events A_i are mutually exclusive and collectively exhaustive, as are the events B_j. However, any A_i event can occur jointly with any B_j event, so that the intersections $A_i \cap B_j$ can occur. These intersections can be regarded as the basic outcomes of the random experiment. Two sets of events, considered jointly in this way, are often called bivariate, and the probabilities are referred to as **bivariate probabilities.** The setup is illustrated in Table 3.1. If probabilities can be attached to all of the events $A_i \cap B_j$, then the whole probability structure of the random experiment is known and other probabilities of interest can be calculated.

As an example, a potential advertiser will want to know not only the likely size of the viewing audience for a particular television show, but also the relevant charac-

TABLE 3.1 Outcomes for bivariate events

	B_1	B_2	$\cdots$	B_k
A_1	$A_1 \cap B_1$	$A_1 \cap B_2$	$\cdots$	$A_1 \cap B_k$
A_2	$A_2 \cap B_1$	$A_2 \cap B_2$	$\cdots$	$A_2 \cap B_k$
$\vdots$	$\vdots$			
A_h	$A_h \cap B_1$	$A_h \cap B_2$	$\cdots$	$A_h \cap B_k$

teristics of that audience. Thus, families may be categorized (corresponding to the A_i categorization) as to whether they regularly, occasionally, or never watch a particular series and also (corresponding to the B_j) according to low, middle, or high income. Then the nine possible cross-classifications can be set out as in Table 3.1, with $h = 3$ and $k = 3$. An alternative way to view this breakdown of the population into nine distinct groups is through the use of a *tree diagram,* as shown in Figure 3.12. Beginning at the left with the whole population of families, there are three distinct branches according to the frequency with which the show is watched. Each of these branches has three sub-branches, corresponding to the three income categories. Therefore, there are nine sub-branches in all, each associated with one of the possible Viewing–Income event intersections.

As a first step, probabilities for the event intersections are required. Values for the television viewing example, obtained from survey results, are given in the body of Table 3.2. For example, it was found that 10% of families in the survey have high incomes and occasionally watch the series. In equating proportions of the survey members and probabilities in this way we are tacitly invoking the relative frequency

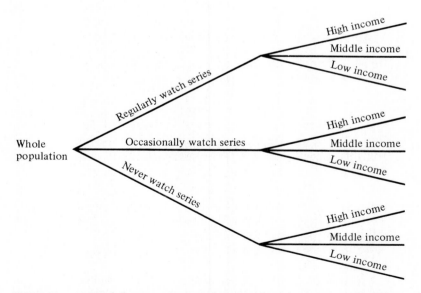

FIGURE 3.12 Tree diagram of events for television viewing–income example

TABLE 3.2 Probabilities for television viewing–income example

VIEWING FREQUENCY	INCOME			
	HIGH	MIDDLE	LOW	TOTALS
REGULAR	.04	.13	.04	.21
OCCASIONAL	.10	.11	.06	.27
NEVER	.13	.17	.22	.52
TOTALS	.27	.41	.32	1.00

concept of probability, assuming that the survey is sufficiently large that proportions can be approximated as probabilities. On this basis, the probability that a family chosen at random from the population has a high income and occasionally watches the show is .10; that is,

$$P(\text{Occasionally watch} \cap \text{High income}) = 0.10$$

Definitions

In the context of bivariate probabilities, the intersection probabilities $P(A_i \cap B_j)$ are called **joint probabilities.** The probabilities for individual events, $P(A_i)$ or $P(B_j)$, are called **marginal probabilities.**

Given the joint probabilities, suppose we require the marginal probabilities. Consider the general setup in Table 3.1, where interest is in the event A_i. Now, A_i is the union of the mutually exclusive events $A_i \cap B_1, A_i \cap B_2, \ldots, A_i \cap B_k$. [This result was formally established in Section 3.2 as result (iii).] Hence, by the addition rule for mutually exclusive events [established in Section 3.4 as result (iii)], the probability of event A_i is just the sum of the intersection probabilities for those intersections involving A_i—that is,

$$P(A_i) = P(A_i \cap B_1) + P(A_i \cap B_2) + \cdots + P(A_i \cap B_k)$$

Thus, when the intersection probabilities are cross-tabulated, the individual event probabilities for the A_i are just the row totals. From an analogous argument, the probabilities for the B_j are the column totals.

For the television viewing–income example, it follows that the probability that a randomly chosen individual occasionally watches the show is

$$P(\text{Occasionally watch}) = P(\text{Occasionally watch} \cap \text{High income})$$
$$+ P(\text{Occasionally watch} \cap \text{Middle income})$$
$$+ P(\text{Occasionally watch} \cap \text{Low income})$$
$$= .10 + .11 + .06 = .27$$

Similarly,

$$P(\text{Regularly watch}) = .21; \qquad P(\text{Never watch}) = .52$$

and

$$P(\text{High income}) = .27; \qquad P(\text{Middle income}) = .41; \qquad P(\text{Low income}) = .32$$

Marginal probabilities can also be deduced directly from tree diagrams, as in Figure 3.13, in which the branches and sub-branches are the same events as in Figure 3.12. Beginning on the right-hand side of the figure, the joint probabilities can be entered immediately on the nine sub-branches. The marginal probabilities for each of the three viewing frequency events can then be entered on the main branches by adding the probabilities on the corresponding sub-branches. This approach is particularly useful when there are more than two attributes of interest. For example, the advertiser might be interested also in the age of the head of household or whether there are children in the household.

Since the events $A_1, A_2, \ldots, A_h$ are mutually exclusive and collectively exhaustive, their marginal probabilities must sum to 1. The same is true of the events $B_1, B_2, \ldots, B_k$. For the television viewing–income example, this is illustrated in Table 3.2, where the row and column totals both sum to 1, and in the tree diagram of Figure 3.13, where the sum of the probabilities on the main branches is also 1. It also follows that the joint probabilities summed over all event combinations add to 1.

In many applications the conditional probabilities are of more interest than the marginal probabilities. For example, an advertiser will be less concerned with the total size of the viewing audience for a show than with the chance that a family, likely to be in the market for a particular product, is watching. The conditional probabilities can

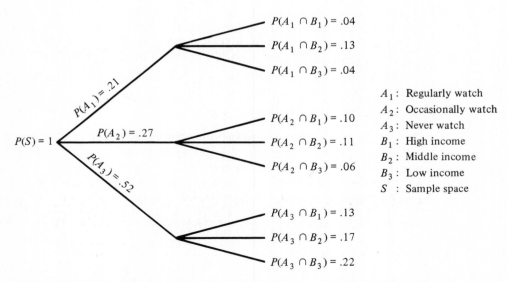

FIGURE 3.13 Tree diagram for television viewing–income example, showing joint and marginal probabilities

be obtained by direct application of the definition of conditional probability introduced in Section 3.6. Thus, the probability of A_i given B_j is

$$P(A_i \mid B_j) = \frac{P(A_i \cap B_j)}{P(B_j)}$$

for any pair of events A_i and B_j. Similarly,

$$P(B_j \mid A_i) = \frac{P(A_i \cap B_j)}{P(A_i)}$$

Once the joint and marginal probabilities are known, the conditional probabilities then follow. For example, the probability that a randomly chosen family occasionally watches the show given that its income is in the middle range is

$$P(\text{Occasionally watch} \mid \text{Middle income}) = \frac{P(\text{Occasionally watch} \cap \text{Middle income})}{P(\text{Middle income})}$$

$$= \frac{.11}{.41} = .27$$

Other conditional probabilities of viewing frequency given income level can be found in the same way, and are given in Table 3.3.

The probabilities for income levels given viewing frequencies can also be found. For example, the probability that a randomly chosen family has middle income given that they occasionally watch the show is

$$P(\text{Middle income} \mid \text{Occasionally watch}) = \frac{P(\text{Occasionally watch} \cap \text{Middle income})}{P(\text{Occasionally watch})}$$

$$= \frac{.11}{.27} = .41$$

The conditional probabilities for income levels given viewing frequencies are displayed in Table 3.4.

We can also check whether or not a pair of events are statistically independent. In general, the events A_i and B_j are independent if and only if their joint probability is the product of their marginal probabilities, that is,

$$P(A_i \cap B_j) = P(A_i)P(B_j)$$

TABLE 3.3 Conditional probabilities of viewing frequencies given income levels

VIEWING FREQUENCY	INCOME		
	HIGH	MIDDLE	LOW
REGULAR	.15	.32	.12
OCCASIONAL	.37	.27	.19
NEVER	.48	.41	.69

TABLE 3.4 Conditional probabilities of income levels given viewing frequencies

VIEWING FREQUENCY	INCOME		
	HIGH	MIDDLE	LOW
REGULAR	.19	.62	.19
OCCASIONAL	.37	.41	.22
NEVER	.25	.33	.42

In our example, for the events "Occasionally watch" and "High income," we have from Table 3.2

$$P(\text{Occasionally watch} \cap \text{High income}) = .10$$

and

$$P(\text{Occasionally watch}) = .27; \qquad P(\text{High income}) = .27$$

The product of the marginal probabilities is .0729, which differs from the joint probability .10. Hence, the two events are not statistically independent.

A case of particular interest arises when every event A_i is independent of every event B_j, in which case the two attributes are said to be **independent.**

Definition

Let A and B be a pair of attributes, each broken down into mutually exclusive and collectively exhaustive event categories, respectively denoted $A_1, A_2, \ldots, A_h$ and B_1, $B_2, \ldots, B_k$. If every event A_i is statistically independent of every event B_j, the attributes A and B are said to be **independent.**

In many practical applications, the joint probabilities will not be known precisely. A problem of considerable importance arises when a sample is available from a population, allowing estimates to be made of the joint probabilities, and it is required to test, on the basis of this sample evidence, whether or not a pair of attributes are independent of one another. A procedure for carrying out such a test will be introduced in Chapter 11.

3.8 BAYES' THEOREM

In this section we introduce a mechanism for the modification of probability assessments when additional information becomes available. To illustrate, suppose that an investor is interested in a particular stock and forms a judgment about the likely profitability of an investment in it. If the investor then learns that the stock is being recommended by an expert analyst, he or she might modify this original judgment, depending on his or her faith in the analyst's ability.

Let A and B be two events with respective probabilities $P(A)$ and $P(B)$. Then the multiplication rule of probabilities gives

$$P(A \cap B) = P(A|B)P(B) \qquad (3.8.1)$$

and also

$$P(A \cap B) = P(B|A)P(A) \qquad (3.8.2)$$

Since the left-hand sides of equations 3.8.1 and 3.8.2 are the same, then so must be the right-hand sides, so that

$$P(B|A)P(A) = P(A|B)P(B)$$

Dividing through this equation by $P(A)$, assuming that probability is not zero, gives Bayes' theorem.[2]

> **Bayes' Theorem**
>
> Let A and B be two events. Then
>
> $$P(B|A) = \frac{P(A|B)P(B)}{P(A)} \qquad (3.8.3)$$

The most interesting interpretation of Bayes' theorem is in terms of subjective probabilities. Suppose that an individual is interested in the event B and forms a subjective view of the probability that B will occur; in this context, the probability $P(B)$ is called a **prior** probability. If then the individual acquires an additional piece of information—namely, that event A has occurred—this may cause a modification of the initial judgment as to the likelihood of the occurrence of B. Since A is known to have happened, the relevant probability for B is now the conditional probability of B given A and is termed the **posterior** probability. Viewed in this way, Bayes' theorem can be thought of as a mechanism for updating a prior probability to a posterior probability when the additional information that event A has occurred becomes avail-

---

[2] The theorem is named for its discoverer, an English clergyman, the Reverend Thomas Bayes (1702–1761).

able. The theorem then states that the updating is accomplished through multiplication of the prior probability by $P(A|B)/P(A)$.

That people form and subsequently modify subjective probability assessments is common experience. An important aspect of an auditor's work is to determine whether or not the account balances are correct. Before examining a particular account the auditor will have formed some view, based on experience from earlier audits, of the probability that it is in error. However, if the balance is found to be substantially different from what might be expected on the basis of the last few years' figures, the auditor will feel that the probability of an error is higher and therefore give the account particularly close attention. Again, the prior probability has been updated in the light of additional information.

<table>
<tr><td>Example
3.25</td><td>In examining past records of a corporation's account balances, an auditor finds that 15% of them have contained errors. Of those balances in error, 60% were regarded as unusual values based on historical figures. Of all the account balances, 20% were unusual values. If the figure for a particular balance appears unusual on this basis, what is the probability that it is in error?</td></tr>
</table>

Denoting the events of interest as "Error" and "Unusual value," we have

$$P(\text{Error}) = 0.15; \qquad P(\text{Unusual value}) = 0.20,$$

and

$$P(\text{Unusual value} | \text{Error}) = 0.60$$

Invoking Bayes' theorem, we obtain

$$P(\text{Error} | \text{Unusual value}) = \frac{P(\text{Unusual value} | \text{Error})P(\text{Error})}{P(\text{Unusual value})}$$

$$= \frac{(0.60)(0.15)}{0.20} = 0.45$$

Thus, given the information that the account balance appears unusual, the probability that it is in error is modified from the prior 0.15 to the posterior 0.45.

Bayes' theorem is often expressed in a different, but equivalent, form. Let E_1, E_2, . . . , E_K be K mutually exclusive and collectively exhaustive events and let A be some other event. For some i, we want to find the conditional probability of E_i given A. This can be obtained directly from Bayes' theorem by setting B in equation 3.8.3 equal to E_i. However, the denominator on the right-hand side of that equation can be expressed in terms of conditional probabilities for A given the E_j and probabilities of the individual E_j. We have seen in Section 3.2 that the events $E_1 \cap A, E_2 \cap A, \ldots,$ $E_K \cap A$ are mutually exclusive and that their union is A. It then follows [see result (iii) of Section 3.4] that the probability of A is

$$P(A) = P(E_1 \cap A) + P(E_2 \cap A) + \cdots + P(E_K \cap A) \qquad (3.8.4)$$

Furthermore, from the multiplication rule of probabilities,

$$P(E_j \cap A) = P(A|E_j)P(E_j) \qquad (j = 1, 2, \ldots, K)$$

and so, substituting into equation 3.8.4 yields

$$P(A) = P(A|E_1)P(E_1) + P(A|E_2)P(E_2) + \cdots + P(A|E_K)P(E_K) \qquad (3.8.5)$$

Finally the restatement of Bayes' theorem is obtained by substituting E_i for B and the right-hand side of equation 3.8.5 for $P(A)$ in equation 3.8.3.

Bayes' Theorem (Alternative Statement)

Let $E_1, E_2, \ldots, E_K$ be K mutually exclusive and collectively exhaustive events and let A be some other event. Then the conditional probability of E_i given A can be expressed as

$$P(E_i|A) = \frac{P(A|E_i)P(E_i)}{P(A|E_1)P(E_1) + P(A|E_2)P(E_2) + \cdots + P(A|E_K)P(E_K)}$$

The advantage of this restatement of the theorem lies in the fact that the probabilities it involves are often precisely those that are directly available.

Example 3.26

An inspector working on an assembly line on which a component is produced has to make quick decisions about the acceptability of the components as they pass through his station. It was found that he rejected 10% of all the components. Of those rejected a further detailed examination showed 80% to in fact be defective, while 5% of the accepted components were also defective. What is the probability that the inspector will accept a defective component?

Given

$$P(\text{Rejects}) = 0.10$$

it follows that

$$P(\text{Accepts}) = 1 - P(\text{Rejects}) = 0.90$$

Also

$$P(\text{Defective}|\text{Rejects}) = 0.80; \qquad P(\text{Defective}|\text{Accepts}) = 0.05$$

Therefore, using the alternative statement of Bayes' theorem, the probability that a component will be accepted, given that it is defective, is

$$P(\text{Accepts} \mid \text{Defective})$$

$$= \frac{P(\text{Defective} \mid \text{Accepts})P(\text{Accepts})}{P(\text{Defective} \mid \text{Accepts})P(\text{Accepts}) + P(\text{Defective} \mid \text{Rejects})P(\text{Rejects})}$$

$$= \frac{(0.05)(0.90)}{(0.05)(0.90) + (0.80)(0.10)} = 0.36$$

Example 3.27

A stock market analyst examined the prospects of the shares of a large number of corporations. When the performance of these stocks was investigated one year later, it turned out that 25% performed much better than the market average, 25% much worse, and the remaining 50% about the same as the average. Forty percent of the stocks that turned out to do much better than the market were rated "good buys" by the analyst, as were 20% of those that did about as well as the market and 10% of those that did much worse. What is the probability that a stock rated a "good buy" by the analyst performed much better than the market average?

Define the following events:

E_1: Stock performs much better than market average.
E_2: Stock performs about the same as market average.
E_3: Stock performs much worse than market average.
A: Stock is rated "good buy" by the analyst.

From the statement of the example, we have the probabilities

$$P(E_1) = 0.25; \qquad P(E_2) = 0.50; \qquad P(E_3) = 0.25$$

and the conditional probabilities

$$P(A \mid E_1) = 0.4; \qquad P(A \mid E_2) = 0.2; \qquad P(A \mid E_3) = 0.1$$

It is required to find the probability that a stock performs much better than the market average, given that it was rated a "good buy" by the analyst. This is the conditional probability $P(E_1 \mid A)$, which is obtained from Bayes' theorem as follows:

$$P(E_1 \mid A) = \frac{P(A \mid E_1)P(E_1)}{P(A \mid E_1)P(E_1) + P(A \mid E_2)P(E_2) + P(A \mid E_3)P(E_3)}$$

$$= \frac{(0.4)(0.25)}{(0.4)(0.25) + (0.2)(0.50) + (0.1)(0.25)} = 0.444$$

In the business area the most common application of Bayes' theorem is in the field of decision-making in an uncertain environment. Problems of this type will be discussed in Chapter 19.

EXERCISES

34. A survey carried out for a supermarket classified customers according to whether their visits to the store are frequent or infrequent and to whether they often, sometimes, or never purchase generic products. The accompanying table gives the proportions of those surveyed in each of the six joint classifications.

FREQUENCY OF VISIT	PURCHASE OF GENERIC PRODUCTS		
	OFTEN	SOMETIMES	NEVER
FREQUENT	.10	.50	.20
INFREQUENT	.05	.05	.10

(a) What is the probability that a customer is both a frequent shopper and often purchases generic products?

(b) What is the probability that a customer who never buys generic products visits the store frequently?

(c) Are the events "Never buys generic products" and "Visits the store frequently" independent?

(d) What is the probability that a customer who infrequently visits the store often buys generic products?

(e) Are the events "Often buys generic products" and "Visits the store infrequently" independent?

(f) What is the probability that a customer frequently visits the store?

(g) What is the probability that a customer never buys generic products?

(h) What is the probability that a customer either frequently visits the store, or never buys generic products, or both?

35. A company test-marketed a new product and classified the responses as favorable or unfavorable, in four different regions of the country. The table given here shows proportions of respondents in eight joint classifications.

RESPONSE	REGION			
	EAST	SOUTH	MIDWEST	WEST
FAVORABLE	.14	.10	.17	.13
UNFAVORABLE	.10	.12	.08	.16

(a) What is the probability that a randomly chosen respondent is both favorable to the product and from the Midwest?

(b) Find the probability that a respondent from the East reacts favorably to the product.

(c) Is reaction to the product independent of region?

36. A company receives deliveries of a particular component from three suppliers. It finds that the proportion of components that are good or defective from its total receipts are as shown in the accompanying table.

COMPONENT	SUPPLIER		
	A	B	C
GOOD	.36	.38	.16
DEFECTIVE	.04	.02	.04

(a) If a component is selected at random from all those received, what is the probability that it is defective?

(b) What is the probability that a component from supplier A is defective?

(c) Is the quality of the component independent of the source of supply?

(d) What is the probability that a randomly selected component is from supplier C?

37. A survey of households in a small town was carried out to determine whether they would or would not subscribe to a cable television service. The households were also classified as low, middle or high income. The table gives the proportions of households in each of the six joint classifications.

INCOME	WILLINGNESS TO SUBSCRIBE	
	YES	NO
LOW	.02	.18
MIDDLE	.15	.45
HIGH	.08	.12

(a) Find the probability that a randomly chosen household would subscribe.

(b) What is the probability that a household that would subscribe has high income?

(c) Are income and willingness to subscribe independent?

(d) Find the probability that a randomly chosen household belongs to at least one of the categories "High income" or "Willing to subscribe."

38. A consulting organization predicts whether corporations' earnings for the coming year will be unusually low, unusually high, or normal. Before deciding whether to continue purchasing these forecasts, a stockbroker compares past predictions with actual outcomes. The accompanying table shows proportions in the nine joint classifications.

OUTCOME	PREDICTION		
	UNUSUALLY HIGH	NORMAL	UNUSUALLY LOW
UNUSUALLY HIGH	.23	.12	.03
NORMAL	.06	.22	.08
UNUSUALLY LOW	.01	.06	.19

(a) What proportion of predictions have been for unusually high earnings?

(b) What proportion of outcomes were for unusually high earnings?

(c) If a firm were to have unusually high earnings, what is the probability that the consulting organization would correctly predict this event?

(d) If the organization predicted unusually high earnings for a corporation, what is the probability that these would materialize?

(e) What is the probability that a corporation for which unusually high earnings had been predicted will have unusually low earnings?

39. The table* shows proportions of common stock investors categorized according to the annual rate of return expected and the level of risk acceptable.

ACCEPTABLE LEVEL OF RISK	EXPECTED ANNUAL RATE OF RETURN		
	LESS THAN 10%	10–15%	MORE THAN 15%
HIGH	.034	.019	.021
MEDIUM	.276	.141	.102
LOW	.230	.092	.046
NONE	.029	.006	.004

*These estimates are given by H. K. Baker, M. B. Hargrove, and J. A. Haslem, "An empirical analysis of the risk-return preferences of individual investors," *Journal of Financial and Quantitative Analysis, 12* (1977), 377–89.

(a) What is the probability that a randomly chosen investor will find a high level of risk acceptable?

(b) What is the probability that a randomly chosen investor will find at least a medium level of risk acceptable?

(c) What is the probability that a randomly chosen investor expects more than 15% annual rate of return?

(d) What is the probability that a randomly chosen investor expects at least 10% annual rate of return?

(e) What is the probability that an investor who accepts a high level of risk expects more than 15% annual rate of return?

(f) What is the probability that an investor who accepts at least a medium level of risk expects at least 10% annual rate of return?

40. In a study of the petrochemical industry, the age of corporate organization was related to the frequency of replacement of its president. The probabilities shown in the table* were estimated.

AGE	FREQUENCY	
	LOW	HIGH
YOUNG	.18	.32
OLD	.35	.15

*Reprinted by permission of the publisher from "Succession: A longitudinal look," by D. L. Helmich, *Journal of Business Research, 3*, 355–64. Copyright 1975 by Elsevier Science Publishing Co., Inc.

(a) What is the probability that an old organization replaces its president with high frequency?

(b) If an organization replaces its president with high frequency, what is the probability that it is old?

(c) Are the age of an organization and the frequency with which its president is replaced independent?

41. Of all the students who take the C.P.A. examination, ¼ have master's degrees and ¾ have bachelor's degrees. Half of the students with bachelor's degrees pass the examination, and ¾ of those with master's degrees pass.
    (a) Find the probability that a student chosen at random from those taking the examination will pass.
    (b) What is the probability that a student both has a bachelor's degree and passes the examination?
    (c) What is the probability that a student with a bachelor's degree will not pass the examination?
    (d) What is the probability that a student who has not passed the examination has a master's degree?
    (e) What is the probability that a student either passes the examination or has a bachelor's degree, or both?

42. In assessing prospects for inflation next year, a group of economic consultants concluded that the probabilities that the inflation rate will be higher next year than in the current year are 0.1 if there is a small rise in energy prices, 0.6 if there is a moderate rise in energy prices, and 0.8 if there is a large rise in energy prices. They also concluded that the probabilities for small, moderate, and large rises in energy prices were, respectively, 0.2, 0.5, and 0.3.
    (a) What is the probability that there will be both a small rise in energy prices and a higher inflation rate next year than in the current year?
    (b) What is the probability that the inflation rate will be higher next year than in the current year?
    (c) If the inflation rate is higher next year than in the current year, what is the probability that the rise in energy prices will have been moderate?

43. A publisher sends advertising material for an accounting text to 80% of all professors teaching the appropriate accounting course. Thirty percent of the professors who received this material adopted the book. Also, 10% of the professors who did not receive the advertising material adopted the book. What is the probability that a professor who adopts the book has received the advertising material?

44. Forty percent of all competitors in a locally organized marathon race were from out of town. Twenty percent of all competitors finished the race. Eighty percent of all those who did finish the race were from out of town.
    (a) What is the probability that a randomly selected competitor both finished the race and was from out of town?
    (b) What is the probability that a randomly selected competitor is in at least one of the categories "Finished the race" or "From out of town"?
    (c) What is the probability that a competitor who was from out of town finished the race?
    (d) What is the probability that a competitor who was not from out of town finished the race?

45. A psychologist has developed a test designed to help predict whether production line workers in a large industry will perform satisfactorily. The test was admistered to all new employees of a corporation. At the end of their first year of work, these employees were rated by their supervisers: 20% were rated excellent, 40% satisfactory, and 40% poor. Fifty percent of the employees rated excellent passed the psychologist's test, as did 25% of those rated satisfactory and 10% of those rated poor.
    (a) What is the probability that a randomly chosen employee will pass the psychologist's test?
    (b) What is the probability that an employee who passes the test will be rated excellent?
    (c) What is the probability that an employee who passes the test will be rated satisfactory or better?

(d) What is the probability that an employee who failed the test will be rated satisfactory or better?

46. Before introducing a new toy to the market in time for Christmas, a toy manufacturer sends prototypes to a panel of children, whose responses are then coded as strongly favorable, weakly favorable, weakly unfavorable, or strongly unfavorable. Historically, 60% of all new toys introduced by the manufacturer have been successful on the market, the remainder proving unsuccessful. Similar panels have examined 1,000 toys in the past. The accompanying table shows their reactions and the resulting market success of the toys.

MARKET SUCCESS	PANEL REACTION			
	STRONGLY FAVORABLE	WEAKLY FAVORABLE	WEAKLY UNFAVORABLE	STRONGLY UNFAVORABLE
SUCCESSFUL	160	220	120	100
UNSUCCESSFUL	20	60	100	220

(a) If the panel reaction is strongly favorable, what is the probability that the toy will be successful?

(b) If the panel reaction is strongly unfavorable, what is the probability that the toy will be unsuccessful?

(c) If the panel reaction is weakly favorable or better, what is the probability that the toy will be successful?

(d) If the toy turns out to be successful, what is the probability that the panel reaction will have been weakly unfavorable or worse?

47. A manufacturer produces boxes of candy, each containing ten pieces. Two machines are used for this purpose. After a large batch has been produced it is discovered that one of the machines, which produces 40% of the total output, has a fault which has led to the introduction of an impurity into 10% of the pieces of candy it makes. From a single box of candy, one piece is selected at random and tested. If that piece contains no impurity, what is the probability that the box from which it came was produced by the faulty machine?

48. A student feels that 80% of his college courses have been enjoyable and the remainder have been boring. He has access to student evaluations of professors and finds that 60% of his enjoyable courses and 25% of his boring courses have been taught by professors who had previously received strong positive evaluations from their students. Next semester the student decides to take three courses, all from professors who have received strongly positive student evaluations. Assume that his reactions to the three courses are independent of one another.

(a) What is the probability that he will find all three courses enjoyable?

(b) What is the probability that he will find at least one of the courses enjoyable?

REVIEW EXERCISES

49. Suppose that you have an intelligent friend who has not studied probability. How would you explain to your friend the distinction between mutually exclusive events and independent events? Illustrate your answer with suitable examples.

50. Explain carefully the meaning of conditional probability. Why is this concept important in discussing the chance of an event's occurrence?

51. "Bayes' theorem is important, as it provides a rule for moving from a prior probability to a posterior probability." Elaborate on this statement so that it would be well understood by a fellow student who has not yet studied probability.

52. State, with reasons, whether each of the following statements is true or false:
    (a) The probability of the union of two events cannot be less than the probability of their intersection.
    (b) The probability of the union of two events cannot be more than the sum of their individual probabilities.
    (c) The probability of the intersection of two events cannot be greater than either of their individual probabilities.
    (d) An event and its complement are mutually exclusive.
    (e) The individual probabilities of a pair of events cannot sum to more than 1.
    (f) If a pair of events are mutually exclusive, they must also be collectively exhaustive.
    (g) If a pair of events are collectively exhaustive, they must also be mutually exclusive.

53. Distinguish among joint probability, marginal probability, and conditional probability. Provide examples to make the distinctions clear.

54. State, giving reasons, whether each of the following claims is true or false:
    (a) The conditional probability of A given B must be at least as large as the probability of A.
    (b) An event must be independent of its complement.
    (c) The probability of A given B must be at least as large as the probability of the intersection of A and B.
    (d) The probability of the intersection of two events cannot exceed the product of their individual probabilities.
    (e) The posterior probability of any event must be at least as large as its prior probability.

55. Show that the probability of the union of the events A and B can be written

$$P(A \cup B) = P(A) + P(B)[1 - P(A \mid B)]$$

56. Parts leaving an assembly line are examined by two inspectors. Each inspector detects 80% of the defective parts, and 70% of the defective parts are detected by both inspectors.
    (a) What is the probability that a defective part will be detected by at least one of the inspectors?
    (b) What is the probability that a defective part which is not detected by the first inspector will be detected by the second?
    (c) A batch of output contains three defective parts. What is the probability that both inspectors will fail to detect at least one of these parts? Assume that detection of one defective part is independent of detection of another.

57. A company places a "rush order" for wire of two thicknesses to a supplier. Consignments of each thickness are to be sent immediately when they are available. Previous experience suggests that the probability is 0.8 that at least one of these consignments will arrive within a week. It is also estimated that, if the thinner wire arrives within a week, the probability is 0.4 that the thicker wire will also arrive within a week. Further, it is estimated that, if the thicker wire arrives within a week, the probability is 0.6 that the thinner wire will also arrive within a week.
    (a) What is the probability that the thicker wire will arrive within a wek?
    (b) What is the probability that the thinner wire will arrive within a week?
    (c) What is the probability that both consignments will arrive within a week?

58. A consultant submits outline proposals to potential clients. There are three possible responses. The proposal may be immediately accepted or rejected, or the consultant may be asked to submit a more detailed proposal. It has been found that 30% of outline proposals are immediately rejected, and for 60% of outline proposals a more detailed proposal is requested. After submitting the detailed proposal, the consultant will receive a final acceptance or rejection from the potential client. It has been found that, of all initial outline proposals, 60% lead to an eventual acceptance, possibly after further submission of a detailed proposal. What is the probability that an acceptance will result, given that a detailed proposal has been requested?

59. A stock market analyst believes that the probability is 0.7 that the market will rise by at least 10% over the next three months if there is a significant drop in interest rates. She also believes that the probability of a significant drop in interest rates is 0.4. According to this analyst, what is the probability that the market will not rise by at least 10% over the next three months, and at the same time there will be a significant drop in interest rates?

60. A jury of twelve people is to be selected from a panel consisting of ten men and ten women.
    (a) How many different jury selections are possible?
    (b) If the choice is made randomly, what is the probability that the jury contains at least four women?

61. A case of twenty cans of food contains one can whose contents are contaminated. Three cans are chosen randomly from this case for testing.
    (a) How many different combinations of three cans could be chosen?
    (b) What is the probability that the can with contaminated contents is among those selected for testing?

62. It is suspected that some "vintage" wines are bottled after the addition of cheap imported wine. An expert claims to be able to detect such "doctoring." He is presented with six bottles, and told that three contain the genuine vintage and that three have been "doctored." If, in fact, he has no power whatever to discriminate, what is the probability of his being able to identify correctly the three genuine wines?

63. A candy bar manufacturer uses the responses of a panel of tasters to help predict the degree of success of new market introductions. The accompanying table shows proportions in nine joint classifications.

MARKET PENETRATION	PANEL RESPONSE		
	POOR	FAIR	GOOD
LOW	.19	.10	.04
MODERATE	.11	.14	.10
HIGH	.06	.10	.16

    (a) What is the probability that the market penetration will be high for a particular candy bar?
    (b) What is the probability that panel response will be poor for a particular candy bar?
    (c) What is the probability of high market penetration if panel response is poor?
    (d) What is the probability of a poor panel response to a candy bar which in fact achieves high market penetration?
    (e) Are panel response and the level of market penetration independent?

64. A survey organization asked respondents' views on the likely future direction of the economy, and also on whether they had voted for the president in the last election. The table given here shows the proportions of respondents in nine joint classifications.

	VIEW ON ECONOMY		
	OPTIMISTIC	PESSIMISTIC	NEUTRAL
VOTED FOR THE PRESIDENT	.20	.08	.12
VOTED AGAINST THE PRESIDENT	.08	.15	.12
DID NOT VOTE	.07	.08	.10

(a) What is the probability that a randomly chosen respondent voted for the president?

(b) What is the probability that a randomly chosen respondent is pessimistic about the economy?

(c) What is the probability that a respondent who voted for the president will be pessimistic about the economy?

(d) What is the probability that a respondent who is pessimistic about the economy voted for the president?

(e) Are views on the economy independent of how respondents voted?

65. The theme music for a new movie is to be released on record. The producers' joint probability assessment for the success of the movie and the record is shown in the table.

SUCCESS OF RECORD	SUCCESS OF MOVIE	
	MODERATE	HIGH
MODERATE	.29	.16
HIGH	.17	.38

(a) What is the probability of high success for the record?

(b) What is the probability of high success for the movie?

(c) What is the probability the record will be highly successful if the movie is highly successful?

(d) What is the probability the movie will be highly successful if the record is highly successful?

(e) Are the levels of success of the record and the movie independent?

66. Subscriptions to *American History Illustrated*[3] are classified as gift, previous renewal, direct mail, or subscription service. In January 1979, 8% of expiring subscriptions were gift; 41%, previous renewal; 6%, direct mail; and 45%, subscription service. The percentages of renewals in these four categories were 81%, 79%, 60%, and 21%, respectively. In February 1979, 10% of expiring subscriptions were gift; 57%, previous

---

[3] This example is adapted from C. H. Wagner, "Simpson's Paradox in real life," *American Statistician, 36* (1982), 46–48.

renewal; 24%, direct mail; and 9%, subscription service. The percentages of renewals were 80%, 76%, 51%, and 14%, respectively.

(a) Find the probability that a randomly chosen subscription expiring in January 1979 was renewed.

(b) Find the probability that a randomly chosen subscription expiring in February 1979 was renewed.

(c) Verify that the probability in part (b) is higher than that in part (a). Do you believe that the editors of *American History Illustrated* should view the change from January to February as a positive or negative development?

67. In a large city, 7% of the inhabitants have contracted a particular disease. A test for this disease is positive in 80% of people who have the disease, and is negative in 80% of people who do not have the disease. What is the probability that a person for whom the test result is positive has the disease?

68. A life insurance salesman finds that, of all the sales he makes, 70% are to people who already own policies. He also finds that, of all contacts for which no sale is made, 50% already own life insurance policies. Furthermore, 40% of all contacts result in sales. What is the probability that a sale will be made to a contact who already owns a policy?

69. A professor finds that she awards a final grade of A to 20% of the students. Of those who obtain a final grade of A, 70% obtained an A in the midterm examination. Also, 10% of students who failed to obtain a final grade of A earned an A in the midterm exam. What is the probability that a student with an A on the midterm examination will obtain a final grade of A?

70. The accompanying table shows, for 1,000 forecasts of earnings per share made by financial analysts, the numbers of forecasts and outcomes in particular categories (compared with the previous year).

OUTCOME	FORECAST		
	IMPROVEMENT	ABOUT THE SAME	WORSE
IMPROVEMENT	220	90	65
ABOUT THE SAME	108	150	75
WORSE	72	80	140

(a) Find the probability that, if the forecast is for a worse performance in earnings, this outcome will result.

(b) If the forecast is for an improvement in earnings, find the probability that this outcome fails to result.

discrete random variables and probability distributions

CHAPTER FOUR

4.1 RANDOM VARIABLES

Suppose that a random experiment is to be carried out and that numerical values can be attached to the possible outcomes. In experiments such as throwing a die or measuring a family's income, the outcomes are naturally in numerical form. When this is not the case it may still be useful and meaningful to attach numbers to the outcomes, particularly in experiments where only two outcomes are possible. For example, a component produced by an industrial process might be classified as "defective" or "not defective." We could attach the value 1 to the former possibility and 0 to the latter.

Before the random experiment is carried out there will be uncertainty as to the outcome and, as we have seen in the previous chapter, this uncertainty can be quantified in terms of probability statements. When the outcomes are numerical values these probabilities can be conveniently summarized through the notion of a **random variable.**

Definition

A **random variable** is a variable that takes on numerical values determined by the outcome of a random experiment.

It is important to distinguish between a random variable and the possible values it can take. Notationally we do this by using capital letters, such as X, to denote the random variable and the corresponding lowercase x to denote a possible value. For example, prior to the result being observed in the throw of a die, the random variable X can be used to denote the outcome. This random variable can take the specific values $x = 1, x = 2, \ldots, x = 6$, each with probability $\frac{1}{6}$.

A further important distinction is between **discrete** and **continuous** random variables. The die throw provides an example of the former; there are only six possible outcomes, and a probability can be attached to each.

Definition

A random variable is **discrete** if it can take on at most countably many values.

It follows from the definition that any random variable that can take on only finitely many values is discrete. For example, the number of heads resulting from ten throws of a coin is a discrete random variable. On the other hand, if the number of possible outcomes is infinite but countable, the random variable is still discrete. An example is the number of throws of a coin needed before a head first appears. The possible outcomes are $1, 2, 3, \ldots$, and a probability can be attached to each. (A discrete random variable that can take a countably infinite number of values will be discussed in Section 4.7.) Other examples of discrete random variables are the following:

1. The number of fatal automobile accidents in a city in a given month
2. The number of customers arriving at a check-out counter in an hour
3. The number of errors detected in a corporation's accounts
4. The number of claims on a medical insurance policy in a particular year

By contrast, suppose we are interested in the day's high temperature. The random variable, temperature, is measured on a continuum, and is said to be **continuous.**

Definition

A random variable is **continuous** if it can take any value in an interval.

For continuous random variables one cannot attach probabilities to specific values. For example, the probability that today's high temperature will be precisely 77.236 degrees Fahrenheit is 0. It will certainly not be *precisely* that figure. However, probabilities can be determined for intervals, so that one could attach a probability to the event "The high temperature today will be between 75 and 80 degrees." Other examples of continuous random variables are

1. The income in a year for a family
2. The amount of oil imported into the United States in a particular month
3. The change in the price of an ounce of gold in a month
4. The time that elapses between the installation of a new component and its failure
5. The percentage of impurity in a batch of chemicals

The distinction that has been made between discrete and continuous random variables may appear rather artificial. After all, rarely is anything actually measured on a continuum. For example, the day's high temperature cannot be reported more precisely than the measuring instrument allows. Moreover, a family's income in a year will be some integer number of cents. However, when measurements can be made on such a fine scale that differences between adjacent values are of no significance, it is convenient to act as if they had truly been made on a continuum. For example, the difference between a family income of \$15,276.21 and \$15,276.22 is of very little significance, and the attachment of probabilities to each would be a tedious and worthless exercise.

For practical purposes we will treat as discrete those random variables for which probability statements about the individual possible outcomes have worthwhile meaning, while other random variables will be regarded as continuous. Because of this distinction it is convenient to treat separately these two classes. In this chapter, discrete random variables are discussed, while continuous random variables will be treated in Chapter 5.

4.2 PROBABILITY DISTRIBUTIONS FOR DISCRETE RANDOM VARIABLES

Suppose that X is a discrete random variable, and x is one of its possible values. The probability that the random variable X takes the specific value x is denoted $P(X = x)$. The **probability distribution** of a random variable is a representation of the probabilities for all the possible outcomes. This representation might be algebraic, graphical, or tabular. For discrete random variables, one simple procedure is to list the probabilities of all possible outcomes, according to the values of x.

Definition

The **probability function,** $P_X(x)$, of a discrete random variable X expresses the probability that X takes the value x, as a function of x. That is,

$$P_X(x) = P(X = x)$$

where the function is evaluated at all possible values of x.

Because the probability function takes only nonzero values at discrete points x, it is sometimes called a **probability mass function.** Once the probabilities have been calculated, the function can easily be graphed.

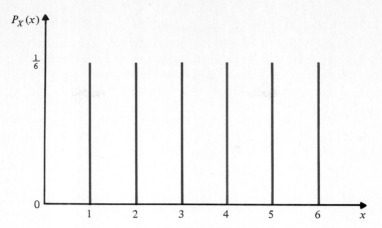

FIGURE 4.1 Probability function for Example 4.1

Example 4.1
A die is rolled. Let the random variable X denote the number resulting. Since $P(X = 1) = P(X = 2) = \cdots = P(X = 6) = \frac{1}{6}$, the probability function is

$$P_X(x) = P(X = x) = \frac{1}{6} \quad \text{for } x = 1, 2, 3, \ldots, 6$$

The function takes the value zero for all other values of x, which cannot occur. The probability function is graphed in Figure 4.1, where spikes of height $\frac{1}{6}$ represent probability masses at the points $x = 1$, $x = 2$, $\ldots$, $x = 6$.

Example 4.2
A roulette wheel contains 38 slots, numbered 1, 2, $\ldots$, 36, 0, and 00. One possible bet is on the event "Odd number results," so that a player making this bet would win in 18 of the 38 possible outcomes. Since there are 18 odd numbers,

$$P(\text{Odd number results}) = \frac{18}{38} = \frac{9}{19}$$

If a player makes a bet in which he or she loses \$1 if an odd number does not result and wins \$1 if it does, we can denote the player's "gain" (in dollars) by the random variable X, where

$$X = \begin{cases} -1 & \text{if odd number does not result} \\ 1 & \text{if odd number results} \end{cases}$$

Then $P(X = 1) = \frac{9}{19}$, and hence $P(X = -1) = \frac{10}{19}$ and so the probability function of the random variable X is

$$P_X(x) = \begin{cases} \dfrac{10}{19} & \text{for } x = -1 \\ \dfrac{9}{19} & \text{for } x = 1 \end{cases}$$

This function is graphed in Figure 4.2.

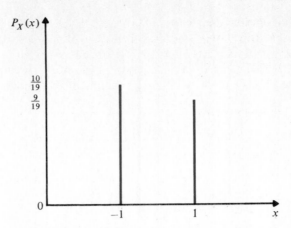

FIGURE 4.2 Probability function for Example 4.2

The probability function of a discrete random variable must satisfy the two conditions given in the box.

Properties of Probability Functions of Discrete Random Variables

Let X be a discrete random variable with probability function $P_X(x)$. Then

(i) $P_X(x) \geq 0$ for any value x
(ii) The individual probabilities sum to 1; that is,

$$\sum_x P_X(x) = 1$$

where the notation indicates summation over all possible values x.

Property (i) merely states that probabilities cannot be negative. Property (ii) follows from the fact that the events "$X = x$," for all possible values x, are mutually exclusive and collectively exhaustive. The probabilities for these events must therefore sum to 1. That this is in fact so for Examples 4.1 and 4.2 can be verified directly.

Definition

The **cumulative probability function**, $F_X(x_0)$ of a random variable X expresses the probability that X does not exceed the value x_0, as a function of x_0. That is,

$$F_X(x_0) = P(X \leq x_0)$$

where the function is evaluated at all values x_0.

For discrete random variables, the cumulative probability function is sometimes called the **cumulative mass function.** It can be seen from the definition that, as x_0 increases, the cumulative probability function will change values only at those points x_0 that can be taken by the random variable with positive probability. Its evaluation at these points can be carried out in terms of the probability function.

Relationship Between Probability Function and Cumulative Probability Function

Let X be a random variable with probability function $P_X(x)$ and cumulative probability function $F_X(x_0)$. Then

$$F_X(x_0) = \sum_{x \leq x_0} P_X(x)$$

where the notation implies that summation is over all possible values x that are less than or equal to x_0.

The result in the box follows since the event "$X \leq x_0$" is the union of the mutually exclusive events "$X = x$" for every x less than or equal to x_0. The probability of the union is then the sum of these individual event probabilities.

Example 4.3

In the die throwing experiment of Example 4.1, where the random variable X denotes the number observed, we have the probability function

$$P_X(x) = \frac{1}{6} \quad \text{for } x = 1, 2, \ldots, 6$$

Now if x_0 is some number less than 1, X cannot be less than x_0, so

$$F_X(x_0) = P(X \leq x_0) = 0 \quad \text{for } x_0 < 1$$

If x_0 is greater than or equal to 1 but strictly less than 2, the only way for X to be less than or equal to x_0 is if $X = 1$. Hence,

$$F_X(x_0) = P(X \leq x_0) = P_X(1) = \frac{1}{6} \quad \text{for } 1 \leq x_0 < 2$$

If x_0 is greater than or equal to 2 but strictly less than 3, then X is less than or equal to x_0 if and only if either $X = 1$ or $X = 2$, so that

$$F_X(x_0) = P(X \leq x_0) = P_X(1) + P_X(2) = \frac{1}{3} \quad \text{for } 2 \leq x_0 < 3$$

Continuing in this way, we see that if x_0 is any number greater than or equal to 6, then X will certainly be less than x_0, so

$$F_X(x_0) = P(X \leq x_0) = \sum_{x=1}^{6} P_X(x) = 1 \quad \text{for } x_0 \geq 6$$

The cumulative probability function may then be written as

$$F_X(x_0) = \begin{cases} 0 & \text{if } x_0 < 1 \\ \dfrac{j}{6} & \text{if } j \le x_0 < j + 1 \qquad (j = 1, 2, \ldots, 5) \\ 1 & \text{if } x_0 \ge 6 \end{cases}$$

This function is plotted in Figure 4.3, from which it can be seen that the cumulative probability function increases in steps until the value 1 is attained.

For discrete random variables, the cumulative probability function is always in the form of a step function beginning at 0 and ending at 1. These properties are expressed formally in the accompanying box.

Properties of Cumulative Probability Functions for Discrete Random Variables

Let X be a discrete random variable with cumulative probability function $F_X(x_0)$. Then

(i) $0 \le F_X(x_0) \le 1$ for every number x_0
(ii) If x_0 and x_1 are two numbers with $x_0 < x_1$, then

$$F_X(x_0) \le F_X(x_1)$$

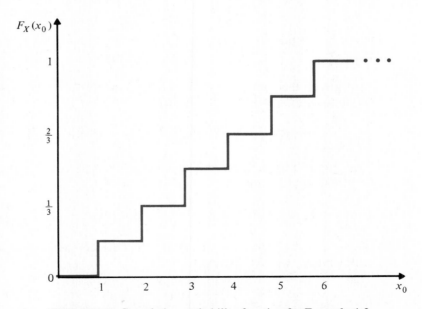

FIGURE 4.3 Cumulative probability function for Example 4.3

Property (i) simply states that a probability cannot be less than 0 or greater than 1. Property (ii) implies that the probability that a random variable does not exceed some number cannot be more than the probability that it does not exceed any larger number.

4.3 EXPECTATIONS FOR DISCRETE RANDOM VARIABLES

The probability distribution contains all the information about the probability properties of a random variable, and graphical inspection of this distribution can certainly be valuable. However, it is frequently desirable to have some numerical summary measures of the distribution's characteristics.

In order to obtain a measure of the center of a probability distribution, we introduce here the notion of the *expectation* of a random variable. In our discussion of sets of numerical observations in Chapter 2, we often found it convenient to compute the mean as a measure of central location. The **expected value** is the corresponding measure of central location for a random variable. Before introducing its definition, it is convenient to dismiss a superficially attractive alternative measure.

Consider the following example: A review of textbooks in a segment of the business area found that 81% of all pages of text were error-free, 17% of all pages contained one error, while the remaining 2% contained two errors. Therefore, if we let the random variable X denote the number of errors in a page chosen at random from one of these books, we see that its possible values are 0, 1, and 2, with probability function

$$P_X(0) = .81; \qquad P_X(1) = .17; \qquad P_X(2) = .02$$

Now, one possible measure of the central location of a random variable might be the simple average of the values it can take. In our example, the possible numbers of errors on a page are 0, 1, and 2. Their average is, then, one error. However, a moment's reflection will convince the reader that this is an absurd measure of central location. In calculating this average, we have paid no attention to the fact that 81% of all pages contain no errors, while only 2% contain two errors. In order to obtain a sensible measure of central location it is desirable to *weight* the various possible outcomes by the probabilities of their occurrence.

Definition

The **expected value**, $E(X)$, of a discrete random variable X is defined as

$$E(X) = \sum_x xP_X(x)$$

where the notation indicates that summation extends over all possible values x.

The expected value of a random variable is called its **mean,** and is denoted μ_X.

The definition of expected value can be motivated in terms of long-run relative frequencies. Suppose that a random experiment is repeated N times and that the event "$X = x$" occurs in N_x of these trials. The average of the values taken by the random variable over all N trials will then be the sum of xN_x/N over all possible values x. Now, as the number of replications N becomes infinitely large, the ratio N_x/N tends to the probability of the occurrence of the event "$X = x$"—that is, to $P_X(x)$. Hence, the quantity xN_x/N tends to $xP_X(x)$. Thus, the expected value can be viewed as the long-run average value that a random variable would take over a very large number of trials. Recall that, in Chapter 2, we used the word *mean* for the average of a set of numerical observations. The foregoing justifies the use of the same term for the expectation of a random variable.

Example 4.4

The probability function for the number of errors, X, in pages from business textbooks is

$$P_X(0) = .81; \qquad P_X(1) = .17; \qquad P_X(2) = .02$$

Find the mean number of errors per page.

We have

$$\mu_X = E(X) = \sum_x xP_X(x)$$

$$= (0)(.81) + (1)(.17) + (2)(.02) = 0.21$$

We thus conclude that, over a large number of pages, we would expect to find an average of 0.21 error per page.

Figure 4.4 shows the probability function, with the location of the mean indicated.

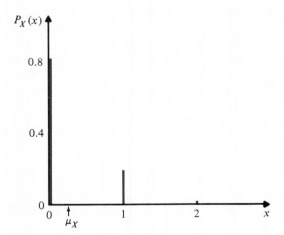

FIGURE 4.4 Probability function for number of errors per page in business textbooks, and location of population mean, μ_X, in Example 4.4

The notion of expectation is not restricted to the random variable itself, but can be applied to any function of the random variable. For example, a contractor may be uncertain of the time required to complete a contract. This uncertainty could be represented by a random variable whose possible values are the numbers of days elapsing from the beginning to the completion of work on the contract. However, the contractor's primary concern is not with the time taken, but rather with the cost of fulfilling the contract. This cost will be a function of the time taken, so that in determining expected cost, it is necessary to find the expectation of a function of the random variable "Time to completion."

Definition

Let X be a discrete random variable with probability function $P_X(x)$, and $g(X)$ be some function of X. Then the **expected value**, $E[g(X)]$, of that function is defined as

$$E[g(X)] = \sum_x g(x)P_X(x)$$

The definition of $E[g(X)]$ can be motivated in precisely the same way as the previous one. That is, the expectation can be thought of as the average value that $g(X)$ would take over a very large number of repeated trials.

In Chapter 2, one useful measure of the spread of a set of numerical observations was found to be the *variance,* the average of the squared discrepancies of the observations from their mean. In the same way, this notion can be used to measure dispersion in the probability distribution of a random variable. In finding the expectation of the squared discrepancy, $(X - \mu_X)^2$, of a random variable from its mean, we are, in forming an average, weighting each possible squared discrepancy by the probability of its occurrence.

Definition

Let X be a discrete random variable. The expectation of the squared discrepancy about the mean, $(X - \mu_X)^2$, is called the **variance,** denoted σ_X^2, and given by

$$\sigma_X^2 = E[(X - \mu_X)^2] = \sum_x (x - \mu_X)^2 P_X(x)$$

The **standard deviation,** σ_X, is the positive square root of the variance.

Taking the square root of the variance to obtain the standard deviation yields a quantity in the original units of measurement, as noted in Chapter 2.

When the probability function is known, the mean and variance of a random variable can be computed through direct application of the definitions.

In some practical applications an alternative, but equivalent, formula for the variance is preferable for computational purposes. That the alternative formula is indeed equivalent to the formula given in the definition can be verified algebraically.[1] It can be conveniently remembered as the expectation of the square less the square of the expectation of X.

Variance of a Discrete Random Variable (Alternative Formula)

The variance of a discrete random variable X can be expressed as

$$\sigma_X^2 = E(X^2) - \mu_X^2$$

$$= \sum_x x^2 P_X(x) - \mu_X^2$$

Example 4.5

Suppose that the probability function for the number of errors, X, in pages from business textbooks is

$$P_X(0) = .81; \qquad P_X(1) = .17; \qquad P_X(2) = .02$$

In Example 4.4, we found the mean number of errors per page to be

$$\mu_X = 0.21$$

To obtain the variance, we first find the expectation of the squares; that is,

$$E(X^2) = \sum_x x^2 P_X(x) = (0)^2(.81) + (1)^2(.17) + (2)^2(.02) = 0.25$$

The variance is then

$$\sigma_X^2 = E(X^2) - \mu_X^2 = 0.25 - (0.21)^2 = 0.2059$$

---

[1]Begin with the original definition of variance:

$$\sigma_X^2 = \sum_x (x - \mu_X)^2 P_X(x) = \sum_x (x^2 - 2\mu_X x + \mu_X^2) P_X(x)$$

$$= \sum_x x^2 P_X(x) - 2\mu_X \sum_x x P_X(x) + \mu_X^2 \sum_x P_X(x)$$

But we have seen that $\sum_x x P_X(x) = \mu_X$ and $\sum_x P_X(x) = 1$, so that

$$\sigma_X^2 = \sum_x x^2 P_X(x) - 2\mu_X^2 + \mu_X^2$$

$$= \sum_x x^2 P_X(x) - \mu_X^2$$

Finally, the standard deviation of the number of errors per page is

$$\sigma_X = \sqrt{\sigma_X^2} = \sqrt{0.2059} = 0.4538$$

The concept of variance can be very useful in comparing the dispersions of probability distributions. Consider, for example, viewing as a random variable the return over a year on an investment. Two investments may have the same expected returns, but will still differ in an important way if the variances of these returns are substantially different. A higher variance indicates that returns substantially different from the mean are more likely than if the variance of returns is small. In this context then, variance of return can be associated with the concept of the riskiness of an investment—the higher the variance, the greater the risk.

We have defined the expectation of a function of a random variable X. The linear function $a + bX$, where a and b are constant fixed numbers, is of particular interest. Let X be a random variable that takes the value x with probability $P_X(x)$, and consider a new random variable Z, defined by

$$Z = a + bX$$

Then, when the random variable X takes the specific value x, Z must take the value $a + bx$. We frequently require the mean and variance of such variables. These quantitites are given in the box.

Let X be a random variable with mean μ_X and variance σ_X^2 and let a and b be any constant fixed numbers. Define the random variable $Z = a + bX$. Then the mean and variance of Z are

$$\mu_Z = E(a + bX) = a + b\mu_X \tag{4.3.1}$$

and

$$\sigma_Z^2 = \mathrm{Var}(a + bX) = b^2\sigma_X^2 \tag{4.3.2}$$

To verify these results, note that it follows from the definition of expectation that if Z takes values $a + bx$ with probabilities $P_X(x)$, its mean is

$$E(Z) = \mu_Z = \sum_x (a + bx)P_X(x)$$

$$= a \sum_x P_X(x) + b \sum_x xP_X(x)$$

Then, since the first summation on the right-hand side of this equation is 1 [by property (ii) of Section 4.2] and the second summation is, by definition, the mean of X, we have

$$E(Z) = a + b\mu_X$$

as in equation 4.3.1. Further, the variance of Z is, by definition,

$$\sigma_Z^2 = E[(Z - \mu_Z)^2] = \sum_x [(a + bx) - \mu_Z]^2 P_X(x)$$

Substituting $a + b\mu_X$ for μ_Z then gives

$$\sigma_Z^2 = \sum_x (bx - b\mu_X)^2 P_X(x) = b^2 \sum_x (x - \mu_X)^2 P_X(x)$$

and, since the summation on the right-hand side of this equation is, by definition, the variance of X, the result 4.3.2 follows.

Example
4.6

A contractor is interested in the total cost of a project on which he intends to bid. He estimates that materials will cost \$5,000 and that his labor costs will be \$300 per day. Then, if the project takes X days to complete, the total labor costs will be 300 X dollars, and the total cost of the project (in dollars) will be

$$C = 5,000 + 300X$$

The contractor forms subjective probability assessments of likely completion times for the project, as indicated in the table.

COMPLETION TIME X (IN DAYS)	10	11	12	13	14
PROBABILITY	.1	.3	.3	.2	.1

The mean and variance for completion time X can then be found directly as

$$\mu_X = E(X) = \sum_x xP_X(x)$$

$$= (10)(.1) + (11)(.3) + (12)(.3) + (13)(.2) + (14)(.1) = 11.9 \text{ days}$$

and

$$\sigma_X^2 = E[(X - \mu_X)^2] = \sum_x (x - \mu_X)^2 P_X(x)$$

$$= (10 - 11.9)^2(.1) + (11 - 11.9)^2(.3) + \cdots + (14 - 11.9)^2(.1) = 1.29$$

The mean and variance of total cost C can now be obtained using equations 4.3.1 and 4.3.2. The expected cost is

$$\mu_C = E(5,000 + 300X) = 5,000 + 300\mu_X$$

$$= 5,000 + (300)(11.9) = 8,570 \text{ dollars}$$

and the variance is

$$\sigma_C^2 = \text{Var}(5,000 + 300X) = (300)^2 \sigma_X^2$$

$$= (90,000)(1.29) = 116,100$$

so that the standard deviation is

$$\sigma_C = \sqrt{\sigma_C^2} = 340.73 \text{ dollars}$$

The following special cases of results 4.3.1 and 4.3.2 are of interest:

(i) By setting $b = 0$ in these equations it follows that, for any constant a,

$$E(a) = a; \qquad \text{Var}(a) = 0$$

Thus, if a random variable always takes the value a, then it will have mean a and variance 0.

(ii) Setting $a = 0$ in these equations, we have, for any constant b,

$$E(bX) = b\mu_X; \qquad \text{Var}(bX) = b^2 \sigma_X^2$$

Thus, if a random variable is multiplied by any constant, the mean is multiplied by the same constant and the variance is multiplied by the square of that constant.

(iii) Setting $a = -\mu_X/\sigma_X$ and $b = 1/\sigma_X$, we have

$$Z = a + bX = \frac{X - \mu_X}{\sigma_X}$$

so that

$$E\left(\frac{X - \mu_X}{\sigma_X}\right) = \frac{-\mu_X}{\sigma_X} + \frac{1}{\sigma_X}\mu_X = 0$$

and

$$\text{Var}\left(\frac{X - \mu_X}{\sigma_X}\right) = \frac{1}{\sigma_X^2}\sigma_X^2 = 1$$

EXERCISES

1. An automobile dealer calculates the proportion of new cars sold that have been returned various numbers of times for the correction of defects during the warranty period. The results are shown in the table.

NUMBER OF RETURNS	0	1	2	3	4
PROPORTION	.26	.35	.30	.06	.03

(a) Draw the probability function.

(b) Calculate and draw the cumulative probability function.

2. A mail-order company examines its records to determine the length of time (in weeks) between sending out an item and receipt of payment. The accompanying table shows the results.

NUMBER OF WEEKS		1	2	3	4	5
PROPORTION		.15	.48	.20	.12	.05

(a) Draw the probability function.

(b) Calculate and draw the cumulative probability function.

(c) For a randomly chosen item, what is the probability that payment will be received within three weeks or less from the time the item is sent out?

3. In a study to investigate energy use, the numbers of occupants in cars on an interstate highway were counted. The accompanying table shows proportions for all cars.

NUMBER OF OCCUPANTS		1	2	3	4	5	6
PROPORTION		.42	.38	.10	.07	.02	.01

(a) Draw the probability function.

(b) Calculate and draw the cumulative probability function.

(c) What is the probability that a randomly chosen car will have at least four occupants?

4. Daily earnings of manufacturing workers at a shoe company are determined, on union scales, in multiples of $5. The proportions of workers at each earnings level are shown in the table.

DOLLAR EARNINGS LEVEL		60	65	70	75	80
PROPORTION		.10	.18	.32	.28	.12

(a) Draw the probability function.

(b) Calculate and draw the cumulative probability function.

(c) What is the probability that a randomly chosen worker will have weekly earnings between $65 and $75, inclusive?

5. A large shipment of lightbulbs contains 20% defectives. Two lightbulbs are chosen at random from the shipment and checked. Let the random variable X denote the number of defectives found. Find the probability function of this random variable.

6. A case of ten lightbulbs contains two defectives. Two lightbulbs are chosen at random from the case and checked. Let the random variable X denote the number of defectives found. Find the probability function of this random variable. Explain why your answer is different from that of Exercise 5.

7. A salesperson finds that the probability that any particular contact will produce an immediate order is 0.25, and that the outcomes from successive contacts are independent of one another. On a particular day, the salesperson decides to continue contacting potential customers until the first order is received. Let the random variable X denote the number of contacts made until this order is secured.

(a) Find the probability function of X.

(b) Find the cumulative probability function of X.

(c) Find the probability that the number of contacts needed is between two and four, inclusive.

8. Refer to the information in Exercise 1. Find the mean and variance of the number of returns of a new car to the dealer during the warranty period.

9. Refer to the information in Exercise 2. Find the mean and standard deviation for the number of weeks between sending out an item and receipt of payment for the mail-order company.

10. Refer to the information in Exercise 4. Find the mean and standard deviation for the weekly earnings of manufacturing workers at the shoe company.

11. A basketball player, who makes 70% of his free throws, comes to the line to shoot a "one-and-one." (That is, if the first shot is successful, he is also allowed a second shot. However, if he misses the first shot, no second shot is taken. One point is scored for each successful shot.) Assume that the outcome of the second shot, if any, is independent of that of the first. Find the expected number of points resulting from the "one-and-one." Compare this with the expected number of points from a "two-shot foul," where a second shot is allowed whatever the outcome of the first.

12. You have been given a sweepstakes ticket. There are five possible mutually exclusive outcomes. You could, with probability $1/20{,}000$, win a grand prize valued at $8,000. You could also, with probability $1/20{,}000$, win a major prize of value $4,000. A minor prize valued at $150 can be won with probability $1/5{,}000$, and a very minor prize of value $3 will be won with probability $1/150$. Finally, you could win nothing (valued at $0). Find the mean and standard deviation of the value of your ticket.

13. A salesperson estimates the probabilities given in the table for the number of weekly sales of a particular photocopying machine.

NUMBER OF SALES	0	1	2	3	4	5
PROBABILITY	.10	.15	.30	.25	.15	.05

(a) Find the mean and standard deviation of the number of sales.

(b) As a weekly income, the salesperson receives $200, plus a commission of $100 for each sale made. Find the mean and standard deviation of his or her weekly income.

14. A company makes packages of candy. The numbers of pieces of candy per package vary, as indicated in the accompanying table.

NUMBER OF PIECES	15	16	17	18
PROPORTION OF PACKAGES	.20	.40	.30	.10

(a) Find the mean and standard deviation of the number of pieces of candy per package.

(b) The cost (in cents) of producing a package of candy is $3 + 1.5X$, where X is the number of pieces of candy in the package. The revenue from selling a package, regardless of how many pieces it contains, is 30 cents. If profit is defined as the difference between revenue and cost, find the mean and standard deviation of profit per package.

15. A store owner stocks an out-of-town newspaper, which is sometimes requested by a small number of customers. Each copy of this newspaper costs him 20 cents, and he sells them for 25 cents each. Any copies left over at the end of the day have no value and are destroyed. Any requests for copies which cannot be met because stocks have been exhausted are considered by the store owner as a loss of 2 cents in goodwill. The probability distribution of the number of requests for the newspaper in a day is shown in the accompanying table. If the store owner defines total daily profit as total revenue from newspaper sales, less total cost of newspapers ordered, less goodwill loss from unsatisfied demand, how many copies per day should he order to maximize expected profit?

NUMBER OF REQUESTS	0	1	2	3	4	5
PROBABILITY	.10	.15	.20	.30	.15	.10

16. A factory manager is considering whether to replace a temperamental machine. A review of past records indicates the accompanying probability distribution for the number of breakdowns of this machine in a week.

NUMBER OF BREAKDOWNS	0	1	2	3	4
PROBABILITY	.10	.30	.40	.15	.05

(a) Find the mean and standard deviation of the number of weekly breakdowns.

(b) It is estimated that each breakdown costs the company $1,000 in lost output. Find the mean and standard deviation of the weekly cost to the company from breakdown of this machine.

17. The probability distribution for the number of machines that might break down in a day has been determined for a machine shop. The probabilities for 0, 1, and 2 breakdowns are, respectively, .3, .6, and .1.

(a) Find the mean and standard deviation of the number of daily breakdowns.

(b) Daily repair costs, R, are found to be (in dollars)

$$R = 300 + 100X^2$$

where X is the number of breakdowns. Calculate expected daily repair cost.

18. An investor is considering three strategies for a $1,000 investment. The probable returns are estimated as follows:

STRATEGY 1 ☐ A profit of $10,000 with probability 0.15 and a loss of $1,000 with probability 0.85

STRATEGY 2 ☐ A profit of $1,000 with probability 0.50, a profit of $500 with a probability of 0.30, and a loss of $500 with a probability of 0.20

STRATEGY 3 ☐ A certain profit of $400

Which strategy has the highest expected profit? Would you necessarily advise the investor to adopt this strategy?

4.4 JOINTLY DISTRIBUTED DISCRETE RANDOM VARIABLES

In Section 3.7 we considered joint probabilities. We now consider the case where we wish to examine two or more, possibly related, discrete random variables. As for a single random variable, the probabilities for all possible outcomes can be summarized in a probability function.

Definition

Let X and Y be a pair of discrete random variables. Their **joint probability function** expresses the probability that simultaneously X takes the specific value x and Y takes the value y, as a function of x and y. The notation used is $P_{X,Y}(x, y)$, so that

$$P_{X,Y}(x, y) = P(X = x \ \cap \ Y = y)$$

More generally, if $X_1, X_2, \ldots, X_K$ are K discrete random variables, their joint probability function is

$$P_{X_1,X_2,\ldots,X_K}(x_1, x_2, \ldots, x_K) = P(X_1 = x_1 \ \cap \ X_2 = x_2 \ \cap \ \cdots \ \cap \ X_K = x_K)$$

To illustrate, consider a pair of random variables X and Y, measuring, respectively, a consumer's satisfaction with food stores in a particular town and the number of years residence in that town. Suppose that X can take values 1, 2, 3, or 4, ranging from low to high satisfaction levels, and that Y takes the value 1 if the consumer has lived in the town less than six years, and 2 otherwise. The main body of Table 4.1 shows the eight joint probabilities for X and Y. These constitute the joint probability function for this pair of random variables.[2] Thus, for instance,

$$P_{X,Y}(1, 1) = .04; \qquad P_{X,Y}(1, 2) = .07$$

Hence, for example, the probability is .07 that a randomly chosen consumer both has satisfaction level 1 and has lived in the town more than six years.

TABLE 4.1 Probabilities for consumer satisfaction (X) and time in residence (Y)

Y	X				
	1	2	3	4	TOTALS
1	.04	.14	.23	.07	.48
2	.07	.17	.23	.05	.52
TOTALS	.11	.31	.46	.12	1.00

---

[2] These probabilities are estimated from results reported in J. A. Miller, "Store satisfaction and aspiration theory," *Journal of Retailing*, 52 (Fall 1976), 65–84.

When dealing with jointly distributed random variables, we are frequently interested in the probability functions for the individual random variables.

Definition

Let X and Y be a pair of jointly distributed random variables. In this context the probability function of the random variable X is called its **marginal probability function,** and is obtained by summing the joint probabilities over all possible y values; that is,

$$P_X(x) = \sum_y P_{X,Y}(x, y)$$

Similarly, the marginal probability function of the random variable Y is

$$P_Y(y) = \sum_x P_{X,Y}(x, y)$$

More generally, if $X_1, X_2, \ldots, X_K$ are K jointly distributed discrete random variables, then the marginal probability function of any one of them is obtained by summing the joint probabilities over all possible combinations of values of all the others.

The marginal probabilities for the random variables X and Y of Table 4.1 can be read from the row and column totals. For consumer satisfaction, the marginal probability function is

$$P_X(1) = .11; \qquad P_X(2) = .31; \qquad P_X(3) = .46; \qquad P_X(4) = .12$$

Thus, for instance, the probability is .11 that a randomly chosen consumer has satisfaction level 1. Similarly, for time in residence in the town,

$$P_Y(1) = .48; \qquad P_Y(2) = .52$$

so that the probability is .48 that a randomly chosen consumer has lived in the town less than six years.

We have already seen that the marginal probabilities must sum to 1; it follows that the joint probabilities necessarily sum to 1. Further, since any probability must be nonnegative, joint probability functions must have the properties given in the box.

Properties of Joint Probability Functions of Discrete Random Variables

Let X and Y be discrete random variables with joint probability function $P_{X,Y}(x, y)$. Then

(i) $P_{X,Y}(x, y) \geq 0$ for any pair of values x and y

(ii) The sum of the joint probabilities $P_{X,Y}(x,y)$ over all possible pairs of values (x,y) must be 1.

The conditional probability function of one random variable, given specified values of another, is the collection of conditional probabilities, which can be found in exactly the manner described in Section 3.7.

Definition

Let X and Y be a pair of jointly distributed discrete random variables. The **conditional probability function** of the random variable Y, given that the random variable X takes the value x, expresses the probability that Y takes the value y, as a function of y, when the value x is specified for X. This is denoted $P_{Y|X}(y|x)$, so that by the definition of conditional probability,

$$P_{Y|X}(y|x) = \frac{P_{X,Y}(x,y)}{P_X(x)}$$

Similarly, the conditional probability function of X, given $Y = y$, is

$$P_{X|Y}(x|y) = \frac{P_{X,Y}(x,y)}{P_Y(y)}$$

To illustrate, consider again the information in Table 4.1. The conditional probability that time in residence Y takes the value 1, given that consumer satisfaction X takes the value 2 is

$$P_{Y|X}(1|2) = P(Y = 1|X = 2) = \frac{P(X = 2 \ \cap \ Y = 1)}{P(X = 2)} = \frac{P_{X,Y}(2,1)}{P_X(2)}$$

$$= \frac{.14}{.31} = .45$$

That is, the probability that a consumer will have resided less than six years in the town given that the satisfaction level is 2 is .45. The conditional probability function of Y given X is displayed in Table 4.2.

Similarly, we can find the conditional probability function for consumer satisfaction given time in residence. The probability that X takes the value 4, given that Y takes the value 2 is

$$P_{X|Y}(4|2) = P(X = 4|Y = 2) = \frac{P(X = 4 \ \cap \ Y = 2)}{P(Y = 2)} = \frac{P_{X,Y}(4,2)}{P_Y(2)}$$

$$= \frac{.05}{.52} = .10$$

TABLE 4.2 Conditional probabilities for time in residence (Y) given consumer satisfaction (X)

Y	X			
	1	2	3	4
1	.36	.45	.50	.58
2	.64	.55	.50	.42

Therefore, the probability that a consumer will have the highest satisfaction level, given that he or she has been resident at least six years in the town, is .10. The conditional probability function of X given Y is shown in Table 4.3.

In Chapter 3 we discussed the notion of independence of events. This concept extends directly to random variables.

Definition

The random variables X and Y are said to be **independent** if and only if their joint probability function is the product of their marginal probability functions, that is, if and only if

$$P_{X,Y}(x, y) = P_X(x)P_Y(y)$$

for all possible pairs of values x and y.

More generally, the K random variables $X_1, X_2, \ldots, X_K$ are independent if and only if

$$P_{X_1,X_2,\ldots,X_K}(x_1, x_2, \ldots, x_K) = P_{X_1}(x_1)P_{X_2}(x_2) \cdots P_{X_K}(x_K)$$

From the definition of conditional probability functions, it follows that if the random variables X and Y are independent, then the conditional probability function of Y given X is the same as the marginal probability function of Y; that is,

$$P_{Y|X}(y|x) = P_Y(y)$$

TABLE 4.3 Conditional probabilities for consumer satisfaction (X) given time in residence (Y)

Y	X			
	1	2	3	4
1	.08	.29	.48	.15
2	.13	.33	.44	.10

Similarly, it follows that

$$P_{X|Y}(x \mid y) = P_X(x)$$

For the example of Table 4.1 we have

$$P_{X|Y}(1, 1) = .04; \qquad P_X(1) = .11; \qquad P_Y(1) = .48$$

so that

$$P_X(1)P_Y(1) = (.11)(.48) = .05$$

Since this differs from $P_{X,Y}(1, 1)$, the two random variables are not independent. It follows that there is some relationship between the time of residence in a town and the level of satisfaction with its food stores. This can be seen from Table 4.3, from which we find, for example, that the probability is .15 that a consumer who has lived less than six years in the town will have the highest satisfaction level. The corresponding probability is only .10 for consumers who have been resident more than six years.

As in the case of a single random variable, the joint probability distribution of a set of random variables can also be represented through their joint cumulative probability function.

Definition

The **joint cumulative probability function,** $F_{X,Y}(x_0, y_0)$ of a pair of discrete random variables X and Y expresses the probability that simultaneously X does not exceed the value x_0 and Y does not exceed the value y_0, as a function of x_0 and y_0. That is,

$$F_{X,Y}(x_0, y_0) = P(X \leq x_0 \ \cap \ Y \leq y_0)$$

where the function is evaluated at all values x_0 and y_0. This can be written

$$F_{X,Y}(x_0, y_0) = \sum_{x \leq x_0} \sum_{y \leq y_0} P_{X,Y}(x, y)$$

where the notation implies that summation extends over all pairs of values of x and y that the random variables can take which simultaneously satisfy $x \leq x_0$ and $y \leq y_0$.

For the example of Table 4.1, consider $F_{X,Y}(2, 1)$—that is, the probability that both $X \leq 2$ and $Y \leq 1$. This is

$$F_{X,Y}(2, 1) = P_{X,Y}(1, 1) + P_{X,Y}(2, 1) = .04 + .14 = .18$$

Hence, the probability is .18 that a randomly chosen consumer will both have lived less than six years in the town and have satisfaction level at most 2.

In the previous section we defined the expectation of a function of a single random variable. This definition can be extended to functions of several random variables.

Definition

Let X and Y be a pair of discrete random variables with joint probability function $P_{X,Y}(x, y)$. The **expectation** of any function $g(X, Y)$ of these random variables is defined as

$$E[g(X, Y)] = \sum_{x}\sum_{y} g(x, y)P_{X,Y}(x, y)$$

More generally, if the K random variables $X_1, X_2, \ldots, X_K$ have joint probability function $P_{X_1, X_2, \ldots, X_K}(x_1, x_2, \ldots, x_K)$, then the expectation of the function $g(X_1, X_2, \ldots, X_K)$ is

$$E[g(X_1, X_2, \ldots, X_K)] = \sum_{x_1}\sum_{x_2} \cdots \sum_{x_K} g(x_1, \ldots, x_K)P_{X_1, \ldots, X_K}(x_1, \ldots, x_K)$$

An important application of the expectation of a function of random variables is to the **covariance.** Suppose that X and Y are a pair of random variables that are not statistically independent. We would like some measure of the strength of the relationship between them. This is rather difficult to achieve, since they could conceivably be related in any number of ways. To simplify matters, we restrict attention to the possiblity of linear association. For example, a high value of X might be associated on the average with a high value of Y and a low value of X with a low value of Y in such a way that, to a good approximation, a straight line might be drawn through the associated values when plotted on a graph. Suppose that the random variable X has mean μ_X and Y has mean μ_Y, and consider the product $(X - \mu_X)(Y - \mu_Y)$. If high values of X tend to be associated with high values of Y and low values of X with low values of Y, then we would expect this product to be positive, and the stronger the association, the larger the expectation of $(X - \mu_X)(Y - \mu_Y)$. On the other hand, if high values of X are associated with low values of Y and low X with high Y, then the expected value for this product would be negative. An expectation of 0 for $(X - \mu_X)(Y - \mu_Y)$ would imply an absence of linear association between X and Y. Thus, as a measure of linear association in the population, we are led to examination of the expected value of $(X - \mu_X)(Y - \mu_Y)$.[3]

Definition

Let X be a random variable with mean μ_X, and Y a random variable with mean μ_Y. The expected value of $(X - \mu_X)(Y - \mu_Y)$ is called the **covariance** between X and Y, and denoted Cov (X, Y). For discrete random variables

---

[3] This measure is not free from interpretational difficulties. In particular, it is not independent of the units in which X and Y are measured. We return to this subject in Chapter 12.

$$\text{Cov}(X, Y) = E[(X - \mu_X)(Y - \mu_Y)] = \sum_x \sum_y (x - \mu_X)(y - \mu_Y)P_{X,Y}(x, y)$$

An equivalent[4] expression is

$$\text{Cov}(X, Y) = E(XY) - \mu_X \mu_Y = \sum_x \sum_y xyP_{X,Y}(x, y) - \mu_X \mu_Y$$

To illustrate, we evaluate the covariance between consumer satisfaction and time in residence, using the joint probabilities in Table 4.1. First, the means of these two random variables are

$$\mu_X = E(X) = \sum_x xP_X(x) = (1)(.11) + (2)(.31) + (3)(.46) + (4)(.12) = 2.59$$

and

$$\mu_Y = E(Y) = \sum_y yP_Y(y) = (1)(.48) + (2)(.52) = 1.52$$

The expectation of the product of X and Y is

$$\begin{aligned}
E(XY) &= \sum_x \sum_y xyP_{X,Y}(x, y) \\
&= (1)(1)(.04) + (1)(2)(.07) + (2)(1)(.14) + (2)(2)(.17) \\
&\quad + (3)(1)(.23) + (3)(2)(.23) + (4)(1)(.07) + (4)(2)(.05) \\
&= 3.89
\end{aligned}$$

Finally, the covariance is

$$\text{Cov}(X, Y) = E(XY) - \mu_X \mu_Y = 3.89 - (2.59)(1.52) = -0.05$$

This negative value for the covariance indicates some tendency for high values of consumer satisfaction to be associated with a low period of time in residence in the town—that is, a *negative association* between this pair of random variables. Again, this property can be seen from the conditional probabilities of Table 4.3. Those who

---

[4] This follows since

$$\begin{aligned}
\sum_x \sum_y (x - \mu_X)(y - \mu_Y)P_{X,Y}(x, y) &= \sum_x \sum_y (xy - \mu_Y x - \mu_X y + \mu_X \mu_Y)P_{X,Y}(x, y) \\
&= \sum_x \sum_y xyP_{X,Y}(x, y) - \mu_Y \sum_x \sum_y xP_{X,Y}(x, y) - \mu_X \sum_x \sum_y yP_{X,Y}(x, y) + \mu_X \mu_Y \\
&= \sum_x \sum_y xyP_{X,Y}(x, y) - \mu_Y \sum_x xP_X(x) - \mu_X \sum_y yP_Y(y) + \mu_X \mu_Y \\
&= \sum_x \sum_y xyP_{X,Y}(x, y) - \mu_X \mu_Y
\end{aligned}$$

have lived longest in the town are more likely than others to have low satisfaction levels, and less likely to have high levels of satisfaction with food stores.

As might be expected, the notions of covariance and statistical independence are not unrelated.

Covariance and Statistical Independence

If a pair of random variables are statistically independent, the covariance between them is 0. However, the converse is not necessarily true.

The reason why a covariance of 0 does not necessarily imply statistical independence is that covariance is designed to measure linear association, and it is possible that this quantity may not detect other types of dependency. As an extreme example, suppose that the random variable X has probability function

$$P_X(-1) = \frac{1}{4}; \qquad P_X(0) = \frac{1}{2}; \qquad P_X(1) = \frac{1}{4}$$

and let the random variable Y be defined as

$$Y = X^2$$

Thus, knowledge of the value taken by X implies knowledge of the value taken by Y and hence, these two random variables are certainly not independent. We know that whenever $X = 0$, then $Y = 0$ and that if X is either -1 or 1, then $Y = 1$. Hence, the joint probability function of X and Y is

$$P_{X,Y}(-1, 1) = \frac{1}{4}; \qquad P_{X,Y}(0, 0) = \frac{1}{2}; \qquad P_{X,Y}(1, 1) = \frac{1}{4}$$

with the probability of any other combination of values being equal to 0. It is then straightforward to verify that

$$E(X) = 0; \qquad E(Y) = \frac{1}{2}; \qquad E(XY) = 0$$

and hence that the covariance between X and Y is 0.

To conclude our discussion of joint distributions, we consider the mean and variance of a random variable that can be written as the sum or difference of other random variables.

Sums and Differences of Random Variables

Let X and Y be a pair of random variables with means μ_X and μ_Y and variances σ_X^2 and σ_Y^2. Then the following properties hold:

$\longrightarrow$

(i) The expected value of their sum is the sum of their expected values:

$$E(X + Y) = \mu_X + \mu_Y$$

(ii) The expected value of their difference is the difference between their expected values:

$$E(X - Y) = \mu_X - \mu_Y$$

(iii) If the covariance between X and Y is 0, then the variance of their sum is the sum of their variances:

$$\text{Var}(X + Y) = \sigma_X^2 + \sigma_Y^2$$

(iv) If the covariance between X and Y is 0, then the variance of their difference is the *sum* of their variances:

$$\text{Var}(X - Y) = \sigma_X^2 + \sigma_Y^2$$

Let $X_1, X_2, \ldots, X_K$ be K random variables with means $\mu_1, \mu_2, \ldots, \mu_K$ and variances $\sigma_1^2, \sigma_2^2, \ldots, \sigma_K^2$. Then the following properties hold:

(v) The expected value of their sum is

$$E(X_1 + X_2 + \cdots + X_K) = \mu_1 + \mu_2 + \cdots + \mu_K$$

(vi) If the covariance between every pair of these random variables is 0, then the variance of their sum is

$$\text{Var}(X_1 + X_2 + \cdots + X_K) = \sigma_1^2 + \sigma_2^2 + \cdots + \sigma_K^2$$

Example 4.7

An investor has $1,000 to invest, and two investment opportunities, each requiring a minimum of $500. The profit per $100 from the first can be represented by a random variable, X, having the following probability function:

x	-5	20
$P_X(x)$	.4	.6

The profit per $100 from the second is given by the random variable Y, whose probability function is

y	0	25
$P_Y(y)$	.6	.4

The random variables X and Y are independent.

The investor has the following three possible strategies:

(a) $1,000 in the first investment
(b) $1,000 in the second investment
(c) $500 in each investment

Find the mean and variance of the profit from each strategy.

The random variable X has mean

$$\mu_X = E(X) = \sum xP_X(x) = (-5)(.4) + (20)(.6) = 10$$

and variance

$$\sigma_X^2 = E[(X - \mu_X)^2] = \sum (x - \mu_X)^2 P_X(x)$$

$$= (-5 - 10)^2(.4) + (20 - 10)^2(.6) = 150$$

Hence, the mean profit from strategy (a) is

$$E(10X) = 10\ E(X) = 100$$

and the variance is

$$\text{Var}(10X) = 100\ \text{Var}(X) = 15{,}000$$

The random variable Y has mean

$$\mu_Y = E(Y) = \sum yP_Y(y) = (0)(.6) + (25)(.4) = 10$$

and variance

$$\sigma_Y^2 = E[(Y - \mu_Y)^2] = \sum (y - \mu_Y)^2 P_Y(y)$$

$$= (0 - 10)^2(.6) + (25 - 10)^2(.4) = 150$$

Therefore, strategy (b) has mean profit

$$E(10Y) = 10\ E(Y) = 100$$

and variance

$$\text{Var}(10Y) = 100\ \text{Var}(Y) = 15{,}000$$

Now, the return from strategy (c) is $5X + 5Y$, which has mean

$$E(5X + 5Y) = E(5X) + E(5Y) = 5E(X) + 5E(Y) = 100$$

Thus, all three strategies have the same expected profit. However, since X and Y are independent, and hence have covariance 0, the variance of the return from strategy (c) is

$$\text{Var}(5X + 5Y) = \text{Var}(5X) + \text{Var}(5Y) = 25\sigma_X^2 + 25\sigma_Y^2 = 7{,}500$$

This is smaller than the variances of the other strategies, reflecting the decrease in risk that follows from diversification.

In the following sections, we will discuss some specific discrete probability distributions that have important applications in the business area.

EXERCISES

19. A personnel officer suspects that job applicants who have had many previous jobs are more likely to leave the company relatively quickly after employment than those with few previous jobs. A review of the records for all employees who stayed four years or less yielded the joint probability function shown in the table.

YEARS BEFORE LEAVING (Y)	NUMBER OF PREVIOUS JOBS (X)			
	1	2	3	4
1	.02	.05	.10	.12
2	.06	.06	.08	.07
3	.07	.06	.06	.02
4	.07	.09	.05	.02

(a) Find the marginal probability function of X, and hence the mean number of previous jobs.

(b) Find the marginal probability function of Y, and hence the mean number of years before leaving.

(c) Find the conditional probability function for Y, given $X = 4$.

(d) Find the covariance between X and Y.

(e) Are number of previous jobs and years before leaving independent of one another?

20. A supermarket manager is interested in the relationship between the number of times a sale item is advertised in the local newspaper during the week, and the volume of demand for the item. Let demand volume be denoted by the random variable X, with the value 0 for low demand, 1 for moderate demand, and 2 for high demand. The manager found the joint probability function shown in the accompanying table.

NUMBER OF ADVERTISEMENTS (Y)	VOLUME OF DEMAND (X)		
	0	1	2
1	.10	.15	.06
2	.05	.24	.16
3	.04	.10	.10

(a) Find the joint cumulative probability function at $X = 1$, $Y = 2$, and interpret your result.

(b) Find the conditional probability function for X, given $Y = 3$.

(c) Find the conditional probability function for Y, given $X = 0$.

(d) Find the covariance between X and Y.

(e) Are number of advertisements and volume of demand independent of one another?

21. The accompanying table shows the joint probabilities for bank card holders of the number of cards owned (X) and the number of credit purchases made per week (Y).

NUMBER OF CARDS (X)	NUMBER OF PURCHASES PER WEEK (Y)			
	0	1	2	3
1	.10	.05	.04	.01
2	.10	.35	.26	.09

(a) If a person in this group owns two cards, what is the probability distribution of the number of purchases made per week by that individual?

(b) Are the number of cards owned and number of credit purchases made per week independent?

22. A market researcher wants to determine whether a new product, which has been advertised on Monday Night Football, has achieved more name-recognition among those who regularly watch that show than among those who do not. Survey information has been collected, and the following random variables are defined:

$$X = \begin{cases} 1 & \text{if regularly watch Monday Night Football} \\ 0 & \text{otherwise} \end{cases}$$

$$Y = \begin{cases} 1 & \text{if respondent correctly identifies the product} \\ 0 & \text{otherwise} \end{cases}$$

The joint probability function of this pair of random variables is estimated as shown in the table.

Y	X	
	0	1
0	.40	.12
1	.30	.18

(a) Find the marginal probability functions of X and Y.

(b) Find the conditional probability function of Y, given $X = 1$.

(c) Are watching Monday Night Football and identification of the product independent of one another?

23. A mechanic suspects that a particular machine is more likely to break down on Monday than on any other day of the week. A review of daily records, one-fifth of which are for

Mondays, indicates that breakdowns occurred on 10% of all Mondays, but only on 5% of all other days. Define the random variables:

$$X = \begin{cases} 1 & \text{if day is Monday} \\ 0 & \text{otherwise} \end{cases}$$

$$Y = \begin{cases} 1 & \text{if machine breaks down during the day} \\ 0 & \text{otherwise} \end{cases}$$

(a) Find the joint probability function of Y and X.
(b) Find the conditional probability function of X, given $Y = 1$.
(c) Are machine breakdowns independent of the day of the week?

24. The customer relations manager at a large department store receives complaints about both the quality of merchandise and of service. The marginal probability functions for the number of daily complaints in each category are shown in the table. If complaints about service and merchandise are independent of one another, find the joint probability function for this pair of random variables.

NUMBER OF MERCHANDISE COMPLAINTS	PROBABILITY	NUMBER OF SERVICE COMPLAINTS	PROBABILITY
0	.05	0	.10
1	.20	1	.25
2	.45	2	.40
3	.30	3	.25

25. Refer to the information in Exercise 24. Find the mean and standard deviation of the total number of complaints received in a day.

26. A publisher finds that the probability distribution for the number of errors per page of text depends on whether the material is technical, nontechnical, or tabular. The three probability distributions are shown in the accompanying table. A book contains 500 pages, of which 150 are nontechnical, 300 are technical, and 50 are tabular. Assuming that the number of errors on any one page is independent of the number on any other, find the mean and standard deviation of the total number of errors in the book.

NUMBER OF ERRORS PER PAGE	0	1	2	3	4	5	6
PROBABILITY FOR TECHNICAL	.25	.20	.15	.15	.10	.10	.05
PROBABILITY FOR NONTECHNICAL	.40	.30	.20	.05	.05	0	0
PROBABILITY FOR TABULAR	.20	.25	.20	.15	.10	.10	0

27. A company has five representatives covering large territories and ten representatives covering smaller territories. The probability distributions for the numbers of orders received by each of these types of representatives in a day are shown in the table. Assuming that the number of orders received by any representative is independent of the

number received by any other, find the mean and standard deviation of the total number of orders received by the company in a day.

NUMBER OF ORDERS (LARGE TERRITORY)	PROBABILITY	NUMBER OF ORDERS (SMALLER TERRITORY)	PROBABILITY
0	.10	0	.20
1	.15	1	.25
2	.25	2	.35
3	.30	3	.15
4	.15	4	.05
5	.05		

4.5 THE BINOMIAL DISTRIBUTION

Suppose that a random experiment can give rise to only two possible mutually exclusive outcomes, which for convenience we will label "success" and "failure." Let p denote the probability of success, so that the probability of failure is $(1 - p)$. Now define the random variable X so that X takes the value 1 if the outcome of the experiment is success and 0 otherwise. The probability function of this random variable is then

$$P_X(0) = (1 - p); \qquad P_X(1) = p$$

This distribution is known as the **Bernoulli distribution.** Its mean and variance can be found by direct application of the definitions in Section 4.3. The mean of a Bernoulli random variable is

$$\mu_X = E(X) = \sum_x x P_X(x) = (0)(1 - p) + (1)(p) = p$$

and the variance is

$$\sigma_X^2 = E[(X - \mu_X)^2] = \sum_x (x - \mu_X)^2 P_X(x)$$

$$= (0 - p)^2(1 - p) + (1 - p)^2 p = p(1 - p)$$

Example 4.8

An insurance broker believes that, for a particular contact, the probability of making a sale is .4. If the random variable X is defined to take the value 1 if a sale is made and 0 otherwise, then X has a Bernoulli distribution with probability of success p equal to .4; that is, the probability function of X is

$$P_X(0) = .6; \qquad P_X(1) = .4$$

The mean of this distribution is $p = .4$ and the variance is

$$\sigma_X^2 = p(1 - p) = (.4)(.6) = .24$$

An important generalization of the Bernoulli distribution concerns the case where a random experiment, with two possible outcomes, is repeated several times.

Suppose again that the probability of a success resulting in a single trial is p, and that n independent trials are carried out, so that the result of any one trial has no influence on the outcome of any other. The number of successes X resulting from these n trials could be any whole number from 0 to n, and we are interested in the probability of obtaining exactly $X = x$ successes in n trials.

We develop the result in two stages. First, observe that the n trials will result in a sequence of n outcomes, each of which must be either success (S) or failure (F). One such sequence is

$$\underbrace{S, S, \ldots, S,}_{x \text{ times}} \quad \underbrace{F, F, \ldots, F}_{(n-x) \text{ times}}$$

In words, the first x trials result in success, while the remainder result in failure. Now, the probability of success in a single trial is p and the probability of failure is $(1 - p)$. Since the n trials are independent of one another, the probability of any particular sequence of outcomes is, by the multiplication rule of probabilities (Section 3.6), equal to the product of the probabilities for the individual events. Thus, the probability of observing the specific sequence of outcomes described above is

$$\underbrace{p \cdot p \cdot \cdots \cdot p \cdot}_{x \text{ times}} \quad \underbrace{(1 - p) \cdot (1 - p) \cdot \cdots \cdot (1 - p)}_{(n-x) \text{ times}} = p^x (1 - p)^{n-x}$$

This line of argument establishes that the probability of observing any specific sequence involving x successes and $(n - x)$ failures is $p^x(1 - p)^{n-x}$.

Our original interest concerned the determination, not of the probability of occurrence of a particular sequence, but of the probability of precisely x successes, regardless of the order of the outcomes. There are several sequences in which x successes could be arranged among $(n - x)$ failures. In fact, the number of such possibilities is just the number of combinations of x objects chosen from n—that is[5] (Section 3.5),

$$\text{Number of sequences involving } x \text{ successes in } n \text{ trials} = {}_nC_x = \frac{n!}{x! \, (n - x)!}$$

Moreover, these sequences are mutually exclusive, since no two of them can occur at the same time.

We have now shown that the event "x successes result from n trials" can occur in ${}_nC_x$ mutually exclusive ways, each with probability $p^x(1 - p)^{n-x}$. Therefore, by the addition rule of probabilities (Section 3.6), the probability required is the sum of these ${}_nC_x$ individual probabilities, that is,

$$P(x \text{ successes in } n \text{ trials}) = \frac{n!}{x! \, (n - x)!} p^x (1 - p)^{n-x}$$

---

[5] In this expression, we define $0! = 1$.

The Binomial Distribution

Suppose that a random experiment can result in two possible mutually exclusive and collectively exhaustive outcomes, "success" and "failure," and that p is the probability of a success resulting in a single trial. If n independent trials are carried out, the distribution of the number of successes X resulting is called the **binomial distribution.** Its probability function is

$$P_X(x) = \frac{n!}{x!\,(n-x)!} p^x (1-p)^{n-x} \quad \text{for } x = 0, 1, 2, \ldots, n$$

Example 4.9

Suppose now that the insurance broker of Example 4.8 has five contacts, and he believes that, for each, the probability of making a sale is .4. The distribution of the number of sales X is then binomial, with $n = 5$ and $p = .4$, that is,

$$P_X(x) = \frac{5!}{x!\,(5-x)!} (.4)^x (.6)^{5-x} \quad \text{for} \quad x = 0, 1, \ldots, 5$$

The probabilities for numbers of successes (sales made) are

$$P(0 \text{ successes}) = P_X(0) = \frac{5!}{0!\,5!} (.4)^0 (.6)^5 = (.6)^5 = .078$$

$$P(1 \text{ success}) = P_X(1) = \frac{5!}{1!\,4!} (.4)^1 (.6)^4 = (5)(.4)(.6)^4 = .259$$

$$P(2 \text{ successes}) = P_X(2) = \frac{5!}{2!\,3!} (.4)^2 (.6)^3 = (10)(.4)^2 (.6)^3 = .346$$

$$P(3 \text{ successes}) = P_X(3) = \frac{5!}{3!\,2!} (.4)^3 (.6)^2 = (10)(.4)^3 (.6)^2 = .230$$

$$P(4 \text{ successes}) = P_X(4) = \frac{5!}{4!\,1!} (.4)^4 (.6)^1 = (5)(.4)^4 (.6) = .077$$

$$P(5 \text{ successes}) = P_X(5) = \frac{5!}{5!\,0!} (.4)^5 (.6)^0 = (.4)^5 = .010$$

The probability function is graphed in Figure 4.5. The shape is rather typical of binomial probabilities when p is neither very large nor very small. At the extremes (0 and 5 successes) the probabilities are quite low, since in either case only one possible sequence of outcomes could give rise to this event. The probabilities peak toward the center of the distribution (the location of the peak depending on p) where the number of possible sequences is higher.

To find the mean and variance of the binomial distribution, it is convenient to return to the Bernoulli distribution. Consider n independent trials, each with probability of success p, and let $X_i = 1$ if the ith trial results in success and 0 otherwise.

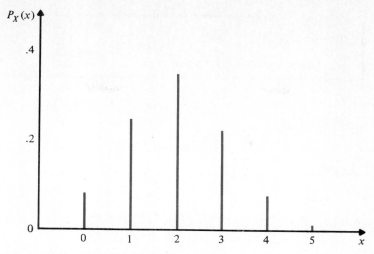

FIGURE 4.5 Binomial probability function for Example 4.9 ($n = 5, p = .4$)

The random variables $X_1, X_2, \ldots, X_n$ are therefore n independent Bernoulli variables, each with probability of success p. Moreover, the total number of successes X is

$$X = X_1 + X_2 + \cdots + X_n$$

That is to say, the binomial random variable can be expressed as the sum of independent Bernoulli random variables. Now, since we have already found the mean and variance for the Bernoulli random variable, the results of Section 4.4 can be used to find the mean and variance of the binomial distribution. Recall that, for a Bernoulli variable,

$$E(X_i) = p; \qquad \text{Var}(X_i) = p(1 - p) \quad \text{for } i = 1, 2, \ldots, n$$

Then, for the binomial distribution,

$$E(X) = E(X_1 + X_2 + \cdots + X_n) = E(X_1) + E(X_2) + \cdots + E(X_n) = np$$

Since the Bernoulli random variables are independent, the covariance between any pair of them is zero, and

$$\text{Var}(X) = \text{Var}(X_1 + X_2 + \cdots + X_n)$$

$$= \text{Var}(X_1) + \text{Var}(X_2) + \cdots + \text{Var}(X_n) = np(1 - p)$$

Mean and Variance of Binomial Distribution

Let X be the number of successes in n independent trials, each with probability of success p. Then X follows a binomial distribution with mean

$$\mu_X = E(X) = np$$

and variance

$$\sigma_X^2 = E[(X - \mu_X)^2] = np(1 - p)$$

Example
4.10

Refer to Example 4.9. For the insurance broker who makes five contacts, each with probability of a sale of .4, the mean (or expected) number of sales is

$$\mu_X = np = (5)(.4) = 2$$

and the variance is

$$\sigma_X^2 = np(1 - p) = (5)(.4)(.6) = 1.2$$

so that the standard deviation for the number of sales is

$$\sigma_X = \sqrt{1.2} = 1.10$$

Given our knowledge of the form of the binomial probability function, it is straightforward to find the probability that the number of successes falls in some specified range. Since the events "0 successes," "1 success," etc., are mutually exclusive, the required probability is just the sum of the probabilities over all numbers of successes in that range.

Example
4.11

Refer to Example 4.9.

(a) What is the probability that the number of successes will be between 2 and 4 (inclusive)? We require

$$P(2 \le X \le 4) = P(X = 2) + P(X = 3) + P(X = 4)$$
$$= P_X(2) + P_X(3) + P_X(4)$$

which, using the results of Example 4.9, is

$$.346 + .230 + .077 = .653$$

(b) What is the probability of at least one success? We could find

$$P(X \ge 1) = P_X(1) + P_X(2) + P_X(3) + P_X(4) + P_X(5)$$

directly, but this is unnecessarily tedious. The probabilities for any discrete distribution sum to 1, so that

$$P(X \ge 1) = 1 - P(X = 0) = 1 - P_X(0) = 1 - .078 = .922$$

Unless the number of trials n is very small, the calculation of binomial probabilities is likely to be extremely burdensome. However, in real-world applications the computations can readily be carried out on an electronic computer. Moreover, as will be seen in Section 4.7 and in the next chapter, when the number of trials n is quite large, the required probabilities can be obtained through convenient approximations to the binomial distribution. In order to facilitate problem-solving, Table 1 in the Appendix lists binomial probabilities for values of n up to 20 and selected values of

p up to and including $p = .5$. For higher values of p one can use the following rule:[6]

☐ $P[x$ successes in n trials when probability of success in a single trial is $p]$
  $= P[(n - x)$ successes in n trials when probability of success in a single trial is $(1 - p)]$

☐ To illustrate, the probability of 7 successes in 12 trials where $p = .6$ is the same as the probability of 5 successes in 12 trials with $p = .4$. That is, setting $x = 7$, $n = 12$, and $p = .6$ in the above formula yields $P[7$ successes in 12 trials when probability of success in a single trial is .6]
  $= P[5$ successes in 12 trials when probability of success in a single trial is .4]

☐ This can be read directly from Table 1 as .2270. This value is obtained by locating, from the first two columns of the table, the row corresponding to $n = 12$ and $x = 5$; we then locate the required probability in the column corresponding to $p = .40$.

Example 4.12

A corporation has a number of vacancies to fill in a particular department. Offers of employment are made to 20 candidates, and it is believed that, for each candidate, the probability that the offer will be accepted is .8. If it is assumed that candidates' decisions are made independently, what is the probability that between 16 and 18 acceptances will be made?

Let the random variable X denote the number of acceptances. Then X has a binomial distribution with $n = 20$ and $p = .8$. We require

$$P(16 \leq X \leq 18) = P_X(16) + P_X(17) + P_X(18)$$

Now, with $n = 20$, the probability of 16 successes for $p = .8$ is the same as the probability of 4 successes for $p = .2$. From Table 1 of the Appendix, we obtain $P_X(16) = .2182$. Similarly, we find $P_X(17) = .2054$ and $P_X(18) = .1369$, so that

$$P(16 \leq X \leq 18) = .2182 + .2054 + .1369 = .5605$$

An important application of the binomial distribution is in **acceptance sampling.** When a firm receives a very large shipment of goods from a manufacturer, it has to decide, based on information about the quality of those goods, whether to accept delivery. Typically, a thorough inspection of the whole consignment would be prohibitively expensive, so a small random sample[7] is selected and these are examined. Based on the results of this examination a decision is made as to whether to accept the shipment. It is possible to calculate, for any particular decision rule of this kind, the probability of accepting a shipment with any given proportion of defectives. This is so, since if p is the proportion of defectives in the shipment, and n is the number sampled[8], then the number of defectives X in the sample follows the binomial distribution with probability function

$$P_X(x) = \frac{n!}{x! \, (n - x)!} p^x (1 - p)^{n-x} \quad \text{for } x = 0, 1, 2, \ldots, n$$

---

[6] The truth of this rule can be seen by reversing x and $(n - x)$ in the binomial probability function.

[7] By a **random sample** of n items, it is meant that the sample is selected in such a way that every set of n items in the shipment is equally likely to be chosen. This concept will be more thoroughly explored in Chapter 6.

[8] We are assuming that the number of items n in the sample is a very small proportion of the total number in the shipment.

The following example illustrates how the probabilities for accepting delivery can be calculated.

Example
4.13

A company receiving a very large shipment of items decides to accept delivery if, in a random sample of twenty items, not more than one is defective. Thus the shipment is accepted if the number of defectives is either zero or one, so that if $P_X(x)$ is the probability function for the number X of defectives in the sample, we have

$$P(\text{Shipment accepted}) = P_X(0) + P_X(1)$$

Suppose that the proportion of defectives in the shipment is $p = .1$. Then, for $n = 20$, we find directly from Table 1 of the Appendix that the probabilities for 0 and 1 defectives in the sample are, respectively, $P_X(0) = .1216$ and $P_X(1) = .2702$. Hence, with this decision rule, the probability that the company accepts delivery is

$$P(\text{Shipment accepted}) = .1216 + .2702 = .3918$$

Similarly, if 20% of the items in the shipment are defective—that is, if $p = .2$, then

$$P(\text{Shipment accepted}) = .0115 + .0576 = .0691$$

and for $p = .3$,

$$P(\text{Shipment accepted}) = .0008 + .0068 = .0076$$

As can be seen from Example 4.13, the higher the proportion of defectives in the shipment, the less likely is the acceptance of the delivery. We can graph the probability of acceptance against the proportion of defectives in the shipment, as in Figure 4.6. This allows acceptance probabilities to be read off for any proportion of defective items.

Of course, any number of curves relating acceptance probabilities to proportion of defectives in a sample can be constructed by choosing different sample sizes n and different rules for determining acceptance. In this way, a company is able to choose

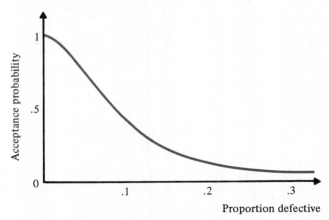

FIGURE 4.6 Acceptance probability as a function of proportion of defectives when shipment is accepted if there are less than two defective items in a random sample of twenty

a scheme with the desired balance between the costs of accepting shipments with particular proportions of defectives, and of carrying out the checks on quality.

4.6 THE HYPERGEOMETRIC DISTRIBUTION

In Example 4.13 we considered a situation in which a sample of goods from a very large consignment was to be checked for defectives. By assuming that the number sampled was extremely small relative to the total number of items in the consignment, we were able to approach the problem through use of the binomial distribution. However, in those cases where the number of sample members is not a very small proportion of the total number of items in the population, the binomial distribution is inappropriate. The reason is that in such circumstances the sample outcomes for individual items will not be independent of one another. To illustrate, suppose that a consignment contains 10 items, 3 of which are defective. If one of these is chosen at random, the probability that it is defective is simply $3/10$. Now, suppose that a second item is to be chosen. If the first was defective, then the remaining 9 would contain 2 defectives, so that the probability that the second item is defective, given that the first was defective, would be $2/9$. On the other hand, the probability that the second item is defective given that the first was not defective would, by a similar argument, be $3/9$. Hence, the events "First item defective" and "Second item defective" are not statistically independent. Recall that, for the distribution of the number of successes in n trials to be binomial, it is necessary that the outcomes of these trials be independent of one another. As we have seen, this will not be the case when sampling from a group, or population, of items. If the total number of items in the population is very large relative to the number to be sampled, the extent of any dependencies will be so trivial that they can be ignored, and the binomial distribution safely used. However, this is not so when the number of population items is not very large.

We can, nevertheless, find the appropriate probability distribution in situations of this kind. In general, suppose that a group of N objects, each of which can be labelled "success" or "failure," contains S successes and $(N - S)$ failures. A random sample of n objects is chosen from this group, and we require the probability that the sample contains x successes. First, the number of possible different samples of n objects that could be drawn from a total of N is the number of combinations

$$_N C_n = \frac{N!}{n!\,(N - n)!}$$

The number of possible ways of getting the x successes in the sample from a total of S successes is

$$_S C_x = \frac{S!}{x!\,(S - x)!}$$

Since the sample contains x successes it must also contain $(n - x)$ failures, and the number of ways of choosing these from a total of $(N - S)$ failures is

$$_{N-S} C_{n-x} = \frac{(N - S)!}{(n - x)!\,(N - S - n + x)!}$$

The total number of samples of n objects containing exactly x successes and $(n - x)$ failures is, therefore,

$$_SC_x \, _{N-S}C_{n-x} = \frac{S!}{x! \, (S - x)!} \frac{(N - S)!}{(n - x)!(N - S - n + x)!}$$

Finally, since the number of possible samples is $_NC_n$, the probability of obtaining x successes in the sample is

$$P(x \text{ successes}) = \frac{_SC_x \, _{N-S}C_{n-x}}{_NC_n} = \frac{\dfrac{S!}{x! \, (S - x)!} \dfrac{(N - S)!}{(n - x)! \, (N - S - n + x)!}}{\dfrac{N!}{n! \, (N - n)!}}$$

The Hypergeometric Distribution

Suppose that a random sample of n objects is chosen from a group of N objects, S of which are successes. The distribution of the number of successes X in the sample is called the **hypergeometric distribution.** Its probability function is

$$P_X(x) = \frac{_SC_x \, _{N-S}C_{n-x}}{_NC_n} = \frac{\dfrac{S!}{x! \, (S - x)!} \dfrac{(N - S)!}{(n - x)! \, (N - S - n + x)!}}{\dfrac{N!}{n! \, (N - n)!}}$$

where x can take integer values ranging from the larger of 0 and $[n - (N - S)]$ to the smaller of n and S.

The mean and variance of this distribution are

$$\mu_X = E(X) = np$$

and

$$\sigma_X^2 = E[(X - \mu_X)^2] = \left(\frac{N - n}{N - 1}\right) np \, (1 - p)$$

where $p = S/N$ is the proportion of successes in the population.

If the sample size n is very small relative to the total number of objects N, the hypergeometric probabilities are very close to binomial probabilities, and the binomial distribution may be used rather than the hypergeometric. In this case, $(N - n)/(N - 1)$ is very close to 1, so that the variance of the hypergeometric distribution is close to $np \, (1 - p)$, the variance of the binomial distribution.

Example 4.14
A company receives a shipment of 20 items. Because inspection of each individual item is expensive, it has a policy of checking a random sample of 6 items from such a shipment, accepting delivery if no more than one sampled item is defective. What is the probability that a shipment with 5 defective items will be accepted?

If we identify "defective" with "success" in this example, the shipment contains $N = 20$ items, $S = 5$ of which are successes. A sample of $n = 6$ items is selected. Then the number of successes X in the sample has a hypergeometric distribution with probability function

$$P_X(x) = \frac{{}_SC_x \; {}_{N-S}C_{n-x}}{{}_NC_n} = \frac{{}_5C_x \; {}_{15}C_{6-x}}{{}_{20}C_6}$$

$$= \frac{\dfrac{5!}{x!\,(5-x)!} \dfrac{15!}{(6-x)!\,(9+x)!}}{\dfrac{20!}{6!\,14!}}$$

The shipment is accepted if the sample contains either 0 or 1 successes (defectives), so the probability of its acceptance is

$$P(\text{Shipment accepted}) = P_X(0) + P_X(1)$$

The probability of no defectives in the sample is

$$P_X(0) = \frac{\dfrac{5!}{0!\,5!} \dfrac{15!}{6!\,9!}}{\dfrac{20!}{6!\,14!}} = .129$$

(Recall that $0! = 1$.) The probability of one defective in the sample is

$$P_X(1) = \frac{\dfrac{5!}{1!\,4!} \dfrac{15!}{5!\,10!}}{\dfrac{20!}{6!\,14!}} = .387$$

Therefore, the probability that the shipment of 20 items containing 5 defectives is accepted using this procedure is

$$P(\text{Shipment accepted}) = P_X(0) + P_X(1) = .129 + .387 = .516$$

4.7 THE POISSON DISTRIBUTION

Consider the following random variables:

(a) The number of fatal traffic accidents in a city in a particular week
(b) The number of telephone calls arriving at a corporation's switchboard in the 15 minutes before noon on a given day
(c) The number of replacement orders for a part received by a firm in a week
(d) The number of times a piece of equipment fails during a 3-month period
(e) The number of strikes at a plant in a year

0 t Time

FIGURE 4.7 Illustration of random occurrences ● of an event over time

Each of these five random variables can be characterized as the number of occurrences of a certain event in a given period of time. Experience indicates that, for a wide range of problems of this kind, the *Poisson probability distribution* well represents the probability structure of the random variable.

Consider the situation illustrated in Figure 4.7, where time is measured along the horizontal line, and we are interested in the period beginning at time 0 and ending at time t. Occurrences of events along the time axis are indicated by ● ● ●, so that in this illustration six events occur in the relevant time period. Suppose that the following assumptions can be made:

(i) For any small time interval, represented by a *small segment* of the time axis between 0 and t in Figure 4.7, the probability that one event will occur in this interval is approximately proportional to the length of the interval.

(ii) The probability of two or more occurrences in such an interval is negligibly small compared to the probability of one occurrence.

(iii) The numbers of occurrences in any nonoverlapping time intervals are independent of one another.

If these assumptions hold, it can be shown that the probability of x occurrences in the interval between 0 and t is

$$P(x \text{ occurrences}) = \frac{e^{-\lambda}\lambda^x}{x!}$$

where λ is the mean number of occurrences in the interval from 0 to t and $e = 2.71828 \ldots$ is the base of the natural logarithms. (Table 2 in the Appendix gives values of $e^{-\lambda}$ for values of λ from 0 to 10.) The distribution with these probabilities is called the **Poisson distribution.**

The Poisson Distribution

The random variable X is said to follow the **Poisson distribution** if it has probability function

$$P_X(x) = \frac{e^{-\lambda}\lambda^x}{x!} \quad \text{for } x = 0, 1, 2, \ldots$$

where λ is any number with $\lambda > 0$.

The mean of this distribution is

$$\mu_X = E(X) = \lambda$$

---

and the variance is

$$\sigma_X^2 = E[(X - \mu_X)^2] = \lambda$$

The shape of the Poisson probability function depends on the mean, λ. Figure 4.8 shows probability functions for $\lambda = .5, 1, 2,$ and 4.

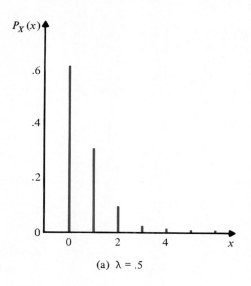

(a) $\lambda = .5$

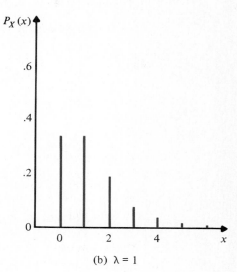

(b) $\lambda = 1$

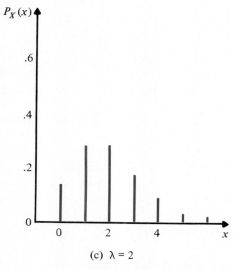

(c) $\lambda = 2$

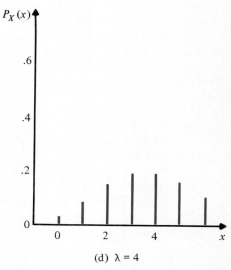

(d) $\lambda = 4$

FIGURE 4.8 Probability functions for the Poisson distribution

Example
4.15

Research has indicated[9] that, for a typical plant with 2,000 employees in Britain, the number of strikes in a year can be represented by a Poisson distribution with mean $\lambda = .4$. The probability function for the number of strikes X in a year is then

$$P_X(x) = \frac{e^{-.4}(.4)^x}{x!} \quad \text{for } x = 0, 1, 2, \ldots$$

We can now calculate probabilities for particular numbers of strikes in a year, using (from Table 2 in the Appendix) $e^{-.4} = .6703$.

The probability of no strikes is

$$P(0 \text{ strikes}) = P_X(0) = \frac{e^{-.4}(.4)^0}{0!} = \frac{(.6703)(1)}{1} = .6703$$

Similarly,

$$P(1 \text{ strike}) = P_X(1) = \frac{e^{-.4}(.4)^1}{1!} = \frac{(.6703)(.4)}{1} = .2681$$

$$P(2 \text{ strikes}) = P_X(2) = \frac{e^{-.4}(.4)^2}{2!} = \frac{(.6703)(.16)}{2} = .0536$$

$$P(3 \text{ strikes}) = P_X(3) = \frac{e^{-.4}(.4)^3}{3!} = \frac{(.6703)(.064)}{6} = .0071$$

$$P(4 \text{ strikes}) = P_X(4) = \frac{e^{-.4}(.4)^4}{4!} = \frac{(.6703)(.0256)}{24} = .0007$$

These probabilities could then be used to find the probability that the number of strikes falls in any given range. For example, the probability of more than one strike in a year is

$$P(\text{More than 1 strike}) = 1 - P(0 \text{ strikes}) - P(1 \text{ strike})$$

$$= 1 - P_X(0) - P_X(1)$$

$$= 1 - .6703 - .2681 = .0616$$

We have seen that the Poisson distribution arises naturally in assessing probabilities for numbers of occurrences of events in a time interval. In certain circumstances it has a further important use—as an approximation to the binomial distribution. In n independent trials, each with probability of success p, the distribution of the number of successes is binomial. However, if the number of trials n is large, the computation of binomial probabilities is a formidable chore. In that case, approximations for the binomial distribution are used. We consider here the case where n is large, but at the same time np is small,[10] so that the probability of a success in any

---

[9] This example is adapted from S. J. Prais, "The strike-proneness of large plants in Britain," *Journal of the Royal Statistical Society A, 141* (1978), 368–84.

[10] An approximation for the binomial distribution that is appropriate when np is not small will be discussed in Chapter 5.

single trial is very small indeed. The following situations would satisfy these conditions:

(a) An insurance company will hold a large number of life policies on individuals of any particular age, and the probability that a single policy will result in a claim during the year is very low. The distribution of the number of claims is binomial with large n and very small p.

(b) A company may have a large number of machines working on a process simultaneously. If the probability that any one of them will break down in a single day is small, the distribution of the number of daily breakdowns is binomial with large n and small p.

In such cases, the binomial distribution can be well approximated by the Poisson distribution with mean $\lambda = np$. That is, the mean, λ, of the approximating Poisson distribution is fixed at the value of the known mean, np, of the binomial distribution being approximated.

Poisson Approximation to the Binomial Distribution

Let X be the number of successes resulting from n independent trials, each with probability of success p. The distribution of the number of successes X is binomial with mean np. However, if the number of trials n is large and np is of only moderate size (preferably $np \leq 7$), this distribution can be well approximated by the Poisson distribution with mean $\lambda = np$. The probability function of the approximating distribution is then

$$P_X(x) = \frac{e^{-np}(np)^x}{x!} \quad \text{for } x = 0, 1, 2, \ldots$$

Example 4.16

A large plant has 50 similar machines operating simultaneously. If the probability that any one of them breaks down in a single day is .04, find probabilities for numbers of breakdowns in one day.

The distribution of the number X of breakdowns is binomial with $n = 50$ and $p = .04$, so that the mean of the distribution is

$$\mu_X = np = (50)(.04) = 2$$

We will use the Poisson approximation to the binomial, with mean $\lambda = 2$. The probability function of X is then

$$P_X(x) = \frac{e^{-2}2^x}{x!} \quad \text{for } x = 0, 1, 2, \ldots$$

Hence, the probability of no breakdowns in a day is

$$P(0 \text{ breakdowns}) = P_X(0) = \frac{e^{-2}(2)^0}{0!} = \frac{(.1353)(1)}{1} = .1353$$

Similarly, for 1 and 2 breakdowns,

$$P(1 \text{ breakdown}) = P_X(1) = \frac{e^{-2}(2)^1}{1!} = \frac{(.1353)(2)}{1} = .2706$$

$$P(2 \text{ breakdowns}) = P_X(2) = \frac{e^{-2}(2)^2}{2!} = \frac{(.1353)(4)}{2} = .2706$$

The probabilities for numbers of breakdowns up to 9 in a day are shown here.

x	0	1	2	3	4	5	6	7	8	9
$P_X(x)$	.1353	.2706	.2706	.1804	.0902	.0361	.0120	.0034	.0009	.0002

This is the probability function graphed in Figure 4.8(c).

EXERCISES

28. A laundromat manager knows that 20% of new washing machines purchased require maintenance during the first year of operation. The manager purchases five new machines, whose performances can be assumed to be independent.
    (a) What is the probability that all of them will require maintenance during the first year of operation?
    (b) What is the probability that none of them will require maintenance during the first year of operation?
    (c) What is the probability that at least two of them will require maintenance during the first year of operation?

29. A salesman knows from past experience that 10% of all calls he makes result eventually in a sale. Assume the salesman makes six calls in a day and that the results of the calls are independent.
    (a) What is the probability that exactly three sales will result?
    (b) What is the probability that no sales will result?
    (c) What is the probability that at least five sales will result?

30. Company records indicate that, in a given day, the probability of an assembly machine breaking down is 0.2. The company has four such machines that operate independently.
    (a) What is the probability that at least one of them will break down in the day?
    (b) What is the probability that three or more of them will break down during a day?

31. A company has determined that the probability that a newly appointed apprentice will remain with them for at least one year is 0.8. Four apprentices are appointed.
    (a) Draw the probability function for the number of apprentices who remain with the company after a year's employment, assuming their decisions to remain are made independently.
    (b) Find the probability that at least three of these apprentices will remain with the company for a year or more.

32. It is known that 40% of all candidates fail the first attempt to pass the test for receipt of a driver's license. Suppose that five of your friends, who can be regarded as a random sample of all candidates, take the test for the first time.
    (a) What is the probability that all of them pass?
    (b) What is the probability that most of them fail?

33. It is estimated that 20% of all new M.B.A. graduates are women. A company advertises a position for an M.B.A. graduate and draws up a list of five candidates for a final interview. Assume that these five individuals can be regarded as a random sample of all M.B.A. graduates.

   (a) What is the probability that there is no woman on the list?
   (b) What is the probability that the majority of those on the list are women?
   (c) What is the probability that the number of women on the list is between one and four (inclusive)?

34. A small commuter airline flies planes that can seat up to eight passengers. The airline has determined that the probability that a ticketed passenger will not show up for a flight is 0.2. For each flight, the airline sells tickets to the first ten people placing orders. The probability distribution for the number of tickets sold per flight is shown in the accompanying table. For what proportion of the airline's flights does the number of ticketed passengers showing up exceed the number of available seats? (Assume independence between number of tickets sold and the probability that a ticketed passenger will show up.)

NUMBER OF TICKETS	6	7	8	9	10
PROBABILITY	.30	.30	.25	.10	.05

35. An instructor sets a test question that he believes will be answered correctly by 70% of all the students in a large class. A random sample of five students is taken. Assume the instructor's belief is correct.

   (a) What is the probability that either all five students answer correctly or none answer correctly?
   (b) The instructor gives 4 points for a correct answer and 0 points for an incorrect answer (no partial credit). Find the mean and standard deviation of the total number of points scored by the five students on this question.

36. A roofing contractor offers a "quick fix" job on leaky roofs. He estimates that, after the job is done, 10% of the roofs will still leak.

   (a) The resident manager of an apartment complex hires this contractor to fix the roofs on six buildings. What is the probability that, after the work has been done, at least two of these roofs will still leak?
   (b) The contractor offers a guarantee that pays $100 if a roof still leaks after he has worked on it. If he works on 81 roofs, find the mean and standard deviation of the amount of money he will have to pay as a result of this guarantee.

37. On a given day, 25% of the cars illegally parked on a college campus are ticketed. Suppose that, on a particular day, ten of your friends park illegally. Assume that these ten people can be regarded as a random sample of all those illegally parked.

   (a) What is the expected number who will be ticketed?
   (b) What is the standard deviation of the number ticketed?
   (c) What is the probability that at least two of your friends will be ticketed?

38. A company produces and distributes promotional literature, the recipients of which are invited to write for a free home trial of a new vacuum cleaner. The company claims that 40% of the recipients will take advantage of this offer. In a trial run the promotional material is sent to 100 randomly selected households. If the company's claim is correct, what are the mean and standard deviation of the number of households in the sample that will take advantage of the offer?

39. The Internal Revenue Service estimates that 7% of all taxpayers filling out the 1040 long forms make mistakes. If a random sample of 10,000 of these forms are checked, find the mean and standard deviation of the number of sampled forms that contain mistakes.

40. A company receives a very large shipment of parts. A random sample of fifteen of these parts are checked and, if less than two are defective, the shipment is accepted. What is the probability of accepting a shipment containing

    (a) 5% defectives?   (b) 10% defectives?   (c) 20% defectives?

41. The following two acceptance rules are being considered for determining whether to take delivery of a large shipment of components:

    (a) A random sample of ten components is checked, and the shipment is accepted only if none of them are defective.

    (b) A random sample of twenty components is checked, and the shipment is accepted only if not more than one of them is defective.

    Which of these acceptance rules has the smaller probability of accepting a shipment containing 20% defectives?

42. A company receives large shipments of parts from two sources. Sixty percent of the shipments come from a supplier whose shipments typically contain 10% defectives, while the remainder are from a supplier whose shipments typically contain 20% defectives. A manager receives a shipment, but does not know the source. A random sample of twenty items from this shipment is tested, and one of the parts is found to be defective. What is the probability that this shipment came from the more reliable supplier?
    [*Hint:* Use Bayes' theorem.]

43. A securities analyst, given a list of ten corporate stocks, nominates (not in order) the five from this group that she believes will perform best in the coming year. The result is that four of the stocks she nominated were among the five best performers. If the five stocks had been chosen at random from the group of ten, what would be the probability of achieving a record as good as, or better than, that of the analyst?

44. A company receives a shipment of sixteen items. A random sample of four items is selected, and the shipment is rejected if any of these items proves to be defective.

    (a) What is the probability of accepting a shipment containing four defective items?

    (b) What is the probability of accepting a shipment containing two defective items?

    (c) What is the probability of rejecting a shipment containing one defective item?

45. A committee of five members is to be formed from a group of six men and four women. If assignments to the committee are made at random, what is the probability that the majority of its members will be women?

46. A supermarket chain has ten stores in an urban area, with five in the inner city and five in the suburbs. Four of these stores are to be used for a new promotional campaign. If these are to be selected at random, what is the probability that at least one inner-city store will be included?

47. Customers arrive at a busy check-out counter at an average rate of two per minute. If the distribution of arrivals is Poisson, find the probability that in any given minute there will be three or fewer arrivals.

48. On the average, 2.5 telephone calls per minute are received at a corporation's switchboard. If the distribution of calls is Poisson, find the probability that in any given minute there will be more than 2 calls.

49. A baker receives an average of six requests per day for a specialty cake. Assume that the distribution of the number of daily requests is Poisson.

    (a) Find the probability that exactly six requests are made in a particular day.

    (b) Find the probability that at least four requests are made in a particular day.

(c) Find the probability that, in a given day, the number of requests is between five and seven.

50. The number of accidents in a manufacturing plant has a Poisson distribution with mean 2.4 per month.
    (a) What is the probability of no accidents in a given month?
    (b) What is the probability of more than three accidents in a given month?

51. An insurance company holds fire insurance policies on 5,000 firms. In any given year the probability that any single policy will result in a claim is 0.001. Find the probability that at least three claims are made in a particular year. Use the Poisson approximation to the binomial distribution.

52. A corporation's office has 200 typewriters. The probability that any one of them will require repair on a given day is 0.02. Find the probability that fewer than four of the typewriters require repair on a particular day. Use the Poisson approximation to the binomial distribution.

53. The Internal Revenue Service reports that 5.5% of all taxpayers filling out the 1040 short forms make mistakes. A random sample of 100 of these forms will be checked.
    (a) What is the probability that none of them contain errors?
    (b) What is the probability that at least three of them contain errors?
    Use the Poisson approximation to the binomial distribution.

54. Recent figures indicate that one out of every 10,000 residents of the United States is murdered in any one year. For a random sample of 15,000 residents, find the probability that at least two will be murdered during the next year.
    Use the Poisson approximation to the binomial distribution.

REVIEW EXERCISES

55. Explain carefully, with an illustrative example, what is meant by the expected value of a random variable. Why is this concept important?

56. As an investment adviser, you tell a client that an investment in a mutual fund has (over the next year) a higher expected return than an investment in the money market. The client then asks the following questions:
    (i) Does that imply that the mutual fund will certainly yield a higher return than the money market?
    (ii) Does it follow that I should invest in the mutual fund rather than in the money market?
    How would you reply?

57. Develop a realistic business example (other than those met in the text and other exercises) in which each of the following probability distributions would be appropriate.
    (a) The binomial distribution
    (b) The hypergeometric distribution
    (c) The Poisson distribution

58. Explain what can be learned from each of the following:
    (a) A graph of the probability function of a random variable
    (b) A graph of the cumulative probability function of a random variable
    (c) The standard deviation of a random variable
    (d) The covariance between a pair of random variables

59. In order to assess the mathematics background of students, an instructor regularly administers, at the beginning of the course, a quiz consisting of ten questions. The accompanying table shows the proportions of students obtaining each possible number of correct answers.

NUMBER OF CORRECT ANSWERS	PROPORTION OF STUDENTS	NUMBER OF CORRECT ANSWERS	PROPORTION OF STUDENTS
0	.02	6	.10
1	.08	7	.06
2	.11	8	.05
3	.13	9	.03
4	.21	10	.01
5	.20		

(a) What is the probability that a randomly chosen student obtains five correct answers?

(b) What is the probability that a randomly chosen student obtains at most five correct answers?

(c) What is the probability that a randomly chosen student obtains more than five correct answers?

(d) Find the mean number of correct answers.

(e) Find the standard deviation of the number of correct answers.

60. A multiple-choice examination consists of twenty questions. For each question, candidates must choose from four possible answers. Five points are awarded for each correct answer, and one point is subtracted for each incorrect answer. Suppose you take this test, without having studied, and guess the answer for each question.

(a) What would be your expected score on the test?

(b) What would be the standard deviation of your score?

61. Develop realistic examples of pairs of random variables for which you would expect to find

(a) A positive covariance

(b) A negative covariance

62. The accompanying table shows the joint probability function for the number of breakdowns in a month, and the age in years, of machines employed in a production process.

AGE	NUMBER OF BREAKDOWNS			
	0	1	2	3
0	.06	.06	.05	.03
1	.05	.04	.04	.04
2	.05	.04	.05	.05
3	.03	.03	.04	.10
4	.02	.03	.06	.13

(a) Find the probability function for the number of breakdowns in a month.

(b) Find the probability function for the machine ages.

(c) Find the expected number of breakdowns per machine in a month.

(d) Find the standard deviation of the number of breakdowns per machine in a month.

(e) Find the mean age of the machines.

(f) Find the standard deviation of the machine ages.

(g) Find, and interpret, the covariance between machine age and number of breakdowns.

(h) What is the probability that a 2-year-old machine will break down at least once a month?

(i) Two machines are chosen at random from all the machines, so that the number of breakdowns of one is independent of the number of breakdowns of the other.

    (i) Find the expected total number of breakdowns for these two machines in a month.

    (ii) Find the standard deviation of the total number of breakdowns for these two machines in a month.

    (iii) What is the probability that at least one of these machines will break down at least once during a month?

63. A long-distance taxi service owns four vehicles. These are of different ages and have different repair records. The probabilities that, on any given day, each vehicle will be available for use are 0.95, 0.90, 0.90, and 0.85. Whether one vehicle is available is independent of whether any other vehicle is available.

(a) Find the probability function for the number of vehicles available for use on a given day.

(b) Find the expected number of vehicles available for use on a given day.

(c) Find the standard deviation of the number of vehicles available for use on a given day.

64. The owner of a small book store estimates that 30% of potential customers who enter the store actually purchase at least one book. At 10:30 AM one day there are seven potential customers in the store. Suppose these individuals can be regarded as a random sample from all potential customers.

(a) What is the probability that at least two of these people purchase at least one book?

(b) What is the probability that no books are purchased by these people?

(c) What is the expected number of these potential customers who will purchase at least one book?

(d) With the information given, can you find the expected number of books purchased by these seven people? Can you say anything about this expected number?

65. A bag contains two black balls and one white ball.

(a) A ball is chosen at random from the bag and is not replaced. A second ball is then chosen at random. Is the probability distribution of the number of black balls chosen binomial? Give reasons for your answer.

(b) A ball is chosen at random from the bag and is then replaced. A second ball is then chosen at random. Is the probability distribution of the number of black balls chosen binomial? Give reasons for your answer.

66. A company sells a set of commemorative stamps, at a price of $100 per set. Orders are solicited by telephone. Potential customers are invited to purchase one set of stamps, and it has been found that 10% of people contacted agree to make this purchase. A random sample of twenty people are contacted.

(a) Find the probability that at least three people purchase a set of stamps.

(b) Find the expected amount of money received from sales of the stamps.

(c) Find the standard deviation of the amount of money received from sales of the stamps.

67. The World Series of baseball is to be played by team A and team B. The first team to win four games wins the series. Suppose that team A is the better team, in the sense that the probability is 0.6 that team A will win any specific game. Assume also that the result of any game is independent of that of any other.

(a) What is the probability that team A will win the series?

(b) What is the probability that a seventh game will be needed to determine the winner?

(c) Suppose that, in fact, each team wins two of the first four games.

(i) What is the probability that team A will win the series?

(ii) What is the probability that a seventh game will be needed to determine the winner?

68. Using detailed cash flow information, a financial analyst claims to be able to spot companies that are likely candidates for bankruptcy. The analyst is presented with information on the past records of fifteen companies, and told that in fact five of these have failed. He selects, as candidates for failure, five companies from the group of fifteen. In fact, three of the five companies selected by the analyst were among those that failed. Evaluate the financial analyst's performance on this test of his ability to detect failed companies.

69. A sales manager has a list of twenty potential clients. Five clients are to be allocated at random to each of four salesmen. If the manager knew which were the five best prospects, he would allocate all of them to his most successful salesman. If the allocation is determined by chance, what is the probability that at least four of the five best prospects are allocated to the most successful salesman?

70. A large answering service receives telephone calls at an average rate of four per minute. If the distribution of the number of phone calls received is Poisson, find the probability that fewer than four calls will be received in any given minute.

71. A company has two assembly lines, each of which stalls an average of 2.4 times per week, according to a Poisson distribution. Assume that the performances of these assembly lines are independent of one another. What is the probability that at least one line stalls at least once in any given week?

72. An analyst predicts that 2.5% of all small companies will file for bankruptcy in the coming year. For a random sample of 200 small companies, estimate the probability that at least two will file for bankruptcy in the next year, assuming the analyst's prediction is correct.

continuous random variables and probability distributions

5.1 CONTINUOUS RANDOM VARIABLES

In this chapter we are concerned with issues of probability statements about random variables that can take any value on a continuum. Measures of time, distance, or temperature fit naturally into this category. In such cases the probability that the random variable takes a single specific value—for instance, the probability that a car will travel precisely 27.236 miles on a gallon of gasoline—is 0. It is also convenient to regard as continuous those essentially discrete random variables that are measured on such a fine grid that the probability of occurrence of any specific value is trivially small. For instance, in a study of the total debt incurred by students while attending four-year colleges, it is certainly true that the total debt for any given student will be some integer number of cents. However, the probability that a randomly chosen student will have debts totalling precisely $5,274.57 is sufficiently small that the random variable of interest can be treated as if it were continuous.

Although the assessment of probabilities for individual values of continuous random variables is meaningless, we may well be interested in the probability that such a variable lies in some given range. For example, the probability that a car travels between 27 and 28 miles on a gallon of gasoline or the probability that a randomly chosen student has incurred debts between $5,000 and $6,000 may be useful quantities to evaluate. Therefore, in characterizing probability distributions for continuous random variables, a natural place to begin (by analogy with our discussion of discrete random variables) is with the idea of cumulative probability.

185

5.2 PROBABILITY DISTRIBUTIONS FOR CONTINUOUS RANDOM VARIABLES

As in the previous chapter we let X be a random variable and x a specific value that it could take. Here we call the probability that X does not exceed x the **cumulative distribution function.** This is analogous to the cumulative probability function of Chapter 4.

Definition

The **cumulative distribution function** $F_X(x)$ of a continuous random variable X expresses the probability that X does not exceed the value x, as a function of x. That is,

$$F_X(x) = P(X \leq x)$$

To illustrate the ideas involved we introduce a random variable that has a particularly simple probability structure. As one example of its use, suppose that a road tunnel is precisely 1 mile long and that we are concerned with vehicle breakdowns in the tunnel. Let the random variable X denote the distance into the tunnel, measured in miles from one of its entrances, that a breakdown occurs, and suppose that for a stretch of the tunnel of fixed length the probability of breakdown is identical to that for any other stretch of the same length. Then the distribution of X, for a particular breakdown, is said to be **uniform** in the range 0 to 1. The cumulative distribution function for this random variable is

$$F_X(x) = \begin{cases} 0 & \text{if} \quad x \leq 0 \\ x & \text{if} \quad 0 < x < 1 \\ 1 & \text{if} \quad x \geq 1 \end{cases}$$

This function, which between $x = 0$ and $x = 1$ is a straight line, is graphed in Figure 5.1. Using this function, we see that the probability that any breakdown occurs in the first ¼ mile of the tunnel is

$$P\left(X \leq \frac{1}{4}\right) = F_X\left(\frac{1}{4}\right) = \frac{1}{4}$$

Now, suppose we want to measure the probability that a continuous random variable falls in a specified range. Let the endpoints of the range of interest be $X = a$ and $X = b$, with $b > a$, so that we require[1] $P(a < X < b)$. If X is less than b, then it is either less than a or it lies between a and b. Moreover, since this latter pair of

---

[1] Note that for continuous random variables it does not matter whether we write "less than," as in $P(X < b)$, or "less than or equal to," as in $P(X \leq b)$, because the probability that X is precisely equal to b is 0.

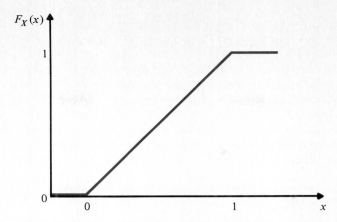

FIGURE 5.1 Cumulative distribution function of a random variable distributed uniformly in the range 0 to 1

events are mutually exclusive,

$$P(X < b) = P(X < a) + P(a < X < b)$$

Hence, from the definition of the cumulative distribution function,

$$F_X(b) = F_X(a) + P(a < X < b)$$

Thus, the probability that a random variable takes a value in a particular range can be obtained from the cumulative distribution function, as described in the box.

Range Probabilities and the Cumulative Distribution Function

Let X be a continuous random variable with cumulative distribution function $F_X(x)$, and let a and b be two possible values of X, with $a < b$. Then the probability that X lies between a and b is

$$P(a < X < b) = F_X(b) - F_X(a) \qquad (5.2.1)$$

For the random variable that is distributed uniformly in the range from 0 to 1, the cumulative distribution function in that range is $F_X(x) = x$. Therefore, if a and b are two numbers between 0 and 1, with $a < b$,

$$P(a < X < b) = F_X(b) - F_X(a) = b - a$$

For example, if a breakdown occurs, the probability that it happens between ¼ and ¾ mile into the tunnel (that is, $a = $ ¼, $b = $ ¾) is

$$P\left(\frac{1}{4} < X < \frac{3}{4}\right) = \frac{3}{4} - \frac{1}{4} = \frac{1}{2}$$

We have seen that the probability that a continuous random variable lies between any two values can be expressed in terms of its cumulative distribution function. This function therefore contains all the information about the probability structure of the random variable. However, for many purposes a different function is more useful. In Chapter 4, we discussed the probability function for discrete random variables, which expresses the probability that a discrete random variable takes any specific value. This concept is not directly relevant in the case of continuous random variables, since here the probability of any specific value arising is 0. However, a related construct called the **probability density function** can be constructed for continuous random variables, and allows ready graphical interpretation of their probability structure.

Probability Density Function

Let X be a continuous random variable, and x any number lying in the range of values this random variable can take. The **probability density function,** $f_X(x)$, of the random variable is a function with the following properties:

(i) $f_X(x) \geq 0$ for all values of x

(ii) Suppose this density function is graphed. Let a and b be two possible values of the random variable X, with $a < b$. Then the probability that X lies between a and b is the area under the density function between these points.

To illustrate property (ii), Figure 5.2 shows the plot of an arbitrary probability density function for some continuous random variable. Two possible values, a and b, are shown, and the shaded area under the curve between these points is the probability that the random variable lies in the interval between them.[2]

To see a specific probability density function, we return to the case of a random variable having a uniform distribution in the range 0 to 1. We introduced this distribution to represent the distance into a 1-mile long tunnel of vehicle breakdowns, where it could be assumed that the probability of a breakdown in any stretch of the tunnel of fixed length is the same as that in any other stretch of the same length. Given this assumption, the shape of the appropriate probability density function can be deduced. Suppose that the complete 1-mile interval is divided into a large number of small subintervals of equal length. Since the probability of a breakdown is the same in each of these subintervals, and since probability corresponds to area under the density function, it follows that the value of the probability density function must be constant

---

[2] Readers with a knowledge of calculus will recognize that the probability that a random variable lies in a given range is the *integral* of the probability density function between the endpoints of the range; that is,

$$P(a < X < b) = \int_a^b f_X(x)\, dx$$

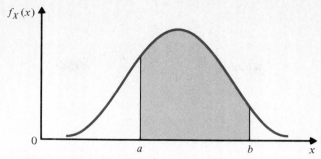

FIGURE 5.2 The shaded area is the probability that the random variable X lies between a and b

throughout the whole range from 0 to 1. Moreover, since, if a breakdown occurs, it must do so between 0 and 1 miles into the tunnel, it follows that the total area under the density function in this range is 1. Hence, the probability density function of this random variable is simply the unit square graphed in Figure 5.3. The uniformity of the density along the range of possible values gives the distribution its name. This probability density function can be written algebraically as

$$f_X(x) = \begin{cases} 1 & \text{for } 0 < x < 1 \\ 0 & \text{for all other values of } x \end{cases}$$

We can now use this density function to find the probability that the random variable falls in any specified range. The calculation of the probability that a given

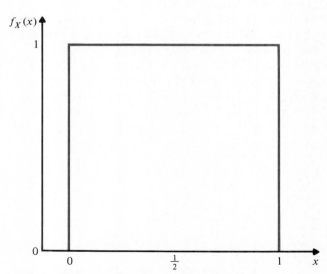

FIGURE 5.3 Probability density function of a random variable distributed uniformly in the range 0 to 1

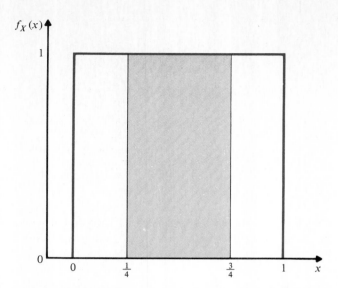

FIGURE 5.4 Probability that a random variable, distributed uniformly in the range 0 to 1, lies between ¼ and ¾; the shaded area is the probability, equal to ½

breakdown occurs between ¼ and ¾ mile into the tunnel is illustrated in Figure 5.4. Since the height of the density function is $f_X(x) = 1$, the area under the curve between these two points is ½, which is the required probability. This is precisely the conclusion reached earlier from consideration of the cumulative distribution function.

We have seen that the probability that a random variable lies between a pair of values is the area under the probability density function between those values. Two special cases of this result are of considerable importance. First, since the random variable must take *some* value—that is, it certainly lies between minus infinity and plus infinity—it follows that the total area under the probability density function is 1. (This is analogous to the requirement for discrete random variables that the individual event probabilities sum to 1.) Second, if we let $F_X(x_0)$ be the cumulative distribution function evaluated at x_0, this is just the probability that the random variable does not exceed x_0; that is,

$$F_X(x_0) = P(X \le x_0)$$

which we now see is the area under the probability density function to the left of x_0.

Areas Under Continuous Probability Density Functions

   Let X be a continuous random variable with probability density function $f_X(x)$ and cumulative distribution function $F_X(x)$. Then the following properties hold:

(i) The total area under the curve $f_X(x)$ is 1.[3]

(ii) The area under the curve $f_X(x)$ to the left of x_0 is $F_X(x_0)$, where x_0 is any value that the random variable can take.[4]

The two results stated in the box are illustrated for the uniform distribution in Figure 5.5. In part (a) of the figure it can be seen that the area under the probability density function is 1, as it is simply the area of a square whose sides are of length 1. Part (b) of the figure shows the cumulative distribution function, evaluated at x_0, as the area of a rectangle of height 1 and width x_0. Hence, we have $F_X(x_0) = x_0$.

Example 5.1

A repair team is responsible for a stretch of oil pipeline 2 miles long. The distance (in miles) along this stretch that any fracture arises can be represented by a uniformly distributed random variable, with probability density function

$$f_X(x) = \begin{cases} 0.5 & \text{for } 0 < x < 2 \\ 0 & \text{for all other values of } x \end{cases}$$

Find the cumulative distribution function, and the probability that any given fracture occurs between 0.5 mile and 1.5 miles along this stretch of pipeline.

Figure 5.6 shows a plot of the probability density function. The cumulative distribution function evaluated at x_0, $F_X(x_0)$, is the probability that the random variable takes a value less than x_0. This is the shaded area shown in the graph—a rectangle of height 0.5 and width x_0. Thus,

$$F_X(x_0) = 0.5x_0$$

To find the probability that a particular fracture arises between 0.5 mile and 1.5 miles along the pipeline, we have

$$P(0.5 < X < 1.5) = F_X(1.5) - F_X(0.5)$$

$$= (0.5)(1.5) - (0.5)(0.5) = 0.5$$

---

[3] Formally, in integral calculus notation,

$$\int_{-\infty}^{\infty} f_X(x)\, dx = 1$$

[4] The cumulative distribution function is thus the integral

$$F_X(x_0) = \int_{-\infty}^{x_0} f_X(x)\, dx$$

It therefore follows that the probability density function is the derivative of the cumulative distribution function; that is,

$$f_X(x) = \frac{dF_X(x)}{dx}$$

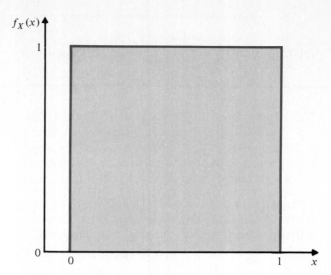

(a) Total area under the uniform probability density function is 1

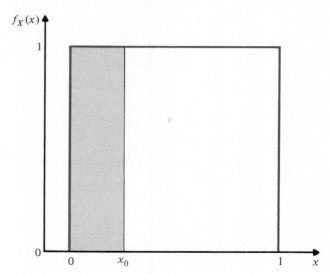

(b) Area under the probability density function to the left of x_0 is $F_X(x_0)$, which is equal to x_0 for this uniform distribution

FIGURE 5.5 Properties of probability density function

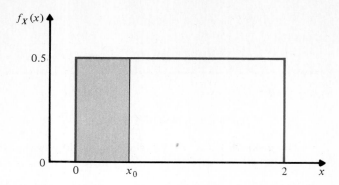

FIGURE 5.6 Probability density function for Example 5.1; the shaded area is the cumulative distribution function evaluated at x_0

5.3 EXPECTATIONS FOR CONTINUOUS RANDOM VARIABLES

In Section 4.3 we introduced the ideas of the expectation of a discrete random variable X and the expectation of a function of that random variable. These concepts extend to the case where the random variable is continuous, though the fact that here the probability of any specific value is 0 necessitates some modification in the method of evaluating expectations, as indicated in the box.

Expectations for Continuous Random Variables

Suppose that a random experiment leads to an outcome that can be represented by a continuous random variable. If N independent replications of this experiment are carried out, then the **expected value** of the random variable is the average of the values taken, as the number N of replications becomes infinitely large. The expected value of a random variable X is denoted $E(X)$.

Similarly, if $g(X)$ is any function of the random variable X, then the expected value of this function is the average value taken by the function over repeated independent trials, as the number of trials becomes infinitely large.[5] This expectation is denoted $E[g(X)]$.

---

[5] Formally, using integral calculus, we express the expected value of the random variable X by

$$E(X) = \int_{-\infty}^{\infty} x f_X(x) \, dx$$

and the expected value of the function $g(X)$ by

$$E[g(X)] = \int_{-\infty}^{\infty} g(x) f(x) \, dx$$

Notice that, in forming these expectations, the integral plays the same role as the summation operator in the discrete case.

In the next box, some important specific expected values are defined in precisely the same manner as for discrete random variables.

Definitions

Let X be a continuous random variable.

(i) The **mean** of X, denoted μ_X, is defined as the expected value of X; that is,

$$\mu_X = E(X)$$

(ii) The **variance** of X, denoted σ_X^2, is defined as the expectation of the squared discrepancy, $(X - \mu_X)^2$, of the random variable from its mean; that is,

$$\sigma_X^2 = E[(X - \mu_X)^2]$$

An alternative, but equivalent, expression is

$$\sigma_X^2 = E(X^2) - \mu_X^2$$

(iii) The **standard deviation** of X, σ_X, is the square root of the variance.

The mean and variance are two pieces of summary information about a probability distribution. The mean provides a measure of the center of the probability distribution. A physical interpretation is as follows: Cut out the graph of a probability density function. The point along the x-axis at which this figure exactly balances on one's finger is the mean of the distribution. For example, the graph of the probability density function of the uniform distribution in Figure 5.3 is perfectly symmetric about $x = \frac{1}{2}$, and so $\mu_X = \frac{1}{2}$ is the mean of this random variable. In the example about vehicle breakdowns in a 1-mile long tunnel, a mean breakdown distance of $\frac{1}{2}$ mile from the tunnel entrance can be interpreted as a statement that, over a very large number of breakdowns, the average distance into the tunnel will be $\frac{1}{2}$ mile.

The variance—or its square root, the standard deviation—gives a measure of the spread (or dispersion) of a probability distribution about its center. To illustrate, consider a uniform distribution in the range 0 to 1 and a second distribution that is uniform on the range $\frac{1}{4}$ to $\frac{3}{4}$. The probability density functions of these two distributions are graphed in Figure 5.7. For the second density, shown in part (b) of the figure, the probability density function is

$$f_X(x) = \begin{cases} 2 & \text{for } \dfrac{1}{4} < x < \dfrac{3}{4} \\ 0 & \text{for all other values of } x \end{cases}$$

This is a proper density function since the area under the curve is 1. Both distributions are centered on $x = \frac{1}{2}$, which is the mean in each case. However, the density of the distribution of part (a) is more disperse about this mean than that of part (b). This is

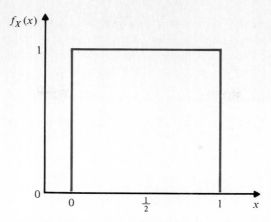

(a) Probability density function of a random variable distributed uniformly in the range 0 to 1

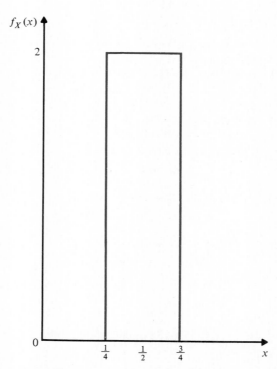

(b) Probability density function of a random variable distributed uniformly in the range 1/4 to 3/4

FIGURE 5.7 Probability density functions for two uniformly distributed random variables

reflected in the fact that the former distribution has the larger variance.[6]

In Section 4.3 we showed how means and variances could be found for linear functions of discrete random variables. In fact, the same results are true for continuous random variables, as summarized in the box.

Let X be a continuous random variable with mean μ_X and variance σ_X^2, and let a and b be any constant fixed numbers. Define the random variable Z as

$$Z = a + bX$$

Then the mean and variance of Z are

$$\mu_Z = E(a + bX) = a + b\mu_X$$

and

$$\sigma_Z^2 = \text{Var}(a + bX) = b^2\sigma_X^2$$

As a special case of these results, the random variable

$$Z = \frac{X - \mu_X}{\sigma_X}$$

has mean 0 and variance 1.

Example 5.2

An author receives from a publisher a contract, according to which she is to be paid a fixed sum of $10,000, plus $1.50 for each copy of her book sold. Her uncertainty about total sales of the book can be represented by a random variable with mean 30,000 and standard deviation 10,000. Find the mean and standard deviation of the total payments she will receive.

Let the random variable X denote total sales. Then X has mean and variance

$$\mu_X = 30,000; \qquad \sigma_X^2 = (10,000)^2 = 100,000,000$$

Total payments (in dollars) can be written as

$$\text{Payment} = 10,000 + 1.5X$$

Therefore, the mean payment is

$$E(\text{Payment}) = 10,000 + 1.5\mu_X$$
$$= 10,000 + (1.5)(30,000) = \$55,000$$

The variance of payments is

$$\text{Var}(\text{Payment}) = (1.5)^2\sigma_X^2 = 225,000,000$$

so that, on taking square roots, we find the standard deviation is $15,000.

---

[6] In fact, the variances are $\frac{1}{12}$ for the distribution of Figure 5.7(a) and $\frac{1}{48}$ for that of (b).

5.4 JOINTLY DISTRIBUTED CONTINUOUS RANDOM VARIABLES

Section 4.4 introduced joint distributions for discrete random variables. Many of the concepts and results discussed there extend quite naturally to the case of continuous random variables.

Definitions

Let $X_1, X_2, \ldots, X_K$ be continuous random variables.

(i) Their **joint cumulative distribution function,** $F_{X_1, X_2, \ldots, X_K}(x_1, x_2, \ldots, x_K)$ expresses the probability that simultaneously X_1 is less than x_1, X_2 is less than x_2, and so on; that is,

$$F_{X_1, X_2, \ldots, X_K}(x_1, x_2, \ldots, x_K) = P(X_1 < x_1 \ \cap \ X_2 < x_2 \ \cap \cdots \cap \ X_K < x_K)$$

(ii) The cumulative distribution functions $F_{X_1}(x_1), F_{X_2}(x_2), \ldots, F_{X_K}(x_K)$ of the individual random variables are called their **marginal distribution functions.** For any i, $F_{X_i}(x_i)$ is the probability that the random variable X_i does not exceed the specific value x_i.

(iii) The random variables are **independent** if and only if

$$F_{X_1, X_2, \ldots, X_K}(x_1, x_2, \ldots, x_K) = F_{X_1}(x_1)F_{X_2}(x_2) \cdots F_{X_K}(x_K)$$

The notion of statistical independence here is precisely the same as in the discrete case. Independence of a set of random variables implies that the probability distribution of any one of them is unaffected by the values taken by the others. Thus, for example, the assertion that consecutive daily changes in the price of a share of common stock are independent of one another implies that information about past price changes is of no value in assessing what is likely to happen tomorrow.

The notion of expectation extends to functions of jointly distributed continuous random variables. As in the case of discrete random variables, an important quantity of this kind is the **covariance,** which is used in assessing the strength of linear association between a pair of random variables.

Definition

Let X and Y be a pair of continuous random variables, with respective means μ_X and μ_Y. The expected value of $(X - \mu_X)(Y - \mu_Y)$ is called the **covariance** between Y and X. That is,

$$\mathrm{Cov}(X, Y) = E[(X - \mu_X)(Y - \mu_Y)]$$

An alternative, but equivalent, expression is

$$\text{Cov}(X, Y) = E(XY) - \mu_X \mu_Y$$

If the random variables X and Y are independent, then the covariance between them is 0. However, the converse is not necessarily true.

The results in Section 4.4 on means and variances of sums and differences of discrete random variables also hold for continuous random variables. For convenience, these are repeated here.

Sums and Differences of Random Variables

Let $X_1, X_2, \ldots, X_K$ be K random variables with means $\mu_1, \mu_2, \ldots, \mu_K$ and variances $\sigma_1^2, \sigma_2^2, \ldots, \sigma_K^2$. Then the following properties hold:

(i) The mean of their sum is the sum of their means; that is,

$$E(X_1 + X_2 + \cdots + X_K) = \mu_1 + \mu_2 + \cdots + \mu_K$$

(ii) If the covariance between every pair of these random variables is 0, then the variance of their sum is the sum of their variances; that is,

$$\text{Var}(X_1 + X_2 + \cdots + X_K) = \sigma_1^2 + \sigma_2^2 + \cdots + \sigma_K^2$$

Let X and Y be a pair of random variables with means μ_X and μ_Y and variances σ_X^2 and σ_Y^2. Then the following properties hold:

(iii) The mean of their difference is the difference of their means; that is,

$$E(X - Y) = \mu_X - \mu_Y$$

(iv) If the covariance between X and Y is 0, then the variance of their difference is the *sum* of their variances; that is,

$$\text{Var}(X - Y) = \sigma_X^2 + \sigma_Y^2$$

Example 5.3

An airline has determined that the weights of its passengers have a distribution with mean 150 pounds and standard deviation 25 pounds. Their baggage weights have a distribution with mean 40 pounds and standard deviation 5 pounds. Assume that the weights for passengers are independent of those for baggage. Then, if passenger weight is denoted by X_1 and baggage weight by X_2, we have $\mu_1 = 150$, $\mu_2 = 40$, $\sigma_1 = 25$, and $\sigma_2 = 5$.

For the total weight of a passenger plus baggage, the mean is

$$\mu_1 + \mu_2 = 150 + 40 = 190 \text{ pounds}$$

and the variance is

$$\sigma_1^2 + \sigma_2^2 = (25)^2 + (5)^2 = 650$$

so that the standard deviation is $\sqrt{650} = 25.5$ pounds.

EXERCISES

1. A car breaks down on an interstate highway. The driver has noticed the frequency of police patrols along the highway, and believes that the time before a patrol car arrives on the scene is a uniformly distributed random variable, between 0 and 1 hour. Therefore, if X denotes the time (in hours) before a patrol car arrives, the probability density function of X is

$$f_X(x) = \begin{cases} 1 & \text{for } 0 < x < 1 \\ 0 & \text{for all other values of } x \end{cases}$$

(a) Draw the probability density function.

(b) Find and draw the cumulative distribution function.

(c) Find the probability that a patrol car arrives within 20 minutes of the breakdown.

(d) Find the probability that a patrol car does not arrive within 45 minutes of the breakdown.

(e) Find the probability that a patrol car arrives between 15 and 30 minutes after the breakdown.

2. A rescue team is responsible for a 5-mile stretch of river. Experience indicates that the distance along this stretch, measured in miles from its northernmost point, that an emergency is likely to arise is a uniformly distributed random variable, between 0 and 5 miles. Thus, if X denotes the distance (in miles) of an emergency from the most northerly point of this stretch of river, its probability density function is

$$f_X(x) = \begin{cases} 0.2 & \text{for } 0 < x < 5 \\ 0 & \text{for all other values of } x \end{cases}$$

(a) Draw the probability density function.

(b) Find and draw the cumulative distribution function.

(c) Find the probability that a given emergency arises within 1 mile of the northernmost point of this stretch of river.

(d) Find the probability that a given emergency arises between 2 and 4 miles from the northernmost point of the stretch.

(e) The rescue team has its headquarters at the midpoint of this stretch of river. Find the probability that a given emergency arises more than 2 miles from the team's headquarters.

3. A contractor is preparing to bid on a project for which he calculates his costs to be $2,000. He believes that the lowest bid of his competitors can be represented by a uniform distribution between $1,500 and $4,000; that is, if X denotes their lowest bid (in thousands of dollars), its probability density function is

$$f_X(x) = \begin{cases} 0.4 & \text{for } 1.5 < x < 4.0 \\ 0 & \text{for all other values of } x \end{cases}$$

The contract is to be awarded to the lowest bidder.

(a) Draw the probability density function of the random variable X.

(b) Find and draw the cumulative distribution function of this random variable.

(c) The contractor generally submits a bid that is 15% higher than his estimated costs. What is the probability that he will obtain the contract if he does this?

(d) Because business is slow, the contractor decides on this occasion to submit a bid that is only 5% above his costs. What is the probability that he will obtain the contract?

4. The probability that a particular brand of spark plug will last less than 36,000 miles is

0.5. The probability that it will last less than 30,000 miles is 0.2. What is the probability that the spark plug will last between 30,000 and 36,000 miles?

5. The incomes of all families in a city can be represented by a continuous probability distribution. For this city, 60% of all family incomes are below $21,000 and 40% are below $15,000. What is the probability that a randomly selected family will have an income between $15,000 and $21,000?

6. A contractor undertakes repair work on an office building. He estimates his materials will cost $100,000, and his labor costs are $1,000 per day. He believes that the number of days needed to complete the contract can be represented by a random variable with mean 80 and standard deviation 10. Find the mean and standard deviation of the total cost (materials plus labor) of fulfilling the contract.

7. A hot dog salesman incurs a daily cost of $100 in the operation of his hot dog stand. For each hot dog he sells, his profit is 25 cents. The number of hot dogs sold in a day is a random variable with mean 600 and standard deivation 120. Find the mean and standard deviation of the salesman's daily net profit (total profit from sales, less operating costs).

8. A salesman receives an annual salary of $5,000 plus 5% of the value of the orders he takes. The annual value of these orders can be represented by a random variable with mean $500,000 and standard deviation $150,000. Find the mean and standard deviation of the salesman's annual income.

9. An investor plans to divide $100,000 between two investments. The first yields a certain profit of 10%, while the second yields a profit with expected value 15% and standard deviation 5%. If the investor divides the money equally between these two investments, find the mean and standard deviation of the total profit.

10. The results of an energy audit indicated that, for an outlay of $5,000, the annual savings on fuel bills to be obtained by a homeowner could be represented by a random variable with mean $1,200 and standard deviation $200. Find the mean and standard deviation of the net savings (total savings less initial cost) after 5 years.

11. An investor has funds in two projects. One has an expected return of $10,000 and standard deviation $3,000, and the other an expected return of $15,000 and standard deviation $5,000. Assuming independence, find the mean and standard deviation of the total return from these investments.

12. A consultant has three sources of income—from teaching short courses, from selling computer software, and from advising on projects. His expected annual incomes from these sources are $10,000, $25,000, $15,000, and the respective standard deviations are $1,000, $5,000, $4,000. Assuming independence, find the mean and standard deviation of his total annual income.

13. A contractor is certain of neither the precise total materials costs nor labor costs for a project. He believes that materials costs can be represented by a random variable with mean $100,000 and standard deviation $10,000. His labor costs are $1,000 per day. If the number of days needed to complete the project can be represented by a random variable with mean 60 and standard deviation 12, find the mean and standard deviation of the total cost (materials plus labor) for the project. Assume that materials costs and labor costs are independent.

14. Road tests on two different makes of automobile showed mean miles per gallon obtained as 18.6 and 19.2, and standard deviations 0.6 and 0.8, respectively. Assuming independence, find the mean and standard deviation of the difference in miles per gallon between these two makes of car.

15. Let X and Y be a pair of random variables, with variances σ_X^2 and σ_Y^2. If the covariance between X and Y is negative, then the variance of $(X + Y)$ is smaller than in the case where X and Y are independent. Explain why this is so.

5.5 THE NORMAL DISTRIBUTION

In this section we introduce a continuous distribution that plays a central role in a very large body of statistical analysis. For example, suppose that a big group of students takes a test. A large proportion of their scores are likely to be concentrated about the mean, and the numbers of scores in ranges of a fixed width are likely to "tail off" away from the mean. If the average score on the test is 60, we would expect to find, for instance, more students with scores in the range 55–65 than in the range 85–95. These considerations suggest a probability density function that peaks at the mean and tails off at its extremeites. One distribution with these properties is the **normal distribution,** whose probability density function is shown in Figure 5.8. As can be seen, this density function is *bell-shaped*.

Probability Density Function of the Normal Distribution

If the random variable X has probability density function

$$f_X(x) = \frac{1}{\sqrt{2\pi\sigma^2}} e^{-(x-\mu)^2/2\sigma^2} \qquad \text{for} \qquad -\infty < x < \infty$$

where μ and σ^2 are any numbers such that $-\infty < \mu < \infty$, $0 < \sigma^2 < \infty$, and $e = 2.71828\ldots$, $\pi = 3.14159\ldots$ are physical constants, then X is said to follow a **normal distribution.**

It can be seen from the definition that there is not a single normal distribution, but a whole family of distributions, resulting from different specifications of μ and σ^2. These two parameters have very convenient interpretations.

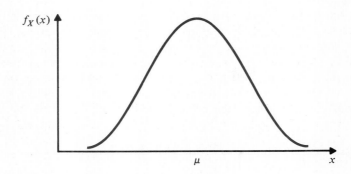

FIGURE 5.8 Probability density function for a normal distribution

Some Properties of the Normal Distribution

Suppose the random variable X follows a normal distibution with parameters μ and σ^2. Then the following properties hold:

(i) The mean of the random variable is μ; that is,

$$E(X) = \mu$$

(ii) The variance of the random variable is σ^2; that is,

$$\text{Var}(X) = E[(X - \mu)^2] = \sigma^2$$

(iii) The shape of the probability density function is a symmetric bell-shaped curve (see Figure 5.8) centered on the mean μ.

It follows from these properties that, given the mean and variance of a normal random variable, an individual member of the family of normal distributions is specified. This allows use of a convenient notation.

Notation

If the random variable X follows a normal distribution with mean μ and variance σ^2, we write

$$X \sim N(\mu, \sigma^2)$$

Now, the mean of any distribution provides a measure of central location, while the variance gives a measure of spread or dispersion about the mean. Thus, the values taken by the parameters μ and σ^2 have different effects on the probability density function of a normal random variable. Figure 5.9(a) shows probability density functions for two normal distributions with a common variance, but different means. It can be seen that increasing the mean, while holding the variance fixed, shifts the density function but does not alter its shape. In Figure 5.9(b) the two density functions are of normal random variables with a common mean, but different variances. Each is symmetric about the common mean, but that with the larger variance is more disperse.

An extremely important practical question concerns the determination of probabilities from a specified normal distribution. As a first step in determining probabilities, we introduce the cumulative distribution function.

Cumulative Distribution Function of the Normal Distribution

Suppose that X is a normal random variable with mean μ and variance σ^2;

that is, $X \sim N(\mu, \sigma^2)$. Then the **cumulative distribution function** $F_X(x_0)$ is

$$F_X(x_0) = P(X \leq x_0)$$

This is the area under the probability density function to the left of x_0, as illustrated in Figure 5.10. As for any proper density function, the total area under the curve is 1; that is,

$$F_X(\infty) = 1$$

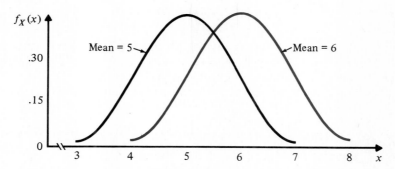

(a) Probability density functions for two normal distributions with means 5 and 6; each distribution has variance 1

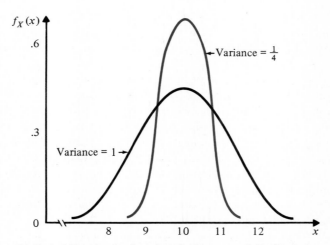

(b) Probability density functions for two normal distributions with variances 1/4 and 1; each distribution has mean 10

FIGURE 5.9 Effects of μ and σ^2 on the probability density function of a normal random variable

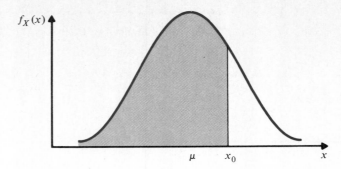

FIGURE 5.10 The shaded area is the probability that X does not exceed x_0 for a normal random variable

There is no simple algebraic expression for calculating the cumulative distribution function of a normally distributed random variable.[7] The general shape of the cumulative distribution function is shown in Figure 5.11.

We have already seen that, for *any* continuous random variable, probabilities can be expressed in terms of the cumulative distribution function.

Range Probabilities for Normal Random Variables

Let X be a normal random variable with cumulative distribution function $F_X(x)$, and let a and b be two possible values of X, with $a < b$. Then

$$P(a < X < b) = F_X(b) - F_X(a)$$

This probability is the area under the corresponding probability density function between a and b, as illustrated in Figure 5.12.

Any required probability can be obtained from the cumulative distribution function. However, a crucial difficulty remains because there does not exist a convenient formula for determining the cumulative distribution function. In principle, for any specific normal distribution, probabilities could be obtained by numerical methods using an electronic computer. However, it would be enormously tedious if we had to carry out such an operation for every normal distribution we encountered. Fortunately, probabilities for *any* normal distribution can always be expressed in terms of probabilities for a *single* normal distribution, for which the cumulative distribution function has been evaluated and tabulated. We now introduce the particular distribution that is used for this purpose.

---

[7] That is to say that the integral

$$F_X(x_0) = \int_{-\infty}^{x_0} \frac{1}{\sqrt{2\pi\sigma^2}} e^{-(x-\mu)^2/2\sigma^2} \, dx$$

does not have a simple algebraic form.

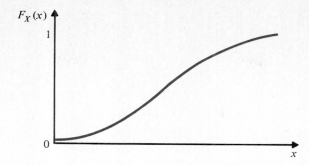

FIGURE 5.11 Cumulative distribution function for a normal random variable

---

The Standard Normal Distribution

Let Z be a normal random variable with mean 0 and variance 1; that is,

$$Z \sim N(0, 1)$$

Then Z is said to follow the **standard normal distribution.**

If the cumulative distribution function of this random variable is denoted $F_Z(z)$, and a^* and b^* are two numbers with $a^* < b^*$, then

$$P(a^* < Z < b^*) = F_Z(b^*) - F_Z(a^*)$$

The cumulative distribution function of the standard normal distribution is tabulated in Table 3 in the Appendix. This table gives values of

$$F_Z(z) = P(Z \leq z)$$

for nonnegative values of z. For example,

$$F_Z(1.25) = .8944$$

Values of the cumulative distribution function for negative values of z can be inferred

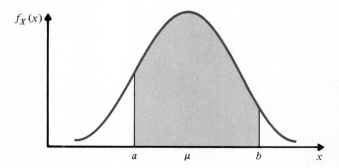

FIGURE 5.12 The shaded area is the probability that X lies between a and b for a normal random variable

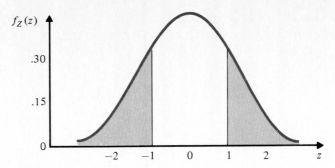

FIGURE 5.13 Probability density function for the standard normal random variable Z; the shaded areas, which are equal, show the probability that Z does not exceed -1 and the probability that Z is greater than 1

from the symmetry of the probability density function. Let z_0 be any positive number and suppose we require

$$F_Z(-z_0) = P(Z \leq -z_0)$$

As illustrated in Figure 5.13, because the density function of the standard normal variable is symmetric about its mean, 0, the area under the curve to the left of $-z_0$ is the same as the area under the curve to the right of z_0; that is,

$$P(Z \leq -z_0) = P(Z \geq z_0)$$

Moreover, since the total area under the curve is 1,

$$P(Z \geq z_0) = 1 - P(Z \leq z_0) = 1 - F_Z(z_0)$$

Hence, it follows that

$$F_Z(-z_0) = 1 - F_Z(z_0)$$

For example,

$$P(Z \leq -1.25) = F_Z(-1.25) = 1 - F_Z(1.25) = 1 - .8944 = .1056$$

Example 5.4

If Z is a standard normal random variable, find $P(-0.50 < Z < 0.75)$. The required probability is

$$P(-0.50 < Z < 0.75) = F_Z(0.75) - F_Z(-0.50)$$
$$= F_Z(0.75) - [1 - F_Z(0.50)]$$

Then, using Table 3 of the Appendix, we obtain

$$P(-0.50 < Z < 0.75) = .7734 - (1 - .6915) = .4649$$

We now show how probabilities for any normal random variable can be expressed in terms of those for the standard normal variable. Let the random variable X be normally distributed with mean μ and variance σ^2. We saw in Section 5.3 that subtracting the mean and dividing by the standard deviation yields a random variable Z that has mean 0 and variance 1. It can also be shown that, if X is normally distributed, so is Z. Hence, Z has a standard normal distribution. Suppose, then, that we require the probability that X lies between the numbers a and b. This is equivalent to $(X - \mu)/\sigma$ lying between $(a - \mu)/\sigma$ and $(b - \mu)/\sigma$, so that the probability of interest is

$$P(a < X < b) = P\left(\frac{a - \mu}{\sigma} < \frac{X - \mu}{\sigma} < \frac{b - \mu}{\sigma}\right)$$

$$= P\left(\frac{a - \mu}{\sigma} < Z < \frac{b - \mu}{\sigma}\right)$$

Finding Range Probabilities for Normal Random Variables

Let X be a normal random variable with mean μ and variance σ^2. Then the random variable $Z = (X - \mu)/\sigma$ has a standard normal distribution; that is, $Z \sim N(0, 1)$.

It follows that, if a and b are any numbers with $a < b$, then

$$P(a < X < b) = P\left(\frac{a - \mu}{\sigma} < Z < \frac{b - \mu}{\sigma}\right)$$

$$= F_Z\left(\frac{b - \mu}{\sigma}\right) - F_Z\left(\frac{a - \mu}{\sigma}\right)$$

where Z is the standard normal variable and $F_Z(z)$ denotes its cumulative distribution function.

The result is illustrated in Figure 5.14. Part (a) of the figure shows the probability density function of a normal random variable X with mean $\mu = 3$ and standard deviation $\sigma = 2$. The shaded area shows the probability that X lies between 4 and 6. This is the same as the probability that a standard normal variable lies between $(4 - \mu)/\sigma$ and $(6 - \mu)/\sigma$, that is, between 0.5 and 1.5. This probability is the shaded area under the standard normal curve in Figure 5.14(b).

Example 5.5

If $X \sim N(15, 16)$, find the probability that X is larger than 18. This probability is

$$P(X > 18) = P\left(Z > \frac{18 - \mu}{\sigma}\right)$$

$$= P\left(Z > \frac{18 - 15}{4}\right)$$

$$= P(Z > 0.75)$$

$$= 1 - P(Z < 0.75)$$

$$= 1 - F_Z(0.75)$$

From Table 3 in the Appendix, $F_Z(0.75)$ is .7734, so that

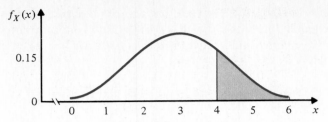

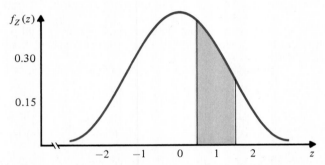

(a) Probability density function for normal random variable X with mean 3 and standard deviation 2; shaded area is probability that X lies between 4 and 6

(b) Probability density function for standard normal random variable Z; shaded area is probability that Z lies between 0.5 and 1.5, and is equal to shaded area in part (a)

FIGURE 5.14 Finding range probabilities for normal random variables

$$P(X > 18) = 1 - .7734 = .2266$$

Example 5.6   If X is normally distributed with mean 3 and standard deviation 2, find $P(4 < X < 6)$. We have

$$P(4 < X < 6) = P\left(\frac{4 - \mu}{\sigma} < Z < \frac{6 - \mu}{\sigma}\right)$$

$$= P\left(\frac{4 - 3}{2} < Z < \frac{6 - 3}{2}\right)$$

$$= P(0.5 < Z < 1.5)$$

$$= F_Z(1.5) - F_Z(0.5)$$

$$= .9332 - .6915 = .2417$$

These are the calculations illustrated in Figure 5.14.

Example 5.7

For a particular model car, it is known that achieved miles per gallon under standard driving patterns can be represented by a normal distribution with mean 28.5 and standard deviation 1.4 miles. What is the probability that such a car will achieve between 28 and 30 miles per gallon?

Let X represent miles per gallon obtained. Then

$$P(28 < X < 30) = P[(28 - \mu)/\sigma < Z < (30 - \mu)/\sigma]$$
$$= P[(28 - 28.5)/1.4 < Z < (30 - 28.5)/1.4]$$
$$= P[-0.36 < Z < 1.07]$$
$$= F_Z(1.07) - F_Z(-0.36)$$
$$= F_Z(1.07) - [1 - F_Z(0.36)]$$
$$= .8577 - [1 - .6406] = .4983$$

Hence the probability of achieving miles per gallon figures between 28 and 30 is very nearly one-half.

Example 5.8

A very large group of students obtains test scores that are normally distributed with mean 60 and standard deviation 15. What proportion of the students obtained scores between 85 and 95?

Let X denote the test score. Then we have

$$P(85 < X < 95) = P\left(\frac{85 - \mu}{\sigma} < Z < \frac{95 - \mu}{\sigma}\right)$$
$$= P\left(\frac{85 - 60}{15} < Z < \frac{95 - 60}{15}\right)$$
$$= P(1.67 < Z < 2.33)$$
$$= F_Z(2.33) - F_Z(1.67)$$
$$= .9901 - .9525 = .0376$$

That is, 3.76% of the students obtained scores in the range 85 to 95.

Example 5.9

For the test scores of Example 5.8, find the cutoff point for the top 10% of all students.

We have previously found probabilities corresponding to cutoff points. Here we need the cutoff point corresponding to a particular probability. The position is illustrated in Figure 5.15, which shows the probability density function of a normally distributed random variable with mean 60 and standard deviation 15. Let the number b denote the minimum score needed to be in the highest 10%. Then, the probability is 0.10 that the score of a randomly chosen student exceeds the number b. This probability is shown as the shaded area in Figure 5.15. If X denotes the test scores, then the probability that X exceeds b is .1, so that

The Normal Distribution   **209**

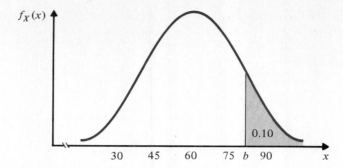

FIGURE 5.15 The probability is 0.10 that the random variable X exceeds the number b; here X is normally distributed, with mean 60 and standard deviation 15

$$.1 = P(X > b)$$

$$= P\left(Z > \frac{b - \mu}{\sigma}\right)$$

$$= P\left(Z > \frac{b - 60}{15}\right)$$

Hence, it follows that

$$.9 = P\left(Z < \frac{b - 60}{15}\right)$$

$$= F_Z\left(\frac{b - 60}{15}\right)$$

Now, from Table 3 of the Appendix, if $F_Z(z) = .9$, then $z = 1.28$. Therefore, we have

$$\frac{b - 60}{15} = 1.28$$

so that

$$b = 79.2$$

The conclusion is that 10% of the students obtain grades higher than 79.2.

In Examples 5.8 and 5.9, if the scores awarded on the test were integers, the distribution of scores would be inherently discrete. Nevertheless, the normal distribution can typically provide an adequate approximation in such circumstances. We will see later in this chapter that the normal distribution can often be employed as an approximation to discrete distributions. As a preliminary we introduce in the next section a result that provides strong justification for the emphasis given to the normal distribution in this book.

16. Let the random variable Z follow a standard normal distribution.
    (a) Find $P(Z < 1.33)$.
    (b) Find $P(Z > 1.20)$.
    (c) Find $P(Z < -1.00)$
    (d) Find $P(Z > -1.80)$.
    (e) Find $P(1.20 < Z < 1.33)$
    (f) Find $P(-1.80 < Z < 1.33)$.
    (g) Find $P(-1.80 < Z < -1.00)$.

17. Let the random variable Z follow a standard normal distribution.
    (a) The probability is 0.8 that Z is less than what number?
    (b) The probability is 0.3 that Z is greater than what number?
    (c) The probability is 0.4 that Z is less than what number?
    (d) The probability is 0.9 that Z is greater than what number?

18. It was found that the selling price of homes in a large city followed a normal distribution with mean $80,000 and standard deviation $15,000. Find the proportion of homes that have selling prices in each of the following ranges:
    (a) Less than $100,000
    (b) More than $90,000
    (c) Less than $75,000
    (d) More than $65,000
    (e) Between $90,000 and $100,000
    (f) Between $65,000 and $100,000
    (g) Between $65,000 and $75,000

19. The tread life of a certain brand of tire has a normal distribution with mean 35,000 miles and standard deviation 4,000 miles. For a randomly selected tire, find the probability that its tread life falls within each of the following ranges:
    (a) Less than 40,000 miles
    (b) More than 42,000 miles
    (c) Less than 32,000 miles
    (d) More than 30,000 miles
    (e) Between 40,000 and 42,000 miles
    (f) Between 30,000 and 42,000 miles
    (g) Between 30,000 and 32,000 miles

20. The diameters of metal rods manufactured by a certain process follow a normal distribution with mean 4 centimeters and standard deviation 0.1 centimeter.
    (a) What proportion of the rods have diameters larger than 4.2 centimeters?
    (b) What proportion have diameters less than 3.8 centimeters?
    (c) What proportion have diameters between 3.8 and 4.2 centimeters?

21. The annual earnings of employees at a large manufacturing plant have a normal distribution with mean $18,000 and standard deviation $3,000.
    (a) What proportion of the employees have earnings over $20,000?
    (b) What proportion of the employees have earnings below $15,000?
    (c) What proportion of the employees have earnings between $18,000 and $20,000?

22. A company produces lightbulbs whose lifetimes follow a normal distribution with mean 1200 hours and standard deviation 250 hours. One lightbulb is to be randomly selected from the company's output.
    (a) Find the probability that the lifetime of the selected bulb is at least 1000 hours.
    (b) Find the probability that its lifetime is less than 1500 hours.
    (c) Find the probability that its lifetime is between 1000 and 1400 hours.
    (d) Without doing the calculations, state in which of the following ranges its lifetime is most likely to be: 1000–1200 hours, 1100–1300 hours, 1200–1400 hours, 1300–1500 hours.

23. The weight of boxes of a particular detergent follows a normal distribution with mean

weight 20 ounces and standard deviation 0.6 ounce. A box of this detergent is randomly selected.

(a) What is the probability that the weight will be less than 19 ounces?

(b) What is the probability that the weight will be between 19.5 and 20.5 ounces?

(c) Without doing the calculations, state in which of the following ranges the weight is most likely to be: 19.2–19.6 ounces, 19.6–20.0 ounces, 19.8–20.2 ounces, 20.0–20.4 ounces.

24. A company services central heating furnaces. A review of its records indicates that the time taken for a service call can be represented by a normal distribution with mean 45 minutes and standard deviation 10 minutes.

(a) What proportion of the service calls take more than 30 minutes?

(b) What proportion of the service calls take less than an hour?

(c) Sketch a graph to illustrate why the answers to (a) and (b) are the same.

(d) An engineer is scheduled to make three calls in a morning. Assume that the times taken for each of these calls are independent of one another. What is the probability that at least one of them will take more than an hour?

25. The breaking strength of plastic bags is normally distributed with a mean of 5 pounds per square inch and standard deviation 1 pound per square inch.

(a) What proportion of the bags have breaking strength at least 3.6 pounds per square inch?

(b) What proportion of the bags have breaking strength between 5.0 and 5.5 pounds per square inch?

(c) What proportion of the bags have breaking strength between 3.2 and 5.2 pounds per square inch?

(d) Without doing the calculations, state which of the following ranges of breaking strength (in pounds per square inch) contains the highest proportion of bags: 4.0–5.0, 4.5–5.5, 5.0–6.0, 5.5–6.5, 6.0–7.0.

26. Seniors at a public school who take a placement test have scores that are normally distributed with a mean of 180 and a standard deviation of 40. Seniors at a private school who take the same test have scores that are normally distributed with a mean of 200 and a standard deviation of 60. A student qualifies for a state honorary society if his or her score exceeds 280.

(a) For a randomly chosen senior from the public school, what is the probability his or her score on the test will qualify the student for the state honorary society?

(b) For a randomly selected senior from the private school, what is the probability his or her score will qualify the student for the state honorary society?

(c) If we randomly and independently select from each school one senior who took the test, what is the probability that at least one of these two students qualifies for the state honorary society?

27. I am considering two alternative investments. In both cases I am unsure about the percentage return, but believe my uncertainty can be represented by normal distributions with the means and standard deviations shown in the accompanying table. I want to make the investment that is more likely to produce a return of at least 10%. Which should I choose?

	MEAN	STANDARD DEVIATION
INVESTMENT A	10.4	1.2
INVESTMENT B	11.0	4.0

28. A company can purchase raw material from either of two suppliers and is concerned about the amounts of impurity the material contains. A review of the records for each supplier indicates that the percentage impurity levels in consignments of the raw material follow normal distributions with the means and standard deviations given in the table. The company is particularly anxious that the impurity level in a consignment not exceed 5%, and wants to purchase from the supplier more likely to meet that specification. Which supplier should be chosen?

	MEAN	STANDARD DEVIATION
SUPPLIER A	4.4	0.4
SUPPLIER B	4.2	0.6

29. The number of days between billing and payment of electricity bills is normally distributed with mean 16 days and standard deviation 5 days.

(a) How many days will have elapsed before 10% of all bills are paid?

(b) After how many days will 25% of all bills remain unpaid?

30. In an automobile race, the length of time required for a crew to complete a routine pit-stop, involving refilling the fuel tank and changing tires, can be represented by a normal random variable with mean 16 seconds and standard deviation 1.5 seconds.

(a) The probability is 0.9 that a pit-stop will take at least how many seconds?

(b) The probability is 0.8 that a pit-stop will take at most how many seconds?

31. An economic forecaster believes that his uncertainty about next year's inflation rate can be represented by a normal random variable with mean 9.2% and standard deviation 1.0%.

(a) According to this forecaster, the probability is 0.05 that inflation will be more than what level?

(b) The probability is 0.10 that inflation will be less than what level?

32. Following an intensive employment interview, a personnel officer assigns to each candidate a score on a scale from 0 to 100. The scores given follow a normal distribution with mean 62 and standard deviation 12. The top 20% of all candidates are invited to attend further interviews, while the remaining candidates are rejected. What score is needed to avoid an immediate rejection?

33. An inspector has determined that the mean weight of the contents of packages of a certain brand of cereal is 19.9 ounces. He also finds that 60% of all packages have contents weighing less than 20 ounces. If the distribution of weights is normal, what is the standard deviation of these weights?

34. An instructor grades a test for a very large class of students. Having completed this task, she determines that the mean score is 60 and the standard deviation is 12. She intends to give grade A to the best 20% of the students. She also wants to adjust the test scores, multiplying each one by the same number, so that an adjusted score of 90 is necessary to obtain an A. By what number should she multiply the scores? (Assume that the original scores are normally distributed. If this is the case, the adjusted scores will also be normally distributed.)

5.6 THE CENTRAL LIMIT THEOREM

Many random variables met in practice can be characterized as either the sum or the average of a fairly large number of independent random variables. Let $X_1, X_2, \ldots ,$

X_n be n independent random variables having identical distributions with mean μ and variance σ^2. Denote their sum by

$$X = X_1 + X_2 + \cdots + X_n$$

We have seen in Sections 4.4 and 5.4 that the mean of a sum is the sum of the means and, for independent random variables, the variance of the sum is the sum of the variances. Hence, the mean and variance of X are

$$E(X) = n\mu \qquad \mathrm{Var}(X) = n\sigma^2$$

Now, for any random variable, subtracting the mean and dividing by the standard deviation yields a random variable with mean 0 and variance 1, so that the random variable

$$Z = \frac{X - E(X)}{\sqrt{\mathrm{Var}(X)}} = \frac{X - n\mu}{\sqrt{n\sigma^2}}$$

has mean 0 and variance 1. Dividing the numerator and denominator of this expression by n yields

$$Z = \frac{\bar{X} - \mu}{\sigma/\sqrt{n}}$$

where

$$\bar{X} = \frac{X_1 + X_2 + \cdots + X_n}{n} = \frac{X}{n}$$

is the average of the X_i.

So far this is not new. The crucial additional information, provided by the **Central Limit Theorem,** is that *whatever the distribution of the X_i* (provided that σ^2 is finite), as the number of terms n in the sum becomes large the distribution of Z tends to the standard normal.

Central Limit Theorem

Let $X_1, X_2, \ldots, X_n$ be n independent random variables having identical distributions with mean μ and variance σ^2. Denote by X and $\bar{X}$, respectively, the sum and average of these random variables. Then as n becomes large, the distribution of

$$Z = \frac{X - n\mu}{\sqrt{n\sigma^2}} = \frac{\bar{X} - \mu}{\sigma/\sqrt{n}}$$

tends to the standard normal.

The Central Limit Theorem has a substantial impact on the practice of statistics. Many practical problems involve sums or averages of random variables, and in these circumstances, by virtue of this theorem, the normal distribution very often provides a satisfactory approximation to the true distribution.

The result has a remarkably wide range of applicability. It states that *whatever* the common distribution of a set of independent random variables, provided that their variance is finite, the sum or average of a moderately large number of them will be a random variable with a distribution close to the normal. To illustrate, in Sections 5.2 and 5.3 we introduced the uniform distribution. The density function of this random variable is certainly very different in shape from that of a normal random variable. Nevertheless, the average of a moderately large number of independent uniformly distributed random variables has a distribution close to the normal. Figure 5.16 shows the shapes of the probability density functions of the averages of one, two, and ten

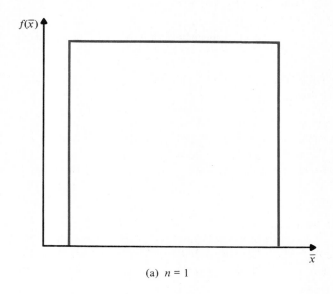

(a) *n* = 1

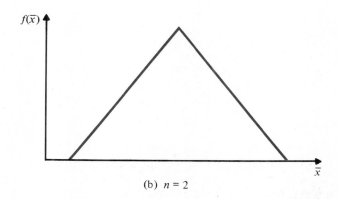

(b) *n* = 2

FIGURE 5.16 Shapes of probability density functions for the average of *n* independent uniformly distributed random variables

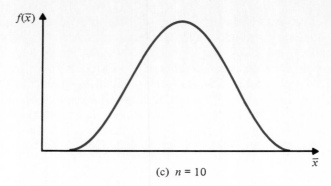

(c) $n = 10$

FIGURE 5.16 Shapes of probability density functions for the average of n independent uniformly distributed random variables (cont.)

independent uniform random variables. In part (a) of the Figure, for $n = 1$, we see the familiar uniform density function. Part (b) shows the density function, which is triangular in shape, for the average of $n = 2$ of these random variables. Then, with as few as $n = 10$ values in the average, we find in part (c) of Figure 5.16 a probability density function whose shape is already very similar to that of a normal distribution.

The uniform distribution is symmetric about its mean. However, the Central Limit Theorem applies also to asymmetric distributions. To illustrate, Figure 5.17(a)

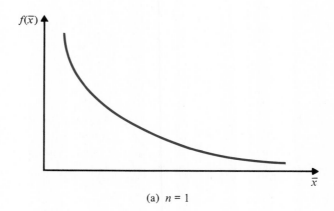

(a) $n = 1$

FIGURE 5.17 Shapes of probability density functions for the average of n independent random variables from the chi-square distribution with one degree of freedom

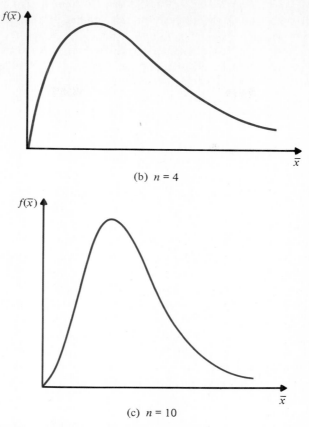

(b) $n = 4$

(c) $n = 10$

FIGURE 5.17 Shapes of probability density functions for the average of n independent random variables from the chi-square distribution with one degree of freedom (cont.)

shows such a distribution.[8] The shape of its probability density function is very different from that of the normal distribution. Parts (b) and (c) of the figure show the probability density functions for the averages of $n = 4$ and $n = 10$ independent observations from this distribution. As can be seen, when there are ten values in the average, the shape of the density function is very close to that of the standard normal random variable.

The validity of the Central Limit Theorem is not restricted to sums of continuous random variables. It extends also to discrete random variables. In the following section we will see how the theorem can be exploited to allow us to obtain good

---

[8] This is the chi-square distribution with one degree of freedom. We will meet this distribution again in Section 6.4.

approximations of range probabilities for random variables following a binomial distribution.

In Chapter 6, we will begin our discussion of the important statistical problem of making inference about a population, based on results obtained from a sample. Many of the quantities calculated from sample data are sums or averages. Thus, the Central Limit Theorem becomes relevant and provides a validity for many of the techniques used for approaching this vital problem.

5.7 THE NORMAL DISTRIBUTION AS AN APPROXIMATION TO THE BINOMIAL AND POISSON DISTRIBUTIONS

In Chapter 4 we introduced the binomial and Poisson distributions and showed how probabilities for these random variables could be calculated. It emerged that, in cases where the number of binomial trials was large and where the mean of the Poisson distribution was large, these computations constituted a formidable burden. Fortunately, as a consequence of the Central Limit Theorem, considerable computational simplification can be achieved through a normal approximation to these distributions. In this section we will discuss and illustrate these approximations.

NORMAL APPROXIMATION TO THE BINOMIAL DISTRIBUTION

If n independent trials, each with probability of success p, are carried out, then the number X of successes resulting has a binomial distribution with mean and variance

$$E(X) = np; \qquad \text{Var}(X) = np(1 - p)$$

We saw in Section 4.5 that the random variable X could be written as the sum of n independent Bernoulli random variables; that is,

$$X = X_1 + X_2 + \cdots + X_n$$

where the random variable X_i takes the value 1 if the outcome of the ith trial is "success," and 0 otherwise, with respective probabilities p and $1 - p$.

Now this is precisely the setup in which the Central Limit Theorem is applicable. It therefore follows that, if the number of trials n is large, the distribution of the random variable

$$Z = \frac{X - E(X)}{\sqrt{\text{Var}(X)}} = \frac{X - np}{\sqrt{np(1 - p)}}$$

is approximately standard normal.

The importance of this result lies in the ease with which it allows us to find, for large n, the probability that the number of successes lies in some given range. Suppose that we want to know the probability that the number of successes will be between a and b, inclusive. We then have

$$P(a \leq X \leq b) = P\left(\frac{a - np}{\sqrt{np(1-p)}} \leq \frac{X - np}{\sqrt{np(1-p)}} \leq \frac{b - np}{\sqrt{np(1-p)}}\right)$$

$$= P\left(\frac{a - np}{\sqrt{np(1-p)}} \leq Z \leq \frac{b - np}{\sqrt{np(1-p)}}\right)$$

If the number of trials n is large, the distribution of Z can be well approximated by the standard normal so that the above probability can be found using the methods of Section 5.5.

If the number of trials n is only of moderate size, a worthwhile improvement to this approximation can be achieved. We are here approximating a discrete distribution by a continuous one. While the binomial random variable can take on only integer values, the normal random variable is defined on a continuum. To allow for this distinction, a *continuity correction* can be applied to the previous formula, replacing a and b by $(a - 0.5)$ and $(b + 0.5)$, respectively. We then have

$$P(a \leq X \leq b) \approx P\left(\frac{a - 0.5 - np}{\sqrt{np(1-p)}} \leq Z \leq \frac{b + 0.5 - np}{\sqrt{np(1-p)}}\right)$$

To see the rationale for this modification, suppose we seek the probability that the number of successes is greater than or equal to 10 and less than or equal to 15. Since the actual number of successes must be an integer, the probability required is the same as the probability that the number of succeses is greater than or equal to 9.001 and less than or equal to 15.999. As a compromise between these two extremes, we employ a range with endpoints 9.5 and 15.5.

Approximating Binomial Probabilities Using the Normal Distribution

Let X be the number of successes resulting from n independent trials, each with probability of success p. If n is large, and provided p is not very small or very large,[9] then to a good approximation

$$P(a \leq X \leq b) \approx P\left(\frac{a - np}{\sqrt{np(1-p)}} \leq Z \leq \frac{b - np}{\sqrt{np(1-p)}}\right)$$

or, using the continuity correction,

$$P(a \leq X \leq b) \approx P\left(\frac{a - 0.5 - np}{\sqrt{np(1-p)}} \leq Z \leq \frac{b + 0.5 - np}{\sqrt{np(1-p)}}\right)$$

where Z has a standard normal distribution.[10]

---

[9] In such a case the Poisson approximation to the binomial distribution should be used (see Section 4.7).

[10] The simpler approximation is generally satisfactory for $n \geq 50$, while for $20 \leq n < 50$, it is preferable to use the continuity correction.

Example
5.10

In Example 4.12 we considered a corporation making offers to 20 candidates for employment, where it was believed that the probability of acceptance was .8 in each case. Letting X denote the number of acceptances, we used the binomial probabilities to calculate the precise probability of between 16 and 18 acceptances:

$$P(16 \leq X \leq 18) = .5605$$

Although one might be rather uncomfortable about using the normal approximation to the binomial distribution based on only $n = 20$ trials, this example provides a good opportunity to evaluate the quality of the approximation for this small number of trials.

If the continuity correction is not used, the approximation is

$$P(16 \leq X \leq 18) \approx P\left[\frac{16 - (20)(.8)}{\sqrt{(20)(.8)(.2)}} \leq Z \leq \frac{18 - (20)(.8)}{\sqrt{(20)(.8)(.2)}}\right]$$

$$= P(0 \leq Z \leq 1.12)$$

$$= F_Z(1.12) - F_Z(0)$$

where $F_Z(z)$ is the cumulative distribution function of the standard normal. Hence, from Table 3 of the Appendix, the approximation yields

$$P(16 \leq X \leq 18) \approx .8686 - .5000 = .3686$$

which is disturbingly far from the true probability of .5605.

Using the continuity correction, we have

$$P(16 \leq X \leq 18) \approx P\left[\frac{15.5 - (20)(.8)}{\sqrt{(20)(.8)(.2)}} \leq Z \leq \frac{18.5 - (20)(.8)}{\sqrt{(20)(.8)(.2)}}\right]$$

$$= P(-0.28 \leq Z \leq 1.40)$$

$$= F_Z(1.40) - F_Z(-0.28)$$

$$= F_Z(1.40) - [1 - F_Z(0.28)]$$

$$= .9192 - (1 - .6103) = .5295$$

This approximation is sufficiently close to the true probability for most practical purposes. The desirability of using the continuity correction when the number of trials is only moderate emerges quite clearly from this example.

Example
5.11

A salesman makes initial contact with potential customers in an effort to assess whether a follow-up visit to their homes is likely to be worthwhile. His experience suggests that 40% of initial contacts lead to follow-up visits. If he contacts 100 people by telephone, what is the probability that between 45 and 50 home visits will result?

Let X be the number of follow-up visits. Then X has a binomial distribution with $n = 100$ and $p = .4$. Approximating the required probability without using the continuity correction gives

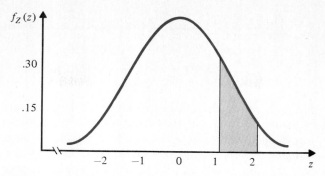

FIGURE 5.18 Probability of 45 to 50 successes in 100 binomial trials each with probability of success .4. This probability is shown as the probability that a standard normal random variable lies between 1.02 and 2.04

$$P(45 \leq X \leq 50) \approx P\left[\frac{45 - (100)(.4)}{\sqrt{(100)(.4)(.6)}} \leq Z \leq \frac{50 - (100)(.4)}{\sqrt{(100)(.4)(.6)}}\right]$$

$$= P(1.02 \leq Z \leq 2.04)$$

$$= F_Z(2.04) - F_Z(1.02)$$

$$= .9793 - .8461 = .1332$$

This probability is shown as an area under the standard normal curve in Figure 5.18.

If the continuity correction is used in approximating this binomial probability, we have

$$P(45 \leq X \leq 50) \approx P\left[\frac{44.5 - (100)(.4)}{\sqrt{(100)(.4)(.6)}} \leq Z \leq \frac{50.5 - (100)(.4)}{\sqrt{(100)(.4)(.6)}}\right]$$

$$= P(0.92 \leq Z \leq 2.14)$$

$$= F_Z(2.14) - F_Z(0.92)$$

$$= .9838 - .8212 = .1626$$

Notice that the two approximations are much closer together here than in Example 5.10, indicating the diminishing importance of the continuity correction as the number of trials n increases.

NORMAL APPROXIMATION TO THE POISSON DISTRIBUTION

Let the random variable X denote the number of occurrences of an event in a particular interval of time and denote by λ the expected, or mean, number of occurrences in that time interval. Then X obeys the Poisson distribution discussed in Section 4.7, with mean and variance

$$E(X) = \lambda; \qquad \text{Var}(X) = \lambda$$

FIGURE 5.19 Occurrences (●) in the interval from 0 to t, broken down into subintervals of equal width

Consider now the situation in which the mean number of occurrences, λ, is large and suppose that the time interval of interest is broken down into subintervals of equal width, as in Figure 5.19. Then the total number of occurrences is the sum of the numbers of occurrences in each subinterval. Thus, we see that when the mean of the Poisson distribution is large, the total number of occurrences can be viewed as the sum of a moderately large number of random variables, each of which represents the number of occurrences in a subinterval of the time period. Hence, invoking the Central Limit Theorem, we conclude that when λ is large, the distribution of the random variable

$$Z = \frac{X - E(X)}{\sqrt{\mathrm{Var}(X)}} = \frac{X - \lambda}{\sqrt{\lambda}}$$

is approximately standard normal.

As in the case of the binomial distribution, this result can be used to approximate probabilities. Here again, if λ is only of moderate size, a continuity correction will be desirable.

Approximating Poisson Probabilities Using the Normal Distribution

Let X be a Poisson random variable with mean λ. If λ is large, then to a good approximation,

$$P(a \leq X \leq b) \approx P\left(\frac{a - \lambda}{\sqrt{\lambda}} \leq Z \leq \frac{b - \lambda}{\sqrt{\lambda}}\right)$$

or, using the continuity correction,

$$P(a \leq X \leq b) \approx P\left(\frac{a - 0.5 - \lambda}{\sqrt{\lambda}} \leq Z \leq \frac{b + 0.5 - \lambda}{\sqrt{\lambda}}\right)$$

where Z has a standard normal distribution.

Example 5.12

A firm receives, on average, 42 replacement orders per week for a particular part. If it begins the week with a stock of 55 of these parts, what is the probability that this stock will be exhausted by the end of the week?

Let X be the number of orders received, so that X has a Poisson distribution with mean $\lambda = 42$. Using the normal approximation to the Poisson (without the continuity correction), we obtain the required probability:

$$P(X > 55) = 1 - P(X \leq 55)$$

$$\approx 1 - P\left(Z \leq \frac{55 - 42}{\sqrt{42}}\right)$$

$$= 1 - P(Z \le 2.01)$$
$$= 1 - F_Z(2.01)$$
$$= 1 - .9778 = .0222$$

Thus, the probability that the stocks will be exhausted is very small.

The reader can verify that, for this high value of λ, use of the continuity correction has very little effect on the final conclusion.

EXERCISES

35. In the 1980 presidential election, 51% of all those voting voted for Ronald Reagan. Suppose we select a random sample of 400 of all those who voted.
    (a) What is the probability that more than 250 of the sample members voted for Ronald Reagan?
    (b) What is the probability that at least 230 of the sample members voted for Mr. Reagan?
    (Use the normal approximation to the binomial distribution, without the continuity correction.)

36. A car rental company has determined that the probability a car will need service work in any given month is 0.2. The company has 900 cars.
    (a) What is the probability that more than 200 will require service work in a particular month?
    (b) What is the probability that less than 175 will need service work in a given month?
    (Use the normal approximation to the binomial distribution, without the continuity correction.)

37. It is known that 10% of all the items produced by a particular manufacturing process are defective. From the very large output of a single day, 400 items are selected at random.
    (a) What is the probability that at least 35 of the selected items are defective?
    (b) What is the probability that between 40 and 50 of the selected items are defective?
    (c) What is the probability that between 34 and 48 of the selected items are defective?
    (d) Without doing the calculations, state which of the following ranges of defectives has the highest probability:

    $$37\text{--}39; \quad 39\text{--}41; \quad 41\text{--}43; \quad 43\text{--}45; \quad 45\text{--}47$$

    (Use the normal approximation to the binomial distribution, without the continuity correction.)

38. Students are examined for anemia. Those with anemia are immediately hospitalized until they recover. It is found that 10% of all students examined have anemia. The length of stay for those hospitalized for anemia is normally distributed, with mean 14 days and standard deviation 2 days.
    (a) Using the normal approximation to the binomial distribution, without the continuity correction, find the probability that at least 14 of 225 randomly chosen students have anemia.
    (b) If one student is randomly selected and examined, what is the probability that he or she will be hospitalized for anemia for at least 17 days?

39. A firm's records indicate that the probability that a machine will break down during a week of operation is 0.4. The firm has 50 of these machines.

(a) What is the probability that less than 15 of these machines break down in a week?

(b) What is the probability that between 18 and 22 of the machines break down in a week?

(Use the normal approximation to the binomial distribution, with and without the continuity correction, and compare your answers.)

40. A gas station owner knows that 30% of his customers pay by credit card. A random sample of 80 customers is selected.

(a) What is the probability that more than 20 pay by credit card?

(b) What is the probability that less than 25 pay by credit card?

(c) What is the probability that between 20 and 25 pay by credit card?

(Use the normal approximation to the binomial distribution, with the continuity correction.)

41. The distribution of the lifetimes of lightbulbs can be represented (as in Exercise 22) by a normal distribution with mean 1200 hours and standard deviation 250 hours. A company purchases 400 of these bulbs.

(a) What is the probability that more than 325 of these bulbs last at least 1000 hours?

(b) What is the probability that more than 325 of these bulbs last less than 1500 hours?

(c) What is the probability that more than 325 of these bulbs last between 1000 and 1400 hours?

(Use the normal approximation to the binomial distribution, without the continuity correction.)

42. In the hour before noon, a busy corporation switchboard receives an average of 60 calls.

(a) What is the probability that more than 70 calls will be received during this hour?

(b) What is the probability that less than 55 calls will be received during this hour?

(c) What is the probability that between 50 and 70 calls will be received during this hour?

(Use the normal approximation to the Poisson distribution, without the continuity correction.)

43. New York City has an average of 36 homicides per week.

(a) Find the probability that there will be less than 30 homicides in a given week.

(b) Find the probability that there will be more than 40 homicides in a given week.

(c) Find the probability that there will be between 30 and 40 homicides in a given week.

(Use the normal approximation to the Poisson distribution, with and without the continuity correction, and compare your answers.)

44. A consumer advice center receives, on average, 25 calls per day.

(a) Find the probability that between 20 and 30 calls will be received on a given day.

(b) Find the probability that between 25 and 40 calls will be received on a given day.

(Use the normal approximation to the Poisson distribution, with and without the continuity correction, and compare your answers.)

REVIEW EXERCISES

45. Explain verbally what can be learned from each of the following:

(a) The cumulative distribution function of a continuous random variable

(b) The probability density function of a continuous random variable

(c) The mean of a continuous random variable

(d) The standard deviation of a continuous random variable

46. "In the real world, measurements on any quantity of interest are almost invariably made on a discrete scale. Therefore, the study of continuous random variables is only of academic interest and has no practical value." Comment on this statement.

47. Answer the following questions:

(a) Why is it necessary to use tables to find probabilities for the normal distribution?

(b) Why is it necessary to tabulate probabilities for only one of the infinite number of normal distributions?

(c) Why is the normal distribution important in the study of statistics?

48. "Many quantities we measure—such as height, weight, and distance—are necessarily positive. It therefore follows that the mean of a sample of observations on such a quantity must also be positive. However, the normal distribution supposes a range of values from minus infinity to plus infinity. Thus, in spite of the Central Limit Theorem, the normal distribution cannot possibly be appropriate for such sample means." Comment on this statement.

49. In practice, we find probabilities for any normal distribution based on probabilities for the standard normal distribution. Suppose that you had available tables of probabilities for the normal distribution with mean 1 and standard deviation 10, rather than tables for the standard normal distribution. Could these tables be used to find probabilities for any other normal distribution? If so, explain how.

50. The percentage impurity content of batches of a particular chemical can be assumed to be uniformly distributed between 1% and 3%. Thus, if X denotes percentage impurity, the probability density function is

$$f_X(x) = \begin{cases} 0.5 & \text{for } 1 < x < 3 \\ 0 & \text{for all other values of } x \end{cases}$$

(a) Draw the probability density function for the random variable X.

(b) Find and draw the cumulative distribution function for this random variable.

(c) Find the probability that a randomly chosen batch has impurity concentration over 2%.

(d) Three batches are chosen at random. Find the probability that at least one of them has impurity concentration over 2%.

51. The ages of managers in a large corporation are uniformly distributed between 30 and 65 years. Therefore, if X denotes age in years, the probability density function is

$$f_X(x) = \begin{cases} \dfrac{1}{35} & \text{for } 30 < x < 65 \\ 0 & \text{for all other values of } x \end{cases}$$

(a) Draw the probability density function for this random variable X.

(b) Find and draw the cumulative distribution function for this random variable.

(c) Managers in this corporation must retire at age 65. What proportion of managers will face compulsory retirement over the next 10 years?

(d) What is the probability that a randomly chosen manager is between 35 and 40 years of age?

(e) Find the mean age of managers in this company.

52. The random variable X has probability density function

$$f_X(x) = \begin{cases} x & \text{for } 0 < x < 1 \\ 2 - x & \text{for } 1 < x < 2 \\ 0 & \text{for all other values of } x \end{cases}$$

(a) Draw the probability density function for this random variable.

(b) Show that the density function has the properties of a proper probability density function.

53. An investor puts $1,000 into a deposit account with a fixed rate of return of 12% per annum. A second sum of $1,000 is invested in a fund with expected rate of return of 15% and standard deviation of 7% per annum.

(a) Find the expected value of the total amount of money this investor will have after a year.

(b) Find the standard deviation of the total amount after a year.

54. A hamburger stand sells burgers for $1.25 each. Daily sales have a distribution with mean 530 and standard deviation 69.

(a) Find the mean daily total revenues from the sale of hamburgers.

(b) Find the standard deviation of total revenues from the sale of hamburgers.

(c) Daily costs (in dollars) are given by

$$C = 100 + 0.75X$$

where X is the number of hamburgers sold. Find the mean and standard deviation of daily profits from sales.

55. An analyst forecasts corporate earnings, and her record is evaluated by comparing actual earnings with predicted earnings. Define

$$\text{Actual earnings} = \text{Predicted earnings} + \text{Forecast error}$$

If the predicted earnings and forecast error are independent of one another, show that the variance of predicted earnings is less than the variance of actual earnings.

56. Let X_1 and X_2 be a pair of random variables. Show that the covariance between the random variables $(X_1 + X_2)$ and $(X_1 - X_2)$ is 0 if and only if X_1 and X_2 have the same variance.

57. The scores of a large group of students on a particular test are normally distributed with mean 70 and standard deviation 10. Between what two numbers do the middle 50% of all scores lie?

58. A management consultant found that the amount of time per day spent by executives performing tasks that could equally well be done by subordinates followed a normal distribution with mean 2.4 hours. It was also found that 10% of executives spent over 3.5 hours per day on such tasks. Find the standard deviation of the distribution of daily time spent by executives on tasks of this type.

59. It was found that 60% of all applicants for a position scored more than 60 points on an aptitude test. The standard deviation of the scores was 15. If the distribution of scores was normal, what was the mean score?

60. The weights of the contents of boxes of a brand of cereal have a normal distribution with mean 24 ounces and standard deviation 0.7 ounce.

(a) What is the probability that the contents of a randomly chosen box weigh less than 23 ounces?

(b) The contents of 10% of all boxes weigh more than how many ounces?

(c) What proportion of boxes have contents weighing between 23.5 and 24.5 ounces?

61. The cereal manufacturer of the previous Exercise wants to adjust the production process so that the mean weight of the contents of the boxes of cereal is still 24 ounces, but so that only 3% of boxes will have contents weighing less than 23 ounces. What standard deviation for the weights of the contents is needed to attain this objective?

62. Lightbulbs of brand A have lifetimes that follow a normal distribution with mean 1200 hours and standard deviation 250 hours. For bulbs of brand B, lifetimes are normally

distributed with mean 1250 hours and standard deviation 400 hours. One lightbulb of each type is chosen at random. Which bulb is the more likely to last at least 1000 hours?

63. What does it mean to say that the normal distribution can be used to approximate the binomial distribution?

64. A survey research organization has found that answers are received from 65% of people sent a particular mail questionnaire. If this questionnaire is sent to 1000 people, what is the probability that at least 500 answers will be received?

65. An organization that gives regular seminars on sales motivation methods determines that 40% of its clients have attended previous seminars. From a sample of 400 clients, what is the probability that more than half have attended previous seminars?

66. An emergency towing service receives an average of 70 calls per day for assistance. For any given day, what is the probability that less than 50 calls will be received?

67. During peak hours, passengers arrive at customs in a large international airport at the rate of 1200 per hour. For a given peak hour, what is the probability of more than 1000 arrivals?

sampling and sampling distributions

CHAPTER SIX

6.1 SAMPLING FROM A POPULATION

In much of the remainder of this volume we will be occupied with a class of problems that involve an attempt to say something about the properties of a large group of objects, given information on a relatively small subset of them. The larger parent group is referred to as a **population** and the subset of population members is called a **sample.**

Examples of populations that might be of interest are

(a) The incomes of all families living in the city of Chicago
(b) The annual yields of all stocks traded on the New York Stock Exchange
(c) The costs of all claims for automobile accident insurance coverage received by a company in a given year
(d) The numbers of miles per gallon achieved by all cars of a particular model

In particular, we might be interested in learning about specific characteristics, or *attributes,* of these populations. For example, we might want to make an *inference* about the mean or variance of the population distribution of family incomes in Chicago, or about the proportion of all families in the city with annual incomes below $12,000.

The major motivation for examining a sample rather than the whole population is that the collection of complete information on the latter would typically be prohibitively expensive. Even in circumstances where sufficient resources are apparently available to contact the whole population, it may well be preferable to devote these resources to just a subset of the population in the hope that such a concentration of effort will produce more accurate measurements. It is well known,[1] for example, that the decennial census of the United States population produces an undercount in which certain groups are seriously underrepresented.

If a sample is to be taken from a population, the eventual aim is to make statements that have some validity for the population at large. Therefore, it is important that the sample be representative of the population. Suppose, for instance, that a marketing manager wants to assess reactions to a new food product. It would be unwise of him to restrict his survey to his circle of friends or to people living in his immediate neighborhood. Such groups are very unlikely to reflect the spectrum of views of the population at large, and may well be heavily weighted toward one end of that spectrum. To avoid problems of this kind, and to allow valid inference about a population based on a sample, it is important that the principle of **randomness** be embodied in the sample selection procedure. The most straightforward way of achieving this is to design the selection mechanism in such a way that every sample of the same size is equally likely to be chosen.

Simple Random Sampling

Suppose that it is required to select a sample of n objects from a population of N objects. A **simple random sampling** procedure is one in which every possible sample of n objects is equally likely to be chosen. This method is in such common use that the adjective *simple* is generally dropped, and the resulting sample is called a **random sample.**

The process of simple random sampling can be thought of as follows: Suppose the N population members are put into a (very large) hat and mixed thoroughly. A random sample is then obtained by pulling out n of them. In practice this is not necessary (even if it is feasible); tables of **random numbers,** such as those given in Table 4 in the Appendix, can be used to achieve the same objective. If the N population members are labelled 1 through N, we can begin at some arbitrary point in the table and read off numbers until a sample of n members has been identified. The tables are constructed in such a way that this process has the properties of simple random sampling. One possible, but very tedious, way of constructing a table of random numbers would be to place ten balls, numbered 0, 1, 2, . . . , 9, in a bag. After shaking thoroughly, draw a ball and record the number on it. Then replace this ball and repeat the process. We can continue in this fashion to obtain random numbers of one, two, three, or as many digits as are required. This process has the properties that

---

[1] See, for instance, N. Keyfitz, "Information and allocation: two uses of the 1980 census," *The American Statistician, 33* (1979), 45–50.

each possible number is equally likely to be chosen and successive choices are independent of one another. In practice, random numbers can be generated much more rapidly using an electronic computer with mechanisms that effectively mimic the process just described.

In this and the next few chapters we will focus on methods for analyzing sample results, with the objective of gaining information about the population. We will concentrate on samples that have been selected through simple random sampling schemes. However, this is by no means the only procedure available for choosing sample members and, in some circumstances, alternative sampling schemes may be preferable. We postpone a more thorough discussion of methods of sample selection until Chapter 18, where the use of random number tables will also be more fully described.

The principle of randomness in the selection of the sample members provides some protection against the sample being unrepresentative of the population, in the sense that, on the average, if the population were repeatedly sampled in this fashion no particular subgroup would be over-represented in the sample. Moreover, through the concept of a *sampling distribution,* it allows us to determine the probability that the particular sample obtained will be, to any specified degree, unrepresentative.

On the basis of sample information, our objective is to make *inference* about the parent population. The distribution of all the values of interest in this population can be represented by a random variable. It would be too ambitious to attempt to describe the whole population distribution based on a small random sample of observations. However, we may well be able to make quite firm inferences about important characteristics of the population distribution. For example, we may want to make statements about its mean and variance. As an illustration, given a random sample of miles per gallon achieved by twenty cars of a particular model, one can make inferential statements about the mean and variance of miles per gallon for all cars of that model. This inference will be based on just the sample information, so we are naturally led to ask such questions as: "If the number of miles per gallon achieved by all cars of a particular model has mean 25 and standard deviation 2, what is the probability that, for a random sample of twenty such cars, the average miles per gallon will be less than 24?" In asking this question, we are implicitly assuming that inference about the population mean will be based on the sample average, or mean.

It is important to distinguish between population attributes and the corresponding sample quantities. In the example of the previous paragraph, the population miles per gallon of all automobiles of a particular model year has a distribution with a specific mean. This mean, which is an attribute of the population, is a fixed (but unknown) number. In attempting to make inference about this attribute, a random sample is drawn from the population, and the sample mean is found. Since, from sample to sample, different values would result for the sample mean, this quantity can be regarded as a random variable, which has a probability distribution. This distribution of possible sample outcomes provides a basis for inferential statements about the population. Our objective in this chapter is to examine the properties of **sampling distributions** of this sort.

Statistics and Sampling Distributions

Suppose that a random sample is drawn from a population, and that inference about some characteristic of the population distribution is to be made. This inference is based on some **statistic,** a particular function of the sample information. The **sampling distribution** of this statistic is the probability distribution of the values it could take over all possible samples of the same number of observations drawn from the population.

To illustrate the important concept of a sampling distribution, let us consider the position of a supervisor with six employees, whose experiences (in terms of years on the job) are

$$2; \quad 4; \quad 6; \quad 6; \quad 7; \quad 8$$

Four of these employees are to be chosen randomly and assigned to a particular work shift. The mean number of years of experience for all six employees is

$$\frac{2 + 4 + 6 + 6 + 7 + 8}{6} = 5.5$$

Our interest now is in the mean number of years of experience of the particular four employees assigned to the work shift. These can be regarded as a simple random sample of four values, chosen from a population of six. There are 15 possible samples that could be selected. Table 6.1 shows the possible samples and associated sample means. Samples such as (2, 4, 6, 7) occur twice because there are two employees in the population with 6 years of work experience.

TABLE 6.1 Possible samples of four observations, and sample means, from the population 2, 4, 6, 6, 7, 8

SAMPLE	SAMPLE MEAN	SAMPLE	SAMPLE MEAN
2, 4, 6, 6	4.50	2, 6, 7, 8	5.75
2, 4, 6, 7	4.75	2, 6, 7, 8	5.75
2, 4, 6, 8	5.00	4, 6, 6, 7	5.75
2, 4, 6, 7	4.75	4, 6, 6, 8	6.00
2, 4, 6, 8	5.00	4, 6, 7, 8	6.25
2, 4, 7, 8	5.25	4, 6, 7, 8	6.25
2, 6, 6, 7	5.25	6, 6, 7, 8	6.75
2, 6, 6, 8	5.50		

Now, since each of the 15 possible samples is equally likely to be selected, the probability is $\frac{1}{15}$ that any specific sample will be selected. Using this information, we can determine the probability that any particular value will result for the sample mean. For example, we see from Table 6.1 that three of the possible samples have mean

5.75. Hence, it follows that the probability is ³⁄₁₅ that the four employees assigned to the work shift have an average of 5.75 years experience. In this way, we can find probabilities for every possible sample mean. The collection of these probabilities constitutes the sampling distribution of the sample mean.

Perhaps the simplest way to describe this sampling distribution is through its probability function. Denoting the sample mean by $\bar{X}$, we have

$$P(\bar{X} = 4.50) = P_{\bar{X}}(4.50) = \frac{1}{15} \qquad P(\bar{X} = 5.75) = P_{\bar{X}}(5.75) = \frac{3}{15}$$

$$P_{\bar{X}}(4.75) = \frac{2}{15} \qquad P_{\bar{X}}(6.00) = \frac{1}{15}$$

$$P_{\bar{X}}(5.00) = \frac{2}{15} \qquad P_{\bar{X}}(6.25) = \frac{2}{15}$$

$$P_{\bar{X}}(5.25) = \frac{2}{15} \qquad P_{\bar{X}}(6.75) = \frac{1}{15}$$

$$P_{\bar{X}}(5.50) = \frac{1}{15}$$

This probability function is graphed in Figure 6.1. Notice that, while the numbers of years of experience for the six workers range from 2 to 8, the possible values for the sample mean have a much more restricted range—from 4.50 to 6.75. Moreover, the preponderance of values lie in the central portion of this range.

In the next section, we consider the sampling distribution of the sample mean.

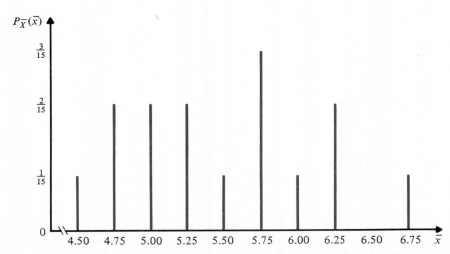

FIGURE 6.1 Probability function of sampling distribution for means of samples of four observations from the population 2, 4, 6, 6, 7, 8

6.2 SAMPLING DISTRIBUTION OF THE SAMPLE MEAN

Suppose that a random sample of n observations is drawn from a population with mean μ_X and variance σ_X^2, and that the sample members are denoted $X_1, X_2, \ldots, X_n$. Then each of these sample members can be regarded as a random variable having mean μ_X and variance σ_X^2. Assume for now that our primary interest is in making inference about the population mean μ_X. An obvious place to start is with the average of the sample values.

Definition

Let $X_1, X_2, \ldots, X_n$ be a random sample from a population. The average value

$$\bar{X} = \frac{1}{n} \sum_{i=1}^{n} X_i$$

of these observations is called the **sample mean.**

We need to consider the sampling distribution of the random variable $\bar{X}$.[2] First we determine the mean of this distribution. In Sections 4.4 and 5.4 it was seen that for discrete and continuous random variables the expectation of a sum is the sum of expectations, so that

$$E\left(\sum_{i=1}^{n} X_i \right) = E(X_1) + E(X_2) + \cdots + E(X_n)$$

Since each random variable X_i has mean μ_X, we can write

$$E\left(\sum_{i=1}^{n} X_i \right) = n\mu_X$$

Now, the sample mean is just the sum of the sample members multiplied by $1/n$, so that its mean is

$$E(\bar{X}) = E\left(\frac{1}{n} \sum_{i=1}^{n} X_i \right) = \frac{1}{n} E\left(\sum_{i=1}^{n} X_i \right) = \frac{n\mu_X}{n} = \mu_X$$

Thus, the mean of the sampling distribution of the sample mean is the population mean. This conclusion states that if samples of n observations are repeatedly and

---

[2] As in Chapters 4 and 5, we distinguish between a random variable and specific values it can take. Thus, when the sample is drawn we might observe the specific values $x_1, x_2, \ldots, x_n$, with mean $\bar{x} = \sum_{i=1}^{n} x_i/n$. In this context $\bar{x}$ is a single specific realization of the random variable $\bar{X}$.

independently drawn from a population, then as the number of samples taken becomes very large, the average of the sample means becomes very close to the true population mean. This result is an important consequence of random sampling, and represents the protection this form of sampling affords against the sample being unrepresentative of the population. Of course, the mean obtained from any *particular* sample could be either much higher or much lower than the population mean. However, *on the average,* there is no reason to expect a value that is either higher or lower than the population value.

Example 6.1	We confirm this result for the example of Table 6.1, in which we considered a population of years of experience figures for six employees:

$$2; \quad 4; \quad 6; \quad 6; \quad 7; \quad 8$$

The mean of this population is simply the average of these six values; that is, $\mu_X = 5.5$.

We found that the probability distribution of the sample mean, for samples of four observations from this population, could be represented by the following probability function:

$$P_{\bar{X}}(4.50) = \frac{1}{15}; \quad P_{\bar{X}}(4.75) = \frac{2}{15}; \quad P_{\bar{X}}(5.00) = \frac{2}{15};$$

$$P_{\bar{X}}(5.25) = \frac{2}{15}; \quad P_{\bar{X}}(5.50) = \frac{1}{15}; \quad P_{\bar{X}}(5.75) = \frac{3}{15};$$

$$P_{\bar{X}}(6.00) = \frac{1}{15}; \quad P_{\bar{X}}(6.25) = \frac{2}{15}; \quad P_{\bar{X}}(6.75) = \frac{1}{15}$$

Thus, the expected value of the sample mean is

$$E(\bar{X}) = \sum \bar{x} P_{\bar{X}}(\bar{x})$$

$$= (4.50)\left(\frac{1}{15}\right) + (4.75)\left(\frac{2}{15}\right) + \cdots + (6.75)\left(\frac{1}{15}\right) = 5.5$$

which is the population mean, μ_X.

It has been established, then, that the distribution of the sample mean is centered on the population mean. It is also of interest to determine how close the sample mean is likely to be to the population mean. For instance, suppose a random sample of twenty cars of a particular model yielded average miles per gallon of 24. How good an approximation might this be to the mean miles per gallon figure for the whole population of such cars? Questions of this kind depend on the spread, or *variance,* in the sampling distribution of $\bar{X}$.

If the number of population members is very large compared to the number in the sample, then an implication of simple random sampling is that *the distributions of the individual sample members are independent of one another.* Recalling from

Sections 4.4 and 5.4 that, in this case, the variance of a sum is the sum of the variances, we have

$$\text{Var}\left(\sum_{i=1}^{n} X_i\right) = \text{Var}(X_1) + \text{Var}(X_2) + \cdots + \text{Var}(X_n)$$

Since each X_i has variance σ_X^2, it follows that

$$\text{Var}\left(\sum_{i=1}^{n} X_i\right) = n\sigma_X^2$$

It then follows that the variance of the sample mean is

$$\text{Var}(\bar{X}) = \text{Var}\left(\frac{1}{n}\sum_{i=1}^{n} X_i\right) = \frac{1}{n^2}\text{Var}\left(\sum_{i=1}^{n} X_i\right) = \frac{n\sigma_X^2}{n^2} = \frac{\sigma_X^2}{n}$$

The variance of the sampling distribution of $\bar{X}$ thus decreases as the sample size n increases. In effect, this says that the more observations in the sample, the more concentrated is the sampling distribution of the sample mean about the population mean. In other words, the larger the sample, the more certain will be our inference about the population mean. This is to be expected: The more information we obtain from a population, the more we are likely to learn about characteristics (such as the mean) of that population. The variance of the sample mean is denoted $\sigma_{\bar{X}}^2$, and the corresponding standard deviation, called the **standard error** of $\bar{X}$, is given by

$$\sigma_{\bar{X}} = \frac{\sigma_X}{\sqrt{n}}$$

If the number of sample members n is not a very small fraction of the number of population members N, it is no longer the case that the individual sample members are distributed independently of one another. For example, since a population member cannot be included more than once in the sample, the probability of any specific population member being the second value chosen in the sample will depend on what was the first chosen sample member. The argument leading to the derivation of the variance of the sample mean given in the previous paragraph then no longer holds. In fact, it can be shown that the appropriate expression[3] is

$$\text{Var}(\bar{X}) = \frac{\sigma_X^2}{n} \cdot \frac{N-n}{N-1}$$

So far we have found expressions for the mean and variance of the sampling distribution of $\bar{X}$. Fortunately, for most applications, this is all that is required to characterize completely that distribution. If the parent population distribution is normal, it is possible to show that the sample mean also has a normal distribution. If the sample size is a small proportion of the population size, then subtracting the mean and dividing by the standard error yields a random variable

---

[3] In Chapter 4 we encountered this phenomenon. The variance of the hypergeometric distribution is $(N-n)/(N-1)$ times that of the binomial distribution.

$$Z = \frac{(\bar{X} - \mu_X)}{\sigma_{\bar{X}}} = \frac{(\bar{X} - \mu_X)}{\sigma_X / \sqrt{n}} \qquad (6.2.1)$$

that has a standard normal distribution. Moreover, by virtue of the Central Limit Theorem, *even if the population distribution is not normal,* but the sample size n is moderately large, then the distribution of $\bar{X}$ will still be very close to normal, so that the quantity 6.2.1 is, to a very close approximation, distributed as standard normal.

The results of this section are summarized in the box.

Sampling Distribution of $\bar{X}$

Let $\bar{X}$ denote the mean of a random sample of n observations from a population with mean μ_X and variance σ_X^2. Then

(i) The sampling distribution of $\bar{X}$ has mean μ_X; that is,

$$E(\bar{X}) = \mu_X$$

(ii) The sampling distribution of $\bar{X}$ has standard deviation

$$\sigma_{\bar{X}} = \frac{\sigma_X}{\sqrt{n}}$$

This quantity is called the **standard error** of $\bar{X}$.

(iii) If the number n of sample members is not a small proportion of the number N of population members, then the standard error of $\bar{X}$ is

$$\sigma_{\bar{X}} = \frac{\sigma_X}{\sqrt{n}} \cdot \sqrt{\frac{N - n}{N - 1}}$$

(iv) If the population distribution is normal, then the random variable

$$Z = \frac{(\bar{X} - \mu_X)}{\sigma_{\bar{X}}}$$

has a standard normal distribution.

(v) If the population distribution is not normal and the sample size n is moderately large, then it follows from the Central Limit Theorem that, to a close approximation, the result (iv) continues to hold.

Figure 6.2 shows the sampling distribution of the sample mean for sample sizes $n = 25$ and $n = 100$ from a normal population. It can be seen that each distribution is centered on the population mean, but that as the sample size increases the distribution becomes more concentrated about that mean, reflecting the fact that the standard error of the sample mean is a decreasing function of the number of observations in the sample. Thus, as we would expect, the probability that the sample mean differs from the population mean by some fixed amount decreases as the sample size increases.

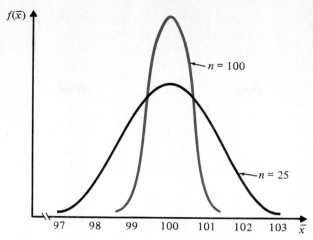

FIGURE 6.2 Probability density functions for the sample means of samples of 25 and 100 observations from a normal distribution with mean 100 and standard deviation 5

We now illustrate the ideas of this section with some specific examples, based on sampling from a normally distributed population.

Example 6.2

If the number of miles per gallon achieved by all cars of a particular model has mean 25 and standard deviation 2, what is the probability that, for a random sample of 20 such cars, average miles per gallon will be less than 24? Assume that the population distribution is normal.

Let $\bar{X}$ denote the sample mean. Then we need to find

$$P(\bar{X} < 24) = P\left[\frac{\bar{X} - \mu_X}{\sigma_{\bar{X}}} < \frac{24 - \mu_X}{\sigma_{\bar{X}}}\right]$$

where $\mu_X = 25$ is the population mean and

$$\sigma_{\bar{X}} = \frac{\sigma_X}{\sqrt{n}} = \frac{2}{\sqrt{20}} = .4472$$

Hence

$$P(\bar{X} < 24) = P\left[Z < \frac{24 - 25}{.4472}\right]$$

$$= P[Z < -2.24]$$

where Z has a standard normal distribution. Therefore

$$P(\bar{X} < 24) = F_Z(-2.24)$$

$$= 1 - F_Z(2.24)$$

$$= 1 - .9875 = .0125$$

using Table 3 of the Appendix.

Example
6.3

A manufacturer claims that the life of its spark plugs is normally distributed with mean 36,000 miles and standard deviation 4,000 miles. For a random sample of sixteen of these plugs, the average life was found to be 34,500 miles. If the manufacturer's claim is correct, what would be the probability of finding a sample mean this small, or smaller?

If $\bar{X}$ denotes the sample mean, then the probability of interest is

$$P(\bar{X} < 34{,}500) = P\left(\frac{\bar{X} - \mu_X}{\sigma_{\bar{X}}} < \frac{34{,}500 - \mu_X}{\sigma_{\bar{X}}}\right)$$

where $\mu_X = 36{,}000$ is the assumed population mean and

$$\sigma_{\bar{X}} = \frac{\sigma_X}{\sqrt{n}} = \frac{4{,}000}{4} = 1{,}000$$

Then

$$P(\bar{X} < 34{,}500) = P\left(Z < \frac{34{,}500 - 36{,}000}{1{,}000}\right)$$

$$= P(Z < -1.5)$$

where Z has a standard normal distribution.

Figure 6.3(a) shows the probability density function of $\bar{X}$, the shaded area being the probability that the sample mean is less than 34,500. In Figure 6.3(b), the same probability is indicated on the graph of the standard normal distribution. The probability is

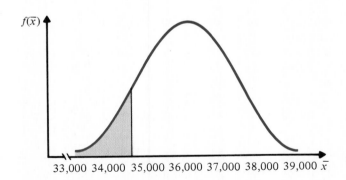

FIGURE 6.3 (a) Probability that sample mean is less than 34,500 in samples of 16 observations from normal distribution with mean 36,000 and standard deviation 4,000; sample mean has normal distribution with mean 36,000 and standard deviation 1,000

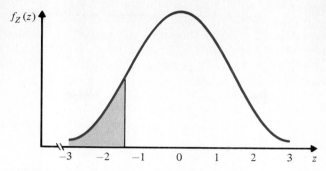

FIGURE 6.3 (b) Probability that a standard normal random variable is less than -1.5

$$P(\bar{X} < 34{,}500) = F_Z(-1.5)$$
$$= 1 - F_Z(1.5)$$
$$= 1 - .9332 = .0668$$

from Table 3 of the Appendix.

This result suggests that, if the manufacturer's claim is correct, the probability of observing such a low value for the sample mean would be quite small. The observed result then casts some doubt on the claim. In Chapter 9, we will discuss a general framework for testing such claims or hypotheses on the basis of sample evidence.

6.3 SAMPLING DISTRIBUTION OF A SAMPLE PROPORTION

We saw in Section 4.5 that if n independent trials, each with probability of success p, are carried out, then the total number of successes, X, obeys a binomial distribution. A common problem arises when the parameter p is unknown. For instance, we may want to determine the proportion of an electorate intending to vote for a particular candidate for office or the proportion of a magazine's readership likely to be in the market for a specific product. In cases of this kind it is natural to base inference on the proportion of successes in the sample.

Definition

Let X be the number of successes in a binomial sample of n observations, where the probability of success is p. (In most applications, the parameter p is the proportion of members of a large population possessing a characteristic of interest.) Then the proportion of successes

$$\hat{p}_X = \frac{X}{n}$$

in the sample is called the **sample proportion.**[4]

The mean and variance of the sampling distribution of the sample proportion can be easily deduced from the mean and variance of the number of successes, which we found in Section 4.5 to be

$$E(X) = np; \qquad \text{Var}(X) = np(1 - p)$$

It then follows that

$$E(\hat{p}_X) = E\left(\frac{X}{n}\right) = \frac{1}{n}E(X) = p$$

That is, the mean of the sample proportion is the proportion p of "successes" in the population. Its variance is

$$\text{Var}(\hat{p}_X) = \text{Var}\left(\frac{X}{n}\right) = \frac{1}{n^2}\text{Var}(X) = \frac{p(1 - p)}{n}$$

Again, the standard deviation of the sample proportion, which is the square root of the variance, is called its **standard error.**

If the number, N, of individuals in the population is not very large compared with the number of sample members, a **finite population correction** is needed in the expression for the variance of the sample proportion. The variance is then

$$\text{Var}(\hat{p}_X) = \frac{p(1 - p)}{n} \cdot \frac{N - n}{N - 1}$$

We saw in Section 5.7 that, as a consequence of the Central Limit Theorem, the distribution of the number of successes is approximately normal for large sample sizes. The same is true of the proportion of successes. It follows that by subtracting from the sample proportion its mean, p, and dividing by its standard error, we obtain a random variable with a standard normal distribution.

Sampling Distribution of the Sample Proportion

Let $\hat{p}_X$ be the proportion of successes in a random sample of n observations from a population in which the proportion of successes is p. Then

---

[4] Again we distinguish between a random variable and its possible specific realizations. We might, for example, observe x successes in a specific sample, in which case the observed sample proportion would be $\hat{p}_x = x/n$. Then $\hat{p}_x$ is a particular realization of the random variable $\hat{p}_X$.

(i) The sampling distribution of $\hat{p}_x$ has mean p; that is,

$$E(\hat{p}_x) = p$$

(ii) The sampling distribution of $\hat{p}_x$ has standard deviation

$$\sigma_{\hat{p}} = \sqrt{\frac{p(1-p)}{n}}$$

The quantity $\sigma_{\hat{p}}$ is called the **standard error** of $\hat{p}_x$.

(iii) If the number n of the sample members is not a small proportion of the number N of population members, then the standard error of $\hat{p}_x$ is

$$\sigma_{\hat{p}} = \sqrt{\frac{p(1-p)}{n}} \sqrt{\frac{N-n}{N-1}}$$

(iv) If the sample size is large,[5] the random variable

$$Z = \frac{(\hat{p}_x - p)}{\sigma_{\hat{p}}}$$

is approximately distributed as standard normal.

Notice that, for fixed p, the standard error of the sample proportion decreases as the sample size increases. This implies that, for increasing sample size, the distribution of $\hat{p}_x$ becomes more concentrated about its mean, as illustrated in Figure 6.4. The implication is that, for any particular population proportion, the probability that the

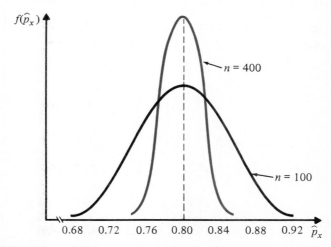

FIGURE 6.4 Probability density functions for the sample proportions in samples of 100 and 400 observations when the population proportion is 0.8

---

[5] In general, the approximation is satisfactory for samples of forty or more observations.

sample and population proportions will differ by any fixed amount decreases as the number of sample members increases. In other words, if we take a bigger sample from a population, our inference about the proportion of population members possessing some particular characteristic becomes more firm.

When the sample size is large, the normal approximation to the binomial distribution provides a very convenient procedure for calculating the probability that a sample proportion lies in some given range. This is illustrated in the following examples.

Example 6.4
A corporation receives 100 applications for a position from recent college graduates in business. Assuming that these applicants can be regarded as a random sample of all such graduates, what is the probability that between 25% and 35% of them are women if 30% of all recent college graduates in business are women?

Denote by $\hat{p}_X$ the proportion of women in the sample of $n = 100$ applicants. The population proportion of women graduates is $p = 0.30$. Then we have to find

$$P(0.25 < \hat{p}_X < 0.35) = P\left(\frac{0.25 - p}{\sigma_{\hat{p}}} < \frac{\hat{p}_X - p}{\sigma_{\hat{p}}} < \frac{0.35 - p}{\sigma_{\hat{p}}}\right)$$

where

$$\sigma_{\hat{p}} = \sqrt{\frac{p(1-p)}{n}} = \sqrt{\frac{(0.30)(0.70)}{100}} = 0.046$$

The required probability is then

$$P\left(\frac{0.25 - 0.30}{0.046} < \frac{\hat{p}_X - p}{\sigma_{\hat{p}}} < \frac{0.35 - 0.30}{0.046}\right) = P(-1.09 < Z < 1.09)$$

where, to a good approximation, the random variable Z has a standard normal distribution. We then have

$$P(0.25 < \hat{p}_X < 0.35) = F_Z(1.09) - F_Z(-1.09)$$
$$= F_Z(1.09) - [1 - F_Z(1.09)]$$
$$= 2F_Z(1.09) - 1$$

where $F_Z(z)$ is the cumulative distribution function of the standard normal random variable. From Table 3 of the Appendix, $F_Z(1.09) = .8621$, so that

$$P(0.25 < \hat{p}_X < 0.35) = 2(.8621) - 1 = .7242$$

Example 6.5
In a random sample[6] of 205 students in adult education classes on a large college campus, it was found that 66% of the sample members were not in the market for cigarettes (that is, do not have and do not expect in the future to have any use for the product). What would be the probability of observing a sample proportion as high or

---

[6] This result was reported by D. Bellenger and B. Greenberg, "A multicategory discrete scale for classifying consumer goods," *Journal of Retailing, 53* (1977), No. 1, 47–60.

higher than this if in fact 60% of all adult education students on the campus are not in the market for cigarettes?

We are interested in the proportion $\hat{p}_X$ of members of a sample of $n = 205$ students who are not in the market for cigarettes, if the population proportion is $p = 0.60$. Specifically, we want to find

$$P(\hat{p}_X > 0.66) = P\left(\frac{\hat{p}_X - p}{\sigma_{\hat{p}}} > \frac{0.66 - p}{\sigma_{\hat{p}}}\right)$$

where

$$\sigma_{\hat{p}} = \sqrt{\frac{p(1 - p)}{n}} = \sqrt{\frac{(0.6)(0.4)}{205}} = 0.034$$

The probability required is then

$$P\left(\frac{\hat{p}_X - p}{\sigma_{\hat{p}}} > \frac{0.66 - 0.60}{0.034}\right) = P(Z > 1.76)$$

where, to a close approximation, Z has a standard normal distribution. This probability is shown as the shaded area under the standard normal density curve in Figure 6.5. We then have

$$P(\hat{p}_X > 0.66) = P(Z > 1.76)$$
$$= 1 - P(Z < 1.76)$$
$$= 1 - F_Z(1.76)$$
$$= 1 - .9608 = .0392$$

from Table 3 of the Appendix. Hence, if 60% of this population is not in the market for cigarettes, the probability of getting a sample proportion .66 or higher would be less than .04—that is, less than one chance in twenty-five.

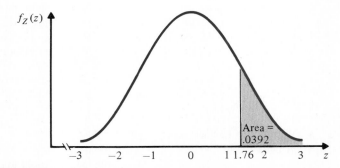

FIGURE 6.5 The probability that a standard normal random variable exceeds 1.76, representing the probability that the sample proportion is bigger than 0.66 for a sample of 205 observations when the population proportion is 0.60

EXERCISES

1. The mean length of all long-distance telephone calls is 220 seconds, and the standard deviation is 50 seconds. Assume that the population distribution is normal. A random sample of sixteen calls was chosen from telephone company records.
   - (a) What is the mean of the sample mean length of call?
   - (b) What is the variance of the sample mean?
   - (c) What is the standard error of the sample mean?
   - (d) What is the probability that the sample mean exceeds 240 seconds?

2. The lifetimes of lightbulbs produced by a particular manufacturer have mean 1,200 hours and standard deviation 300 hours. The population distribution is normal. Suppose you purchase nine bulbs, which can be regarded as a random sample from the manufacturer's output.
   - (a) What is the mean of the sample mean lifetime?
   - (b) What is the variance of the sample mean?
   - (c) What is the standard error of the sample mean?
   - (d) What is the probability that, on average, these nine lightbulbs have lifetime less than 1,050 hours?

3. A company produces baked beans. The true mean weight of the contents of their cans is 16 ounces, and the standard deviation is 0.6 ounce. The population distribution of weights is normal. Suppose you purchase four cans, which can be regarded as a random sample of all those produced.
   - (a) What is the standard error of the sample mean weight?
   - (b) What is the probability that, on average, the contents of these four cans will weigh less than 15.8 ounces?
   - (c) What is the probability that, on average, the contents of the four cans weigh more than 16.3 ounces?
   - (d) What is the probability that, on average, the contents of the four cans weigh more than 15.9 ounces?

4. A random sample of sixteen people employed in a large city was taken to determine how far they travel to work each day. Suppose that the population distribution is normal with mean 6.4 miles and standard deviation 2.4 miles.
   - (a) What is the standard error of the sample mean distance travelled?
   - (b) What is the probability that the sample mean is less than 6 miles?
   - (c) What is the probability that the sample mean is more than 7 miles?
   - (d) What is the probability that the sample mean is less than 7.5 miles?
   - (e) Suppose that a second (independent) random sample, of fifty people, is taken. Without doing the calculations, state whether the probabilities in parts (b)–(d) would be higher, lower, or the same for the second sample compared with the first.

5. A corporation gives a standard aptitude test to all applicants for employment. Historically the scores on this test have been normally distributed with mean 70 and standard deviation 12. The test is taken by twenty-five candidates, who are viewed as a random sample of all candidates for employment.
   - (a) What is the standard error of the sample mean score?
   - (b) What is the probability that the sample mean score is less than 72?
   - (c) What is the probability that the sample mean is more than 69?
   - (d) What is the probability that the sample mean is between 70 and 72?
   - (e) What is the probability that the sample mean is between 69 and 72?

6. It is known that, for students graduating from a large secretarial school, the average

typing speed is 64 words per minute and the standard deviation is 8 words per minute. A random sample of 100 graduates is selected.

   (a) What is the probability that the sample mean typing speed is more than 63 words per minute?
   (b) What is the probability that the sample mean is less than 66 words per minute?
   (c) What is the probability that the sample mean is between 62 and 66 words per minute?
   (d) What is the probability that the sample mean is between 63 and 67 words per minute?

7. The mean income for families in a particular town is $22,500, and the standard deviation is $4,200. Suppose we select a random sample of 400 families from this town.

   (a) What is the probability that the sample mean income is more than $22,000?
   (b) What is the probability that the sample mean income is between $22,200 and $22,300?
   (c) What is the probability that the sample mean income is between $22,600 and $22,700?
   (d) What is the probability that the sample mean income is between $22,000 and $23,000?
   (e) What is the probability that the sample mean income is between $22,250 and $23,000?
   (f) Without doing the calculations, state in which of the following ranges the sample mean income is most likely to lie: $22,000–$22,200, $22,200–$22,400, $22,400–$22,600, $22,600–$22,800.

8. Assume that the standard deviation of the amount of money spent in a semester on textbooks by students on a large campus is $20. A random sample of 100 students is taken in order to estimate the mean expenditure for the whole student population.

   (a) What is the standard error of the sample mean expenditure?
   (b) What is the probability that the sample mean expenditure exceeds the population mean by more than $2?
   (c) What is the probability that the sample mean is more than $3 below the population mean?
   (d) What is the probability that the sample mean differs from the population mean by more than $4?

9. The dividend yields for all companies whose shares were traded on the New York Stock Exchange last year obeyed a normal distribution with standard deviation 2.4%. A random sample of sixteen of these companies is taken in order to estimate the population mean dividend yield.

   (a) What is the probability that the sample mean dividend yield exceeds the population mean by more than 0.8%?
   (b) What is the probability that the sample mean is below the population mean by more than 1.0%?
   (c) What is the probability that the sample mean differs from the population mean by more than 1.2%?
   (d) Suppose that a second (independent) random sample of forty companies is taken. Without doing the calculations, state whether the probabilities in parts (a)–(c) would be higher, lower, or the same for the second sample compared with the first.

10. An industrial process produces batches of chemicals whose impurity levels follow a normal distribution with standard deviation 1.2 grams per hundred grams of chemical. A random sample of 100 batches is selected in order to estimate the population mean impurity level.

   (a) The probability is 0.05 that the sample mean impurity level exceeds the population mean by how much?

(b) The probability is 0.10 that the sample mean impurity level is below the population mean by how much?

(c) The probability is 0.15 that the sample mean impurity level differs from the population mean by how much?

11. The weekly expenditures on food by families in a particular neighborhood follow a normal distribution with standard deviation $20. A random sample of families is selected in order to estimate the population mean weekly food expenditure. How large a sample is necessary in order to ensure that the probability that the sample mean differs from the population mean by more than $1 is less than 0.10?

12. The annual value of all items charged by customers of a major credit card company has a normal distribution with standard deviation $1,100. A random sample of customer records is drawn in order to estimate the mean annual value of items charged per customer.

(a) How large a sample is necessary to ensure that the probability that the sample mean differs from the population mean by more than $100 is less than 0.05?

(b) Without doing the calculations, state whether a larger or smaller sample than that in part (a) would be required to guarantee that the probability that the sample mean differs from the population mean by more than $100 is less than 0.10.

(c) Without doing the calculations, state whether a larger or smaller sample than that in part (a) would be required to ensure that the probability that the sample mean differs from the population mean by more than $50 is less than 0.05.

13. A town has 500 real estate agents. The mean value of the properties sold in a year by these agents is $800,000, and the standard deviation is $200,000. A random sample of 100 agents is selected and the value of the properties they sold in a year is recorded.

(a) What is the standard error of the sample mean?

(b) What is the probability that the sample mean exceeds $825,000?

(c) What is the probability that the sample mean exceeds $780,000?

(d) What is the probability that the sample mean is between $790,000 and $820,000?

14. For an audience of 800 people attending a concert, the average distance travelled was 5 miles, and the standard deviation of distance travelled was 2 miles. A random sample of 160 members of the audience is selected.

(a) What is the probability that the sample mean distance travelled is less than 5.5 miles?

(b) What is the probability that the sample mean distance travelled is more than 4.75 miles?

(c) What is the probability that the sample mean distance travelled is between 4.8 and 5.2 miles?

(d) What is the probability that the sample mean distance travelled is not between 4.6 and 5.4 miles?

15. A business statistics course is taken by 400 students. Each member of a random sample of 120 of these students was asked to estimate the number of hours per week, outside of class, he or she spent working on this course. Suppose that the population standard deviation is 3.2 hours.

(a) What is the probability that the sample mean exceeds the population mean by more than 0.4 hour?

(b) What is the probability that the sample mean is lower than the population mean by more than 0.5 hour?

(c) What is the probability that the sample mean differs from the population mean by more than 0.6 hour?

16. The heights of a large population of individuals are known to be normally distributed with mean 70 inches and standard deviation 3 inches.

(a) What is the probability that the height of a randomly chosen individual from this population is between 69 and 71 inches?

(b) What is the probability that the average height of four randomly chosen people from this population is between 69 and 71 inches?

(c) Without doing the calculations, state whether the probability that the average height of fifty randomly chosen people from this population is between 69 and 71 inches is higher or lower than the probability found in part (b). Sketch a graph to illustrate your reasoning.

17. In a statistics examination taken by a large group of students, the scores were normally distributed with mean 70 and standard deviation 10.

(a) What is the probability that the score of a randomly chosen student is higher than 80?

(b) What is the probability that the average score of four randomly chosen students is higher than 80?

(c) Without doing the calculations, state whether the probability that the average score of ten randomly chosen students is higher than 80 is larger or smaller than the probability found in part (b). Sketch a graph to illustrate your reasoning.

18. Bare Mountain Nature Preserve has a large number of bears. It has been determined that, during porridge season, their daily consumption of porridge is normally distributed with mean 5.5 bowls per bear and standard deviation 2 bowls. If four bears are chosen at random, what is the probability that their total consumption of porridge in a day will be between 12 and 24 bowls?

19. According to the Internal Revenue Service, 75% of all tax returns lead to a refund. A random sample of 100 tax returns is taken.

(a) What is the mean of the sample proportion of returns leading to refunds?

(b) What is the variance of the sample proportion?

(c) What is the standard error of the sample proportion?

(d) What is the probability that the sample proportion exceeds 0.8?

20. A record store owner finds that 20% of customers entering her store make a purchase. One morning 120 people, who can be regarded as a random sample of all customers, enter the store.

(a) What is the mean of the sample proportion of customers making a purchase?

(b) What is the variance of the sample proportion?

(c) What is the standard error of the sample proportion?

(d) What is the probability that the sample proportion is less than 0.15?

21. A random sample of 150 homes is taken from a large population of older homes to estimate the proportion of such homes in which the electrical wiring is unsafe. Suppose that, in fact, 30% of all homes in this population have unsafe wiring.

(a) What is the standard error of the sample proportion of homes with unsafe wiring?

(b) What is the probability that the sample proportion is less than 0.25?

(c) What is the probability that the sample proportion is more than 0.33?

(d) What is the probability that the sample proportion is less than 0.35?

22. A company has determined that 10% of the parts from a particular supplier are defective. The company receives a shipment of 400 parts, which can be regarded as a random sample of all parts produced by this supplier.

(a) What is the standard error of the sample proportion of defectives?

(b) What is the probability that the sample proportion is more than 0.13?

(c) What is the probability that the sample proportion is less than 0.08?

(d) What is the probability that the sample proportion is more than 0.09?

(e) The company receives a second shipment, which contains 200 parts. Without doing the calculations, state whether the probabilities in parts (b)–(d) are higher, lower, or the same for the second shipment compared with the first.

23. A department store accepts 60% of all applications for credit. A random sample of 100 applications is selected.
    (a) What is the probability that the sample proportion of applications accepted is less than 0.67?
    (b) What is the probability that the sample proportion is between 0.55 and 0.65?
    (c) What is the probability that the sample proportion is between 0.60 and 0.70?
    (d) What is the probability that the sample proportion is between 0.50 and 0.70?

24. Last week's Nielsen ratings estimated that a particular program was watched in 30% of all homes. (Assume this estimate is correct.) A random sample of 200 homes is selected.
    (a) What is the probability that the sample proportion of homes in which the show was watched is between 0.28 and 0.32?
    (b) What is the probability that the sample proportion is between 0.32 and 0.36?
    (c) What is the probability that the sample proportion is between 0.29 and 0.33?

25. A new record club company mails advertising to 400 randomly selected homes. The company believes that, for the population at large, 20% of all recipients of this material will accept the introductory membership offer. Assume the company's belief is correct.
    (a) What is the probability that the sample proportion of recipients accepting the introductory offer is between 0.18 and 0.21?
    (b) What is the probability that the sample proportion is between 0.21 and 0.24?
    (c) What is the probability that the sample proportion is between 0.15 and 0.18?
    (d) Without doing the calculations, state in which of the following ranges the sample proportion is most likely to lie: 0.17–0.19, 0.19–0.21, 0.21–0.23, 0.23–0.25.

26. A random sample of 100 voters is taken to estimate the proportion of a state's electorate in favor of an increase in the level of gasoline tax to provide additional revenue for highway repairs. What is the largest value that the standard error of the sample proportion in favor of this measure can take?

27. In Exercise 26, suppose it is decided that a sample of 100 voters is too small to provide a sufficiently reliable estimate of the population proportion. It is required instead that the probability that the sample proportion differs from the population proportion (whatever its value) by more than 0.03 should not exceed 0.05. How large a sample is needed to guarantee that this requirement be met?

28. A brewing company wants to estimate the proportion of all beer drinkers who watch the nationally telecast Saturday afternoon baseball game. A random sample obtained information from 120 beer drinkers. Suppose that the proportion of beer drinkers in the population who watch the telecast is 0.20.
    (a) The probability is 0.10 that the sample proportion exceeds the population proportion by how much?
    (b) The probability is 0.05 that the sample proportion is lower than the population proportion by how much?
    (c) The probability is 0.25 that the sample proportion differs from the population proportion by how much?

29. In a random sample[7] of 205 adult education students it was found that 53% expected to buy automobile tires within the next year. What is the probability of observing a sample proportion as high or higher than this if, in fact, 50% of all the adult education students on the campus expect to buy tires?

---

[7] Reported in D. Bellenger and B. Greenberg, "A multicategory discrete scale for classifying consumer goods," *Journal of Retailing, 53* (1977), No. 1, 47–60.

248   Sampling and Sampling Distributions

30. For the same sample as in Exercise 29 it was found that 21% of the sample members expected to purchase life insurance within the next year. What is the probability of observing a sample proportion as low or lower than this if the population proportion is 0.25?

31. A company is considering introducing a stock option into the bonus plan for 1,000 of its employees. To estimate the proportion of employees in favor of this move, the company management solicits reactions from a random sample of 400 of those eligible. Suppose that, in fact, 60% of the employees are in favor of the option.
    (a) What is the standard error of the sample proportion in favor of the option?
    (b) What is the probability that the sample proportion exceeds 0.67?
    (c) What is the probability that the sample proportion is larger than 0.55?
    (d) What is the probability that the sample proportion is between 0.58 and 0.62?

32. A supermarket chain has 400 stores in the Midwest. In 320 of these stores, shoplifting has increased by more than 10% in the past year. A random sample of 100 of these stores is selected.
    (a) What is the standard error of the sample proportion of stores in which shoplifting has increased by more than 10% in the past year?
    (b) What is the probability that the sample proportion is less than 0.84?
    (c) What is the probability that the sample proportion is between 0.77 and 0.83?
    (d) What is the probability that the sample proportion is between 0.76 and 0.82?

33. A course is taken by 500 students. The instructor takes a random sample of 200 students to determine the proportion in favor of a take-home final exam. Suppose that, in fact, 275 of the students in the class are in favor. What is the probability that less than half of the students in the sample will be in favor of a take-home final?

34. The lifetimes of lightbulbs produced by a particular manufacturer have a normal distribution with mean 1,200 hours and standard deviation 300 hours. If a random sample of 144 of these bulbs is selected, what is the probability that more than 20% of the sampled bulbs will last less than 900 hours?

6.4 SAMPLING DISTRIBUTION OF THE SAMPLE VARIANCE

In Section 6.2 we considered the problem of making inference about the mean of a population, based on sample information. We now turn our attention to the population variance.

Suppose that a random sample of n observations is drawn from a population with unknown mean μ_X and unknown variance σ_X^2, the sample members being denoted as $X_1, X_2, \ldots, X_n$. Now, the population variance is the expectation

$$\sigma_X^2 = E[(X - \mu_X)^2]$$

so that an obvious quantity to look at would be the average of $(X_i - \mu_X)^2$ over the n sample members. However, the population mean μ_X is unknown, so that in practice this quantity cannot be calculated. It is natural then to replace the unknown μ_X by the sample mean $\bar{X}$, and to consider the average of $(X_i - \bar{X})^2$. In fact, as we noted in Chapter 2, the sample variance is defined as

$$s_X^2 = \frac{1}{n-1} \sum_{i=1}^{n} (X_i - \bar{X})^2$$

Definition

Let $X_1, X_2, \ldots, X_n$ be a random sample from a population. The quantity

$$s_X^2 = \frac{1}{n-1} \sum_{i=1}^{n} (X_i - \bar{X})^2$$

is called the **sample variance**.[8] Its square root, s_X, is called the **sample standard deviation.**

At first sight, the use of $(n-1)$ rather than n as the divisor in our definition of the sample variance may be rather surprising. The motivation for this formulation is that, if the sample variance is defined in this way, it can be shown that [9] the mean of its sampling distribution is the true population variance; that is,

$$E(s_X^2) = \sigma_X^2$$

The conclusion that the expected value of the sample variance is the population variance is quite general. However, in order to characterize further the sampling distribution we need to know more about the underlying population distribution. In many practical applications the assumption that the population distribution is normal is not unreasonable. In this case it can be shown that the random variable

$$\frac{(n-1)s_X^2}{\sigma_X^2} = \frac{\sum_{i=1}^{n} (X_i - \bar{X})^2}{\sigma_X^2}$$

has a distribution known as the **χ^2 distribution (chi-square distribution) with $(n-1)$ degrees of freedom.**[10]

The chi-square family of distributions is frequently employed in statistical analysis. The distributions are defined only for positive values of a random variable, which is appropriate in the present context since a sample variance cannot be negative. The density function, which is illustrated in Figure 6.6, is asymmetric. A specific member of the chi-square family is characterized by a single parameter, referred to as the number of *degrees of freedom,* for which the symbol ν is typically used. If a random variable has a χ^2 distribution with ν degrees of freedom, it will be denoted χ_ν^2. The mean and variance of this distribution are equal to the number of degrees of freedom and twice the number of degrees of freedom, respectively; that is,

---

[8] Again we distinguish between the random variable s_X^2 and specific values it can take. Thus, if the actual sample observed is $x_1, x_2, \ldots, x_n$, then the realization of s_X^2 is

$$s_x^2 = \frac{1}{n-1} \sum_{i=1}^{n} (x_i - \bar{x})^2$$

[9] The result is established in Appendix A6.1 at the end of this chapter.

[10] The chi-square distribution with ν degrees of freedom is the distribution of the sum of squares of ν independent standard normal random variables.

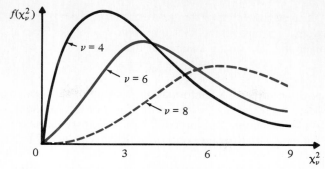

FIGURE 6.6 Probability density functions of the chi-square distribution with $\nu = 4$, 6, and 8 degrees of freedom

$$E(\chi_\nu^2) = \nu; \qquad \mathrm{Var}(\chi_\nu^2) = 2\nu$$

In the present context, the random variable $(n - 1)s_X^2/\sigma_X^2$ has a $\chi_{(n-1)}^2$ distribution, so that its mean is

$$E\left[\frac{(n - 1)s_X^2}{\sigma_X^2}\right] = (n - 1)$$

Hence, we have

$$\frac{(n - 1)}{\sigma_X^2}E(s_X^2) = (n - 1)$$

so that

$$E(s_X^2) = \sigma_X^2$$

as before. To get the variance of s_X^2, we have

$$\mathrm{Var}\left[\frac{(n - 1)s_X^2}{\sigma_X^2}\right] = 2(n - 1)$$

Hence,

$$\frac{(n - 1)^2}{\sigma_X^4}\mathrm{Var}(s_X^2) = 2(n - 1)$$

so that

$$\mathrm{Var}(s_X^2) = \frac{2\sigma_X^4}{(n - 1)}$$

The properties of the χ^2 distribution can therefore be used to find the variance of the sampling distribution of the sample variance.[11]

---

[11] It should be repeated that this result holds only when the parent population is normal.

The parameter ν of the χ^2 distribution is known as the number of **degrees of freedom.** To understand this terminology, let us look at the sample variance. It involves the sum of the squares of the quantities

$$(X_1 - \bar{X}); \qquad (X_2 - \bar{X}); \qquad \ldots ; \qquad (X_n - \bar{X})$$

Thus, these n pieces of information are employed to calculate the sample variance. However, they are not *independent* pieces of information, since they must sum to 0 (as follows from the definition of $\bar{X}$). Hence, if we know any $(n - 1)$ of the $(X_i - \bar{X})$, we can calculate the other one from the first $(n - 1)$. For example, since

$$\sum_{i=1}^{n} (X_i - \bar{X}) = 0$$

it follows that

$$X_n - \bar{X} = -\sum_{i=1}^{n-1} (X_i - \bar{X})$$

The n quantities $(X_i - \bar{X})$ are equivalent to a set of $(n - 1)$ independent pieces of information. The situation can be thought of as follows: We want to make inference about the unknown σ_X^2. If the population mean μ_X were known, this inference could be based on the sum of squares of

$$(X_1 - \mu_X); \qquad (X_2 - \mu_X); \qquad \ldots ; \qquad (X_n - \mu_X)$$

These quantities are independent of one another, and we would say that we have n degrees of freedom for the estimation of σ_X^2. However, since the unknown population mean must be replaced in practice by its estimate $\bar{X}$, one of these degrees of freedom is used up and we are left with the equivalent of $(n - 1)$ independent observations for use in making inference about the population variance. It is then said that $(n - 1)$ degrees of freedom are available.

We frequently need to find values of the cumulative distribution function for a χ^2 random variable. Such problems are often phrased in terms of the determination of cutoff points corresponding to particular specified probabilities. For instance, if a random variable has a χ_{10}^2 distribution we may require the number K for which

$$P(\chi_{10}^2 < K) = 0.90$$

or, equivalently,

$$P(\chi_{10}^2 > K) = 0.10$$

The distribution function of the chi-square random variable is tabulated in Table 5 in the Appendix in such a way that these cutoff points can be read directly. For the χ_{10}^2 random variable, it can be seen from Table 5 that if $P(\chi_{10}^2 > K) = 0.10$, then $K = 15.99$. This probability is shown as an area under the density function of the random variable in Figure 6.7.

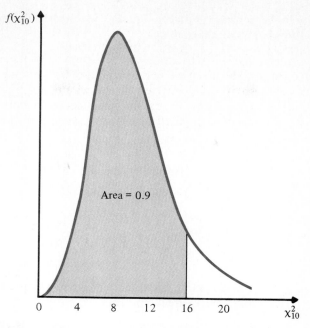

FIGURE 6.7 Probability (0.9) that a chi-square random variable with 10 degrees of freedom is less than 15.99

Sampling Distribution of the Sample Variance

Let s_X^2 denote the sample variance for a random sample of n observations from a population with variance σ_X^2. Then

(i) The sampling distribution of s_X^2 has mean σ_X^2; that is,

$$E(s_X^2) = \sigma_X^2$$

(ii) The variance of the sampling distribution of s_X^2 depends on the underlying population distribution. If that distribution is normal, then

$$\operatorname{Var}(s_X^2) = \frac{2\sigma_X^4}{n-1}$$

(iii) If the population distribution is normal, then

$$\frac{(n-1)s_X^2}{\sigma_X^2} \text{ is distributed as } \chi_{(n-1)}^2$$

Suppose that we take a random sample from a population and want to make some inferential statements about the population variance. Given an assumption of normality in the underlying population, the chi-square distribution can be used, as illustrated in the following example.

Example
6.6

A manufacturer of canned peas is concerned that the mean weight of the product be close to the advertised weight. In addition, he does not want too much variability in the weights of the cans of peas; otherwise, a large proportion will differ markedly from the advertised weight. Assume that the population distribution of weights is normal. If a random sample of twenty cans is checked, find the numbers K_1, and K_2 such that

$$P\left(\frac{s_X^2}{\sigma_X^2} < K_1\right) = 0.05; \qquad P\left(\frac{s_X^2}{\sigma_X^2} > K_2\right) = 0.05$$

We have

$$0.05 = P\left(\frac{s_X^2}{\sigma_X^2} < K_1\right) = P\left[\frac{(n-1)s_X^2}{\sigma_X^2} < (n-1)K_1\right]$$

$$= P[\chi_{(n-1)}^2 < (n-1)K_1]$$

where $n = 20$ is the sample size and $\chi_{(n-1)}^2$ is a chi-square random variable with $(n-1) = 19$ degrees of freedom. Then

$$0.05 = P(\chi_{19}^2 < 19K_1) \qquad \text{or} \qquad 0.95 = P(\chi_{19}^2 > 19K_1)$$

From Table 5 in the Appendix, we therefore have

$$19K_1 = 10.12$$

so that

$$K_1 = 0.533$$

The conclusion then is that the probability is .05 that the sample variance will be less than 53.3% of the population variance.

We also require the number K_2 such that

$$0.05 = P\left(\frac{s_X^2}{\sigma_X^2} > K_2\right)$$

Equivalently, we have

$$0.05 = P\left[\frac{(n-1)s_X^2}{\sigma_X^2} > (n-1)K_2\right]$$

$$= P[\chi_{(n-1)}^2 > (n-1)K_2]$$

Hence, since $n = 20$,

$$0.05 = P(\chi_{19}^2 > 19K_2)$$

Then, from Table 5, it follows that

$$19K_2 = 30.14$$

so that

$$K_2 = 1.586$$

This implies that the probability is .05 that the sample variance will be more than 58.6% bigger than the population variance.

These probabilities are shown in Figure 6.8 as areas under the probability density function of the χ^2_{19} distribution.

It should be emphasized that the technique of Example 6.6 is less universally applicable in practice than are those of earlier sections of this chapter. The assumption that the distribution of the population being sampled is normal is more critical here. We have seen how probability statements can be made about both the sample mean and sample variance when sampling from a normal distribution. However, the latter will typically be far more affected than the former by any departures from normality in the distribution being sampled. In making probability statements about the sample mean, the Central Limit Theorem ensures that, for moderately large samples, modest departures from normality in the sampled population have only minor influence on the validity of any derived probability statements. As a result, we say that inference based on the sample mean is *robust* to departures from assumed normality in the parent population, whereas inference based on the sample variance is not.

Nevertheless, it often happens in practice that the population variance is of direct interest to an investigator. It must be kept in mind that, if only a moderate number of sample observations are available, serious departures from normality in the parent population can severely invalidate the conclusions of analyses based on the technique described in this section. The cautious analyst will, therefore, be rather tentative in making inference in these circumstances.

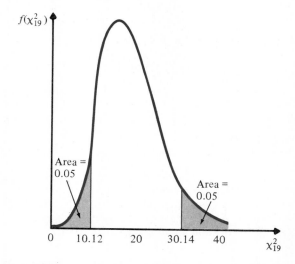

FIGURE 6.8 Probability is 0.05 that a chi-square random variable with 19 degrees of freedom is less than 10.12, and also that this random variable is bigger than 30.14

35. The diameters of metal rods produced by a particular process have a normal distribution with variance (in squared centimeters) of 0.01. A random sample of ten rods is selected.
    (a) Is the probability more than 0.95 that the sample variance is bigger than 0.005?
    (b) Is the probability more than 0.95 that the sample variance is less than 0.02?

36. The percentages of impurity in batches of chemicals produced by a certain process have a normal distribution with variance 1.50. A random sample of twenty batches is selected.
    (a) Is the probability less than 0.10 that the sample variance is bigger than 2?
    (b) Is the probability less than 0.10 that the sample variance is below 1?

37. The scores on a test taken by a very large group of students have a normal distribution with standard deviation 12. A random sample of fifteen tests is selected.
    (a) Is the probability less than 0.05 that the sample variance is bigger than 230?
    (b) Is the probability less than 0.10 that the sample variance is below 75?

38. Weekly expenditures on nonfood items in supermarkets by families in a particular neighborhood have a normal distribution with standard deviation $30. A random sample of nine families is selected.
    (a) The probability is 0.10 that the sample variance is bigger than what number?
    (b) The probability is 0.05 that the sample variance is smaller than what number?

39. A credit card company has determined that for all its customers, the elapsed time between sending out bills and receipt of payment is normally distributed with standard deviation 1.2 weeks. A random sample of twelve customers is selected.
    (a) The probability is 0.01 that the sample variance is bigger than what number?
    (b) The probability is 0.025 that the sample variance is less than what number?
    (c) Find any pair of numbers such that the probability that the sample variance lies between these numbers is 0.90.

40. In a particular town, family incomes have a normal distribution with standard deviation $5,200. A random sample of twenty-five families is selected from this town.
    (a) The probability is 0.05 that the sample standard deviation is less than what number?
    (b) The probability is 0.05 that the sample standard deviation is bigger than what number?

41. Each member of a random sample of twenty business economists was asked to predict the rate of inflation for the coming year. Assume that the predictions for the whole population of business economists follow a normal distribution with standard deviation 1.8%.
    (a) The probability is 0.01 that the sample standard deviation is bigger than what number?
    (b) The probability is 0.025 that the sample standard deviation is smaller than what number?
    (c) Find any pair of numbers such that the probability that the sample standard deviation lies between these numbers is 0.90.

42. A production process produces ball bearings. The process is periodically halted and a random sample of fifteen ball bearings is selected and their diameters measured. The population distribution of diameters is normal.
    (a) The probability is 0.05 that the sample variance is more than what percentage of the population variance?
    (b) The probability is 0.05 that the sample variance is less than what percentage of the population variance?

43. Weekly incomes were measured for a random sample of twenty outlets of a fast-food chain. The population distribution of weekly incomes is normal.

(a) The probability is 0.10 that the sample variance is more than what percentage of the population variance?

(b) The probability is 0.01 that the sample variance is less than what percentage of the population variance?

44. A random sample of eight financial analysts were asked to predict the earnings of General Motors for the coming year. Suppose the distribution of predictions for the population of financial analysts is normal.

(a) The probability is 0.10 that the sample variance is more than what percentage of the population variance?

(b) Determine any pair of appropriate numbers a and b to complete the following sentence: The probability is 0.95 that the sample variance is between $a\%$ and $b\%$ of the population variance.

45. A precision instrument is checked by making twelve readings on the same quantity. The population distribution of readings is normal.

(a) The probability is 0.90 that the sample variance is more than what percentage of the population variance?

(b) The probability is 0.99 that the sample variance is less than what percentage of the population variance?

(c) Determine any pair of appropriate numbers a and b to complete the following sentence: The probability is 0.95 that the sample variance is between $a\%$ and $b\%$ of the population variance.

46. A drug company produces pills containing an active ingredient. The company is concerned about the mean weight of this ingredient per pill, but it also requires that the variance (in squared milligrams) be no more than 1.5. A random sample of twenty pills is selected and the sample variance is found to be 2.05. How likely is it that a sample variance this high or higher would be found if the population variance is in fact 1.5? Assume that the population distribution is normal.

47. A manufacturer has been purchasing raw materials from a supplier whose consignments have a variance of 15.4 (in squared pounds) in impurity levels. A rival supplier claims that he can supply consignments of this raw material with the same mean impurity level, but with lower variance. For a random sample of twenty-five consignments from the second supplier, the variance in impurity levels was found to be 12.2. What is the probability of observing a value this low or lower for the sample variance if, in fact, the true population variance is 15.4? Assume that the population distribution is normal.

REVIEW EXERCISES

48. What is meant by the statement that the sample mean has a sampling distribution?

49. An investor is considering six different money market funds. The average days to maturity for these funds are:

$$31; \quad 39; \quad 35; \quad 35; \quad 33; \quad 38$$

Two of these funds are to be chosen at random.

(a) How many possible samples of two funds are there?

(b) List all possible samples.

(c) Find the probability function of the sampling distribution of the sample mean.

50. Of what relevance is the Central Limit Theorem to the sampling distribution of the sample mean?

51. Refer to Exercise 49. Find the probability function of the sampling distribution of the sample proportion of funds with average maturity more than 36 days, for samples of two observations.

52. A test is taken by a large group of students. The test scores are normally distributed with mean 60 and standard deviation 12. Is the probability higher or lower that the mean score for a random sample of four students will exceed 55 than the probability that the mean score for a random sample of sixteen students will exceed 55? Demonstrate your conclusion graphically.

53. It is known that the durations of long-distance telephone calls follow a normal distribution with mean 5.7 minutes. For a random sample of twenty-five such calls, the probability that the sample mean duration exceeds 6 minutes is 0.2. Find the population standard deviation.

54. The percentage increase in earnings per share of large corporations over the past year followed a normal distribution with standard deviation 3.2%. A random sample of thirty-six companies is taken to estimate the population mean percentage increase in earnings per share.

    (a) What is the probability that the sample mean percentage increase exceeds the population mean by more than 1.0%?

    (b) What is the probability that the sample mean percentage increase is more than 0.9% below the population mean?

    (c) What is the probability that the sample mean percentage increase differs from the population mean by more than 0.8%?

    (d) Suppose that a second random sample, of twenty observations, is taken. Without doing the calculations, state whether the probabilities in parts (a)–(c) would be higher, lower, or the same for the second sample compared with the first.

55. The selling prices of houses in a particular area follow a normal distribution with standard deviation $30,000. A random sample of twenty-five houses is taken to estimate the population mean selling price.

    (a) What is the probability that the sample mean selling price exceeds the population mean by more than $10,000?

    (b) What is the probability that the sample mean selling price is more than $9,000 less than the population mean?

    (c) What is the probability that the sample mean selling price differs from the population mean by more than $8,000?

    (d) Suppose that a second random sample, of fifty houses, is taken. Without doing the calculations, state whether the probabilities in parts (a)–(c) would be higher, lower, or the same for the second sample compared with the first.

56. As in Exercise 55, the selling price of houses is normally distributed with standard deviation $30,000, and a random sample of twenty-five observations is to be taken. Denote by $\bar{X}$ the sample mean. What is the probability that the interval $(\bar{X} - 11,760)$ to $(\bar{X} + 11,760)$ contains the true population mean?

A manufacturer of liquid detergent claims that the mean weight of liquid in containers sold is 30 ounces. It is known that the population distribution of weights is normal, with standard deviation 1.2 ounces. In order to check the manufacturer's claim, a random sample of nine containers of detergent is examined. The claim will be questioned if the sample mean weight is less than 29.2 ounces. What is the probability that the claim will be questioned if, in fact, the population mean weight is 30 ounces?

It has been found that 12% of all utility bills are not paid by the required date. A random sample of 500 bills is selected.

(a) What is the probability that the percentage not paid by the required date is more than 10%?

(b) What is the probability that the percentage not paid by the required date is less than 13%?

(c) What is the probability that the percentage not paid by the required date is between 10% and 12.5%?

59. A political pressure group wants to propose a referendum on a property tax reform proposal in statewide elections. A random sample of 300 voters is taken. Suppose that, in the population at large, 60% of voters favor the proposal.

    (a) Find the probability that the sample proportion in favor of the proposal is less than 0.5.

    (b) Find the probability that the sample proportion in favor is between 0.6 and 0.7.

    (c) Find the probability that the sample proportion in favor is between 0.55 and 0.65.

60. In a particular year, 60% of home sales were partially financed by the seller. A random sample of 250 sales is examined.

    (a) The probability is 0.8 that the sample proportion is bigger than what amount?

    (b) The probability is 0.9 that the sample proportion is smaller than what amount?

    (c) The probability is 0.7 that the sample proportion differs from the population proportion by how much?

61. A candidate for office intends to campaign in a state if his initial support level exceeds 40% of the voters. A random sample of 300 voters is taken, and it is decided to campaign if the sample proportion supporting the candidate exceeds 0.36.

    (a) What is the probability of a decision to campaign if in fact the initial support level is 30%?

    (b) What is the probability of a decision not to campaign if in fact the initial support level is 40%?

62. The scores on an aptitude test taken by a very large group of candidates have a normal distribution with variance 230. Suppose we select a random sample of twenty candidates.

    (a) What is the probability that the sample variance of scores is less than 150?

    (b) What is the probability that the sample variance of scores is bigger than 300?

63. It is known that the incomes of subscribers to a particular magazine have a normal distribution with standard deviation $6,400. A random sample of twenty-five subscribers is taken.

    (a) What is the probability that the sample standard deviation of their incomes is bigger than $4,000?

    (b) What is the probability that the sample standard deviation of their incomes is less than $8,000?

64. Batches of chemical are manufactured by a production process. Samples of twenty batches from a production run are selected for testing. If the standard deviation of the percentage impurity contents in the sampled batches exceeds 2.5%, the production process is thoroughly checked. Assume that the population distribution of percentage impurity concentrations is normal. What is the probability that the production process will be thoroughly checked if the population standard deviation of percentage impurity concentrations is 2%?

APPENDIX A6.1

In this appendix, we will show that the mean of the sampling distribution of the sample variance is the population variance. We begin by finding the expectation of the sum of squares of the sample members about their mean; that is, the expectation of

$$\sum_{i=1}^{n} (X_i - \bar{X})^2 = \sum_{i=1}^{n} [(X_i - \mu_X) - (\bar{X} - \mu_X)]^2$$

$$= \sum_{i=1}^{n} [(X_i - \mu_X)^2 - 2(\bar{X} - \mu_X)(X_i - \mu_X) + (\bar{X} - \mu_X)^2]$$

$$= \sum_{i=1}^{n} (X_i - \mu_X)^2 - 2(\bar{X} - \mu_X) \sum_{i=1}^{n} (X_i - \mu_X) + \sum_{i=1}^{n} (\bar{X} - \mu_X)^2$$

$$= \sum_{i=1}^{n} (X_i - \mu_X)^2 - 2n(\bar{X} - \mu_X)^2 + n(\bar{X} - \mu_X)^2$$

$$= \sum_{i=1}^{n} (X_i - \mu_X)^2 - n(\bar{X} - \mu_X)^2$$

Taking expectations then gives

$$E\left[\sum_{i=1}^{n} (X_i - \bar{X})^2 \right] = E\left[\sum_{i=1}^{n} (X_i - \mu_X)^2 \right] - nE[(\bar{X} - \mu_X)^2]$$

$$= \sum_{i=1}^{n} E[(X_i - \mu_X)^2] - nE[(\bar{X} - \mu_X)^2]$$

Now, the expectation of each $(X_i - \mu_X)^2$ is the population variance σ_X^2, and the expectation of $(\bar{X} - \mu_X)^2$ is the variance of the sample mean—that is, σ_X^2/n. Hence, we have

$$E\left[\sum_{i=1}^{n} (X_i - \bar{X})^2 \right] = n\sigma_X^2 - \frac{n\sigma_X^2}{n} = (n - 1)\sigma_X^2$$

Finally, for the expected value of the sample variance, we have

$$E(s_X^2) = E\left[\frac{1}{n-1} \sum_{i=1}^{n} (X_i - \bar{X})^2 \right]$$

$$= \frac{1}{n-1} E\left[\sum_{i=1}^{n} (X_i - \bar{X})^2 \right]$$

$$= \frac{1}{n-1} \cdot (n - 1)\sigma_X^2 = \sigma_X^2$$

This is the result we set out to establish.

point estimation

CHAPTER SEVEN

7.1 INTRODUCTION

In this chapter we begin to explore the possibility of making inferential statements about a population, based on the information contained in a random sample. We will focus attention on specific characteristics, or **parameters,** of the population. Parameters of interest might include the population mean or variance, or the proportion of population members possessing some specific attribute. For example, we may want to make inference about

(a) The mean income of all families in a particular neighborhood
(b) The variation in the impurity levels in batches of a manufactured chemical
(c) The proportion of a corporation's employees favoring the introduction of a modified bonus scheme

Any inference drawn about the population will, of necessity, be based on sample **statistics**—that is, on functions of the sample information. The choice of appropriate statistics will depend on which population parameter is of interest. The true parameter will be unknown, and one objective of sampling could be to estimate its value.

Definitions

An **estimator** of a population parameter is a random variable that depends on the sample information, and whose realizations provide approximations to this unknown parameter. A specific realization of that random variable is called an **estimate.**

261

To clarify the distinction between the terms *estimator* and *estimate,* consider the estimation of the mean income of all families in a neighborhood, based on a random sample of twenty families. It seems reasonable to base our conclusions on the sample mean income, so we say that *the estimator of the population mean is the sample mean.* Suppose that, having obtained the sample, we find that the average income of the families in the sample is $29,356. Then *the estimate of the population mean family income is $29,356.* Notationally, we have already made this distinction, using $\bar{X}$ to denote the random variable, and $\bar{x}$ a specific realization.

In discussing the estimation of an unknown population parameter, two possibilities must be considered. First, we could compute from the sample a single number as "representative," or perhaps "most representative." The estimate $29,356 for the neighborhood mean family income above, is an example of such an estimate. Alternatively, we could try to find an interval, or range, which we are fairly sure contains the true parameter. In this chapter we consider the first type of estimation problem, postponing until Chapter 8 a discussion of interval estimation.

Definitions

A **point estimator** of a population parameter is a function of the sample information that yields a single number. The corresponding realization is called the **point estimate** of the parameter.

In the neighborhood family income example, the parameter to be estimated is the population mean family income. The point estimator used is the sample mean, and the resulting point estimate is $29,356.

For purposes of illustration, we will discuss in this chapter four point estimators, all of which were met in Chapter 6. These are the sample mean, variance, standard deviation, and proportion. Table 7.1 summarizes the notation we have used.

Example 7.1

Price-earnings ratios for a random sample of ten stocks traded on the New York Stock Exchange on December 27, 1980 were:

$$10; \quad 16; \quad 5; \quad 10; \quad 12; \quad 8; \quad 4; \quad 6; \quad 5; \quad 4$$

Find point estimates of the population mean, variance, and standard deviation, and of the proportion of stocks in the population for which the price-earnings ratio exceeded 8.5.

TABLE 7.1 Notation for population parameters, point estimators, and estimates

POPULATION PARAMETER	ESTIMATOR	ESTIMATE
Mean (μ_X)	$\bar{X}$	$\bar{x}$
Variance (σ_X^2)	s_X^2	s_x^2
Standard deviation (σ_X)	s_X	s_x
Proportion (p)	$\hat{p}_X$	$\hat{p}_x$

To find the first three of these sample quantities, we show the calculations in tabular form:

i	x_i	x_i^2
1	10	100
2	16	256
3	5	25
4	10	100
5	12	144
6	8	64
7	4	16
8	6	36
9	5	25
10	4	16
Sums	80	782

We then have

$$n = 10; \qquad \Sigma x_i = 80; \qquad \Sigma x_i^2 = 782$$

Hence, the sample mean is

$$\bar{x} = \frac{1}{n}\Sigma x_i = \frac{80}{10} = 8.0$$

which is our point estimate of the population mean.

A point estimate of the population variance is provided by

$$s_x^2 = \frac{1}{n-1}(\Sigma x_i^2 - n\bar{x}^2)$$

$$= \frac{782 - (10)(8.0)^2}{9} = 15.78$$

For the population standard deviation, the point estimate is

$$s_x = \sqrt{s_x^2} = \sqrt{15.78} = 3.97$$

Finally, in the sample, the number of stocks for which the price-earnings ratio exceeds 8.5 is $x = 4$. Hence, our point estimate of the population proportion is

$$\hat{p}_x = \frac{x}{n} = \frac{4}{10} = 0.4$$

For the specific estimation problems discussed in this section, the choice of point estimator has been based mainly on intuitive plausibility. In the remainder of this chapter we consider various desirable properties of point estimators. This provides a framework within which a particular choice can be evaluated and alternatives examined. At the outset it must be stated that no single mechanism exists for the determination of a uniquely "best" point estimator in all circumstances. What is available instead is a set of criteria under which particular estimators can be evaluated. We will see that for most general purposes the sample mean, variance, standard deviation, and proportion provide satisfactory estimators of the corresponding population quantities. However, in subsequent chapters we will meet estimation problems for which the choice of an appropriate point estimator is rather less obvious.

7.2 UNBIASED ESTIMATORS AND THEIR EFFICIENCY

In this discussion, we will denote by θ a parameter to be estimated, and by $\hat{\theta}$ the corresponding point estimator. As we saw in the previous chapter, it is sometimes possible to find the sampling distribution of the random variable $\hat{\theta}$. That estimator is said to be **unbiased** if the mean of its sampling distribution is the unknown parameter θ.

Definitions

The estimator $\hat{\theta}$ is said to be an **unbiased estimator** of the parameter θ if the mean of the sampling distribution of $\hat{\theta}$ is θ, that is,

$$E(\hat{\theta}) = \theta$$

We say that the corresponding point estimate is obtained through an **unbiased estimation procedure.**

It follows from the notion of expectation that if the sampling procedure is repeated many times, then *on the average* the value obtained for an unbiased estimator will be equal to the population parameter. It seems reasonable to assert that, all other things being equal, unbiasedness is a desirable property in a point estimator. Figure 7.1 shows the sampling distributions of two estimators, one unbiased and one not.

For three of the estimators being considered, we saw in Chapter 6 that

$$E(\overline{X}) = \mu_X; \qquad E(s_X^2) = \sigma_X^2; \qquad E(\hat{p}_X) = p$$

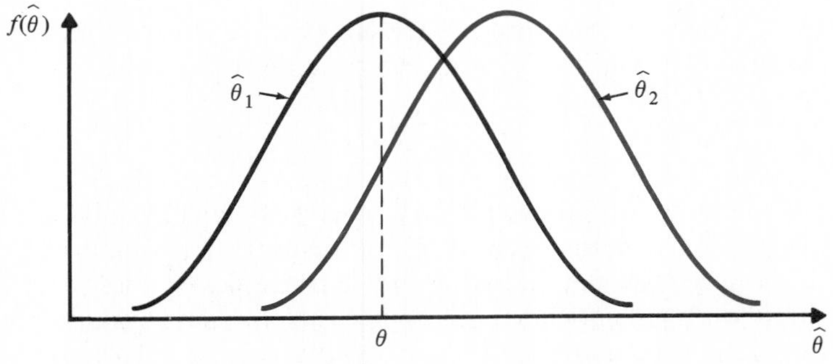

FIGURE 7.1 Probability density functions for the estimators $\hat{\theta}_1$ and $\hat{\theta}_2$: $\hat{\theta}_1$ is an unbiased estimator of θ, $\hat{\theta}_2$ is not

Thus, we can say that the sample mean, variance, and proportion are unbiased estimators of the corresponding population parameters. It is for this reason that, in defining the sample variance, we divided the sum of squared discrepancies from the sample mean by $(n - 1)$ rather than n. The former produces an unbiased estimator; the latter does not. The mean of the sampling distribution of the sample standard deviation is *not* equal to the population standard deviation. Hence, the sample standard deviation is not an unbiased estimator of the population standard deviation.

Unbiasedness of Some Estimators

(i) The sample mean, variance, and proportion are unbiased estimators of the corresponding population quantities.

(ii) In general, the sample standard deviation is *not* an unbiased estimator of the population standard deviation.

Example 7.2

Refer to the results in Example 7.1. We can now state that the estimates of the population mean, variance, and proportion of stocks for which the price-earnings ratio exceeded 8.5,

$$\bar{x} = 8.0; \qquad s_x^2 = 15.78; \qquad \hat{p}_x = 0.4$$

are obtained through unbiased estimation procedures. However, the estimate of the population standard deviation, $s_x = 3.97$, is not obtained through an unbiased estimation procedure.

An estimator that is not unbiased is said to be **biased.** The extent of the **bias** is the difference between the mean of the estimator and the true parameter.

Definition

Let $\hat{\theta}$ be an estimator of θ. Then the **bias** in $\hat{\theta}$ is defined as the difference between its mean and θ; that is,

$$\text{Bias}(\hat{\theta}) = E(\hat{\theta}) - \theta$$

It follows that the bias of an unbiased estimator is 0.

In many practical problems several different unbiased estimators can be obtained, and some method of choosing among them needs to be found. In this situation, it is natural to prefer the estimator whose distribution is most closely concentrated about the population parameter being estimated. Values of such an estimator are less likely to differ, by any fixed amount, from the parameter being estimated than are those of its competitors. Using variance as a measure of concentration, we introduce the concept of **efficiency** of an estimator as a criterion for preferring one estimator over another.

Figure 7.2 shows the sampling distributions of two unbiased estimators. Clearly, $\hat{\theta}_1$ is the more efficient.

Example 7.3

Let $X_1, X_2, \ldots, X_n$ be a random sample from a normal distribution with mean μ_X and variance σ_X^2. The sample mean, $\overline{X}$, is an unbiased estimator of the population mean, with variance

$$\text{Var}(\overline{X}) = \frac{\sigma_X^2}{n}$$

As an alternative estimator, we could use the median of the sample observations. It can be shown that this estimator is also unbiased for μ_X, and that its variance is

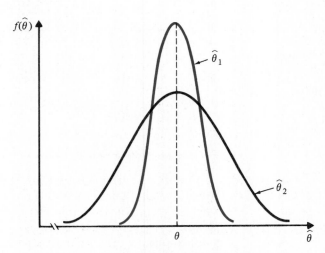

FIGURE 7.2 Probability density functions of two unbiased estimators, $\hat{\theta}_1$ and $\hat{\theta}_2$; $\hat{\theta}_1$ is the more efficient

$$\text{Var(Median)} = \frac{\pi}{2} \cdot \frac{\sigma_{\bar{X}}^2}{n} \approx \frac{1.57\sigma_{\bar{X}}^2}{n}$$

The sample mean is more efficient than the median, the relative efficiency of the mean with respect to the median being

$$\text{Relative efficiency} = \frac{\text{Var(Median)}}{\text{Var}(\bar{X})} = 1.57$$

The variance of the sample median is 57% higher than that of the sample mean. Thus, in order for the sample median to have as small a variance as the sample mean, it would have to be based on 57% more observations. In Chapter 2 we noted that one advantage of the median over the mean is that it gives far less weight to extreme observations. We now see, in terms of its relative inefficiency, a potential disadvantage in using the sample median as a measure of central location.

In some estimation problems[1] the point estimator with the smallest variance among a group of unbiased estimators is sought. In a few relatively simple cases it is possible to find the most efficient of *all* unbiased estimators for a parameter.

Definition

If $\hat{\theta}$ is an unbiased estimator of θ, and no other unbiased estimator has smaller variance, then $\hat{\theta}$ is said to be the **most efficient** or **minimum variance unbiased estimator** of θ.

Specific examples of minimum variance unbiased estimators include

(a) The sample mean when sampling from a normal distribution
(b) The sample variance when sampling from a normal distribution
(c) The binomial sample proportion

The use of minimum variance unbiased estimators is appealing. However, it is not always possible to find such estimators.

EXERCISES

1. A random sample of eight batches of chemical were tested for impurity concentration. The percentage impurity levels found in this sample were

3.2; 4.3; 2.1; 2.8; 3.2; 3.6; 4.0; 2.7

(a) Find the sample mean, variance, and standard deviation. Find the sample proportion of batches with impurity level greater than 3.75%.

---

[1] The most important examples of this type will be discussed in Chapters 12 and 13.

(b) For what population quantities have estimates based on unbiased estimation procedures been found in part (a)?

2. A random sample of six securities analysts gave the following forecasts for the percentage increase in the Dow-Jones average over the next 12 months:

$$12; \quad 8; \quad 6; \quad -2; \quad 0; \quad 5$$

(a) Find the sample mean, variance, and standard deviation.

(b) For what population quantities have estimates based on unbiased estimation procedures been found in part (a)?

(c) Use an unbiased estimation procedure to find a point estimate of the variance of the sample mean.

3. A random sample of ten gas stations in a large city yielded the following prices (in cents) for a gallon of regular gasoline:

$$136; \quad 140; \quad 145; \quad 149; \quad 136; \quad 141; \quad 143; \quad 152; \quad 151; \quad 144$$

(a) Find the sample mean, variance, and standard deviation. Find the sample proportion of stations with prices below 139.5 cents.

(b) For what population quantities have estimates based on unbiased estimation procedures been found in part (a)?

(c) Use an unbiased estimation procedure to find a point estimate of the variance of the sample mean.

4. Let X_1, X_2 be a random sample of two observations from a population with mean μ and variance σ^2. Consider the following two point estimators of μ:

$$\bar{X} = \frac{1}{2}X_1 + \frac{1}{2}X_2 \quad \text{and} \quad \hat{\mu} = \frac{1}{4}X_1 + \frac{3}{4}X_2$$

(a) Show that both estimators are unbiased.

(b) Which estimator is the more efficient?

(c) Find their relative efficiency.

5. Let X_1, X_2 be a random sample from a population with mean μ and variance σ^2. Consider the following two point estimators of μ:

$$\hat{\mu}_1 = \frac{X_1 + 4X_2}{5} \quad \text{and} \quad \hat{\mu}_2 = \frac{2X_1 + 3X_2}{5}$$

(a) Show that both estimators are unbiased.

(b) Which estimator is the more efficient?

(c) Find their relative efficiency.

(d) Find an unbiased estimator of the population mean that is more efficient than either of these estimators.

6. A random sample of fifteen recent graduates in business from college A showed that, in their first job, the mean starting salary was $21,500 and the sample standard deviation was $1,200. An independent random sample of twelve recent business graduates from college B found a mean starting salary of $23,000 and a sample standard deviation of $1,500. Let μ_1 and σ_1^2 denote the population mean and variance of starting salaries for business graduates of college A, and μ_2 and σ_2^2 the population mean and variance for college B.

(a) Use an unbiased estimation procedure to find a point estimate of $(\mu_2 - \mu_1)$, the difference between the population means.

(b) If $\bar{X}_1$ and $\bar{X}_2$ denote the two sample means, find the variance of $(\bar{X}_2 - \bar{X}_1)$.

(c) Use an unbiased estimation procedure to find a point estimate of the variance of $(\bar{X}_2 - \bar{X}_1)$.

7. A random sample of sixteen airline passengers had a mean weight of 162 pounds and a sample standard deviation of 21 pounds. An independent random sample of twenty observations of passengers' baggage produced a sample mean of 48 pounds per passenger and a sample standard deviation of 9 pounds. Let μ_1 and σ_1^2 denote the population mean and variance for passenger weights, and μ_2 and σ_2^2 the population mean and variance for baggage weights.

   (a) Use an unbiased estimation procedure to find a point estimate of $(\mu_1 + \mu_2)$, the population mean weight of a passenger plus baggage.

   (b) If $\bar{X}_1$ and $\bar{X}_2$ denote the two sample means, find the variance of $(\bar{X}_1 + \bar{X}_2)$.

   (c) Use an unbiased estimation procedure to find a point estimate of the variance of $(\bar{X}_1 + \bar{X}_2)$.

8. A random sample of 100 voters in a city found 57 in favor of an immediate reduction in property taxes. An independent random sample of 200 voters from the same population found 102 in favor of extending the services of the public library system. Let p_1 denote the population proportion favoring property tax reduction and p_2 the proportion favoring extending public library services.

   (a) Use an unbiased estimation procedure to find a point estimate of $(p_1 - p_2)$, the difference between the two population proportions.

   (b) If $\hat{p}_1$ and $\hat{p}_2$ denote the two sample proportions, find the variance of $(\hat{p}_1 - \hat{p}_2)$.

9. A random sample of sixteen investment analysts contained twelve who believe that oil company stocks represent a sound investment for the coming year. An independent random sample of ten analysts contained six favoring investment in precious metals. Let p_1 denote the population proportion of investment analysts supporting oil stocks and p_2 the population proportion supporting precious metals.

   (a) Use an unbiased estimation procedure to find a point estimate of $(p_1 - p_2)$, the difference between the two population proportions.

   (b) If $\hat{p}_1$ and $\hat{p}_2$ denote the two sample proportions, find the variance of $(\hat{p}_1 - \hat{p}_2)$.

10. Two adjacent cities are considering a merger. The first city has a population of 80,000. A random sample of 100 of its citizens found 48 in favor of merger. A random sample of 100 people in the second city, which has a population of 20,000, found 36 in favor of merger. Use an unbiased estimation procedure to find a point estimate of the proportion of the combined population of the two cities favoring merger.

11. A company with 200 managers and 1,000 nonmanagerial employees is considering a modification to its employees' bonus scheme. A random sample of 80 managers found 50 favoring the modification, while an independent random sample of 100 nonmanagerial employees found 42 favoring the modification.

    (a) Use an unbiased estimation procedure to find a point estimate of the proportion of all the employees who are in favor of the bonus scheme modification.

    (b) If p_1 denotes the population proportion of managers favoring modification of the scheme, and p_2 the corresponding population proportion of nonmanagerial employees, find the variance of the estimator used in part (a).

12. A random sample of 100 voters found 65 in favor of a federal tax reduction and 55 in favor of sharply increased military spending. Fifty of the sample members were in favor of both these policies. Use an unbiased estimation procedure to find a point estimate of the proportion of all voters favoring at least one of these policies.

13. A random sample of eighty account balances contained four with at least one serious error and ten with at least one minor error. Two of the balances contained at least one serious and one minor error. Use an unbiased estimation procedure to find a point estimate of the proportion of all balances in the population containing no errors.

14. Let X be the number of successes in n independent trials, each with probability of success p. Find an unbiased estimator of the variance of the sample proportion of successes.

7.3 CONSISTENT ESTIMATORS[2]

Up to this point we have considered properties (such as the mean and variance) of point estimators based on a random sample of n observations, for any sample size n. This section deals with **asymptotic** properties of estimators—that is, their behavior as the sample size becomes very large. There are two major motivations for such an inquiry. First, for many estimation problems met in practice it is not possible to obtain even basic characteristics of an estimator's sampling distribution. However, we can often determine the behavior of that distribution when the sample size becomes very large. Second, if an estimator behaves poorly when a very large amount of sample information is available, then its worth may be suspect for any sample size.

In Section 6.2 we saw that, when sampling from a normal population, the distribution of the sample mean becomes more and more concentrated about the true population mean as the sample size becomes large. We now formalize this property through the notion of **consistency.** As before, let θ be a population parameter to be estimated. To denote an estimator based on a sample of n observations, we use $\hat{\theta}_n$, where the subscript is employed to make clear the dependence on the sample size. Our concern is with the "closeness" of the estimator to the population parameter, so we consider the probability that $\hat{\theta}_n$ differs from θ by less than some positive amount ϵ, that is,

$$P[\,|\,\hat{\theta}_n - \theta\,| < \epsilon\,]$$

If, for any positive ϵ, however small, this probability approaches 1 as the sample size becomes very large, then the probability density function of the estimator becomes more concentrated about the true parameter. The estimator is then said to be **consistent.** This notion is illustrated in Figure 7.3.

Definition

    Let $\hat{\theta}_n$ be an estimator of θ based on n observations. If for every positive amount ϵ, however small, the probability

$$P[\,|\,\hat{\theta}_n - \theta\,| < \epsilon\,]$$

approaches 1 as the sample size n becomes infinitely large, then $\hat{\theta}_n$ is said to be a **consistent estimator** of θ.

    Loosely speaking, the use of a consistent estimator with an infinite amount of sample information would yield the correct result. Conversely, an inconsistent estimator would not yield the correct result, even if it were based on an infinite amount of sample information. Viewed in this light, consistency certainly seems to be a

---

[2] This section is rather technical, and can be omitted without loss of continuity. The property of consistency will be briefly mentioned in Section 14.8.

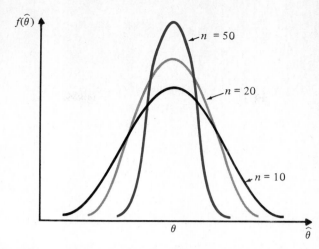

FIGURE 7.3 Illustration of consistency: As sample size increases, the distribution of the estimator becomes more tightly concentrated around the true parameter value, θ

desirable property of an estimator. However, if we knew that for a particular small sample size an inconsistent estimator had superior properties (unbiased, with smaller variance, for example) to a consistent estimator, then we would prefer to use the former in analyzing samples of that size.

The definition of consistency is not very useful in establishing whether a particular estimator is consistent. The two procedures stated in the box are frequently applicable.

Establishing Consistency of an Estimator

(i) Let $\hat{\theta}_n$ be an estimator of θ. If both the bias and variance of $\hat{\theta}_n$ approach 0 as n becomes infinitely large, then $\hat{\theta}_n$ is a consistent estimator.

(ii) If $\hat{\theta}_n$ is a consistent estimator of θ, then under very general conditions the function $g(\hat{\theta}_n)$ is a consistent estimator of $g(\theta)$.

Notice that a biased estimator can be consistent; all that is necessary is that the bias decrease to 0 as the sample size increases. Moreover, unbiasedness alone is not enough to establish consistency; the variance of the estimator must also decrease to 0 as the sample size increases.

Example 7.4

Let $X_1, X_2, \ldots, X_n$ be a random sample from a distribution with mean μ_X and variance σ_X^2. Show that the sample mean $\bar{X}$ is a consistent estimator of the population mean.

Since

$$E(\bar{X}) = \mu_X$$

the bias is 0 for *any* sample size. The variance of $\bar{X}$ is

$$\text{Var}(\bar{X}) = \frac{\sigma_X^2}{n}$$

which approaches 0 as n becomes increasingly large. Hence, consistency is established.

Example 7.5 Let $X_1, X_2, \ldots, X_n$ be a random sample from a normal distribution with variance σ_X^2. Show that the sample variance s_X^2 is a consistent estimator of the population variance.

Since

$$E(s_X^2) = \sigma_X^2$$

the bias is 0 for *any* sample size. In Section 6.4, we saw that the variance of this estimator is

$$\text{Var}(s_X^2) = \frac{2\sigma_X^4}{n-1}$$

Since this quantity approaches 0 as n becomes infinitely large, the sample variance is consistent. (In fact, consistency of the sample variance can be established for a very wide class of population distributions.)

Example 7.6 Let $X_1, X_2, \ldots, X_n$ be a random sample from a normal distribution with variance σ_X^2.

Since s_X^2 is a consistent estimator of σ_X^2, it follows that the sample standard deviation s_X is a consistent estimator of the population standard deviation σ_X.

Example 7.7 Let X be the number of successes in n independent trials, each with probability of success p. Show that the sample proportion

$$\hat{p}_X = \frac{X}{n}$$

is a consistent estimator of p.

Since

$$E(\hat{p}_X) = p$$

the bias is 0 for *any* sample size. In Section 6.3, we saw that

$$\text{Var}(\hat{p}_X) = \frac{p(1-p)}{n}$$

which approaches 0 as n becomes infinitely large. Hence, consistency is established.

EXERCISES

15. Let $X_1, X_2, \ldots, X_n$ be a random sample from a distribution with mean μ_X and variance σ_X^2. Define the following point estimator of the population mean:

$$\hat{\mu}_X = \frac{(n + 1)\bar{X}}{n}$$

where $\bar{X}$ is the sample mean. Show that this estimator is consistent.

16. Let X be the number of successes in n independent trials, each with probability of success p. Let

$$\hat{p} = \frac{X + 2}{n + 2}$$

be a point estimator of the population proportion. Show that this estimator is consistent.

17. Let X be the number of successes in n independent trials, each with probability of success p. Consider the following two estimators of p:

$$\hat{p}_X = \frac{X}{n} \quad \text{and} \quad \hat{p}_X^* = \frac{2n}{2n + 1}\hat{p}_X + \frac{1}{2n + 1}$$

(a) Are both estimators unbiased?
(b) Are both estimators consistent?

REVIEW EXERCISES

18. Explain carefully the distinction between the properties of unbiasedness and consistency in a point estimator.

19. Of what practical value is the concept of efficiency in evaluating the merits of a point estimator?

20. Let $\hat{\theta}_1$ be an unbiased estimator of θ_1, and $\hat{\theta}_2$ an unbiased estimator of θ_2.
    (a) Show that $(\hat{\theta}_1 + \hat{\theta}_2)$ is an unbiased estimator of $(\theta_1 + \theta_2)$.
    (b) Show that $(\hat{\theta}_1 - \hat{\theta}_2)$ is an unbiased estimator of $(\theta_1 - \theta_2)$.

21. Let $\hat{\theta}_1$ be an unbiased estimator of θ_1, and $\hat{\theta}_2$ an unbiased estimator of θ_2.
    (a) Is $\hat{\theta}_1^2$ necessarily an unbiased estimator of θ_1^2?
    (b) Is $\hat{\theta}_1 \hat{\theta}_2$ necessarily an unbiased estimator of $\theta_1 \theta_2$?

22. A random sample of ten money market funds showed the following average days to maturity:

29	35	36	31	40	33	36	32	37	28

Use unbiased estimation procedures to find point estimates for each of the following:
(a) The population mean
(b) The population variance
(c) The variance of the sample mean
(d) The population proportion of funds with average days to maturity less than 35
(e) The variance of the sample proportion of funds with average days to maturity less than 35

23. A random sample of twelve employees of a large corporation yielded the following figures for the numbers of sick days taken during the past year:

12	8	4	7	8	9
15	10	10	6	5	3

Use unbiased estimation procedures to find point estimates for each of the following:
- **(a)** The population mean
- **(b)** The population variance
- **(c)** The variance of the sample mean
- **(d)** The population proportion of employees taking less than 6 sick days
- **(e)** The variance of the sample proportion of employees taking less than 6 sick days

24. A population has mean μ (different from 0) and variance σ^2. A random sample of two observations from this population has mean $\bar{X}_1$ and an independent random sample of three observations from the same population has mean $\bar{X}_2$.

- **(a)** Let a and b be any two numbers. Show that the estimator

$$\hat{\mu} = a\bar{X}_1 + b\bar{X}_2$$

  is unbiased for μ if and only if $a + b = 1$.
- **(b)** Consider the class of unbiased estimators

$$\hat{\mu} = a\bar{X}_1 + (1 - a)\bar{X}_2$$

  where a is any number. Show that the most efficient of these estimators has $a = \frac{2}{5}$.

25. Consider a population with a proportion $p \, (0 < p < 1)$ of members possessing a particular attribute. Let $\hat{p}_1$ be the sample proportion possessing this attribute in a random sample of three observations, and $\hat{p}_2$ the sample proportion possessing this attribute in an independent random sample of four observations from the same population.

- **(a)** Show that $\hat{p}_1$ and $\hat{p}_2$ are both unbiased estimators of p, and determine which is the more efficient.
- **(b)** Let a and b be any two numbers. Show that the estimator

$$\hat{p} = a\hat{p}_1 + b\hat{p}_2$$

  is unbiased for p if and only if $a + b = 1$.
- **(c)** Consider the class of unbiased estimators

$$\hat{p} = a\hat{p}_1 + (1 - a)\hat{p}_2$$

  where a is any number. Show that the most efficient of these estimators has $a = \frac{3}{7}$.

26. A random sample of ten X-cars achieved the following miles per gallon figures:

27.2	26.8	25.3	26.4	27.1
27.2	26.9	26.0	25.7	25.7

An independent random sample of twelve Y-cars achieved the following miles per gallon results:

24.2	25.3	25.1	24.9	26.0	26.0
24.3	24.8	25.0	23.9	26.1	25.7

(a) Use an unbiased estimation procedure to obtain a point estimate of the difference in population mean miles per gallon between X-cars and Y-cars.

(b) Use an unbiased estimation procedure to obtain a point estimate of the difference between the population proportion of X-cars achieving more than 25.5 miles per gallon and the population proportion of Y-cars achieving more than 25.5 miles per gallon.

27. A random sample of nine seniors taking a business statistics course achieved the following scores on the final examination:

62	57	85	64	59	71	63	58	62

An independent random sample of eight sophomores taking the same exam obtained the following scores:

73	79	62	60	51	57	59	78

(a) Use an unbiased estimation procedure to obtain a point estimate of the difference in population mean scores between seniors and sophomores.

(b) A score of at least 70 was required on this examination for grade A. Use an unbiased estimation procedure to find a point estimate of the difference between the population proportions of seniors and sophomores obtaining grade A.

28. A random sample, $X_1, X_2, \ldots, X_n$, of n observations is taken from a population with mean μ and variance σ^2. Consider the following estimator of μ:

$$\hat{\mu} = \frac{2}{n(n + 1)}(X_1 + 2X_2 + 3X_3 + \cdots + nX_n)$$

(a) Show that $\hat{\mu}$ is an unbiased estimator of μ.
(b) Find the efficiency of $\hat{\mu}$ relative to $\bar{X}$, the sample mean.
(c) Show that $\hat{\mu}$ is a consistent estimator of μ.

$$\left[Hint: \sum_{i=1}^{n} i = \frac{n(n + 1)}{2}; \quad \sum_{i=1}^{n} i^2 = \frac{n(n + 1)(2n + 1)}{6} \right]$$

interval estimation

8.1 CONFIDENCE INTERVALS

In the previous chapter we considered the point estimation of an unknown population parameter—that is, the production of a single number which in some sense is a "good bet." For most practical problems, a point estimate alone is inadequate. For instance, suppose that a check on a random sample of parts from a large shipment leads to the estimate that 10% of all the parts are defective. Faced with this figure, a manager is likely to ask such questions as, "Can I be fairly sure that the true percentage of defectives is between 5% and 15%?" or, "Does it then seem very likely that between 9% and 11% of all the parts are defective?" Questions of this kind seek information beyond that contained in a single point estimate; they are asking about the reliability of that estimate. More directly, the quest is for an **interval estimate,** a range of values in which the quantity to be estimated appears likely to lie.

In sampling from a population, we would expect (all other things equal) to obtain more secure knowledge about that population from a relatively large sample than we would from a small sample. However, this factor is not reflected in point estimates. For example, our point estimate of the proportion of defective parts in a shipment would be the same if we observed one defective in a sample of ten parts as if we observed one hundred defectives in a sample of one thousand parts. As we will see, increased precision in our information about population parameters *is* reflected in interval estimates.

So far we have described interval estimators as being "likely" or "very likely" to contain the true, but unknown, population parameter. In order to make our dis-

276

> **Definitions**
>
> An **interval estimator** for a population parameter is a rule for determining (based on sample information) a range, or interval, in which the parameter is likely to fall. The corresponding estimate is called an **interval estimate.**

cussion more precise, it is necessary to phrase such terms as probability statements. Let us denote by θ the parameter to be estimated. Suppose that a random sample has been taken and that, based on the sample information, it is possible to find two random variables, A and B, with A less than B. If these random variables have the property that the probability that both A is less than θ and B is bigger than θ is 0.9, we can write

$$P(A < \theta < B) = 0.9$$

The interval from A to B is then said to be a 90% confidence interval estimator for θ. If the specific sample realizations of the random variables A and B are denoted a and b, then the interval from a to b is called a 90% **confidence interval** for θ. According to the frequency concept of probability, we can interpret such intervals as follows: If the population is repeatedly sampled and intervals calculated in this fashion, then 90% of the intervals would contain the unknown parameter.

More generally, we can define confidence intervals of any required probability content less than 1. Suppose that the random variables A and B are such that

$$P(A < \theta < B) = 1 - \alpha$$

where α is any number between 0 and 1. (In the example of the previous paragraph, $1 - \alpha = 0.9$, so that $\alpha = 0.1$.) Then, if the population is repeatedly sampled, and this interval calculated, a proportion $1 - \alpha$, or a percentage $100(1 - \alpha)\%$, of those intervals will contain θ. A confidence interval calculated in this way is called a $100(1 - \alpha)\%$ confidence interval for θ.

> **Definitions**
>
> Let θ be an unknown parameter and suppose that, on the basis of sample information, we can find random variables A and B such that
> $$P(A < \theta < B) = 1 - \alpha$$
> If the specific sample realizations of A and B are denoted a and b, then the interval from a to b is called a **$100(1 - \alpha)\%$ confidence interval** for θ. The quantity $(1 - \alpha)$ is called the **probability content**, or **level of confidence,** of the interval.

> If the population was repeatedly sampled a very large number of times, the parameter θ would be contained in $100(1 - \alpha)\%$ of intervals calculated this way. The confidence interval calculated in this manner is written
>
> $$a < \theta < b$$

In the remainder of this chapter we will develop and illustrate procedures for finding confidence intervals in several common types of estimation problems. Further examples of interval estimation will also be met in subsequent chapters.

8.2 CONFIDENCE INTERVALS FOR THE MEAN OF A NORMAL DISTRIBUTION: POPULATION VARIANCE KNOWN

Let us assume that a random sample is taken from a normal distribution with an unknown mean and a *known* variance, and that the objective is to find a confidence interval for the population mean. This problem is somewhat unrealistic, because rarely (if ever) will a population variance be precisely known and yet the mean be unknown. It does sometimes happen, however, that similar populations have been sampled so often in the past that the variance of the population of interest can be assumed known to a very close approximation on the basis of past experience. Further, as we will see later in this section, if a sufficiently large sample is available, the procedures developed for the case where the population variance is known can be used if that variance has to be estimated from the sample. Nevertheless, the chief virtue in beginning with this problem is that it allows a fairly straightforward development of the procedures involved in finding confidence intervals.

Denote by $X_1, X_2, \ldots, X_n$ a random sample of n observations from a normal population with unknown mean μ and known variance σ^2, and let $\overline{X}$ be the sample mean. Then confidence intervals for the population mean are based on the result that the random variable

$$Z = \frac{\overline{X} - \mu}{\sigma/\sqrt{n}}$$

has a standard normal distribution.

Suppose that we want to find a 90% confidence interval for the population mean. Now, from the cumulative distribution function of the standard normal random variable, given in Table 3 of the Appendix, we find that

$$P(Z < 1.645) = F_Z(1.645) = 0.95$$

It therefore follows that

$$P(Z > 1.645) = 0.05$$

Also, since the density function of the standard normal random variable is symmetric about its mean, 0, it follows that

$$P(Z < -1.645) = 0.05$$

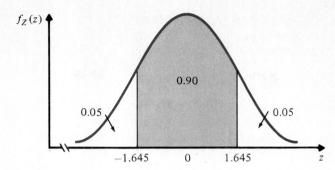

FIGURE 8.1 $P(-1.645 < Z < 1.645) = 0.90$, where Z is a standard normal random variable

The probability that Z is between -1.645 and 1.645 is therefore

$$P(-1.645 < Z < 1.645) = 1 - P(Z > 1.645) - P(Z < -1.645)$$
$$= 1 - 0.05 - 0.05 = 0.90$$

These probability calculations are illustrated in Figure 8.1.

So far then, we have shown that the probability is 0.9 that a standard normal random variable lies between the numbers -1.645 and 1.645. We now convert this probability statement into a confidence interval for the population mean as follows:

$$0.90 = P(-1.645 < Z < 1.645)$$

$$= P\left(-1.645 < \frac{\bar{X} - \mu}{\sigma/\sqrt{n}} < 1.645\right)$$

$$= P\left(\frac{-1.645\sigma}{\sqrt{n}} < \bar{X} - \mu < \frac{1.645\sigma}{\sqrt{n}}\right)$$

$$= P\left(\bar{X} - \frac{1.645\sigma}{\sqrt{n}} < \mu < \bar{X} + \frac{1.645\sigma}{\sqrt{n}}\right)$$

Therefore, the probability is 0.9 that the random interval from $(\bar{X} - 1.645\sigma/\sqrt{n})$ to $(\bar{X} + 1.645\sigma/\sqrt{n})$ contains the population mean μ. Thus, from our definition of confidence intervals, it follows that the interval from $(\bar{x} - 1.645\sigma/\sqrt{n})$ to $(\bar{x} + 1.645\ \sigma/\sqrt{n})$ is a 90% confidence interval for μ, where $\bar{x}$ is the specific value observed for the sample mean. For simplicty, this interval is written

$$\bar{x} - \frac{1.645\sigma}{\sqrt{n}} < \mu < \bar{x} + \frac{1.645\sigma}{\sqrt{n}}$$

Example 8.1   A random sample of sixteen observations from a normal population with standard deviation 6 had mean 25. Find a 90% confidence interval for the population mean, μ.

We have

$$\bar{x} = 25; \qquad \sigma = 6; \qquad n = 16$$

The 90% confidence interval is given by

$$\bar{x} - \frac{1.645\sigma}{\sqrt{n}} < \mu < \bar{x} + \frac{1.645\sigma}{\sqrt{n}}$$

so that we have

$$25 - \frac{(1.645)(6)}{\sqrt{16}} < \mu < 25 + \frac{(1.645)(6)}{\sqrt{16}}$$

or

$$22.5325 < \mu < 27.4675$$

We now pause to consider the proper interpretation of confidence intervals in terms of Example 8.1. In that example we found, based on a sample of sixteen observations, a 90% confidence interval for the unknown population mean running from 22.5325 to 27.4675. Now, this particular sample is just one of many that might have been drawn from the population. Suppose that we were to start over again, taking a second sample of sixteen observations. It is virtually certain that the mean of the second sample will differ from that of the first. Accordingly, if a 90% confidence interval is calculated from the results of the second sample, it will differ from that found in Example 8.1. We can imagine taking a very large number of independent random samples of sixteen observations from this population and, from each sample result, calculating a 90% confidence interval. The probability content of the interval implies that 90% of intervals found in this manner contain the true population mean. It is in this sense that we can say that we have "90% confidence" in our interval estimate.

This position is illustrated in Figure 8.2. Part (a) of the figure shows the sampling distribution of the sample mean of n observations from a normal population with mean μ and standard deviation σ. This sampling distribution is normal, with mean μ and standard deviation $\sigma/\sqrt{n}$. A confidence interval for the population mean will be based on the observed value of the sample mean—that is, on an observation drawn from our sampling distribution. Part (b) of Figure 8.2 shows a sequence of 90% confidence intervals, obtained from independent samples taken from the population. The centers of these intervals, which are just the observed sample means, will often be quite close to the population mean, μ. However, some may differ quite substantially. We are assured that 90% of a large number of these intervals will contain the population mean.

We now turn to the general case of finding confidence intervals with any required probability content. In order to do so, we introduce a new notation.

Notation

Let Z be a standard normal random variable and α be any number such that $0 < \alpha < 1$. Then we denote by z_α that number for which

$$P(Z > z_\alpha) = \alpha$$

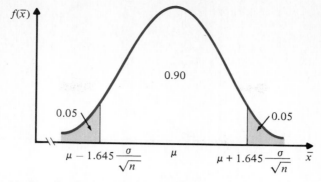

(a) Sampling distribution of sample mean of n observations
from a normal distribution with mean μ and variance σ^2

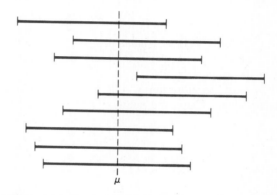

(b) Some 90% confidence intervals for the population mean

FIGURE 8.2 Interpretation of a 90% confidence interval

This notation is illustrated in Figure 8.3, from which it is clear how z_α can be found from tables of the cumulative distribution function of the standard normal random variable. Since $P(Z > z_\alpha) = \alpha$, it follows that

$$F_Z(z_\alpha) = P(Z < z_\alpha) = 1 - \alpha$$

as is illustrated in Figure 8.3. Therefore, for any specified α, z_α can be determined from Table 3 in the Appendix. For example, if $\alpha = .025$, then

$$1 - \alpha = .975$$

so that

$$F_Z(z_\alpha) = F_Z(z_{.025}) = .975$$

and from the table

$$z_{.025} = 1.96$$

Confidence Intervals for the Mean of a Normal Distribution: Population Variance Known

281

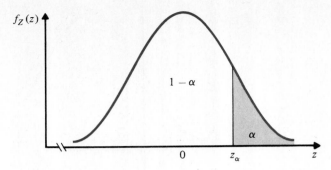

FIGURE 8.3 $P(Z > z_\alpha) = \alpha$, where Z is a standard normal random variable

Therefore,

$$P(Z > 1.96) = .025$$

Table 8.1 shows values of z_α corresponding to some α values that are frequently used in finding confidence intervals. The reader can verify these quantities from Table 3 of the Appendix.

Now, suppose that a $100(1 - \alpha)\%$ confidence interval is required for the population mean. Using the notation just established, we have

$$P(Z > z_{\alpha/2}) = \frac{\alpha}{2}$$

and so, by the symmetry of the standard normal density function about its mean of 0,

$$P(Z < -z_{\alpha/2}) = \frac{\alpha}{2}$$

It therefore follows that

$$P(-z_{\alpha/2} < Z < z_{\alpha/2}) = 1 - \frac{\alpha}{2} - \frac{\alpha}{2} = 1 - \alpha$$

These probability calculations are illustrated in Figure 8.4.

We have found a range of values of the standard normal random variable with a specified probability content. This information can now be used to develop a

TABLE 8.1 Values of z_α, from table of the standard normal cumulative distribution function

α	z_α
.005	2.575
.01	2.33
.025	1.96
.05	1.645
.10	1.28

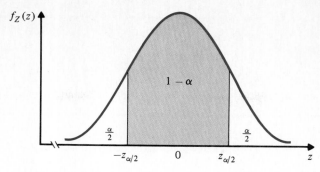

FIGURE 8.4 $P(-z_{\alpha/2} < Z < z_{\alpha/2}) = 1 - \alpha$ where Z is a standard normal random variable

confidence interval of the same probability content for the population mean. Using exactly the same line of argument as employed in developing the 90% interval, we have

$$1 - \alpha = P(-z_{\alpha/2} < Z < z_{\alpha/2})$$

$$= P\left(-z_{\alpha/2} < \frac{\bar{X} - \mu}{\sigma/\sqrt{n}} < z_{\alpha/2}\right)$$

$$= P\left(\frac{-z_{\alpha/2}\sigma}{\sqrt{n}} < \bar{X} - \mu < \frac{z_{\alpha/2}\sigma}{\sqrt{n}}\right)$$

$$= P\left(\bar{X} - \frac{z_{\alpha/2}\sigma}{\sqrt{n}} < \mu < \bar{X} + \frac{z_{\alpha/2}\sigma}{\sqrt{n}}\right)$$

It follows from the definition of confidence intervals that, if $\bar{x}$ is the specific value observed for the sample mean, then a $100(1 - \alpha)\%$ confidence interval for the population mean is given by

$$\bar{x} - \frac{z_{\alpha/2}\sigma}{\sqrt{n}} < \mu < \bar{x} + \frac{z_{\alpha/2}\sigma}{\sqrt{n}}$$

Confidence Intervals for the Mean of a Normal Population: Population Variance Known

Suppose we have a random sample of n observations from a normal distribution with mean μ and variance σ^2. If σ^2 is known, and the observed sample mean is $\bar{x}$, then a $100(1 - \alpha)\%$ confidence interval for the population mean is given by

$$\bar{x} - \frac{z_{\alpha/2}\sigma}{\sqrt{n}} < \mu < \bar{x} + \frac{z_{\alpha/2}\sigma}{\sqrt{n}}$$

---

where $z_{\alpha/2}$ is that number for which

$$P(Z > z_{\alpha/2}) = \frac{\alpha}{2}$$

and the random variable Z has a standard normal distribution.

The interpretation of these general confidence intervals corresponds to that for the specific 90% interval. If samples of n observations are drawn repeatedly and independently from the population, and $100(1 - \alpha)\%$ confidence intervals are calculated using the formula given in the box, then over a very large number of repeated trials, $100(1 - \alpha)\%$ of these intervals will contain the true population mean.

Example 8.2

A process produces bags of refined sugar. The weights of the contents of these bags are normally distributed with standard deviation 1.2 ounces. The contents of a random sample of twenty-five bags had mean weight 19.8 ounces. Find a 95% confidence interval for the true mean weight for all bags of sugar produced by the process.

Since a 95% confidence interval is required, we have

$$100(1 - \alpha) = 95$$

so that

$$\alpha = 0.05$$

Hence,

$$z_{\alpha/2} = z_{.025}$$

and we require that number $z_{.025}$ for which

$$P(Z > z_{.025}) = .025$$

Hence,

$$P(Z < z_{.025}) = F_Z(z_{.025}) = .975$$

where $F_Z(z)$ is the cumulative distribution function of the standard normal random variable. Then, from Table 3 in the Appendix, it follows that

$$z_{.025} = 1.96$$

The 95% confidence interval for the population mean μ is

$$\bar{x} - \frac{z_{\alpha/2}\sigma}{\sqrt{n}} < \mu < \bar{x} + \frac{z_{\alpha/2}\sigma}{\sqrt{n}}$$

where

$$\bar{x} = 19.8; \qquad z_{\alpha/2} = 1.96; \qquad \sigma = 1.2; \qquad n = 25$$

Hence, the required confidence interval is

$$19.8 - \frac{(1.96)(1.2)}{\sqrt{25}} < \mu < 19.8 + \frac{(1.96)(1.2)}{\sqrt{25}}$$

or

$$19.33 < \mu < 20.27$$

Hence, the 95% confidence interval for the true mean weight ranges from 19.33 to 20.27 ounces.

We now pause to consider the general nature of confidence intervals for the mean of a normal population, when the variance is known. The results that follow provide some insight into the structure of interval estimation. Many extend also to other interval estimation problems.

We have seen that a $100(1 - \alpha)\%$ confidence interval for the population mean is

$$\bar{x} - \frac{z_{\alpha/2}\sigma}{\sqrt{n}} < \mu < \bar{x} + \frac{z_{\alpha/2}\sigma}{\sqrt{n}}$$

As might be expected, the sample mean $\bar{x}$ is at the center of this interval. The width, w, of the interval—that is, the distance between its endpoints—is

$$w = \frac{2z_{\alpha/2}\sigma}{\sqrt{n}}$$

It can be seen, then, that the width of the confidence interval is determined by its probability content, the population standard deviation, and the number of sample observations. In particular, the following results hold:

(a) For a given probability content and sample size, *the bigger the population standard deviation σ the wider the confidence interval for the population mean*. This is intuitively plausible since, all other things equal, the more disperse is the population distribution about its mean, the more uncertain will be our inference about that mean. This extra uncertainty is reflected in wider confidence intervals.

(b) For a given probability content and population standard deviation, *the bigger the sample size n, the narrower the confidence interval for the population mean*. Again, this conclusion accords with intuition. The more information we obtain from a population, the more precise should be our inference about its mean. This extra precision is reflected in narrower confidence intervals.

(c) For a given population standard deviation and sample size, *the bigger the probability content $1 - \alpha$, the wider the confidence interval for the population mean*. For example, a 99% confidence interval will be wider than a 95% confidence interval based on the same information. This result follows since, the larger is $1 - \alpha$, the smaller will be α, and hence the bigger will be $z_{\alpha/2}$. Once again, we pay for an increased certainty in probability statements through a reduced definitiveness in those statements. This is reflected in wider confidence intervals for population parameters.

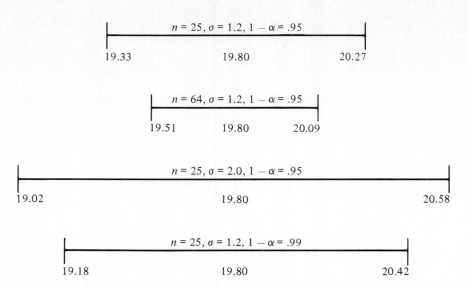

FIGURE 8.5 The effects of sample size n, population standard deviation σ, and probability content $1 - \alpha$ on confidence intervals for the mean of a normal distribution; in each case the sample mean is 19.80

These results are illustrated in Figure 8.5 for the data of Example 8.2. This figure illustrates, for the same sample mean, the effect on the confidence interval of increases in the sample size, the population standard deviation, and the probability content of the interval.

In developing confidence intervals for the mean in this section, we introduced two requirements that limit the range of applicability of these interval estimates in practical problems—namely, that the population distribution is normal and that the variance of this population is known. However, if the sample size is large, neither requirement is critical. In this case, by virtue of the Central Limit Theorem, the confidence intervals described in this section remain appropriate to a sufficiently good approximation even when the population distribution is not normal. Moreover, when the sample size is large, the sample standard deviation will be a sufficiently good estimator of the population standard deviation to allow us to use the former in place of the latter without affecting very much the probability content of the intervals. Thus, we can compute intervals in exactly the manner described previously, substituting the sample standard deviation for the population standard deviation.

Confidence Intervals for the Population Mean: Large Sample Sizes

Suppose we have a sample of n observations from a distribution with mean μ. Denote the observed sample mean and standard deviation by $\bar{x}$ and s_x. Then if n is large, to a good approximation, a $100(1 - \alpha)\%$ confidence interval for μ is given by[1]

$$\bar{x} - \frac{z_{\alpha/2} s_x}{\sqrt{n}} < \mu < \bar{x} + \frac{z_{\alpha/2} s_x}{\sqrt{n}}$$

This approximation will typically remain adequate even if the population distribution is not normal.

Thus, if the sample size is large, we can relax two of the assumptions made earlier, and consequently broaden considerably the range of applicability of our results, as illustrated in the following example.

Example 8.3

The cloze readability procedure is designed to measure the effectiveness of a written communication. Research has indicated that a score of 57% or more on the cloze test demonstrates adequate understandability of the written material. A random sample[2] of 352 certified public accountants were asked to read financial report messages. The sample mean cloze score was 60.41% and the sample standard deviation was 11.28%. Find a 95% confidence interval for the population mean score.

Since the sample size is large, we can use as an interval estimate of the population mean

$$\bar{x} - \frac{z_{\alpha/2} s_x}{\sqrt{n}} < \mu < \bar{x} + \frac{z_{\alpha/2} s_x}{\sqrt{n}}$$

where

$$\bar{x} = 60.41; \qquad s_x = 11.28; \qquad n = 352$$

and $z_{\alpha/2}$ is that number for which

$$P(Z > z_{\alpha/2}) = \frac{\alpha}{2}$$

Hence, for a 95% confidence interval, we have

$$\frac{\alpha}{2} = .025 \qquad \text{and} \qquad z_{\alpha/2} = 1.96$$

The required 95% confidence interval is then

---

[1] As a rule of thumb, we will consider $n = 30$ observations or more to constitute a "large" sample. However, it should not be inferred that the approximation is virtually perfect for sample size 30, and absolutely awful for sample size 29. Rather, the quality of this approximation gradually improves with increasing sample size.

[2] Reported in A. H. Adelberg, "A methodology for measuring the understandability of financial report messages," *Journal of Accounting Research, 17* (1979), 565–92.

$$60.41 - \frac{(1.96)(11.28)}{\sqrt{352}} < \mu < 60.41 + \frac{(1.96)(11.28)}{\sqrt{352}}$$

or

$$59.23 < \mu < 61.59$$

Thus, our 95% confidence interval for the mean cloze score for the population of certified public accountants reading financial report messages runs from 59.23% to 61.59%. Note that the score 57%, generally taken to represent adequate understandability, lies well below the lower endpoint of this interval.

8.3 THE STUDENT'S *t* DISTRIBUTION

In the previous section we derived confidence intervals for the mean of a normal population when the population variance was known. It was noted that the assumption of known population variance could be relaxed when the sample size was large. We will now deal with the case, of considerable practical importance, where the population variance is unknown and the sample size is not large. In order to do so, it is necessary to introduce a new class of probability distributions.

The development of Section 8.2 was based on the fact that the random variable

$$Z = \frac{\bar{X} - \mu}{\sigma/\sqrt{n}} \qquad (8.3.1)$$

has a standard normal distribution. In the case where the population standard deviation is unknown this result cannot be used directly. It is natural in such circumstances to consider the random variable obtained by replacing the unknown σ in equation 8.3.1 by the sample standard deviation, s_X, giving

$$t = \frac{\bar{X} - \mu}{s_X/\sqrt{n}}$$

This random variable does not follow a standard normal distribution. However, its distribution is known, and is in fact a member of a family of distributions called **Student's** *t*.[3]

Student's *t* Distribution

> Given a random sample of n observations, with mean $\bar{X}$ and standard deviation s_X, from a normal population with mean μ, the random variable

---

[3] This result was published in 1908 by W. S. Gosset. Gosset was employed by Guinness Breweries in Ireland, which forbade the publication of scientific research by their employees. His research appeared under the pseudonym "Student," and this name is now given to the distribution.

$$t = \frac{\bar{X} - \mu}{s_X/\sqrt{n}}$$

follows the **Student's t distribution with $(n-1)$ degrees of freedom.**[4]

A specific member of the family of Student's t distributions is characterized by the number of degrees of freedom, for which we will use the parameter ν. A Student's t random variable with ν degrees of freedom will be denoted t_ν. The shape of the Student's t distribution is rather similar to that of the standard normal. Both distributions have mean 0, and the probability density functions of both are symmetric about their means. However, the density function of the Student's t has larger dispersion (reflected in a larger variance) than the standard normal. This can be seen in Figure 8.6, which shows density functions for the standard normal and the Student's t distribution with 3 degrees of freedom. The additional dispersion in the Student's t distribution arises as a result of the extra uncertainty caused by replacing the known population standard deviation by its sample estimator. As the number of degrees of freedom increases, the Student's t becomes increasingly similar to the standard normal. For large degrees of freedom, the two distributions are virtually identical. This is intuitively reasonable, and follows from the fact that, for a large sample, the sample standard deviation is a very precise estimator of the population standard deviation.

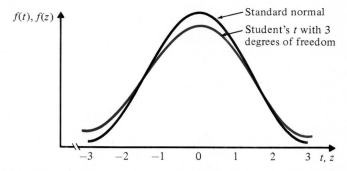

FIGURE 8.6 Probability density functions of the standard normal and the Student's t distribution with 3 degrees of freedom

---

[4] Note that the number of degrees of freedom here corresponds to that quantity for the chi-square distribution based on the same sample size, as in Section 6.4. Formally, the Student's t random variable with ν degrees of freedom is defined as

$$t_\nu = \frac{Z}{\sqrt{\chi_\nu^2/\nu}}$$

where Z is a standard normal random variable, χ_ν^2 is a chi-square random variable with ν degrees of freedom, and Z and χ_ν^2 are independent.

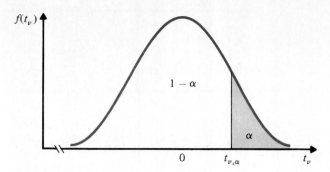

FIGURE 8.7 $P(t_\nu > t_{\nu,\alpha}) = \alpha$, where t_ν is a Student's t random variable with ν degrees of freedom

In order to base inference about a population mean on the Student's t distribution, we need probabilities associated with that distribution. To describe these probabilities, some further notation is required, as presented in the box and illustrated in Figure 8.7.

Notation

A random variable having the Student's t distribution with ν degrees of freedom will be denoted t_ν. We define as $t_{\nu,\alpha}$ that number for which

$$P(t_\nu > t_{\nu,\alpha}) = \alpha$$

In the typical application we will need to find the value $t_{\nu,\alpha}$ corresponding to a specified probability α. These quantities can be read directly from Table 6 in the Appendix. To illustrate, suppose we want to find the number that is exceeded with probability 0.1 by a Student's t random variable with 15 degrees of freedom. That is,

$$P(t_{15} > t_{15,.1}) = 0.1$$

Then, reading directly from the table, we have

$$t_{15,.1} = 1.341$$

We will now use the Student's t distribution to compute confidence intervals for the mean of a normal population. Paralleling our discussion of the standard normal distribution in the previous section, we have

$$P(t_\nu > t_{\nu,\alpha/2}) = \frac{\alpha}{2}$$

Furthermore, as a result of the symmetry of the density function of the Student's t distribution about its mean of 0,

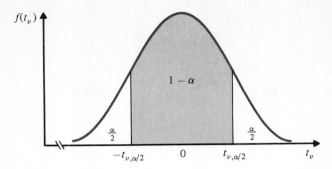

FIGURE 8.8 $P(-t_{\nu,\alpha/2} < t_\nu < t_{\nu,\alpha/2}) = 1 - \alpha$, where t_ν is a Student's t random variable with ν degrees of freedom

$$P(t_\nu < -t_{\nu,\alpha/2}) = \frac{\alpha}{2}$$

Finally, since probabilities for mutually exclusive and collectively exhaustive events must sum to 1, it follows that

$$P(-t_{\nu,\alpha/2} < t_\nu < t_{\nu,\alpha/2}) = 1 - P(t_\nu > t_{\nu,\alpha/2}) - P(t_\nu < -t_{\nu,\alpha/2})$$

$$= 1 - \frac{\alpha}{2} - \frac{\alpha}{2} = 1 - \alpha$$

These probabilities are shown in Figure 8.8, which corresponds to Figure 8.4 for the standard normal distribution.

8.4 CONFIDENCE INTERVALS FOR THE MEAN OF A NORMAL POPULATION: POPULATION VARIANCE UNKNOWN

We can now use the Student's t distribution to derive confidence intervals for the mean of a normal population when the variance is unknown, using an argument similar to that of Section 8.2.

Assume that a random sample of n observations is available from a normal population with mean μ and variance σ^2, and that confidence intervals for the population mean are required.

Let $\bar{X}$ and s_X^2 denote the sample mean and variance. Then, from Section 8.3, we know that the random variable

$$t_{n-1} = \frac{\bar{X} - \mu}{s_X/\sqrt{n}}$$

follows a Student's t distribution with $(n - 1)$ degrees of freedom.

Suppose that a $100(1 - \alpha)\%$ confidence interval for the population mean is required. Then, following precisely the same line of reasoning used to obtain the confidence intervals of Section 8.2, we have

$$1 - \alpha = P(-t_{n-1,\alpha/2} < t_{n-1} < t_{n-1,\alpha/2})$$

$$= P\left(-t_{n-1,\alpha/2} < \frac{\bar{X} - \mu}{s_X/\sqrt{n}} < t_{n-1,\alpha/2}\right)$$

$$= P\left(\frac{-t_{n-1,\alpha/2}s_X}{\sqrt{n}} < \bar{X} - \mu < \frac{t_{n-1,\alpha/2}s_X}{\sqrt{n}}\right)$$

$$= P\left(\bar{X} - \frac{t_{n-1,\alpha/2}s_X}{\sqrt{n}} < \mu < \bar{X} + \frac{t_{n-1,\alpha/2}s_X}{\sqrt{n}}\right)$$

Therefore, from our definition of confidence intervals, it follows that if $\bar{x}$ and s_x are the specific values observed for the sample mean and standard deviation, then a $100(1 - \alpha)\%$ confidence interval for the population mean is given by

$$\bar{x} - \frac{t_{n-1,\alpha/2}s_x}{\sqrt{n}} < \mu < \bar{x} + \frac{t_{n-1,\alpha/2}s_x}{\sqrt{n}}$$

Confidence Intervals for the Mean of a Normal Population: Population Variance Unknown

Suppose we have a random sample of n observations from a normal distribution with mean μ and unknown variance. If the observed sample mean and standard deviation are, respectively, $\bar{x}$ and s_x, then a $100(1 - \alpha)\%$ confidence interval for the population mean is given by

$$\bar{x} - \frac{t_{n-1,\alpha/2}s_x}{\sqrt{n}} < \mu < \bar{x} + \frac{t_{n-1,\alpha/2}s_x}{\sqrt{n}}$$

where $t_{n-1,\alpha/2}$ is that number for which

$$P(t_{n-1} > t_{n-1,\alpha/2}) = \frac{\alpha}{2}$$

and the random variable t_{n-1} has a Student's t distribution with $(n - 1)$ degrees of freedom.

We now illustrate the use of the Student's t distribution in finding confidence intervals for the mean of a normal population when only a moderate number of sample observations is available.

Example
8.4

A random sample of six cars from a particular model year achieved the following miles per gallon figures:

$$18.6; \quad 18.4; \quad 19.2; \quad 20.8; \quad 19.4; \quad 20.5$$

Find a 90% confidence interval for the population mean miles per gallon for cars of this model year, assuming that the population distribution is normal.

As a first step, we must find the sample mean and variance, which can be conveniently calculated from the accompanying table.

i	x_i	x_i^2
1	18.6	345.96
2	18.4	338.56
3	19.2	368.64
4	20.8	432.64
5	19.4	376.36
6	20.5	420.25
Sums	116.9	2,282.41

The sample mean is then

$$\bar{x} = \frac{1}{n} \sum x_i = \frac{1}{6}(116.9) = 19.48$$

and the sample variance is

$$s_x^2 = \frac{1}{(n-1)}\left(\sum x_i^2 - n\bar{x}^2\right)$$

$$= \frac{1}{5}[2,282.41 - (6)(19.48)^2] = 1.12$$

so that the sample standard deviation is

$$s_x = \sqrt{1.12} = 1.06$$

Given the assumption that the population distribution is normal, we can construct confidence intervals for the population mean as

$$\bar{x} - \frac{t_{n-1,\alpha/2}s_x}{\sqrt{n}} < \mu < \bar{x} + \frac{t_{n-1,\alpha/2}s_x}{\sqrt{n}}$$

where

$$\bar{x} = 19.48; \qquad s_x = 1.06; \qquad n = 6$$

Also, since $(n - 1) = 5$ and a 90% confidence interval is needed, we have $\alpha = 0.1$, so that $\alpha/2 = 0.05$, and

$$t_{n-1,\alpha/2} = t_{5,.05} = 2.015$$

from Table 6 in the Appendix. The 90% confidence interval for the population mean miles per gallon is then

$$19.48 - \frac{(2.015)(1.06)}{\sqrt{6}} < \mu < 19.48 + \frac{(2.015)(1.06)}{\sqrt{6}}$$

or

$$18.61 < \mu < 20.35$$

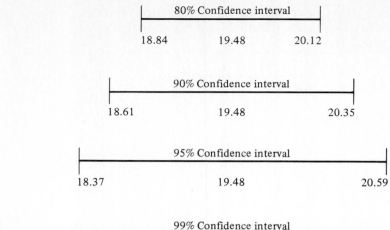

FIGURE 8.9 80%, 90%, 95%, and 99% confidence intervals for population mean miles per gallon from data of Example 8.4

Thus, our 90% confidence interval for the population mean miles per gallon for these cars ranges from 18.61 to 20.35.

Rather than presenting just a single confidence interval, it is sometimes convenient to calculate a sequence of confidence intervals of differing probability contents. Taken together, these allow us to form a clearer picture about likely values of the population parameter being estimated. Figure 8.9 shows also the 80%, 95%, and 99% confidence intervals calculated from the data of Example 8.4. Notice that the intervals increase in width with increasing probability content, as we would expect.

Example 8.5

In a study[5] of the effects of mergers in the motor carrier industry, a random sample of seventeen merged carriers was examined, and their increase in the growth rate of truckload tons of revenue freight in the postmerger period, compared with the premerger period, was measured. For each of these sample carriers, a nonmerged firm with similar location and size characteristics was examined for comparison. The difference between the increased growth rate for the merged firm and the corresponding nonmerged firm was calculated. The sample values had mean 0.105 and standard deviation 0.440. Find a 95% confidence interval for the population mean, assuming the population distribution is normal.

Confidence intervals for the population mean may be found from

$$\bar{x} - \frac{t_{n-1,\alpha/2} s_x}{\sqrt{n}} < \mu < \bar{x} + \frac{t_{n-1,\alpha/2} s_x}{\sqrt{n}}$$

---

[5] This research is described in R. P. Boisjoly and T. M. Corsi, "The economic implications of less-than-truckload motor carrier mergers," *Journal of Economics and Business, 33* (1980), 13–20.

where, in this example,

$$\bar{x} = 0.105; \qquad s_x = 0.440; \qquad n = 17$$

Since $(n - 1) = 16$ and we require a 95% confidence interval, then $\alpha = .05$ and we have

$$t_{n-1,\alpha/2} = t_{16,.025} = 2.120$$

from Table 6 of the Appendix. Thus, the 95% confidence interval for the population mean is

$$0.105 - \frac{(2.120)(0.440)}{\sqrt{17}} < \mu < 0.105 + \frac{(2.120)(0.440)}{\sqrt{17}}$$

or

$$-0.121 < \mu < 0.331$$

The 95% confidence interval for the population mean difference in increased growth rates between merged and nonmerged firms ranges from -0.121 to 0.331. Notice that this interval for the population mean difference includes 0, suggesting that the data do not strongly contradict the conclusion of no difference between merged and nonmerged firms, on the average.

EXERCISES

1. A university admissions officer has determined that, historically, applicants to the business school have a grade point average that is normally distributed with standard deviation 0.40. He takes a random sample of thirty-six of the current year's applications, and obtains a sample mean grade point average of 2.95. Find a 90% confidence interval for the mean grade point average of all business school applicants to the university in the current year.

2. A process producing bricks is known to give an output whose weights are normally distributed with standard deviation 0.16 pound. A random sample of sixteen bricks from a day's output had mean weight 4.24 pounds.
   (a) Find a 95% confidence interval for the mean weight of all bricks produced that day.
   (b) Without doing the calculations, state whether a 99% confidence interval for the population mean would be wider or narrower than that calculated in part (a).

3. A supermarket manager finds from reviewing his records that the dollar purchases of individual customers on Saturdays are normally distributed with standard deviation $18.20. A random sample of nine customers on a particular Saturday gave the following purchase amounts (in dollars)

   19.41; 9.27; 45.23; 70.24; 65.23; 50.47; 43.19; 40.84; 29.53

   (a) Find a 99% confidence interval for the mean dollar amount of customer purchases on that day.
   (b) Without doing the calculations, state whether a 90% confidence interval for the population mean would be wider or narrower than that in part (a).
   (c) For the following Saturday, the manager decides to calculate a 95% confidence interval, based on a random sample of twenty customers. Will that interval be wider or narrower than the one found in part (a)?

4. A process produces cans of baked beans. It is known from past production records that the weights of the contents of these cans each day follow a normal distribution with standard deviation 0.25 ounce. However, due to minor adjustments made to the process prior to the beginning of a production run, the daily mean weights vary somewhat. A random sample of twenty cans is taken from a day's output, and the sample mean weight is found to be 19.91 ounces.

(a) Find a 95% confidence interval for the mean weight of the contents of all cans of baked beans produced that day.

(b) In fact, the adjustments made to the production process on this particular day had the effect of increasing the population standard deviation to 0.30 ounce. Would a properly calculated 95% confidence interval for the population mean be wider or narrower than that found in part (a)?

5. It was found that a 99% confidence interval for the population mean level of satisfaction of students with their automobiles was given by

$$75.7 < \mu < 82.5$$

Would it therefore be proper to say that the probability is 0.99 that the population mean satisfaction level is between 75.7 and 82.5? If not, replace this statement by a valid probability statement.

6. A random sample of 100 restaurants in a large city showed a sample mean number of employees of 12.28, and a sample standard deviation of 3.68. Find a 99% confidence interval for the mean number of employees in all restaurants in that city.

7. A political scientist is interested in the amount of time voters spend in the voting booth. A random sample of sixty-four voters showed an average of 8.24 minutes in the booth, and a standard deviation of 2.14 minutes. Find an 80% confidence interval for the mean amount of time spent by all voters in the voting booth.

8. A publisher contracts for the first time with a printer to produce several books. An editor is concerned about the number of errors that might be made by this printer. She chooses at random 100 pages from the books when they are returned, and proofreads them carefully. For this sample, the number of errors per page had an average of 1.24 and a standard deviation of 0.60. Find 90% and 95% confidence intervals for the population mean number of errors per page for this printer.

9. A large university is concerned about the amount of time its students spend studying each week. A random sample of sixteen students had mean weekly study time 18.36 hours and a sample standard deviation of 3.42 hours. Assume that study times are normally distributed.

(a) Find a 90% confidence interval for the average amount of study time per week for all students at this university.

(b) State, without doing the calculations, whether the interval would be wider or narrower under each of the following conditions:

(i) The sample contained thirty students, with everything else the same as in part (a).

(ii) The sample standard deviation was 4.15 hours, with everything else the same as in part (a).

(iii) An 80% confidence interval was required, with everything else the same as in part (a).

10. A clinic offers a weight reduction program. A review of its records shows that, for a random sample of nine patients, the mean weight loss after following this program was 18.15 pounds, and the sample standard deviation was 7.85 pounds. Find a 95% confidence interval for the mean weight loss of all the clinic's patients, assuming the population distribution is normal.

11. A homeowner, intending to put her house on the market, obtained valuations from four independently selected real estate appraisers. These valuations (in dollars) were as follows:

$$90,250; \quad 86,200; \quad 91,400; \quad 88,750$$

Assuming the population distribution is normal, find an 80% confidence interval for the mean valuation of all appraisers for this home.

12. A random sample of four students is chosen to estimate the mean number of miles walked in a week by a student. The numbers of miles walked in the week by these four students were as follows:

$$18; \quad 20; \quad 16; \quad 24$$

Find a 98% confidence interval for the population mean number of miles walked by students in a week. What assumption, if any, have you made in calculating this interval estimate?

13. A random sample of twenty-five cars from a particular model was checked to measure their depreciation in value during the first year. The sample mean was $1,900 and the sample standard deviation was $700. Assuming the population distribution is normal, find an 80% confidence interval for the average depreciation of all cars of this model during the first year.

14. The number of mushrooms per pizza on a large Flying Cucumber Pizza can be assumed to have a normal distribution. For a random sample of twenty such pizzas, the following sample statistics were found:

$$\sum x_i = 510; \qquad \sum (x_i - \bar{x})^2 = 179$$

Find a 90% confidence interval for the population mean number of mushrooms per pizza.

15. A chief of police is concerned about the speed of cars traveling over a stretch of highway that has experienced a heavy accident rate. A random sample of twenty-five automobiles showed an average speed of 61.4 miles per hour on this stretch, and a standard deviation of 6.8 miles per hour. Find 90% and 99% confidence intervals for the mean speed of all cars traveling on this stretch of the highway.

16. A car rental company is interested in the amount of time its vehicles are out of operation for repair work. A random sample of nine cars showed that, over the past year, the numbers of days each had been inoperative were as follows:

$$15; \quad 11; \quad 19; \quad 24; \quad 6; \quad 18; \quad 20; \quad 15; \quad 18$$

Assuming that the population distribution is normal, find a 95% confidence interval for the mean number of days (in a year) that all vehicles in the company's fleet are out of operation.

17. A new lightbulb is advertised as being "long-lasting." A random sample of twenty of these lightbulbs was checked. The sample mean lifetime was 4,120 hours, and the sample standard deviation was 345 hours. Find 80% and 90% confidence intervals for the mean lifetime of these "long-lasting" lightbulbs, assuming a normal population distribution.

8.5 CONFIDENCE INTERVALS FOR THE POPULATION PROPORTION (LARGE SAMPLES)

Suppose now that we are interested in the proportion of population members possessing some specific attribute. For example, we might want to estimate the proportion of all adult Americans in favor of handgun control legislation. If a random sample is taken from the population, a natural point estimator of the population proportion is provided by the sample proportion. In this section we derive confidence intervals for the population proportion.

Using the binomial set-up, we let $\hat{p}_X$ denote the proportion of "successes" in n independent trials, each with probability of success p. In Section 6.3, we saw that if the number n of sample members is large, then the random variable

$$Z = \frac{\hat{p}_X - p}{\sqrt{p(1 - p)/n}} \qquad (8.5.1)$$

has, to a close approximation, a standard normal distribution. Unfortunately, this result is not quite sufficient to allow us to find confidence intervals for the population proportion, as the denominator of equation 8.5.1 involves the unknown p. However, if the sample size is large, we can obtain a good approximation if we replace p by its point estimator $\hat{p}_X$ in this denominator; that is,

$$\sqrt{\frac{p(1 - p)}{n}} \approx \sqrt{\frac{\hat{p}_X(1 - \hat{p}_X)}{n}}$$

Hence, for large sample sizes, the distribution of the random variable

$$Z = \frac{\hat{p}_X - p}{\sqrt{\hat{p}_X(1 - \hat{p}_X)/n}}$$

is approximately standard normal. This result can then be used to obtain confidence intervals for the population proportion.

As before, we define $z_{\alpha/2}$ as that number for which

$$P(Z > z_{\alpha/2}) = \frac{\alpha}{2}$$

where the random variable Z follows a standard normal distribution. Then

$$1 - \alpha = P(-z_{\alpha/2} < Z < z_{\alpha/2})$$

$$= P\left(-z_{\alpha/2} < \frac{\hat{p}_X - p}{\sqrt{\hat{p}_X(1 - \hat{p}_X)/n}} < z_{\alpha/2}\right)$$

$$= P\left(-z_{\alpha/2}\sqrt{\frac{\hat{p}_X(1 - \hat{p}_X)}{n}} < \hat{p}_X - p < z_{\alpha/2}\sqrt{\frac{\hat{p}_X(1 - \hat{p}_X)}{n}}\right)$$

$$= P\left(\hat{p}_X - z_{\alpha/2}\sqrt{\frac{\hat{p}_X(1 - \hat{p}_X)}{n}} < p < \hat{p}_X + z_{\alpha/2}\sqrt{\frac{\hat{p}_X(1 - \hat{p}_X)}{n}}\right)$$

It therefore follows that, if the observed sample proportion is $\hat{p}_x$, an approximate $100(1 - \alpha)\%$ confidence interval for the population proportion is

$$\hat{p}_x - z_{\alpha/2}\sqrt{\frac{\hat{p}_x(1 - \hat{p}_x)}{n}} < p < \hat{p}_x + z_{\alpha/2}\sqrt{\frac{\hat{p}_x(1 - \hat{p}_x)}{n}}$$

> ## Confidence Intervals for the Population Proportion (Large Samples)
>
> Let $\hat{p}_x$ denote the observed proportion of "successes" in a random sample of n observations from a population with a proportion p of successes. Then, if n is large, a $100(1 - \alpha)\%$ confidence interval for the population proportion is given by[6]
>
> $$\hat{p}_x - z_{\alpha/2} \sqrt{\frac{\hat{p}_x(1 - \hat{p}_x)}{n}} < p < \hat{p}_x + z_{\alpha/2} \sqrt{\frac{\hat{p}_x(1 - \hat{p}_x)}{n}}$$
>
> where $z_{\alpha/2}$ is that number for which
>
> $$P(Z > z_{\alpha/2}) = \frac{\alpha}{2}$$
>
> and the random variable Z has a standard normal distribution.

Confidence intervals for the population proportion are centered on the sample proportion. Also, it can be seen that (all other things equal) the larger the sample size n, the narrower the confidence interval. This reflects the increasing precision of the information about the population proportion obtained as the sample size becomes larger.

Example 8.6

In a study[7] of attitudes toward shoplifting and devices for its prevention, a random sample of 403 shopping center patrons were questioned. Of this sample, 24.0% indicated that they were both aware of, and uncomfortable with, the use of television cameras as a shoplifting prevention device. Find a 90% confidence interval for the population proportion with this attitude.

If p denotes the true population proportion, and $\hat{p}_x$ the sample proportion, then confidence intervals for the population proportion are given by

$$\hat{p}_x - z_{\alpha/2} \sqrt{\frac{\hat{p}_x(1 - \hat{p}_x)}{n}} < p < \hat{p}_x + z_{\alpha/2} \sqrt{\frac{\hat{p}_x(1 - \hat{p}_x)}{n}}$$

where, for a 90% confidence interval, $\alpha = 0.10$, so that

$$\frac{\alpha}{2} = .05 \quad \text{and} \quad z_{\alpha/2} = z_{.05} = 1.645$$

from Table 3 of the Appendix. We then have

$$\hat{p}_x = 0.240; \quad n = 403; \quad z_{\alpha/2} = 1.645$$

so that a 90% confidence interval for the population proportion is

---

[6] Confidence intervals formed in this way are generally quite reliable when based on samples of $n = 40$ or more observations.

[7] H. J. Guffey, J. R. Harris, and J. F. Laumer, "Shopper attitudes toward shoplifting and shoplifting prevention devices," *Journal of Retailing, 55* (1979), No. 3, 75–89.

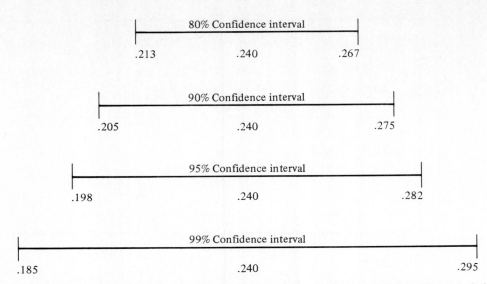

FIGURE 8.10 80%, 90%, 95%, and 99% confidence intervals based on data from Example 8.6

$$0.240 - (1.645) \sqrt{\frac{(0.240)(0.760)}{403}} < p < 0.240 + (1.645) \sqrt{\frac{(0.240)(0.760)}{403}}$$

or

$$0.205 < p < 0.275$$

Hence, the interval ranges from 20.5% to 27.5% of the population. Figure 8.10 also shows the 80%, 95%, and 99% confidence intervals for the population proportion. Note again that as the probability content increases, so does the width of the confidence interval.

Example
8.7

For a random sample[8] of 89 new product market failures, managers associated with the products assessed "Price set higher than customers would pay" as the main cause of failure in 18.2% of these cases. Find a 95% confidence interval for the proportion of all market failures for which high price is held to be the main cause.

For a 95% confidence interval, $\alpha = .05$, so that $z_{\alpha/2} = z_{.025} = 1.96$ from Table 3 of the Appendix. Thus, we have

$$\hat{p}_x = 0.182; \qquad n = 89; \qquad z_{\alpha/2} = 1.96$$

Substitution of these values into the general formula for a 95% confidence interval yields

$$0.182 - (1.96) \sqrt{\frac{(0.182)(0.818)}{89}} < p < 0.182 + (1.96) \sqrt{\frac{(0.182)(0.818)}{89}}$$

---

[8] This example is taken from R. J. Calantine and R. G. Cooper, "A discriminant model for identifying scenarios of industrial new product failure," *Journal of the Academy of Marketing Science, 7* (1979), 163–83.

or

$$0.102 < p < 0.262$$

Hence, a 95% confidence interval for the proportion of all market failures for which high price is held to be the main cause ranges from 0.102 to 0.262.

The reader may find this interval disturbingly wide, mirroring the lack of precision in our information about the true proportion of market failures primarily attributable to high price. Of course, an obvious way to achieve narrower interval estimates is to take a larger sample. In Section 8.9 we will see how to determine the number of sample observations necessary to ensure confidence intervals of any desired width.

8.6 CONFIDENCE INTERVALS FOR THE VARIANCE OF A NORMAL POPULATION

In Sections 8.2 and 8.4 we saw how to obtain interval estimates for the population mean. On occasion, interval estimates are also required for the variance of a population. As might be expected, such estimates are based on the sample variance.

Suppose we have a sample of n observations from a normal population with variance σ^2, and that the sample variance is denoted s_X^2. In Section 6.4, we saw that the random variable

$$\chi_{n-1}^2 = \frac{(n-1)s_X^2}{\sigma^2}$$

follows a chi-square distribution with $(n-1)$ degrees of freedom. This result forms the basis for the derivation of confidence intervals for the population variance when sampling from a normal distribution.

In order to develop the formula for calculating confidence intervals for the variance, an additional notation is needed, as described in the box and illustrated in Figure 8.11.

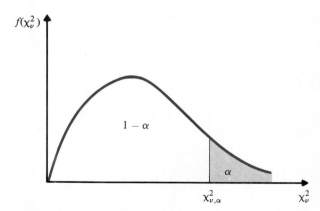

FIGURE 8.11 $P(\chi_\nu^2 > \chi_{\nu,\alpha}^2) = \alpha$, where χ_ν^2 is a chi-square random variable with ν degrees of freedom

A random variable having the chi-square distribution with v degrees of freedom will be denoted χ_v^2. We define as $\chi_{v,\alpha}^2$ that number for which

$$P(\chi_v^2 > \chi_{v,\alpha}^2) = \alpha$$

For a specified probability α, we will need to find the corresponding number $\chi_{v,\alpha}^2$. This can be achieved from the values of the cumulative distribution function of the chi-square random variable given in Table 5 in the Appendix. For instance, suppose we need the number that is exceeded with probability 0.05 by a chi-square random variable with 6 degrees of freedom; that is,

$$P(\chi_6^2 > \chi_{6,.05}^2) = .05$$

then, we have from Table 5

$$\chi_{6,.05}^2 = 12.59$$

Using the notation just defined, we can write

$$P(\chi_v^2 > \chi_{v,\alpha/2}^2) = \frac{\alpha}{2}$$

Similarly, we define $\chi_{v,1-\alpha/2}^2$ so that

$$P(\chi_v^2 > \chi_{v,1-\alpha/2}^2) = 1 - \frac{\alpha}{2}$$

and hence,

$$P(\chi_v^2 < \chi_{v,1-\alpha/2}^2) = \frac{\alpha}{2}$$

It therefore follows that

$$P(\chi_{v,1-\alpha/2}^2 < \chi_v^2 < \chi_{v,\alpha/2}^2) = 1 - \frac{\alpha}{2} - \frac{\alpha}{2} = 1 - \alpha$$

This probability is illustrated in Figure 8.12.

To illustrate, suppose we want to find a pair of numbers such that the probability that a chi-square random variable with 6 degrees of freedom lies between these numbers is 0.9. Then,

$$1 - \alpha = 0.9 \qquad \text{and} \qquad \alpha = 0.1$$

so that

$$P(\chi_{6,.95}^2 < \chi_6^2 < \chi_{6,.05}^2) = 0.9$$

We determined previously that $\chi_{6,.05}^2 = 12.59$. From Table 5 of the Appendix we see that

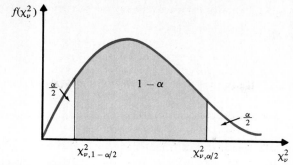

FIGURE 8.12 $P(\chi^2_{\nu,1-\alpha/2} < \chi^2_\nu < \chi^2_{\nu,\alpha/2}) = 1 - \alpha$, where χ^2_ν is a chi-square random variable with ν degrees of freedom

$$\chi^2_{6,.95} = 1.64$$

Thus, the probability is 0.9 that this chi-square random variable lies between 1.64 and 12.59.

To find confidence intervals for the population variance, we have for the chi-square distribution with $(n - 1)$ degrees of freedom

$$1 - \alpha = P(\chi^2_{n-1,1-\alpha/2} < \chi^2_{n-1} < \chi^2_{n-1,\alpha/2})$$

$$= P\left(\chi^2_{n-1,1-\alpha/2} < \frac{(n-1)s_X^2}{\sigma^2} < \chi^2_{n-1,\alpha/2}\right)$$

$$= P\left[\frac{(n-1)s_X^2}{\chi^2_{n-1,\alpha/2}} < \sigma^2 < \frac{(n-1)s_X^2}{\chi^2_{n-1,1-\alpha/2}}\right]$$

Therefore, if s_x^2 is the specific value observed for the sample variance, it follows that a $100(1 - \alpha)\%$ confidence interval for the population variance is given by

$$\frac{(n-1)s_x^2}{\chi^2_{n-1,\alpha/2}} < \sigma^2 < \frac{(n-1)s_x^2}{\chi^2_{n-1,1-\alpha/2}}$$

Confidence Intervals for the Variance of a Normal Population

Suppose we have a random sample of n observations from a normal population with variance σ^2. If the observed sample variance is s_x^2, then a $100(1 - \alpha)\%$ confidence interval for the population variance is given by

$$\frac{(n-1)s_x^2}{\chi^2_{n-1,\alpha/2}} < \sigma^2 < \frac{(n-1)s_x^2}{\chi^2_{n-1,1-\alpha/2}}$$

where $\chi^2_{n-1,\alpha/2}$ is that number for which

$$P(\chi^2_{n-1} > \chi^2_{n-1,\alpha/2}) = \frac{\alpha}{2}$$

---

and $\chi^2_{n-1, 1-\alpha/2}$ is that number for which

$$P(\chi^2_{n-1} < \chi^2_{n-1, 1-\alpha/2}) = \frac{\alpha}{2}$$

and the random variable χ^2_{n-1} obeys a chi-square distribution with $(n - 1)$ degrees of freedom.

Example 8.8

A random sample of fifteen pills for headache relief showed a standard deviation of 0.8% in the concentration of the active ingredient. Find a 90% confidence interval for the population variance for these pills.

We have

$$n = 15; \qquad s_x^2 = (0.8)^2 = 0.64$$

Since a 90% confidence interval is wanted, we have $\alpha = 0.1$, and from Table 5 of the Appendix,

$$\chi^2_{n-1, \alpha/2} = \chi^2_{14, .05} = 23.68 \qquad \text{and} \qquad \chi^2_{n-1, 1-\alpha/2} = \chi^2_{14, .95} = 6.57$$

The 90% confidence interval for the population variance is given by

$$\frac{(n - 1)s_x^2}{\chi^2_{n-1, \alpha/2}} < \sigma^2 < \frac{(n - 1)s_x^2}{\chi^2_{n-1, 1-\alpha/2}}$$

Substitution yields

$$\frac{(14)(0.64)}{23.68} < \sigma^2 < \frac{(14)(0.64)}{6.57}$$

or

$$0.378 < \sigma^2 < 1.364$$

Our 90% confidence interval for the population variance in concentration of the active ingredient therefore ranges from 0.378 to 1.364.

Since the standard deviation is the square root of the variance, we can take square roots to obtain a 90% confidence interval for the population standard deviation. We have, then, for this interval

$$0.61 < \sigma < 1.17$$

Hence, our 90% confidence interval for the population standard deviation in the percentage concentration of the active ingredient in these pills runs from 0.61% to 1.17%.

We conclude this section with the warning that it is dangerous to follow the above procedure when the population distribution is not normal. The validity of the interval estimator for the variance depends far more critically on the assumption of normality than does that of the interval estimator for the population mean, developed in Section 8.4.

EXERCISES

18. In a commercial shown during half time of the telecast of the 1981 Super Bowl game, 100 "loyal Michelob drinkers" were asked to compare the taste of Michelob and Schlitz beers. Fifty of them indicated a preference for Schlitz. Assuming that these 100 people were a random sample of all "loyal Michelob drinkers," find a 90% confidence interval for the proportion of this population preferring Schlitz.

19. The campaign organization of a candidate for the presidency must decide in which states most effort should be concentrated during the primaries. It is important not to spend too much time in those states in which the chance of eventual success is very slim. A random sample of 900 Democrats in a state indicated an initial support level of 28.4% for a candidate. Find a 95% confidence interval for the proportion of all Democrats in the state supporting the candidate at this stage.

20. In a study[9] designed to assess consumers' views of new product warranties, 64.9% of a random sample of 237 consumers disagreed with the statement: "Products today are built to last a long time before needing service or repair." Find a 99% confidence interval for the population proportion of all consumers disagreeing with this statement.

21. In the study described in Exercise 20, 72.6% of the sample members agreed with the statement: "Product guarantees seem to be worded more for lawyers to understand than to be easily understood by customers."

   (a) Find a 90% confidence interval for the population proportion of all consumers agreeing with this statement.

   (b) Would a 95% confidence interval for the population proportion be wider or narrower than that found in part (a)?

22. The president of a state Medical Librarians' Association seeks information about the makeup of the national organization. She takes a random sample of 150 members. From this sample she constructed, for the population proportion of women members, the confidence interval

$$0.73 < p < 0.87$$

Find the probability content of this interval.

23. A random sample of 400 stockholders of a large corporation contained 120 stockholders who believed that the corporation should curtail its investments in South Africa. On the basis of this sample information, an analyst calculated the confidence interval

$$0.25 < p < 0.35$$

for the population proportion of stockholders favoring curtailment in South Africa. What is the probability content of this interval?

24. A random sample[10] of 202 vice presidents of marketing in corporations from the manufacturing sectors were given a list of eleven marketing areas, and asked to nominate the area they believed to be most important. "Pricing" was selected by 16.1% of the sample members. Find 80% and 90% confidence intervals for the population proportion of all vice presidents of marketing in the manufacturing sectors viewing pricing as the most important marketing area.

25. Using the data of Exercise 10, find a 90% confidence interval for the variance in weight loss of all the participants in the clinic's program.

---

[9] See R. E. Wilkes and J. B. Wilcox, "Consumer perceptions of product warranties and their implications for retail strategy," *Journal of Business Research, 4* (1976), 35–43.

[10] These results are reported by S. Samiee, "Elements of marketing strategy: How important are they from the executive viewpoint?", *Journal of the Academy of Marketing Science, 8* (1980), 40–50.

26. The confidence intervals for the population variance described in this section are *not* centered on the sample variance. Explain graphically why this is so.

27. Given the same sample information, it is possible (in principle) to obtain narrower confidence intervals, of the same probability content, for the population variance than those found in this section. Explain graphically why this is so.

28. Using the data of Exercise 11, find an 80% confidence interval for the population variance of real estate appraisers' valuations of this home.

29. A psychologist wants to estimate the variance of employee test scores. A random sample of eighteen scores had sample standard deviation 10.4. Find a 90% confidence interval for the population variance. What assumption, if any, have you made in calculating this interval estimate?

30. A manufacturer is concerned about the variability of the levels of impurity contained in consignments of raw material from a supplier. A random sample of fifteen consignments showed a standard deviation of 2.36% in the concentration of impurity levels. Assume a normal population distribution.

    (a) Find a 95% confidence interval for the population variance.

    (b) Would a 99% confidence interval for this variance be wider or narrower than that found in part (a)?

31. Using the data of Exercise 15, find a 90% confidence interval for the variance of the speeds of all automobiles travelling on this stretch of highway.

32. A manufacturer bonds a plastic coating to a metal surface. A random sample of nine observations on the thicknesses of this coating is taken from a week's output. The sample thicknesses (in millimeters) were as follows:

$$20.5; \quad 21.2; \quad 18.6; \quad 20.4; \quad 21.6; \quad 19.8; \quad 19.9; \quad 20.3; \quad 20.8$$

Assuming that the population distribution is normal, find a 90% confidence interval for the population variance.

33. Using the data of Exercise 17, find an 80% confidence interval for the population variance of lifetimes for these lightbulbs.

8.7 CONFIDENCE INTERVALS FOR THE DIFFERENCE BETWEEN THE MEANS OF TWO NORMAL POPULATIONS

An important problem in statistical inference deals with the comparison of two population means. As one illustration, a company might receive shipments of a chemical from two suppliers and be concerned about the difference between the mean levels of impurity present in the chemicals from the two sources of supply. As another example, a farmer may consider the use of two alternative fertilizers, his interest being in the difference between the resulting mean crop yields per acre.

In order to compare population means, a random sample is drawn from the two populations, and inference about the difference between population means is based on the sample results. The appropriate method for analyzing this information depends on the procedure used in selecting the samples. We will consider the following two very common sampling schemes:

(a) MATCHED PAIRS ☐ In this scheme, the sample members are chosen in pairs, one from each population. The idea is that, apart from the factor under study, the members of these pairs should resemble one another as closely as possible, so that the comparison of interest can be directly made. For instance, suppose we want to measure the effectiveness of a speed-reading course. One possible approach would be to record the number of words per minute read by a sample of students before taking the course, and compare with results *for the same students* after completing the course. In this case each pair of observations consists of "before" and "after" measurements on a single student.

(b) INDEPENDENT SAMPLES ☐ In this scheme, samples are drawn independently from the two populations of interest, so that the membership of one sample is not influenced by that of another. In the example of the company that receives shipments of chemical from two suppliers, we might choose independent random samples of batches from each supplier and measure the impurity levels of each batch sampled.

Whichever sampling method is used, our objective in this section is to obtain confidence intervals for the difference between the two population means.

CONFIDENCE INTERVALS BASED ON MATCHED PAIRS

Suppose, in general, that we obtain a random sample of n matched pairs of observations, denoted $(x_1, y_1), (x_2, y_2), \ldots, (x_n, y_n)$, from populations with means μ_X and μ_Y. Thus, $x_1, x_2, \ldots, x_n$ denote the observations from the population with mean μ_X, and $y_1, y_2, \ldots, y_n$ the matched sampled values from the population with mean μ_Y. Table 8.2 shows miles per gallon figures obtained for a random sample of eight cars from each of two different models. The sample cars were paired, and each member of a particular pair was driven over the same route by the same driver, so that variability between drivers and routes could be eliminated from the comparisons. The table also shows the differences, d_i, between these miles per gallon figures. These differences represent a random sample from a population whose mean is $(\mu_X - \mu_Y)$, the difference between the population means for the two car models.

From the information in the table, the sample mean and variance of the differences in achieved miles per gallon can be calculated. We have, for the mean,

$$\bar{d} = \frac{1}{n} \sum_{i=1}^{n} d_i = \frac{1}{8}(6.2) = 0.775$$

and, for the variance,

TABLE 8.2 Miles per gallon achieved by matched pairs of cars

	X-CARS	Y-CARS	DIFFERENCES	
i	x_i	y_i	d_i	d_i^2
1	19.4	19.6	−0.2	0.04
2	18.8	17.5	1.3	1.69
3	20.6	18.4	2.2	4.84
4	17.6	17.5	0.1	0.01
5	19.2	18.0	1.2	1.44
6	20.9	20.0	0.9	0.81
7	18.3	18.8	−0.5	0.25
8	20.4	19.2	1.2	1.44
			Sums 6.2	10.52

$$s_d^2 = \frac{1}{n-1}\left(\sum_{i=1}^{n} d_i^2 - n\bar{d}^2\right)$$

$$= \frac{1}{7}[10.52 - (8)(0.775)^2] = 0.816$$

so that the observed sample standard deviation is

$$s_d = \sqrt{0.816} = 0.903$$

We are now in the position of requiring a confidence interval for a population mean $(\mu_X - \mu_Y)$, given a random sample (the values of the differences d_i) from that population. If the population distribution is assumed to be normal, the procedure developed in Section 8.4 is immediately applicable because the differences in matched pairs constitute a random sample from a population whose mean is the quantity we are trying to estimate.

Confidence Intervals for Difference Between Means: Matched Pairs

Suppose we have a random sample of n matched pairs of observations from distributions with means μ_X and μ_Y. Let $\bar{d}$ and s_d denote the observed sample mean and standard deviation for the n differences $d_i = x_i - y_i$. If the population distribution of the differences is assumed to be normal, then a $100(1-\alpha)\%$ confidence interval for $(\mu_X - \mu_Y)$ is given by

$$\bar{d} - \frac{t_{n-1,\alpha/2}s_d}{\sqrt{n}} < \mu_X - \mu_Y < \bar{d} + \frac{t_{n-1,\alpha/2}s_d}{\sqrt{n}}$$

where $t_{n-1,\alpha/2}$ is that number for which

$$P(t_{n-1} > t_{n-1,\alpha/2}) = \frac{\alpha}{2}$$

and the random variable t_{n-1} has a Student's t distribution with $(n-1)$ degrees of freedom.

In the automobile mileage example, we have found

$$\bar{d} = 0.775; \qquad s_d = 0.903; \qquad n = 8$$

For a 99% confidence interval, $\alpha = 0.01$, so that

$$t_{n-1,\alpha/2} = t_{7,.005} = 3.499$$

from Table 6 in the Appendix. Hence, we substitute to obtain the 99% confidence interval for the difference between the population means:

$$\bar{d} - \frac{t_{n-1,\alpha/2}s_d}{\sqrt{n}} < \mu_X - \mu_Y < \bar{d} + \frac{t_{n-1,\alpha/2}s_d}{\sqrt{n}}$$

that is,

$$0.775 - \frac{(3.499)(0.903)}{\sqrt{8}} < \mu_X - \mu_Y < 0.775 + \frac{(3.499)(0.903)}{\sqrt{8}}$$

or

$$-0.342 < \mu_X - \mu_Y < 1.892$$

We therefore find that, based on the data of Table 8.2, a 99% confidence interval for the difference in population mean miles per gallon for these two types of automobile ranges from -0.342 to 1.892. Since the interval includes 0, the sample evidence against the conjecture that the population means are the same is not very strong.

CONFIDENCE INTERVALS BASED ON INDEPENDENT SAMPLES

We now consider the case where independent samples, not necessarily of equal size, are taken from the two populations of interest. Suppose we have a random sample of n_x observations from a population with mean μ_X and variance σ_X^2, and an independent random sample of n_y observations from a population with mean μ_Y and variance σ_Y^2. Let the respective sample means be $\bar{X}$ and $\bar{Y}$.

As a first step, we examine the situation where the two population distributions are normal with known variances. Since the object of interest is the difference between the two population means, it is natural to base inference on the difference between the corresponding sample means. This random variable has mean

$$E(\bar{X} - \bar{Y}) = E(\bar{X}) - E(\bar{Y}) = \mu_X - \mu_Y$$

and, since the samples are independent, variance

$$\text{Var}(\bar{X} - \bar{Y}) = \text{Var}(\bar{X}) + \text{Var}(\bar{Y}) = \frac{\sigma_X^2}{n_x} + \frac{\sigma_Y^2}{n_y}$$

Furthermore, it can be shown that its distribution is normal. It therefore follows that the random variable

$$Z = \frac{(\bar{X} - \bar{Y}) - (\mu_X - \mu_Y)}{\sqrt{\dfrac{\sigma_X^2}{n_x} + \dfrac{\sigma_Y^2}{n_y}}}$$

has a standard normal distribution. An argument parallel to that of Section 8.2 can then be used to obtain confidence intervals for the difference between the population means. Since this interval requires knowledge of the true population variances, it is rarely of much direct use. However, as was the case in Section 8.2, its range of applicability is greatly extended when the sample sizes are large, as indicated in the box.

Confidence Intervals for Difference Between Means: Independent Samples (Known Variances or Large Sample Sizes)

Suppose we have independent random samples of n_x and n_y observations from normal distributions with means μ_X and μ_Y and variances σ_X^2 and σ_Y^2. If the observed sample means are $\bar{x}$ and $\bar{y}$, then a $100(1 - \alpha)\%$ confidence interval for $(\mu_X - \mu_Y)$ is given by

$$(\bar{x} - \bar{y}) - z_{\alpha/2}\sqrt{\frac{\sigma_X^2}{n_x} + \frac{\sigma_Y^2}{n_y}} < \mu_X - \mu_Y < (\bar{x} - \bar{y}) + z_{\alpha/2}\sqrt{\frac{\sigma_X^2}{n_x} + \frac{\sigma_Y^2}{n_y}}$$

where $z_{\alpha/2}$ is that number for which

$$P(Z > z_{\alpha/2}) = \frac{\alpha}{2}$$

and the random variable Z has a standard normal distribution.

If the sample sizes n_x and n_y are large,[11] then to a good approximation, a $100(1 - \alpha)\%$ confidence interval for $(\mu_X - \mu_Y)$ is obtained by replacing the population variances in the previous expression by the corresponding observed sample variances s_x^2 and s_y^2. For large sample sizes, this approximation will typically remain adequate even if the population distributions are not normal.

Example 8.9

A professor,[12] teaching two sections of an introductory financial accounting course, organized quizzes differently in the two sections. In one section quizzes, based on pre-assigned reading, were given on the first day of discussion of a topic. In the other section quizzes were given after each chapter was completed. A common final examination was set for the two sections, which each contained 40 students. For the first group, the mean score was 143.7, and the standard deviation was 21.2, while for the second the mean and standard deviation were 131.7 and 20.9. If we can regard these students as independent random samples from the populations of all students who might be exposed to these two approaches to quizzes, find a 99% confidence interval for the difference between the population mean scores.

For the pre-quiz group we have

$$\bar{x} = 143.7; \qquad n_x = 40; \qquad s_x^2 = (21.2)^2 = 449.44$$

and for the post-quiz group

$$\bar{y} = 131.7; \qquad n_y = 40; \qquad s_y^2 = (20.9)^2 = 436.81$$

Since the sample sizes are quite large, we can use the sample variances in place of the population variances, in the formula given above, to find confidence intervals

---

[11] Thirty observations in each sample is generally adequate for this approximation.

[12] See B. A. Baldwin, "On positioning the quiz: an empirical analysis," *Accounting Review, 55* (1980), 666–71.

for the difference between the two population means. These intervals then take the form

$$(\bar{x} - \bar{y}) - z_{\alpha/2}\sqrt{s_x^2/n_x + s_y^2/n_y} < \mu_X - \mu_Y < (\bar{x} - \bar{y}) + z_{\alpha/2}\sqrt{s_x^2/n_x + s_y^2/n_y}$$

where, for a 99% interval,

$$z_{\alpha/2} = z_{.005} = 2.575$$

The interval needed is then

$$(143.7 - 131.7) - (2.575)\sqrt{449.44/40 + 436.81/40} < \mu_X - \mu_Y$$
$$< (143.7 - 131.7) + (2.575)\sqrt{449.44/40 + 436.81/40}$$

which is

$$-0.1 < \mu_X - \mu_Y < 24.1$$

This 99% confidence interval for the difference in population mean scores just includes zero, suggesting that the evidence in the data against the conjecture that the two population means are the same is not overwhelmingly strong.

Example 8.10

Researchers interested in verbal responses to survey questions have found that not only the responses themselves, but also subjects' judgment time in responding to the question, contain useful information. Typically the data analyzed are the reciprocals of judgment time, and are called the *certainty values*. Thus, the larger the time taken to produce a response, the smaller the certainty value. A random sample[13] of 143 people were contacted by telephone and asked: "Assuming that it's convenient for you, do you expect to get a swine flu shot?" One hundred of these subjects answered "yes." Their mean certainty value was 0.76, and the standard deviation was 0.50. For the 43 subjects answering "no," the mean certainty value was 1.41 and the standard deviation was 0.72. Denoting by μ_X the population mean of those answering "yes" and by μ_Y the population mean for the "no" group, find a 95% confidence interval for $(\mu_X - \mu_Y)$.

Again, since the sample sizes are large, we can use the sample variances in place of the population variances and obtain intervals from

$$(\bar{x} - \bar{y}) - z_{\alpha/2}\sqrt{\frac{s_x^2}{n_x} + \frac{s_y^2}{n_y}} < \mu_X - \mu_Y < (\bar{x} - \bar{y}) + z_{\alpha/2}\sqrt{\frac{s_x^2}{n_x} + \frac{s_y^2}{n_y}}$$

where

$$\bar{x} = 0.76; \qquad n_x = 100; \qquad s_x = 0.50$$
$$\bar{y} = 1.41; \qquad n_y = 43; \qquad s_y = 0.72$$

and for a 95% confidence interval,

$$z_{\alpha/2} = z_{.025} = 1.96$$

---

[13] These results are given by P. A. LaBarbera and J. M. MacLachlan, "Response latency in telephone interviews," *Journal of Advertising Research, 19* (1979), No. 3, 49–55.

The interval is then

$$(0.76 - 1.41) - (1.96)\sqrt{\frac{(0.50)^2}{100} + \frac{(0.72)^2}{43}}$$

$$< \mu_X - \mu_Y < (0.76 - 1.41) + (1.96)\sqrt{\frac{(0.50)^2}{100} + \frac{(0.72)^2}{43}}$$

or

$$-0.89 < \mu_X - \mu_Y < -0.41$$

This interval includes only negative values, indicating that those who say they will get a shot are less certain of their answers, on the average, than those who say they will not, in the population at large. Figure 8.13 shows this confidence interval, together with 80%, 90%, and 99% confidence intervals for the difference in the population means.

We now have to consider the case where the sample sizes are not large, and a confidence interval is needed for the difference between the means of two normal populations based on independent random samples from the two populations. In fact, when the population variances are unknown, there is considerable difficulty in attacking this general problem. However, in one special case, where it can be assumed

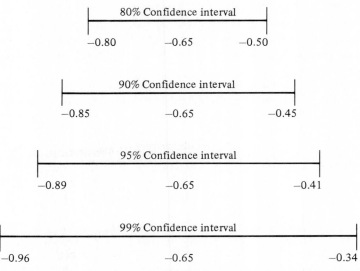

FIGURE 8.13 80%, 90%, 95%, and 99% confidence intervals for difference in population means based on data of Example 8.10

that the two population variances are equal,[14] a fairly straightforward method is available.

Suppose again that we have independent random samples of n_x and n_y observations from normal populations with means μ_X and μ_Y, and that the populations have a common (unknown) variance σ^2. Inference about the population means is, as before, based on the difference $(\bar{X} - \bar{Y})$ between the two sample means. This random variable has a normal distribution with mean $(\mu_X - \mu_Y)$ and variance

$$\text{Var}(\bar{X} - \bar{Y}) = \text{Var}(\bar{X}) + \text{Var}(\bar{Y})$$

$$= \frac{\sigma^2}{n_x} + \frac{\sigma^2}{n_y}$$

$$= \sigma^2\left(\frac{1}{n_x} + \frac{1}{n_y}\right)$$

$$= \sigma^2\left(\frac{n_x + n_y}{n_x n_y}\right)$$

It therefore follows that the random variable

$$Z = \frac{(\bar{X} - \bar{Y}) - (\mu_X - \mu_Y)}{\sqrt{\sigma^2\left(\dfrac{n_x + n_y}{n_x n_y}\right)}} \tag{8.7.1}$$

has a standard normal distribution. However, this result cannot be used as it stands because the unknown population variance is involved. Since this variance is common to the two populations, the two sets of sample information can be pooled together to estimate it. The estimator used is

$$s^2 = \frac{(n_x - 1)s_X^2 + (n_y - 1)s_Y^2}{(n_x + n_y - 2)}$$

where s_X^2 and s_Y^2 are the two sample variances.

Replacing the unknown σ^2 by its estimator s^2 in equation 8.7.1 gives the random variable

$$t = \frac{(\bar{X} - \bar{Y}) - (\mu_X - \mu_Y)}{s\sqrt{\dfrac{n_x + n_y}{n_x n_y}}}$$

It can be shown that this random variable obeys the Student's t distribution with

---

[14]We will see in Chapter 9 how the data can be used to check this assumption.

$(n_x + n_y - 2)$ degrees of freedom. Given this result, confidence intervals for the difference between the population means can be obtained through an argument similar to that used in Section 8.4.

Confidence Intervals for Difference Between the Means of Two Normal Populations: Independent Samples, Population Variances Equal

Suppose we have independent random samples of n_x and n_y observations from normal distributions with means μ_X and μ_Y and a common variance. If the observed sample means and variances are $\bar{x}$, $\bar{y}$ and s_x^2, s_y^2, then a $100\,(1 - \alpha)\%$ confidence interval for $\mu_X - \mu_Y$ is given by

$$(\bar{x} - \bar{y}) - t_{n_x+n_y-2,\alpha/2}s\sqrt{\frac{n_x + n_y}{n_x n_y}}$$

$$< \mu_X - \mu_Y < (\bar{x} - \bar{y}) + t_{n_x+n_y-2,\,\alpha/2}s\sqrt{\frac{n_x + n_y}{n_x n_y}}$$

where

$$s^2 = \frac{(n_x - 1)s_x^2 + (n_y - 1)s_y^2}{(n_x + n_y - 2)}$$

and $t_{n_x+n_y-2,\alpha/2}$ is that number for which

$$P(t_{n_x+n_y-2} > t_{n_x+n_y-2,\alpha/2}) = \frac{\alpha}{2}$$

where the random variable $t_{n_x+n_y-2}$ has a Student's t distribution with $(n_x + n_y - 2)$ degrees of freedom

---

Example 8.11

In a study[15] of the effects of planning on the financial performance of banks, a random sample of six "partial formal planners" showed mean annual percentage increase in net income of 9.972 and a standard deviation of 7.470. An independent random sample of nine banks with no formal planning system had a mean annual percentage increase in net income of 2.098 and a standard deviation of 10.834. Assuming the two population distributions are normal with the same variance, find a 90% confidence interval for the difference between their means.

We have, with x referring to the "partial formal planners," and y to those with no formal planning,

$$n_x = 6; \quad \bar{x} = 9.972; \quad s_x = 7.470$$

$$n_y = 9; \quad \bar{y} = 2.098; \quad s_y = 10.834$$

The estimate of the common population variance is then

---

[15] These results are reported in D. R. Wood and R. L. LaForge, "The impact of comprehensive planning on financial performance," *Academy of Management Journal, 22* (1979), 516–26.

$$s^2 = \frac{(n_x - 1)s_x^2 + (n_y - 1)s_y^2}{(n_x + n_y - 2)}$$

$$= \frac{(5)\,(7.470)^2 + (8)(10.834)^2}{13} = 93.693$$

so that

$$s = \sqrt{93.693} = 9.680$$

The interval required is of the form

$$(\bar{x} - \bar{y}) - t_{n_x + n_y - 2, \alpha/2} s\sqrt{\frac{n_x + n_y}{n_x n_y}} < \mu_X - \mu_Y < (\bar{x} - \bar{y}) + t_{n_x + n_y - 2, \alpha/2} s\sqrt{\frac{n_x + n_y}{n_x n_y}}$$

where, for a 90% confidence interval, $\alpha = 0.10$, so that

$$t_{n_x + n_y - 2, \alpha/2} = t_{13, .05} = 1.771$$

from Table 6 of the Appendix.

Hence, the 90% confidence interval for the difference between the population mean percentage increases in net incomes is

$$(9.972 - 2.098) - (1.771)(9.680)\sqrt{\frac{6 + 9}{54}}$$

$$< \mu_X - \mu_Y < (9.972 - 2.098) + (1.771)(9.680)\sqrt{\frac{6 + 9}{54}}$$

or

$$-1.161 < \mu_X - \mu_Y < 16.909$$

Our 90% confidence interval for the difference between population mean annual percentage increases in net income for these two groups of banks includes 0. This suggests that the evidence in the data against the conjecture that the two population means are the same is not strong.

8.8 CONFIDENCE INTERVALS FOR THE DIFFERENCE BETWEEN TWO POPULATION PROPORTIONS (LARGE SAMPLES)

In Section 8.5, we derived confidence intervals for a single population proportion. Often we are interested in comparing two proportions. For instance, we might want to compare the proportion of football players who succeed in graduating with the proportion of nonathletes who graduate. In this section we show how to obtain confidence intervals for the difference between population proportions when independent large samples are taken from the two populations.

Suppose that a random sample of n_x observations from a population with proportion p_X of "successes" yields sample proportion $\hat{p}_X$, and that an independent random sample of n_y observations from a population with proportion p_Y of "successes" produces sample proportion $\hat{p}_Y$. Since our concern is with the population difference $(p_X - p_Y)$, it is natural to examine the random variable $(\hat{p}_X - \hat{p}_Y)$. This has mean

$$E(\hat{p}_X - \hat{p}_Y) = E(\hat{p}_X) - E(\hat{p}_Y) = p_X - p_Y$$

and, since the samples are taken independently, variance

$$\begin{array}{c} \text{Var}(\hat{p}_X - \hat{p}_Y) = \text{Var}(\hat{p}_X) + \text{Var}(\hat{p}_Y) \\ \\ = \dfrac{p_X(1 - p_X)}{n_x} + \dfrac{p_Y(1 - p_Y)}{n_y} \end{array} \qquad (8.8.1)$$

Furthermore, if the sample sizes are large, the distribution of this random variable is approximately normal, so that subtracting its mean and dividing by its standard deviation gives a standard normal random variable. Moreover, for large sample sizes, this approximation remains good when the unknown population proportions in equation 8.8.1 are replaced by the corresponding sample quantities. Thus, to a good approximation, the random variable

$$Z = \frac{(\hat{p}_X - \hat{p}_Y) - (p_X - p_Y)}{\sqrt{\left[\dfrac{\hat{p}_X(1 - \hat{p}_X)}{n_x} + \dfrac{\hat{p}_Y(1 - \hat{p}_Y)}{n_y}\right]}}$$

has a standard normal distribution.

This result allows the derivation of confidence intervals for the difference between the two population proportions when the sample sizes are large, as shown in the box.

Confidence Intervals for the Difference Between Proportions (Large Samples)

Let $\hat{p}_x$ denote the observed proportion of successes in a random sample of n_x observations from a population with proportion p_X successes, and $\hat{p}_y$ the proportion of successes observed in an independent random sample from a population with proportion p_Y successes. Then, if the sample sizes are large,[16] a $100(1 - \alpha)\%$ confidence interval for $(p_X - p_Y)$ is given by

$$(\hat{p}_x - \hat{p}_y) - z_{\alpha/2}\sqrt{\frac{\hat{p}_x(1 - \hat{p}_x)}{n_x} + \frac{\hat{p}_y(1 - \hat{p}_y)}{n_y}}$$

$$< p_X - p_Y < (\hat{p}_x - \hat{p}_y) + z_{\alpha/2}\sqrt{\frac{\hat{p}_x(1 - \hat{p}_x)}{n_x} + \frac{\hat{p}_y(1 - \hat{p}_y)}{n_y}}$$

---

[16] This approximation is generally adequate if there are at least 40 observations in each sample.

where $z_{\alpha/2}$ is that number for which

$$P(Z > z_{\alpha/2}) = \frac{\alpha}{2}$$

and the random variable Z has a standard normal distribution.

We now illustrate this procedure with two examples.

Example 8.12

In a study[17] of the effectiveness of premiums given as an inducement to open bank accounts, a random sample of 200 accounts attracted by premiums (either cookware or a calculator) contained 79% that were retained over a 6-month period. An independent random sample of 200 accounts not attracted by premiums contained 89% that were retained over a 6-month period. If the respective population proportions are denoted p_X and p_Y, find a 90% confidence interval for $(p_X - p_Y)$.

From the sample information, we have

$$n_x = 200; \qquad \hat{p}_x = 0.79; \qquad n_y = 200; \qquad \hat{p}_y = 0.89$$

For a 90% confidence interval, $\alpha = 0.10$, so that

$$z_{\alpha/2} = z_{.05} = 1.645$$

Substituting these values into the formula for the confidence interval gives

$$(0.79 - 0.89) - (1.645)\sqrt{\frac{(0.79)(0.21)}{200} + \frac{(0.89)(0.11)}{200}}$$

$$< p_X - p_Y < (0.79 - 0.89) + (1.645)\sqrt{\frac{(0.79)(0.21)}{200} + \frac{(0.89)(0.11)}{200}}$$

or

$$-0.16 < p_X - p_Y < -0.04$$

Figure 8.14 shows also 80%, 95%, and 99% confidence intervals for the difference between the two population proportions. Notice that none of these intervals contains the difference 0 in population proportions. Thus, the data strongly suggest that accounts attracted by premiums are less likely to be held for as long as 6 months than those not attracted by premiums.

Example 8.13

Frequently populations are surveyed by mail questionnaires, and the investigators are anxious to obtain as high a response rate as possible. A questionnaire,[18] printed on a single sheet, front and back, was sent to a random sample of 220 households, of

---

[17] See R. H. Preston, F. R. Dwyer, and W. Rudelius, "The effectiveness of bank premiums," *Journal of Marketing, 42* (1978), No. 3, 96–101.

[18] This study is due to T. L. Childers and O. C. Ferrell, "Response rates and perceived questionnaire length in mail surveys," *Journal of Marketing Research, 16* (1979), 429–31.

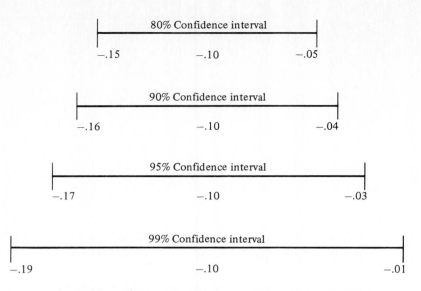

FIGURE 8.14 80%, 90%, 95%, and 99% confidence intervals for difference in population proportions, using data of Example 8.12

which 36% responded. The same questionnaire, printed on two sheets, front only, was sent to an independent random sample of 220 households, and the achieved response rate was 30%. Find a 95% confidence interval for the difference between the two population proportions responding.

The sample values are

$$n_x = 220; \quad \hat{p}_x = 0.36; \quad n_y = 220; \quad \hat{p}_y = 0.30$$

For a 95% confidence interval, $\alpha = 0.05$, and so

$$z_{\alpha/2} = z_{.025} = 1.96$$

Substituting into the formula above then yields the interval

$$(0.36 - 0.30) - (1.96)\sqrt{\frac{(0.36)(0.64)}{220} + \frac{(0.30)(0.70)}{220}}$$

$$< p_X - p_Y < (0.36 - 0.30) + (1.96)\sqrt{\frac{(0.36)(0.64)}{220} + \frac{(0.30)(0.70)}{220}}$$

that is

$$-0.03 < p_X - p_Y < 0.15$$

The 95% confidence interval for the difference between these two population proportions contains zero. This indicates that the evidence in the data against the proposition that the two population proportions are the same is not very strong.

EXERCISES

34. A random sample of ten pairs of identical houses was chosen in a large midwestern city, and a passive solar heating system was installed in one member of each pair. The total fuel bills (in dollars) for three winter months for these homes were then determined as shown in the accompanying table. Assuming normal population distributions, find a 90% confidence interval for the difference between the two population means.

PAIR	WITHOUT PASSIVE SOLAR	WITH PASSIVE SOLAR	PAIR	WITHOUT PASSIVE SOLAR	WITH PASSIVE SOLAR
1	497	452	6	386	380
2	423	386	7	426	395
3	515	502	8	473	411
4	425	376	9	454	420
5	653	605	10	496	441

35. A company claims to have developed a course that will lead to an increase in the typing speeds of experienced typists. A random sample of six typists who took this course were assessed before and after its completion. Their typing speeds, in words per minute, are recorded in the table. Assuming that the population distributions are normal, find an 80% confidence interval for the difference between the two population means.

TYPIST	AFTER COURSE	BEFORE COURSE
1	61.2	57.8
2	60.3	60.5
3	64.6	60.2
4	58.5	57.4
5	56.2	55.9
6	62.8	58.3

36. A random sample of 150 students in dormitories on a large college campus were asked to rate, on a scale from 1 to 10, their level of satisfaction with their rooms as a place to study, and with the food in the dormitories. The differences in their scores (room rating minus food rating) had mean 1.67 and standard deviation 3.82. Find a 95% confidence interval for the difference between the two population means.

37. Managers frequently conduct performance appraisal interviews with their subordinates in an effort to improve the subsequent quality of work performance. However, this intention can be frustrated by subordinates' negative reactions to the manner in which the interview is carried out. A study[19] was designed to assess the extent to which feedback from previous interviews, presented in the form of the subordinates' aggregate reactions,

---

[19] These results are given by W. F. Nemeroff and J. Cosentino, "Utilizing feedback and goal setting to increase performance appraisal interviewer skills of managers," *Academy of Management Journal, 22* (1979), 516–26.

would lead managers to conduct interviews with which their subordinates were more satisfied. For a random sample of 45 interviews conducted by managers who had received this feedback, the subordinates' mean satisfaction level was 13.98 and the standard deviation was 3.32. For an independent random sample of 41 interviews carried out by managers who had received no feedback, the mean subordinate satisfaction level was 13.07 and the standard deviation was 3.65. Find a 90% confidence interval for the difference between the two population means.

38. A department store sends out monthly statements to its credit customers. In the past it had not enclosed with these statements a preaddressed envelope for returning payments. From a random sample of 100 of its accounts for that period, it was found that the mean time to payment was 10.4 days and the sample standard deviation was 4.3 days. As an experiment, it decided to send preaddressed envelopes with the accounts of 150 randomly and independently selected customers. For this sample, the mean time to payment was found to be 9.2 days and the standard deviation was 5.1 days. Find a 95% confidence interval for the difference between the two population means.

39. The Energy Policy and Conservation Act of 1976 banned large oil companies from bidding jointly with one another in federal Outer Continental Shelf lease sales. For bids[20] on 2,870 tracts before the ban, the mean number of bids per tract was 3.315 and the standard deviation was 3.043. In the postban period, 589 tracts attracted an average of 2.849 bids and the standard deviation was 2.262. Assuming independence, find a 99% confidence interval for the difference in the mean number of bids per tract before and after the ban.

40. A researcher intends to estimate the effect of a drug on the scores of human subjects performing a task of psychomotor coordination. A random sample of nine subjects were given the drug prior to testing. Their mean score was 9.78 and the sample variance was 17.64. An independent random sample of ten subjects was used as a control group, and given a placebo prior to testing. Their mean score was 15.10 and the sample variance was 27.21. Assuming that the population distributions are normal with equal variances, find a 90% confidence interval for the difference between the population mean scores.

41. A company sends a random sample of twelve of its salespeople to a course designed to increase their motivation and hence, presumably, their effectiveness. In the following year these people generated sales with an average value of $425,000 and a sample standard deviation of $56,000. During the same period, an independently chosen random sample of fifteen salespeople who had not attended the course obtained sales with average value $408,000 and standard deviation $42,000. Assuming the two population distributions are normal and have the same variance, find a 95% confidence interval for the difference between their means.

42. A toy manufacturer is considering two different model trains for possible production. Prototypes are built and presented to two independently chosen random samples of ten children. For the first model, the times the children played with the train before moving on to another activity had an average of 25.3 minutes and a sample standard deviation of 6.4 minutes. For the second model, the sample mean and standard deviation were 19.6 minutes and 8.2 minutes, respectively. If the population distributions are normal with a common variance, find an 80% confidence interval for the difference between the population means.

43. A study[21] was designed to estimate the improvement in the response rate to mail surveys when people are first contacted by phone to solicit cooperation. For a particular survey, a random sample of 239 people were contacted by telephone. The purpose of the survey

[20] The data are given in B. Sullivan and P. Kolrin, "The joint bidding ban: pro- and anti-competitive theories of joint bidding in OCS lease sales," *Journal of Economics and Business, 33* (1980), 1–12.

[21] C. T. Allen, C. S. Schewe, and G. Wijk, "More on self-perception theory's foot technique in the pre-call/mail survey setting," *Journal of Marketing Research, 17* (1980), 498–502.

was explained and their cooperation requested. Of these people, 136 both agreed to fill out the mailed questionnaire and did in fact do so. The same questionnaire was sent to an independent random sample of 836 individuals, accompanied by a covering letter explaining its purpose. Of this group, who had not been first contacted by phone, 186 responded. Find a 99% confidence interval for the difference between the proportions of people responding to this questionnaire in the two populations.

44. A random sample of 100 managers at a large manufacturing plant contained 58 opposed to the unionization of the work force. An independent random sample of 200 production line workers contained 64 opposed to unionization. Find a 95% confidence interval for the difference between the two population proportions.

45. A random sample[22] was taken of shoppers in a Sacramento, California, supermarket that had used unit pricing of goods for 6 years. It was found that 532 of the sample members had received some college education and, of these, 86.3% were aware of unit pricing. The remaining 213 sample members had no college education, and 74.4% of them were aware of unit pricing. Find a 95% confidence interval for the difference in the population proportions aware of unit pricing between those with and without some college education.

46. A random sample of 100 men contained 61 in favor of a state constitutional amendment to retard the rate of growth of property taxes. An independent random sample of 100 women contained 54 in favor of this amendment. The confidence interval

$$0.03 < p_X - p_Y < 0.11$$

was calculated for the difference between the population proportions. What is the probability content of this interval?

47. A survey[23] was carried out of senior citizens' attitudes and desires about retailing. For a random sample of 232 shoppers aged 65 or older, 25.0% indicated a desire for more courteous, dignified, and patient treatment of older customers, in response to the question: "What might companies and stores do to help customers who are senior citizens?" For a random sample of 106 shoppers aged 55–64, the corresponding percentage was 19.8%. Find a 90% confidence interval for the difference between the two population proportions.

8.9 ESTIMATING THE SAMPLE SIZE

Thus far we have developed methods for finding confidence intervals for a population parameter on the basis of the information contained in a given sample. Following such a process, an investigator may believe that the resulting confidence interval is too wide, reflecting an undesirable amount of uncertainty about the parameter being estimated. Typically, the only way to obtain a narrower interval with a given probability content is to take a larger sample.

In some circumstances, the investigator may be able to fix in advance the width of the confidence interval, choosing a sample size big enough to guarantee that width. In this section we show how the sample size can be chosen in this way for two interval

---

[22] These results are reported by B. F. McElroy and D. A. Aaker, "Unit pricing six years after introduction," *Journal of Retailing, 55* (1979), No. 3, 44–57.

[23] See Z. V. Lambert, "An investigation of older consumers' unmet needs at the retail level," *Journal of Retailing, 55* (1979), No. 4, 35–57.

estimation problems. Similar procedures can be employed in other interval estimation approaches.

INTERVALS FOR THE MEAN OF A NORMAL DISTRIBUTION: POPULATION VARIANCE KNOWN

If a random sample of n observations is taken from a normal population with mean μ and known variance σ^2, it was seen in Section 8.2 that a $100(1 - \alpha)\%$ confidence interval for the population mean is provided by

$$\bar{x} - \frac{z_{\alpha/2}\sigma}{\sqrt{n}} < \mu < \bar{x} + \frac{z_{\alpha/2}\sigma}{\sqrt{n}}$$

where $\bar{x}$ is the observed sample mean and $z_{\alpha/2}$ is the appropriate cutoff point of the standard normal distribution. This interval is centered on the sample mean and extends a distance

$$L = \frac{z_{\alpha/2}\sigma}{\sqrt{n}} \qquad\qquad (8.9.1)$$

on each side of the sample mean, so that L is half the width of the interval. Suppose, now, that the investigator wants to fix L in advance. From equation 8.9.1 we have

$$\sqrt{n} = \frac{z_{\alpha/2}\sigma}{L}$$

and by squaring both sides of this equation, we obtain

$$n = \frac{z_{\alpha/2}^2\sigma^2}{L^2}$$

This choice of the sample size ensures that the confidence interval extends a distance L on each side of the sample mean.

Sample Size for Confidence Intervals for the Mean of a Normal Distribution: Population Variance Known

Suppose we take a random sample from a normal population with known variance σ^2. Then a $100(1 - \alpha)\%$ confidence interval for the population mean extends a distance L on each side of the sample mean, if the number of observations is

$$n = \frac{z_{\alpha/2}^2\sigma^2}{L^2}$$

---

where $z_{\alpha/2}$ is that number for which

$$P(Z > z_{\alpha/2}) = \frac{\alpha}{2}$$

and Z has a standard normal distribution.

Of course, the number of sample observations must necessarily be an integer. If the number n resulting from the sample size formula is not an integer, we *round up* to the next whole number in order to guarantee that our confidence interval does not exceed the required width.

Example 8.14
The lengths of metal rods produced by an industrial process are normally distributed with standard deviation 1.8 millimeters. Based on a random sample of nine observations from this population, the 99% confidence interval

$$194.65 < \mu < 197.75$$

was found for the population mean length. Suppose that a production manager believes that the interval is too wide for practical use, and instead requires a 99% confidence interval extending no further than 0.50 millimeter on each side of the sample mean. How large a sample is needed to achieve such an interval?

We have

$$L = 0.50; \qquad \sigma = 1.8; \qquad z_{\alpha/2} = z_{.005} = 2.575$$

Hence, the required sample size is

$$n = \frac{z_{\alpha/2}^2 \sigma^2}{L^2}$$

$$= \frac{(2.575)^2(1.8)^2}{(0.5)^2} = 85.93$$

Therefore, to satisfy the manager's requirement, a sample of at least 86 observations is needed. This large increase in the sample size represents the additional cost of achieving the higher precision in the estimate of the true mean, reflected in a narrower confidence interval.

INTERVALS FOR THE POPULATION PROPORTION

In Section 8.5, we saw that, based on a random sample of n observations, a $100(1 - \alpha)\%$ confidence interval for the population proportion p is given by

$$\hat{p}_x - z_{\alpha/2}\sqrt{\frac{\hat{p}_x(1 - \hat{p}_x)}{n}} < p < \hat{p}_x + z_{\alpha/2}\sqrt{\frac{\hat{p}_x(1 - \hat{p}_x)}{n}}$$

where $\hat{p}_x$ is the observed sample proportion. This interval is centered on the sample proportion, and extends a distance

$$L = z_{\alpha/2} \sqrt{\frac{\hat{p}_x(1 - \hat{p}_x)}{n}}$$

on each side of the sample proportion. Now, this result cannot be used directly to determine the sample size necessary to obtain a confidence interval of some specified width, since it involves the sample proportion, which will not be known at the outset. However, whatever the outcome, $\hat{p}_x(1 - \hat{p}_x)$ cannot be bigger than 0.25, its value when the sample proportion is 0.5. Thus, the largest possible value for L is

$$L_* = z_{\alpha/2} \sqrt{\frac{0.25}{n}} = \frac{0.5z_{\alpha/2}}{\sqrt{n}} \qquad (8.9.2)$$

Suppose, then, that an investigator wants to choose a sufficiently large sample size to *guarantee* that the confidence interval extends no more than L_* on each side of the sample proportion. From equation 8.9.2, we have

$$\sqrt{n} = \frac{0.5z_{\alpha/2}}{L_*}$$

and squaring yields

$$n = \frac{0.25z_{\alpha/2}^2}{L_*^2}$$

This provides the required sample size.

Sample Size for Confidence Intervals for the Population Proportion

Suppose we take a random sample from a population. Then a $100(1 - \alpha)\%$ confidence interval for the population proportion, extending a distance of *at most* L_* on each side of the sample proportion, can be guaranteed if the number of sample observations is

$$n = \frac{0.25z_{\alpha/2}^2}{L_*^2}$$

Example 8.15

In Example 8.7 we calculated a 95% confidence interval for the proportion of managers viewing high price as the major cause of new product market failures. This interval, based on a random sample of 89 observations, was

$$0.102 < p < 0.262$$

Suppose that we want to ensure that a 95% confidence interval for the population proportion extends no further than 0.04 on each side of the sample proportion. How large a sample must be taken?

We have

$$L_* = 0.04 \quad \text{and} \quad z_{\alpha/2} = z_{.025} = 1.96$$

Thus, the sample size needed is

$$n = \frac{0.25 z_{\alpha/2}^2}{L_*^2}$$

$$= \frac{0.25(1.96)^2}{(0.04)^2} = 600.25$$

In order to achieve this shorter confidence interval, we see that a minimum of 601 sample observations are required.

EXERCISES

48. Using the data in Exercise 2 find the number of sample observations necessary in order to obtain a 95% confidence interval for the mean weight of bricks that extends a distance 0.02 pound on each side of the sample mean.

49. Using the data in Exercise 3 find the sample size needed so that a 99% confidence interval for the mean dollar value of the supermarket's customer purchases extends a distance $10 on each side of the sample mean.

50. Using the data in Exercise 20 find the number of sample observations necessary in order to guarantee that a 99% confidence interval for the proportion of people disagreeing with the statement extends a distance at most 0.05 on each side of the sample proportion.

51. A congressman wants to determine the reactions of voters in his constituency to a proposal on modifying inflation adjustments to Social Security payments. He wants a 95% confidence interval for the proportion of electors favoring the proposal to extend no further than 0.05 on each side of the sample proportion. How large a sample is needed in order to be sure that this requirement will be met?

52. A company wants to estimate the proportion of flashbulbs that will give proper illumination. Suppose that a 99% confidence interval that extends at most 0.04 on each side of the sample proportion is required. How many sample observations are needed?

53. An investigator wants to compare two population proportions, and intends to take independent random samples of the same size from each population. She wants to be sure that a 90% confidence interval for the difference between the two population proportions extends no further than 0.05 on each side of the difference between the sample proportions. How large a sample should she take from each population?

REVIEW EXERCISES

54. Based on independent random samples from the two populations, an investigator found a 95% confidence interval for the difference between mean miles per gallon for X-cars and mean miles per gallon for Y-cars running from 0.6 to 1.9. Does this imply that the probability is 0.95 that the true difference in mean miles per gallon lies between 0.6 and 1.9? If not, provide a valid interpretation of the interval estimate.

55. Explain the relevance of the Central Limit Theorem to interval estimation for a population mean.

56. We discussed two different procedures for finding confidence intervals for the difference between population means. Explain why two different procedures are needed.

57. A random sample of 100 houses in a city had a mean selling price of $78,300 and a sample standard deviation of $12,760.

(a) Find a 95% confidence interval for the population mean selling price.

(b) Without doing the calculations, state whether a 90% confidence interval for the population mean would be wider or narrower than that found in part (a).

(c) Based on these results, a statistician produces a confidence interval for the population mean running from $77,800 to $78,800. What is the probability content of this interval?

58. For a random sample of 150 of the employees of a large corporation, it was found that the mean number of days lost in a year through sickness was 9.37. The sample standard deviation was 4.82 days.

(a) Find an 80% confidence interval for the population mean number of days lost through sickness.

(b) Suppose these sample results had been based on a random sample of 100 employees. Would an 80% confidence interval for the population mean be narrower or wider than that found in part (a)?

59. In order to estimate the mean value of the purchases of card holders in a month, a credit card company takes a random sample of twelve monthly statements and obtains the following amounts in dollars:

79.21	143.62	95.08	34.27	211.87	139.53
98.26	65.93	159.11	127.26	53.91	87.80

Assuming the population distribution is normal, find a 90% confidence interval for the mean monthly value of purchases of all card holders.

60. A manufacturer of electronic games is considering their installation in campus bars. In a pilot study on the potential profitability of this enterprise, games were placed for one week in ten randomly chosen college bars. The weekly profits in dollars were as follows:

100.80	63.90	151.20	93.60	75.80
201.30	102.40	77.80	94.10	98.00

Assuming the population distribution is normal, find a 99% confidence interval for the mean weekly profit for these games in all college bars.

61. A random sample of 83 managers were asked if they would like more opportunities for choice in their jobs. Of these sample members, 47 answered "no" to this question.[24] Find a 90% confidence interval for the population proportion of all managers not wanting more choice in their jobs.

---

[24] This information is reported in J. Marshall and R. Stewart, "Managers' job perceptions, part II: Opportunities for and attitudes to choice," *Journal of Management Studies, 18* (1981), 263–75.

62. In a random sample of 1,158 newly promoted executives, 47.9% rated a statistics course as very important or somewhat important as part of the preparation for a career in general management.[25]

   (a) Find a 99% confidence interval for the population proportion of all newly promoted executives holding this view.

   (b) Based on the sample information, a statistician computed a confidence interval for the population proportion running from 0.458 to 0.500. What is the probability content of this interval?

63. A random sample of 54 union stewards were asked how often they talk employees out of filing grievances. Of these sample members, 25.9% answered "never" to this question.[26]

   (a) Find a 95% confidence interval for the population proportion of union stewards who never talk employees out of filing grievances.

   (b) Based on this information, a statistician calculated a confidence interval for the population proportion ranging from 0.228 to 0.290. What is the probability content of this interval?

64. Of a random sample of 182 Scottish accountants,[27] 90 agreed that "Cash flow accounting should not be introduced as a complete substitute for, or addition to, the present system of allocation based accounting data."

   (a) Find a 90% confidence interval for the population proportion of Scottish accountants agreeing with this statement.

   (b) Without doing the calculations, state whether a 99% confidence interval for the population proportion would be narrower or wider than that found in part (a).

65. Of a random sample of 174 new car buyers in the St. Louis metropolitan area, 129 bought a brand they had previously owned.[28]

   (a) Find a 95% confidence interval for the population proportion.

   (b) Without doing the calculations, state whether a 90% confidence interval would be narrower or wider than that found in part (a).

66. A random sample of twelve financial analysts were asked to predict earnings per share over the next year for a particular corporation. Their forecasts in dollars were as follows:

10.20	7.50	8.30	11.10	9.20	10.00
9.70	8.90	9.40	9.10	10.70	10.30

Find a 90% confidence interval for the population standard deviation of financial

---

[25] If you don't believe this, see H. W. Hildebrandt, F. A. Bond, E. L. Miller, and A. W. Swinyard, "An executive appraisal of courses which best prepare one for general management," *Journal of Business Communication, 19* (1982), No. 1, 5–15.

[26] This information is given in D. R. Dalton and W. D. Todor, "Antecedents of grievance filing behavor: Attitude/behavioral consistency and the union steward," *Academy of Management Journal, 25* (1982), 158–69.

[27] See T. A. Lee, "A survey of accountants' opinions on cash flow reporting," *Abacus, 17* (1981), 130–44.

[28] Reported by J. Arndt and F. E. May, "The hypothesis of a dominance hierarchy of information sources," *Journal of Academy of Marketing Science, 9* (1981), 337–50.

analysts' predictions of earnings per share for this corporation. State any assumptions that you have made in obtaining this interval.

67. The performances of a random sample of ten stocks traded on the New York Stock Exchange were examined. The following figures are the rates of returns for these stocks over the last year:

| 10.51 | 16.28 | 3.10 | 7.20 | −1.89 |
| 8.20 | 9.73 | 13.12 | 6.35 | 4.20 |

   (a) Find a 95% confidence interval for the population standard deviation, and state any assumptions that you have made.
   (b) Without doing the calculations, state whether a 99% confidence interval for the population standard deviation would be narrower or wider than that found in part (a).

68. A random sample of twelve middle managers was sent to an intensive business school summer session on modern techniques. The job performance appraisal scores given by their immediate superiors in the year before and the year after this session are shown in the accompanying table. Stating any assumptions you need to make, find a 95% confidence interval for the difference between the population mean appraisal scores after and before attendance at this business school summer session.

BEFORE	AFTER	BEFORE	AFTER
67	73	75	74
54	50	78	87
82	83	64	69
67	78	72	72
60	56	70	77
73	74	63	80

69. Candidates for employment in a large corporation must take a written aptitude test and complete an interview with a personnel manager. After the interview, the personnel manager grades each candidate on a scale from 0 to 100. To check the consistency, in the aggregate, of the scores of personnel managers, a random sample of ten pairs of candidates was selected. The pairing was arranged so that, within each pair, the two candidates obtained identical scores on the written aptitude test. Candidates in each pair were then interviewed by one of two personnel managers, John Doe and Jean Ray. The grades awarded following these interviews are shown in the accompanying table.

JOHN DOE	JEAN RAY
80	72
65	63
87	91
64	65
73	64
78	71
83	69
91	90
84	79
85	87

(a) Stating any assumptions you have made, find a 90% confidence interval for the difference between the population mean scores awarded by Doe and Ray.

(b) Without doing the calculations, state whether a 99% confidence interval for the difference in the population means would be narrower or wider than that found in part (a).

70. For a random sample[29] of 40 accounting students in a class using group learning techniques, the mean examination score was 322.12, and the sample standard deviation was 54.53. For an independent random sample of 61 students in the same course, but in a class not using group learning techniques, the sample mean and standard deviation of the scores were 304.61 and 62.61, respectively. Find 90% and 95% confidence intervals for the difference in the population mean scores.

71. A random sample of ten department store credit accounts of customers who had these accounts for over a year showed the following charges (in dollars) at the end of a month:

37.21	65.18	17.97	63.21	51.70
43.97	62.39	47.62	75.91	18.23

For an independent random sample of eight new credit customers, charges at the end of the same month were as follows:

18.73	29.21	37.83	16.25
43.81	27.96	23.64	31.93

Stating carefully any assumptions you make, find a 90% confidence interval for the difference between the two population means.

72. A random sample of eight presidents of manufacturing corporations yielded the following results for number of months in the position:

83	71	144	62	31	18	56	92

For an independent random sample of six bank presidents, the corresponding figures were as follows:

61	173	94	111	135	87

Stating carefully any assumptions you make, find a 95% confidence interval for the difference between the two population means.

---

[29] Results reported by S. M. Lightner, "Accounting education and participatory group dynamics," *Collegiate News and Views, 35* (1981), No. 1, 5–9.

73. In a survey of the approaches to forecasting used by commercial banks,[30] it was found that 28 of a random sample of 78 very large banks used multiple regression,[31] while 21 of an independent random sample of 79 large banks used this technique.
    (a) Find a 90% confidence interval for the difference between the two population proportions of commercial banks using multiple regression in forecasting.
    (b) Without doing the calculations, state whether a 95% confidence interval for the difference in the population proportions would be wider or narrower than that found in part (a).

74. A large company currently operates two shifts per day. At the present time, employees remain in the same shift each week. However, management is considering the introduction of weekly rotation of shifts. For a random sample of 70 workers in the first shift, 32 were in favor of this move, while for an independent random sample of 80 workers from the second shift, 28 were in favor. Find a 99% confidence interval for the difference between the two population proportions of workers in favor of the change to rotating shifts.

75. In Exercise 57 we found a confidence interval for the population mean selling price of houses in a city. Suppose now that it is known that the population standard deviation of selling prices is $12,760. How many sample observations would be needed in order to obtain a 95% confidence interval for the population mean that extends $1,000 on each side of the sample mean?

76. In Exercise 61 we found a confidence interval for the proportion of managers not wanting more choice in their jobs. How many sample observations would be needed to ensure that a 95% confidence interval for the population proportion extends at most 0.04 on each side of the sample proportion?

---

[30]G. A. Giroux, "A survey of forecasting techniques used by commercial banks," *Journal of Bank Research, 11* (1980), 51–53.

[31]We will discuss multiple regression in Chapters 13 and 14.

hypothesis
testing

CHAPTER NINE

9.1 CONCEPTS OF HYPOTHESIS TESTING

When a sample is drawn from a population, the evidence obtained can be used to make inferential statements about the characteristics of the population. As we have seen, one possibility is to estimate the unknown population parameters through the calculation of point estimates or confidence intervals. Alternatively, the sample information can be employed to assess the validity of some conjecture, or **hypothesis,** that an investigator has formed about the population. Examples of situations of this kind are as follows:

1. A manufacturer who produces boxes of cereal claims that, on average, the contents weigh at least 20 ounces. In order to check this claim, the contents of a random sample of boxes can be weighed, and inference based on the sample results.
2. A company receiving a large shipment of parts may want to accept delivery only if no more than 5% of the parts are defective. The decision on whether to take delivery might be based on a check of a random sample of these parts.
3. An instructor is interested in the value of regularly administered quizzes in a statistics course. She uses these quizzes in one section of the course, but not in another. At the end of the course she compares the average performances of students in the two sections on the final examination, in order to check her hypothesis that the quizzes raise average performance.
4. A political scientist wants to know if a tax reform proposal appeals equally to men and women. In order to check whether this is so he obtains the opinions of randomly selected samples of males and females.

The examples given here have a common theme. A hypothesis is formed about some population, and conclusions about the merits of this hypothesis are to be formed on the basis of sample information. In this section we introduce a general framework for approaching such problems. Specific procedures are then developed in the following sections.

To keep our discussion quite general, let us denote the population parameter of interest (for example, the population mean, variance, or proportion) by θ. Suppose that some hypothesis has been formed about this parameter, and that this hypothesis will be believed unless sufficient contrary evidence is produced. This can be thought of as a *maintained hypothesis*. In the language of statistical hypothesis testing, it is called a **null hypothesis.** For example, we might, in the absence of evidence to dispute it, believe the manufacturer's claim that, on average, the contents of its boxes of cereal weigh at least 20 ounces. When sample information is collected, this hypothesis is put in jeopardy, or *tested*. If the hypothesis is not true, then some alternative must be true and, in carrying out a hypothesis test, the investigator formulates an **alternative hypothesis** against which the null hypothesis is tested. For the cereal manufacturer we could test the null hypothesis that the mean contents weight is at least 20 ounces against the alternative hypothesis that the mean weight is less than 20 ounces. The null hypothesis will be denoted H_0 and the alternative hypothesis H_1.

A hypothesis, whether null or alternative, might specify just a single value, say θ_0, for the population parameter θ. In that case the hypothesis is said to be **simple.** A convenient shorthand notation would read, for example,

$$H_0: \quad \theta = \theta_0$$

for "The null hypothesis is that the population parameter θ is equal to the specific value θ_0." For instance, in the fourth example listed previously, the political scientist might begin his investigation with the simple null hypothesis that the difference between the proportions of men and women in the population who favor the tax reform proposal is 0.

On the other hand, a hypothesis could specify a *range* of values for the unknown population parameter. Such a hypothesis is said to be **composite**, and will hold true for more than one value of the population parameter. For instance, the null hypothesis that the mean weight of boxes of cereal is at least 20 ounces is composite. The hypothesis is true for *any* population mean weight greater than or equal to 20 ounces.

In many applications, a simple null hypothesis, say $H_0: \quad \theta = \theta_0$, is tested against a composite alternative. In some cases, only alternatives on one side of the null hypothesis are of interest. For example, we might want to test this null hypothesis against the alternative hypothesis that the true value of θ is bigger than θ_0, which we can write

$$H_1: \quad \theta > \theta_0$$

Conversely, the alternative of interest might be

$$H_1: \quad \theta < \theta_0$$

Such alternative hypotheses are called **one-sided alternatives.** Another possibility is that we want to test this simple null hypothesis against the very general alternative that the true value of θ is something other than θ_0, that is,

$$H_1: \quad \theta \neq \theta_0$$

This is referred to as a **two-sided alternative.**

The specification of appropriate null and alternative hypotheses is problem-specific. To illustrate, we return to our earlier examples:

1. Let θ denote the population mean weight (in ounces) of cereal per box. The null hypothesis is that this mean is at least 20 ounces, so we have the composite null hypothesis

$$H_0: \quad \theta \geq 20$$

   The obvious alternative is that the true mean weight is less than 20 ounces, that is,

$$H_1: \quad \theta < 20$$

2. A company intends to accept delivery of parts unless it has evidence to suspect that more than 5% are defective. Let θ denote the population proportion of defectives. The null hypothesis here is that this proportion is at most 0.05, that is,

$$H_0: \quad \theta \leq 0.05$$

   On the basis of sample information, this hypothesis is tested against the alternative

$$H_1: \quad \theta > 0.05$$

   The null hypothesis, then, is that the shipment of parts is of adequate quality overall, while the alternative is that it is not.

3. Suppose an instructor conjectures that the regular administration of quizzes in class makes no difference to the average scores on the final examination. Let θ denote the difference between the population mean scores for sections with and without regular quizzes. The null hypothesis is then the simple null hypothesis

$$H_0: \quad \theta = 0$$

   However, she may suspect the possibility that quizzes lead to an increase in average performance, and thus would want to test the null hypothesis against the alternative hypothesis

$$H_1: \quad \theta > 0$$

4. A political scientist might hold, as a working hypothesis, the view that the tax reform proposal is equally appealing to men and women. If θ is the difference between the two population proportions in favor of the proposal, then the null hypothesis is

$$H_0: \quad \theta = 0$$

   If the political scientist has no good reason to suspect the bulk of support comes from one population rather than the other, this null hypothesis would be tested against the two-sided alternative hypothesis

$$H_1: \quad \theta \neq 0$$

Having specified a null and alternative hypothesis and collected sample information, a decision concerning the null hypothesis must be made. The two possibilities are to **accept** the null hypothesis, or **reject** it in favor of the alternative. In order to reach one of these conclusions, some **decision rule,** based on the sample evidence, has to be formulated. In subsequent sections we will discuss specific decision rules, noting for now only that their general form is often fairly obvious. Suppose, for instance, that a random sample of ten boxes of cereal is taken and their contents weighed. If the sample mean weight is much less than 20 ounces we might suspect the validity of the null hypothesis that the population mean is at least 20 ounces. The decision rule would then involve the rejection of this null hypothesis if the sample mean was "too low." Thus, in testing a null hypothesis about a population mean, it is plausible that our conclusion will be based on the value observed for the sample mean. All other things equal, the greater the difference between the sample mean and the values postulated by the null hypothesis for the population mean, the more suspicious would we be of the truth of that hypothesis.

If all that is available is a sample from a population, then the population parameters will not be precisely known. Accordingly, it *cannot be known for sure* whether a null hypothesis is true or false. Therefore, whatever decision rule is adopted, there is some chance of reaching an erroneous conclusion about the population parameter of interest. In fact, as indicated in Table 9.1, either of two possible kinds of error could be made. There are two possible states of nature—either the null hypothesis is true or it is false. One error that could be made, called a **Type I error,** is the rejection of a true null hypothesis. If the decision rule is such that the probability of rejecting the null hypothesis when it is true is α, then α is said to be the **significance level** of the test. Since the null hypothesis must either be accepted or rejected, it follows that the probability of accepting the null hypothesis when it is true is $(1-\alpha)$. The other possible error, called a **Type II error,** arises when a false null hypothesis is accepted. Suppose that, for a particular decision rule, the probability of making such an error when the null hypothesis is false is denoted β. Then, the probability of rejecting a false null hypothesis is $(1-\beta)$, which is called the **power** of the test.

We will illustrate these ideas by reference to one of our earlier examples. Consider, again, the problem of the political scientist trying to determine whether a tax reform proposal appeals equally to men and women. The null hypothesis is that, *in the population,* the proportion of men in favor of this proposal is the same as the proportion of women. This null hypothesis is to be tested against the alternative that the two population proportions differ. In order to test the null hypothesis, independent

TABLE 9.1 States of nature and decisions on null hypothesis, with associated probabilities of making the decisions, given the particular states of nature

		STATES OF NATURE	
		NULL HYPOTHESIS TRUE	NULL HYPOTHESIS FALSE
DECISION ON NULL HYPOTHESIS	ACCEPT	Correct decision Probability = $1 - \alpha$	Type II error Probability = β
	REJECT	Type I error Probability = α (α is called **significance level**)	Correct decision Probability = $1 - \beta$ ($1 - \beta$ is called **power**)

random samples of men and women are taken, and the views of the sample members are solicited. It is natural to base inference about the null hypothesis on the difference between the *sample proportions* of men and women in favor of the proposal. If this difference is large, the null hypothesis of equality of the *population proportions* would be rejected; otherwise, this null hypothesis would be accepted. Let $\hat{p}_x$ denote the sample proportion of men, and $\hat{p}_y$ the sample proportion of women, in favor of the tax reform proposal. Then a possible decision rule is

$$\text{Reject } H_0 \text{ if } \quad (\hat{p}_x - \hat{p}_y) > 0.05 \qquad \text{or} \qquad (\hat{p}_x - \hat{p}_y) < -0.05$$

Now suppose that, in fact, the null hypothesis that the two population proportions favoring the proposal are equal is true. It nevertheless could happen that the sample proportions differ by more than 0.05 so that, according to our decision rule, the null hypothesis would be rejected. In that case, a Type I error would have been made. The probability of this occurring (when the null hypothesis is true) is the significance level α. On the other hand, suppose that the null hypothesis is false and that, in fact, the population proportions of men and women in favor of the proposal are not the same. It may still be the case that the two sample proportions differ by less than 0.05. Then according to our decision rule, the null hypothesis would be accepted, and a Type II error would have been made. The probability of making such an error will depend on just how different are the two population proportions. We would be less likely, for given sample sizes, to accept the null hypothesis if 80% of men and 20% of women favored the proposal than if these percentages were 55% and 45%.

Ideally, of course, we would like to have the probabilities of both types of error be as small as possible. However, there is clearly a trade-off between the two. Once a sample has been taken, any adjustment to the decision rule that makes it less likely to reject a true null hypothesis will inevitably render it more likely to accept this hypothesis when it is false. Specifically, suppose we want to test, on the basis of a random sample, the null hypothesis that the true mean weight of the contents of boxes of cereal is at least 20 ounces. Given a specific sample size—say, $n = 30$ observations—we might adopt the decision rule that the null hypothesis is rejected if the sample mean weight is less than 18.5 ounces. Now, it is easy to find a decision rule for which the probability of Type I error is lower. If we modify our decision rule to "Reject null hypothesis if sample mean weight is less than 18 ounces" this objective will have been achieved. However, there is a price to be paid. Using the modified decision rule, we will be more likely to accept the null hypothesis, whether it is true or false. Thus, in decreasing the Type I error probability, we have increased the Type II error probability. The only way of simultaneously lowering both error probabilities would be to obtain more information about the true population mean, by taking a larger sample. Typically what is done in practice is to fix at some desired level the probability of making a Type I error; that is, the significance level is fixed. This then determines the appropriate decision rule, which in turn determines the probability of a Type II error. This sequence is illustrated in Figure 9.1.

To illustrate this sequence, consider again the problem of testing, based on a sample of thirty observations, whether the true mean weight of boxes of cereal is at least 20 ounces. Given a decision rule, we could determine the probabilities of Type

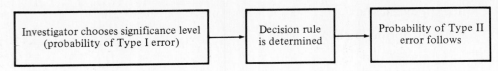

FIGURE 9.1 Consequences of fixing the significance level of a test

I and Type II errors associated with the test. However, in fact, we proceed by first fixing the Type I error probability. Suppose, for example, that we want to ensure that the probability of rejecting the null hypothesis when it is true is at most 0.05. We can do this by choosing an appropriate number, K, in the decision rule, "Reject the null hypothesis if the sample mean is less than K ounces." (We will discuss in the next section how this can be done.) Once the number K is chosen, the Type II error probabilities can be computed, using procedures to be discussed in Section 9.9.

We have seen that, since the decision rule is determined by the particular significance level chosen, the concept of power plays no direct part in the decision as to whether to reject a null hypothesis. However, calculations of power, stemming from particular significance level choices, provide the investigator with valuable information about the properties of the decision rule. Often an investigator has some flexibility in the choice of the number of sample observations to take. For a given significance level, the bigger the sample size, the higher will be the power of the test. In deciding how big the sample should be, the analyst must balance the benefits from increased power against the costs of acquiring additional sample information. Another important use of power calculations arises when we have available two or more possible tests for analyzing the same problem. If the decision rules associated with these tests are determined so that each has the same significance level, for a given sample size, then it is natural to prefer the procedure with the smallest probability of Type II error—that is, the procedure with the highest power.

In Sections 9.2–9.8, we show how, for given significance levels, decision rules can be formulated for some important classes of hypothesis testing problems. We will return in Section 9.9 to a consideration of the power of a test.

For convenience, the new terminology introduced in this section is summarized in the accompanying box.

Some Hypothesis Testing Terminology

NULL HYPOTHESIS (H_0) □ A maintained hypothesis that is held to be true until sufficient evidence to the contrary is obtained

ALTERNATIVE HYPOTHESIS (H_1) □ A hypothesis against which the null hypothesis is tested, and which will be held to be true if the null is held false

SIMPLE HYPOTHESIS □ A hypothesis that specifies a single value for a population parameter

---

COMPOSITE HYPOTHESIS □ A hypothesis that specifies a range of values for a population parameter

ONE-SIDED ALTERNATIVE □ An alternative hypothesis involving all possible values of a population parameter on either one side or the other of (that is, either greater than or less than) the value specified by a simple null hypothesis

TWO-SIDED ALTERNATIVE □ An alternative hypothesis involving all possible values of a population parameter other than the value specified by a simple null hypothesis

HYPOTHESIS TEST DECISIONS □ A decision rule is formulated, leading the investigator to either *accept* or *reject* the null hypothesis on the basis of sample evidence

TYPE I ERROR □ The rejection of a true null hypothesis

TYPE II ERROR □ The acceptance of a false null hypothesis

SIGNIFICANCE LEVEL □ The probability of rejecting a null hypothesis which is true (This probability is sometimes expressed as a percentage, so that a test of significance level α is referred to as a $100\alpha\%$ level test.)

POWER □ The probability of rejecting a null hypothesis which is false

The terminology "accept" and "reject" for the possible decisions about a null hypothesis is commonly used in formal summaries of the outcomes of particular tests. However, these terms do not adequately reflect the asymmetry of the status of the null and alternative hypotheses, or the consequences of a procedure in which the significance level is fixed, and the probability of a Type II error is not controlled. As we have already noted, the null hypothesis has the status of a maintained hypothesis— a hypothesis that will be held true, *unless the data contain sufficient contrary evidence*. Moreover, in fixing a significance level, generally at some small probability, we are ensuring that the chance is low that a true null hypothesis will be rejected. In such a setup, we are not likely with only a modest amount of data to be in a position to reject a null hypothesis, unless it is wildly in error. As we have seen, as the number of sample observations increases, so does the chance of our being able to detect a false null hypothesis. Thus, in "accepting" a null hypothesis, we are not necessarily saying a great deal in its favor. A more accurate, though more pedantic, statement of the position might be "The data available do not provide enough evidence for rejection of the null hypothesis, given that we want to fix at α the probability of rejecting a null hypothesis that is true." For this reason, some writers prefer the phrase "The null hypothesis is not rejected" rather than "The null hypothesis is accepted." We will continue to use "accept," as an efficient way of expressing this idea, but it is important that this interpretation of the phrase be kept in mind. The position is rather similar to that prevailing in a court of law, where the defendant is, at the outset, deemed innocent, and the burden is on the prosecution to present sufficiently strong contrary evidence to secure a verdict of "guilty." In the classical hypothesis testing framework, the null hypothesis is, in the same sense, initially held to be true. The burden of persuading us otherwise rests on the sample data.

In the following sections of this chapter we will present tests of a number of specific hypotheses.

9.2 TESTS OF THE MEAN OF A NORMAL DISTRIBUTION: POPULATION VARIANCE KNOWN

We introduce the methodology of classical hypothesis testing by considering the case where a random sample of n observations, $X_1, X_2, \ldots, X_n$, from a normal distribution with mean μ and variance σ^2, is available. The objective is to test hypotheses about the unknown population mean. Initially it will be assumed that the population variance is known. Later we will see that this assumption and that of normality can be relaxed when the number of sample observations is large.

We begin with the problem of testing the simple null hypothesis that the population mean is equal to some specified value, μ_0. This hypothesis is denoted

$$H_0: \quad \mu = \mu_0$$

Suppose that the alternative hypothesis of interest is that the population mean exceeds this specified value; that is,

$$H_1: \quad \mu > \mu_0$$

It is natural to base tests of the population mean on the sample mean $\bar{X}$. In particular, one would doubt the truth of the null hypothesis, as opposed to this alternative, if the observed sample mean was greatly in excess of μ_0. We require the format of a test with some preassigned significance level α. That is, we want a decision rule such that the probability of rejecting the null hypothesis, when it is in fact true, is α. The basis for such a test lies in the fact that the random variable

$$Z = \frac{\bar{X} - \mu}{\sigma/\sqrt{n}}$$

follows a standard normal distribution; that is, the sampling distribution of the sample mean is normal, with mean μ and standard deviation $\sigma/\sqrt{n}$. *When the null hypothesis is true,* μ is equal to μ_0, so that the random variable

$$Z = \frac{\bar{X} - \mu_0}{\sigma/\sqrt{n}} \qquad (9.2.1)$$

then has a standard normal distribution. Now, the null hypothesis is to be rejected if the sample mean greatly exceeds the value μ_0 hypothesized for the population mean. Thus, H_0 will be rejected if a high value for the random variable 9.2.1 is observed. We want to fix at α the probability of rejecting the null hypothesis when it is true. As in the previous chapter, we denote by z_α that number for which

$$P(Z > z_\alpha) = \alpha$$

It then follows that, when the null hypothesis is true, the probability that the random variable 9.2.1 is bigger than z_α is α. Hence, denoting the observed sample mean by $\bar{x}$, suppose we adopt the following decision rule:

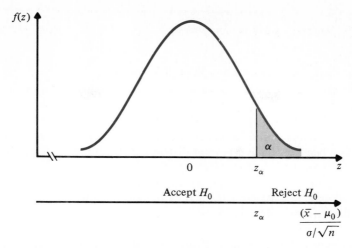

FIGURE 9.2 The probability density function of $Z = (\bar{X} - \mu_0)/(\sigma/\sqrt{n})$ when the null hypothesis H_0: $\mu = \mu_0$ is true, and the decision rule for testing H_0 against the alternative H_1: $\mu > \mu_0$ at significance level α

$$\text{Reject } H_0 \text{ if } \quad \frac{\bar{x} - \mu_0}{\sigma/\sqrt{n}} > z_\alpha$$

Then the probability of rejecting H_0 when it is true will be α, so that α is the significance level of the test based on this decision rule. This situation is illustrated in Figure 9.2, which shows the sampling distribution of the random variable 9.2.1 when the null hypothesis is true, through a graph of its probability density function. The figure shows the value z_α, which is such that the probability of its being exceeded, when the null hypothesis is true, is the significance level α of the test. It follows that the probability of a sample result in the corresponding rejection region, shown as the shaded area in the figure, must be α when the null hypothesis is correct.

A Test of the Mean of a Normal Population: Population Variance Known

Suppose we have a random sample of n observations from a normal population with mean μ and known variance σ^2. If the observed sample mean is $\bar{x}$, then a test with significance level α of the null hypothesis

$$H_0: \quad \mu = \mu_0$$

against the alternative

$$H_1: \quad \mu > \mu_0$$

is obtained from the decision rule

$$\text{Reject } H_0 \text{ if } \quad \frac{\bar{x} - \mu_0}{\sigma/\sqrt{n}} > z_\alpha$$

---

where z_α is that number for which

$$P(Z > z_\alpha) = \alpha$$

and Z is a standard normal random variable.

Example
9.1
When a process producing ball bearings is operating correctly, the weights of the ball bearings have a normal distribution with mean 5 ounces and standard deviation 0.1 ounce. An adjustment has been made to the process, and the plant manager suspects this has raised the mean weight of ball bearings produced, leaving the standard deviation unchanged. A random sample of sixteen ball bearings is taken, and their mean weight is found to be 5.038 ounces. Test at significance levels 0.05 and 0.10 (that is, at 5% and 10% levels) the null hypothesis that the population mean weight is 5 ounces against the alternative that it is bigger.

Denoting by μ the population mean weight (in ounces), we want to test

$$H_0: \quad \mu = \mu_0 = 5$$

against

$$H_1: \quad \mu > 5$$

The decision rule is to reject H_0 in favor of H_1 if

$$\frac{\bar{x} - \mu_0}{\sigma/\sqrt{n}} > z_\alpha$$

From the statement of the example, we have

$$\bar{x} = 5.038; \quad \mu_0 = 5; \quad \sigma = 0.1; \quad n = 16$$

so that

$$\frac{\bar{x} - \mu_0}{\sigma/\sqrt{n}} = \frac{5.038 - 5}{0.1/\sqrt{16}} = 1.52$$

For a 5% level test, we find from Table 3 in the Appendix,

$$z_{.05} = 1.645$$

Since 1.52 does not exceed 1.645, we fail to reject the null hypothesis at the 5% level of significance; that is, the null hypothesis is accepted at this significance level. In other words, if we use a test that ensures that the probability of rejecting the null hypothesis when it is in fact true is 0.05, the sample data do not contain enough evidence to allow rejection of that hypothesis.

For a 10% level test, we have

$$z_{.10} = 1.28$$

Since 1.52 is bigger than 1.28, the null hypothesis is rejected at the 10% level of significance. To this extent, then, there is some evidence in the data to suggest that the true mean weight exceeds 5 ounces.

Let us pause to consider what is meant by the rejection of a null hypothesis. In Example 9.1 the hypothesis that the population mean weight is 5 ounces was rejected by a test with significance level 0.1. This certainly does not mean that we have *proved* that the true mean exceeds 5 ounces. Given only sample information, it will never be possible to be *certain* about a population parameter. Rather, we might view the data as having cast some doubt on the truth of the null hypothesis. If that hypothesis were true, then the observed value

$$\frac{\bar{x} - \mu_0}{\sigma/\sqrt{n}} = 1.52$$

would represent a single observation drawn from a standard normal population. In testing hypotheses, we are really asking how likely it would be to observe such an extreme value if the null hypothesis were in fact true. In Example 9.1, we saw that the probability of observing a value bigger than 1.28 is 0.1. Hence, in rejecting the null hypothesis we are saying either that the null hypothesis is false, or that we have observed an unlikely event—one that would occur only with the probability specified by the significance level. This is the sense in which the sample information has aroused doubt about the null hypothesis.

Notice that in Example 9.1 the null hypothesis was rejected at significance level 0.10, but was not rejected at the lower level 0.05. In lowering the significance level, we are reducing the probability of rejecting a true null hypothesis, and therefore modifying the decision rule to make it less likely that the null hypothesis will be rejected, whether or not it is true. Obviously the lower the significance level at which a null hypothesis can be rejected, the greater the doubt cast on its truth. Rather than testing hypotheses at preassigned levels of significance, investigators often determine the smallest level of significance at which a null hypothesis can be rejected.

Definition

The smallest significance level at which a null hypothesis can be rejected is called the **probability-value**, or ***p*-value**, of the test.

In Example 9.1 we found

$$\frac{\bar{x} - \mu_0}{\sigma/\sqrt{n}} = 1.52$$

Therefore, according to our decision rule, the null hypothesis is rejected for any significance level α for which z_α is less than 1.52. From Table 3 in the Appendix we find that, when z_α is 1.52, α is equal to 0.0643. This, then, is the *p*-value of the test. The implication is that the null hypothesis can be rejected at all levels of significance

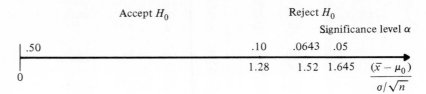

FIGURE 9.3 Rejection regions for testing H_0: $\mu = \mu_0$ against H_1: $\mu > \mu_0$ for significance levels .10, .0643, .05

higher than 6.43%. This is illustrated in Figure 9.3, which shows the correspondence between the significance level of the test, α, and the corresponding value z_α, which enters the decision rule.

Suppose that, in place of the simple null hypothesis, we had wanted to test the composite null hypothesis

$$H_0: \quad \mu \leq \mu_0$$

against the alternative

$$H_1: \quad \mu > \mu_0$$

at significance level α. For the decision rule developed in the case of the simple null hypothesis, we saw that if the population mean is precisely μ_0, then the probability of rejecting the null hypothesis is α. For this same decision rule, if the true population mean is anything less than μ_0 we would be even less likely to reject the null hypothesis. Hence, use of this decision rule in the present context *guarantees* a probability of *at most* α of rejecting the composite null hypothesis when it is true.

A Test of the Mean of a Normal Distribution (Variance Known): Composite Null and Alternative Hypotheses

The appropriate procedure for testing, at significance level α, the null hypothesis

$$H_0: \quad \mu \leq \mu_0$$

against the alternative hypothesis

$$H_1: \quad \mu > \mu_0$$

is precisely the same as when the null hypothesis is H_0: $\mu = \mu_0$

Consider, now, the problem of testing the simple null hypothesis

$$H_0: \quad \mu = \mu_0$$

against the composite alternative that the true mean is *less than* μ_0, that is,

$$H_1: \quad \mu < \mu_0$$

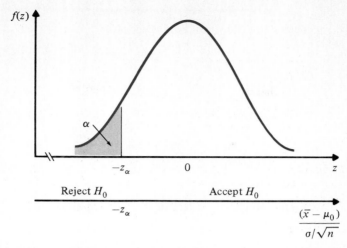

FIGURE 9.4 The probability density function of $Z = (\bar{X} - \mu_0)/(\sigma/\sqrt{n})$ when the null hypothesis H_0: $\mu = \mu_0$ is true, and the decision rule for testing H_0 against the alternative H_1: $\mu < \mu_0$ at significance level α

In this circumstance, doubt would be cast on the null hypothesis if the sample mean were a good deal *lower* than the hypothesized population mean. Once again, if the null hypothesis were true, the random variable 9.2.1 would follow a standard normal distribution. To achieve a test with significance level α, we only need to note that

$$P(Z < -z_\alpha) = \alpha$$

if Z is a standard normal random variable. Hence, if $\bar{x}$ is the observed sample mean, the appropriate decision rule is:

$$\text{Reject } H_0 \text{ if } \quad \frac{\bar{x} - \mu_0}{\sigma/\sqrt{n}} < -z_\alpha$$

This is illustrated in Figure 9.4, which should be compared with Figure 9.2. Clearly, the former is simply the mirror image of the latter.

Using an analogous argument to that developed earlier, we can see that this decision rule continues to be appropriate if, in place of the simple null hypothesis, we have the composite hypothesis

$$H_0: \quad \mu \geq \mu_0$$

with the same alternative hypothesis.

Tests of the Mean of a Normal Distribution: Population Variance Known

Suppose we have a random sample of n observations from a normal population with mean μ and known variance σ^2. If the observed sample mean is $\bar{x}$, a test with

---

significance level α of either null hypothesis

$$H_0: \quad \mu = \mu_0 \quad \text{or} \quad H_0: \quad \mu \geq \mu_0$$

against the alternative

$$H_1: \quad \mu < \mu_0$$

is obtained from the decision rule

$$\text{Reject } H_0 \text{ if} \quad \frac{\bar{x} - \mu_0}{\sigma/\sqrt{n}} < -z_\alpha$$

We now consider the test of the null hypothesis

$$H_0: \quad \mu = \mu_0$$

against the two-sided alternative

$$H_1: \quad \mu \neq \mu_0$$

It is assumed here that the investigator has no strong reason for suspecting departures on one side rather than the other side of the hypothesized population mean. The null hypothesis would then be doubted if the observed sample mean were either much higher *or* much lower than μ_0. Once again, if the null hypothesis is true, the random variable 9.2.1 has a standard normal distribution. To obtain a test with significance level α, note that, under the null hypothesis,

$$P(Z > z_{\alpha/2}) = \frac{\alpha}{2} \quad \text{and} \quad P(Z < -z_{\alpha/2}) = \frac{\alpha}{2}$$

Hence, the probability that Z either exceeds $z_{\alpha/2}$ or is less than $-z_{\alpha/2}$ is α. It therefore follows that a test of level α is obtained from the decision rule:

$$\text{Reject } H_0 \text{ if} \quad \frac{\bar{x} - \mu_0}{\sigma/\sqrt{n}} \text{ is either bigger than } z_{\alpha/2} \text{ or less than } -z_{\alpha/2}$$

This is illustrated in Figure 9.5, from which we see that the region of sample outcomes for which the null hypothesis is rejected is divided into two parts. The upper part of the region corresponds to observed values of the sample mean greatly in excess of the hypothesized population mean, and the lower part to values of the sample mean that are substantially below μ_0.

Test for the Mean of a Normal Distribution Against Two-Sided Alternative: Population Variance Known

Suppose we have a random sample of n observations from a normal population

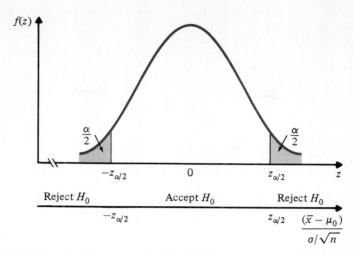

FIGURE 9.5 The probability density function of $Z = (\bar{X} - \mu_0)/(\sigma/\sqrt{n})$ when the null hypothesis H_0: $\mu = \mu_0$ is true, and the decision rule for testing H_0 against the alternative H_1: $\mu \neq \mu_0$ at significance level α

with mean μ and known variance σ^2. If the observed sample mean is $\bar{x}$, then a test with significance level α of the null hypothesis

$$H_0: \quad \mu = \mu_0$$

against the two-sided alternative

$$H_1: \quad \mu \neq \mu_0$$

is obtained from the decision rule

$$\text{Reject } H_0 \text{ if } \quad \frac{\bar{x} - \mu_0}{\sigma/\sqrt{n}} > z_{\alpha/2} \quad \text{or} \quad \frac{\bar{x} - \mu_0}{\sigma/\sqrt{n}} < -z_{\alpha/2}$$

The reader has probably noticed the similarity between the developments of procedures for determining confidence intervals and testing hypotheses. A review of the material in Section 8.2 will clarify the relationship: *The null hypothesis $H_0: \mu = \mu_0$ is rejected against the two-sided alternative H_1: $\mu \neq \mu_0$ at significance level α if and only if the $100(1 - \alpha)\%$ confidence interval for μ does not contain μ_0. This relationship holds for most such tests developed in subsequent sections of this chapter.*

Example 9.2

A drill, as part of an assembly line operation, is used to drill holes in sheet metal. When the drill is functioning properly, the diameters of these holes have a normal distribution with mean 2 inches and standard deviation 0.06 inch. Periodically, to check that the drill is functioning properly, the diameters of a random sample of holes are measured. Assume that the standard deviation does not vary. A random sample of nine measurements yielded mean diameter 1.95 inches. Test the null hypothesis that the population mean is 2 inches against the alternative that it is not. Use a 5% significance level and also find the p-value of the test.

Let μ denote the population mean diameter (in inches). Then we require to test

$$H_0: \quad \mu = \mu_0 = 2$$

against

$$H_1: \quad \mu \neq 2$$

The decision rule is to reject H_0 in favor of H_1 if

$$\frac{\bar{x} - \mu_0}{\sigma/\sqrt{n}} > z_{\alpha/2} \qquad \text{or} \qquad \frac{\bar{x} - \mu_0}{\sigma/\sqrt{n}} < -z_{\alpha/2}$$

We have

$$\bar{x} = 1.95; \qquad \mu_0 = 2; \qquad \sigma = 0.06; \qquad n = 9$$

and so

$$\frac{\bar{x} - \mu_0}{\sigma/\sqrt{n}} = \frac{1.95 - 2}{0.06/\sqrt{9}} = -2.50$$

For a 5% level test, $\alpha = 0.05$, and $z_{\alpha/2} = z_{.025} = 1.96$. Then, since -2.50 is less than -1.96, the null hypothesis is rejected at the 5% significance level.

In fact, according to the decision rule, the null hypothesis will be rejected for any significance level α for which $-z_{\alpha/2}$ is bigger than -2.50. From Table 3 in the Appendix, we see that when $z_{\alpha/2}$ is 2.50, $\alpha/2$ is equal to 0.0062. Hence, $\alpha = 0.0124$. This is the p-value of the test, implying that the null hypothesis can be rejected against the two-sided alternative at any level of significance greater than 1.24%. This certainly casts substantial doubt on the hypothesis that the drill is functioning correctly.

Until now we have dealt only with the (generally unrealistic) case where the population variance is known. However, if the available number of sample observations is large, the tests can readily be modified to deal with an important class of practical problems. As in Section 8.2, we are indebted to the Central Limit Theorem, which allows us to conclude that, for large samples, the sampling distribution of the sample mean will be approximately normal, even though the population distribution is not normal.

Tests for the Mean: Large Sample Sizes

Suppose we have a random sample of n observations from a population with mean μ and variance σ^2. If the sample size n is large,[1] the test procedures developed for the

---

[1] This approximation is generally satisfactory for samples of thirty or more observations.

case where the population variance is known can be employed when it is unknown, replacing σ^2 by the observed sample variance s_x^2. Moreover, these procedures remain approximately valid even if the population distribution is not normal.

Example 9.3

It might be suspected that the firms most likely to attract take-over bids are those that have been achieving relatively poor returns. One measure of such performance is through "abnormal returns," which average 0 over all firms. A random sample[2] of 88 firms for which cash tender offers had been made showed abnormal returns with a mean of -0.0029 and a standard deviation of 0.0169 in the period from 24 months to 4 months prior to the take-over bids. Test the null hypothesis that the mean abnormal returns for this population is 0 against the alternative that it is negative.

Let μ denote the population mean abnormal returns. Then we want to test

$$H_0: \quad \mu = \mu_0 = 0$$

against

$$H_1: \quad \mu < 0$$

The decision rule is to reject H_0 in favor of H_1 if

$$\frac{\bar{x} - \mu_0}{s_x/\sqrt{n}} < -z_\alpha$$

Here we have

$$\bar{x} = -0.0029; \qquad \mu_0 = 0; \qquad s_x = 0.0169; \qquad n = 88$$

and hence,

$$\frac{\bar{x} - \mu_0}{s_x/\sqrt{n}} = \frac{-0.0029}{0.0169/\sqrt{88}} = -1.61$$

According to our decision rule, the null hypothesis is rejected for any significance level α for which $-z_\alpha$ is bigger than -1.61. From Table 3 in the Appendix, we see that when z_α is 1.61, α is equal to 0.0537. Hence, the null hypothesis is rejected at any significance level bigger than 5.37%. Thus, the probability of observing sample mean abnormal returns as low as or lower than those actually observed would be .0537 if the true mean abnormal returns for all firms attracting take-over bids were 0. The data suggest quite strongly that, on average, abnormal returns are lower for such firms.

The conclusion is illustrated in Figure 9.6, which shows the distribution of the decision rule criterion under the null hypothesis, with the lower-tail area probability corresponding to the observed value.

---

[2] See D. R. Kummer and J. R. Hoffmeister, "Valuation consequences of cash tender offers," *Journal of Finance, 33* (1978), 505–16.

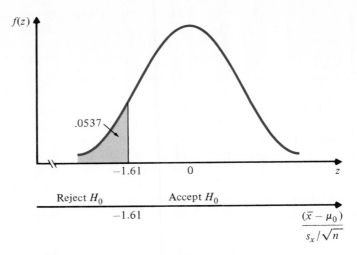

FIGURE 9.6 Conclusion of the test in Example 9.3; the null hypothesis H_0: $\mu = \mu_0$ is rejected against the alternative H_1: $\mu < \mu_0$ at significance levels greater than .0537

9.3 TESTS OF THE MEAN OF A NORMAL DISTRIBUTION: POPULATION VARIANCE UNKNOWN

In this section we again consider the problem where a random sample of n observations is taken from a normal population, and it is required to test hypotheses about the population mean μ. However, the population variance is no longer assumed known. If the sample size is not large, the procedures discussed at the end of Section 9.2 are no longer appropriate. Nevertheless, valid tests can be derived, based on a result discussed in Sections 8.3 and 8.4. It was seen there that, if the sample mean and variance are denoted $\bar{X}$ and s_X^2, the random variable

$$t_{n-1} = \frac{\bar{X} - \mu}{s_X / \sqrt{n}}$$

follows a Student's t distribution with $(n-1)$ degrees of freedom. Following precisely the line of argument adopted in Section 9.2, with the Student's t distribution now playing the same role as the standard normal distribution, we can obtain valid tests, as indicated in the box.

Tests of the Mean of a Normal Distribution: Population Variance Unknown

Suppose we have a random sample of n observations from a normal population with mean μ. If the observed sample mean and standard deviation are $\bar{x}$ and s_x, then the following tests have significance level α:

(i) To test either null hypothesis

$$H_0: \quad \mu = \mu_0 \qquad \text{or} \qquad H_0: \quad \mu \leq \mu_0$$

against the alternative

$$H_1: \quad \mu > \mu_0$$

the decision rule is

$$\text{Reject } H_0 \text{ if } \quad \frac{\bar{x} - \mu_0}{s_x/\sqrt{n}} > t_{n-1,\alpha}$$

(ii) To test either null hypothesis

$$H_0: \quad \mu = \mu_0 \qquad \text{or} \qquad H_0: \quad \mu \geq \mu_0$$

against the alternative

$$H_1: \quad \mu < \mu_0$$

the decision rule is

$$\text{Reject } H_0 \text{ if } \quad \frac{\bar{x} - \mu_0}{s_x/\sqrt{n}} < -t_{n-1,\alpha}$$

(iii) To test the null hypothesis

$$H_0: \quad \mu = \mu_0$$

against the two-sided alternative

$$H_1: \quad \mu \neq \mu_0$$

the decision rule is

$$\text{Reject } H_0 \text{ if } \quad \frac{\bar{x} - \mu_0}{s_x/\sqrt{n}} > t_{n-1,\alpha/2} \qquad \text{or} \qquad \frac{\bar{x} - \mu_0}{s_x/\sqrt{n}} < -t_{n-1,\alpha/2}$$

Here, $t_{n-1,\alpha}$ is that number for which

$$P(t_{n-1} > t_{n-1,\alpha}) = \alpha$$

where the random variable t_{n-1} follows a Student's t distribution with $(n-1)$ degrees of freedom.

Figure 9.7 illustrates the setup of the test against a two-sided alternative. The probability density function is now that of the Student's t distribution, and this figure is the analogue of Figure 9.5, which related to the case where the population variance was known. In an obvious way, tests against one-sided alternative hypotheses can be viewed pictorially in a manner analogous to Figures 9.2 and 9.4.

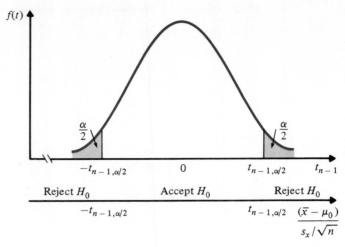

FIGURE 9.7 The probability density function of $t_{n-1} = \dfrac{(\bar{X} - \mu_0)}{s_X/\sqrt{n}}$ when the null hypothesis H_0: $\mu = \mu_0$ is true, and the decision rule for testing H_0 against the alternative H_1: $\mu \neq \mu_0$ at significance level α

Example
9.4

A retail chain knows that, on average, sales in its stores are 20% higher in December than in November. For the current year, a random sample of six stores was selected. Their percentage December sales increases were found to be

$$19.2; \qquad 18.4; \qquad 19.8; \qquad 20.2; \qquad 20.4; \qquad 19.0$$

Assuming a normal population distribution, test the null hypothesis that the true mean percentage sales increase is 20, against the two-sided alternative, at the 10% significance level.

Letting μ denote the population mean percentage increase in sales in December, we want to test the null hypothesis

$$H_0: \quad \mu = \mu_0 = 20$$

against the alternative

$$H_1: \quad \mu \neq 20$$

The decision rule is to reject H_0 in favor of H_1 if

$$\frac{\bar{x} - \mu_0}{s_x/\sqrt{n}} > t_{n-1,\alpha/2} \qquad \text{or} \qquad \frac{\bar{x} - \mu_0}{s_x/\sqrt{n}} < -t_{n-1,\alpha/2}$$

The sample mean and variance are obtained by using the computations in the accompanying table.

We have for the sample mean,

$$\bar{x} = \frac{\Sigma x_i}{n} = \frac{117}{6} = 19.5$$

	x_i	x_i^2
	19.2	368.64
	18.4	338.56
	19.8	392.04
	20.2	408.04
	20.4	416.16
	19.0	361.00
Sums	117	2,284.44

and for the sample variance,

$$s_x^2 = \frac{\Sigma x_i^2 - n\bar{x}^2}{n-1} = \frac{2,284.44 - (6)(19.5)^2}{5} = 0.588$$

so that the sample standard deviation is

$$s_x = \sqrt{0.588} = 0.767$$

We then have

$$\frac{\bar{x} - \mu_0}{s_x/\sqrt{n}} = \frac{19.5 - 20}{0.767/\sqrt{6}} = -1.597$$

Since a test of significance level $\alpha = 0.10$ is required, we have from Table 6 of the Appendix,

$$t_{n-1,\alpha/2} = t_{5,.05} = 2.015$$

Thus, since -1.597 lies between -2.015 and 2.015, the null hypothesis that the true mean percentage increase is 20 is accepted at the 10% level. The evidence in the data against this hypothesis is not terribly strong.

EXERCISES

1. An automobile manufacturer claims that a new model car achieves on average 31.5 miles per gallon in highway driving. The distribution is known to be normal, with standard deviation 2.4 miles per gallon. A random sample of sixteen of the automobiles obtained an average of 30.6 miles per gallon in highway trials. Test at the 5% level of significance the null hypothesis that the population mean is at least 31.5.

2. A breakfast cereal manufacturer is told by an advertising agency that the use of a new advertising display in supermarkets will lead to an average increase of at least 10% in the sales of a brand of cereal. A random sample of nine supermarkets was chosen for a pilot study. On the basis of past experience, it was believed that the population distribution for percentage sales increases would be normal, with standard deviation 3.8%. For this sample, the mean percentage increase in sales was 8.1%. Find the lowest level of significance at which it is possible to reject the null hypothesis that the population mean sales increase is 10% against the alternative that it is less.

3. A chemical manufacturer is concerned about the level of impurities in batches of chemicals sent out to customers. It is known that, from a particular production run, the impurity concentration of these batches will follow a normal distribution with standard

deviation 2.2%. A random sample of twenty-five batches from a production run is tested for impurity content, and the total output of the run is labeled substandard, and sold at a lower price, if the sample mean impurity level exceeds 11.25%. What is the probability, under this decision rule, that a production run for which the population mean impurity level is 10% will be classified as substandard?

4. A professor of finance has administered a standard examination to his students in an introductory course over a period of several years. He has found that the scores follow a normal distribution with mean 75.2 and standard deviation 10.4. The professor takes a year's leave of absence to teach at another university, where, at the end of his course, he administers the same examination to twenty students, who can be regarded as a random sample of all that university's finance students. These students achieved a sample mean score of 73.5. Assuming a normal population distribution with standard deviation 10.4, test at the 5% level of significance, against a two-sided alternative, the null hypothesis that the population mean score is 75.2. Find the p-value for this test.

5. The cloze readability test was given to a random sample[3] of 328 commercial bank loan officer trainees to check their understanding of financial report messages. Their mean score on the test was 55.3177 and the sample standard deviation was 10.8324. Test the null hypothesis that the population mean is at least 57, the minimum score necessary to demonstrate adequate understanding of written material.

6. In Example 9.3 we found that the null hypothesis that mean abnormal returns were 0, for firms attracting take-over bids, could be rejected against the alternative that the population mean was negative at all levels of significance higher than 5.37%. Would it therefore be correct to say that the probability that the null hypothesis is true is 0.0537? If not, replace this statement by a valid probability statement.

7. A book club advertises an introductory offer, through which new members are entitled to receive an initial package of books at a nominal price, with no obligation to purchase additional books. The club estimates that this promotion will be profitable if the new members buy, on average, at least 2.8 additional books during the following year. A random sample of 100 records of members attracted by this offer showed a mean purchase in the next year of 2.61 books, and a sample standard deviation of 0.9 book. Test the null hypothesis that the promotion was profitable.

8. The accounts of a company show that, on average, accounts receivable are $94.27. An auditor checks a random sample of 150 of these accounts, finding a sample mean of $88.61 and standard deviation $40.56. Find the lowest level of significance at which the null hypothesis that the population mean is $94.27 can be rejected against a two-sided alternative.

9. A fertilizer company claims that its product improves average corn yields by at least 8 bushels per acre. A random sample of sixteen observations shows an average increase of 6.8 bushels per acre, and sample standard deviation 2.2 bushels per acre. Assuming that yield increases are normally distributed, test the company's claim at the 5% significance level.

10. A magazine claims that the average annual income of its readers is at least $36,000. A random sample of six readers yielded the following annual incomes (in dollars):

    28,400;   20,900;   39,300;   38,400;   30,100;   24,200

    Assuming that readers' incomes are normally distributed, test the magazine's claim.

11. An assembly line operation produces on average 650 finished components per day. At the insistence of a government inspector, a new safety device is installed, which management fears will lead to a reduction in average daily output. After the installation

---

[3] See A. H. Adelberg, "A methodology for measuring the understandability of financial report messages," *Journal of Accounting Research, 17* (1979), 565–92.

of this device, a random sample of eight days' output gave the following results:

620; 662; 587; 641; 623; 567; 598; 639

Test the null hypothesis that mean daily output is unchanged against the alternative that it has fallen since the safety device was installed.

12. On the basis of a random sample, the null hypothesis

$$H_0: \quad \mu = \mu_0$$

is tested against the alternative

$$H_1: \quad \mu > \mu_0$$

and the null hypothesis is accepted at the 5% significance level.

(a) Does this necessarily imply that μ_0 is contained in the 95% confidence interval for μ?

(b) Does this necessarily imply that μ_0 is contained in the 90% confidence interval for μ, if the observed sample mean is bigger than μ_0?

13. A company selling franchises advertises that operators obtain, on average during the first year, a yield of 10% on their initial investments. A random sample of ten of these franchises produced the following yields for the first year of operation:

6.1; 9.2; 11.5; 8.6; 12.1;

3.9; 8.4; 10.1; 9.4; 8.7

Assuming that population yields are normally distributed, test the company's claim.

14. A process that produces bags of sugar, when operating correctly, yields bags whose contents weigh, on average, 20 ounces. A random sample of nine bags produced the following weights (in ounces):

21.3; 19.8; 19.6; 20.5; 20.9;

20.2; 19.9; 20.7; 20.3

Test, against a two-sided alternative, the null hypothesis that the process is functioning correctly, assuming the population distribution is normal.

15. In contract negotiations, a company claims that a new incentive scheme has resulted in average weekly earnings of at least $400 for all production workers. A union representative takes a random sample of fifteen workers, and finds their weekly earnings have an average of $381.25 and a standard deviation of $48.62.

(a) Test the company's claim.

(b) If the same sample results had been obtained from a random sample of fifty employees, could the company's claim be rejected at a lower significance level than in part (a)?

9.4 TESTS OF THE VARIANCE OF A NORMAL DISTRIBUTION

In this section, we develop procedures for testing the population variance σ^2, based on a random sample of n observations from a normal population.

It is natural to base these tests on the sample variance s_X^2. The basis for developing particular tests lies in the fact that the random variable

$$\chi^2_{n-1} = \frac{(n-1)s^2_X}{\sigma^2}$$

follows a chi-square distribution with $(n-1)$ degrees of freedom.[4] If the null hypothesis is that the population variance is equal to some specified value σ^2_0, that is,

$$H_0: \quad \sigma^2 = \sigma^2_0$$

then when this hypothesis is true the random variable

$$\chi^2_{n-1} = \frac{(n-1)s^2_X}{\sigma^2_0} \qquad (9.4.1)$$

obeys a chi-square distribution with $(n-1)$ degrees of freedom. Tests of hypotheses about the variance of a normal population are then based on the sample value observed for the quantity 9.4.1. If the alternative hypothesis is that the true variance exceeds σ^2_0, we would be suspicious of the null hypothesis if the observed sample variance was much bigger than σ^2_0. Hence, the null hypothesis would be rejected if a high value of 9.4.1 was observed. Conversely, if the alternative is that the population variance is less than the value specified by the null hypothesis, the null hypothesis would be rejected for low values of 9.4.1. Finally, for the two-sided alternative that the population variance differs from σ^2_0, we would want to reject the null hypothesis on observing either unusually high or unusually low values of 9.4.1.

The rationale for the development of appropriate tests now follows the same pattern as in Section 9.2. As a preliminary, we recall a notation introduced in Section

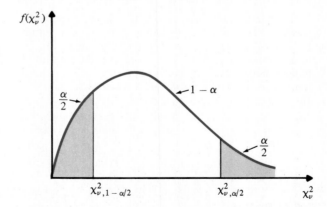

FIGURE 9.8 Some probabilities for the chi-square distribution

---

[4] Once again, we caution the reader that the validity of this result rests crucially on the assumption that the population distribution is normal.

8.5. We denote by $\chi^2_{\nu,\alpha}$ that number which is exceeded with probability α by a chi-square random variable with v degrees of freedom. Hence,

$$P(\chi^2_\nu > \chi^2_{\nu,\alpha}) = \alpha$$

It then follows that

$$P(\chi^2_\nu < \chi^2_{\nu,1-\alpha}) = \alpha$$

and that

$$P(\chi^2_\nu > \chi^2_{\nu,\alpha/2}) \quad \text{or} \quad \chi^2_\nu < \chi^2_{\nu,1-\alpha/2}) = \alpha$$

These probabilities are shown in Figure 9.8. Tests for the normal variance then follow as indicated in the box.

Tests of the Variance of a Normal Population

Suppose we have a random sample of n observations from a normal population with variance σ^2. If the observed sample variance is s_x^2, then the following tests have significance level α:

(i) To test either null hypothesis

$$H_0: \quad \sigma^2 = \sigma_0^2 \quad \text{or} \quad H_0: \quad \sigma^2 \leq \sigma_0^2$$

against the alternative

$$H_1: \quad \sigma^2 > \sigma_0^2$$

the decision rule is

$$\text{Reject } H_0 \text{ if } \quad \frac{(n-1)s_x^2}{\sigma_0^2} > \chi^2_{n-1,\alpha}$$

(ii) To test either null hypothesis

$$H_0: \quad \sigma^2 = \sigma_0^2 \quad \text{or} \quad H_0: \quad \sigma^2 \geq \sigma_0^2$$

against the alternative

$$H_1: \quad \sigma^2 < \sigma_0^2$$

the decision rule is

$$\text{Reject } H_0 \text{ if } \quad \frac{(n-1)s_x^2}{\sigma_0^2} < \chi^2_{n-1,1-\alpha}$$

(iii) To test the null hypothesis

$$H_0: \quad \sigma^2 = \sigma_0^2$$

against the two-sided alternative

$$H_1: \quad \sigma^2 \neq \sigma_0^2$$

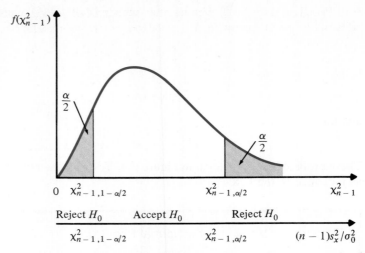

$f(\chi^2_{n-1})$

$\dfrac{\alpha}{2}$

$\dfrac{\alpha}{2}$

0    $\chi^2_{n-1,1-\alpha/2}$ $\chi^2_{n-1,\alpha/2}$ χ^2_{n-1}

Reject H_0 Accept H_0 Reject H_0

$\chi^2_{n-1,1-\alpha/2}$ $\chi^2_{n-1,\alpha/2}$ $(n-1)s^2_x/\sigma^2_0$

FIGURE 9.9 The probability density function of $\chi^2_{n-1} = (n-1)s^2_x/\sigma^2_0$ when the null hypothesis H_0: $\sigma^2 = \sigma^2_0$ is true, and the decision rule for testing H_0 against the alternative H_1: $\sigma^2 \neq \sigma^2_0$ at significance level α

---

the decision rule is

Reject H_0 if $\quad \dfrac{(n-1)s^2_x}{\sigma^2_0} > \chi^2_{n-1,\alpha/2} \quad$ or $\quad \dfrac{(n-1)s^2_x}{\sigma^2_0} < \chi^2_{n-1,1-\alpha/2}$

Here $\chi^2_{n-1,\alpha}$ is that number for which

$$P(\chi^2_{n-1} > \chi^2_{n-1,\alpha}) = \alpha$$

where the random variable χ^2_{n-1} follows a chi-square distribution with $(n-1)$ degrees of freedom.

The decision rule for the test against the two-sided alternative is illustrated in Figure 9.9.

Example
9.5

In order to meet established standards, it is important that the variance of the percentage impurity levels in consignments of a chemical not exceed 4.0. A random sample of twenty consignments had a sample variance of 5.62 in impurity level percentage. Test the null hypothesis that the population variance is no more than 4.0.

Let σ^2 denote the population variance of impurity concentrations. The null hypothesis

$$H_0: \quad \sigma^2 \leq \sigma^2_0 = 4.0$$

is to be tested against

$$H_1: \quad \sigma^2 > 4.0$$

Based on the assumption that the population distribution is normal, the decision rule, for a test of significance level α, is to reject H_0 in favor of H_1 if

$$\frac{(n-1)s_x^2}{\sigma_0^2} > \chi_{n-1,\alpha}^2$$

From the statement of the example we have

$$s_x^2 = 5.62; \qquad n = 20; \qquad \sigma_0^2 = 4.0$$

Hence,

$$\frac{(n-1)s_x^2}{\sigma_0^2} = \frac{(19)(5.62)}{4.0} = 26.695$$

For a 10% level test, $\alpha = 0.10$ and we see from Table 5 in the Appendix that the corresponding cutoff point of the chi-square distribution with $(n-1) = 19$ degrees of freedom is

$$\chi_{19,.10}^2 = 27.20$$

Therefore, since 26.695 is not bigger than 27.20, the null hypothesis cannot be rejected at the 10% level. Hence, the data do not contain terribly strong evidence against the hypothesis that the population variance in impurity level percentages is at most 4.0.

9.5 TESTS OF THE POPULATION PROPORTION (LARGE SAMPLES)

In many practical problems we want to test hypotheses about the proportion of members of a large population possessing some particular attribute. Inference about the population proportion is based on the proportion of individuals in a random sample who possess the attribute of interest.

Denoting by p the population proportion, and by $\hat{p}_X$ the proportion in a random sample of n observations, we know that if the sample size is large, then to a good approximation the random variable

$$Z = \frac{\hat{p}_X - p}{\sqrt{p(1-p)/n}}$$

has a standard normal distribution. That the sampling distribution for the sample proportion is approximately normal when the sample size is large follows, as we have noted in previous chapters, as a result of the Central Limit Theorem.

If the null hypothesis is that the population proportion is equal to some specific value p_0, it follows that when this hypothesis is true, the random variable

$$Z = \frac{\hat{p}_X - p_0}{\sqrt{p_0(1-p_0)/n}}$$

follows a standard normal distribution. We can now deduce appropriate tests for the population proportion, as described in the box.

Tests of the Population Proportion (Large Sample Sizes)

Suppose we have a sample of n observations from a population, a proportion p of whose members possess a particular attribute. Then, if the number of sample observations is large[5] and the observed sample proportion is $\hat{p}_x$, the following tests have significance level α:

(i) To test either null hypothesis

$$H_0: \quad p = p_0 \quad \text{or} \quad H_0: \quad p \leq p_0$$

against the alternative

$$H_1: \quad p > p_0$$

the decision rule is

$$\text{Reject } H_0 \text{ if } \quad \frac{\hat{p}_x - p_0}{\sqrt{p_0(1 - p_0)/n}} > z_\alpha$$

(ii) To test either null hypothesis

$$H_0: \quad p = p_0 \quad \text{or} \quad H_0: \quad p \geq p_0$$

against the alternative

$$H_1: \quad p < p_0$$

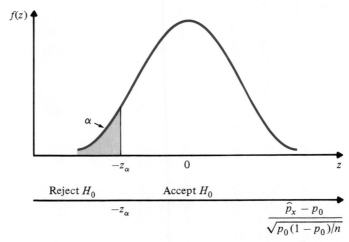

FIGURE 9.10 The probability density function of $Z = \dfrac{\hat{p}_x - p_0}{\sqrt{p_0(1 - p_0)/n}}$ when the null hypothesis $H_0: \quad p = p_0$ is true, and the decision rule for testing H_0 against the alternative $H_1: \quad p < p_0$ at significance level α

---

[5] The approximations implied here are generally adequate for samples of forty or more observations.

the decision rule is

$$\text{Reject } H_0 \text{ if } \quad \frac{\hat{p}_x - p_0}{\sqrt{p_0(1 - p_0)/n}} < -z_\alpha$$

(iii) To test the null hypothesis

$$H_0: \quad p = p_0$$

against the two-sided alternative

$$H_1: \quad p \neq p_0$$

the decision rule is

$$\text{Reject } H_0 \text{ if } \quad \frac{\hat{p}_x - p_0}{\sqrt{p_0(1 - p_0)/n}} > z_{\alpha/2} \quad \text{or} \quad \frac{\hat{p}_x - p_0}{\sqrt{p_0(1 - p_0)/n}} < -z_{\alpha/2}$$

Here, as previously, z_α is that number for which

$$P(Z > z_\alpha) = \alpha$$

where the random variable Z has a standard normal distribution.

The decision rule for the second of these tests is illustrated in Figure 9.10.

Example 9.6

Forecasts of corporate earnings per share are made on a regular basis by many financial analysts. In a random sample[6] of 600 forecasts, it was found that 382 of these forecasts exceeded the actual outcome for earnings. Test against a two-sided alternative the null hypothesis that the population proportion of forecasts that are higher than actual outcomes is 0.50. (This is the hypothesis we would expect to be true if there were no overall tendency for financial analysts to be either unduly optimistic or unduly pessimistic about earnings prospects.)

Let p denote the population proportion of forecasts that are above actual outcomes. We want to test

$$H_0: \quad p = p_0 = 0.50$$

against

$$H_1: \quad p \neq 0.50$$

The decision rule is to reject H_0 in favor of the alternative if

$$\frac{\hat{p}_x - p_0}{\sqrt{p_0(1 - p_0)/n}} > z_{\alpha/2} \quad \text{or} \quad \frac{\hat{p}_x - p_0}{\sqrt{p_0(1 - p_0)/n}} < -z_{\alpha/2}$$

We have

$$p_0 = 0.50; \quad n = 600; \quad \hat{p}_x = \frac{382}{600} = 0.637$$

---

[6] These results are reported by R. M. Barefield and E. E. Comiskey, "The accuracy of analysts' forecasts of earnings per share," *Journal of Business Research, 3* (1975), 241–52.

Then

$$\frac{\hat{p}_x - p_0}{\sqrt{p_0(1 - p_0)/n}} = \frac{0.637 - 0.50}{\sqrt{(0.50)(0.50)/600}} = 6.71$$

Referring to Table 3 of the Appendix, we see that such an outcome would be extremely unlikely if the null hypothesis were indeed true. For instance, for a 0.2% level test, $\alpha = 0.002$, so that

$$z_{\alpha/2} = z_{.001} = 3.09$$

Since 6.71 is much bigger than 3.09, the null hypothesis is clearly rejected even at such a low significance level. The evidence against the hypothesis that the population proportion of forecasts that exceed outcome is 0.50 is really overwhelming. It is clear from these data that there is an overall tendency for the proportion of financial analysts' forecasts of future corporate earnings that exceed the actual values to differ from 0.50.

EXERCISES

16. Refer to Exercise 14. The manufacturer requires that the variance of weights of bags of sugar not exceed 0.25. Test at the 5% significance level the null hypothesis that the population variance is no more than 0.25.

17. A manufacturer claims that the variance of the tensile strengths of the wire it produces is no more than 120. A random sample of twenty wires has a sample variance of 210.4. Assuming a normal population distribution, test the manufacturer's claim.

18. Refer to Exercise 11. Management is concerned also about the variance of daily output, and regards as undesirable any value exceeding 500. Test the null hypothesis that the population variance for daily output does not exceed 500.

19. A company routinely administers a standard test to determine the mathematical aptitude of candidates for management positions. Historically, the variance of the test scores has been 400. A personnel officer suspects that the test scores of the current year's candidates are more variable than those in the past. A random sample of fifteen scores from the current year had variance 656. Test the null hypothesis that the variance for this population is 400, against the alternative that it is bigger.

20. One way to evaluate the effectiveness of a teaching assistant is to examine the scores achieved by his or her students in an examination at the end of the course. Obviously, the mean score is of interest. However, the variance also contains useful information— some teachers have a style that works very well with more able students, but is unsuccessful with less able or poorly motivated students. A professor sets a standard examination at the end of each semester for all sections of a course. The variance of the scores on this test is typically very close to 300. A new teaching assistant has a class of thirty students, whose test scores had variance 483. Regarding these students' test scores as a random sample from a normal population, test the null hypothesis that the population variance of their scores is 300, against a two-sided alternative.

21. In a random sample of 1,549 consumers taken during the first weeks of the Reagan administration, 53% said they believed that, if the president's plans to cut taxes and government spending were approved, he would also be able to control inflation. Test the null hypothesis that the proportion of population members with this view is no more than 0.5.

22. An investigator is interested in the sex composition of the audience for a horror movie.

A random sample of 100 audience members contained 45 females. Test, against a two-sided alternative, the null hypothesis that half the audience members were female.

23. A manufacturer contemplating a bulk mailing advertising campaign has determined that, in order for the campaign to be successful, a response rate of at least 20% is needed. In a pilot study involving 600 randomly selected households, the actual number responding was 102. Test the null hypothesis that the population proportion that would respond is no less than 0.2.

24. A manufacturer claims that at least 90% of the parts supplied to a customer conform to required specifications. A random sample of 200 parts contained 162 that in fact conformed to these specifications. Test the manufacturer's claim.

25. A mayor in a large city claims that, in one particularly depressed neighborhood, at least 20% of all males between ages 18 and 65 are unemployed. A random sample of 120 people from this population contained 20 unemployed. Test the mayor's claim.

26. In a random sample[7] of 1,119 shoppers in a Sacramento, California, supermarket that had used unit pricing for 6 years, 71.4% understood it. Test the null hypothesis that at least 75% of this population understand unit pricing.

27. It was found that 60.7% of the members of the sample in Exercise 26 claimed to make use of unit pricing in deciding on purchases. Test the null hypothesis that no more than half of this population uses unit pricing.

9.6 TESTS FOR THE DIFFERENCE BETWEEN TWO MEANS

In this section we examine the case where random samples are available from two populations, and the quantity of interest is the difference between the two population means. In developing procedures for testing hypotheses, the appropriate methodology once again depends on the manner in which the samples are taken. As in Section 8.7, we need to consider separately the cases of matched pairs and independent samples.

TESTS BASED ON MATCHED PAIRS

Here we assume that a random sample of n matched pairs of observations is obtained from populations with means μ_X and μ_Y. The actual sample observations will be denoted $(x_1, y_1), (x_2, y_2), \ldots, (x_n, y_n)$.

In a study[8] aimed at assessing the relationship between a subject's brain activity while watching a television commercial, and the subject's subsequent ability to recall the contents of the commercial, the data in Table 9.2 were obtained. Subjects were shown commercials for two brands of each of ten products. For each commercial, the ability to recall 24 hours later was measured, and each member of a pair of commercials was then designated "high recall" or "low recall." The table shows an index of the total amount of brain activity of subjects while watching these commercials.

If μ_X denotes the population mean brain activity for the high recall commercials

---

[7] B. F. McElroy and D. A. Aaker, "Unit pricing six years after introduction," *Journal of Retailing,* 55 (1979), No. 3, 44–57.

[8] V. Appel, S. Weinstein, and C. Weinstein, "Brain activity and recall of TV advertising," *Journal of Advertising Research,* 19 (1979), No. 4, 7–15.

TABLE 9.2 Brain activities of subjects watching ten pairs of television commercials

PRODUCT	HIGH RECALL	LOW RECALL	DIFFERENCES	
i	x_i	y_i	d_i	d_i^2
1	137	53	84	7,056
2	135	114	21	441
3	83	81	2	4
4	125	86	39	1,521
5	47	34	13	169
6	46	66	−20	400
7	114	89	25	625
8	157	113	44	1,936
9	57	88	−31	961
10	144	111	33	1,089
		Sums	210	14,202

and μ_Y the population mean for the low recall commercials, then the differences d_i represent a random sample of ten observations from a population with mean $(\mu_X - \mu_Y)$. If the population distribution of these differences can be assumed normal, the methods of Section 9.3 are immediately applicable for testing hypotheses about $(\mu_X - \mu_Y)$. For this particular example, a natural test is of the null hypothesis of no difference in mean brain activity levels,

$$H_0: \quad \mu_X - \mu_Y = 0$$

against the alternative that, on average, brain activity is greater for the high recall commercials, that is,

$$H_1: \quad \mu_X - \mu_Y > 0$$

More generally, we can test the hypothesis that the difference $(\mu_X - \mu_Y)$ is equal to any specified value D_0 using the procedure given in the box.

Tests of the Difference Between Population Means: Matched Pairs

Suppose we have a random sample of n matched pairs of observations from distributions with means μ_X and μ_Y. Let $\bar{d}$ and s_d denote the observed sample mean and standard deviation for the n differences $(x_i - y_i)$. If the population distribution of the differences is normal, then the following tests have significance level α:

(i) To test either null hypothesis

$$H_0: \quad \mu_X - \mu_Y = D_0 \quad \text{or} \quad H_0: \quad \mu_X - \mu_Y \leq D_0$$

against the alternative

$$H_1: \quad \mu_X - \mu_Y > D_0$$

the decision rule is

$$\text{Reject } H_0 \text{ if } \quad \frac{\bar{d} - D_0}{s_d/\sqrt{n}} > t_{n-1,\alpha}$$

(ii) To test either null hypothesis

$$H_0: \quad \mu_X - \mu_Y = D_0 \qquad \text{or} \qquad H_0: \quad \mu_X - \mu_Y \geq D_0$$

against the alternative

$$H_1: \quad \mu_X - \mu_Y < D_0$$

the decision rule is

$$\text{Reject } H_0 \text{ if } \quad \frac{\bar{d} - D_0}{s_d/\sqrt{n}} < -t_{n-1,\alpha}$$

(iii) To test the null hypothesis

$$H_0: \quad \mu_X - \mu_Y = D_0$$

against the alternative

$$H_1: \quad \mu_X - \mu_Y \neq D_0$$

the decision rule is

$$\text{Reject } H_0 \text{ if } \quad \frac{\bar{d} - D_0}{s_d/\sqrt{n}} < -t_{n-1,\alpha/2} \qquad \text{or} \qquad \frac{\bar{d} - D_0}{s_d/\sqrt{n}} > t_{n-1,\alpha/2}$$

Here $t_{n-1,\alpha}$ is that number for which

$$P(t_{n-1} > t_{n-1,\alpha}) = \alpha$$

where the random variable t_{n-1} follows a Student's t distribution with $(n-1)$ degrees of freedom.

When we want to test the null hypothesis that the two population means are equal, we set $D_0 = 0$ in the formulas.

We now return to our example of brain activity of subjects watching television commercials; from Table 9.2, the sample mean of the differences is

$$\bar{d} = \frac{1}{n} \sum_{i=1}^{n} d_i = \frac{1}{10}(210) = 21.0$$

The sample variance is

$$s_d^2 = \frac{1}{n-1} \left(\sum_{i=1}^{n} d_i^2 - n\bar{d}^2 \right)$$

$$= \frac{1}{9}[14{,}202 - (10)(21.0)^2] = 1{,}088$$

so that the sample standard deviation is

$$s_d = \sqrt{1{,}088} = 32.98$$

We want to test the null hypothesis

$$H_0: \quad \mu_X - \mu_Y = D_0 = 0$$

against the alternative

$$H_1: \quad \mu_X - \mu_Y > 0$$

The test is based on

$$\frac{\bar{d} - D_0}{s_d/\sqrt{n}} = \frac{21.0}{32.98/\sqrt{10}} = 2.014$$

This quantity must be compared with tabulated values of the Student's t distribution with $(n - 1) = 9$ degrees of freedom. From Table 6 of the Appendix, we have for 5% level and 2.5% level tests,

$$t_{9,\,.05} = 1.833 \qquad \text{and} \qquad t_{9,\,.025} = 2.262$$

Hence, the null hypothesis of equality of the population means can be rejected at the 5% level, but not at the 2.5% level of significance. We see then that the data of Table 9.2 contain much evidence suggesting that, on the average, brain activity is higher for the high recall than for the low recall group. If, in fact, the mean brain activity were the same for these two groups, then the probability of finding a sample result as extreme or more extreme than that actually obtained would be between .025 and .05.

TESTS BASED ON INDEPENDENT SAMPLES

Suppose now that we have a random sample of n_x observations from a normal population with mean μ_X and variance σ_X^2, and an independent random sample of n_y observations from a normal population with mean μ_Y and variance σ_Y^2.

In Section 8.7 we saw that, if the sample means are denoted $\bar{X}$ and $\bar{Y}$, then the random variable

$$Z = \frac{(\bar{X} - \bar{Y}) - (\mu_X - \mu_Y)}{\sqrt{\dfrac{\sigma_X^2}{n_x} + \dfrac{\sigma_Y^2}{n_y}}}$$

has a standard normal distribution. If the two population variances are known, tests for the difference between the population means can be based on this result, using the same arguments as before. Moreover (thanks to the Central Limit Theorem), if the two sample sizes are large, the result continues to hold to a good approximation when the sample variances replace the population variances, even when the population distributions are not normal. This allows the derivation of tests of wide applicability, as summarized in the box.

Tests for the Difference Between Population Means: Independent Samples (Known Variances or Large Sample Sizes)

Suppose we have independent random samples of n_x and n_y observations from normal distributions with means μ_X and μ_Y and variances σ_X^2 and σ_Y^2. If the observed sample means are $\bar{x}$ and $\bar{y}$, then the following tests have significance level α:

(i) To test either null hypothesis

$$H_0: \quad \mu_X - \mu_Y = D_0 \qquad \text{or} \qquad H_0: \quad \mu_X - \mu_Y \leq D_0$$

against the alternative

$$H_1: \quad \mu_X - \mu_Y > D_0$$

the decision rule is

$$\text{Reject } H_0 \text{ if } \quad \frac{\bar{x} - \bar{y} - D_0}{\sqrt{\dfrac{\sigma_X^2}{n_x} + \dfrac{\sigma_Y^2}{n_y}}} > z_\alpha$$

(ii) To test either null hypothesis

$$H_0: \quad \mu_X - \mu_Y = D_0 \qquad \text{or} \qquad H_0: \quad \mu_X - \mu_Y \geq D_0$$

against the alternative

$$H_1: \quad \mu_X - \mu_Y < D_0$$

the decision rule is

$$\text{Reject } H_0 \text{ if } \quad \frac{\bar{x} - \bar{y} - D_0}{\sqrt{\dfrac{\sigma_X^2}{n_x} + \dfrac{\sigma_Y^2}{n_y}}} < -z_\alpha$$

(iii) To test the null hypothesis

$$H_0: \quad \mu_X - \mu_Y = D_0$$

against the alternative

$$H_1: \quad \mu_X - \mu_Y \neq D_0$$

The decision rule is

$$\text{Reject } H_0 \text{ if } \quad \frac{\bar{x} - \bar{y} - D_0}{\sqrt{\dfrac{\sigma_X^2}{n_x} + \dfrac{\sigma_Y^2}{n_y}}} < -z_{\alpha/2} \qquad \text{or} \qquad \frac{\bar{x} - \bar{y} - D_0}{\sqrt{\dfrac{\sigma_X^2}{n_x} + \dfrac{\sigma_Y^2}{n_y}}} > z_{\alpha/2}$$

If the sample sizes n_x and n_y are large,[9] then to a good approximation, tests of significance level α for the difference between population means are obtained by replacing the population variances by the observed sample variances, s_x^2 and s_y^2. For large sample sizes, these approximations remain good even when the population distributions are not normal.

---

[9] This approximation is generally satisfactory when each sample contains at least thirty observations.

Example 9.7

The international banking crisis of 1974, involving the failure of the Franklin National Bank in New York, led the Federal Reserve System to guarantee the international as well as the domestic deposits of the bank. It might be hypothesized[10] that this "Franklin Message" would lead to a decrease in the risk premium attached to large American banks' deposits. (*Risk premium* here is taken to be measured by the excess of secondary market Certificate of Deposit rates over Treasury bill yields.) For 48 months before the "Franklin Message," the mean risk premium was 0.899, and the variance was 0.247. For 48 months after the message, the mean and variance were 0.703 and 0.320. If μ_X and μ_Y denote respectively the means before and after the message, test the null hypothesis

$$H_0: \quad \mu_X - \mu_Y = 0$$

against the alternative

$$H_1: \quad \mu_X - \mu_Y > 0$$

Assume that the data can be regarded as independent random samples from the two populations.

The decision rule is to reject H_0 in favor of H_1 if

$$\frac{\bar{x} - \bar{y}}{\sqrt{\dfrac{s_x^2}{n_x} + \dfrac{s_y^2}{n_y}}} > z_\alpha$$

In this example,

$$\bar{x} = 0.899; \quad s_x^2 = 0.247; \quad n_x = 48; \quad \bar{y} = 0.703; \quad s_y^2 = 0.320; \quad n_y = 48$$

so that

$$\frac{\bar{x} - \bar{y}}{\sqrt{\dfrac{s_x^2}{n_x} + \dfrac{s_y^2}{n_y}}} = \frac{0.899 - 0.703}{\sqrt{\dfrac{0.247}{48} + \dfrac{0.320}{48}}} = 1.80$$

From Table 3 of the Appendix, we find that the value of α corresponding to $z_\alpha = 1.80$ is 0.0359. Hence, the null hypothesis can be rejected at all levels of significance greater than 3.59%. Hence, were the null hypothesis of equality of population means true, the probability of observing a sample result as extreme or more extreme than that found would be .0359. This represents pretty strong evidence against the null hypothesis of equality of these means, suggesting rather a decrease in the mean risk premium after the "Franklin Message."

We will now treat the case where the sample sizes are not large. If it can be assumed that the two population variances are equal, then tests can be based on the result (given in Section 8.7) that the random variable

---

[10] This example is taken from a study by I. H. Giddy, "Moral hazard and Central Bank rescues in an international context," *The Financial Review, 15* (1980), No. 2, 50–56.

$$t = \frac{(\bar{X} - \bar{Y}) - (\mu_X - \mu_Y)}{s\sqrt{\dfrac{n_x + n_y}{n_x n_y}}}$$

has a Student's t distribution with $(n_x + n_y - 2)$ degrees of freedom, where

$$s^2 = \frac{(n_x - 1)s_X^2 + (n_y - 1)s_Y^2}{(n_x + n_y - 2)}$$

is an estimator of the common variance. The appropriate hypothesis tests on the difference between the population means are given in the box.

Tests for the Difference Between the Means of Two Normal Populations: Independent Samples, Population Variances Equal

Suppose we have independent random samples of n_x and n_y observations from normal distributions with means μ_X and μ_Y and a common variance. If the observed sample variances are s_x^2 and s_y^2, an estimate of the common population variance is provided by

$$s^2 = \frac{(n_x - 1)s_x^2 + (n_y - 1)s_y^2}{n_x + n_y - 2}$$

Then, if the observed sample means are $\bar{x}$ and $\bar{y}$, the following tests have significance level α:

(i) To test either null hypothesis

$$H_0: \quad \mu_X - \mu_Y = D_0 \qquad \text{or} \qquad H_0: \quad \mu_X - \mu_Y \leq D_0$$

against the alternative

$$H_1: \quad \mu_X - \mu_Y > D_0$$

the decision rule is

$$\text{Reject } H_0 \text{ if } \quad \frac{\bar{x} - \bar{y} - D_0}{s\sqrt{\dfrac{n_x + n_y}{n_x n_y}}} > t_{n_x + n_y - 2, \alpha}$$

(ii) To test either null hypothesis

$$H_0: \quad \mu_X - \mu_Y = D_0 \qquad \text{or} \qquad H_0: \quad \mu_X - \mu_Y \geq D_0$$

against the alternative

$$H_1: \quad \mu_X - \mu_Y < D_0$$

the decision rule is

$$\text{Reject } H_0 \text{ if } \quad \frac{\bar{x} - \bar{y} - D_0}{s\sqrt{\dfrac{n_x + n_y}{n_x n_y}}} < -t_{n_x + n_y - 2, \alpha}$$

(iii) To test the null hypothesis

$$H_0: \quad \mu_X - \mu_Y = D_0$$

against the alternative

$$H_1: \quad \mu_X - \mu_Y \neq D_0$$

the decision rule is

$$\text{Reject } H_0 \text{ if } \quad \frac{\bar{x} - \bar{y} - D_0}{s\sqrt{\dfrac{n_x + n_y}{n_x n_y}}} < -t_{n_x + n_y - 2, \alpha/2} \quad \text{or} \quad \frac{\bar{x} - \bar{y} - D_0}{s\sqrt{\dfrac{n_x + n_y}{n_x n_y}}} > t_{n_x + n_y - 2, \alpha/2}$$

Here $t_{n_x + n_y - 2, \alpha}$ is that number for which

$$P\left(t_{n_x + n_y - 2} > t_{n_x + n_y - 2, \alpha}\right) = \alpha$$

where $t_{n_x + n_y - 2}$ has a Student's t distribution with $(n_x + n_y - 2)$ degrees of freedom.

Example 9.8

In a study[11] aimed at finding early warning signals of business failure, a random sample of twenty-three failed retail firms showed mean return on assets 3 years previously of 0.058, and sample standard deviation 0.055. An independent random sample of twenty-three nonfailed retail firms showed mean return of 0.146 and standard deviation 0.058 for the same period. If μ_X and μ_Y denote the population means for failed and nonfailed firms, respectively, test the null hypothesis

$$H_0: \quad \mu_X - \mu_Y = 0$$

against the alternative

$$H_1: \quad \mu_X - \mu_Y < 0$$

Assume the two population distributions are normal and have the same variance.
The decision rule is to reject H_0 in favor of H_1 if

$$\frac{\bar{x} - \bar{y}}{s\sqrt{\dfrac{n_x + n_y}{n_x n_y}}} < -t_{n_x + n_y - 2, \alpha}$$

---

[11] S. Sharma and V. Mahajan, "Early warning indicators of business failure," *Journal of Marketing*, *44* (1980), No. 4, 80–89.

For these data, we have

$$\bar{x} = 0.058; \quad s_x = 0.055; \quad n_x = 23; \quad \bar{y} = 0.146; \quad s_y = 0.058; \quad n_y = 23$$

Hence,

$$s^2 = \frac{(n_x - 1)s_x^2 + (n_y - 1)s_y^2}{n_x + n_y - 2}$$

$$= \frac{(22)(0.055)^2 + (22)(0.058)^2}{23 + 23 - 2} = 0.0031945$$

so that

$$s = \sqrt{0.0031945} = 0.0565$$

Then

$$\frac{\bar{x} - \bar{y}}{s\sqrt{\dfrac{n_x + n_y}{n_x n_y}}} = \frac{0.058 - 0.146}{0.0565 \sqrt{\dfrac{23 + 23}{23^2}}} = -5.282$$

For a 0.5% level test, we have by interpolation from Table 6, for the Student's t distribution with $(n_x + n_y - 2) = 44$ degrees of freedom,

$$t_{44,\,.005} = 2.695$$

Then, since -5.282 is much less than -2.695, the null hypothesis is overwhelmingly rejected even at this low level of significance. The data cast considerable doubt on the hypothesis that the two population means are equal. Rather, they suggest very strongly that the population mean return on assets is lower for failed than for nonfailed retail firms.

The test just discussed and illustrated is based on an assumption that the two population variances are equal. In fact, it is possible to develop tests that are valid when this assumption does not hold. However, these will not be discussed further here.

9.7 TESTS FOR THE DIFFERENCE BETWEEN TWO POPULATION PROPORTIONS (LARGE SAMPLES)

We turn now to the problem of comparing two population proportions. As in Section 8.8, suppose that a random sample of n_x observations from a population with proportion p_X "successes" gives a sample proportion $\hat{p}_X$, and that an independent random sample of n_y observations from a population with proportion p_Y "successes" yields sample proportion $\hat{p}_Y$.

In Section 8.8 we saw that, if the numbers of sample observations are large, then to a very good approximation, the random variable

$$Z = \frac{(\hat{p}_X - \hat{p}_Y) - (p_X - p_Y)}{\sqrt{\dfrac{p_X(1 - p_X)}{n_x} + \dfrac{p_Y(1 - p_Y)}{n_y}}}$$

has a standard normal distribution, by virtue of the Central Limit Theorem.

Suppose that we want to test the hypothesis that the population proportions p_X and p_Y are equal. If their common value is denoted p_0, then we have under this hypothesis that

$$
\begin{aligned}
Z &= \frac{(\hat{p}_X - \hat{p}_Y)}{\sqrt{\dfrac{p_0(1 - p_0)}{n_x} + \dfrac{p_0(1 - p_0)}{n_y}}} \\[2em]
&= \frac{\hat{p}_X - \hat{p}_Y}{\sqrt{p_0(1 - p_0)\left(\dfrac{n_x + n_y}{n_x n_y}\right)}}
\end{aligned}
\tag{9.7.1}
$$

follows to a good approximation a standard normal distribution.

Finally, the unknown common proportion p_0 in expression 9.7.1 can be estimated by the pooled estimator $\hat{p}_0$, given by

$$\hat{p}_0 = \frac{n_x \hat{p}_X + n_y \hat{p}_Y}{n_x + n_y}$$

Replacing the unknown p_0 by $\hat{p}_0$ in 9.7.1 gives a random variable that has a distribution close to the standard normal, provided the sample sizes are large. This result forms the basis for our tests, as indicated in the box.

Testing the Equality of Two Population Proportions (Large Samples)

Let $\hat{p}_x$ denote the proportion of successes in a random sample of n_x observations from a population with proportion p_X successes, and $\hat{p}_y$ the proportion of successes observed in an independent random sample of n_y observations from a population with proportion p_Y successes. If it is hypothesized that the population proportions are equal, an estimate of the common proportion is given by

$$\hat{p}_0 = \frac{n_x \hat{p}_x + n_y \hat{p}_y}{n_x + n_y}$$

Then, if the sample sizes are large,[12] the following tests have significance level α:

---

[12] The approximation is generally adequate if each sample contains at least forty observations.

(i) To test either null hypothesis

$$H_0: \quad p_X - p_Y = 0 \qquad \text{or} \qquad H_0: \quad p_X - p_Y \leq 0$$

against the alternative

$$H_1: \quad p_X - p_Y > 0$$

the decision rule is

$$\text{Reject } H_0 \text{ if } \quad \frac{\hat{p}_x - \hat{p}_y}{\sqrt{\hat{p}_0(1 - \hat{p}_0)\left(\dfrac{n_x + n_y}{n_x n_y}\right)}} > z_\alpha$$

(ii) To test either null hypothesis

$$H_0: \quad p_X - p_Y = 0 \qquad \text{or} \qquad H_0: \quad p_X - p_Y \geq 0$$

against the alternative

$$H_1: \quad p_X - p_Y < 0$$

the decision rule is

$$\text{Reject } H_0 \text{ if } \quad \frac{\hat{p}_x - \hat{p}_y}{\sqrt{\hat{p}_0(1 - \hat{p}_0)\left(\dfrac{n_x + n_y}{n_x n_y}\right)}} < -z_\alpha$$

(iii) To test the null hypothesis

$$H_0: \quad p_X - p_Y = 0$$

against the alternative

$$H_1: \quad p_X - p_Y \neq 0$$

the decision rule is

$$\text{Reject } H_0 \text{ if } \quad \frac{\hat{p}_x - \hat{p}_y}{\sqrt{\hat{p}_0(1 - \hat{p}_0)\left(\dfrac{n_x + n_y}{n_x n_y}\right)}} < -z_{\alpha/2}$$

$$\text{or} \quad \frac{\hat{p}_x - \hat{p}_y}{\sqrt{\hat{p}_0(1 - \hat{p}_0)\left(\dfrac{n_x + n_y}{n_x n_y}\right)}} > z_{\alpha/2}$$

Example 9.9

In a study[13] designed to determine important criteria in selecting a retail pharmacy, a random sample of 200 people, aged 18–30, contained 36 who said price was the most important criterion. An independent random sample of 118 people, aged 50

---

[13] These data are taken from P. Nickel and A. I. Wertheimer, "Factors affecting consumers' images and choices of drugstores," *Journal of Retailing, 55* (1979), No. 2, 71–78.

years or older, showed 29 indicating price as the most important criterion. Denoting by p_X and p_Y the respective population proportions, test the null hypothesis

$$H_0: \quad p_X - p_Y = 0$$

against the alternative

$$H_1: \quad p_X - p_Y \neq 0$$

The decision rule is to reject H_0 in favor of H_1 if

$$\frac{\hat{p}_x - \hat{p}_y}{\sqrt{\hat{p}_0(1 - \hat{p}_0)\left(\dfrac{n_x + n_y}{n_x n_y}\right)}} < -z_{\alpha/2} \quad \text{or} \quad \frac{\hat{p}_x - \hat{p}_y}{\sqrt{\hat{p}_0(1 - \hat{p}_0)\left(\dfrac{n_x + n_y}{n_x n_y}\right)}} > z_{\alpha/2}$$

For these data, we have

$$\hat{p}_x = \frac{36}{200} = 0.180; \quad n_x = 200; \quad \hat{p}_y = \frac{29}{118} = 0.246; \quad n_y = 118$$

Hence,

$$\hat{p}_0 = \frac{n_x \hat{p}_x + n_y \hat{p}_y}{n_x + n_y} = \frac{(200)(0.180) + (118)(0.246)}{200 + 118} = 0.204$$

Then

$$\frac{\hat{p}_x - \hat{p}_y}{\sqrt{\hat{p}_0(1 - \hat{p}_0)\left(\dfrac{n_x + n_y}{n_x n_y}\right)}} = \frac{0.180 - 0.246}{\sqrt{(0.204)(0.796)\left[\dfrac{200 + 118}{(200)(118)}\right]}} = -1.41$$

The value of $\alpha/2$ corresponding to $z_{\alpha/2} = 1.41$ is, from Table 3 of the Appendix, $\alpha/2 = .0793$, so that $\alpha = .1586$. Hence, the null hypothesis can be rejected only at significance levels higher than 15.86%. The evidence against the hypothesis that the population proportions viewing price as the most important criterion are the same in these two age groups is not very strong.

EXERCISES

28. A college placement officer wants to determine whether male and female business graduates receive, on average, different salary offers for their first position after graduation. The placement officer randomly selected eight pairs of business graduates in such a way that the qualifications, interests, and backgrounds of the members of any pair were as similar as possible. The major difference was that one member of each pair was male and one was female. The accompanying table shows the highest salary offer received by each sample member at the end of the recruiting round. Assuming that the distributions are normal, test the null hypothesis that the population means are equal against the alternative that the true mean for males is higher than for females.

HIGHEST SALARY OFFER (IN DOLLARS)		
PAIR	MALE	FEMALE
1	26,200	21,400
2	24,700	23,600
3	28,400	29,300
4	21,700	22,300
5	28,600	26,200
6	29,300	25,900
7	28,300	28,500
8	24,300	20,900

29. An investigator wants to determine whether, on average, automobiles achieve higher mileage using gasohol rather than unleaded gasoline. A random sample of six compact cars was selected, and each was driven under identical conditions with the two types of fuel. The table shows the resulting miles per gallon figures. Assuming normality, test the null hypothesis that the population means are equal against the alternative that the true mean is higher when gasohol is used.

AUTOMOBILE	GASOHOL	UNLEADED GASOLINE
1	24.3	23.9
2	26.2	26.1
3	23.8	24.0
4	27.9	27.3
5	21.8	21.1
6	23.9	24.1

30. A random sample of 100 production workers in a large corporation were asked to rate, on a scale from 1 to 10, their levels of satisfaction with two supervisors. The differences in their scores for the two supervisors had mean 0.24 and standard deviation 1.43. Test, against a two-sided alternative, the null hypothesis that the population mean ratings for the two supervisors are the same.

31. A charity experiments with two procedures for contacting regular donors. In the first procedure, a phone call is made, the charity's objectives for the coming year are discussed, and any questions are answered. The second procedure is similar, except that an agent of the charity arranges to meet with donors in their homes. For a random sample of eighty donors contacted by telephone, the mean donation that resulted was $115 and the standard deviation was $71. For an independent random sample of sixty donors receiving a home visit, the mean and standard deviation of donations received were $131 and $83, respectively. Test the null hypothesis that the two population means are equal against the alternative that the mean is higher for donors contacted at home.

32. A professor[14] taught two sections of an introductory marketing course using very differ-

---

[14] These results are reported in R. W. Cook, "An investigation of the relationship between student attitudes and student achievement," *Journal of the Academy of Marketing Science*, 7 (1979), 71–79.

ent styles. In the first section the approach was extremely formal and rigid, while in the second section a more relaxed and informal attitude was adopted. At the end of the course, a common final examination was administered. In the first section the seventy-two students obtained a mean score of 214.03, and the sample standard deviation was 22.91. In the second section there were sixty-four students, with mean score 215.22 and standard deviation 23.11. Assume that these two groups of students can be regarded as independent random samples from the populations of all students who might be exposed to these teaching methods. Test against a two-sided alternative the null hypothesis that the population means are equal.

33. A political science professor is interested in comparing the characteristics of students who do and do not vote in national elections. For a random sample of 114 students who claimed to have voted in the last presidential election, she found a mean grade point average of 2.71, and standard deviation 0.61. For an independent random sample of 123 students who did not vote, the mean grade point average was 2.79, and the standard deviation was 0.56. Test, against a two-sided alternative, the null hypothesis that the population means are equal.

34. A random sample of fifteen manufacturing concerns with total assets under $1 million showed an average after-tax profit of 2.2% of sales, and sample standard deviation 0.5%. An independent random sample of twelve manufacturing concerns with total assets over $1 million yielded an average after-tax profit of 2.5% of sales, and sample standard deviation 0.7%. Assuming that the two population distributions are normal with the same variance, test, against a two-sided alternative, the null hypothesis that the population means are equal.

35. The levels of product familiarity of students with two types of product were compared through the administration of a questionnaire.[15] For a random sample of thirty-two students the mean familiarity level for floor wax was found to be 4.355, and the sample standard deviation was 2.524. For an independent random sample of thirty-two students, the mean familiarity level for shampoo was 9.129, and the sample standard deviation was 2.291. Assuming that the two population distributions are normal and have the same variance, test against a two-sided alternative the null hypothesis that the population means are equal.

36. A publisher is interested in the effects on sales of college texts of expensive three-color cover designs. The publisher is planning to bring out twenty texts in the business area, and randomly chooses ten of them to have expensive cover designs. The remaining ten are produced with plain covers. For those with expensive cover designs, first year sales averaged 9,254, and the sample standard deviation was 2,127. For the books with plain cover designs, average first year sales were 8,167, and the sample standard deviation was 1,631. Assuming the two population distributions are normal with the same variance, test the null hypothesis that the population means are equal against the alternative that the true mean is higher for books with expensive cover designs.

37. It is possible to time-compress a television commercial by electronically speeding it up in such a way that the voice pitch does not change at all, provided the compression is not too severe. An Allstate Insurance commercial was shown in both a normal (30 seconds) version and a time-compressed (24 seconds) version. For a random sample[16] of fifty-seven people who had seen the normal version, 26.3% could recall the brand 2 days

---

[15] See P. S. Raju and M. D. Reilly, "Product familiarity and information processing strategies: an exploratory investigation," *Journal of Business Research, 8* (1980), 187–212.

[16] These results are given by J. MacLachlan and M. H. Siegel, "Reducing the costs of TV commercials by use of time compressions," *Journal of Marketing Research, 17* (1980), 52–57.

later. For an independent random sample of seventy-four people who had seen the compressed version, the recall rate was 43.2%. Test the null hypothesis that the two population proportions are equal against the alternative that the true proportion is higher for viewers of the compressed version of the commercial.

38. A random sample of 120 middle- and high-income voters contained 69 in favor of reducing the number of members in a state legislature. For an independent random sample of 100 low-income voters, 48 were in favor of this move. Test against a two-sided alternative the null hypothesis that the two population proportions are equal.

39. Manufacturers frequently offer deals, such as media-distributed coupons, to encourage brand switching. One concern is that consumers who have switched in this way may be less likely than others to stay with the brand. In a study[17] designed to test this view, people who had switched to a particular brand of margarine were surveyed. For a random sample of 23,794 who had switched without the inducement of a deal, the repeat-purchase rate was 28%. For an independent random sample of 671 consumers who had been induced to switch through media-distributed coupons, repeat purchases were made by 13%. Test the null hypothesis that the two population proportions are equal against the alternative that the true proportion is higher for those who switched without a deal.

40. Populations are frequently surveyed by mail questionnaires. In order to obtain as representative a sample as possible, the investigator must aim at a high level of response. One fear is that, if a question on the survey form is felt to be offensive, a lower response rate will result. Perhaps for this reason, many investigators have been reluctant to include questions about race, even when they would have liked to compare attitudes of different racial groups. A questionnaire[18] which did not ask about race was sent to a random sample of 600 people, of whom 75.3% responded. An otherwise identical questionnaire, but with a question asking the subject's racial group, was sent to an independent random sample of 600 people; the response rate was 74.9%. Test the null hypothesis that the two population response rates are equal against the alternative that the response rate is higher when the race question is not asked.

41. Frequently, in mail surveys, it is even more important to obtain information about income than about race. It is known that a nontrivial proportion of the population is reluctant to provide this information. It might be feared that the inclusion of a question about race would provide an additional irritant, thereby rendering it less likely that income information would be supplied. In the study described in Exercise 40, 432 usable responses were provided when the race question was not asked and 9.1% of these respondents failed to give income information. When the race question was asked, 431 usable responses were received, of which 10.4% did not contain information about income. Test the null hypothesis that the two population proportions are equal against the alternative that the true proportion is higher when the race question is asked.

42. Needham, Harper and Steers, a major advertising agency, periodically surveys heads of household on the issue of brand loyalty, asking whether they agree with the statement: "I try to stick to well-known brand names." In the 1981 survey, 58% of the 2,000 women and 65% of the 2,000 men questioned agreed with this statement. Test the null hypothesis that there is no difference between the population proportions of men and women who would agree.

---

[17] See J. A. Dodson, A. M. Tybout, and B. Sternthal, "Impact of deals and deal retraction on brand switching," *Journal of Marketing Research, 15* (1978), 72–81.

[18] These results are taken from J. N. Sheth, A. LeClaire, and D. Wachspress, "Impact of asking race information in mail surveys," *Journal of Marketing, 44* (1980), No. 1, 67–70.

9.8 TESTING THE EQUALITY OF THE VARIANCES OF TWO NORMAL POPULATIONS

One of the tests developed in Section 9.6 for the comparison of population means depends on an assumption of equality of the two population variances. While in many practical applications such an assumption is reasonable, it is prudent to use the available data to test its validity. In addition, it sometimes happens when comparing population distributions that the population variances are of interest in their own right, and that we wish to compare them.

In this section, we consider the problem where independent random samples from two normal populations are available, and it is required to test the equality of the population variances. To develop such a test, another probability distribution must be introduced. Let s_X^2 be the sample variance for a random sample of n_x observations from a normal population with variance σ_X^2, and s_Y^2 be the sample variance from an independent random sample of n_y observations from a normal population with variance σ_Y^2. Then the random variable

$$F = \frac{s_X^2/\sigma_X^2}{s_Y^2/\sigma_Y^2} \qquad (9.8.1)$$

follows a distribution known as the **F distribution**.[19] This family of distributions is widely used in statistical analysis. A particular member of the family is distinguished by two values—the degrees of freedom associated with the numerator and with the denominator. In the present context, recall that the degrees of freedom associated with the sample variance s_X^2 is $(n_x - 1)$, and that with s_Y^2 is $(n_y - 1)$. The random variable 9.8.1 then has an F distribution with numerator degrees of freedom $(n_x - 1)$ and denominator degrees of freedom $(n_y - 1)$.

The F distribution has an asymmetric probability density function, defined only for nonnegative values. This density function is illustrated in Figure 9.11.

The F Distribution

Suppose that independent random samples of n_x and n_y observations are taken from two normal populations with variances σ_X^2 and σ_Y^2. If the sample variances are s_X^2 and s_Y^2, then the random variable

$$F = \frac{s_X^2/\sigma_X^2}{s_Y^2/\sigma_Y^2}$$

has an **F distribution** with numerator degrees of freedom $(n_x - 1)$ and denominator degrees of freedom $(n_y - 1)$.

An F distribution with numerator degrees of freedom ν_1 and denominator degrees of freedom ν_2 will be denoted F_{ν_1, ν_2}. We denote by $F_{\nu_1, \nu_2, \alpha}$ that number for which

$$P(F_{\nu_1, \nu_2} > F_{\nu_1, \nu_2, \alpha}) = \alpha$$

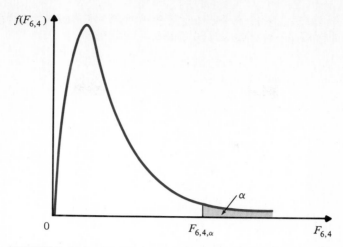

FIGURE 9.11 Probability density function of the F distribution with 6 numerator degrees of freedom and 4 denominator degrees of freedom; the probability is α that $F_{6,4}$ exceeds $F_{6,4,\alpha}$

The cutoff points $F_{\nu_1,\nu_2,\alpha}$, for α equal to 0.05 and 0.01, are provided in Table 7 of the Appendix. For example, for 10 numerator degrees of freedom and 20 denominator degrees of freedom, we see from the table

$$F_{10,20,.05} = 2.35 \quad \text{and} \quad F_{10,20,.01} = 3.37$$

Hence,

$$P(F_{10,20} > 2.35) = 0.05 \quad \text{and} \quad P(F_{10,20} > 3.37) = 0.01$$

In practical applications, provided we arrange matters so that the larger sample variance appears in the numerator, these are the only cutoff points necessary to test the hypothesis of equality of population variances. When the population variances are equal, it follows from 9.8.1 that the random variable

$$F = \frac{s_X^2}{s_Y^2}$$

obeys an F_{n_x-1,n_y-1} distribution. Appropriate hypothesis tests are described in the box.

Tests for the Equality of the Variances of Two Normal Populations

Let s_x^2 and s_y^2 be observed sample variances from independent random samples of n_x and n_y observations from normal populations with variances σ_X^2 and σ_Y^2. If s_x^2 is bigger than s_y^2, then the following tests have significance level α:

$\longrightarrow$

---

[19] Formally, the F distribution is defined as the distribution followed by the ratio of two independent chi-square random variables, each divided by its associated degrees of freedom.

(i) To test either null hypothesis

$$H_0: \quad \sigma_X^2 = \sigma_Y^2 \quad \text{or} \quad H_0: \quad \sigma_X^2 \leq \sigma_Y^2$$

against the alternative

$$H_1: \quad \sigma_X^2 > \sigma_Y^2$$

the decision rule is

$$\text{Reject } H_0 \text{ if } \quad \frac{s_x^2}{s_y^2} > F_{n_x-1, n_y-1, \alpha}$$

(ii) To test the null hypothesis

$$H_0: \quad \sigma_X^2 = \sigma_Y^2$$

against the alternative

$$H_1: \quad \sigma_X^2 \neq \sigma_Y^2$$

the decision rule is

$$\text{Reject } H_0 \text{ if } \quad \frac{s_x^2}{s_y^2} > F_{n_x-1, n_y-1, \alpha/2}$$

Here, $F_{n_x-1, n_y-1, \alpha}$ is that number for which

$$P(F_{n_x-1, n_y-1} > F_{n_x-1, n_y-1, \alpha}) = \alpha$$

where F_{n_x-1, n_y-1} has an F distribution with numerator degrees of freedom $(n_x - 1)$ and denominator degrees of freedom $(n_y - 1)$.

Example 9.10

It is hypothesized that the market share of a corporation should vary more in an industry with active price competition than in one with duopoly and tacit collusion. In a study[20] of the steam turbine generator industry, it was found that in 4 years of active price competition, the variance of General Electric's market share was 114.0895. In the following 7 years, in which there was duopoly and tacit collusion, this variance was 16.0780. If the two population variances are denoted σ_X^2 and σ_Y^2, test

$$H_0: \quad \sigma_X^2 = \sigma_Y^2$$

against

$$H_1: \quad \sigma_X^2 > \sigma_Y^2$$

Assume that the data can be regarded as independent random samples from two normal distributions.

The decision rule is to reject H_0 in favor of H_1 if

[20] B. T. Allen, "Tacit collusion and market sharing: The case of steam turbine generators," *Industrial Organization Review, 4* (1976), 48–57.

$$\frac{s_x^2}{s_y^2} > F_{n_x-1, n_y-1, \alpha}$$

From the statement of the example, we have

$$n_x = 4; \qquad s_x^2 = 114.0895; \qquad n_y = 7; \qquad s_y^2 = 16.0780$$

and so

$$\frac{s_x^2}{s_y^2} = \frac{114.0895}{16.0780} = 7.10$$

Since the degrees of freedom are $(n_x - 1) = 3$ and $(n_y - 1) = 6$, the cutoff points for 5% level and 1% level tests are

$$F_{3,6,.05} = 4.76 \qquad \text{and} \qquad F_{3,6,.01} = 9.78$$

from Table 7 of the Appendix. Hence, the null hypothesis can be rejected at the 5% level, but not at the 1% level of significance. The evidence is pretty strong against the hypothesis that variability in market shares is the same in an industry with price competition as in one with duopoly and tacit collusion. It appears that the data strongly support the alternative hypothesis of more variablity in market shares under price competition.

9.9 MEASURING THE POWER OF A TEST

In Sections 9.2–9.8, we have concentrated on the development of tests at a particular significance level. That is, we have developed decision rules for which the probability of making a Type I error—rejecting the null hypothesis when it is true—is fixed at some preassigned value. As noted in Section 9.1, a decision rule of this kind will necessarily imply some probability of making a Type II error—accepting a null hypothesis which is false. Moreover, it is often important to know what are the probabilities of making this kind of error, so that, if a null hypothesis is accepted, we will have an assessment of how likely such a decision would be when that hypothesis is false.

In this section we consider for the first time the characteristics of some of our tests when the null hypothesis is not true. In particular, we show how the power can be calculated for tests of the mean of a normal distribution when the variance is known, and for tests of the population proportion.

TESTS OF THE MEAN OF A NORMAL DISTRIBUTION: POPULATION VARIANCE KNOWN

Suppose we have a random sample of n observations from a normal population with unknown mean μ and known variance σ^2, and we want to test the null hypothesis

$$H_0: \quad \mu = \mu_0$$

against the alternative

$$H_1: \quad \mu > \mu_0$$

We saw in Section 9.2 that a test of significance level α is obtained from the decision rule

$$\text{Reject } H_0 \text{ if } \quad \frac{\overline{x} - \mu_0}{\sigma/\sqrt{n}} > z_\alpha$$

Hence, the probability that the null hypothesis will be rejected is

$$P\left(\frac{\overline{X} - \mu_0}{\sigma/\sqrt{n}} > z_\alpha\right)$$

This probability is, of course, equal to the significance level α when the null hypothesis is true. We now wish to determine its value when that hypothesis is false.

Assume, then, that the null hypothesis is false, and that the true population mean is in fact some number μ_1, bigger than μ_0. In that case, the random variable

$$Z = \frac{\overline{X} - \mu_1}{\sigma/\sqrt{n}}$$

has a standard normal distribution. This fact allows us to calculate the power of the test. We have, when the population mean is μ_1,

$$\text{Power} = 1 - \beta$$

$$= P(\text{Null hypothesis rejected when it is false})$$

$$= P\left(\frac{\overline{X} - \mu_0}{\sigma/\sqrt{n}} > z_\alpha\right)$$

$$= P\left(\overline{X} > \mu_0 + \frac{z_\alpha \sigma}{\sqrt{n}}\right)$$

$$= P\left(\frac{\overline{X} - \mu_1}{\sigma/\sqrt{n}} > \frac{\mu_0 - \mu_1}{\sigma/\sqrt{n}} + z_\alpha\right)$$

$$= P\left(Z > \frac{\mu_0 - \mu_1}{\sigma/\sqrt{n}} + z_\alpha\right)$$

where Z is a standard normal random variable. This probability can be calculated using tables of the standard normal cumulative distribution function, once μ_1, σ, n, and z_α are specified.

Example 9.11

In Example 9.1 we tested the null hypothesis that the population mean weight (in ounces) of ball bearings was

$$H_0: \quad \mu = \mu_0 = 5$$

against the alternative

$$H_1: \quad \mu > 5$$

The population standard deviation was $\sigma = 0.1$, and the test was based on $n = 16$ observations. The test was carried out at significance level $\alpha = 0.05$, so that

$$z_\alpha = z_{.05} = 1.645$$

We now determine the probability that our decision rule will reject the null hypothesis when the true mean weight is $\mu_1 = 5.02$. The power is then

$$1 - \beta = P\left(Z > \frac{\mu_0 - \mu_1}{\sigma/\sqrt{n}} + z_\alpha\right)$$

$$= P\left(Z > \frac{5.00 - 5.02}{0.1/\sqrt{16}} + 1.645\right)$$

$$= P(Z > 0.845) = 0.1991$$

from Table 3 of the Appendix.[21] Thus, the probability is 0.1991 that our decision rule will reject the null hypothesis when the true mean weight is 5.02 ounces.

These power calculations are illustrated in Figure 9.12, which shows the proba-

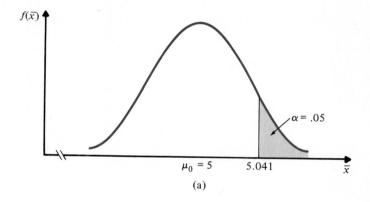

(a)

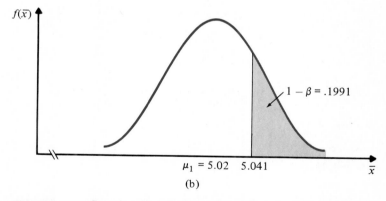

(b)

FIGURE 9.12 Sampling distributions of sample mean in Example 9.11 for population means $\mu_0 = 5$, $\mu_1 = 5.02$. Figures show calculation of power, $1 - \beta$, corresponding to significance level $\alpha = .05$ for testing H_0: $\mu = 5$ against H_1: $\mu > 5$; power is evaluated at $\mu = 5.02$

---

[21] This probability was obtained by interpolating between the z values 0.84 and 0.85 in the table.

bility density functions of the sample mean when the population mean is 5 and 5.02. Recall that a 5% level test rejects the null hypothesis when

$$\frac{\bar{x} - \mu_0}{\sigma/\sqrt{n}} > 1.645$$

or, with $\mu_0 = 5$, $\sigma = 0.1$, and $n = 16$, when

$$\frac{\bar{x} - 5}{0.1/4} > 1.645$$

This is equivalent to requiring

$$\bar{x} > 5.041$$

Thus, when the null hypothesis is true, the probability that the sample mean exceeds 5.041 is 0.05. This is shown in part (a) of Figure 9.12. Part (b) of the figure shows the density function of the sampling distribution of the sample mean when the population mean is 5.02. It differs from part (a) of the figure in being shifted to the right by an amount 0.02—the difference between the means 5.02 and 5. The shaded area in this figure shows the probability that the sample mean exceeds 5.041 when the population mean is 5.02. This is the power, evaluated at that point, as calculated previously.

In a similar manner, such probabilities can be calculated for any value of μ_1. Specific powers are shown in the table and graphed in Figure 9.13.

μ_1	5.00	5.01	5.02	5.03	5.04	5.05
POWER	.05	.1066	.1991	.3282	.4820	.6387

μ_1	5.06	5.07	5.08	5.09	5.10	5.11
POWER	.7749	.8760	.9400	.9747	.9908	.9971

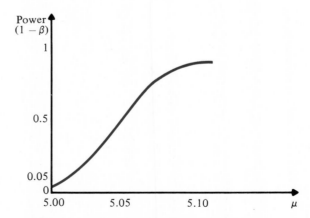

FIGURE 9.13 Power function for Example 9.11; test of H_0: $\mu = 5.00$ against H_1: $\mu > 5.00 (\alpha = 0.05$, $\sigma = 0.1$, $n = 16)$

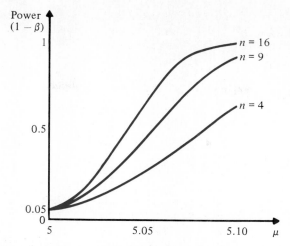

FIGURE 9.14 Power functions for test of H_0: $\mu = 5.00$ against H_1: $\mu > 5.00 (\alpha = 0.05, \sigma = 0.1)$, shown for sample sizes 4, 9, 16

We note that the power function has the following features:

(i) Everything else equal, the further is the true mean μ_1 from the hypothesized mean μ_0, the greater the power of the test. This is illustrated in Figure 9.13, and means that we are more likely to detect large than small discrepancies from the hypothesized mean.

(ii) Everything else equal, the smaller the significance level of the test, the smaller the power. In other words, reducing the probability of a Type I error will increase the probability of a Type II error.

(iii) Everything else equal, the larger the population variance, the lower the power of the test. We are less likely to detect small departures from the hypothesized mean when there is greater variability in the population.

(iv) Everything else equal, the larger the sample size, the greater the power of the test. Again, this is intuitively plausible. The more information obtained from the population, the greater the chance of detecting any departure from the null hypothesis. This is illustrated in Figure 9.14. Together with the power function derived in Example 9.11 for samples of $n = 16$ observations, this figure shows also the power functions for the same test based on $n = 4$ and $n = 9$ observations. Note that, for every value of the population mean higher than the hypothesized mean, the greater the number of observations, the greater the probability of rejecting the null hypothesis.

Using arguments similar to those given previously, we can also calculate the power of a test against the two-sided alternative hypothesis. However, rather than pursuing that case here, we will illustrate these calculations for tests on a population proportion.

TESTS OF THE POPULATION PROPORTION (LARGE SAMPLES)

Assume that a random sample of n observations is taken from a population containing a proportion p of "successes," and that the null hypothesis

$$H_0: \quad p = p_0$$

is to be tested against the two-sided alternative

$$H_1: \quad p \neq p_0$$

It was seen in Section 9.5 that a test of significance level α is obtained from the decision rule

Reject H_0 if $\quad \dfrac{\hat{p}_x - p_0}{\sqrt{p_0(1 - p_0)/n}} > z_{\alpha/2} \quad$ or $\quad \dfrac{\hat{p}_x - p_0}{\sqrt{p_0(1 - p_0)/n}} < -z_{\alpha/2}$

Therefore, the probability that the null hypothesis will be rejected is

$$P\left[\frac{\hat{p}_x - p_0}{\sqrt{p_0(1 - p_0)/n}} > z_{\alpha/2} \quad \text{or} \quad \frac{\hat{p}_x - p_0}{\sqrt{p_0(1 - p_0)/n}} < -z_{\alpha/2}\right]$$

$$= P\left[\frac{\hat{p}_x - p_0}{\sqrt{p_0(1 - p_0)/n}} < -z_{\alpha/2}\right] + P\left[\frac{\hat{p}_x - p_0}{\sqrt{p_0(1 - p_0)/n}} > z_{\alpha/2}\right]$$

Again, this probability is equal to the significance level α when the null hypothesis is true.

Suppose now that the null hypothesis is false and that the population proportion of successes is p_1, which differs from p_0. In that case, provided the sample size is large, we know that to a good approximation, the random variable

$$Z = \frac{\hat{p}_x - p_1}{\sqrt{p_1(1 - p_1)/n}}$$

has a standard normal distribution. The power of the test can then be found as

$$\text{Power} = 1 - \beta = P\left[\frac{\hat{p}_x - p_0}{\sqrt{p_0(1 - p_0)/n}} < -z_{\alpha/2}\right] + P\left[\frac{\hat{p}_x - p_0}{\sqrt{p_0(1 - p_0)/n}} > z_{\alpha/2}\right]$$

$$= P\left[\hat{p}_x < p_0 - z_{\alpha/2}\sqrt{\frac{p_0(1 - p_0)}{n}}\right] + P\left[\hat{p}_x > p_0 + z_{\alpha/2}\sqrt{\frac{p_0(1 - p_0)}{n}}\right]$$

$$= P\left[\frac{\hat{p}_x - p_1}{\sqrt{p_1(1 - p_1)/n}} < \frac{p_0 - p_1}{\sqrt{p_1(1 - p_1)/n}} - z_{\alpha/2}\frac{\sqrt{p_0(1 - p_0)/n}}{\sqrt{p_1(1 - p_1)/n}}\right]$$

$$+ P\left[\frac{\hat{p}_x - p_1}{\sqrt{p_1(1 - p_1)/n}} > \frac{p_0 - p_1}{\sqrt{p_1(1 - p_1)/n}} + z_{\alpha/2}\frac{\sqrt{p_0(1 - p_0)/n}}{\sqrt{p_1(1 - p_1)/n}}\right]$$

$$= P\left[Z < \frac{p_0 - p_1}{\sqrt{p_1(1 - p_1)/n}} - z_{\alpha/2}\frac{\sqrt{p_0(1 - p_0)/n}}{\sqrt{p_1(1 - p_1)/n}}\right]$$

$$+ P\left[Z > \frac{p_0 - p_1}{\sqrt{p_1(1 - p_1)/n}} + z_{\alpha/2}\frac{\sqrt{p_0(1 - p_0)/n}}{\sqrt{p_1(1 - p_1)/n}}\right]$$

where Z is a standard normal random variable. This result allows the calculation of power as a function of p_1 and n.

<table>
<tr><td>Example
9.12</td><td>

In Example 9.6 we tested the null hypothesis that the population proportion of earnings forecasts by financial analysts that exceed the actual outcome is

</td></tr>
</table>

$$H_0: \quad p = p_0 = 0.50$$

against the alternative

$$H_1: \quad p \neq 0.50$$

The test was based on a random sample of $n = 600$ observations. For a test at significance level $\alpha = 0.05$, we have

$$z_{a/2} = z_{.025} = 1.96$$

We now determine the probability that, for a 5% level test, the null hypothesis will be rejected when the true population proportion is $p_1 = 0.52$.

Substitution into the formula gives the power

$$1 - \beta = P\left[Z < \frac{0.50 - 0.52}{\sqrt{(0.52)(0.48)/600}} - 1.96 \frac{\sqrt{(0.50)(0.50)/600}}{\sqrt{(0.52)(0.48)/600}} \right]$$

$$+ P\left[Z > \frac{0.50 - 0.52}{\sqrt{(0.52)(0.48)/600}} + 1.96 \frac{\sqrt{(0.50)(0.50)/600}}{\sqrt{(0.52)(0.48)/600}} \right]$$

$$= P(Z < -2.94) + P(Z > 0.98)$$

$$= 0.0016 + 0.1635 = 0.1651$$

from Table 3 of the Appendix. So, the probability is 0.1651 that this decision rule will reject the null hypothesis when the population proportion is 0.52.

Similarly, this probability can be calculated for any population proportion p_1. Figure 9.15 shows the power function for this example. Because the alternative hypothesis is two-sided, the power function differs in shape from that of Figure 9.13.

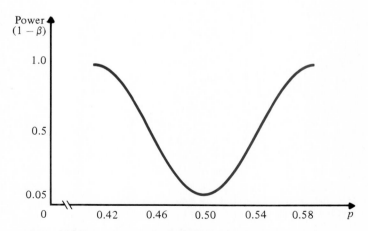

FIGURE 9.15 Power function for Example 9.12; test of
$H_0: \quad p = 0.50$ against $H_1: \quad p \neq 0.50 (\alpha = 0.05, n = 600)$

Here we are considering possible values of the population proportion on either side of the hypothesized value, 0.50. As we see, the probability of rejecting the null hypothesis when it is false increases the further is the true population proportion from the hypothesized value.

9.10 SOME COMMENTS ON HYPOTHESIS TESTING

Classical hypothesis testing methods are probably used more frequently in empirical investigations than any other statistical technique. The relative simplicity of the procedures, together with the frequency of real-world problems for which competing hypotheses exist, render hypothesis testing extremely appealing. Nevertheless, some caution is necessary in interpreting test results.

Perhaps of paramount concern is the crucial role played by the null hypothesis in the hypothesis testing framework. In a typical investigation, the significance level—that is, the probability of rejecting the null hypothesis when it is true—will be set at some low level. Evidence is then gathered from the population to put this hypothesis in jeopardy. However, we might fail to reject a drastically false null hypothesis simply because very little sample information is available, or because the test employed has low power. This may be perfectly appropriate if the null hypothesis is indeed special and the investigator is, for some reason, reluctant to abandon it. Such might be the case, for example, if rejection of a null hypothesis would lead to costly modifications in a production process. However, the special status of the null hypothesis is very often neither warranted nor appropriate. In such circumstances a more symmetric treatment of competing hypotheses would be desirable. One obvious possibility would be to take into account the actual costs (if they could be quantified) of making both Type I errors and Type II errors in deciding between competing hypotheses. Considerations of this sort are embedded in statistical decision theory, which will be discussed in Chapter 19.

On some occasions, very large amounts of sample information are available, and the opposite problem arises—the null hypothesis is put in very considerable jeopardy indeed. It is then important to distinguish between the use of the word *significant,* as it is used in statistical significance testing, compared with its dictionary definition. Suppose very large numbers of households are sampled from two cities. One occasionally meets such statements as: "The difference between the sample average annual incomes for the two cities is $2.37; this is significant." Presumably what is meant by this statement is that the null hypothesis of equality of the population means can, on the basis of such large samples, be rejected at the usual significance levels. Nevertheless, the best estimate available of the difference between the population mean annual incomes is a mere $2.37, which is of no practical significance whatever!

The tests developed in this chapter have typically been based on an assumption about the underlying population distribution. Often, we have assumed that distribution to be normal. When such an assumption fails to hold, the procedures are not strictly valid. Specifically, the true significance levels will differ from those derived on the basis of the normality assumption. Since it is difficult to believe that any population distribution is *precisely* normal, it is important to ask how badly the significance levels might be affected by nonnormality in the parent population. If the effect is relatively small, the tests are said to be *robust* to departures from normality. Generally speaking, it is known that tests on population means are robust.[22] However, tests (such as those of Section 9.8) that compare population variances are not. In the following chapter we will discuss some tests that do not depend for their validity on specific assumptions about the population distribution.

EXERCISES

43. In Example 9.8 we tested the hypothesis of the equality of population mean returns on assets for failed and nonfailed business firms 3 years prior to failure. This test was based on the assumption that the two population variances were equal. Test this assumption against a two-sided alternative.

44. In Exercise 34, it was assumed that the population variances of after-tax profit rates for large and small firms were identical. Test this assumption against a two-sided alternative.

45. It is hypothesized that the more expert a group of people examining a corporation's financial report, the more variable will be their predictions about its future. Independent random samples,[23] each of thirty individuals, from groups of different levels of expertise were chosen. The "low expertise" group consisted of people who had just completed their first intermediate accounting course. Members of the "high expertise" group had completed undergraduate studies and were employed by reputable C.P.A. firms. The sample members were asked to predict next period's operating net cash flow of a company on the basis of its annual financial report. For the low expertise group the sample variance was 451.770, while for the high expertise group it was 1,614.208. Test the null hypothesis that the two population variances are equal against the alternative that the true variance is higher for the high expertise group.

46. In Example 8.11, a confidence interval was found for the difference between mean annual percentage increases in net income for "partial formal planners" and banks with no formal planning system. The calculation of this interval was based on the assumption that the two population variances were equal. Test this assumption against a two-sided alternative.

47. In Exercise 36, it was assumed that the population variances of sales of textbooks, with expensive and plain cover designs, were equal. Test this assumption against a two-sided alternative.

48. In Exercise 1, find the power of a 5% level test when the true population mean miles per gallon is 30.5

49. Refer to Exercise 5. Find the probability, for a 10% level test, of rejecting the null hypothesis when the true population mean cloze score is 55.

50. Refer to Exercise 32. Find the probability, for a 1% level test, of rejecting the null hypothesis when the true difference between the population mean scores is 5—that is, the mean score for the second group exceeds that of the first by 5 points.

51. In Exercise 25, find the power of a 10% level test when the true population proportion of unemployed males is 0.15.

52. Refer to Exercise 26. For a 1% level test, find the probability of accepting the null hypothesis if, in fact, 72.5% of the population understood unit pricing.

53. In Exercise 40, find the power of a 5% level test when the true response rates are 80% for questionnaires not asking race information and 75% for those asking about race.

---

[22] However, the tests for equality of population means based on independent samples, discussed in Section 9.6, become rather less robust when the two sample sizes are not equal.

[23] See D. Snowball, "Some effects of accounting expertise and information load: An empirical study," *Accounting, Organizations and Society, 5* (1980), 323–38.

54. A fast-food chain tests each day that the average weight of their "two-pounders" is at least 32 ounces. The alternative hypothesis is that the average weight is less than 32 ounces, so that new processing procedures are needed. The weights of "two-pounders" can be assumed to be normally distributed, with a standard deviation of 3 ounces. The decision rule adopted is to reject the null hypothesis if the sample mean weight is less than 30.7 ounces.

    (a) If random samples of $n = 36$ two-pounders are selected, what is the probability of a Type I error, using this decision rule?

    (b) If random samples of $n = 9$ two-pounders are selected, what is the probability of a Type I error, using this decision rule? Explain why your answer differs from that in part (a).

    (c) Suppose the true mean weight is 31 ounces. If random samples of 36 two-pounders are selected, what is the probability of a Type II error, using this decision rule?

55. A wine producer claims that the proportion of its customers who cannot distinguish its product from frozen grape juice is at most 0.10. The producer decides to test this null hypothesis against the alternative that the true proportion is more than 0.10. The decision rule adopted is to reject the null hypothesis if the sample proportion who cannot distinguish between these two flavors exceeds 0.145.

    (a) If a random sample of 100 customers is chosen, what is the probability of a Type I error, using this decision rule?

    (b) If a random sample of 400 customers is selected, what is the probability of a Type I error, using this decision rule? Explain, verbally and graphically, why your answer differs from that in part (a).

    (c) Suppose the true proportion of customers who cannot distinguish between these flavors is 0.20. If a random sample of 100 customers is selected, what is the probability of a Type II error?

    (d) Suppose that, instead of the given decision rule, it is decided to reject the null hypothesis if the sample proportion of customers who cannot distinguish between the two flavors exceeds 0.16. A random sample of 100 customers is selected.

        (i) Without doing the calculations, state whether the probability of a Type I error will be higher than, lower than, or the same as that in part (a).

        (ii) If the true proportion is 0.20, will the probability of a Type II error be higher than, lower than, or the same as that in part (c)?

REVIEW EXERCISES

56. Explain carefully the distinction between each of the following pairs of terms:

    (a) Null and alternative hypotheses

    (b) Simple and composite hypotheses

    (c) One-sided and two-sided alternatives

    (d) Type I and Type II errors

    (e) Significance level and power

57. A statistician tests the null hypothesis that the proportion of men favoring a tax reform proposal is the same as the proportion of women. Based on sample data, the null hypothesis is rejected at the 5% significance level. Does this imply that the probability is at least 0.95 that the null hypothesis is false? If not, provide a valid probability statement.

58. Carefully explain what is meant by the p-value of a test, and discuss the use of this concept in hypothesis testing.

59. A secretarial school claims that, after completion of its advanced course, typing speed is increased (in terms of words per minute) by an average of 15%. A random sample of 100 graduates of this course found a mean percentage increase of 14.14% in words per minute. The sample standard deviation was 3.61%. Test the secretarial school's claim at the 5% significance level.

60. A manufacturer of television sets claims that its picture tubes last, on average, 5 years. A random sample of eighty of these sets found a mean tube lifetime of 4.7 years, and a sample standard deviation of 1.9 years. Test the manufacturer's claim.

61. When operating normally, a manufacturing process produces tablets for which the mean weight of the active ingredient is 5 grams, and the standard deviation is 0.1 gram. For a random sample of twelve tablets, the following weights of active ingredient (in grams) were found:

5.01	5.03	4.98	5.00	5.03	5.04
4.96	4.98	4.95	5.00	5.01	4.95

(a) Without assuming the population variance known, test the null hypothesis that the population mean weight of active ingredient per tablet is 5 grams. Use a two-sided alternative and a 5% significance level. State any assumptions that you make.

(b) Stating any assumptions that you make, test the null hypothesis that the population standard deviation is 0.1 gram against the alternative hypothesis that the population standard deviation exceeds 0.1 gram. Use a 5% significance level.

62. An insurance company employs agents on a commission basis. It claims that, during their first year, agents will earn a mean commission of $35,000 and that the population standard deviation is $5,000. For a random sample of ten agents, first-year commissions (in dollars) were as follows:

36,200	33,400	20,700	28,100	18,300
34,900	20,300	19,600	25,200	30,700

(a) Without assuming the population variance known, test the null hypothesis that the population mean commission is $35,000. Use a 10% significance level and state any assumptions you make.

(b) Stating any assumptions that you make, test at the 10% significance level, against a two-sided alternative, the null hypothesis that the population standard deviation is $5,000.

63. An assembly line has typically produced an average of 500 finished components in a day, the population standard deviation being 35. In preliminary contract negotiations, union representatives proposed modified work practices. These were tried for a period of ten days, and the numbers of components produced were as follows:

510	462	493	482	436
523	499	506	430	421

(a) Without assuming the population variance known, test the null hypothesis that the population mean output is at least 500 finished components against the alternative that the true mean is less than 500. State any assumptions you make.

(b) Stating the assumptions needed, test against a two-sided alternative the null hypothesis that the population standard deviation of output under the modified work practices is 35 finished components.

64. A manufacturer of video discs believes that, under adequate quality control, no more than 10% of the discs should be returned as faulty. For a random sample of 250 sales of these discs, it was found that 32 were returned as faulty. Test, at the 10% significance level, the null hypothesis that the population percentage of discs returned as faulty is at most 10%.

65. Of a random sample of 200 students attending a campus rock concert, 112 were women. Test against a two-sided alternative the null hypothesis that half the audience members were women.

66. I want to test the null hypothesis that a population proportion is 0.5, against a two-sided alternative. One dilemma faced in such circumstances is that for any given significance level, the larger the number of sample observations taken, the more likely is the null hypothesis to be rejected. Why is this so, and why might the fact present a dilemma to an investigator planning to use standard hypothesis testing techniques in these circumstances?

67. In an agricultural experiment, two expensive, high-yield, varieties of corn are to be tested, and yield improvements measured. The experiment is arranged so that each variety is planted in one of each of ten pairs of similar plots. The data shown in the accompanying table are the percentage yield increases obtained for these two varieties. Stating any assumptions you make, test at the 10% significance level the null hypothesis that the two population mean percentage yield increases are the same. Use a two-sided alternative hypothesis.

PLOT	VARIETY A	VARIETY B
1	12.6	10.3
2	9.2	8.1
3	6.4	6.2
4	9.8	10.1
5	15.3	12.2
6	16.1	14.0
7	12.3	10.1
8	11.3	13.6
9	12.2	9.8
10	14.3	13.4

68. Two financial analysts were asked to predict earnings per share for a random sample of twelve corporations over the coming year. The quality of their forecasts was evaluated in terms of absolute percentage forecast error, defined as

$$100 \cdot \frac{|\text{Actual} - \text{Predicted}|}{\text{Actual}}$$

The absolute percentage forecast errors made are shown in the table. Stating any assumptions you make, test against a two-sided alternative the null hypothesis that the population mean absolute percentage forecast errors are the same for these two financial analysts.

CORPORATION	ANALYST A	ANALYST B
1	12.3	6.2
2	15.4	12.1
3	5.3	7.4
4	9.2	8.1
5	8.6	11.3
6	14.2	12.3
7	5.2	3.1
8	4.1	0.6
9	5.3	5.5
10	4.1	2.8
11	3.6	4.3
12	5.2	1.7

69. A study[24] was made of the job performance characteristics of managerial, professional, and technical employees of a large oil company. The objective was to compare employees who left the company after less than 15 years of service with those who stayed. For a random sample of 355 employees who left, the mean initial performance rating was 3.24 and the standard deviation was 0.52. For an independent random sample of 174 employees who stayed, the mean initial performance rating was 3.51 and the standard deviation was 0.51. Test at the 5% level, against a two-sided alternative, the null hypothesis that the population mean initial performance ratings are the same for these two groups.

70. In the study cited in Exercise 69, for a random sample of 310 employees who left, the mean number of promotions per years was 0.31 and the standard deviation was 0.31. (Those employees who left within the first 2 years of service were excluded from the sample.) For an independent random sample of 174 employees who stayed, the mean number of promotions per year was 0.43 and the standard deviation was 0.20. What is the lowest level of significance at which the null hypothesis, that the population mean number of promotions per year for these two groups is the same, can be rejected against the alternative that the mean number of promotions is higher for employees who remain with the company?

71. A recent study[25] was aimed at assessing the effects of the size and characteristics of groups on the generation of ideas. To assess the influence of group size, groups of four and eight members were compared. For a random sample of four 4-member groups, the mean number of ideas generated per group was 78.0, and the sample standard deviation was 24.4. For an independent random sample of four 8-member groups, the mean number of ideas generated per group was 114.7, and the sample standard deviation was 14.6. (In each case the groups had a moderator.) Stating any assumptions you need to make, test at the 1% level the null hypothesis that the population means are the same against the alternative that the mean is higher for 8-member groups.

72. In the study described in Exercise 71, an attempt was made to assess the effect of the presence of a moderator on the number of ideas generated. Groups of four members, with or without moderators, were examined. For a random sample of four groups with a

---

[24] G. F. Dreher, "The role of performance in the turnover process," *Academy of Management Journal, 25* (1982), 137–47.

[25] E. F. Fern, "The use of focus groups for idea generation: The effects of group size, acquaintanceship, and moderator on response quantity and quality," *Journal of Marketing Research, 19* (1982), 1–13.

moderator, the mean number of ideas generated per group was 78.0, and the sample standard deviation was 24.4. For an independent random sample of four groups without a moderator, the mean number of ideas generated was 63.5, and the sample standard deviation was 20.2. Stating any assumptions you need to make, test against a two-sided alternative the null hypothesis that the population means are the same.

73. The *fog index* is used to measure the reading difficulty of a written text. The index is calculated through the following steps:

   (i) Find the average number of words per sentence.
   (ii) Find the percentage of words with three or more syllables.
   (iii) The fog index is then 40% of the sum of (i) and (ii).

   A random sample of six advertisements[26] taken from *Scientific American* had the following fog indices:

15.75	11.55	11.16	9.92	9.23	8.20

   An independent random sample of six advertisements from *Sports Illustrated* had the following fog indices:

9.17	8.44	6.10	5.78	5.58	5.36

   Stating any assumptions you need to make, test at the 5% level the null hypothesis that the population mean fog indices are the same, against the alternative that the true mean is higher for *Scientific American* than for *Sports Illustrated*.

74. From the report described in Exercise 73, the fog indices for a random sample of six advertisements in *People Weekly* were as follows:

9.50	8.60	8.59	6.50	4.79	4.29

   For an independent random sample of six advertisements in *Newsweek* the fog indices were as follows:

10.21	9.66	7.67	5.12	4.88	3.12

   Stating any assumptions you need to make, test against a two-sided alternative the null hypothesis that the two population mean fog indices are the same.

---

[26] These results are reported by F. K. Shuptrine and D. D. McVicker, "Readability levels of magazine advertisements," *Journal of Advertising Research, 21* (1981), No. 5, 45–50.

75. An attempt was made to compare the characteristics of price-sensitive and non–price-sensitive consumers. A sample [27] of customers for automobile insurance were asked if they would be willing to switch companies, given a price reduction in premium. Those indicating a willingness to switch were classified as price-sensitive. A random sample of 773 price-sensitive customers were asked if they agreed with the statement: "Agents that sell for only one company usually provide the best service." Of these sample members, 44% agreed. For an independent random sample of 326 non–price-sensitive customers, 51% agreed with this statement. Test at the 1% level the null hypothesis that the population proportions agreeing are the same against the alternative that the true proportion is higher for the non–price-sensitive customers.

76. In a random sample[28] of 70 academicians, 26 nominated "ease of interpretation" as an important criterion in a forecasting procedure. For an independent random sample of 75 practitioners, 29 viewed this quality as an important criterion. What is the smallest significance level at which the null hypothesis of equality of the two population proportions can be rejected against a two-sided alternative?

77. Using the data of Exercise 71, test against a two-sided alternative the null hypothesis that the population standard deviation of the number of ideas generated by 4-member groups is the same as the population standard deviation for 8-member groups. State any assumptions you make.

78. Based on the material of Section 9.8, can you use the data of Exercise 67 to test the null hypothesis of equality of population variances for percentage yield increases for the two varieties of corn?

79. An inspector takes a random sample of ninety components for checking. The decision rule is to reject the null hypothesis that the population proportion of defectives is at most 0.1, against the alternative that more than 10% are defective, if the sample proportion of defectives exceeds 0.125.

(a) What is the significance level of this test?

(b) What is the power of the test if the population proportion of defectives is 0.15?

[27] Results reported by S. J. Skinner, T. L. Childers, and W. H. Jones, "Consumer responsiveness to price differentials: A case for insurance industry deregulation," *Journal of Business Research, 9* (1981), 381–95.

[28] See R. Carbone and J. S. Armstrong, "Evaluation of extrapolative forecasting methods: Results of a survey of academicians and practitioners," *Journal of Forecasting, 1* (1982), 215–17.

some nonparametric tests

CHAPTER TEN

10.1   INTRODUCTION

In Chapter 9, several hypothesis tests that depended on the assumption of normality for population distributions were introduced. Frequently, the assumption of normality is reasonable. Moreover, by virtue of the Central Limit Theorem, many of these test procedures remain approximately valid when applied to large samples even if the population distribution is not normal. However, it is often the case in practical applications that a normality assumption is not tenable. In these circumstances, it is desirable to base inference on tests that are valid over a wide range of distributions of the parent population. Such tests are called **nonparametric,** or **distribution-free.**

In this chapter, we describe some nonparametric tests that are appropriate for analyzing some of the problems we have already met. Additional nonparametric tests will be discussed in subsequent chapters. Although they do require certain assumptions, such as independent sample observations, nonparametric tests are generally valid *whatever* the population distribution. That is to say, tests can be developed that have the required significance levels, no matter what the distribution of the population members. It is not our intention here to attempt to describe the wide array of such tests that are available.[1] Rather, our objective is the more modest one of providing a flavor of the methods used. In this chapter we will discuss nonparametric procedures for testing the equality of the centers of two population distributions. These tests, then, are nonparametric alternatives to the tests discussed in Section 9.6.

---

[1] A great many procedures of this type are discussed in M. Hollander and D. A. Wolfe, *Nonparametric Statistical Methods,* Wiley (1973).

10.2 THE SIGN TEST

The simplest nonparametric test to carry out is the **sign test.** It is used for testing hypotheses about the central location of a population distribution, and is most frequently employed in analyzing data from matched pairs.

Table 10.1 shows the results of a taste-comparison experiment. A manufacturer of baked beans was contemplating a new recipe for the sauce used in its product. A random sample of eight individuals was chosen, and each asked to rate on a scale from 1 to 10 the taste of the original and the proposed new product. The scores are shown in the table. Also shown are the differences in the scores for every taster, and the signs of these differences. Thus a "+" is assigned if the original product is preferred, a "−" if the new product is preferred, and 0 if the two products are rated equally. In this particular experiment, two tasters preferred the original product, five the new, and one rated them equal.

The null hypothesis of interest is that, in the population at large, there is no overall tendency to prefer one product over the other. In assessing this hypothesis, we compare the numbers expressing a preference for each product, discarding those who rated the products equally. In the present example then, the effective sample size is reduced to 7, and the only sample information on which our test is based is that two of the seven tasters preferred the original product.

The null hypothesis can be viewed as the hypothesis that the population median of the differences is 0 (which would be true, for example, if the differences came from a population whose distribution was symmetric about a mean of 0). If this hypothesis were true, our sequence of + and − differences could be regarded as a random sample from a population in which the probabilities for + and − were each 0.5. In that case, the observations would constitute a random sample from a binomial population in which the probability of + was 0.5. Thus, if p denotes the true proportion of +s in the population, then the null hypothesis is simply

$$H_0: \quad p = 0.5$$

We may want to test this hypothesis against either a one-sided or a two-sided alternative. Suppose that, in the taste-preference example, the alternative of interest

TABLE 10.1 Taster ratings for baked beans

	RATING			
TASTER	ORIGINAL PRODUCT	NEW PRODUCT	DIFFERENCE	SIGN OF DIFFERENCE
A	6	8	−2	−
B	4	9	−5	−
C	5	4	1	+
D	8	7	1	+
E	3	9	−6	−
F	6	9	−3	−
G	7	7	0	0
H	5	9	−4	−

is that, in the population, the majority of preferences are for the new product. This alternative is expressed as

$$H_1: \quad p < 0.5$$

In testing the null hypothesis against this alternative, we ask what is the probability of observing a sample result as extreme as or more extreme than that found if the null hypothesis were in fact true. If we denote by $P(x)$ the probability of observing x "successes" (+s) in $n = 7$ binomial trials, each with probability of success 0.5, then the probability of observing 2 or fewer +s is

$$P(0) + P(1) + P(2) = .0078 + .0547 + .1641$$
$$= .2266$$

from Table 1 in the Appendix. Therefore, if we adopt the decision rule, "Reject H_0 if 2 or fewer +s occur in the sample," then the probability is .2266 that the null hypothesis will be rejected when it is in fact true. Hence, such a test has significance level 22.66% and in the present example the null hypothesis can be rejected at this level. It is also important to ask at what level we fail to reject the null hypothesis. Had the decision rule required zero or one +s for rejection, then H_0 would not have been rejected. The significance level for this test is

$$P(0) + P(1) = .0625$$

Thus, the null hypothesis is not rejected by a 6.25% level test.

We now summarize our findings about these data. The null hypothesis that, in the population at large, as many people prefer the original product as the new, is rejected against the alternative that the majority of the population prefer the new product, using a test with significance level 22.66%. However, the null hypothesis cannot be rejected using a test with significance level 6.25%. Thus, these data contain a modest amount of evidence against the hypothesis of population equality of preferences, but by no means an overwhelming amount. This could be a consequence of our having such a small number of sample observations.

Finally, we need to consider the case where the alternative hypothesis is two-sided, that is,

$$H_1: \quad p \neq 0.5$$

In our example, this is the hypothesis that there is an overall preference in the population for either one of the products. If alternatives on each side of the null hypothesized value are treated symmetrically, then a decision rule that would lead to rejection of the null hypothesis for these data is: "Reject H_0 if either 2 or fewer or 5 or more +s occur in the sample." The significance level for this test is

$$P(0) + P(1) + P(2) + P(5) + P(6) + P(7) = 2[P(0) + P(1) + P(2)] = .4532$$

since the probability function of the binomial distribution is symmetric for $p = 0.5$.

The null hypothesis will not be rejected by the rule, "Reject H_0 if either 1 or fewer or 6 or more +s occur in the sample." This test has significance level

$$P(0) + P(1) + P(6) + P(7) = 2[P(0) + P(1)] = .1250$$

Hence, for a 12.5% level test, the null hypothesis that half those population members with a preference prefer the new product is not rejected against a two-sided alternative.

The Sign Test

The **sign test** can be used to test the null hypothesis that a population median is 0. Suppose a random sample is taken from the population, and the observations equal to 0 are discarded, leaving n observations. The null hypothesis to be tested is that the proportion p of nonzero observations in the population that are positive is 0.5; that is,

$$H_0: \quad p = 0.5$$

The test is then based on the fact that the number of positive observations in the sample has a binomial distribution (with $p = 0.5$ under the null hypothesis).

If the number of sample observations is large, the normal approximation to the bionomial distribution can be used to carry out the sign test. Once again, this is a consequence of the Central Limit Theorem.

The Sign Test: Large Samples

If the number n of nonzero sample observations is large,[2] the sign test is based on the normal approximation to the binomial. The test of

$$H_0: \quad p = 0.5$$

is then precisely that described in Section 9.5.

Example 10.1

A random sample of 100 children was asked to compare two new ice cream flavors—peanut butter ripple and bubblegum surprise. Fifty-six sample members preferred peanut butter ripple, forty preferred bubblegum surprise, and four expressed no preference. Test, against a two-sided alternative, the null hypothesis that there is no overall preference in this population for one flavor over the other.

---

[2] This approximation is adequate for samples of more than twenty observations. (This sample size is generally sufficient for using the normal approximation when testing the null hypothesis that the binomial probability is 0.5; however, as we noted in Chapter 9, larger sample sizes are desirable for more extreme values of this parameter.)

If p is the population proportion of all those expressing a preference who favor peanut butter ripple, we want to test

$$H_0: \quad p = 0.5$$

against

$$H_1: \quad p \neq 0.5$$

Since four children gave no preference, we are left with $n = 96$ sample members. The sample proportion preferring peanut butter ripple is

$$\hat{p}_x = \frac{56}{96} = 0.583$$

For a test of significance level α, the decision rule (see Section 9.5) is

$$\text{Reject } H_0 \text{ if} \quad \frac{\hat{p}_x - 0.5}{\sqrt{(0.5)(0.5)/n}} < -z_{\alpha/2} \quad \text{or} \quad \frac{\hat{p}_x - 0.5}{\sqrt{(0.5)(0.5)/n}} > z_{\alpha/2}$$

Here we have

$$\frac{\hat{p}_x - 0.5}{\sqrt{(0.5)(0.5)/n}} = \frac{0.583 - 0.5}{\sqrt{(0.5)(0.5)/96}} = 1.63$$

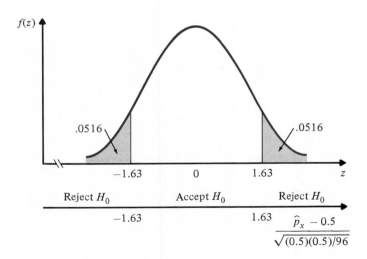

FIGURE 10.1 The distribution of $Z = \dfrac{\hat{p}_x - 0.5}{\sqrt{(0.5)(0.5)/96}}$ when the null hypothesis H_0: $p = 0.5$ is true, and the decision rule for testing H_0 against the alternative H_1: $p \neq 0.5$ at significance level 0.1032 (sample size $n = 96$)

From Table 3 of the Appendix, we find that if $z_{\alpha/2} = 1.63$, then $\alpha/2 = .0516$, so that $\alpha = .1032$. Hence, the null hypothesis can be rejected at all significance levels greater than 10.32%. If the null hypothesis that as many children prefer peanut butter ripple as prefer bubblegum surprise were true, the chance of observing a sample outcome as extreme or more extreme than that actually obtained would be a little higher than 1 in 10. These data then contain a modest amount of evidence against that hypothesis.

This test is illustrated in Figure 10.1, which shows tail area probabilities of the standard normal distribution corresponding to the upper and lower 5.16% of the total area under the probability density function.

EXERCISES

1. A random sample of twelve financial analysts was asked to predict the percentage increases in the prices of two common stocks over the next year. The results obtained are shown in the table. Use the sign test to test the null hypothesis that, for the population of analysts, there is no overall preference for one stock over the other.

ANALYST	A	B	C	D	E	F
STOCK 1	6.8	9.8	2.1	6.2	7.1	6.5
STOCK 2	7.1	12.3	5.3	6.8	7.2	6.2

ANALYST	G	H	I	J	K	L
STOCK 1	9.3	1.0	−0.2	9.6	12.1	6.3
STOCK 2	10.1	2.7	1.3	9.8	12.2	8.9

2. An organization offers a program designed to increase the level of comprehension achieved by students when quickly reading technical material. Each member of a random sample of ten students was given 30 minutes to read an article. A test of the level of comprehension achieved was then administered. This process was repeated after these students had completed the program. The accompanying table shows comprehension scores before and after completion of the program. Use the sign test to test the null hypothesis that, for this population, there is no overall improvement in comprehension levels following completion of the program.

STUDENT	A	B	C	D	E	F	G	H	I	J
BEFORE	62	63	84	72	60	53	49	58	83	92
AFTER	69	72	80	70	69	61	63	59	87	98

3. A random sample of fourteen voters was asked to rate, on a scale from 1 to 10, both the likeability of a political candidate and their view of the candidate's competence. The results shown in the table were obtained. Use the sign test to test the null hypothesis that, overall, the candidate is rated equally on these two attributes.

VOTER	A	B	C	D	E	F	G
LIKEABILITY	9	8	5	7	6	9	8
COMPETENCE	6	8	1	7	7	4	6

VOTER	H	I	J	K	L	M	N
LIKEABILITY	8	7	6	9	6	7	9
COMPETENCE	8	8	4	3	7	7	5

4. A company introduced canned background music into its assembly shops. The personnel department found that, in a random sample of eighty-five workers, forty preferred to have the music, thirty-three preferred not to have it, and twelve were indifferent. Test the null hypothesis that, for this population, there is no overall preference, one way or the other, against the alternative that the balance of preference is in favor of having the canned music.

5. For a random sample of 120 adults, a survey organization found that lawyers were more highly respected than politicians by fifty-seven, that politicians were the more highly respected by forty-nine, and that the two groups were equally respected by the remaining fourteen sample members. Test the null hypothesis that, overall, politicians and lawyers are equally respected in the adult population at large.

6. A random sample of sixty professional economists was asked to predict whether next year's inflation rate would be higher, lower, or about the same as in the current year. The results are shown in the following table:

HIGHER	22
LOWER	28
ABOUT THE SAME	10

Test the null hypothesis that the profession is evenly divided on the question.

10.3 THE WILCOXON TEST

We saw in Section 10.2 that the sign test provides an easy test procedure for comparing populations when a sample of matched pairs is available. One difficulty with the test is that it takes account of only a very limited amount of information—namely, the signs of the differences. For example, in Table 10.1, the sign test takes account only of which product is preferred, ignoring the strengths of the preferences. When the sample size is small, it might be suspected therefore that the test would not be very powerful. In fact, it is possible to take some account, not only of the signs, but also of the magnitudes of the differences between matched pairs, and still achieve a test that is distribution-free. The **Wilcoxon test** provides a method of incorporating information about the relative sizes of the differences. Like very many nonparametric tests, it is based on *ranks*.

To illustrate, Table 10.2 extends the calculations of Table 10.1 for the data on the ratings of two baked bean products.

TABLE 10.2 Calculation of Wilcoxon test statistic for taste-preference data

TASTER	DIFFERENCE	RANK (+)	RANK (−)
A	−2		3
B	−5		6
C	1	1.5	
D	1	1.5	
E	−6		7
F	−3		4
G	0		
H	−4		5
	Sums	3	25

The test statistic for the Wilcoxon test is calculated as follows:

(i) As in the sign test, differences of 0 are ignored, so that in this example the effective number of sample observations is 7.

(ii) The nonzero absolute differences are then ranked in ascending order of magnitude. If two or more values are equal they are assigned the average of the next available ranks. In our example, the two smallest absolute differences are equal. The rank assigned them is therefore the average of ranks 1 and 2—that is, 1.5. The next value is assigned rank 3, and so on. Proceeding in this way, we rank all differences.

(iii) The ranks for positive and negative differences are summed separately. The smaller of these sums is the Wilcoxon statistic T. From Table 10.2, the two sums in this case are 3 and 25, so that $T = 3$.

We will now suppose, as is often reasonable, that the population distribution of the paired differences is symmetric. The null hypothesis to be tested is that the center of this distribution is 0. In our example, then, we are assuming that differences in the ratings of the two products have a symmetric distribution, and want to test whether that distribution is centered on 0—that is, no difference between ratings. We would be suspicious of the null hypothesis if the sum of the ranks for positive differences was very different from that for negative differences. Hence, the null hypothesis will be rejected for low values of the statistic T. Cutoff points for the distribution of this random variable are given in Table 8 of the Appendix for tests against a one-sided alternative, that the population distribution of the paired differences is specified either to be centered on some number bigger than 0 or to be centered on some number less than 0. In our example, we might want to take the alternative hypothesis that the new product tends to be preferred over the original. This would imply that the distribution of the paired differences in ratings is centered on some number less than 0. From Table 8, we see that for a sample size of $n = 7$, the cutoff points are 3 for a 2.5% level test and 4 for a 5% level test. Hence, the null hypothesis is rejected against the one-sided alternative at the 5% level, and is just rejected at the 2.5% level. Therefore, the evidence against the hypothesis that the population differences in ratings for these products is centered on 0 is quite strong. It appears likely that, overall, ratings are higher for the new product.

If the alternative hypothesis is two-sided—that is, that the two population differences are centered on some number other than 0—then the appropriate significance levels are twice those for the one-sided alternative. Hence, for these data, the null hypothesis is rejected against a two-sided alternative at the 10% level, and is just rejected at the 5% level of significance.

Notice that using the additional information provided by the ranks allows the rejection of the null hypothesis at a much lower level of significance than was possible for the sign test.

The Wilcoxon Test

The **Wilcoxon test** can be employed when a random sample of matched pairs of observations is available. We assume that the population distribution of the differences in matched pairs is symmetric, and want to test the null hypothesis that this distribution is centered on 0. Discarding those pairs for which the difference is 0, we rank the remaining n absolute differences in ascending order. The sums of the ranks corresponding to positive and negative differences are calculated, and the smaller of these sums is the Wilcoxon test statistic T. The null hypothesis is rejected if T is smaller than the tabulated value in Table 8 of the Appendix.

When the number n of nonzero differences in the sample is large,[3] the normal distribution provides a good approximation to the distribution of the Wilcoxon statistic T under the null hypothesis. It can be shown that, when the null hypothesis that the population differences are centered on 0 is true, the mean and variance of this distribution are

$$E(T) = \mu_T = \frac{n(n + 1)}{4}$$

and

$$\text{Var}(T) = \sigma_T^2 = \frac{n(n + 1)(2n + 1)}{24}$$

Then, for large n, the distribution of the random variable

$$Z = \frac{T - \mu_T}{\sigma_T}$$

is approximately standard normal, and tests can be based on this result, as indicated in the box.

The Wilcoxon Test: Large Samples

If the number n of nonzero differences is large and T is the observed value of the Wilcoxon statistic, then the following tests have significance level α:

$\longrightarrow$

---

[3] The approximation is adequate for twenty or more observations.

(i) If the alternative hypothesis is one-sided, reject the null hypothesis if

$$\frac{T - \mu_T}{\sigma_T} < -z_\alpha$$

(ii) If the alternative hypothesis is two-sided, reject the null hypothesis if

$$\frac{T - \mu_T}{\sigma_T} < -z_{\alpha/2}$$

Example 10.2

It was hypothesized[4] that (for Australian companies) the time from year-end to the final publication of the annual report would be longer for a company in a year when auditors had reported a "subject to" qualification than in previous normal years. Times to publication of annual reports for a random sample of eighty-five companies with "subject to" qualifications were compared to the times for these same companies in normal years. The eighty-five differences ("subject to" years minus normal years reporting times) were calculated, and the absolute differences ranked. The smaller of the rank sums, 1,195, was for negative differences. Test the null hypothesis that the distribution of the population differences is centered on 0 against the alternative that reporting times tend to be longer in the "subject to" years.

Given a sample of $n = 85$ observations, the Wilcoxon statistic has (under the null hypothesis) mean

$$\mu_T = \frac{n(n+1)}{4} = \frac{(85)(86)}{4} = 1,827.5$$

and variance

$$\sigma_T^2 = \frac{n(n+1)(2n+1)}{24} = \frac{(85)(86)(171)}{24} = 52,083.75$$

If T is the observed value of the statistic, the null hypothesis is rejected against the one-sided alternative if

$$\frac{T - \mu_T}{\sigma_T} < -z_\alpha$$

Here, $T = 1,195$, so that

$$\frac{T - \mu_T}{\sigma_T} = \frac{1,195 - 1,827.5}{\sqrt{52,083.75}} = -2.77$$

Now, the value of α corresponding to $z_\alpha = 2.77$ is, from Table 3 of the Appendix, .0028. Hence, the null hypothesis can be rejected at all levels higher than 0.28%.

---

[4] This example is adapted from results given by G. P. Whittred, "Audit qualification and the timeliness of corporate annual reports," *Accounting Review, 55* (1980), 563–77.

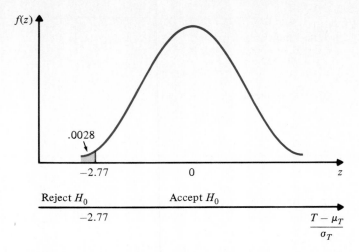

FIGURE 10.2 The distribution of $Z = \dfrac{T - \mu_T}{\sigma_T}$ when the null hypothesis that the distribution of the paired differences is centered on zero is true, and the decision rule for testing this hypothesis against the alternative that the center of this distribution is less than zero, at significance level 0.0028

The data, then, strongly suggest the truth of the alternative hypothesis, that reporting times tend to be longer in the "subject to" years. The test is illustrated in Figure 10.2.

10.4 THE MANN-WHITNEY TEST

In Section 10.3 we saw how the central locations of two population distributions could be compared when a random sample of matched pairs was available. In this section we introduce a test for the same problem when *independent random samples* are taken from the two populations.

Table 10.3 shows the numbers of hours per week students claim to spend studying for introductory finance and accounting courses. The data are from independent random samples of ten finance students and twelve accounting students.

Our null hypothesis is that the central locations of the two population distributions are identical. As a first step in testing this hypothesis, the two samples are pooled together and the observations are ranked in ascending order, ties being treated in the same way as previously. These ranks are shown in Table 10.3. Now, if the null hypothesis were true, we would expect the average ranks for the two samples to be quite close. In this particular example, the average rank for the finance students is 9.35, while that for the accounting students is 13.29. As usual, when testing hypotheses, we want to know how likely would be a discrepancy of this magnitude if the null hypothesis were true.

TABLE 10.3 Numbers of hours per week spent studying for introductory finance and accounting courses

FINANCE	(RANK)	ACCOUNTING	(RANK)
10	(10)	13	(17.5)
6	(2)	17	(22)
8	(4.5)	14	(19)
10	(10)	12	(15.5)
12	(15.5)	10	(10)
13	(17.5)	9	(7)
11	(13)	15	(20)
9	(7)	16	(21)
5	(1)	11	(13)
11	(13)	8	(4.5)
		9	(7)
		7	(3)
Rank sum	93.5	Rank sum	159.5

We note here that it is not necessary to calculate both rank sums, for if we know one we can deduce the other. In our example, for instance, the ranks must sum to the sum of the integers 1 through 22—that is, to 253. Thus, any test of our hypothesis can be based on just one of the rank sums.

In general, suppose that n_1 observations are available from the first population, and n_2 from the second, and let R_1 denote the sum of the ranks of the observations from the first population. The **Mann-Whitney test statistic** is then defined as

$$U = n_1 n_2 + \frac{n_1(n_1 + 1)}{2} - R_1$$

In testing the null hypothesis that the central locations of the two population distributions are the same, we assume that, apart from any possible differences in central location, the two population distributions are identical. It can be shown then that, if the null hypothesis is true, the random variable U has mean and variance

$$E(U) = \mu_U = \frac{n_1 n_2}{2}$$

and

$$\text{Var}(U) = \sigma_U^2 = \frac{n_1 n_2 (n_1 + n_2 + 1)}{12}$$

Furthermore, under the null hypothesis, the distribution of U approaches the normal quite rapidly as the number of sample observations increases.[5] Hence, for moderately large sample sizes, the distribution of the random variable

$$Z = \frac{U - \mu_U}{\sigma_U}$$

---

[5] The approximation is adequate if each sample contains ten or more observations.

is well-approximated by the standard normal. This allows tests to be carried out in a straightforward manner, as described in the box.

The Mann-Whitney Test

Suppose we have independent random samples of n_1 and n_2 observations from two populations. If the sample observations are pooled together and ranked, with R_1 denoting the sum of the ranks for the first population, then the Mann-Whitney test statistic is

$$U = n_1 n_2 + \frac{n_1(n_1 + 1)}{2} - R_1$$

It is assumed that the two population distributions are identical, apart from any possible differences in central location. In testing the null hypothesis that the two population distributions have the same central location, the following tests have significance level α:

(i) If the alternative is the one-sided hypothesis that the location of population 1 is higher than that of population 2, the decision rule is

$$\text{Reject } H_0 \text{ if } \quad \frac{U - \mu_U}{\sigma_U} < -z_\alpha$$

(ii) If the alternative is the one-sided hypothesis that the location of population 1 is lower than that of population 2, the decision rule is

$$\text{Reject } H_0 \text{ if } \quad \frac{U - \mu_U}{\sigma_U} > z_\alpha$$

(iii) If the alternative is the two-sided hypothesis that the two population distributions differ, the decision rule is

$$\text{Reject } H_0 \text{ if } \quad \frac{U - \mu_U}{\sigma_U} < -z_{\alpha/2} \quad \text{or} \quad \frac{U - \mu_U}{\sigma_U} > z_{\alpha/2}$$

We now return to the example introduced in this section. Suppose we want to test the null hypothesis that the central locations of the distributions of study times are identical against the two-sided alternative. For these data, we have

$$n_1 = 10; \qquad n_2 = 12; \qquad R_1 = 93.5$$

so that the value observed for the Mann-Whitney statistic is

$$U = n_1 n_2 + \frac{n_1(n_1 + 1)}{2} - R_1$$

$$= (10)(12) + \frac{(10)(11)}{2} - 93.5 = 81.5$$

Under the null hypothesis, the distribution of the statistic has mean

$$\mu_U = \frac{n_1 n_2}{2} = \frac{(10)(12)}{2} = 60$$

and variance

$$\sigma_U^2 = \frac{n_1 n_2 (n_1 + n_2 + 1)}{12} = \frac{(10)(12)(23)}{12} = 230$$

The decision rule is to reject the null hypothesis if

$$\frac{U - \mu_U}{\sigma_U} < -z_{\alpha/2} \qquad \text{or} \qquad \frac{U - \mu_U}{\sigma_U} > z_{\alpha/2}$$

Here,

$$\frac{U - \mu_U}{\sigma_U} = \frac{81.5 - 60}{\sqrt{230}} = 1.42$$

Now from Table 3 of the Appendix, the value of $\alpha/2$ corresponding to a value 1.42 for $z_{\alpha/2}$ is .0778, so that the corresponding α is .1556. Hence, the null hypothesis can be rejected against the two-sided alternative at levels higher than 15.56%, as indicated in Figure 10.3. Thus, these data do not contain terribly strong evidence against the hypothesis that the central locations of the distributions of study times in

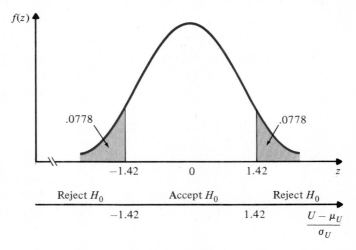

FIGURE 10.3 The distribution of $Z = \dfrac{U - \mu_U}{\sigma_U}$ when the null hypothesis that the central locations of the two population distributions are identical is true, and the decision rule for testing this hypothesis against the two-sided alternative at significance level 0.1556

accounting and finance are the same. There is not very strong support for the conclusion that, on the whole, students spend more time studying for one of these subjects rather than the other.

Example 10.3

In a study[6] designed to compare the performance of firms that give management forecasts of earnings with those that do not, independent random samples of 80 firms from each of the populations were taken. The variability in the growth rate of earnings over the previous ten periods was measured for each of the 160 firms, and these variabilities were ranked. The sum of the ranks for firms not disclosing management earnings forecasts was 7,287. Test, against a two-sided alternative, the null hypothesis that the central locations of the population distributions of earnings variabilities are the same for these two types of firms.

Since we have $n_1 = 80$, $n_2 = 80$, and $R_1 = 7,287$, the calculated value of the Mann-Whitney statistic is

$$U = n_1 n_2 + \frac{n_1(n_1 + 1)}{2} - R_1 = (80)(80) + \frac{(80)(81)}{2} - 7,287 = 2,353$$

Under the null hypothesis, the Mann-Whitney statistic has mean

$$\mu_U = \frac{n_1 n_2}{2} = \frac{(80)(80)}{2} = 3,200$$

and variance

$$\sigma_U^2 = \frac{n_1 n_2 (n_1 + n_2 + 1)}{12} = \frac{(80)(80)(161)}{12} = 85,867$$

The decision rule is to reject the null hypothesis, against a two-sided alternative, if

$$\frac{U - \mu_U}{\sigma_U} < -z_{\alpha/2} \qquad \text{or} \qquad \frac{U - \mu_U}{\sigma_U} > z_{\alpha/2}$$

Here, we have

$$\frac{U - \mu_U}{\sigma_U} = \frac{2,353 - 3,200}{\sqrt{85,867}} = -2.89$$

From Table 3 of the Appendix, we see that the value of $\alpha/2$ corresponding to a value 2.89 for $z_{\alpha/2}$ is .0019, so that α is .0038. Hence, the null hypothesis can be rejected at all levels higher than 0.38%. These data, then, present very strong evidence against the hypothesis that the central locations of the distributions of population variabilities in earnings growth rates are the same for those firms that give management earnings forecasts as for those that do not. On the contrary, it appears that, in

---

[6] This example is adapted from results given by B. Jaggi and P. Grier, "A comparative analysis of forecast disclosing and nondisclosing firms," *Financial Management, 9* (1980), No. 2, 38–43.

the aggregate, there is more variability in earnings for one group of firms than for the other.

Now, if we had been given the actual data rather than just the ranks, we could have carried out a test of the null hypothesis using the methods of Section 9.6. However, using the Mann-Whitney test, we have found that the null hypothesis can be rejected *without our having to make the assumption of population normality*.

10.5 DISCUSSION

The nonparametric tests discussed in this chapter represent a very small subset of the nonparametric procedures in current use. In later chapters we will meet some other distribution-free tests.

It is instructive to compare the tests of this chapter with those of Section 9.6, where we considered the problem of testing the equality of two population means, *assuming the population distributions are normal*. The tests developed in this chapter can be regarded also as tests of this null hypothesis, but assuming only that the two population distributions have the same shape. This brings out the major advantage of nonparametric methods—they are appropriate under a wide range of assumptions about the underlying populations.

Nonparametric methods also have the following advantages:

(i) They are often computationally more straightforward, so that the tests can be carried out quite rapidly. This is particularly true of the sign test.

(ii) The data available to the investigator may be of the natural form for these tests. For example, if all that was known in a product comparison study was which product is preferred, the sign test would be immediately applicable. In many practical situations, data are available only in the form of ranks, leading naturally to such procedures as the Wilcoxon test or the Mann-Whitney test.

(iii) Just as the mean is greatly susceptible to influence by extreme outlying observations, so are inferences based on the *t* tests of Section 9.6. On the other hand, tests based on ranks give far less weight to odd outlying sample values.

Of course, when the assumption of normality in the population is at least approximately true, one would expect tests based on that assumption to be more powerful than tests based on rankings, since the latter discard some of the information in the data. In fact, however, at least in samples of moderate size, tests such as the Wilcoxon and Mann-Whitney are only a little less powerful than the competing *t* tests when the population distributions are normal. For this reason, together with their broader applicability, these nonparametric tests are very popular. Moreover, when the population distribution differs markedly from the normal, nonparametric tests can have much more power than the corresponding normal-theory tests.

On the other hand, nonparametric methods are rather difficult to extend to problems (such as those to be discussed in chapters 13 and 14) that involve complex model building. For this reason, the traditional procedures of Chapter 9, whose development is far more straightforward, remain in the mainstream of statistical analysis.

EXERCISES

7. Use the Wilcoxon test to analyze the data of Exercise 1. Discuss your findings.

8. The Cola-Cola Company recently ran a national advertising campaign based on the slogan: "Twice the cola, twice the fun." To test whether this campaign had improved brand awareness, random samples of 500 people in each of ten cities were asked to name five soft drinks, both before and after the campaign had run. The accompanying table shows the numbers naming Cola-Cola. Use the Wilcoxon test to test the hypothesis that the distribution of the paired differences is centered on 0 against the alternative that the brand is better known after the campaign.

CITY	BEFORE	AFTER	CITY	BEFORE	AFTER
Atlanta	86	123	Miami	215	193
Boston	151	160	New Orleans	254	311
Chicago	192	180	New York	123	112
Denver	71	93	Philadelphia	97	131
Los Angeles	86	99	St. Louis	153	169

9. Two cars from each of eight different models were chosen to assess the influence of a gasoline additive on achieved miles per gallon. One car from each pair was tested with the additive and one without. The resulting miles per gallon figures are shown in the table. Use the Wilcoxon test to analyze the data. Discuss the implications of the test results.

MODEL	A	B	C	D	E	F	G	H
WITHOUT ADDITIVE	21.6	20.5	26.3	27.2	19.5	20.3	25.2	19.8
WITH ADDITIVE	22.1	20.5	26.0	27.3	19.7	21.4	26.4	19.9

10. A random sample of fifty farms was chosen for the comparison of the effects of two fertilizers on corn yields. Each fertilizer—Wonder-grow and Barn buster—was used on half the acreage of each of the farms. The fifty yield differences (Wonder-grow minus Barn buster) were calculated and the absolute differences ranked. The smaller of the rank sums was 622. Test the null hypothesis that the two fertilizers are equally effective.

11. A business school dean is interested in the comparison of verbal and mathematical Scholastic Aptitude Test scores of applicants to the school. For a random sample of sixty applicants, the differences between the two test scores were calculated, and the absolute differences ranked. The smaller of the rank sums was 684. What can you conclude from these results?

12. A random sample of eighty automobiles was selected to compare two brands of tires. One tire from each brand was tested on the front wheels of each automobile. The number of miles obtained from each tire before it became dangerously worn was recorded. The eighty differences were then calculated and their absolute values ranked. The smaller of the rank sums was 1,565. Discuss the implications of these sample results.

13. A corporation interviews both liberal arts and business graduates for certain managerial

positions. A random sample of twelve liberal arts applicants and an independent random sample of twelve business graduates were subjected to intensive interviewing and testing by a team of the corporation's senior managers. The candidates were then ranked from 1 (most suitable for employment) to 24, as indicated in the accompanying table. Test the null hypothesis that, overall, the corporation's senior management has no preference between liberal arts and business graduates, against the alternative that business graduates are preferred.

1	business	9	liberal arts	17	liberal arts
2	business	10	liberal arts	18	liberal arts
3	liberal arts	11	liberal arts	19	business
4	business	12	business	20	business
5	business	13	liberal arts	21	liberal arts
6	liberal arts	14	business	22	liberal arts
7	business	15	business	23	liberal arts
8	liberal arts	16	business	24	business

14. A random sample of twelve of a company's Midwestern sales representatives and an independent random sample of ten of the company's Southern representatives showed the following figures for the percentage increase in orders received this year over last year:

Midwest: 8.2, 6.5, 7.4, 2.1, −0.8, 5.3,
9.4, 6.1, 8.2, 7.3, 2.4, −1.3

South: 7.4, 8.6, 9.2, 11.8, 12.3,
6.4, 13.1, 5.0, 7.1, 2.8

Test the null hypothesis of no difference between the central locations of population distributions of rates of sales increase in the two regions.

15. A bank executive is presented with a random sample of fifteen single female managers and an independent random sample of fifteen single male managers, and asked to rank the sample members in order of creditworthiness from 1 (highest) to 30. He produced the rankings:

Female: 18, 12, 8, 15, 26, 28, 19, 11,
6, 29, 16, 24, 7, 20, 13

Male: 3, 5, 14, 17, 22, 23, 1, 9,
30, 21, 25, 2, 10, 27, 4

Test the hypothesis that this executive considers the sexes equally creditworthy against the alternative that males are preferred.

16. A random sample of 50 people who voted Republican and an independent random sample of 50 people who voted Democratic in the last presidential election were asked their annual family income. The 100 income figures were then pooled together and ranked in order (with rank 1 assigned to the lowest income). The sum of the ranks for Democratic voters was 2,062. Test the null hypothesis that there is no difference between the central locations of the distributions of incomes of Republican and Democratic voters against the alternative that, on the whole, Republican voters have higher incomes.

17. The time taken in days from year-end for a random sample[7] of 120 Australian companies with clean audit reports to release a preliminary profit report was compared with the time taken for an independent random sample of 86 companies whose reports had a "subject to" qualification. The times taken for the 206 companies were pooled together and ranked, with shortest time assigned rank 1. The sum of the ranks for companies with a "subject to" qualification was 9,686. Test the null hypothesis that the central locations of the two population distributions are identical against the alternative that companies with "subject to" qualifications tend to take longer to produce their preliminary profit reports.

18. Starting salaries of M.B.A. graduates from two leading business schools were compared. Independent random samples of thirty students from each school were taken, and the sixty starting salaries were pooled and ranked. The sum of the ranks for students from one of these schools was 1,243. Test the null hypothesis that the central locations of the population distributions are identical.

REVIEW EXERCISES

19. What does it mean to say that a test is nonparametric? What are the relative advantages of such tests?

20. Construct a realistic example of a statistical problem in the business area where you would prefer the use of a nonparametric test to the alternative parametric test.

21. In a random sample of twelve analysts, seven believed that automobile sales in the United States were likely to be significantly higher next year than in the present year, two believed that sales would be significantly lower, and the others anticipated that next year's sales would be roughly the same as those in the current year. What can you conclude from these data?

22. From a random sample of sixteen international bankers, five believed there was need for additional caution in making loans to East European countries, eight believed there was no such need, and the others were undecided. What conclusions can be drawn from these data?

23. A large supermarket chain switched to electronic check-out. Of a random sample of 100 customers, 46 preferred the new system, 39 preferred the old system, and the others were indifferent between the two. Do these data present strong evidence that the new check-out system is more popular than the old?

24. In a random sample of 120 corporations, it was found that new orders were running at a noticeably higher level in the current quarter than in the same quarter of the previous year for 67 corporations, while new orders were noticeably lower for 39 corporations. For the remaining corporations in the sample there had been little noticeable change. Evaluate the strength of the sample evidence in support of the contention that, over the population of all corporations, new orders are more likely to be higher than lower in this quarter, compared with the previous year.

25. A random sample of ten corporate analysts were asked to rate, on a scale from 1 (very poor) to 10 (very high) the prospects for their own corporations, and for the economy at large, in the current year. The results obtained are shown in the accompanying table. Using the Wilcoxon test, discuss the proposition that, in the aggregate, corporate analysts

---

[7] Results reported by G. P. Whittred, "Audit qualification and the timeliness of corporate annual reports," *Accounting Review, 55* (1980), 563–77.

are more optimistic about the prospects for their own companies than for the economy at large.

ANALYST	OWN CORPORATION	ECONOMY AT LARGE
1	9	8
2	7	5
3	6	7
4	5	4
5	8	4
6	6	9
7	7	7
8	5	2
9	4	6
10	9	7

26. Nine pairs of hypothetical profiles were constructed for business school graduates applying for general management positions. Within each pair, these profiles were identical, except that one candidate was male and the other female. For interviews for employment of these graduates, evaluations on a scale of 1 (low) to 10 (high) were made of the suitability for employment of the candidates. The results are shown in the table. Analyze these data using the Wilcoxon test.

INTERVIEW	MALE	FEMALE
1	8	7
2	9	10
3	7	5
4	4	7
5	8	8
6	9	9
7	5	3
8	4	5
9	6	3

27. A random sample of fifty schools was chosen to compare the effectiveness of two different introductory economics texts. At the end of the course, a common examination was administered in each school, and the average scores of students recorded. The fifty differences in average scores were calculated and their absolute values ranked. The smaller of the rank sums was 453. Interpret these results.

28. It is claimed that a computer-based evaluation system can pick out those companies most, or least, likely to enjoy high earnings growth in the year ahead. The system selects 10% of companies as "best prospects" and 10% as "worst prospects." For a random sample of ten best prospects, it emerged that percentage growth rates in earnings were as follows:

9.23	14.71	8.26	3.15	6.24
1.09	18.30	21.97	4.17	2.30

For an independent random sample of ten worst prospects, the corresponding results were as follows:

−6.20	3.91	6.23	1.49	−6.01
−1.30	0.21	4.62	0.27	−1.32

Use the Mann-Whitney test to evaluate these data.

29. For a random sample of twelve graduates in business from a technical college, the starting salaries accepted for employment on graduation (in thousands of dollars) were the following:

26.2	31.3	27.4	26.0	27.5	33.5
29.3	28.7	25.1	27.2	29.8	34.6

For an independent random sample of ten graduates from a state university, the corresponding figures were as follows:

25.3	29.2	26.8	30.7	24.9
28.2	27.1	26.5	31.3	24.9

Analyze the data using the Mann-Whitney test, and comment on the results.

30. The data of Exercise 29 could have been analyzed using one of the tests discussed in Chapter 9. Which one? Which test would you prefer to use in this particular case? Why?

goodness
of fit tests
and contingency
tables

CHAPTER ELEVEN

11.1 GOODNESS OF FIT TESTS

In this chapter, we discuss some tests based on the chi-square distribution. We begin with a very simple numerical example. In a study[1] of the reasons why United States firms adopt exporting as a marketing strategy, the process leading to a decision to export was categorized according to two possibilities. The first, called a *problem-oriented adoption process* (POAP), is where exporting occurs as a result of a combination of problems such as mature product, increased competition in the domestic market, decreasing sales, and a desire for market expansion. The second, called an *innovation-oriented adoption process* (IOAP), is where the initiating force for the decision to export is the knowledge of the existence of a marketing opportunity in a foreign market, or the desire to learn about such opportunities. The results obtained for a random sample of thirty-five firms are shown in Table 11.1.

More generally, we may have a random sample of n observations that can be

TABLE 11.1 Motivations for exporting as a marketing strategy

	CATEGORY		
	IOAP	POAP	TOTAL
Number of Firms	24	11	35

[1] W.Y. Lee and J. J. Brasch, "The adoption of export as an innovative strategy," *Journal of International Business Studies, 9* (1978), No. 1, 85–93.

TABLE 11.2 Classification of n observations into K categories

	CATEGORY			TOTAL
	1	2 ...	K	
Number of Observations	O_1	O_2 ...	O_K	n

classified according to K categories. If the numbers of observations falling into each category are $O_1, O_2, \ldots, O_K$, the setup is as shown in Table 11.2.

The sample data are to be used to test some hypothesis about the underlying population distribution. Here we consider the case where a null hypothesis specifies the probabilities that an observation falls in each of the categories. To illustrate, suppose our null hypothesis (H_0) for the exporting decision example is that, in the population, IOAP and POAP are equally likely motivations—that is, the probability is 0.5 that a sample observation falls into either one of these categories. In order to check this hypothesis, it is natural to compare what is actually *observed* in the sample with what would be *expected* if the null hypothesis were true. For our example, given a total sample size of 35 observations, the expected number of firms in each category under the null hypothesis would be $35(0.5) = 17.5$. This is summarized in Table 11.3.

More generally, where there are K categories, suppose that the null hypothesis specifies $p_1, p_2, \ldots, p_K$ for the probabilities that an observation falls into the categories. We assume these possibilities are mutually exclusive and collectively exhaustive—that is, each sample observation must belong to one of the categories, and cannot belong to more than one. In that case, the hypothesized probabilities must sum to 1, that is,

$$p_1 + p_2 + \cdots + p_K = 1$$

Then, if there are n sample observations, the expected numbers in each category, under the null hypothesis, will be

$$E_i = np_i \qquad (i = 1, 2, \ldots, K)$$

This is shown in Table 11.4.

The setup, then, is that we have a null hypothesis about the population which

TABLE 11.3 Comparison of observed and expected number of firms for exporting data

	CATEGORY		TOTAL
	IOAP	POAP	
Observed Number of Firms	24	11	35
Probability (under H_0)	0.5	0.5	1
Expected Number of Firms (under H_0)	17.5	17.5	35

TABLE 11.4 Observed and expected numbers for n observations and K categories

	1	2	...	K	TOTAL
Observed Number	O_1	O_2	...	O_K	n
Probability (under H_0)	p_1	p_2	...	p_K	1
Expected Number (under H_0)	$E_1 = np_1$	$E_2 = np_2$	... $E_K = np_K$		n

specifies the probabilities that a sample observation will fall into each possible category. The sample observations are to be used to check this hypothesis. If the numbers of sample values observed in each category are very close to those expected if the null hypothesis were true, this fact would lend support to that hypothesis. We might, in such circumstances, say that the data provide a close *fit* to the assumed population distribution of probabilities. Our tests of the null hypothesis are based on an assessment of the closeness of this fit and are generally referred to as **goodness of fit tests.**

Now, in order to test the null hypothesis, it is natural to look at the magnitudes of the discrepancies between what is observed and what is expected. The larger are these discrepancies in absolute value, the more suspicious we are of the null hypothesis. It can be shown that, when the null hypothesis is true and the sample size is moderately large,[2] the random variable associated with

$$\chi^2 = \sum_{i=1}^{K} \frac{(O_i - E_i)^2}{E_i} \qquad (11.1.1)$$

has, to a good approximation, a chi-square distribution with $(K - 1)$ degrees of freedom. Intuitively, the number of degrees of freedom follows from the fact that the O_i must sum to n. Hence, if we know the number of sample members n, and the numbers of observations falling in any $K - 1$ of the categories, then we necessarily know the number in the Kth category. We will want to reject the null hypothesis when the observed numbers differ substantially from the expected numbers—that is, for unusually large values of the statistic 11.1.1. The appropriate test is given in the box.

A Goodness of Fit Test

Suppose we are given a random sample of n observations, each of which can be classified into exactly one of K categories. Denote the observed numbers in each category by $O_1, O_2, \ldots, O_K$. If a null hypothesis (H_0) specifies probabilities $p_1, p_2,$

---

[2] The approximation works well if each of the expected values (E_i) is at least 5.

$\ldots, p_K$ for an observation falling into each of these categories, the expected numbers in the categories, under H_0, would be

$$E_i = np_i \qquad (i = 1, 2 \ldots, K)$$

A test, of significance level α, of H_0 against the alternative that the specified probabilities are not correct, is based on the decision rule

$$\text{Reject } H_0 \text{ if } \sum_{i=1}^{K} \frac{(O_i - E_i)^2}{E_i} > \chi^2_{K-1,\alpha}$$

where $\chi^2_{K-1,\alpha}$ is that number for which

$$P(\chi^2_{K-1} > \chi^2_{K-1,\alpha}) = \alpha$$

and the random variable χ^2_{K-1} follows a chi-square distribution with $(K - 1)$ degrees of freedom.

To illustrate, we return to the data on exporting firms, given in Table 11.3. Our null hypothesis is that the probabilities are the same for the two categories IOAP and POAP. The test of this hypothesis is based on

$$\chi^2 = \sum_{i=1}^{2} \frac{(O_i - E_i)^2}{E_i}$$

$$= \frac{(24 - 17.5)^2}{17.5} + \frac{(11 - 17.5)^2}{17.5} = 4.829$$

There are $K = 2$ categories, so that the degrees of freedom associated with the chi-square distribution are $(K - 1) = 1$. From Table 5 of the Appendix, we find

$$\chi^2_{1,.05} = 3.84 \qquad \text{and} \qquad \chi^2_{1,.025} = 5.02$$

Hence, according to our decision rule, which is illustrated in Figure 11.1, the null hypothesis can be rejected at the 5% level, but not at the 2.5% level of significance. Our data then contain very strong evidence against the hypothesis that, for the population of exporting firms, decisions to export were equally likely to be based on the problem-oriented adoption process and the innovation-oriented adoption process. It appears, rather, that decisions to export based on these two possible motivations were not equally likely.

This example is rather simple, in the sense that there are only two categories and the hypothesized probabilities are the same for each. Example 11.1 illustrates the more general case.

Example 11.1

A gas company has determined from past experience that, at the end of winter, 80% of its accounts are fully paid, 10% are 1 month in arrears, 6% are 2 months in arrears, and 4% are more than 2 months in arrears. At the end of this winter the company checked a random sample of 400 of its accounts, finding 287 to be fully paid, 49 to be 1 month in arrears, 30 to be 2 months in arrears, and 34 to be more than 2 months in arrears. Do these data suggest that the pattern of previous years is still being followed this winter?

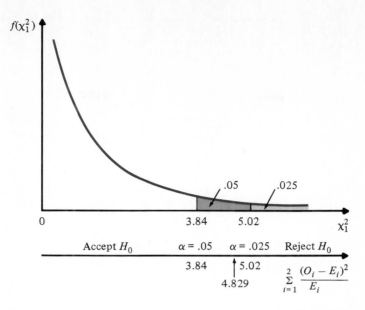

FIGURE 11.1 The distribution of $\chi_1^2 = \Sigma_{i=1}^{2}\,(O_i - E_i)^2/E_i$ when the null hypothesis, H_0, of goodness of fit is true, and the decision rules for testing H_0 at the 2.5% and 5% significance levels

Under the null hypothesis that the proportions in the present winter conform to the historical record, the respective probabilities for the four categories are 0.8, 0.1, 0.06, 0.04. Under that hypothesis, the expected numbers of accounts in each category, for a random sample of 400 accounts, would be

$$400(0.8) = 320; \qquad 400(0.1) = 40; \qquad 400(0.06) = 24; \qquad 400(0.04) = 16$$

Thus, we have the observed and expected numbers shown in the table.

	NUMBER OF MONTHS IN ARREARS				
	0	1	2	MORE THAN 2	TOTAL
Observed Number	287	49	30	34	400
Probability (under H_0)	0.80	0.10	0.06	0.04	1
Expected Number (under H_0)	320	40	24	16	400

The test of the null hypothesis (H_0) is based on

$$\chi^2 = \sum_{i=1}^{4} \frac{(O_i - E_i)^2}{E_i}$$

$$= \frac{(287 - 320)^2}{320} + \frac{(49 - 40)^2}{40} + \frac{(30 - 24)^2}{24} + \frac{(34 - 16)^2}{16} = 27.178$$

Here there are $K = 4$ categories, so that the degrees of freedom are $(K - 1) = 3$. From Table 5 of the Appendix, we have

$$\chi^2_{3, .005} = 12.84$$

Since 27.178 is much bigger than 12.84 the null hypothesis is very clearly rejected, even at the 0.5% significance level. Certainly these data provide considerable evidence to suspect that the pattern of payments of gas bills this year differs from the historical norm. Inspection of the numbers in the table shows that more accounts are in arrears over a longer time period than is usually the case.

We conclude this section with a word of caution. *The figures used in calculating the test statistic 11.1.1 must be the observed and expected numbers in each category. It is not correct, for example, to use instead the percentages of sample members in each category.*

EXERCISES

1. A professor is planning to use a new book for a financial accounting course, and is considering three possibilities: *Financial Accounting Made Easy, Financial Accounting Without Tears,* and *Financial Accounting for Profit and Pleasure*. He contacted a random sample of sixty students who had already taken his course, and asked each to review the three books, indicating a first preference. The results obtained are shown in the table. Test the null hypothesis that, for this population, first preferences are evenly distributed over the three books.

	MADE EASY	WITHOUT TEARS	PROFIT AND PLEASURE
Number of First Preferences	16	26	18

2. A small company sets up hot dog stands at five different locations in the downtown area of a large city. The accompanying table shows the numbers of sales made in the first day. Test the null hypothesis that the five locations are equally successful in attracting sales.

	LOCATION				
	A	B	C	D	E
Number of Sales	127	92	164	131	106

3. A company is preparing to adopt some form of dental insurance scheme for its employees. A random sample of 160 employees indicated their preferences among the four schemes under consideration, as shown in the table. Test the null hypothesis that, for this population, the distribution of preferences is even over the four schemes.

	SCHEME			
	A	B	C	D
Number of Employees	31	46	29	54

4. A small commuter airline runs several daily flights between two cities. Historically, it has found that 28% of these flights are full, 31% have one empty seat, 17% have two empty seats, 10% have three empty seats, 8% have four empty seats, and 6% have more than four empty seats. The airline mounts a radio advertising campaign, and finds, for the 200 flights following the campaign, 64 have no empty seats, 81 have one empty seat, 40 have two empty seats, 6 have three empty seats, 5 have four empty seats, and 4 have more than four empty seats. Test the null hypothesis that the population distribution of empty seats has not changed since the launching of the advertising campaign.

5. A securities analyst claims that, in the year following his recommendations, 50% of his "buy now" stocks have beaten the market average by more than 10%, 25% have beaten the average by 0%–10%, and 25% have failed to beat the market average. For a random sample of 60 stocks that had been recommended by the analyst, it was found that 22 beat the market average by more than 10%, 18 beat the average by 0%–10%, and 20 failed to beat the average. Test the analyst's claim.

6. In a large business statistics course it has been found historically that 30% of all students receive either an A or B, 40% receive a C, and 30% a D or lower. A student from a section containing 40 students complains that his section was harshly graded, as 10 students received A or B, 17 received a C, and 13 received a D or lower. What do you think of the student's claim?

11.2 GOODNESS OF FIT TESTS: POPULATION PARAMETERS UNKNOWN

The goodness of fit test discussed in Section 11.1 is directly applicable when the category probabilities are completely specified by the null hypothesis. However, it is often the case that we want to test the hypothesis that a sample is from some distribution whose unknown parameters we are unable to specify, but must instead estimate from the sample data.

To illustrate, we consider a situation where the hypothesized distribution is binomial, but with the probability of success unknown. A company receiving large shipments of parts from a supplier uses an acceptance sampling scheme of the sort described in Section 4.5. A random sample of twenty parts from a shipment are checked, and delivery is accepted only if no more than three of the sample parts are defective. The supplier protests that all shipments sent out are of equal quality, so that the acceptance sampling scheme is redundant, and the returned shipments do not differ substantially from those that are accepted. This is equivalent to the hypothesis that every shipment contains the same proportion p of defectives, where p is an unknown parameter. It is this hypothesis that we want to test.

To continue our example, suppose that the receiving company reviews the

TABLE 11.5 Results from 200 acceptance sampling tests

	NUMBER OF DEFECTIVES IN SAMPLE OF 20						
	0	1	2	3	4	5 OR MORE	TOTAL
Frequency of Observation	29	61	46	29	23	12	200

outcomes of its acceptance sampling scheme for 200 shipments from this supplier. The company selected samples of 20 from each of 200 shipments and recorded the number of defectives in each sample. The results are reported in Table 11.5, where we include an open-ended category "5 or more," rather than the number of samples with 5, the number with 6, the number with 7, and so on. Following this latter strategy would lead to a large number of categories containing either zero or a very small number of observations. Besides making the computation of the test statistic unduly cumbersome, having a large number of categories of this sort leads to a statistic whose distribution under the null hypothesis is not well approximated by the chi-square. However, the actual numbers of defectives in each sample were used to determine that the sample mean proportion of defectives, over the 200 shipments, was 0.10. We thus have

Number of shipments: 200

Mean proportion of defectives in samples
of 20 parts from these shipments: 0.10

This information can now be used to check the claim of homogeneity of quality of the shipments. The null hypothesis is that the true proportion of defective parts is the same in every shipment. If that hypothesis were true, our best estimate of the common proportion of defectives per shipment would be $\hat{p} = 0.10$, the average proportion of defectives in the 200 samples. Suppose, then, we set out to test the hypothesis that the data of Table 11.5 represent a sample from a binomial distribution with parameter 0.10. As before, we compare the observed category frequencies with the expected frequencies under the null hypothesis. The calculations are shown in Table 11.6.

For each category, p_i is the probability of the corresponding number of successes in 20 binomial trials, each with probability of success 0.10. These probabilities are obtained from Table 1 of the Appendix. The expected frequencies are then

$$E_i = np_i = 200p_i$$

For example, from Table 1 of the Appendix, we find that the probability of 0 "successes" (defectives) in 20 trials, each with probability of success 0.10, is 0.1216. Therefore, in 200 such samples we would expect to find

$$200(0.1216) = 24.32$$

samples with no defectives.

TABLE 11.6 Observed and expected frequencies for acceptance sampling data

| | NUMBER OF DEFECTIVES IN SAMPLE | | | | | | |
	0	1	2	3	4	5 OR MORE	TOTAL
Observed Frequencies (O_i)	29	61	46	29	23	12	200
Probabilities (p_i) under H_0	.1216	.2702	.2852	.1901	.0898	.0431	1
Expected Frequencies (E_i) under H_0	24.32	54.04	57.04	38.02	17.96	8.62	200

Exactly as in the previous section, the appropriate test is based on the quantity 11.1.1. For these data we have

$$\sum_{i=1}^{6} \frac{(O_i - E_i)^2}{E_i} = \frac{(29 - 24.32)^2}{24.32} + \cdots + \frac{(12 - 8.62)^2}{8.62} = 8.813$$

The test is performed precisely as before, except that the number of degrees of freedom for the chi-square random variable is reduced by 1 for each population parameter that is estimated. The general result is given in the box.

Goodness of Fit Tests When Population Parameters Are Estimated

Suppose a null hypothesis specifies category probabilities that depend on the estimation (from the data) of m unknown population parameters. The appropriate test of the null hypothesis is precisely as in Section 11.1, except that the number of degrees of freedom for the chi-square random variable is $(K - m - 1)$, where K is the number of categories.

In the acceptance sampling example, there are six categories and one population parameter has been estimated from the data. Thus, the appropriate number of degrees of freedom is 4. From Table 5 of the Appendix we find

$$\chi^2_{4,.10} = 7.78 \quad \text{and} \quad \chi^2_{4,.05} = 9.49$$

Therefore, the null hypothesis of homogeneity of shipment quality can be rejected at the 10% level, but not at the 5% level of significance. Our null hypothesis for this problem was the claim of the supplier that every shipment is of the same quality—that is, the true proportion of defectives is the same for each shipment of parts. As a result of our test procedure, we see that the sample data contain strong, but not overwhelming, evidence against this claim.

The following example illustrates this procedure for the Poisson population distribution.

Example
11.2

In Section 4.7 we discussed the Poisson distribution and noted that, under certain conditions, this distribution could be used to describe the probabilities for the number of occurrences of an event over a period of time. One means of deciding whether the Poisson distribution is appropriate for a particular problem is to make a subjective judgment about the reasonableness of the required conditions in that instance. An alternative, which is in many ways preferable, is to collect sample information and test the hypothesis that the Poisson is the appropriate parent distribution.

The numbers of breakdowns per week of a machine were observed over a 100-week period, giving the results shown in the table.

NUMBER OF BREAKDOWNS IN WEEK						
	0	1	2	3	4	5 OR MORE
Observed Frequency	10	24	32	23	6	5

It was also found that the average number of breakdowns per week over this period was 2.1. Test the null hypothesis that the population distribution of breakdowns is Poisson.

Recall from Section 4.7 that, if the Poisson distribution is appropriate, the probability of x breakdowns in a given week is

$$P(x) = \frac{e^{-\lambda}\lambda^x}{x!} \qquad (11.2.1)$$

where λ is the mean number of breakdowns per week. In this case, we estimate the unknown population mean λ by the sample mean 2.1, and then calculate expected frequencies. The results are shown in the next table.

NUMBER OF BREAKDOWNS IN WEEK							
	0	1	2	3	4	5 OR MORE	TOTAL
Observed Frequencies (O_i)	10	24	32	23	6	5	100
Probabilities (p_i) under H_0	.122	.256	.270	.189	.099	.064	1
Expected Frequencies (E_i) under H_0	12.2	25.6	27.0	18.9	9.9	6.4	100

The probabilities under the null hypothesis are obtained by substituting the estimate $\hat{\lambda} = 2.1$ in 11.2.1. Thus, for example, for three breakdowns

$$P(3) = \frac{e^{-2.1}(2.1)^3}{3!} = .189$$

The expected frequencies are then

$$E_i = np_i = 100p_i$$

To carry out the test, we calculate

$$\sum_{i=1}^{6} \frac{(O_i - E_i)^2}{E_i} = \frac{(10 - 12.2)^2}{12.2} + \cdots + \frac{(5 - 6.4)^2}{6.4} = 4.155$$

Since there are six categories, and one parameter has been estimated, the appropriate number of degrees of freedom is 4. From Table 5 of the Appendix, we obtain

$$\chi^2_{4, .10} = 7.78$$

Hence, the null hypothesis that the population distribution is Poisson cannot be rejected at the 10% significance level. These data, then, contain scant grounds for suspicion of the hypothesis that the weekly distribution of breakdowns is Poisson.

EXERCISES

7. A psychic has asserted that some people are luckier than others, so that when a game of chance is played, some will tend to be successful and others will be unsuccessful. The contrary view is that observed differences in success can be attributed to random variation. A random sample of 100 people each played a game of chance three times, so that the possible number of successes for each person is 0, 1, 2, or 3. The numbers of people who achieved various numbers of successes are shown in the table. Test the null hypothesis that the population distribution is binomial, implying a common probability of success for each individual in every game.

NUMBER OF SUCCESSES	0	1	2	3
NUMBER OF PEOPLE	12	37	35	16

8. A company employs 200 salespeople. If they are all equally effective, the probability of making a sale from any contact would be the same for all of them, and the binomial distribution could be used to calculate probabilities for numbers of sales. On a particular day, the results of the first two contacts made by these salespeople were recorded, with the results shown in the accompanying table. Test the null hypothesis of a common binomial distribution for the success rates of these salespeople.

NUMBER OF SALES	0	1	2
NUMBER OF SALESPEOPLE	110	64	26

9. The number of telephone calls received each minute over a 200-minute period by a corporation's switchboard was recorded, giving the results shown in the table. The average number of calls per minute received over this period was 2.50. Test the null hypothesis that the population distribution is Poisson.

NUMBER OF CALLS IN MINUTE	0	1	2	3	4	5	6 or more
OBSERVED FREQUENCY	11	46	50	47	28	11	7

10. In a 100-minute period there were a total of 180 arrivals at a supermarket check-out counter. The accompanying table shows the frequency of arrivals per minute over this period. Test the hypothesis that the population distribution is Poisson.

NUMBER OF ARRIVALS IN MINUTE	0	1	2	3	4 or more
OBSERVED FREQUENCY	14	26	32	22	6

11. The chief of police in a large city finds that, over a 100-day period, there have been 220 homicides in the city. The table shows the frequencies of homicides per day over that period. Test the hypothesis that the distribution is Poisson.

NUMBER OF HOMICIDES IN DAY	0	1	2	3	4	5 or more
OBSERVED FREQUENCY	16	29	19	13	12	11

11.3 CONTINGENCY TABLES

In this section we show how the test of Section 11.1 can be extended to deal with a problem to which we alluded briefly in Section 3.7. Suppose that a sample is taken from a population, each of whose members can be uniquely cross-classified according to a pair of attributes. The hypothesis to be tested is of no association or dependence, in the population, between possessions of these attributes.

To illustrate, Table 11.7 contains information on a sample of 1,140 loans.[3] The loan accounts were classified as either good or bad, and the ages of the loan holders

TABLE 11.7 Cross-classification of loan accounts according to status and age of holder

AGE	TYPE OF LOAN ACCOUNT		
	GOOD	BAD	TOTALS
25 OR LESS	92	172	264
26–35	166	211	377
36–45	114	97	211
46 OR MORE	198	90	288
TOTALS	570	570	1,140

[3] The data are taken from J. R. Nevin and G. A. Churchill, "The equal credit opportunity act: an evaluation," *Journal of Marketing*, *43* (1979), No. 2, 95–104, published by the American Marketing Association.

were recorded. For convenience, row and column totals are also recorded in the table. In what follows, it makes no difference to the analysis whether a random sample of population members is first taken and then cross-classified, or whether the row or column totals are fixed in advance and independent random samples taken. In some circumstances, the latter scheme (where possible) may be preferable, as it can increase the chances of having a reasonable number of observations in each joint classification. This is the procedure followed in the present example: Independent random samples of 570 good and 570 bad accounts were selected, and the age of each of the account holders was then recorded.

The null hypothesis to be tested is of no association between the attributes—in this case, that there is no association between type of account and age. The implication of this null hypothesis is that, for the population, the proportion of good accounts belonging to any particular age group is the same as the proportion of bad accounts belonging to the same age group.

More generally, suppose we are concerned with two attributes, A and B. We assume that there are r categories for A and c categories for B, so that a total of rc cross-classifications are possible. The number of sample observations belonging to both the ith category of A and the jth category of B will be denoted O_{ij}, as shown in Table 11.8.

This tabulation is called an $r \times c$ **contingency table.** In this terminology, Table 11.7 is a 4×2 contingency table. For convenience, we have added to Table 11.8 the row and column totals, denoted respectively $R_1, R_2, \ldots, R_r$, and $C_1, C_2, \ldots, C_c$.

In order to test the null hypothesis of no association between the attributes, we again ask how many observations we would expect to find in each cross-classification if that hypothesis were true. In fact, this question becomes meaningful when the row and column totals are taken to be *fixed*. Consider, then, the joint classification corresponding to the ith row and jth column of the table. There are a total of C_j observations in the jth column, and given no association we would expect these to be distributed among the rows in proportion to the total number of observations in each row. Thus, we would expect a proportion R_i/n of these C_j observations to be in the ith row. Hence, the estimated expected numbers of observations in the cross-classifications are

$$\hat{E}_{ij} = \frac{R_i C_j}{n} \qquad (i = 1, 2, \ldots, r; \quad j = 1, 2, \ldots, c)$$

TABLE 11.8 Cross-classification of n observations in an $r \times c$ contingency table

ATTRIBUTE A	ATTRIBUTE B				
	1	2	$\cdots$	c	TOTALS
1	O_{11}	O_{12}	$\cdots$	O_{1c}	R_1
2	O_{21}	O_{22}	$\cdots$	O_{2c}	R_2
.	.	.	$\cdots$	.	.
.	.	.		.	.
r	O_{r1}	O_{r2}	$\cdots$	O_{rc}	R_r
TOTALS	C_1	C_2	$\cdots$	C_c	n

TABLE 11.9 Observed (and estimated expected) numbers in each cross-classification for status of loan account–age data

AGE	TYPE OF LOAN ACCOUNT	
	GOOD	BAD
25 OR LESS	92 (132)	172 (132)
26–35	166 (188.5)	211 (188.5)
36–45	114 (105.5)	97 (105.5)
46 OR MORE	198 (144)	90 (144)

For example, if there were no association between age and status of account in Table 11.7, then since 264 of the 1,140 sample members are aged 25 or less, we would expect a proportion $264/1,140$ of the 570 good accounts to be those of persons aged 25 or less; that is,

$$\hat{E}_{11} = \frac{(264)(570)}{1,140} = 132$$

The other estimated expected numbers are calculated in the same way, and are shown in Table 11.9, together with the corresponding observed values.

Our test of the null hypothesis of no association is based on the magnitudes of the discrepancies between the observed numbers and those that would be expected if that hypothesis were true. The test is similar to that of Section 11.1. It can be shown that, under the null hypothesis, for moderately large sample sizes,[4] the random variable associated with

$$\chi^2 = \sum_{i=1}^{r} \sum_{j=1}^{c} \frac{(O_{ij} - \hat{E}_{ij})^2}{\hat{E}_{ij}} \qquad (11.3.1)$$

has, to a good approximation, a chi-square distribution with $(r - 1)(c - 1)$ degrees of freedom. The double summation in expression 11.3.1 implies that summation extends over all rc cells of the table. The number of degrees of freedom follows from regarding the row and column totals as fixed. If these are known, and the $(r - 1)(c - 1)$ entries corresponding to the first $(r - 1)$ rows and $(c - 1)$ columns are also known, then the remaining entries in the table can be deduced. Clearly, the null hypothesis of no association will be rejected for large absolute discrepancies between observed and expected numbers—that is, for high values of 11.3.1. The test procedure is given in the box.

---

[4] The approximation works well if each of the estimated expected number ($\hat{E}_{ij}$) is at least 5.

A Test of Association in Contingency Tables

Suppose that a sample of n observations is cross-classified, according to two attributes, in an $r \times c$ contingency table. Denote by O_{ij} the number of observations in the cell which is in the ith row and jth column. If the null hypothesis is

H_0: No association between the two attributes in the population

the estimated expected number of observations in this cell, under H_0, is

$$\hat{E}_{ij} = \frac{R_i C_j}{n}$$

where R_i and C_j are the corresponding row and column totals. A test of significance level α is based on the following decision rule:

$$\text{Reject } H_0 \text{ if } \sum_{i=1}^{r}\sum_{j=1}^{c} \frac{(O_{ij} - \hat{E}_{ij})^2}{\hat{E}_{ij}} > \chi^2_{(r-1)(c-1),\alpha}$$

Returning to our example, we find

$$\sum_{i=1}^{4}\sum_{j=1}^{2} \frac{(O_{ij} - \hat{E}_{ij})^2}{\hat{E}_{ij}} = \frac{(92 - 132)^2}{132} + \frac{(172 - 132)^2}{132} + \cdots$$

$$+ \frac{(198 - 144)^2}{144} + \frac{(90 - 144)^2}{144} = 71.5$$

Here there are $r = 4$ rows and $c = 2$ columns in the table, so the appropriate number of degrees of freedom is

$$(r - 1)(c - 1) = (4 - 1)(2 - 1) = 3$$

From Table 5 of the Appendix we find

$$\chi^2_{3,.005} = 12.84$$

Thus, the null hypothesis of no association is very clearly rejected, even at the 0.5% significance level. The evidence against that hypothesis is overwhelming. The cause of rejection can be seen from Table 11.9: The youngest age group has far more bad accounts and the oldest age group far fewer than would be compatible with the hypothesis of independence of age and account status.

Example 11.3

In a study[5] to check for association between the social class of a supermarket shopper and choice of soft drink brand—national or store—a random sample of 300 supermarket shoppers was questioned, yielding the results shown in the table. Test the null hypothesis of no association between brand choice and social class.

---

[5] P. E. Murphy, "The effect of social class on brand and price consciousness for supermarket products," *Journal of Retailing, 54* (1978), No. 2, 33–42.

SOCIAL CLASS	BRAND CHOSEN	
	NATIONAL	STORE
LOWER	70	24
MIDDLE	71	39
UPPER	55	41

The necessary calculations are summarized in the following table. The expected numbers are shown in brackets, following the corresponding observed numbers.

SOCIAL CLASS	BRAND CHOSEN		TOTALS
	NATIONAL	STORE	
LOWER	70 (61.4)	24 (32.6)	94
MIDDLE	71 (71.9)	39 (38.1)	110
UPPER	55 (62.7)	41 (33.3)	96
TOTALS	196	104	300

Given the row and column totals, the expected values, under the null hypothesis, are found from

$$\hat{E}_{ij} = \frac{R_i C_j}{n}$$

For instance, for the middle income–national brand cell, we have

$$\hat{E}_{21} = \frac{R_2 C_1}{n} = \frac{(110)(196)}{300} = 71.9$$

The test is then based on

$$\sum_{i=1}^{3}\sum_{j=1}^{2} \frac{(O_{ij} - \hat{E}_{ij})^2}{\hat{E}_{ij}} = \frac{(70 - 61.4)^2}{61.4} + \frac{(24 - 32.6)^2}{32.6} + \cdots$$
$$+ \frac{(55 - 62.7)^2}{62.7} + \frac{(41 - 33.3)^2}{33.3} = 6.232$$

Since there are $r = 3$ rows and $c = 2$ columns in the table, the number of degrees of freedom associated with the test is

$$(r - 1)(c - 1) = (3 - 1)(2 - 1) = 2$$

From Table 5 of the Appendix, we obtain

$$\chi^2_{2,.05} = 5.99 \qquad \text{and} \qquad \chi^2_{2,.025} = 7.38$$

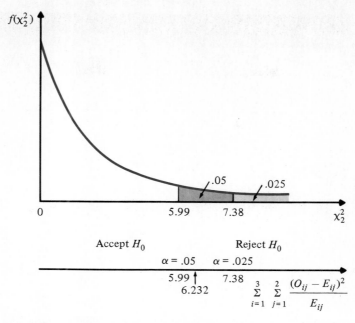

FIGURE 11.2 The distribution of $\chi_2^2 = \Sigma_{i=1}^3 \Sigma_{j=1}^2 (O_{ij} - \hat{E}_{ij})^2/\hat{E}_{ij}$ when the null hypothesis, H_0, of no association is true, and the decision rules for testing H_0 at the 2.5% and 5% significance levels

Hence, the null hypothesis of no association between social class and brand choice can be rejected at the 5% level, but not at the 2.5% level of significance. The test is illustrated in Figure 11.2. The reason for the rejection of the null hypothesis emerges from inspection of the previous table. Upper social class members appear more likely to choose the store brand product than do lower social class members.

Example 11.4

In a comparison[6] of corporate disclosure practices of United Kingdom and continental European Economic Community (EEC) firms, it was found that the reports of 11 of a sample of 45 United Kingdom companies and 15 of a sample of 55 continental EEC companies contained assets analyses. Test the null hypothesis of no association between company location and the reporting of assets analyses. (Notice that the hypothesis being tested is of the equality of two population proportions. A test of this hypothesis was discussed in Section 9.7. Here we use the contingency table approach).

The calculations are summarized in the table. Given the row and column totals, the expected numbers are found as before. For instance, for continental EEC companies disclosing assets analyses, we would expect

$$\hat{E}_{12} = \frac{R_1 C_2}{n} = \frac{(26)(55)}{100} = 14.3$$

---

[6] See S. J. Gray, "Segment reporting and the EEC multinationals," *Journal of Accounting Research, 16* (1978), 242–53.

ASSETS ANALYSES DISCLOSED	COMPANY LOCATION		
	U.K.	CONTINENTAL EEC	TOTALS
YES	11 (11.7)	15 (14.3)	26
NO	34 (33.3)	40 (40.7)	74
TOTALS	45	55	100

Then the value of the test statistic is

$$\sum_{i=1}^{2}\sum_{j=1}^{2}\frac{(O_{ij} - \hat{E}_{ij})^2}{\hat{E}_{ij}} = \frac{(11 - 11.7)^2}{11.7} + \frac{(15 - 14.3)^2}{14.3}$$
$$+ \frac{(34 - 33.3)^2}{33.3} + \frac{(40 - 40.7)^2}{40.7} = 0.103$$

Since the table is 2×2, the number of degrees of freedom is

$$(r - 1)(c - 1) = (2 - 1)(2 - 1) = 1$$

Comparison of the test statistic with tabulated values of the chi-square distribution with 1 degree of freedom gives no grounds on which to question the null hypothesis of no association between company location and assets analysis reporting.

It should be noted, as was the case for the goodness of fit tests of the previous sections, that *the figures used in calculating the statistic 11.3.1 must be the actual numbers observed and not, for example, percentages of the total.*

EXERCISES

12. In a study[7] aimed at comparing the importance attached by shoppers to various factors for specialty stores and department stores, patrons of each were asked whether they attached little, moderate, or great importance to convenience of location. The results are shown in the accompanying table. Test the null hypothesis that shoppers regard convenience of location as equally important for these two types of stores.

IMPORTANCE OF CONVENIENCE OF LOCATION	TYPE OF STORE	
	SPECIALTY	DEPARTMENT
LITTLE	156	61
MODERATE	69	68
GREAT	35	24

[7] L. G. Schiffman, J. F. Dash, and W. R. Dillon, "The contribution of store-image characteristics to store-type choice," *Journal of Retailing, 53* (1977), No. 2, 3–22.

13. In the study described in Exercise 12, shoppers were also asked how much importance they attached to best price or "deal." The results obtained are shown in the table. Test the null hypothesis that, in the population, there is no difference in the level of importance attached to best price or "deal" for these two types of stores.

IMPORTANCE OF BEST PRICE OR "DEAL"	TYPE OF STORE	
	SPECIALTY	DEPARTMENT
LITTLE	9	15
MODERATE	61	21
GREAT	194	118

14. In the study described in Example 11.4, 3 of the 45 United Kingdom companies and 25 of the 55 continental EEC companies disclosed production analyses in their reports. Test the hypothesis of no association between company location and disclosure of production analyses.

15. A professor[8] compared two methods for teaching elementary managerial accounting classes—the personalized system of instruction (PSI), and traditional lecture/discussion classes. At the end of the term students in the two sections were asked: "Would you recommend that a friend take this course?" Their responses are shown in the accompanying table. Test the null hypothesis of no association between recommendation and teaching method.

TEACHING METHOD	RECOMMENDED COURSE	
	YES	NO
LECTURE	43	30
PSI	51	9

16. Suppose that, in a 2×2 contingency table, the numbers observed are denoted $a, b, c, d,$ as indicated in the following table:

			TOTALS
	a	b	$a + b$
	c	d	$c + d$
TOTALS	$a + c$	$b + d$	n

---

[8] See J. C. Bailes, "Lectures versus personalized instruction: An experimental study in elementary managerial accounting," *Accounting Review, 54* (1979), 147–54.

Show that the test statistic of equation 11.3.1 can be written

$$\chi^2 = \frac{n(ad - bc)^2}{(a + b)(a + c)(b + d)(c + d)}$$

17. In a study[9] of differences in satisfaction levels of male and female pharmaceutical salespeople with various job factors, the results shown in the table were observed on the question of pay. Test the null hypothesis of no association between the salesperson's sex and satisfaction level with pay.

	SATISFACTION WITH PAY		
	LOW	MEDIUM	HIGH
MALE	46	61	53
FEMALE	8	9	12

18. In the study of Exercise 17, the sample members were also asked about levels of satisfaction with promotion, yielding the results given in the table. Test the null hypothesis of no association between the salesperson's sex and satisfaction level with promotion.

	SATISFACTION WITH PROMOTION		
	LOW	MEDIUM	HIGH
MALE	29	77	54
FEMALE	8	11	10

19. In a comparison[10] of recent buyers of a major appliance (those who had purchased within the past 2 months) and prospective buyers (those who planned to buy within the next 2 months), samples from both groups were asked about their satisfaction with prepurchase information. The results are shown in the accompanying table. Test the hypothesis that the two groups are equally satisfied with prepurchase information.

---

[9] J. E. Swan, C. M. Futrell, and J. T. Todd, "Some job-different views: Women and men in industrial sales," *Journal of Marketing, 42* (1978), No. 1, 92–98., published by the American Marketing Association.

[10] These data are taken from R. A. Westbrook, J. W. Newman, and J. R. Taylor, "Satisfaction/dissatisfaction in the purchase decision process," *Journal of Marketing, 42* (1978), No. 4, 54–60., published by the American Marketing Association.

	RECENT BUYERS	PROSPECTIVE BUYERS
Available Information Met Needs	215	105
Needed Additional Information	21	53

20. In an attempt[11] to assess the "political congeniality" factor in importers' evaluation of United States exporters' business practices, in comparison with exporters of other countries, importers in Spain and Greece were chosen for study. This choice was made because, at the time, political congeniality between the United States and these countries differed substantially. Concerning the "right type of sales literature," the results shown in the table were found. Test the null hypothesis of no association between importing country and view of United States exporters' practices.

IMPORTING COUNTRY	U.S. EXPORTERS' PRACTICES SAID TO BE		
	BETTER	SAME	WORSE
GREECE	25	10	11
SPAIN	20	21	6

21. In an experiment[12] to determine the effects of an incentive on the response rate to mail questionnaires, the same survey form was sent to three groups of people. One group was offered no incentive to return the form, the second group was offered twenty-five cents, and the third group was offered a pen (valued at twenty-five cents). The results obtained are shown in the table.

	INCENTIVE		
	TWENTY-FIVE CENTS	PEN	NONE
Response	308	177	114
Nonresponse	502	627	697

(a) Test the null hypothesis of no association between response rate and incentive.

---

[11] C. P. Rao, "An assessment of 'political congeniality' factor in international business transactions," *Journal of Academy of Marketing Science, 7* (1979), 48–60; © 1979, reprinted by permission of the Academy of Marketing Science.

[12] See R. A. Hansen, "A self-perception interpretation of monetary and non-monetary incentives on mail survey respondent behavior," *Journal of Marketing Research, 17* (1980), 77–83, published by the American Marketing Association.

(b) Do these data indicate a significant difference between response rates when the incentive is twenty-five cents, rather than a pen?

22. A random sample of thirty sportswriters was asked to nominate which of three graduating college senior guards had the best professional prospects. These guards came respectively from a "big power" college program, a small state school, and a small private school. The following table shows the number of nominations:

	BIG POWER	STATE SCHOOL	PRIVATE SCHOOL
Number of Nominations	14	10	6

Test the null hypothesis that, for the population of sportswriters, nominations are evenly distributed over these three players.

23. It is known that, in a state gubernatorial election, 42% of eligible voters voted for the Democratic candidate, 36% for the Republican candidate, and 22% did not vote. In a random sample of 40 eligible voters in one town in this state, 21 voted for the Democrat, 11 for the Republican, and 8 did not vote. Test the null hypothesis that, for this town, voting patterns were the same as for the state as a whole.

24. Last year it was noted by an admissions dean for a college that 72% of all the applicants for an engineering program were male. It was also found that 50% of male applicants and 62% of female applicants were in the top 10% of their high school classes. From a random sample of 50 applicants for the next class, this admissions dean found 32 males, of whom 18 were in the top 10% of their high school classes. Of the women in this sample, 12 were in the top 10% of their classes. Test the null hypothesis that the population profile of applicants for the next class is the same as in the previous year.

25. Over a period of 200 5-day trading weeks, the numbers of daily rises per week in the Dow-Jones industrials index were measured. The results are shown in the table. Test the null hypothesis that the number of daily rises per week is binomial and state any assumptions you need to make.

NUMBER OF INCREASES IN WEEK	0	1	2	3	4	5
NUMBER OF WEEKS	8	37	54	51	38	12

26. One procedure for attempting to resolve questions of disputed authorship is to count the number of occurrences of particular words in blocks of text. These can then be compared with results from passages whose authorship is known. An example of this type of research involves the study of the Federalist Papers.[13] For a sample of 262 blocks of text

---

[13] The data of this exercise are taken from F. Mosteller and D. L. Wallace, *Inference and Disputed Authorship: The Federalist,* © 1964. Addison-Wesley, Reading, MA. Table 2.3-4. Reprinted with permission.

(each approximately 200 words in length) from this source, the mean number of occurrences of the word *may* was 0.66. The frequencies of occurrence are given in the table. Test the null hypothesis that the population distribution is Poisson.

NUMBER OF OCCURRENCES	0	1	2	3 or more
FREQUENCY	156	63	29	14

27. A study[14] was aimed at assessing the reactions of firms to qualified audit opinions. The accompanying table shows, for samples of firms that did and did not receive qualified opinions, the numbers that switched auditors in the following year. Test at the 1% significance level the null hypothesis that the switching of auditors is independent of whether a qualified opinion has been given.

	OPINION RECEIVED	
	QUALIFIED	UNQUALIFIED
SWITCHED	141	227
DID NOT SWITCH	991	8,051

28. In the study of Exercise 27, a random sample of 273 firms that had received qualified opinions was taken. The table shows the numbers of qualified opinions received by these firms in the following year, for firms that did and did not switch auditors. What can be learned from these data?

	OPINION RECEIVED NEXT YEAR	
	QUALIFIED	UNQUALIFIED
SWITCHED	103	29
DID NOT SWITCH	100	41

29. The accompanying table shows frequencies of funds statements for samples of nonpublic companies in Canada and the United States.[15] Test at the 5% level the null hypothesis of no relation between company nationality and frequency of funds statements.

---

[14] C. W. Chow and S. J. Rice, "Qualified audit opinions and auditor switching," *Accounting Review, 57* (1982), 326–35.

[15] These data are given in G.R. Chesley and J. H. Scheiner, "The statement of changes in financial position: An empirical investigation of Canadian and U.S. users in nonpublic companies," *International Journal of Accounting, Education and Research, 17* (1982), No. 2, 49–58.

FREQUENCY OF STATEMENTS	NATIONALITY	
	CANADA	UNITED STATES
BIWEEKLY OR MONTHLY	35	50
QUARTERLY OR SEMIANNUALLY	9	13
ANNUALLY	51	13

30. The table[16] shows, for independent random samples of men and women, the numbers who watch television for more or less than 2.5 hours per day. Test at the 10% level the null hypothesis of no relationship between a person's sex and the amount of television watched.

	HOURS TELEVISION PER DAY	
	LESS THAN 2.5	2.5 OR MORE
MEN	18	10
WOMEN	17	13

31. In the study of Exercise 30, respondents were cross-classified according to the amounts of time spent reading newspapers and watching television. The sample results are shown in the accompanying table. What can you conclude from these data?

HOURS TELEVISION PER DAY	HOURS READING NEWSPAPER PER DAY		
	LESS THAN 0.5	0.5	MORE THAN 0.5
LESS THAN 2.5	10	16	9
2.5 OR MORE	8	10	5

32. In a study of newspaper readership in nonmetropolitan markets,[17] sample members from urban, rural, and farm communities were classified as readers or nonreaders. The results given in the table were obtained. Test at the 5% significance level the null hypothesis that readership is independent of community type.

[16] Data taken from B. W. Becker and P. E. Connor, "Personal values of the heavy user of mass media," *Journal of Advertising Research, 21* (1981), No. 5, 37–43.

[17] See J. R. Lynn, "Newspaper ad. impact in nonmetropolitan markets," *Journal of Advertising Research, 21* (1981), No. 4, 13–19.

COMMUNITY		
	READERS	NONREADERS
URBAN	529	121
RURAL	373	137
FARM	237	89

33. In the study of Exercise 32, readers and nonreaders of newspapers were asked if they voted in the last election. The results obtained are summarized in the table. What can be learned from these results?

VOTED		
	READERS	NONREADERS
YES	834	174
NO	185	140

34. The table given here shows numbers of sample members[18] in different income ranges who view bank salaries as high, above average, and average or below. Test at the 1% significance level the null hypothesis of no association between income and attitude toward bank salaries.

INCOME	ATTITUDE TOWARD BANK SALARIES		
	HIGH	ABOVE AVERAGE	AVERAGE OR BELOW
ABOVE AVERAGE	175	124	92
AVERAGE	118	110	126
BELOW AVERAGE	127	82	147

---

[18] This information is obtained from L. Mandell, R. Lachman, and Y. Orgler, "Interpreting the image of banking," *Journal of Bank Research, 12* (1981), 96–104.

linear correlation and regression

CHAPTER TWELVE

12.1 CORRELATION

To this point, we have dealt almost exclusively with problems of inference about a single random variable or independent random variables. Very often, however, in business and economic applications, interest is focused on the *relationships* between two or more random variables. In this chapter we will consider the case where a pair of random variables is under study, extending our results to the more general case in Chapter 13.

Now, in principle, there are any number of ways in which a pair of random variables might be related to one another, and before much progress can be made, it is helpful to postulate some functional form for their relationship. It is often reasonable to conjecture, as a good approximation, that the association is linear. Thus, if the pair of random variables X and Y are being considered, joint observations on this pair will tend to be clustered around a straight line, as in the graph of Figure 12.1. We do not mean to imply that the relationships studied here need necessarily be very strong, as would be the case if observations on a pair of random variables were very tightly clustered around a straight line. In practice, many relationships that we want to analyze will be rather weak, as is the one depicted in Figure 12.1. Then, in order to learn about such relations, we will need more sophisticated techniques than graphical inspection. Nevertheless, the plotting of a graph is a desirable preliminary to the more detailed analysis that follows. In particular, while the assumption that any association that might exist between random variables can be characterized as linear is very convenient, and frequently realistic, it is only an assumption. One benefit from

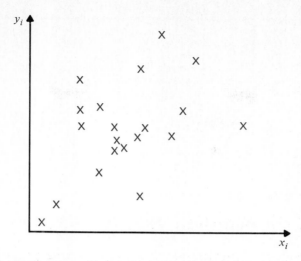

FIGURE 12.1 Observations on a pair of linearly related random variables

making plots is that they might provide an indication of any serious departures from linearity in the association between random variables. Looked at in this light, Figure 12.1 provides no indication that the association between these two particular random variables is anything other than linear.

In this section we are concerned with measuring the *strength* of the linear association between a pair of random variables. Let X and Y be a pair of random variables, with means μ_X and μ_Y and variances σ_X^2 and σ_Y^2. As a measure of the association between these quantities, we introduced in Sections 4.4 and 5.4 the **covariance**, defined as

$$\text{Cov}(X, Y) = E[(X - \mu_X)(Y - \mu_Y)]$$

If there is positive association between the random variables, so that high values of X tend to be associated with high values of Y and low X with low Y (as in Figure 12.1), then the covariance is positive. When there is negative association, so that high values of X are associated with low values of Y, and low X with high Y, the covariance is negative. If there is no linear association between X and Y, their covariance is 0.

As it stands, however, the covariance is of little direct use in assessing the strength of the relation between a pair of random variables, as its value depends on the units in which they are measured. Ideally, we would like a pure, scale-free measure. Fortunately, such a measure is easily obtained by dividing the covariance by the product of the individual standard deviations. The quantity resulting is called the **correlation coefficient.**

The Correlation Coefficient

Let X and Y be a pair of random variables, with means μ_X and μ_Y and variances σ_X^2 and σ_Y^2. A measure of the strength of their linear association is provided by the **correlation coefficient, ρ,** defined as

$$\rho = \operatorname{Corr}(X, Y) = \frac{\operatorname{Cov}(X, Y)}{\sigma_X \sigma_Y} = \frac{E[(X - \mu_X)(Y - \mu_Y)]}{\sqrt{E[(X - \mu_X)^2]E[(Y - \mu_Y)^2]}} \quad (12.1.1)$$

It can be shown that the correlation must lie between -1 and 1, that is,

$$-1 \leq \rho \leq 1$$

with the following interpretations:

(i) A correlation of -1 implies perfect negative linear association.
(ii) A correlation of 1 implies perfect positive linear association.
(iii) A correlation of 0 implies no linear association.
(iv) The larger in absolute value is the correlation, the stronger is the linear association between the random variables.

The implication of the value of the correlation coefficient is illustrated in Figure 12.2, which displays plots of random samples of observations from joint distributions for which the correlations are -1, -0.8, -0.4, 0, 0.4, 0.8, 1. The two extreme cases exhibit perfect linear association, while for the other cases, the higher in absolute value is the correlation, the more closely are the observations clustered about a straight line.

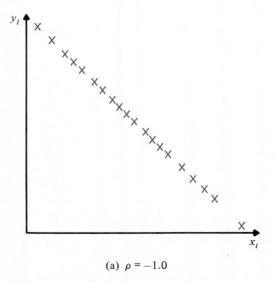

(a) $\rho = -1.0$

FIGURE 12.2 Samples of observations from joint distributions with different correlations

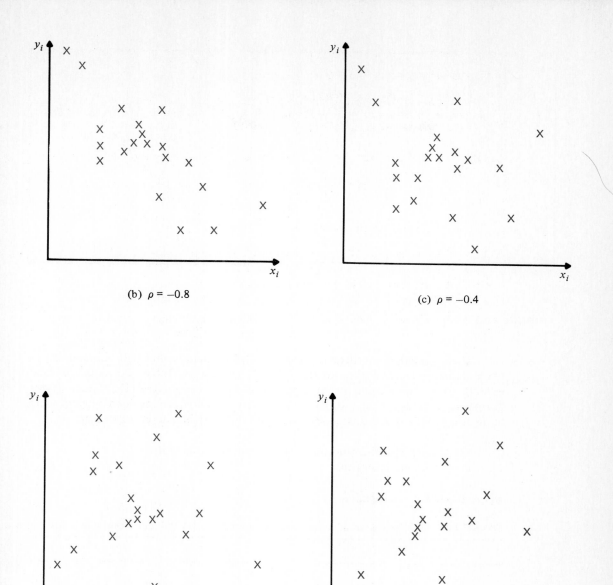

FIGURE 12.2 Samples of observations from joint distributions with different correlations (cont.)

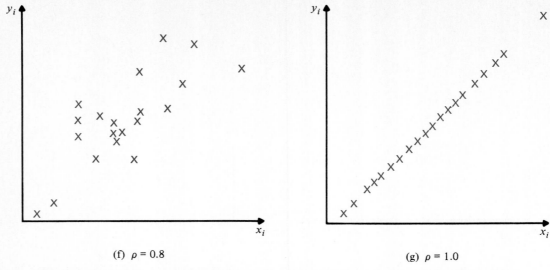

(f) $\rho = 0.8$ (g) $\rho = 1.0$

FIGURE 12.2 Samples of observations from joint distributions with different correlations (cont.)

The correlation coefficient ρ is a population quantity that will be unknown in practice, and must be estimated from data. To illustrate, Table 12.1 contains data[1] relating to an advertising promotion run in seventeen magazines. These advertisements were aimed at promoting tourism in South Carolina, and readers were invited to write for additional literature. The two variables to be related are

☐ X: Cost of advertisement and circulation (in thousands of dollars)
☐ Y: Return-on-inquiry cost

where the latter is defined as

TABLE 12.1 Data on cost of advertisement and circulation (x) and return-on-inquiry cost (y)

x_i	y_i	x_i	y_i
4.07	17.41	1.50	78.74
2.51	22.25	1.68	66.42
1.25	106.84	2.72	121.95
14.67	14.41	1.61	21.93
16.02	24.18	1.52	31.29
3.81	29.73	3.10	88.31
9.87	35.95	3.32	92.70
1.27	61.81	3.07	59.06
1.80	48.36		

[1] These data are reprinted by permission of the publisher from "Is CPM related to the advertising effectiveness of magazines?" by A. G. Woodside and D. M. Reid, *Journal of Business Research, 3,* 323–34. Copyright 1975 by Elsevier Science Publishing Co., Inc.

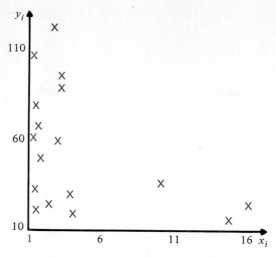

FIGURE 12.3 Plot of cost–return advertising data

$$Y = \frac{\text{Estimated revenue from inquiries} - \text{Cost of advertisement}}{\text{Cost of advertisement}}$$

The data are plotted in Figure 12.3, from which it appears that the relationship between cost and return is negative, with high values of one tending to be associated with low values of the other.

In general, suppose that a sample contains n pairs of observations, denoted (x_1, y_1), (x, y_2), . . . , (x_n, y_n), and that these data are to be used to estimate the population correlation ρ. As usual, estimates of population standard deviations are provided by the corresponding sample standard deviations

$$s_x = \sqrt{\frac{1}{n-1} \sum_1^n (x_i - \bar{x})^2} \qquad \text{and} \qquad s_y = \sqrt{\frac{1}{n-1} \sum_1^n (y_i - \bar{y})^2}$$

Similarly, an estimate of the covariance is given by

$$\text{Estimated Cov}(X, Y) = \frac{1}{n-1} \sum_1^n (x_i - \bar{x})(y_i - \bar{y})$$

The population correlation is then estimated by substituting sample estimates for the corresponding population values in formula 12.1.1.

The Sample Correlation Coefficient

Let (x_1, y_1), (x_2, y_2), . . . , (x_n, y_n) denote a random sample of n pairs of observations on the random variables X and Y. Then an estimate of the population correlation between X and Y is provided by the **sample correlation coefficient**

$$r = \dfrac{\dfrac{1}{n-1} \sum\limits_{1}^{n} (x_i - \overline{x})(y_i - \overline{y})}{s_x s_y} = \dfrac{\sum\limits_{1}^{n} (x_i - \overline{x})(y_i - \overline{y})}{\sqrt{\sum\limits_{1}^{n} (x_i - \overline{x})^2 \sum\limits_{1}^{n} (y_i - \overline{y})^2}}$$

An equivalent expression, which is often computationally simpler, is

$$r = \dfrac{\sum\limits_{1}^{n} x_i y_i - n\overline{x}\,\overline{y}}{\sqrt{\left(\sum\limits_{1}^{n} x_i^2 - n\overline{x}^2 \right)\left(\sum\limits_{1}^{n} y_i^2 - n\overline{y}^2 \right)}}$$

The calculation of the sample correlation for the advertising data is set out in Table 12.2. The sample means are

$$\overline{x} = \frac{73.79}{17} = 4.3406 \qquad \text{and} \qquad \overline{y} = \frac{921.34}{17} = 54.1965$$

Hence, the sample correlation is

$$
\begin{aligned}
r &= \frac{\sum\limits_{1}^{n} x_i y_i - n\overline{x}\,\overline{y}}{\sqrt{\left(\sum\limits_{1}^{n} x_i^2 - n\overline{x}^2 \right)\left(\sum\limits_{1}^{n} y_i^2 - n\overline{y}^2 \right)}} \\[2mm]
&= \frac{2{,}899.7659 - (17)(4.3406)(54.1965)}{\sqrt{[660.4933 - (17)(4.3406)^2][68{,}164.143 - (17)(54.1965)^2]}} \\[2mm]
&= -0.441
\end{aligned}
$$

The sample correlation, -0.441, indicates a mild negative relationship between cost and return for the magazine advertisements. The fact that the estimated correlation coefficient is negative indicates that high values of cost tend to be associated with low values of return. However, since a correlation of 0 implies no linear association, while a value of -1 is equivalent to perfect negative linear association, the value found for the sample correlation here does not suggest a terribly strong association between cost and return.

The sample correlation coefficient is useful as a descriptive measure of the strength of linear association in a sample. It can also be used as the basis of a test of the hypothesis of no linear association, in the population, between a pair of random variables; that is,

$$H_0: \quad \rho = 0$$

This particular null hypothesis, of no (linear) relationship between a pair of random variables, is often of great interest to an investigator. Of course, if we estimate

TABLE 12.2 Calculation of sample correlation for data of Table 12.1

x_i	y_i	x_i^2	y_i^2	$x_i y_i$
4.07	17.41	16.5649	303.1081	70.8587
2.51	22.25	6.3001	495.0625	55.8475
1.25	106.84	1.5625	11,414.7856	133.5500
14.67	14.41	215.2089	207.6481	211.3947
16.02	24.18	256.6404	584.6724	387.3636
3.81	29.73	14.5161	883.8729	113.2713
9.87	35.95	97.4169	1,292.4025	354.8265
1.27	61.81	1.6129	3,820.4761	78.4987
1.80	48.36	3.2400	2,338.6896	87.0480
1.50	78.74	2.2500	6,199.9876	118.1100
1.68	66.42	2.8224	4,411.6164	111.5856
2.72	121.95	7.3984	14,871.8025	331.7040
1.61	21.93	2.5921	480.9249	35.3073
1.52	31.29	2.3104	979.0641	47.5608
3.10	88.31	9.6100	7,798.6561	273.7610
3.32	92.70	11.0224	8,593.2900	307.7640
3.07	59.06	9.4249	3,488.0836	181.3142
73.79	921.34	660.4933	68,164.1430	2,899.7659

the correlation, we will almost inevitably obtain some value other than 0, whether or not a relationship exists. What we are asking is how likely we are to find sample correlations that differ by particular amounts from 0 when in fact there is no linear association between the random variables under study.

It can be shown that, when this null hypothesis is true, and the random variables have a joint normal distribution[2], the random variable corresponding to

$$t = \frac{r}{\sqrt{(1 - r^2)/(n - 2)}}$$

follows a Student's t distribution with $(n - 2)$ degrees of freedom. The appropriate tests are then derived as indicated in the box.

Test for Population Correlation

Let r be the sample correlation coefficient, calculated from a random sample of n pairs of observations from a joint normal distribution. The following tests of the null hypothesis

$$H_0: \quad \rho = 0$$

have significance level α:

(i) To test H_0 against the alternative

---

[2] This is equivalent to requiring that every linear combination of the random variables X and Y has a normal distribution.

$$H_1: \quad \rho > 0$$

the decision rule is

$$\text{Reject } H_0 \text{ if } \quad \frac{r}{\sqrt{(1 - r^2)/(n - 2)}} > t_{n-2,\alpha}$$

(ii) To test H_0 against the alternative

$$H_1: \quad \rho < 0$$

the decision rule is

$$\text{Reject } H_0 \text{ if } \quad \frac{r}{\sqrt{(1 - r^2)/(n - 2)}} < -t_{n-2,\alpha}$$

(iii) To test H_0 against the two-sided alternative

$$H_1: \quad \rho \neq 0$$

the decision rule is

$$\text{Reject } H_0 \text{ if } \quad \frac{r}{\sqrt{(1 - r^2)/(n - 2)}} < -t_{n-2,\alpha/2} \quad \text{or} \quad \frac{r}{\sqrt{(1 - r^2)/(n - 2)}} > t_{n-2,\alpha/2}$$

Here, $t_{n-2,\alpha}$ is that number for which

$$P(t_{n-2} > t_{n-2,\alpha}) = \alpha$$

where the random variable t_{n-2} follows a Student's t distribution with $(n - 2)$ degrees of freedom.

Returning to the advertising cost–return example, we will test the hypothesis of no population correlation against the two-sided alternative. The test is based on

$$\frac{r}{\sqrt{(1 - r^2)/(n - 2)}} = \frac{-0.441}{\sqrt{[1 - (-0.441)^2]/(17 - 2)}} = -1.903$$

Since $n = 17$, there are $(n - 2) = 15$ degrees of freedom, so that, from Table 6 of the Appendix, appropriate points of comparison for 10% and 5% level tests against the two-sided alternative are, respectively,

$$t_{15,.05} = 1.753 \qquad \text{and} \qquad t_{15,.025} = 2.131$$

Hence, according to our decision rule, these data allow rejection of the null hypothesis of no population correlation, against the two-sided alternative, at the 10% but not at the 5% level of significance. Although the data contain some evidence suggesting that there is association between costs and returns, the indications against the hypothesis of no (linear) association between these variables are only moderately strong.

Example
12.1

For a random sample of thirty organizations, the sample correlation between average worker tenure and organization age[3] was found to be 0.33. Test the null hypothesis that these two quantities are not correlated in the population, against the alternative that the population correlation is positive.

Denoting by ρ the population correlation, we want to test

$$H_0: \quad \rho = 0$$

against

$$H_1: \quad \rho > 0$$

using the sample information

$$n = 30; \qquad r = 0.33$$

The test is based on

$$\frac{r}{\sqrt{(1 - r^2)/(n - 2)}} = \frac{0.33}{\sqrt{[1 - (0.33)^2]/(30 - 2)}} = 1.850$$

Since there are $(n - 2) = 28$ degrees of freedom, we have from Table 6 of the Appendix,

$$t_{28,\,.05} = 1.701 \qquad \text{and} \qquad t_{28,\,.025} = 2.048$$

Therefore, the null hypothesis of no population correlation can be rejected, against the alternative that the true correlation is positive, at the 5% but not at the 2.5% significance level. These data, then, contain fairly strong evidence supporting the hypothesis of positive (linear) association between average worker tenure and organization age, as opposed to the hypothesis of no association.

12.2 RANK CORRELATION

The sample correlation coefficient of Section 12.1 can be seriously affected by odd extreme observations. Moreover, tests based on it rely for their validity on an assumption of normality. A measure of correlation that is not susceptible to serious influence by extreme values, and on which valid tests can be based for very general population distributions, is obtained through the use of *ranks*, as in Chapter 10. The resulting test will then be nonparametric. In Table 12.3, we illustrate the calculation of this coefficient for the advertising cost–return data of the previous section.

The x_i and y_i observations are first ranked in ascending order. **Spearman's rank correlation coefficient** is then calculated as the sample correlation between the *ranks* of x_i and y_i, using the formula of Section 12.1. However, *if there are no ties in the*

---

[3] Result reported in C. A. Glisson and P. Y. Martin, "Productivity and efficiency in human service organizations as related to structure, size, and age," *Academy of Management Journal, 23* (1980), 21–37.

TABLE 12.3 Rank correlation calculations for data of Table 12.1

x_i	RANK(x_i)	y_i	RANK (y_i)	d_i = RANK (x_i) − RANK (y_i)	d_i^2
4.07	14	17.41	2	12	144
2.51	8	22.25	4	4	16
1.25	1	106.84	16	−15	225
14.67	16	14.41	1	15	225
16.02	17	24.18	5	12	144
3.81	13	29.73	6	7	49
9.87	15	35.95	8	7	49
1.27	2	61.81	11	−9	81
1.80	7	48.36	9	−2	4
1.50	3	78.74	13	−10	100
1.68	6	66.42	12	−6	36
2.72	9	121.95	17	−8	64
1.61	5	21.93	3	2	4
1.52	4	31.29	7	−3	9
3.10	11	88.31	14	−3	9
3.32	12	92.70	15	−3	9
3.07	10	59.06	10	0	0
				Sum	1,168

rankings of the x_i and no ties in the rankings of the y_i, it can be shown that the computationally simpler expression

$$r_s = 1 - \frac{6 \sum_{1}^{n} d_i^2}{n(n^2 - 1)}$$

is equivalent, where the d_i are the differences in ranks. Thus, since there are no ties in our advertising data, we have from Table 12.3,

$$r_s = 1 - \frac{(6)(1,168)}{(17)[(17)^2 - 1]} = -0.431$$

In this particular case, note that the rank correlation coefficient is remarkably close to the ordinary sample correlation coefficient found for these data in the previous section.

Spearman's rank correlation coefficient can be used to test the null hypothesis of no association between a pair of random variables. Its distribution under that hypothesis is known, and cutoff points are given in Table 9 of the Appendix. Specifically, for various sample sizes n, the tabulated values are those numbers $r_{s,\alpha}$ that are exceeded with probability α by the rank correlation when the null hypothesis is true.

Spearman's Rank Correlation Test

Suppose we have a random sample $(x_1, y_1), \ldots, (x_n, y_n)$ of n pairs of observations. If the x_i and y_i are each ranked in ascending order, and the sample correlation of these ranks is calculated, the resulting coefficient is called **Spearman's rank correlation coefficient**. If there are no tied ranks, an equivalent formula for computing this coefficient is

$\longrightarrow$

$$r_s = 1 - \frac{6 \sum\limits_{1}^{n} d_i^2}{n(n^2 - 1)}$$

where the d_i are the differences of the ranked pairs.

The following tests of the null hypothesis H_0 of no association in the population have significance level α:

(i) To test against the alternative of positive association, the decision rule is

$$\text{Reject } H_0 \text{ if } r_s > r_{s,\alpha}$$

(ii) To test against the alternative of negative association, the decision rule is

$$\text{Reject } H_0 \text{ if } r_s < -r_{s,\alpha}$$

(iii) To test against the two-sided alternative of some association, the decision rule is

$$\text{Reject } H_0 \text{ if } r_s < -r_{s,\alpha/2} \quad \text{or} \quad r_s > r_{s,\alpha/2}$$

Here, $r_{s,\alpha}$ is the cutoff point of the distribution of the Spearman coefficient, given in Table 9 of the Appendix.

For the advertising cost–return example, we will test the null hypothesis of no association in the population against the two-sided alternative. Since there are 17 pairs of observations, we find from Table 9 of the Appendix that cutoff points for 10% and 5% level tests are, respectively,

$$r_{s,.05} = 0.412 \quad \text{and} \quad r_{s,.025} = 0.490$$

Since the calculated value of the Spearman coefficient is -0.431, the null hypothesis of no association can be rejected against the two-sided alternative, according to the decision rule, at the 10% but not at the 5% significance level. We reached the same conclusion for these data in the previous section. However, we have now been able to do so without an assumption of population normality. There is, then, some gain in using the rank correlations here, since our conclusions are no longer conditioned on an assumption about the population distribution.

EXERCISES

1. An instructor in a statistics course set a final examination and also required the students to do a data analysis project. For a random sample of ten students, the scores obtained are shown in the table. Find the sample correlation between the examination and project scores.

EXAMINATION	81	62	74	78	93	69	72	83	90	85
PROJECT	74	71	69	76	87	62	80	75	92	79

2. A random sample of eight introductory accounting texts yielded the figures shown in the table for annual sales (in thousands) and price (in dollars).

SALES	12.4	18.6	29.2	15.7	25.4	35.2	14.7	11.1
PRICE	19.2	20.5	19.7	21.3	20.8	19.9	17.8	17.2

(a) Find the sample correlation between sales and price.

(b) Test, at the 5% significance level against a two-sided alternative, the null hypothesis that the population correlation coefficient is 0.

3. A company routinely administers an aptitude test to all new management trainees. At the end of their first year with the company, these trainees are graded by their immediate supervisors. For a random sample of twelve trainees, the results given in the accompanying table were obtained.

APTITUDE SCORE	86	79	92	53	65	82
SUPERVISOR'S GRADE	76	75	81	72	68	88

APTITUDE SCORE	87	75	77	76	69	88
SUPERVISOR'S GRADE	79	81	70	83	62	84

(a) Find the sample correlation between the aptitude scores and supervisor's grades.

(b) Test, at the 5% and 10% significance levels, the null hypothesis that the population correlation coefficient is 0 against the alternative that it is positive.

4. In the advertising study discussed in Section 12.1, the following results were also found:

x_i	y_i	x_i	y_i
7.70	141.77	4.65	171.81
4.17	96.97	2.97	200.23
1.52	163.92	0.98	120.49
10.04	154.70	4.18	95.83
6.02	151.61	6.09	196.67
4.81	147.82	3.09	275.97
1.57	98.61	3.08	289.59
3.63	179.18	1.76	105.71
1.57	125.19		

x_i = Cost of advertisement ÷ Number of inquiries received

y_i = Revenue from inquiries ÷ Number of inquiries received

Find the sample correlation, and test against a two-sided alternative the null hypothesis that the population correlation is 0.

5. In the study of thirty human service organizations described in Example 12.1, the authors found a sample correlation of 0.07 between productivity and organization age. Test against a two-sided alternative the null hypothesis that the population correlation is 0.

6. For the thirty organizations of Exercise 5, the sample correlation between average worker tenure and productivity was 0.15. Test the null hypothesis of no linear association, in the population, between average worker tenure and productivity, against the alternative of positive linear association.

7. The accompanying table shows total returns on common stocks and on long-term government bonds over a period of 20 years. Calculate Spearman's rank correlation coefficient, and use it to test, against a two-sided alternative, the null hypothesis of no association in the population.

COMMON STOCKS	GOVERNMENT BONDS	COMMON STOCKS	GOVERNMENT BONDS
−.1078	.0745	.2398	−.0919
.4336	−.0610	.1106	−.0026
.1195	−.0226	−.0850	−.0508
.0047	.1378	.0401	.1210
.2689	.0097	.1431	.1323
−.0873	.0689	.1898	.0568
.2280	.0121	−.1466	−.0111
.1648	.0351	−.2648	.0435
.1245	.0071	.3720	.0919
−.1006	.0365	.2384	.1675

8. Two procedures are commonly used to elicit brand purchase information from shoppers. One possibility is to carry out a survey, in which questions about brands purchased are asked. Alternatively, a sample of families are asked to keep diary records of actual purchases over a particular time period. Market researchers are naturally interested in the extent to which conclusions drawn from these two methods of inquiry agree. The accompanying table[4] shows rankings for seven brands of margarine, obtained from the percentage of households buying the brands according to a diary study over a 6-month period, and the percentage of households who recalled buying the brands when questioned in a survey. Calculate Spearman's rank correlation coefficient, and test the null hypothesis of no association against the alternative of positive association.

	PERCENTAGE HOUSEHOLDS BUYING	
BRAND	DIARY STUDY RANKS	SURVEY RANKS
Blue Bonnet	2	1
Chiffon	6	6
Fleischmann	3	4
Imperial	4	3
Mazola	7	5
Mrs. Filbert's	5	7
Parkay	1	2

---

[4] Results given by Y. Wind and D. Lerner, "On the measurement of purchase data: Surveys versus purchase diaries," *Journal of Marketing Research*, *16* (1979), 39–47, published by the American Marketing Association.

9. Some products or services cause more shopping difficulty than others. The table[5] given here shows rankings of twenty-four products and services, in order of shopping difficulty as perceived by men and women. Find Spearman's rank correlation coefficient, and test the null hypothesis of no association, against the alternative of positive association.

	ORDER OF SHOPPING DIFFICULTY	
PRODUCT OR SERVICE	MEN	WOMEN
Auto repairs	2	1
Automobiles	4	2
Home improvement	1	3
Clothing and footwear	13	4
Household furnishings	5	5
Household appliances	7	6
Housing and real estate	6	7
Home entertainment	8	8
Life insurance	3	9
Household moving	9	10
Sporting goods	17	11
Groceries	19	12
Photographic equipment	12	13
Auto insurance	10	14
Children's toys	15	15
Home gardening supplies	18	16
Legal services	11	17
Dry cleaning	23	18
Stationery supplies	20	19
Travel agency services	16	20
Drugs and pharmaceuticals	21	21
Jewelry	14	22
Financial services	22	23
Personal care	24	24

10. Calculate Spearman's rank correlation coefficient for the advertising cost–revenue data of Exercise 4. Then carry out a nonparametric test of the null hypothesis of no association in the population, against a two-sided alternative.

11. Multinational corporations frequently transfer, from one division to another, goods and services across international boundaries. They have much flexibility in the pricing of the transferred items. A study[6] of large United States and Japanese multinational corporations was aimed at determining the relative importance of twenty variables that might be considered in formulating international transfer pricing policies. Rankings for these variables are shown in the table.

   (a) Compute Spearman's rank correlation coefficient. (Notice that there are tied ranks.)
   (b) Test the hypothesis of no association against the alternative of positive association.

---

[5] See J. D. Claxton and J. R. B. Ritchie, "Consumer prepurchase shopping problems: A focus on the retailing component," *Journal of Retailing, 55* (1979), No. 3, 24–43.

[6] R. Y. W. Tang and K. H. Chan, "Environmental variables of international transfer pricing: A Japan–United States comparison," *Abacus, 15* (1979), 3–12, published by Sydney University Press.

VARIABLE	RANKING UNITED STATES	JAPANESE
Overall profit to company	1	1
Restrictions imposed by foreign countries on repatriation of profits or dividends	2	4
Competitive position of subsidiaries in foreign countries	3	2
Differentials in tax rates and tax legislation among countries	4	14
Performance evaluation of foreign subsidiaries	5	5
Rate of customs duties and customs legislation	6	9.5
Import restrictions imposed by foreign countries	7	12
Need to maintain adequate cash flows in foreign subsidiaries	8.5	7
Maintaining good relations with host governments	8.5	9.5
Restrictions on royalty or management fees that can be charged against foreign subsidiaries	10	11
Rules of financial reporting for foreign subsidiaries	11	15
Devaluation and revaluation in countries where company has operations	12	3
Rates of inflation in foreign countries	13	8
Volume of industrial transfers	14	18
Need of subsidiaries in foreign countries to seek local funds	15	16
Antidumping legislation in foreign countries	16	13
Interests of local partners of foreign subsidiaries	17	6
Domestic government requirements on foreign investments	18	20
Risk of expropriation in foreign countries	19	17
Antitrust legislation of foreign countries	20	19

12. The accompanying table shows total returns on United States treasury bills and the rate of inflation of the Consumer Price Index over a 20-year period. Calculate Spearman's rank correlation coefficient, and use it to test the null hypothesis of no association in the population against the alternative of positive association.

TREASURY BILLS	CONSUMER PRICE INDEX	TREASURY BILLS	CONSUMER PRICE INDEX
.0314	.0302	.0421	.0304
.0154	.0176	.0521	.0472
.0295	.0150	.0658	.0611
.0266	.0148	.0653	.0549
.0213	.0067	.0439	.0336
.0273	.0122	.0384	.0341
.0312	.0165	.0693	.0880
.0354	.0119	.0800	.1220
.0393	.0192	.0580	.0701
.0476	.0335	.0508	.0481

12.3 THE LINEAR REGRESSION MODEL

We use correlation to provide a measure of the strength of any linear association between a pair of random variables. The random variables are treated perfectly symmetrically, and it is a matter of indifference whether we speak of "the correlation between X and Y" or "the correlation between Y and X." In the remainder of this chapter, we will continue to discuss the linear relationship between a pair of variables, but in terms of the *dependency* of one on the other. The symmetry of our previous discussion is now removed. Rather, the concept here is that, given that the random variable X takes a specific value, we expect a response in the random variable Y. That is, the value taken by X influences the value of Y. This can be thought of as a dependence of Y on X.

To illustrate,[7] Table 12.4 and Figure 12.4 show twenty-two annual values for retail sales per household and disposable income per household (data in 1972 U.S. dollars). We would expect expenditures on retail sales to depend on available income. Specifically, in years when income is relatively high, then high retail sales would be expected. That this pattern generally emerges is clear from the plot of the data in Figure 12.4.

The objective of regression analysis is to *model* this relationship. At the outset, two problems must be faced. First, in typical business and economic applications, the precise functional form of any underlying relation will not be known. In the context of our example, we expect high income to lead to high sales, but know very little more than that. In such circumstances, it is sensible to begin by postulating as simple a structure as seems plausible. For a great many problems it is reasonable to posit a **linear** model,[8] at least in the range of interest.

Our concern, then, is with the value taken by the random variable Y, when the

TABLE 12.4 Data on disposable income per household (x) and retail sales per household (y); data in 1972 U.S. dollars

YEAR	x_i	y_i	YEAR	x_i	y_i
1	9,098	5,492	12	11,307	5,907
2	9,138	5,540	13	11,432	6,124
3	9,094	5,305	14	11,449	6,186
4	9,282	5,507	15	11,697	6,224
5	9,229	5,418	16	11,871	6,496
6	9,347	5,320	17	12,018	6,718
7	9,525	5,538	18	12,523	6,921
8	9,756	5,692	19	12,053	6,471
9	10,282	5,871	20	12,088	6,394
10	10,662	6,157	21	12,215	6,555
11	11,019	6,342	22	12,494	6,755

[7] These data are taken from a study by N. K. Dhalla, "Short-term forecasts of advertising expenditures," *Journal of Advertising Research, 19* (1979), No. 1, 7–14.

[8] The analyst should, however, be prepared to abandon this assumption if it is strongly contradicted by the data.

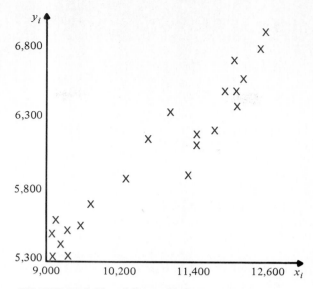

FIGURE 12.4 Plot of disposable income–retail sales data

value of retail sales per household in a year in which disposable income per household is \$12,000. Now, we do not know, and will not know, *precisely* what value would result for sales in a particular instance in which disposable income was \$12,000. In reality, the relationship between the quantities will not be exact, so that it is not reasonable to think of just a single possible sales level resulting from a particular value for disposable income. Rather, it is more realistic to conceive of a *distribution* of possible sales levels resulting from each possible level of disposable income. In the terminology of Chapters 4 and 5, we might reasonably think of the *conditional distribution* of sales, when disposable income has some specific value such as \$12,000. A crucial characteristic of this distribution, in the present context, is its mean, or expected value. Therefore, we can reasonably ask what would be the **expected value,** or average level, of retail sales per household in years in which disposable income per household was \$12,000. In general, we will denote by $E(Y|X = x)$ the expected value of the random variable Y, when the random variable X takes the specific value x. Our assumption of linearity is the assumption that this conditional expectation depends linearly on x.

We are concerned with the expected value of the random variable Y, when X takes a specific value x. The assumption of linearity then implies that this expectation can be written

$$E(Y|X = x) = \alpha + \beta x \qquad (12.3.1)$$

where the numbers α and β determine a specific straight line. Each of these numbers has a simple interpretation—one extremely important, the other typically much less so. Suppose that retail sales are related to disposable income through the equation

$$E(Y|X = x) = 2{,}000 + 0.4x \qquad (12.3.2)$$

so that, in equation 12.3.1, $\alpha = 2{,}000$ and $\beta = 0.4$. Now, substituting $x = 0$ in 12.3.1 gives

$$E(Y|X = 0) = \alpha$$

so that α emerges as the expected value of the **dependent variable** Y when the **independent variable** X takes the value 0. Although theoretically correct, this interpretation may not always be taken seriously. The implication for our example would be that, if disposable income were \$0, retail sales per household would be expected to be \$2,000. However, we have no interest in the case of \$0 disposable income, since it will never arise. It is important to remember that the assumption of linearity should not be stretched too far. We have observations on disposable income in a range from roughly \$9,000 to \$12,500 per household and, while linearity in this range seems not unreasonable, it would be dangerous to extrapolate our conclusions very far outside that range. Thus, the assertion that expected retail sales per household would be \$2,000 in a year in which disposable income per household was \$0 would imply a belief that our assumption of linearity held true well beyond the boundaries of the \$9,000 to \$12,500 income range. Since we have no observational experience outside of this range, such a belief would be unsupported by our data.

We now return to equation 12.3.1. Suppose that X is increased by 1 unit, from x to $(x + 1)$. Then we have

$$E(Y|X = x + 1) = \alpha + \beta(x + 1)$$

so that

$$E(Y|X = x + 1) - E(Y|X = x) = \alpha + \beta(x + 1) - (\alpha + \beta x) = \beta$$

Thus, β, the **slope** of the line, is the expected increase in Y for a 1 unit increase in X. If the relation 12.3.2 held for this disposable income–retail sales data, the implication would be that, for each \$1 increase in disposable income per household, a \$0.40 increase in retail sales per household would be expected. We have already seen that the objective of regression is to describe, or model, the dependence of one variable on another. One way to think of such dependence is in terms of the change in the dependent variable, Y, resulting from a change in the independent variable, X. As we have just seen, each unit increase in X results in an expected change β in Y. Thus, if β is positive, an increase in X leads to an expected increase in Y, while if β is negative, an increase in X leads to an expected decrease in Y. In each case, the magnitude of the expected change in the dependent variable is a multiple of β of the change in the independent variable.

The second problem is that no hypothesized theoretical relation will hold *exactly* in the real world, as we have already suggested. The data points in Figure 12.4 do not all lie on a single straight line (or any other convenient simple curve we might draw). Suppose that the independent variable takes the value x_i. Let Y_i denote the corresponding dependent variable, whose expected value would be

$$E(Y_i \mid X = x_i) = \alpha + \beta x_i \qquad (12.3.3)$$

In practice, the observed value of Y_i would almost invariably deviate somewhat from this expectation. If the discrepancy is denoted by the random variable ε_i (which should, by virtue of 12.3.3, have mean 0), then we can write

$$\varepsilon_i = Y_i - E(Y_i \mid X = x_i) = Y_i - (\alpha + \beta x_i)$$

or

$$Y_i = \alpha + \beta x_i + \varepsilon_i \qquad (12.3.4)$$

Equation 12.3.4 is called the **population (or true) regression line** of Y on X.

The Population Regression Line

Suppose we are interested in the relation of a dependent variable Y to an independent variable X. If the random variable X takes specific values x_i, the **population regression line** expresses the corresponding values Y_i as

$$Y_i = \alpha + \beta x_i + \varepsilon_i$$

where α and β are constants, and ε_i is a random variable with mean 0.

Thus, the response of retail sales to a particular value, say x_i, of disposable income will be the sum of two parts—an expectation $(\alpha + \beta x_i)$ reflecting their systematic relationship, and a discrepancy (ε_i) from that expectation. One can think of the discrepancy, or **error term** ε_i, as embodying the multitude of factors *other than disposable income* that influence retail sales.

The regression model we have just described is illustrated in Figure 12.5, which shows a line representing the linear relationship between the expected value of the dependent variable and the value taken by the independent variable. For each different possible value of the independent variable, the value of the dependent variable can be represented by a random variable, whose mean lies on the regression line. We represent this in the figure by drawing a series of probability density functions for the dependent variable, given values of the independent variable. The regression line traces out the means of these distributions. For any given value x_i, the deviation of the

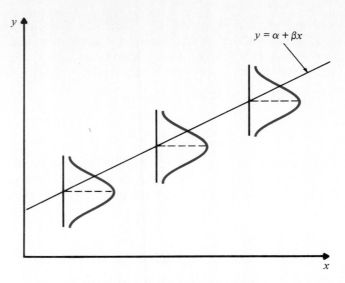

FIGURE 12.5 Illustration of the population regression model; shown are probability density functions of the dependent variable for given values, x, of the independent variable

dependent variable Y_i from the regression line is the discrepancy, or error, term ε_i. The density functions drawn in Figure 12.5 can be regarded as probability density functions of the random variables ε_i, when each of these density functions is centered on 0.

12.4 LEAST SQUARES ESTIMATION

The population regression line introduced in the previous section is a valuable theoretical construct. However, in practical applications, one will never be able to determine it *precisely*. Instead, it will be necessary to obtain an *estimate* from whatever data are available.

Suppose that we have n pairs of observations (x_1, y_1), (x_2, y_2), . . . , (x_n, y_n). We would like to find the straight line that best *fits* these points, in some sense. In other words, we would like to find estimates of the unknown coefficients α and β of the population regression line. An obvious approach would be through visual inspection of the plotted points. One might, after a little trial and error, be able to draw a line that passes reasonably near every point. However, a more formal approach that has attractive features is available.

Consider, as possible estimates of α and β, the numbers a and b, so that the estimated line is

$$y = a + bx$$

To determine how good an estimate this is, we need some measure of the distance of the points (x_i, y_i) from this line. Figure 12.6 shows, for a single point, how this

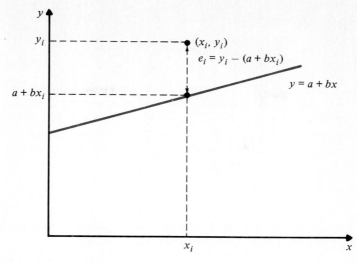

FIGURE 12.6 Distance $e_i = y_i - (a + bx_i)$ from point (x_i, y_i) to the line $y = a + bx$

distance is measured. For the value x_i the corresponding y value on our line is $(a + bx_i)$, while the value actually observed for the dependent variable is y_i. The discrepancy between the two is therefore

$$e_i = y_i - (a + bx_i) \qquad (12.4.1)$$

At first sight, it may seem surprising that we did not take the shortest distance from the point to the line. Recall that our objective is to use the independent variable to explain the behavior of the dependent variable. The discrepancy e_i therefore reflects departure of the dependent variable from the value $(a + bx_i)$ predicted by the postulated line. To illustrate, consider the first observations on disposable income and retail sales in Table 12.4. These are

$$x_1 = 9{,}098; \qquad y_1 = 5{,}492$$

Consider also the possible line

$$y = 2{,}000 + 0.4x$$

that is, the line with values

$$a = 2{,}000 \qquad \text{and} \qquad b = 0.4$$

for the intercept and slope. The value predicted by this line for retail sales, when disposable income is 9,098, is

$$a + bx_1 = 2{,}000 + (0.4)(9{,}098) = 5{,}639.2$$

The discrepancy e_1 between the actual value y_1 and the value predicted by this line is

$$e_1 = y_1 - (a + bx_1)$$
$$= 5{,}492 - 5{,}639.2 = -147.2$$

Now, for any reasonable candidate as an estimate of the true regression line, some of the observed data points will be above the estimated line and some below. Hence, some of the e_i of 12.4.1 will be positive and some negative. If we want to penalize equally positive and negative values of the same magnitude, one way to achieve this is to work with the *squares* of the e_i. The sum of squared discrepancies from the points to the line is

$$SS = \sum_{i=1}^{n} e_i^2 = \sum_{i=1}^{n} (y_i - a - bx_i)^2$$

The method of **least squares** selects, as an estimate of the population regression line, that line for which this sum of squares is smallest.

Least Squares Estimation and the Sample Regression Line

Let (x_1, y_1), (x_2, y_2), . . . , (x_n, y_n) be a sample of n pairs of observations on a process with population regression line

$$Y_i = \alpha + \beta x_i + \varepsilon_i$$

The **least squares estimates** of the coefficients α and β are those values a and b for which the sum of squared discrepancies

$$SS = \sum_{i=1}^{n} (y_i - a - bx_i)^2$$

is a minimum. It can be shown[9] that the resulting estimates are

$$b = \frac{\sum_{1}^{n} (x_i - \bar{x})(y_i - \bar{y})}{\sum_{1}^{n} (x_i - \bar{x})^2} = \frac{\sum_{1}^{n} x_i y_i - n\bar{x}\bar{y}}{\sum_{1}^{n} x_i^2 - n\bar{x}^2}$$

and

$$a = \bar{y} - b\bar{x}$$

where $\bar{x}$ and $\bar{y}$ are the respective sample means.
The line

$$y = a + bx$$

is called the **sample regression line** of Y on X.

---

[9] The result is most easily derived using calculus. The interested reader is referred to Appendix A12.1 at the end of this chapter.

TABLE 12.5 Calculations for the sample regression of retail sales per household on disposable income per household (x_iy_i and x_i^2 rounded to nearest thousand)

	x_i	y_i	x_iy_i	x_i^2
	9,098	5,492	49,966,000	82,774,000
	9,138	5,540	50,625,000	83,503,000
	9,094	5,305	48,244,000	82,701,000
	9,282	5,507	51,116,000	86,156,000
	9,229	5,418	50,003,000	85,174,000
	9,347	5,320	49,726,000	87,366,000
	9,525	5,538	52,749,000	90,726,000
	9,756	5,692	55,531,000	95,180,000
	10,282	5,871	60,366,000	105,720,000
	10,662	6,157	65,646,000	113,678,000
	11,019	6,342	69,882,000	121,418,000
	11,307	5,907	66,790,000	127,848,000
	11,432	6,124	70,010,000	130,691,000
	11,449	6,186	70,824,000	131,080,000
	11,697	6,224	72,802,000	136,820,000
	11,871	6,496	77,114,000	140,921,000
	12,018	6,718	80,737,000	144,432,000
	12,523	6,921	86,672,000	156,826,000
	12,053	6,471	77,995,000	145,275,000
	12,088	6,394	77,291,000	146,120,000
	12,215	6,555	80,069,000	149,206,000
	12,494	6,755	84,397,000	156,100,000
Sums	237,579	132,933	1,448,555,000	2,599,715,000

The calculation of the sample regression line for the retail sales–disposable income data follows from the computations of Table 12.5 from which we have

$$\sum x_i = 237,579; \quad \sum y_i = 132,933; \quad \sum x_iy_i = 1,448,555,000;$$

$$\sum x_i^2 = 2,599,715,000$$

Hence, the sample means are

$$\bar{x} = \frac{\sum x_i}{n} = \frac{237,579}{22} = 10,799.0 \quad \text{and} \quad \bar{y} = \frac{\sum y_i}{n} = \frac{132,933}{22} = 6,042.4$$

The least squares estimates of the coefficients of the population regression line are therefore

$$b = \frac{\sum x_iy_i - n\bar{x}\bar{y}}{\sum x_i^2 - n\bar{x}^2}$$

$$= \frac{1,448,555,000 - (22)(10,799.0)(6,042.4)}{2,599,715,000 - (22)(10,799.0)^2} = 0.3815$$

and

$$a = \bar{y} - b\bar{x} = 6,042.4 - (0.3815)(10,799.0) = 1,923$$

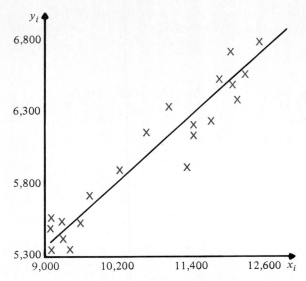

FIGURE 12.7 Disposable income–retail sales data and least squares estimated regression line, $y = 1,923 + 0.3815x$

The sample, or estimated, regression line is thus

$$y = 1,923 + 0.3815x$$

Recalling the interpretation of the slope of the regression line, we are estimating that each \$1 increase in disposable income per household leads, on average, to an increase of \$0.3815 in retail sales per household. Figure 12.7 shows the least squares estimated regression line, together with the twenty-two data points. The estimated regression line seems to give a good visual fit to these data points.

The arithmetic calculations involved in obtaining these estimates are already rather tedious. However, for more complicated regression problems, the computational burden can be enormously higher. In practice, this presents no great problem, since computer packages have been developed to perform regression calculations. We will discuss such packages in Chapter 13.

12.5 STANDARD ASSUMPTIONS FOR THE LINEAR REGRESSION MODEL

The least squares method was presented in Section 12.4 as a procedure to use to estimate a population regression line. It is not invariably the case that the least squares estimates are the most appropriate. However, given certain assumptions, it can be shown that least squares estimates possess certain desirable properties. These standard assumptions are generally taken to hold, unless the available data or theoretical subject matter considerations suggest otherwise. In this section we introduce these assumptions, while an important consequence of them is discussed in the following section.

Standard Assumptions of Linear Regression Model

Denote the population regression line by

$$Y_i = \alpha + \beta x_i + \varepsilon_i$$

and assume that n pairs of observations are available. The following standard assumptions are often made:

1. Either the x_i are fixed numbers (set, for example, by an experimenter) or they are realizations of random variables X_i that are independent of the error terms ε_i. In the latter case, inference is carried out conditionally on the observed x_i.
2. The error terms ε_i are random variables with mean 0, that is,

$$E(\varepsilon_i) = 0 \qquad (i = 1, 2, \ldots, n)$$

3. The random variables ε_i all have the same variance, say σ_ε^2, so that

$$E(\varepsilon_i^2) = \sigma_\varepsilon^2 \qquad (i = 1, 2, \ldots, n)$$

4. The random variables ε_i are not correlated with one another, so that

$$E(\varepsilon_i \varepsilon_j) = 0 \text{ for all } i \neq j$$

The first of these assumptions is generally, with justification, taken to be true, although in some advanced econometric work it is untenable. (The assumption fails to hold, for example, when the x_i cannot be measured precisely or when the regression equation is part of a system of interdependent equations.) Henceforth, however, we will take this assumption as given.

Assumptions 2–4 concern the error terms ε_i in the regression equation—that is, the differences between the Y_i and their conditional expectations $(\alpha + \beta x_i)$. The expected discrepancy is 0, and all discrepancies are assumed to have the same variance. Thus, we do not expect the magnitudes of the error terms to be higher for some observations than for others. Finally, it is assumed that the discrepancies are not correlated with one another. Thus, for example, the occurrence of a large positive discrepancy at one observation point does not help us predict the values of any of the other error terms.

In the remainder of this chapter, it will be taken that these assumptions hold. In Chapter 14 the possibility of their breakdown will be considered.

12.6 THE GAUSS-MARKOV THEOREM

In this section we present a justification for the use of least squares in the estimation of a population regression line. Suppose that we have available n pairs of sample observations $(x_1, y_1), (x_2, y_2), \ldots, (x_n, y_n)$. Any number of possible estimates of the parameters α and β could be constructed. One possibility is to restrict attention to those quantities that are linear functions of the y_i—that is, estimates of the general form

$$c_1 y_1 + c_2 y_2 + \cdots + c_n y_n$$

where the c_i are numbers that do not depend on the y_i. There is a sense in which, from this broad range of potential estimates, the least squares estimates are the most appealing, as discussed in the box.

Gauss-Markov Theorem

Denote the population regression line by

$$Y_i = \alpha + \beta x_i + \varepsilon_i$$

and assume that n pairs of observations are available. Suppose, further, that assumptions 1–4 of Section 12.5 hold.

Then, of all possible estimators of α and β that are linear in the Y_i and unbiased, the least squares estimators (that is, the random variables corresponding to the least squares estimates a and b of Section 12.4) have the smallest variances.

Further, if d_0 and d_1 are any fixed numbers, and we want to estimate

$$d_0 \alpha + d_1 \beta$$

then the estimator corresponding to

$$d_0 a + d_1 b$$

has the smallest variance in the class of all estimators that are linear in the Y_i and unbiased. (This result is useful when using the regression line to obtain predictions of the dependent variable.)

By virtue of this theorem, least squares estimators are said to be **best linear unbiased estimators** (BLUE).

The Gauss-Markov theorem provides a powerful motivation for estimating the parameters of a regression model by least squares. Recalling from Chapter 7 the definition of the efficiency of a point estimator, we see that the method of least squares yields the most efficient estimators from a wide class of unbiased estimators. It should be emphasized, however, that this result depends on the assumptions of the previous section and does not necessarily hold if these assumptions break down. In the remainder of this chapter we continue to explore the properties of the least squares estimation method when those assumptions are true.

The technique of regression analysis is perhaps the most commonly applied statistical tool in business and economics. The result of this section, though briefly stated, is therefore of great practical importance. Regression models are, in practice, generally estimated by the method of least squares. In Section 12.4 we introduced this technique as a sensible procedure for fitting a straight line to a set of data points. However, it is not difficult to think of other sensible ways of accomplishing the same objective. What distinguishes the method of least squares is the fact, following from the Gauss-Markov theorem, that the resulting estimators have desirable statistical properties.

12.7 THE EXPLANATORY POWER OF A LINEAR REGRESSION EQUATION

A regression equation can be viewed as an attempt to employ information on an independent variable, X, to *explain* the behavior of a dependent variable, Y. In this section we develop a measure of the degree to which this attempt has been successful for the sample data. The observations on the dependent variable will exhibit a certain amount of *variability* within the sample. In essence, we are asking here what *proportion* of that variability can be explained by the linear dependence of Y on X.

For the sample values, the estimated regression can be written

$$y_i = a + bx_i + e_i$$

or

$$y_i = \hat{y}_i + e_i \qquad (12.7.1)$$

where

$$\hat{y}_i = a + bx_i$$

The quantity $\hat{y}_i$ is the value of the dependent variable predicted by the regression line and the **residual** e_i is the difference between the observed and predicted values. Therefore, the residual represents that part of the behavior of the dependent variable, in the sample, that cannot be explained by its linear relationship with the independent variable. For the disposable income–retail sales data, the three quantities in equation 12.7.1 are tabulated in the first three columns of Table 12.6.

For our present purpose, a small modification of equation 12.7.1 is useful. We can think of the sample variability of the dependent variable in terms of deviations from its sample mean. Subtracting $\bar{y}$ from each side of equation 12.7.1, we can write

$$(y_i - \bar{y}) = (\hat{y}_i - \bar{y}) + e_i \qquad (12.7.2)$$

or

Observed deviation from sample mean

$$= \text{Predicted deviation from sample mean} + \text{Residual}$$

The first two quantities in equation 12.7.2 are tabulated, for our data, in columns 4 and 5 of Table 12.6.

Now, it can be shown that, if we square both sides of equation 12.7.2 and sum over the sample index i, the result is

TABLE 12.6 Actual and predicted values for retail sales per household, and residuals from its linear regression on income per household

y_i	$\hat{y}_i = a + bx_i$ $= 1,923 + 0.3815x_i$	$e_i = y_i - \hat{y}_i$	$y_i - \bar{y}$ $= y_i - 6,042.4$	$\hat{y}_i - \bar{y}$ $= \hat{y}_i - 6,042.4$
5,492	5,394	98	−550.4	−648.4
5,540	5,409	131	−502.4	−633.4
5,305	5,392	−87	−737.4	−650.4
5,507	5,464	43	−535.4	−578.4
5,418	5,444	−26	−624.4	−598.4
5,320	5,489	−169	−722.4	−553.4
5,538	5,557	−19	−504.4	−485.4
5,692	5,645	47	−350.4	−397.4
5,871	5,846	25	−171.4	−196.4
6,157	5,991	166	114.6	−51.4
6,342	6,127	215	299.6	84.6
5,907	6,237	−330	−135.4	194.6
6,124	6,284	−160	81.6	241.6
6,186	6,291	−105	143.6	248.6
6,224	6,385	−161	181.6	342.6
6,496	6,452	44	453.6	409.6
6,718	6,508	210	675.6	465.6
6,921	6,701	220	878.6	658.6
6,471	6,521	−50	428.6	478.6
6,394	6,535	−141	351.6	492.6
6,555	6,583	−28	512.6	540.6
6,755	6,689	66	712.6	646.6

$$\sum_{i=1}^{n} (y_i - \bar{y})^2 = \sum_{i=1}^{n} (\hat{y}_i - \bar{y})^2 + \sum_{i=1}^{n} e_i^2 \qquad (12.7.3)$$

Equation 12.7.3 has a valuable interpretation. The quantity on its left-hand side represents the total variability, in the sample, of the dependent variable about its mean. This is decomposed into two parts. The first term on the right-hand side of 12.7.3 represents that part of the variability that is explained by the regression, while the second term represents unexplained variability. The equation can be expressed therefore as

Total sample variability = Explained variability + Unexplained variability

In a sense, the higher the proportion of the sample variability explained, the stronger is the explanatory power of the regression.

Sum of Squares Decomposition and the Coefficient of Determination

Suppose that a linear regression equation is fitted by least squares to n pairs of observations, yielding

---

$$y_i = a + bx_i + e_i = \hat{y}_i + e_i \qquad (i = 1, 2, \ldots, n)$$

where a and b are the least squares estimates of the intercept and slope of the population regression, and e_i are the **residuals** from the fitted regression line.

We define the following quantities (where $\bar{y}$ is the sample mean for the dependent variable):

TOTAL SUM OF SQUARES:
$$\text{SST} = \sum_{i=1}^{n} (y_i - \bar{y})^2$$

REGRESSION SUM OF SQUARES:
$$\text{SSR} = \sum_{i=1}^{n} (\hat{y}_i - \bar{y})^2$$

ERROR SUM OF SQUARES:
$$\text{SSE} = \sum_{i=1}^{n} e_i^2$$

It can be shown that

$$\text{SST} = \text{SSR} + \text{SSE}$$

The **coefficient of determination, R^2**, of the fitted regression is defined as

$$R^2 = \frac{\text{SSR}}{\text{SST}} = 1 - \frac{\text{SSE}}{\text{SST}}$$

This is the proportion of the sample variability of the dependent variable explained by its linear relationship with the independent variable. It can further be shown that R^2 is the square of the sample correlation coefficient[10] defined in Section 12.1.

It follows from its definition as the proportion of sample variability explained that

$$0 \le R^2 \le 1$$

and that, the higher is R^2, the greater is the explanatory power of the regression.

For our example, the sums of squares of the elements in columns 3 and 4 of Table 12.6 are, respectively,

$$\text{SSE} = \sum_{i=1}^{n} e_i^2 = 435{,}799 \qquad \text{and} \qquad \text{SST} = \sum_{i=1}^{n}(y_i - \bar{y})^2 = 5{,}397{,}560$$

The coefficient of determination is, therefore

$$R^2 = 1 - \frac{\text{SSE}}{\text{SST}} = 1 - \frac{435{,}799}{5{,}397{,}560} = 0.92$$

This result implies that 92% of the sample variability in retail sales per household is explained by its linear dependence on income per household. We could therefore conclude that, in using disposable income per household as the independent variable, we have been rather successful in explaining variability in retail sales per household.

---

[10] Here we use the notation R^2, rather than the equivalent r^2, as the use of the coefficient of determination extends naturally to the case of multiple regression, to be discussed in the next chapter.

13. A company sets different prices for a particular stereo system in eight different regions of the country. The accompanying table shows the numbers of units sold and the corresponding prices (in hundreds of dollars).

SALES	410	380	350	400	440	380	450	420
PRICE	5.5	6.0	6.5	6.0	5.0	6.5	4.5	5.0

(a) Plot these data, and estimate the linear regression of sales on price.

(b) What effect would you expect a $100 increase in price to have on sales?

14. A random sample of ten students on the University of Illinois campus were asked their grade point averages and the number of hours per week, outside of class, they spent studying. The table shows the results (on a scale where A = 5.0).

GRADE POINT AVERAGE	4.8	4.3	3.8	3.8	4.2	4.3	3.8	4.0	3.1	3.9
STUDY TIME	25	22	9	15	15	30	20	30	10	18

(a) Plot these data and estimate the linear regression of grade point average on study time.

(b) What effect would you expect an increase of 1 hour per week in study time to have on grade point average?

15. The accompanying table shows motor vehicle death rates per 100,000 population and the percentage of vehicles on highways exceeding 60 miles per hour over a 10-year period.

DEATH RATE	22.0	23.1	24.3	26.1	26.5
PERCENTAGE EXCEEDING 60 MPH	18	21	29	32	34

DEATH RATE	28.3	27.8	28.8	27.6	25.3
PERCENTAGE EXCEEDING 60 MPH	40	44	45	46	47

(a) Plot these data and estimate the linear regression of death rates on percentage of vehicles exceeding 60 miles per hour.

(b) What effect would you expect an increase of 1% in the percentage of vehicles exceeding 60 miles per hour to have on the motor vehicle death rate?

16. It is sometimes argued that corporations, in making salary offers to students, take little

or no account of their grade point averages. The results given in the table were obtained for a random sample of twelve graduating M.B.A. students from the same business school.

GRADE POINT AVERAGE	3.0	3.2	3.6	3.5	2.8	2.9
STARTING SALARY (IN DOLLARS)	28,000	34,000	35,000	32,000	27,000	31,000
GRADE POINT AVERAGE	3.8	3.9	3.1	2.7	3.6	3.5
STARTING SALARY (IN DOLLARS)	36,000	35,000	29,000	25,000	30,000	35,000

(a) Estimate the linear regression of starting salary on grade point average.

(b) Interpret the estimated slope of the regression line.

17. A mail-order company is investigating the relationship between its sales revenues (Y) and the price per gallon of gasoline (X). For a 10-month period, the following information was found:

$$\sum_{i=1}^{10} y_i = 120; \qquad \sum_{i=1}^{10} x_i = 1,280; \qquad \sum_{i=1}^{10} x_i^2 = 165,200; \qquad \sum_{i=1}^{10} x_i y_i = 15,480$$

where y_i is measured in millions of dollars, and x_i in cents.

(a) Estimate the linear regression of sales revenues on gasoline price.

(b) Interpret the estimated slope of the regression line.

18. A corporation administers an aptitude test to all new sales representatives. Management is interested in the extent to which this test is able to predict their eventual success. The accompanying table records average weekly sales (in thousands of dollars) and aptitude test scores for a random sample of eight representatives.

WEEKLY SALES	8	12	28	24	18	16	15	12
TEST SCORE	55	60	85	75	80	85	65	55

(a) Estimate the linear regression of weekly sales on aptitude test scores.

(b) Interpret the estimated slope of the regression line.

19. A union official hypothesizes that hourly wage rates (Y) of unskilled workers depend linearly on the number of years (X) that a worker has been with a particular firm. A random sample of fifty workers was taken. Let

y_i = Hourly wage rate (in dollars)
x_i = Years with firm

The following sample statistics were obtained:

$$n = 50, \qquad \bar{x} = 4.0, \qquad \bar{y} = 6.2$$

$$\frac{\sum\limits_{i=1}^{n} (x_i - \bar{x})^2}{n - 1} = 800, \qquad \frac{\sum\limits_{i=1}^{n} (x_i - \bar{x})(y_i - \bar{y})}{n - 1} = 480$$

(a) Estimate the linear regression of wage rates on number of years with firm.

(b) Interpret the estimated slope of the regression line.

20. Let the sample regression line be

$$y_i = a + bx_i + e_i = \hat{y}_i + e_i \qquad (i = 1, 2, \ldots, n)$$

and let $\bar{x}$ and $\bar{y}$ denote the sample means for the independent and dependent variables, respectively.

(a) Show that

$$e_i = y_i - \bar{y} - b(x_i - \bar{x})$$

(b) Using the result in part (a), show that

$$\sum_{i=1}^{n} e_i = 0$$

(c) Using the result in part (a), show that

$$\sum_{i=1}^{n} e_i^2 = \sum_{i=1}^{n} (y_i - \bar{y})^2 - b^2 \sum_{i=1}^{n} (x_i - \bar{x})^2$$

(d) Show that

$$\hat{y}_i - \bar{y} = b(x_i - \bar{x})$$

(e) Using the results in parts (c) and (d), show that

$$SST = SSR + SSE$$

(f) Using the result in part (a), show that

$$\sum_{i=1}^{n} e_i(x_i - \bar{x}) = 0$$

21. Let

$$R^2 = \frac{SSR}{SST}$$

denote the coefficient of determination for the sample regression line.

(a) Using part (d) of Exercise 20, show that

$$R^2 = b^2 \frac{\sum\limits_{i=1}^{n} (x_i - \bar{x})^2}{\sum\limits_{i=1}^{n} (y_i - \bar{y})^2}$$

(b) Using the result in part (a), show that the coefficient of determination is equal to the square of the sample correlation between X and Y.

(c) Let b be the slope of the least squares regression of Y on X, b^* the slope of the least squares regression of X on Y, and r the sample correlation between X and Y. Show that

$$b \cdot b^* = r^2$$

22. Find and interpret the coefficient of determination for the regression of stereo system sales on price, using the data of Exercise 13.

23. Find and interpret the coefficient of determination for the regression of grade point average on hours per week of study time, using the data of Exercise 14.

24. Find the proportion of the sample variability in death rates explained by their linear dependence on the percentage of vehicles on highways exceeding 60 miles per hour for the data of Exercise 15.

25. Refer to the data on starting salaries and grade point averages in Exercise 16.

(a) Find the predicted values, $\hat{y}_i$, and residuals, e_i, for the least squares regression of starting salary on grade point average.

(b) Find the sums of squares SST, SSR, and SSE, and verify that

$$\text{SST} = \text{SSR} + \text{SSE}$$

(c) Using the results in part (b), find and interpret the coefficient of determination.

26. For the problem in Exercise 17, use

$$\sum_{i=1}^{10} y_i^2 = 1{,}475$$

to find the coefficient of determination for the regression of the mail-order company's sales revenue on the price of gasoline. [*Hint*: Use the result in part (a) of Exercise 21.]

27. Refer to the data on weekly sales and aptitude test scores achieved by sales representatives, given in Exercise 18.

(a) Find the predicted values, $\hat{y}_i$, and residuals, e_i, for the least squares regression of weekly sales on aptitude test scores.

(b) Find the sums of squares SST, SSR, and SSE, and verify that

$$\text{SST} = \text{SSR} + \text{SSE}$$

(c) Using the results in part (b), find and interpret the coefficient of determination.

(d) Find directly the sample correlation coefficient between sales and aptitude test scores, and verify that its square is equal to the coefficient of determination.

28. The table lists observations on number of days absent from work during a year and age for a random sample of eight production workers.

DAYS ABSENT	6	12	8	9	9	7	4	15
AGE	49	40	41	35	42	49	54	28

(a) Find the sample correlation between days absent and age.

(b) Using the result in part (a), find the coefficient of determination for the regression of days absent on age.

12.8 CONFIDENCE INTERVALS AND HYPOTHESIS TESTS

In studying the population regression line

$$Y_i = \alpha + \beta x_i + \varepsilon_i$$

we have produced point estimates, through the method of least squares, of the unknown parameters α and β. In addition, we have seen that, given certain assumptions, the least squares estimators have desirable properties by virtue of the Gauss-Markov theorem. However, point estimation is rarely, by itself, sufficient for a thorough analysis of data. It is natural, for example, in the present context to ask how reliable are the estimates that have been obtained. In this section we consider the problems of finding confidence intervals for, and testing hypotheses about, the population regression parameters.

To begin, we note in the box an important use of the quantity SSE, introduced in Section 12.7.

Estimation of the Error Variance

Suppose the population regression line is

$$Y_i = \alpha + \beta x_i + \varepsilon_i$$

and that assumptions 1–4 of Section 12.5 hold. Let σ_ε^2 denote the common variance of the error terms ε_i. Then an unbiased estimate of σ_ε^2 is provided by [11]

$$s_e^2 = \frac{\sum\limits_{i=1}^{n} e_i^2}{(n-2)} = \frac{\text{SSE}}{(n-2)}$$

In the next box we consider the sampling distribution of the least squares estimator of the slope of the population regression line.

Sampling Distribution of the Least Squares Estimator

Let b denote the least squares estimate of the slope β of the population regession line. If assumptions 1–4 of Section 12.5 hold, then the estimator corresponding to b is unbiased for β and has variance

---

[11] It is natural to use the least squares residuals, e_i, as proxies for the unknown ε_i. Intuitively, the reason for division by $(n-2)$ is that 2 degrees of freedom are "lost" through the necessity to estimate the unknown parameters α and β.

$$\sigma_b^2 = \frac{\sigma_\varepsilon^2}{\sum\limits_{i=1}^{n}(x_i - \overline{x})^2} = \frac{\sigma_\varepsilon^2}{\sum x_i^2 - n\overline{x}^2}$$

An unbiased estimate of σ_b^2 is provided by

$$s_b^2 = \frac{s_e^2}{\sum\limits_{i=1}^{n}(x_i - \overline{x})^2} = \frac{s_e^2}{\sum x_i^2 - n\overline{x}^2}$$

In the disposable income–retail sales example, we have

$$s_e^2 = \frac{\text{SSE}}{n-2} = \frac{435,799}{22-2} = 21,789.95$$

Also, we found in Section 12.4,

$$\overline{x} = 10,799 \quad \text{and} \quad \sum_{i=1}^{n} x_i^2 = 2,599,715,000$$

Hence,

$$s_b^2 = \frac{21,789.95}{[2,599,715,000 - (22)(10,799)^2]} = 0.0006388$$

so that the estimated standard deviation of the least squares estimator of the slope of the population regression line is

$$s_b = \sqrt{s_b^2} = 0.0253$$

In the overwhelming majority of practical applications the major focus of interest is on the slope rather than the intercept of the regression line. Accordingly, we will concentrate on that quantity in what follows, noting here that inference about the intercept can be carried out in an analogous fashion by substituting for β, b, and s_b^2, the quantities α, a, and s_a^2, where

$$s_a^2 = s_e^2 \left[\frac{1}{n} + \frac{\overline{x}^2}{(\sum x_i^2 - n\overline{x}^2)} \right]$$

Up to this point we have not required specific distributional assumptions about the population errors ε_i. However, in order to take inference further, something more definite has to be assumed. Almost invariably, unless strong contrary evidence is available, these errors are taken to obey a normal distribution. Given this additional assumption, we can develop confidence intervals and hypothesis tests. Moreover, as a result of the Central Limit Theorem, the procedures remain approximately valid for a wide range of nonnormal error distributions. The main result, from which appropriate confidence intervals and tests immediately follow, is stated in the box.

Confidence intervals for the slope, β, of the population regression line can then be derived through a line of argument used repeatedly in Chapter 8, as summarized in the next box.

For the regression of retail sales on disposable income, we have already found

$$n = 22; \qquad b = 0.3815; \qquad s_b = 0.0253$$

If a 99% confidence interval for β is required, we have $(1 - \alpha) = 0.99$, and so, from Table 6 of the Appendix,

$$t_{n-2,\alpha/2} = t_{20,.005} = 2.845$$

Hence, the 99% confidence interval is

$$0.3815 - (2.845)(0.0253) < \beta < 0.3815 + (2.845)(0.0253)$$

or

$$0.3095 < \beta < 0.4535$$

Thus, in the context of our problem, the 99% confidence interval for the expected

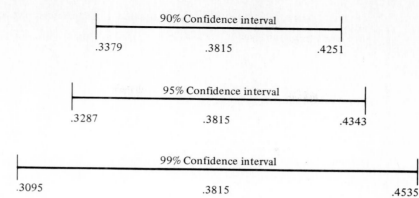

FIGURE 12.8 90%, 95%, and 99% confidence intervals for the population regression slope for regression of retail sales on disposable income

increase in retail sales per household, resulting from a $1 increase in disposable income per household, runs from $0.3095 to $0.4535. Figure 12.8 shows also the 90% and 95% confidence intervals for the population regression slope, calculated from these data.

Following the same kind of reasoning used frequently in Chapter 9, we can readily develop tests of the hypothesis that the population regression slope β is equal to some specified value β_0. The procedure is described in the box.

Tests of the Population Regression Slope

If the regression errors ε_i are normally distributed and assumptions 1–4 of Section 12.5 hold, then the following tests have significance level α:

(i) To test either null hypothesis

$$H_0: \quad \beta = \beta_0 \quad \text{or} \quad H_0: \quad \beta \leq \beta_0$$

against the alternative

$$H_1: \quad \beta > \beta_0$$

the decision rule is

$$\text{Reject } H_0 \text{ if } \quad \frac{b - \beta_0}{s_b} > t_{n-2,\alpha}$$

(ii) To test either null hypothesis

$$H_0: \quad \beta = \beta_0 \quad \text{or} \quad H_0: \quad \beta \geq \beta_0$$

against the alternative

$$H_1: \quad \beta < \beta_0$$

---

the decision rule is

$$\text{Reject } H_0 \text{ if } \quad \frac{b - \beta_0}{s_b} < -t_{n-2,\alpha}$$

(iii) To test the null hypothesis

$$H_0: \quad \beta = \beta_0$$

against the two-sided alternative

$$H_1: \quad \beta \neq \beta_0$$

the decision rule is

$$\text{Reject } H_0 \text{ if } \quad \frac{b - \beta_0}{s_b} > t_{n-2,\alpha/2} \quad \text{ or } \quad \frac{b - \beta_0}{s_b} < -t_{n-2,\alpha/2}$$

A special case of some practical interest arises when the value hypothesized for the regression slope is 0. The population regression line is

$$Y_i = \alpha + \beta x_i + \varepsilon_i$$

so that, on setting β equal to 0, we have

$$Y_i = \alpha + \varepsilon_i$$

This implies that, *irrespective of the value taken by the independent variable,* the dependent variable will be a random variable with mean α and variance σ_ε^2. Thus, the expected value of the dependent variable will not be (linearly) affected by the value of the independent variable. In other words, variability in the dependent variable cannot at all be explained by a linear relation with the independent variable.

To illustrate, we test for the retail sales–disposable income data the null hypothesis

$$H_0: \quad \beta = 0$$

This is the hypothesis that income does not (linearly) influence sales. The alternative of interest is that an increase in income leads to an increase in sales; that is,

$$H_1: \quad \beta > 0$$

We have

$$n = 22; \quad b = 0.3815; \quad s_b = 0.0253; \quad \beta_0 = 0$$

so that

$$\frac{b - \beta_0}{s_b} = \frac{0.3815 - 0}{0.0253} = 15.08$$

From Table 6 of the Appendix, for $(n - 2) = 20$ degrees of freedom, we find

$$t_{20,.005} = 2.845$$

Hence, the null hypothesis that the population regression slope is 0 is very clearly rejected, against the alternative that it is positive, at the 0.5% significance level. Thus, we see that the evidence in these data against the hypothesis that a change in disposable income per household does not induce a change in expected retail sales per household is overwhelming. The data point very strongly to the conclusion that an increase in disposable income leads to an expected increase in retail sales.

We have now developed, under normality assumptions, two tests of the hypothesis of no linear association between a pair of random variables. In Section 12.1, such a test was based on the sample correlation coefficient, while the test just illustrated is based on the least squares estimate of the slope of the regression line. In fact, no contradiction in conclusions can arise from these two tests. It is possible to show that precisely the same test statistic emerges whichever route is followed, so that the two tests are equivalent. The distinction between linear regression and correlation analysis lies not in the formal procedures for testing for lack of association, but in the interpretation of any association. The correlation model does not postulate a *direction* for any dependence between random variables. On the other hand, in estimating the regression model for our data in this chapter, we have implicitly assumed that changes in disposable income per household lead to changes in retail sales per household, rather than the converse.

12.9 PREDICTION

One important use of regression is in the computation of *predictions*, or forecasts, for the dependent variable, conditional on an assumed value for the independent variable. Suppose that the independent variable is equal to some specified value, x_{n+1}, and that the linear relationship between dependent and independent variables continues to hold. The corresponding value of the dependent variable will then be

$$Y_{n+1} = \alpha + \beta x_{n+1} + \varepsilon_{n+1} \qquad (12.9.1)$$

which, given x_{n+1}, has expectation

$$E(Y_{n+1}|x_{n+1}) = \alpha + \beta x_{n+1} \qquad (12.9.2)$$

Two distinct prediction problems are of interest:

(i) We may want to estimate the *actual value* that will result for Y_{n+1} in equation 12.9.1.
(ii) We might want to estimate the conditional expectation $E(Y_{n+1}|x_{n+1})$ of 12.9.2—that is, the *average value* of the dependent variable when the independent variable is fixed at x_{n+1}.

Provided the assumptions of Section 12.5 continue to hold, the same point estimate results for either problem. It is natural to replace the unknown α and β by their least squares estimates, a and b. Hence, $(\alpha + \beta x_{n+1})$ is estimated by $(a + bx_{n+1})$. We know from the Gauss-Markov theorem that the corresponding estimator is best linear unbiased. Thus, for both problems, an appropriate point estimate under our assumptions is

$$\hat{Y}_{n+1} = a + bx_{n+1}$$

This follows since we know nothing useful, in the present context, about the random variable ε_{n+1} of 12.9.1, except that its mean is 0. In the absence of any other relevant information, then, the best we can do is to use 0 as its point estimate.

However, confidence intervals are usually wanted along with the point estimates, and at this point the two problems become distinct. This is because there will be uncertainty about the value to be taken by the random variable ε_{n+1}, which appears in 12.9.1, but not in 12.9.2. The appropriate procedures are summarized in the box.

Confidence Intervals for Predictions

Suppose that the population regression model

$$Y_i = \alpha + \beta x_i + \varepsilon_i \qquad (i = 1, \ldots, n + 1)$$

holds, that the assumptions of Section 12.5 hold, and that the ε_i are normally distributed. Let a and b be the least squares estimates of α and β, based on (x_1, y_1), (x_2, y_2), . . . , (x_n, y_n). Then it can be shown that the following are $100(1 - \alpha)\%$ confidence intervals:

(i) For the prediction of the actual value resulting for Y_{n+1}, the interval is

$$\hat{Y}_{n+1} \pm t_{n-2, \alpha/2} \sqrt{\left[1 + \frac{1}{n} + \frac{(x_{n+1} - \overline{x})^2}{\left(\sum_{i=1}^{n} x_i^2 - n\overline{x}^2 \right)} \right] s_e^2} \qquad (12.9.3)$$

(ii) For the prediction of the conditional expectation, $E(Y_{n+1} \mid x_{n+1})$, the interval is

$$\hat{Y}_{n+1} \pm t_{n-2, \alpha/2} \sqrt{\left[\frac{1}{n} + \frac{(x_{n+1} - \overline{x})^2}{\left(\sum_{i=1}^{n} x_i^2 - n\overline{x}^2 \right)} \right] s_e^2} \qquad (12.9.4)$$

Here,

$$\overline{x} = \frac{\sum_{i=1}^{n} x_i}{n} \qquad \text{and} \qquad \hat{Y}_{n+1} = a + bx_{n+1}$$

To illustrate these procedures, consider again the retail sales–disposable income example, and suppose we are interested in the prediction of retail sales per household in a year in which disposable income per household is $12,000. Thus

$$x_{n+1} = 12,000$$

For point prediction, we then have

$$\hat{Y}_{n+1} = a + bx_{n+1}$$

$$= 1,923 + (0.3815)(12,000) = 6,501$$

Thus, we estimate sales of $6,501 when income is $12,000. We have earlier found

$$n = 22; \quad \bar{x} = 10,799; \quad \sum_{i=1}^{n} x_i^2 = 2,599,715,000; \quad s_e^2 = 21,789.95$$

Hence,

$$\sqrt{\left[1 + \frac{1}{n} + \frac{(x_{n+1} - \bar{x})^2}{\left(\sum_{i=1}^{n} x_i^2 - n\bar{x}^2\right)}\right] s_e^2}$$

$$= \sqrt{\left[1 + \frac{1}{22} + \frac{(12,000 - 10,799)^2}{[2,599,715,000 - (22)(10,799)^2]}\right](21,789.95)}$$

$$= 153.954$$

Similarly, we find

$$\sqrt{\left[\frac{1}{n} + \frac{(x_{n+1} - \bar{x})^2}{\left(\sum_{i=1}^{n} x_i^2 - n\bar{x}^2\right)}\right] s_e^2} = 43.725$$

Suppose that 95% confidence intervals are required for the predictions, so that $\alpha = .05$ and

$$t_{n-2,\alpha/2} = t_{20,.025} = 2.086$$

Then, for a prediction of the actual value resulting for retail sales in a year when disposable income is $12,000, we have the 95% interval

$$6,501 \pm (2.086)(153.954)$$

or

$$6,501 \pm 321$$

Thus the 95% confidence interval for sales in a year in which income is $12,000 runs from $6,180 to $6,822.

For the confidence interval for the expected value of retail sales when disposable income is $12,000, we have

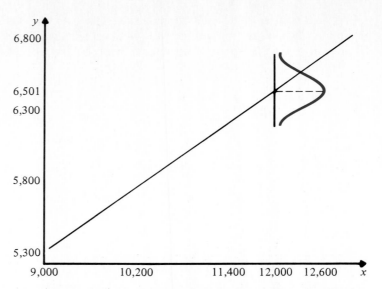

FIGURE 12.9 Least squares estimated regression line of retail sales on disposable income: The probability density function represents our uncertainty about actual retail sales when disposable income is $12,000

$$6{,}501 \pm (2.086)(43.725)$$

or

$$6{,}501 \pm 91$$

Hence, the 95% confidence interval runs from $6,410 to $6,592.

The distinction between these two interval estimation problems is illustrated in Figures 12.9 and 12.10. Each figure shows the estimated regression line for our retail sales–disposable income data. Also shown in Figure 12.9 is a probability density function representing our uncertainty about the value that retail sales will take in any specific year in which disposable income is $12,000. The probability density function in Figure 12.10 represents our uncertainty about *expected,* or average, retail sales in years when disposable income is $12,000. Now, we would be far more uncertain about sales in a single specific year than about average sales, and this is reflected in the shapes of the two density functions. Although both are centered on a retail sales figure of $6,501, the density function in Figure 12.9 is far more dispersed about this value. This additional uncertainty is reflected in wider confidence intervals for a specific value of retail sales than for expected retail sales.

A study of the general forms of the confidence intervals 12.9.3 and 12.9.4 provides some insight. Keeping in mind that, the wider the confidence interval, the greater the uncertainty surrounding the point forecast, we see from these formulas that:

(a) All other things equal, the larger the sample size n, the narrower the confidence interval. This reflects the fact that, the more sample information available, the more sure will be our inference.

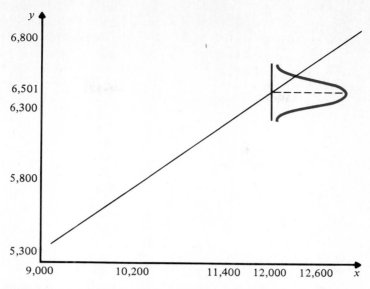

FIGURE 12.10 Least squares estimated regression line of retail sales on disposable income: The probability density function represents our uncertainty about expected, or average, retail sales when disposable income is $12,000

(b) All other things equal, the larger is s_e^2, the wider the confidence interval. Again this is to be expected, since s_e^2 is an estimate of σ_ε^2, the variance of the regression errors ε_i. Since these errors

$$\varepsilon_i = Y_i - \alpha - \beta x_i$$

represent the discrepancy between the observed values of the dependent variables and their expectations given the independent variables, the bigger the magnitude of this discrepancy the more imprecise will be our inference.

(c) Consider now the quantity $(\Sigma\, x_i^2 - n\bar{x}^2)$. This is simply a multiple of the sample variance of the observations on the independent variable. A large variance implies that we have information for a wide range of values of this variable, which allows more precise estimates of the population regression line, and correspondingly narrower confidence intervals.

(d) Finally, we see that, the larger is $(x_{n+1} - \bar{x})^2$, the wider the confidence intervals for the predictions. That is, our inference becomes more uncertain the farther is x_{n+1} from the sample mean of the independent variable. This, too, is a plausible conclusion. If our sample data are centered on $\bar{x}$, we would expect to be able to be more definitive about our inference when the independent variable is relatively close to this central value than when it is some distance away.

Suppose, now, given the data of Table 12.4, you are asked to predict retail sales per household in a year in which disposable income per household is $30,000. In principle, you could routinely follow the procedures of this section, and produce point and interval forecasts. However, to do so would be extremely rash. The available data do suggest, *within the income range observed,* a linear relationship between expected sales and income. However, we have no observed experience of what happens when income is as high as $30,000. It would be an act of pure faith to assume that our linear relationship continues to hold for such high income levels. It might, of course, but this certainly should not be inferred from the data. If, indeed, the assumed relation does break down, forecasts based on the assumption that it holds can be wildly in error. The point is that it is unwise to extrapolate an estimated linear regression very far outside the range of the independent variable for which we have sample data.

EXERCISES

29. Consider the linear regression of stereo system sales on price, based on the data of Exercise 13.
    (a) Use an unbiased estimation procedure to find an unbiased estimate of the variance of the error terms in the population regression.
    (b) Use an unbiased estimation procedure to find an unbiased estimate of the variance of the least squares estimator of the slope of the population regression line.
    (c) Find a 90% confidence interval for the slope of the population regression line.
30. Consider the linear regression of grade point average on hours per week of study time, based on the data of Exercise 14.
    (a) Use an unbiased estimation procedure to find a point estimate of the variance of the error terms in the population regression.
    (b) Use an unbiased estimation procedure to find a point estimate of the variance of the least squares estimator of the slope of the population regression line.
    (c) Find a 95% confidence interval for the expected increase in grade point average resulting from an increase of 1 hour per week in study time.
31. Consider the linear regression of motor vehicle death rates on the percentage of vehicles on highways exceeding 60 miles per hour, based on the data of Exercise 15.
    (a) Use an unbiased estimation procedure to obtain a point estimate of the variance of the error terms in the population regression.
    (b) Use an unbiased estimation procedure to obtain a point estimate of the variance of the least squares estimator of the slope of the population regression line.
    (c) Find 90%, 95%, and 99% confidence intervals for the slope of the population regression line.
32. A fast-food franchise chain decided to carry out an experiment in order to assess the influence of advertising expenditure on sales. Different relative changes in advertising expenditure, compared to the previous year, were made in eight regions of the country, and resulting changes in sales levels were observed. The accompanying table shows the results.

PERCENTAGE INCREASE IN ADVERTISING EXPENDITURE	0.0	4.0	14.0	10.0	9.0	8.0	6.0	2.0
PERCENTAGE INCREASE IN SALES	2.5	7.2	10.3	9.1	10.2	4.1	7.6	3.5

    (a) Estimate by least squares the linear regression of increase in sales on increase in advertising expenditure.
    (b) Find a 90% confidence interval for the slope of the population regression line.
33. A random sample of twelve families gave the figures shown in the table for their incomes and savings (in thousands of dollars) during a 1-year period. Find a 95% confidence interval for the expected increase in savings resulting from a $1,000 increase in income.

INCOME	30.5	26.0	18.0	42.5	30.0	28.0
SAVINGS	2.6	2.2	1.5	4.0	2.7	2.9

INCOME	27.5	32.5	35.0	26.0	27.5	39.0
SAVINGS	2.6	3.0	3.2	2.7	2.2	3.4

34. Using the data of Exercise 16, test the null hypothesis that starting salaries for M.B.A. students are not linearly affected by grade point averages against the alternative that there is a positive relationship.

35. For the mail-order company of Exercise 17, use

$$\sum_{i=1}^{10} y_i^2 = 1,475$$

to test the null hypothesis that the slope of the population regression line of sales revenues on the price of gasoline is 0 against the alternative that the slope is positive.

36. Using the data on Exercise 18, test the null hypothesis that representatives' weekly sales are not linearly related to their aptitude test scores against the alternative that there is positive association.

37. A study of the dependence of annual output of corn, y (in bushels per acre), on the amount of insecticide used, x (in gallons per acre), yielded for a random sample of thirty farms the estimated regression

$$y = 200 + 0.8x$$

Also calculated was $s_b = 0.42$. Test the null hypothesis that the quantity of insecticide used does not linearly affect the yield of corn.

38. Denote by r the sample correlation between a pair of random variables.
   (a) Show that

$$\frac{1 - r^2}{n - 2} = \frac{s_e^2}{\text{SST}}$$

   (b) Using the result in part (a), show that

$$\frac{r}{\sqrt{(1 - r^2)/(n - 2)}} = \frac{b}{s_e/\sqrt{\Sigma(x_i - \overline{x})^2}}$$

   (c) Using the result in part (b), deduce that the test of the null hypothesis of 0 population correlation, given in Section 12.1, is the same as the test of 0 population regression slope, given in Section 12.8.

39. In the regression of Exercise 19, the coefficient of determination of the fitted model was found to be 0.50. Test the null hypothesis that number of years with a firm do not linearly influence wage rates of unskilled workers.

40. In a study of building societies in the United Kingdom, one focus of interest was the possibility of economies of scale in expenses. It was believed that management expenses for a fixed amount of assets might be lower for larger than for smaller societies. For a sample of 43 building societies, the least squares regression[12]

$$y = 0.652 - 0.0000207x$$

was estimated, where

---

[12] Roughly speaking, British building societies are the equivalents of United States savings and loan associations. These data were analyzed by T. J. Gough, "Building society mergers and the size-efficiency relationship," *Applied Economics, 11* (1979), 185–94.

y = Management expenses per hundred pounds of assets of the society

x = Total assets (in millions of pounds) of the society

The estimated standard deviation of the estimator of the slope of the population regression line was 0.0000292. Test the null hypothesis that the slope of the population regression line is 0 against the alternative that the slope is negative.

41. Refer to the data of Exercise 32. Test at the 5% significance level, against the appropriate one-sided alternative, the null hypothesis that sales do not depend linearly on advertising expenditure.

42. Refer to the data of Exercise 13.

   (a) Find a point estimate for the value of sales when the price of the stereo system is $480 in a given region.

   (b) If the price of the system is set at $480, find 95% confidence intervals for each of the following:

      (i) The actual volume of sales in a particular region

      (ii) The expected number of sales in that region

43. Refer to the data in Exercise 14. For a student who studies 22 hours per week, find 90% confidence intervals for each of the following:

   (a) Actual grade point average

   (b) Expected grade point average

   Explain verbally why these intervals differ.

44. Refer to the data in Exercise 15. For a year in which 42% of vehicles on highways exceed 60 miles per hour, find 99% confidence intervals for each of the following:

   (a) Actual death rate from motor vehicle accidents

   (b) Expected death rate

45. Suppose you have a friend who is an M.B.A. student. Based on the data of Exercise 16, find 95% and 99% confidence intervals for her starting salary if her grade point average is 3.6.

46. Using the data of Exercises 17 and 26, find 90% and 95% confidence intervals for the mail-order company's expected sales revenues in a month in which the price of gasoline is $1.45 per gallon.

47. A new sales representative for the corporation of Exercise 18 scores 70 on the aptitude test. Find 80% and 90% confidence intervals for the value of weekly sales he will achieve.

48. Refer to the data of Exercise 28.

   (a) Find a point estimate for the number of days absent in a year for a 36-year-old production worker.

   (b) For a 36-year-old worker find 90% confidence intervals for each of the following:

      (i) The actual number of days absent during a year

      (ii) The expected number of days absent during a year

REVIEW EXERCISES

49. (a) What is meant by the statement that a pair of random variables are positively correlated?

   (b) Give examples of pairs of random variables for which you would expect

(i) Positive correlation

(ii) Negative correlation

(iii) Zero correlation

50. In this chapter we have discussed two measures of correlation—the sample correlation coefficient r and Spearman's rank correlation coefficient. Discuss the circumstances in which each measure might be appropriate.

51. A random sample of five sets of observations on a pair of random variables yielded the observations given in the table.

y_i	4	1	0	1	4
x_i	-2	-1	0	1	2

(a) Find the sample correlation coefficient.

(b) In light of the fact that each y_i value is the square of the corresponding x_i value, comment on your answer in part (a).

52. For a random sample of ten students who were taking introductory courses in business statistics and accounting, the final examination scores shown in the table were obtained.

BUSINESS STATISTICS	79	93	86	62	74	87	91	62	75	73
ACCOUNTING	82	83	74	69	79	81	94	70	82	77

(a) Find the sample correlation between the two sets of scores.

(b) Test at the 5% level the null hypothesis of no population correlation between scores in business statistics and accounting, against the alternative of positive population correlation.

53. For a random sample of seventy standard metropolitan statistical areas, it was found[13] that the sample correlation between the economic dimension of quality of life and the health–educational dimension of quality of life was 0.25. Test at the 5% significance level, against a two-sided alternative, the null hypothesis that the population correlation is 0.

54. For a period of thirty quarters, it was found[14] that the sample correlation between the percentage change in the price of gold bullion and the percentage change in the price of silver bullion was 0.690. Stating any assumptions you need to make, test at the 1% significance level, against the alternative of positive correlation, the null hypothesis that the population correlation is 0.

55. In the study of Exercise 54, for a period of thirty quarters, a sample correlation of -0.056 between the percentage change in the price of gold bullion and the percentage change in the value of the Standard and Poor's industrials stock price index was found. Test at the

---

[13] See J. M. Pennings, "The urban quality of life and entrepreneurship," *Academy of Management Journal*, 25 (1982), 63–79.

[14] See A. Renshaw and E. Renshaw, "Does gold have a role in investment portfolios?", *Journal of Portfolio Management*, 8 (1982), No. 3, 28–31.

10% significance level, against a two-sided alternative, the null hypothesis that the population correlation is 0.

56. In a study[15] of the motives for patronage of a used merchandise outlet, the rankings of motives were obtained from samples of light and heavy shoppers, as indicated in the accompanying table. Calculate Spearman's rank correlation coefficient and test, against the alternative of positive association, the null hypothesis of no association between patronage motives of light and heavy shoppers.

	LIGHT SHOPPERS	HEAVY SHOPPERS
Price	1	1
Adventure–Treasure hunting	2	4
Quality	3	2
Location	4	3
Salespeople	5	5
Cleanliness	6	6

57. (a) Calculate Spearman's rank correlation coefficient for the examination scores of Exercise 52.

(b) Carry out a nonparametric test of the null hypothesis of no association between scores in business statistics and accounting, against the alternative of positive association.

58. A study[16] was carried out on the perceptions of problems arising in relationships between advertising agencies and their clients. Agencies were asked the extent to which various agency characteristics caused problems. The table given here shows average responses on a scale from 1 (seldom a problem) to 4 for agency-client relationships lasting less than, and more than, 10 years.

	LENGTH OF RELATIONSHIP	
	LESS THAN 10 YEARS	MORE THAN 10 YEARS
High personnel turnover	2.21	1.80
Failure to meet deadlines	2.14	2.04
Poor communications	2.14	2.05
Poor follow-through	2.07	1.81
Understaffing	2.00	1.81
Lack of cost-consciousness	2.00	2.10
Tendency not to listen	1.95	1.86
Inexperienced account personnel	1.93	1.76
Tendency to be defensive	1.79	1.62
Unstructured procedures	1.50	2.00
Inflexible procedures	1.43	1.57
Too much politics	1.20	1.80

[15] Reprinted by permission of the publisher from "Heavy, medium, light shoppers, and nonshoppers of a used merchandise outlet," by U. Yavas and G. Riecken, *Journal of Business Research, 9,* 243–53. Copyright 1981 by Elsevier Science Publishing Co., Inc.

[16] See M. R. Hotz, J. K. Ryans, and W. L. Shanklin, "Agency-client relationships as seen by influentials on both sides," *Journal of Advertising, 11* (1982), No. 1, 37–44.

(a) Calculate Spearman's rank correlation coefficient.

(b) Carry out, against a two-sided alternative, a nonparametric test of the null hypothesis of no association.

59. Explain carefully what is meant by the population linear regression of a dependent variable on an independent variable.

60. Based on a sample of n observations, (x_1, y_1), (x_2, y_2), . . . , (x_n, y_n), the sample regression of y on x is calculated. Show that the sample regression line passes through the point $(x = \bar{x}, y = \bar{y})$, where $\bar{x}$ and $\bar{y}$ are the sample means.

61. For a random sample of twelve business executives who had been with their firms for at least 4 years, the information shown in the table was obtained on number of years with the firm and number of promotions per year.

NUMBER OF YEARS WITH FIRM	NUMBER OF PROMOTIONS PER YEAR
4	0.25
7	0.29
9	0.33
16	0.31
8	0.38
5	0.40
6	0.50
7	0.43
15	0.47
18	0.33
5	0.60
6	0.50

(a) Estimate the regression of number of promotions per year on number of years with firm.

(b) Interpret the slope of the estimated regression line.

(c) Find and interpret the coefficient of determination.

(d) Test, against a two-sided alternative, the null hypothesis that the slope of the population regression line is 0 and interpret the test result.

(e) Find a 95% confidence interval for the number of promotions per year for a randomly chosen executive who has been with the firm 8 years.

62. An attempt was made[17] to evaluate the forward rate as a predictor of the spot rate in the Canadian treasury bill market. For a sample of 79 quarterly observations, the estimated linear regression

$$y = -0.00027 + 0.7916x$$

was obtained, where

y = Actual change in the spot rate
x = Change in the spot rate predicted by the forward rate

[17] Reported by S. B. Park, "Spot and forward rates in the Canadian treasury bill market," *Journal of Financial Economics, 10* (1982), 107–14.

The coefficient of determination was 0.097, and the estimated standard deviation of the estimator of the slope of the population regression line was 0.2759.

(a) Interpret the slope of the estimated regression line.

(b) Interpret the coefficient of determination.

(c) Test the null hypothesis that the slope of the population regression line is 0, against the alternative that the true slope is positive, and interpret your result.

(d) Test, against a two-sided alternative, the null hypothesis that the slope of the population regression line is 1, and interpret your result.

63. A corporation imposes different percentage price increases on its product in eight different regions of the country, and measures over the next year the percentage changes in sales volume. The results are shown in the accompanying table.

PERCENTAGE CHANGE IN PRICE	6.0	5.0	4.0	7.0	7.0	6.0	10.0	8.0
PERCENTAGE CHANGE IN SALES VOLUME	5.2	7.3	7.4	4.6	5.3	5.0	−1.0	2.7

(a) Estimate the regression of percentage change in sales volume on percentage change in price.

(b) Interpret the slope of the estimated regression line.

(c) Find and interpret the coefficient of determination.

(d) Find and interpret a 90% confidence interval for the slope of the population regression line.

(e) Find a 90% confidence interval for the expected percentage change in sales volume in the year following an 8% price increase.

64. In a study on the determinants of absenteeism,[18] the sample regression model

$$y = -0.0003 + 0.32x$$

was fitted based on a sample of sixty observations. Here

y = Change in rate of absence

x = Change in the compensation ratio, where compensation ratio is defined as the proportion of total compensation distributed as fringe benefits

The coefficient of determination was 0.15, and the estimated standard deviation of the estimator of the slope of the population regression line was 0.10.

(a) Interpret the slope of the estimated regression line.

(b) Interpret the coefficient of determination.

(c) Test against a two-sided alternative, at the 5% level, the null hypothesis that the slope of the population regression line is 0. Interpret your result.

65. The accompanying table shows the performance ratings of a sample of ten employees one year after initial employment, together with the evaluation scores made for these people at the time they were interviewed for their positions.

---

[18] See J.R. Chelias, "Understanding absenteeism: The potential contribution of economic theory," *Journal of Business Research*, 9 (1981), 409–18.

EMPLOYEE	PERFORMANCE RATING	EVALUATION SCORE
1	6	15
2	4	13
3	7	14
4	7	12
5	9	17
6	3	12
7	8	15
8	4	14
9	9	16
10	8	13

(a) Estimate the linear regression of performance ratings on evaluation scores.

(b) Interpret the slope of the estimated regression line.

(c) Find and interpret the coefficient of determination.

(d) Find and interpret a 95% confidence interval for the slope of the population regression line.

(e) Find a 95% confidence interval for the performance rating of a randomly chosen employee whose evaluation score was 15.

66. Comment on the following statement: "If a regression of the yield per acre of corn on the quantity of fertilizer used were estimated, using fertilizer quantities in the range typically used by farmers, then the slope of the estimated regression line would certainly be positive. However, it is well known that, if an enormously high amount of fertilizer were to be used, corn yield would be very low. Therefore, regression equations are not of much use in forecasting."

APPENDIX A12.1

In this appendix we derive the least squares estimates of the population regression parameters. We want to find those values a and b for which the sum of squared discrepancies

$$SS = \sum_{i=1}^{n} (y_i - a - bx_i)^2 \qquad (A12.1.1)$$

is as small as possible.

As a first step, we keep b in equation A12.1.1 constant and differentiate[19] with respect to a, giving

$$\frac{\partial SS}{\partial a} = -2 \sum_{i=1}^{n} (y_i - a - bx_i)$$

$$= -2(\Sigma y_i - na - b\Sigma x_i)$$

Since this derivative must be 0 for a minimum, we have

$$\Sigma y_i - na - b\Sigma x_i = 0$$

Hence, dividing through by n yields

$$a = \bar{y} - b\bar{x}$$

Substituting this expression for a in equation A12.1.1 gives

$$SS = \sum_{i=1}^{n} [(y_i - \bar{y}) - b(x_i - \bar{x})]^2$$

Differentiating this expression with respect to b then gives

$$\frac{dSS}{db} = -2 \sum_{i=1}^{n} (x_i - \bar{x})[(y_i - \bar{y}) - b(x_i - \bar{x})]$$

$$= -2[\Sigma(x_i - \bar{x})(y_i - \bar{y}) - b\Sigma(x_i - \bar{x})^2]$$

---

[19] Here we are using the concept of **partial differentiation.** The partial derivative of SS with respect to a is denoted $\partial SS/\partial a$ and is obtained by differentiating SS with respect to a, treating other variables as constant. The sum of squares SS is a minimum with respect to a and b when both partial derivatives, $\partial SS/\partial a$ and $\partial SS/\partial b$, are 0.

This derivative must be 0 for a minimum, and so we have

$$\Sigma(x_i - \overline{x})(y_i - \overline{y}) = b\Sigma(x_i - \overline{x})^2$$

Hence,

$$b = \frac{\Sigma(x_i - \overline{x})(y_i - \overline{y})}{\Sigma(x_i - \overline{x})^2}$$

multiple regression

13.1 THE MULTIPLE REGRESSION MODEL

In the linear regression model of Chapter 12, the behavior of a *single* independent variable was employed to explain the behavior of a dependent variable. In this chapter, our objective remains the construction of a model to explain, as much as possible, variability in some dependent variable in which we are interested. However, we now admit the possibility of several relevant independent variables, or *multiple influences*. To illustrate, suppose we want to explain the variability through the years of profit margins of savings and loan associations. It is reasonable to conjecture that, all other things equal, profit margins will be positively related to net revenues per deposit dollar; that is, the higher net revenues, the higher will be profit margins. Another possibility is that profit margins will fall due to increased competition, all other things equal, as the number of savings and loan offices increases. We therefore seek a model in which the dependent variable, profit margin (Y), is related to the pair of independent variables, net revenues (X_1) and number of savings and loan offices (X_2). In order to build a model of this relationship, we require data on these three quantities. Table 13.1 shows 25 annual sets of observations.[1]

At the outset, we are again faced with the difficulty that an infinite number of possible functional forms are available to describe the relationship of interest. Once more it is convenient to consider the appropriateness of the linear form. In the previous chapter, when relating a dependent variable to a single independent variable, we

---

[1] These data are taken from L. J. Spellman, "Entry and profitability in a rate-free savings and loan market," *Quarterly Review of Economics and Business, 18* (1978), No. 2, 87–95.

TABLE 13.1 Data on percentage net revenues per deposit dollar (x_1), number of offices (x_2), and percentage profit margin (y) for savings and loan associations

YEAR	x_{1i}	x_{2i}	y_i	YEAR	x_{1i}	x_{2i}	y_i
1	3.92	7,298	0.75	14	3.78	6,672	0.84
2	3.61	6,855	0.71	15	3.82	6,890	0.79
3	3.32	6,636	0.66	16	3.97	7,115	0.70
4	3.07	6,506	0.61	17	4.07	7,327	0.68
5	3.06	6,450	0.70	18	4.25	7,546	0.72
6	3.11	6,402	0.72	19	4.41	7,931	0.55
7	3.21	6,368	0.77	20	4.49	8,097	0.63
8	3.26	6,340	0.74	21	4.70	8,468	0.56
9	3.42	6,349	0.90	22	4.58	8,717	0.41
10	3.42	6,352	0.82	23	4.69	8,991	0.51
11	3.45	6,361	0.75	24	4.71	9,179	0.47
12	3.58	6,369	0.77	25	4.78	9,318	0.32
13	3.66	6,546	0.78				

constructed a formulation in which the expected value of the dependent variable was related linearly to the value of the independent variable. Once again, our interest is in the expected value of the dependent variable, but this is now conditioned by the values taken by all the independent variables. For example, in the context of our savings and loan problem, we might ask what value would be expected for percentage profit margin in a year in which percentage net revenues per deposit dollar were 4.0 and there were 8,000 offices. Again, we need a convenient notation for such a concept. For the case where a dependent variable, Y, is related to a pair of independent variables, X_1 and X_2, we will use $E(Y|X_1 = x_1, X_2 = x_2)$ to represent the expected value of the dependent variable when the independent variables take the respective values x_1 and x_2. Our assumption of linearity, within this context, implies that this conditional expected value is of the form

$$E(Y|X_1 = x_1, X_2 = x_2) = \alpha + \beta_1 x_1 + \beta_2 x_2 \qquad (13.1.1)$$

where the numbers α, β_1, and β_2 must be estimated from data.

More generally, we may want to relate a dependent variable, Y, to K independent variables, $X_1, X_2, \ldots, X_K$. Then, if X_1 takes the value x_1, X_2 takes the value x_2, and so on, the generalization of equation 13.1.1 gives the expected value of the dependent variable as

$$\begin{aligned} E(Y|X_1 = x_1, X_2 = x_2, \ldots, X_K = x_K) \\ = \alpha + \beta_1 x_1 + \beta_2 x_2 + \cdots + \beta_K x_K \end{aligned} \qquad (13.1.2)$$

where $E(Y|X_1 = x_1, X_2 = x_2, \ldots, X_K = x_K)$ denotes the expected value of the dependent variable when the independent variables take the specific values $x_1, x_2, \ldots, x_K$, and where the fixed numbers $\alpha, \beta_1, \beta_2, \ldots, \beta_K$ determine the nature of the relationship. These numbers have straightforward interpretations. First, if each of the independent variables is set to 0, then it follows from 13.1.2 that

$$E(Y|X_1 = 0, X_2 = 0, \ldots, X_K = 0) = \alpha$$

Thus, α is the expected value of the dependent variable when every independent variable takes the value 0. Frequently this interpretation is of no practical interest since the point at which all independent variables are 0 is of no concern, and indeed may be meaningless. (For example, we have no interest in the case where the number of savings and loan offices is 0.) Moreover, while the assumed model form may be reasonable in the region of observed values of the independent variables, it would be unduly optimistic to assume validity of the model very far outside that region.

The interpretation of the coefficients $\beta_1, \beta_2, \ldots, \beta_K$ is extremely important. Referring to equation 13.1.2, suppose that one of the independent variables, say X_1, is increased by 1 unit from x_1 to $(x_1 + 1)$, *while the values of the other independent variables are held constant.* Then we have

$$E(Y|X_1 = x_1 + 1, X_2 = x_2, \ldots, X_K = x_K)$$
$$= \alpha + \beta_1(x_1 + 1) + \beta_2 x_2 + \cdots + \beta_K x_K$$

Hence, using equation 13.1.2, we have

$$E(Y|X_1 = x_1 + 1, X_2 = x_2, \ldots, X_K = x_K) - E(Y|X_1 = x_1, X_2 = x_2, \ldots, X_K = x_K)$$
$$= \alpha + \beta_1(x_1 + 1) + \beta_2 x_2 + \cdots + \beta_K x_K - (\alpha + \beta_1 x_1 + \beta_2 x_2 + \cdots + \beta_K x_K)$$
$$= \beta_1$$

It therefore follows that β_1 is the expected increase in Y resulting from a 1 unit increase in X_1, when the values of the other independent variables remain constant. In general, the coefficient β_j is the expected increase in the dependent variable resulting from a 1 unit increase in the independent variable X_j, *when the values of the other independent variables are held constant.* The quantities β_j, called **partial regression coefficients,** provide separate measures of the influences of the independent variables on the dependent variable, when all other relevant factors remain unchanged.

Returning to our savings and loan example, suppose that the true relationship is

$$E(Y|X_1 = x_1, X_2 = x_2) = 1.5 + 0.2x_1 - 0.00025x_2$$

The coefficient on x_1 implies that an increase of 1 unit in net revenues leads to an expected increase of 0.2 in the percentage profit margin of savings and loan associations, *when the number of savings and loan offices remains fixed.* Similarly, the coefficient on x_2 implies that, *holding net revenues fixed,* an increase of 1 in the number of savings and loan offices leads to an expected increase of -0.00025—that is, to an expected *decrease* of 0.00025—in percentage profit margin. To use more realistic numbers, an increase of 1,000 in the number of savings and loan offices, with net revenues held fixed, leads to an expected decrease of 0.25 in percentage profit margins. These partial regression coefficients are illustrated in Figure 13.1.

Figure 13.1(a) shows the relationship between expected profit margin and net revenues, when the number of savings and loan offices is fixed at 8,000. This relation slopes upward, indicating that an increase in net revenues leads to an expected increase in profit margins when the number of offices remains unchanged. Part (b) of the figure depicts the relation between expected profit margin and the number of offices, when percentage net revenues per deposit dollar are fixed at 4.00. The downward-sloping relation indicates that an increase in the number of offices leads to an expected decrease in profit margins when net revenues per deposit dollar remain unchanged.

We extend the graphical illustration of this population regression in Figure 13.2. Part (a) of the figure shows the dependence of expected profit margin on net revenues for three different numbers of savings and loan offices—7,000, 7,500, and 8,000.

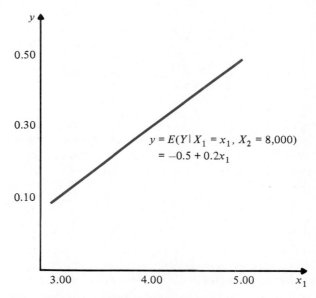

FIGURE 13.1 Postulated model $E(Y|X_1 = x_1,$ $X_2 = x_2) = 1.5 + 0.2x_1 - 0.00025x_2$, where

$$Y = \text{Percentage profit margin for savings and loan associations}$$
$$X_1 = \text{Percentage net revenues per deposit dollar}$$
$$X_2 = \text{Number of savings and loan offices}$$

(a) When number of offices is fixed at 8,000, substituting $x_2 = 8,000$ gives

$$E(Y|X_1 = x_1, X_2 = 8,000) = -0.5 + 0.2x_1$$

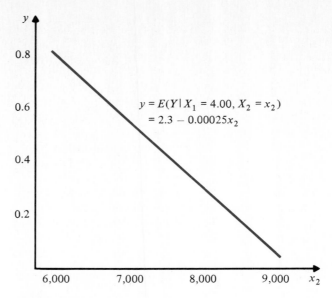

FIGURE 13.1 (cont.)
(b) When net revenues per deposit dollar are fixed at 4%, substituting $x_1 = 4.00$ gives

$$E(Y \mid X_1 = 4.00, X_2 = x_2) = 2.3 - 0.00025x_2$$

Notice that the straight lines depicting these relations are parallel to one another. This reflects the fact that, according to our model, an increase in net revenues, with the number of offices held constant, leads to the same increase in expected profit margin, whatever the fixed value of the number of offices. Notice also that the higher the number of offices, the lower is the line relating expected profit margin to net revenues. This says that, for any given value of net revenues, the bigger the number of offices, the smaller expected profit margins. This is precisely the interpretation of the partial regression coefficient on number of offices that we noted earlier.

Part (b) of Figure 13.2 shows the relationship between expected profit margins and the number of savings and loan offices for three different values of percentage net revenues per deposit dollar—3.5, 4.0, and 4.5. Once again the three lines are parallel. Also, we see that the higher net revenues, the higher the line relating expected profit margin to the number of offices. This is to be expected since, as we have already seen, for any fixed number of offices, an increase in net revenues leads to an expected increase in profit margin.

When, as in the present example, the multiple regression involves only two independent variables, a further graphical description of the model is available. The relationship can be viewed as three-dimensional, as in Figure 13.3. This figure shows the regression function as a plane in three-dimensional space, and is the multiple regression analogue of the line in two-dimensional space used to represent the relation between a dependent variable and a single independent variable. For each pair of possible values of the independent variables, the expected value of the dependent variable is a point on this plane. As we have drawn it, Figure 13.3 shows the situation in our savings and loan example. An increase in X_1 leads, all other things equal, to an expected increase in the dependent variable, while an increase in X_2 leads, all else the same, to an expected decrease in the dependent variable.

In order to complete our model, it is necessary to add an error term, in acknowledgment of the fact that, in the real world, no postulated relationship will hold precisely. Suppose that the K independent variables take the specific values x_{1i}, x_{2i}, ..., x_{Ki}. Then, if the corresponding value of the dependent variable is denoted Y_i, its expectation is

$$E(Y_i \mid X_1 = x_{1i}, X_2 = x_{2i}, \ldots, X_K = x_{Ki}) = \alpha + \beta_1 x_{1i} + \beta_2 x_{2i} + \cdots + \beta_K x_{Ki}$$

Now, let the random variable ε_i denote the discrepancy between Y_i and its expected value given the independent variables, so that

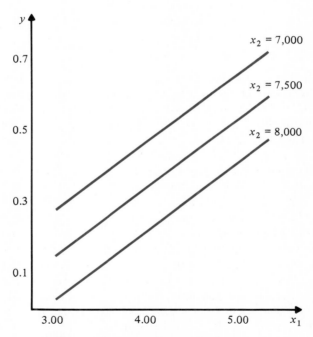

FIGURE 13.2 Postulated model $E(Y \mid X_1 = x_1,$ $X_2 = x_2) = 1.5 + 0.2x_1 - 0.00025x_2$
(a) Relationship between expected Y and x_1, with x_2 fixed at 7,000, 7,500, and 8,000

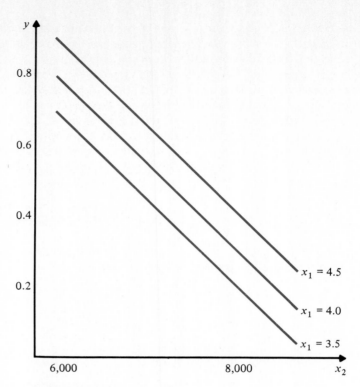

FIGURE 13.2 (cont.)
(b) Relationship between expected Y and x_2, with x_1 fixed at 3.5, 4.0, and 4.5

$$\varepsilon_i = Y_i - E(Y_i \mid X_1 = x_{1i}, X_2 = x_{2i}, \ldots, X_K = x_{Ki})$$
$$= Y_i - (\alpha + \beta_1 x_{1i} + \beta_2 x_{2i} + \cdots + \beta_K x_{Ki})$$

Hence, we can write

$$Y_i = \alpha + \beta_1 x_{1i} + \beta_2 x_{2i} + \cdots + \beta_K x_{Ki} + \varepsilon_i$$

The Population Multiple Regression

Suppose we are interested in the relation of a dependent variable Y to K independent variables, $X_1, X_2, \ldots, X_K$. If the independent variables take the specific values x_{1i}, $x_{2i}, \ldots, x_{Ki}$, then the **population multiple regression** expresses the corresponding value of the dependent variable, Y_i, as

$$Y_i = \alpha + \beta_1 x_{1i} + \beta_2 x_{2i} + \cdots + \beta_K x_{Ki} + \varepsilon_i$$

where $\alpha, \beta_1, \beta_2, \ldots, \beta_K$ are constants, and ε_i is a random variable with mean 0.

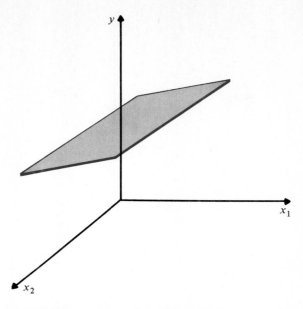

FIGURE 13.3 Representation by a plane, in three dimensions, of the relation of a dependent variable to a pair of independent variables

For our savings and loan example, where there are two independent variables, the population multiple regression is

$$Y_i = \alpha + \beta_1 x_{1i} + \beta_2 x_{2i} + \varepsilon_i$$

Thus, for particular values, x_{1i} and x_{2i}, of net revenues and number of savings and loan offices, the corresponding profit margin is the sum of two parts—an expectation $(\alpha + \beta_1 x_{1i} + \beta_2 x_{2i})$ and a discrepancy, or error term, ε_i. The error term can be regarded as the amalgamation of the influences of the multitude of factors, *other than net revenues and number of offices,* that affect profit margins. We illustrate the population regression model for the case of a pair of independent variables in Figure 13.4. This figure shows the regression plane relating the expected value of the dependent variable to the values of the independent variables, exactly as in Figure 13.3. Also, marked by a cross, is a possible observation point, representing the value of the dependent variable that actually occurs for given values of the independent variables. Now, such observations will not lie precisely on the regression plane. The difference between observed and expected values of the dependent variable is represented by the error term, ε_i. If, as drawn in Figure 13.4, the observation point is above the regression plane, this error term will be positive. For observations below the regression plane, the error term is negative. On average, these error terms are 0.

We note that the simple linear regression model developed in Chapter 12 is

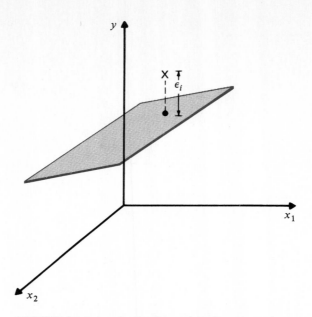

FIGURE 13.4 An observation point, distance ε_i along the y-axis, from the regression plane showing the relation between the expected value of the dependent variable and the values taken by a pair of independent variables

merely the special case of the multiple regression model with just a single independent variable. Our analysis of the multiple regression model will, accordingly, run along lines similar to those of the previous chapter.

13.2 **LEAST SQUARES ESTIMATION**

The principle of least squares estimation of the parameters of the population multiple regression is identical to that in the simple linear regression case. Once again we seek an estimated equation that is as close as possible to the observed data, in the sense of minimizing the sum of squared discrepancies.

Suppose that we have n sets of joint observations on the K independent variables and the dependent variable. These observation points can be denoted as follows:

$$(x_{11}, x_{21}, \ldots, x_{K1}, y_1)$$
$$(x_{12}, x_{22}, \ldots, x_{K2}, y_2)$$
$$\vdots$$
$$(x_{1n}, x_{2n}, \ldots, x_{Kn}, y_n)$$

Referring to Table 13.1, we see for the savings and loan example that there are two independent variables and $n = 25$ observation points which, in the specified format, would be depicted

$$(x_{11}, x_{21}, y_1) = (3.92, 7{,}298, 0.75)$$

$$(x_{12}, x_{22}, y_2) = (3.61, 6{,}855, 0.71)$$

$$\vdots$$

$$(x_{1n}, x_{2n}, y_n) = (4.78, 9{,}318, 0.32)$$

Given these data, our problem is to find estimates of the unknown parameters α, β_1, $\beta_2, \ldots, \beta_K$ of the population multiple regression model.

Consider, as possible estimates of these unknown parameters, the numbers a, b_1, $b_2, \ldots, b_K$, so that the estimated model is

$$\hat{y} = a + b_1 x_1 + b_2 x_2 + \cdots + b_K x_K$$

Then, for the observed values of the independent variables, the values predicted for the dependent variable by this estimated model are $(a + b_1 x_{1i} + b_2 x_{2i} + \cdots + b_K x_{Ki})$, whereas the values actually observed are y_i. Hence, the discrepancies between the observed and predicted outcomes for the dependent variable are

$$e_i = y_i - (a + b_1 x_{1i} + b_2 x_{2i} + \cdots + b_K x_{Ki}) \qquad (i = 1, 2, \ldots, n)$$

Precisely as for the linear regression model, the least squares estimates are those numbers a, b_1, $b_2, \ldots, b_K$ for which the sum of squared discrepancies

$$SS = \sum_{i=1}^{n} e_i^2 = \sum_{i=1}^{n} (y_i - a - b_1 x_{1i} - b_2 x_{2i} - \cdots - b_K x_{Ki})^2$$

is as small as possible.

Least Squares Estimation and the Sample Multiple Regression

Let $(x_{11}, x_{21}, \ldots, x_{K1}, y_1), (x_{12}, x_{22}, \ldots, x_{K2}, y_2), \ldots, (x_{1n}, x_{2n}, \ldots, x_{Kn}, y_n)$ be a sample of n sets of observations on a process with population multiple regression

$$Y_i = \alpha + \beta_1 x_{1i} + \beta_2 x_{2i} + \cdots + \beta_K x_{Ki} + \varepsilon_i$$

The **least squares estimates** of the coefficients α, β_1, $\beta_2, \ldots, \beta_K$ are those values, a, b_1, $b_2, \ldots, b_K$, for which the sum of squared discrepancies

$$SS = \sum_{i=1}^{n} (y_i - a - b_1 x_{1i} - b_2 x_{2i} - \cdots - b_K x_{Ki})^2 \qquad (13.2.1)$$

is a minimum.

The equation

$$y = a + b_1 x_1 + b_2 x_2 + \cdots + b_K x_K$$

represents the **sample multiple regression** of Y on $X_1, X_2, \ldots, X_K$.

Unless the number, K, of independent variables is very small, expressions for the computation of the least squares estimates can be extremely cumbersome.[2] For the interested reader, some algebraic details are included in Appendix A13.1 at the end of this chapter. Fortunately, from a practical point of view, this matter need not detain us here. The fitting of regression models is such a commonly used tool in so many branches of applied statistics that computer packages for performing the necessary arithmetic have long been widely available. A fuller discussion of such packages will be postponed until additional concepts have been developed. It is sufficient to note for the present that the operation involved in computing the least squares estimates is the minimization of the sum of squared discrepancies in equation 13.2.1.

Based on the data in Table 13.1, the least squares estimates of the parameters of the regression of savings and loan profit margins on net revenues and number of offices were found to be

$$a = 1.565; \qquad b_1 = 0.237; \qquad b_2 = -0.000249$$

The sample multiple regression is therefore

$$y = 1.565 + 0.237x_1 - 0.000249x_2$$

We therefore estimate that, for a fixed number of offices, an increase of 1% in net revenues leads to an expected increase of 0.237 in percentage profit margin. Similarly, we estimate that, for a fixed level of net revenues, an increase of 1,000 in the number of offices leads to an expected decrease of 0.249 in percentage profit margin.

When, as in the present example, the regression model contains two independent variables, the least squares estimates can be computed through a moderate amount of arithmetic effort, without recourse to a computer program. The details are set out in Appendix A13.1.

13.3 STANDARD ASSUMPTIONS FOR THE MULTIPLE REGRESSION MODEL

Before continuing our discussion of the practice of regression analysis, we introduce in the box some assumptions that are reasonable in many applications. When these assumptions do hold, a powerful justification for the use of least squares estimation exists.

Standard Assumptions

Denote the population multiple regression by

$$Y_i = \alpha + \beta_1 x_{1i} + \beta_2 x_{2i} + \cdots + \beta_K x_{Ki} + \varepsilon_i$$

---

[2] In fact, algebraically manageable expressions can be achieved through the use of matrix algebra. This approach, however, is beyond the scope of this text.

and assume that n sets of observations are available. The following standard assumptions are often made:

1. Either the $x_{1i}, x_{2i}, \ldots, x_{Ki}$ are fixed numbers or they are realizations of random variables $X_{1i}, X_{2i}, \ldots, X_{Ki}$, which are independent of the error terms ε_i. In the latter case, inference is carried out conditionally on the observed $x_{1i}, x_{2i}, \ldots, x_{Ki}$.

2. The error terms ε_i are random variables with mean 0, that is

$$E(\varepsilon_i) = 0 \qquad (i = 1, 2, \ldots, n)$$

3. The random variables ε_i all have the same variance, say σ_ε^2, so that

$$E(\varepsilon_i^2) = \sigma_\varepsilon^2 \qquad (i = 1, 2, \ldots, n)$$

4. The random variables ε_i are not correlated with one another, so that

$$E(\varepsilon_i \varepsilon_j) = 0 \text{ for all } i \neq j$$

5. It is not possible to find a set of numbers, $c_0, c_1, c_2, \ldots, c_K$, such that

$$c_0 + c_1 x_{1i} + c_2 x_{2i} + \cdots + c_K x_{Ki} = 0$$

for every $i = 1, 2, \ldots, n$.

The first four assumptions are essentially the same as those made in Section 12.5 for the linear regression model. The purpose of assumption 5 is to exclude from consideration certain pathological cases in which unique least squares estimates do not exist. To illustrate, suppose we are interested in explaining the variability in rates charged for shipping corn. One obvious explanatory variable would be the distance that the shipment is to travel. We can measure distance in any convenient units. However, there is clearly no point in including in the regression two separate measures of distance, such as distance in miles (x_1) and distance in kilometers (x_2). These two independent variables are really just alternative measures of the same quantity—distance—and it would be foolish to try to assess their *separate* effects. It is this kind of situation that is excluded by the fifth assumption. As a practical matter, it will be safe to assume (provided the regression model is sensibly specified) that assumption 5 is not violated.

In the remainder of this chapter, we will take it that all of the standard assumptions hold and in the next chapter we will deal with the consequences of their breakdown.

13.4 THE GAUSS-MARKOV THEOREM

In this section we extend the result of Section 12.6. Suppose that we want to use the observations $(x_{11}, x_{21}, \ldots, x_{K1}, y_1), (x_{12}, x_{22}, \ldots, x_{K2}, y_2), \ldots, (x_{1n}, x_{2n}, \ldots, x_{Kn}, y_n)$ to estimate the parameters of the multiple regression model. As in Section

12.6, we restrict attention to those estimates that are linear functions of the y_i—that is, to estimates of the form

$$c_1 y_1 + c_2 y_2 + \cdots + c_n y_n$$

where the c_i are numbers that do not depend on the y_i. The Gauss-Markov theorem asserts that of all possible estimates of this type, the least squares estimates are best, in a sense, as described in the box.

Gauss-Markov Theorem

Denote the population multiple regression by

$$Y_i = \alpha + \beta_1 x_{1i} + \beta_2 x_{2i} + \cdots + \beta_K x_{Ki} + \varepsilon_i$$

and assume that n sets of observations are available. Suppose, further, that assumptions 1–5 of Section 13.3 hold.

Then, of all possible estimators of $\alpha, \beta_1, \beta_2, \ldots, \beta_K$, which are linear in the Y_i and unbiased, the least squares estimators (that is, the random variables corresponding to the least squares estimates $a, b_1, b_2, \ldots, b_K$ of Section 13.2) have smallest variances.

Further, if $d_0, d_1, d_2, \ldots, d_K$ are any fixed numbers, and we want to estimate

$$d_0 \alpha + d_1 \beta_1 + d_2 \beta_2 + \cdots + d_K \beta_K$$

the estimator corresponding to

$$d_0 a + d_1 b_1 + d_2 b_2 + \cdots + d_K b_K$$

has smallest variance in the class of all estimators that are linear in the Y_i and unbiased.

By virtue of this theorem, least squares estimators are said to be **best linear unbiased estimators** (BLUE).

Now, in certain circumstances it may be entirely reasonable to prefer a biased estimator to the best available unbiased estimator. Nevertheless, the Gauss-Markov theorem provides a strong justification for the use of the least squares method of estimation, and provides the bulk of the theoretical support for the broad popularity of this procedure in practical applications. The Gauss-Markov theorem asserts that, from a very wide class of unbiased estimators, the least squares estimators are the most efficient. Thus, exactly as in our discussion of simple linear regression in the previous chapter, the Gauss-Markov theorem elevates the status of the method of least squares from an apparently sensible method for obtaining point estimates to a procedure for generating point estimators with optimal statistical properties.

We hasten to add that the method of least squares is not invariably the best approach to the problem of estimating the coefficients of a regression model. We have seen that there is strong justification for using this approach when the standard assumptions of the previous section hold. However, if one or more of these assumptions break down, alternative estimators may (as will be discussed in the next chapter) be preferable.

13.5 THE EXPLANATORY POWER OF A MULTIPLE REGRESSION EQUATION

The purpose of regression analysis is to use the independent variables to explain the behavior of the dependent variable. Variability in the dependent variable can, in part, be explained by its linear association with the independent variables. In this section we discuss a measure of the proportion of the variability, in the sample, of the dependent variable that is explained by the estimated multiple regression.

For the sample, the estimated regression is written

$$y_i = a + b_1 x_{1i} + b_2 x_{2i} + \cdots + b_K x_{Ki} + e_i$$

Alternatively, we may write

$$y_i = \hat{y}_i + e_i \qquad (13.5.1)$$

where

$$\hat{y}_i = a + b_1 x_{1i} + b_2 x_{2i} + \cdots + b_K x_{Ki}$$

is the value of the dependent variable predicted by the regression and the **residual** e_i is the difference between the observed and predicted value. For the savings and loan data, the three quantities in equation 13.5.1 are given in the first three columns of Table 13.2.

As in Section 12.7, we subtract the sample mean of the dependent variable from each side of equation 13.5.1, giving

$$(y_i - \bar{y}) = (\hat{y}_i - \bar{y}) + e_i \qquad (13.5.2)$$

which can be expressed as

$$\text{Observed deviation from sample mean}$$
$$= \text{Predicted deviation from sample mean} + \text{Residual}$$

For our example, the sample mean for the dependent variable is

$$\bar{y} = \frac{\Sigma y_i}{n} = \frac{16.86}{25} = 0.674$$

Using this value, we have computed the first two terms of equation 13.5.2 for the savings and loan data in the final two columns of Table 13.2.

It can be shown that squaring both sides of equation 13.5.2 and summing over the index i yields

TABLE 13.2 Actual values, predicted values, and residuals for savings and loan regression

y_i	$\hat{y}_i = a + b_1x_{1i} + b_2x_{2i}$ $= 1.565 + 0.237x_{1i} - 0.000249x_{2i}$	e_i $= y_i - \hat{y}_i$	$y_i - \bar{y}$ $= y_i - 0.674$	$\hat{y}_i - \bar{y}$ $= \hat{y}_i - 0.674$
0.75	0.677	0.073	0.076	0.003
0.71	0.714	−0.004	0.036	0.040
0.66	0.699	−0.039	−0.014	0.025
0.61	0.673	−0.063	−0.064	−0.001
0.70	0.684	0.016	0.026	0.010
0.72	0.708	0.012	0.046	0.034
0.77	0.740	0.030	0.096	0.066
0.74	0.759	−0.019	0.066	0.085
0.90	0.795	0.105	0.226	0.121
0.82	0.794	0.026	0.146	0.120
0.75	0.799	−0.049	0.076	0.125
0.77	0.828	−0.058	0.096	0.154
0.78	0.802	−0.022	0.106	0.128
0.84	0.800	0.040	0.166	0.126
0.79	0.755	0.035	0.116	0.081
0.70	0.734	−0.034	0.026	0.060
0.68	0.705	−0.025	0.006	0.031
0.72	0.693	0.027	0.046	0.019
0.55	0.635	−0.085	−0.124	−0.039
0.63	0.613	0.017	−0.044	−0.061
0.56	0.570	−0.010	−0.114	−0.104
0.41	0.480	−0.070	−0.264	−0.194
0.51	0.438	0.072	−0.164	−0.236
0.47	0.396	0.074	−0.204	−0.278
0.32	0.378	−0.058	−0.354	−0.296
Sum 16.86				

$$\sum_{i=1}^{n} (y_i - \bar{y})^2 = \sum_{i=1}^{n} (\hat{y}_i - \bar{y})^2 + \sum_{i=1}^{n} e_i^2$$

This is precisely the sum of squares decomposition obtained in Section 12.7, and it is interpreted here in an analogous fashion, as indicated in the box.

Sum of Squares Decomposition and the Coefficient of Determination

Suppose that a multiple regression model is fitted by least squares to n sets of observations, yielding

$$y_i = a + b_1x_{1i} + b_2x_{2i} + \cdots + b_Kx_{Ki} + e_i = \hat{y}_i + e_i$$

where $a, b_1, b_2, \ldots, b_K$ are the least squares estimates of the parameters of the population regression model, and e_i are the **residuals** from the estimated regression.

We define the following quantities (where $\bar{y}$ is the sample mean for the dependent variable):

TOTAL SUM OF SQUARES: $\quad \text{SST} = \sum_{i=1}^{n} (y_i - \bar{y})^2$

$$\text{REGRESSION SUM OF SQUARES:} \quad SSR = \sum_{i=1}^{n} (\hat{y}_i - \bar{y})^2$$

$$\text{ERROR SUM OF SQUARES:} \quad SSE = \sum_{i=1}^{n} e_i^2$$

It can then be shown that

$$SST = SSR + SSE$$

In terms of the observed values of the dependent variable, this decomposition can be interpreted as

$$\text{Total sample variability} = \text{Explained variability} + \text{Unexplained variability}$$

The **coefficient of determination, R^2**, of the fitted regression is defined as the proportion of the total sample variability explained, that is,

$$R^2 = \frac{SSR}{SST} = 1 - \frac{SSE}{SST}$$

It necessarily follows from this definition that

$$0 \leq R^2 \leq 1$$

From Table 13.2, we find (on summing the squares of the elements of the third and fourth columns) that, for our savings and loan example,

$$SSE = \sum_{i=1}^{n} e_i^2 = 0.0623 \quad \text{and} \quad SST = \sum_{i=1}^{n} (y_i - \bar{y})^2 = 0.4640$$

The coefficient of determination is, therefore,

$$R^2 = 1 - \frac{SSE}{SST} = 1 - \frac{0.0623}{0.4640} = 0.87$$

Thus, in this sample, 87% of the variability in savings and loan associations' profit margins is explained by their linear association with net revenues and number of offices.

If the regression sum of squares is needed, we do not have to go to the trouble of summing the squares of the terms in the final column of Table 13.2. Rather, we can obtain this quantity through

$$SSR = SST - SSE$$

so that, for our example,

$$SSR = 0.4640 - 0.0623 = 0.4017$$

The sum of squared errors is also useful in estimating the variance of the error terms in the population regression model, as described in the box.

Estimation of the Error Variance

Suppose that the population regression model is

$$Y_i = \alpha + \beta_1 x_{1i} + \beta_2 x_{2i} + \cdots + \beta_K x_{Ki} + \varepsilon_i$$

and that assumptions 1–5 of Section 13.3 hold. Let σ_ε^2 denote the common variance of the error terms ε_i. Then an unbiased estimate of σ_ε^2 is provided by[3]

$$s_e^2 = \frac{\sum_{i=1}^{n} e_i^2}{(n - K - 1)} = \frac{\text{SSE}}{(n - K - 1)}$$

The major use of the coefficient of determination is as a descriptive statistic that provides a measure of the success of the independent variables in explaining the behavior of the dependent variable. While it can certainly be valuable in this context, its use can be criticized in circumstances where the number, K, of independent variables is not a small proportion of the number, n, of data points. In this case the model can appear to fit the data rather well, even when in truth the independent variables are not strongly linked to the dependent variable. As a trivial example, we can fit perfectly a straight line to just two points. In order to alleviate this problem somewhat, a modified measure of the strength of the regression relation is sometimes calculated. Essentially, the idea is that, in the expression

$$R^2 = 1 - \frac{\text{SSE}}{\text{SST}}$$

some compensation for the inevitable reduction in the sum of squared errors occasioned by the addition of more independent variables (relevant or not) to the regression equation can be achieved by dividing each of the sums of squares by their appropriate number of degrees of freedom. We have already seen that, in order to obtain an unbiased estimate of the error variance, SSE should be divided by $(n - K - 1)$. Further, division of SST by $(n - 1)$ provides an unbiased estimate of the variance of the dependent variable when the observations constitute a random sample. The introduction of these modifications produces an **adjusted, or corrected, coefficient of determination.**

Definition

Suppose that a multiple regression relates a dependent variable to K independent variables. The **adjusted (or corrected) coefficient of determination, $\bar{R}^2$,** is defined as

---

[3] Intuitively, the reason for division by $(n - K - 1)$ is that $(K + 1)$ degrees of freedom are "lost" in the estimation of the unknown population parameters $\alpha, \beta_1, \beta_2, \ldots, \beta_K$.

$$\bar{R}^2 = 1 - \frac{\text{SSE}/(n - K - 1)}{\text{SST}/(n - 1)}$$

For our example,

$$n = 25; \quad K = 2; \quad \text{SSE} = 0.0623; \quad \text{SST} = 0.4640$$

The adjusted coefficient of determination is therefore

$$\bar{R}^2 = 1 - \frac{0.0623/22}{0.4640/24} = 0.85$$

In this particular case the adjustment is very minor. More severe adjustments result when the number of independent variables is a larger proportion of the number of sample points.

EXERCISES

1. In a study[4] of the influence of financial institutions on share prices in the United Kingdom, quarterly data for a 12-year period were analyzed. The postulated model was

$$Y_i = \alpha + \beta_1 x_{1i} + \beta_2 x_{2i} + \varepsilon_i$$

where

Y_i = Change over the quarter in the Financial Times stock price index
x_{1i} = Change over the quarter in equity purchases by financial institutions
x_{2i} = Change over the quarter in equity sales by financial institutions

The estimated partial regression coefficients were

$$b_1 = 0.057 \quad \text{and} \quad b_2 = -0.065$$

Interpret these estimates.

2. The following model was formulated to explain sales of imported cars in the United States:

$$Y_i = \alpha + \beta_1 x_{1i} + \beta_2 x_{2i} + \beta_3 x_{3i} + \beta_4 x_{4i} + \varepsilon_i$$

where

Y_i = Sales of imported cars as a percentage of total car sales in the United States
x_{1i} = Average real price of gasoline (in dollars per gallon)
x_{2i} = Ratio of average United States car prices to average foreign car prices
x_{3i} = Ratio of productivity of foreign to United States car workers
x_{4i} = Difference between average gas mileage rating of United States and imported cars

The estimated partial regression coefficients were

$$b_1 = 10.9; \quad b_2 = 3.6; \quad b_3 = 0.2; \quad b_4 = 3.9$$

Interpret these estimates.

---

[4] R. Dobbins and S. F. Witt, "Stock market prices and sector activity," *Journal of Business, Finance and Accounting, 7* (1980), 261–76.

3. The following model was fitted to a sample of supermarkets in order to explain profit levels:

$$Y_i = \alpha + \beta_1 x_{1i} + \beta_2 x_{2i} + \beta_3 x_{3i} + \varepsilon_i$$

where

Y_i = Profits (in thousands of dollars)
x_{1i} = Food sales (in tens of thousands of dollars)
x_{2i} = Nonfood sales (in tens of thousands of dollars)
x_{3i} = Store size (in thousands of square feet)

The estimated partial regression coefficients were

$$b_1 = 0.027; \qquad b_2 = 0.097; \qquad b_3 = 0.525$$

Interpret these estimates.

4. A small commuter airline formulates the following model to explain ridership on its flights:

$$Y_i = \alpha + \beta_1 x_{1i} + \beta_2 x_{2i} + \beta_3 x_{3i} + \varepsilon_i$$

where

Y_i = Number of passengers on flight
x_{1i} = Price of ticket (in dollars)
x_{2i} = Price of gasoline (in dollars per gallon)
x_{3i} = Length of flight (in minutes)

The estimated partial regression coefficients were

$$b_1 = -0.92; \qquad b_2 = 22.66; \qquad b_3 = 5.80$$

Interpret these estimates.

5. In the study of Exercise 2, twenty observations were used to calculate the least squares estimates. The total sum of squares and error sum of squares were found to be

$$\text{SST} = 980 \quad \text{and} \quad \text{SSE} = 49$$

(a) Find and interpret the coefficient of determination.
(b) Find the corrected coefficient of determination.

6. In the study of Exercise 3, twenty-five observations were used to calculate the least squares estimates. The total sum of squares and regression sum of squares were found to be

$$\text{SST} = 656.98 \quad \text{and} \quad \text{SSR} = 610.48$$

(a) Find and interpret the coefficient of determination.
(b) Find the corrected coefficient of determination.

7. In the study of Exercise 4, thirty observations were used to calculate the least squares estimates. The regression sum of squares and error sum of squares were found to be

$$\text{SSR} = 620.91 \quad \text{and} \quad \text{SSE} = 56.09$$

(a) Find and interpret the coefficient of determination.
(b) Find the corrected coefficient of determination.

13.6 CONFIDENCE INTERVALS AND HYPOTHESIS TESTS FOR INDIVIDUAL REGRESSION PARAMETERS

Up to this stage we have discussed point estimation of the parameters of the multiple regression model

$$Y_i = \alpha + \beta_1 x_{1i} + \beta_2 x_{2i} + \cdots + \beta_K x_{Ki} + \varepsilon_i$$

In this section we extend our results to develop confidence intervals and tests of hypotheses.

The least squares estimators are unbiased for the corresponding population parameters. Expressions for the variances of these estimators can be derived, but their precise algebraic forms are rather complicated. However, multiple regression computer packages routinely calculate unbiased estimates, s_a^2, $s_{b_1}^2$, $s_{b_2}^2$, . . . , $s_{b_K}^2$, for the variances of the least squares estimators of the population parameters. In fact, it is more usual to report the square roots of these quantities, the estimated standard deviations, or **standard errors.** For the savings and loan data, the values

$$s_a = 0.079; \qquad s_{b_1} = 0.0555; \qquad s_{b_2} = 0.0000320$$

were obtained. Thus, for example, the standard deviation of the sampling distribution of the least squares estimator of β_1 is estimated by 0.0555.

In reporting estimated regressions, it is common practice to include these standard deviations in parentheses beneath the corresponding estimated parameters, and also to include either the coefficient of determination or the adjusted coefficient. Thus, the savings and loan example results would be reported as

$$y = \underset{(0.079)}{1.565} + \underset{(0.0555)}{0.237 x_1} - \underset{(0.0000320)}{0.000249 x_2}; \qquad R^2 = 0.87$$

The reader is cautioned, however, that this convention is not universal. Some authors report the ratios of estimated coefficients to their estimated standard deviations instead of the estimated standard deviations. Also, the estimated standard deviation, s_a, of the intercept term is frequently omitted.

At this point, in order to make further progress, an assumption about the distribution of the population error terms, ε_i, is needed. It is common practice to assume their distribution is normal, and to base inference on this assumption. In fact, by virtue of the Central Limit Theorem, the resulting confidence intervals and hypothesis tests are not very seriously affected by moderate departures from normality in the distribution of the error terms. Moreover, the Central Limit Theorem provides, in many circumstances, some justification for the assumption that the error terms are normal. We might think of the error term in the population regression model as representing the sum of the influences on the dependent variable of a multitude of factors not specifically accounted for in the list of independent variables. Now, individually, none of these factors should exert a strong influence, as such influential factors should be included among the independent variables. However, their joint effect may be nontrivial. Thus, since the error term is made up of the sum of a large number of components of this sort, the Central Limit Theorem suggests that its distribution will often be close to normal. The main result, on which inference about the parameters of the population multiple regression model is based, is stated in the box.

Basis for Inference About the Population Regression Parameters

Let the population regression model be

$$Y_i = \alpha + \beta_1 x_{1i} + \beta_2 x_{2i} + \cdots + \beta_K x_{Ki} + \varepsilon_i$$

Let $a, b_1, b_2, \ldots, b_K$ be the least squares estimates of the population parameters and $s_a, s_{b_1}, s_{b_2}, \ldots, s_{b_K}$ the estimated standard deviations of the least squares estimators. Then, if assumptions 1–5 of Section 13.3 hold, and if the error terms ε_i are normally distributed, the random variables corresponding to

$$t_a = \frac{a - \alpha}{s_a}$$

and

$$t_{b_i} = \frac{b_i - \beta_i}{s_{b_i}} \qquad (i = 1, 2, \ldots, K)$$

are distributed as Student's t with $(n - K - 1)$ degrees of freedom.

Typically, interest is in the partial regression coefficients, β_i, rather than the intercept, α. Accordingly, we will concentrate on the former, noting that inference about the latter proceeds along similar lines.

Confidence intervals for the β_i can be derived using familiar arguments. The procedure is summarized in the next box.

Confidence Intervals for the Partial Regression Coefficients

If the population regression errors, ε_i, are normally distributed and assumptions 1–5 of Section 13.3 hold, then $100(1 - \alpha)\%$ confidence intervals for the partial regression coefficients, β_i, are given by

$$b_i - t_{n-K-1,\alpha/2} s_{b_i} < \beta_i < b_i + t_{n-K-1,\alpha/2} s_{b_i}$$

where $t_{n-K-1,\alpha/2}$ is that number for which

$$P(t_{n-K-1} > t_{n-K-1,\alpha/2}) = \frac{\alpha}{2}$$

and the random variable t_{n-K-1} follows a Student's t distribution with $(n - K - 1)$ degrees of freedom.

For the savings and loan regression, we have found

$$n = 25; \quad b_1 = 0.237; \quad s_{b_1} = 0.0555; \quad b_2 = -0.000249 ; \quad s_{b_2} = 0.0000320$$

To obtain 99% confidence intervals for β_1 and β_2, we have from Table 6 of the Appendix

$$t_{n-K-1,\alpha/2} = t_{22,.005} = 2.819$$

Hence, the 99% confidence interval for β_1 is

$$0.237 - (2.819)(0.0555) < \beta_1 < 0.237 + (2.819)(0.0555)$$

or

$$0.081 < \beta_1 < 0.393$$

Thus, the 99% confidence interval for the expected increase in savings and loan profit margins resulting from a 1 unit increase in net revenues, given a fixed number of offices, runs from 0.081 to 0.393. The 99% confidence interval for β_2 is

$$-0.000249 - (2.819)(0.0000320) < \beta_2 < -0.000249 + (2.819)(0.0000320)$$

or

$$-0.000339 < \beta_2 < -0.000159$$

Therefore, the 99% confidence interval for the expected *decrease* in savings and loan profit margins resulting from an increase of 1,000 offices, for a fixed level of net revenues, runs from 0.159 to 0.339.

Tests of hypotheses about individual regression parameters can also be obtained. We consider in the box the hypothesis that the parameter β_i is equal to some specific value $\beta_{i,0}$.

Tests of Hypotheses for the Partial Regression Coefficients

If the regression errors, ε_i, are normally distributed and assumptions 1–5 of Section 13.3 hold, then the following tests have significance level α:

(i) To test either null hypothesis

$$H_0: \quad \beta_i = \beta_{i,0} \qquad \text{or} \qquad H_0: \quad \beta_i \leq \beta_{i,0}$$

against the alternative

$$H_1: \quad \beta_i > \beta_{i,0}$$

the decision rule is

$$\text{Reject } H_0 \text{ if } \quad \frac{b_i - \beta_{i,0}}{s_{b_i}} > t_{n-K-1,\alpha}$$

(ii) To test either null hypothesis

$$H_0: \quad \beta_i = \beta_{i,0} \qquad \text{or} \qquad H_0: \quad \beta_i \geq \beta_{i,0}$$

---

against the alternative

$$H_1: \quad \beta_i < \beta_{i,0}$$

the decision rule is

$$\text{Reject } H_0 \text{ if } \quad \frac{b_i - \beta_{i,0}}{s_{b_i}} < -t_{n-K-1,\alpha}$$

(iii) To test the null hypothesis

$$H_0: \quad \beta_i = \beta_{i,0}$$

against the two-sided alternative

$$H_1: \quad \beta_i \neq \beta_{i,0}$$

the decision rule is

$$\text{Reject } H_0 \text{ if } \quad \frac{b_i - \beta_{i,0}}{s_{b_i}} > t_{n-K-1,\alpha/2} \quad \text{or} \quad \frac{b_i - \beta_{i,0}}{s_{b_i}} < -t_{n-K-1,\alpha/2}$$

An important special case arises when the hypothesized value for an individual parameter is 0. In this case, all else equal, the expected value of the dependent variable will not be affected by a change in the corresponding independent variable. For example, suppose that in the regression

$$Y_i = \alpha + \beta_1 x_{1i} + \beta_2 x_{2i} + \cdots + \beta_K x_{Ki} + \varepsilon_i$$

the true value of the parameter β_1 is 0. The implication is that, given that $X_2, \ldots, X_K$ are also to be used, information on X_1 contributes nothing further, at least within a linear framework, toward explaining the behavior of the dependent variable.

We will test, for the savings and loan data, the hypothesis

$$H_0: \quad \beta_1 = 0$$

It has already been noted that a reasonable conjecture is that, all else equal, an increase in net revenues might be expected to lead to an increase in profit margins. Hence, the appropriate alternative hypothesis is

$$H_1: \quad \beta_1 > 0$$

The test is based on

$$\frac{b_1 - \beta_{1,0}}{s_{b_1}} = \frac{0.237 - 0}{0.0555} = 4.27$$

From Table 6 of the Appendix, we have for $(n - K - 1) = 22$ degrees of freedom,

$$t_{22,.005} = 2.819$$

Thus, the null hypothesis that net revenues do not contribute toward explaining the behavior of profit margins, given that number of offices is also used as an explanatory variable, is clearly rejected, even at the 0.5% significance level.

Next we test, using these data, the hypothesis

$$H_0: \quad \beta_2 = 0$$

In fact, it is suspected that, all other things the same, an increase in the number of offices will lead, on the average, to a decrease in the profit margins of savings and loan associations. Thus, as an alternative hypothesis, we employ

$$H_1: \quad \beta_2 < 0$$

The test is based on

$$\frac{b_2 - \beta_{2,0}}{s_{b_2}} = \frac{-0.000249 - 0}{0.0000320} = -7.78$$

Comparing this with tabulated values of Student's t for 22 degrees of freedom, we see that the null hypothesis that the number of offices does not contribute toward explaining the behavior of profit margins, given that net revenues are also used as an explanatory variable, is very clearly rejected at the 0.5% significance level.

Example 13.1
A study was aimed at determining whether certain of their features could be used to explain variability in the prices of air conditioners. For a sample of nineteen air conditioners, the following regression was estimated:[5]

$$y = -68.236 + 0.023 x_1 + 19.729 x_2 + 7.653 x_3; \quad R^2 = 0.84$$
$$\qquad\qquad\quad (0.005) \qquad (8.992) \qquad (3.082)$$

where

$y =$ Price (in dollars)
$x_1 =$ BTU per hour rating of air conditioner
$x_2 =$ Energy efficiency ratio
$x_3 =$ Number of settings

The figures in parentheses beneath the least squares parameter estimates are the corresponding estimated standard deviations.

The estimates have the following implications:

(a) All else equal, an increase of 1 unit in BTU per hour rating leads to an expected increase of $0.023 in price.
(b) All else equal, an increase of 1 unit in the energy efficiency ratio leads to an expected increase of $19.729 in price.
(c) All else equal, an increase of 1 in the number of settings leads to an expected increase of $7.653 in price.
(d) Taken together, these three independent variables explain 84% of the variability in prices of air conditioners in this sample.

---

[5] See B. T. Ratchford, "The value of information for selected appliances," *Journal of Marketing Research, 17* (1980), 14–25.

We will now find confidence intervals for β_3, the expected increase in price for an additional setting, when the values of the other independent variables are held fixed. The appropriate intervals, based on the usual assumptions, are of the form

$$b_3 - t_{n-K-1,\alpha/2}s_{b_3} < \beta_3 < b_3 + t_{n-K-1,\alpha/2}s_{b_3}$$

where

$$n = 19; \quad K = 3; \quad b_3 = 7.653, \quad s_{b_3} = 3.082$$

For a 95% confidence interval, $\alpha/2 = .025$, so from Table 6 of the Appendix,

$$t_{n-K-1,\alpha/2} = t_{15,.025} = 2.131$$

The interval is, therefore

$$7.653 - (2.131)(3.082) < \beta_3 < 7.653 + (2.131)(3.082)$$

or

$$1.085 < \beta_3 < 14.221$$

Thus, our 95% confidence interval for the expected increase in price resulting from an additional setting, when the values of the BTU per hour rating and the energy efficiency ratio remain fixed, runs from \$1.085 to \$14.221. Figure 13.5 shows also 80%, 90%, and 99% confidence intervals.

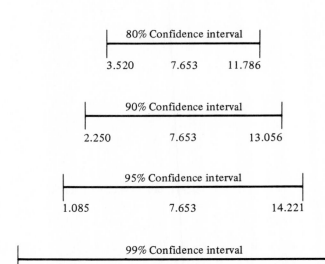

FIGURE 13.5 80%, 90%, 95%, and 99% confidence intervals for β_3, the slope of the partial regression of price of air conditioners on number of settings, using the data of Example 13.1

Example
13.2

The following regression model was fitted[6] to data on thirty-one stock repurchase cash tender offers—that is, offers by corporations repurchasing large blocks of their own outstanding shares:

$$y = 0.69 + 4.70x_1 + 0.00041x_2 - 0.72x_3 + 0.023x_4; \qquad R^2 = 0.72$$
$$\quad (0.651) \quad (0.00128) \quad\;\; (0.353) \quad (0.0185)$$

where

y = Tender premium, as percentage of closing market price 1 week prior to offer date
x_1 = Soliciting fee per share, as percentage of closing market price 1 week prior to offer date (This provides a measure of the perceived degree of difficulty in obtaining the shares)
x_2 = Percentage of shares sought
x_3 = Relative monthly change in the Dow Jones industrial average
x_4 = Volume of shares traded as percentage of those outstanding

The figures in brackets under the least squares parameter estimates are the corresponding estimated standard deviations.

The reported estimates imply that:

(a) All else equal, an increase of one unit in the soliciting fee per share leads to an expected increase of 4.70 in the tender premium
(b) All else equal, an increase of one in the percent of shares sought leads to an expected increase of 0.00041 in the tender premium
(c) All else equal, an increase of one unit in the relative monthly change of the Dow Jones industrial average leads to an expected decrease of 0.72 in the tender premium
(d) All else equal, an increase of one unit in the volume of shares traded leads to an expected increase of 0.023 in the tender premium. The authors of the study note that this result is surprising since, if trading volume is high, the corporation would be able to purchase more of the required shares on the open market, and hence presumably would not need to offer such a high tender premium. This would suggest, contrary to the above findings, that, all else equal, the higher the volume of shares traded, the lower the tender premium.
(e) Taken together, these four independent variables explain 72% of the variability in tender premiums in this sample.

We want to test the null hypothesis

$$H_0: \quad \beta_3 = 0$$

This is the hypothesis that, given that the other independent variables are included in the regression, the relative monthly change in the Dow Jones industrial average contributes nothing further toward explaining variability in tender premiums. The authors of the study note that, for this variable, a case could be made for either a positive or negative relationship with cash tender premium. Therefore, we will test the null hypothesis against the two-sided alternative

$$H_1: \quad \beta_3 \neq 0$$

---

[6]These regression results were reported by K. R. Ferris, A. Melnik, and A. Rappaport, "Factors influencing the pricing of stock repurchase tenders," *Quarterly Review of Economics and Business, 18* (1978), No. 1, 31–39.

The test is based on

$$\frac{b_3 - \beta_{3,0}}{s_{b_3}} = \frac{-0.72 - 0}{0.353} = -2.040$$

Here we have $(n - K - 1) = (31 - 4 - 1) = 26$ degrees of freedom. From Table 6 in the Appendix,

$$t_{26, .05} = 1.706 \qquad \text{and} \qquad t_{26, .025} = 2.056$$

Thus, the null hypothesis can be rejected against a two-sided alternative at the 10% but not at the 5% significance level. These data, then, contain moderately strong evidence against the hypothesis that, all else equal, the relative monthly change in the Dow Jones industrial average does not influence the tender premium.

We now test the null hypothesis

$$H_0: \quad \beta_4 = 0$$

that, given that the other independent variables are used, the inclusion of volume of shares traded in the regression contributes nothing more toward explaining variability in tender premiums. Again we test against the two-sided alternative, that is,

$$H_1: \quad \beta_4 \neq 0$$

The test is based on

$$\frac{b_4 - \beta_{4,0}}{s_{b_4}} = \frac{0.023 - 0}{0.0185} = 1.243$$

For $(n - K - 1) = 26$ degrees of freedom, we find from Table 6 of the Appendix,

$$t_{26, .10} = 1.315$$

Hence, the null hypothesis cannot be rejected against a two-sided alternative at the 20% significance level. This test is illustrated in Figure 13.6. We noted earlier that the sign of the least squares estimate b_4 was the opposite of what one would have expected on the basis of subject matter theory. However, it now emerges that the data contain little evidence to contradict the hypothesis that the population parameter β_4 is in fact 0. In such circumstances, common practice is to omit variables such as x_4 from the regression model and to reestimate the other parameters.

The hypothesis tests discussed in this section are tests on the *individual* regression parameters. We may also want to test the hypothesis that, *simultaneously,* two or more of these parameters are 0. Tests of such hypotheses will be discussed in the next section.

13.7 TESTS ON SETS OF REGRESSION PARAMETERS

In the previous section we showed how to perform tests on individual regression parameters. Often it is the case that one requires to test hypotheses that specify values for some or all of these parameters. Consider, once again, the model

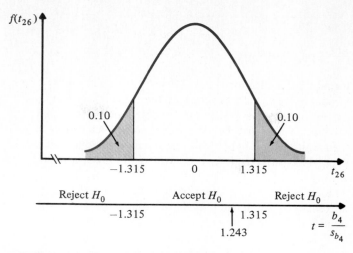

FIGURE 13.6 The probability density function of $t = b_4/s_{b_4}$ when the null hypothesis H_0: $\beta_4 = 0$ is true. (The distribution is the Student't t with 26 degrees of freedom in this case.) Illustrated is the test of the null hypothesis that β_4, the slope of the partial regression of tender premium on volume of shares traded, is 0, based on the data of Example 13.2. The alternative hypothesis is $\beta_4 \neq 0$, and the significance level is 0.20

$$Y_i = \alpha + \beta_1 x_{1i} + \beta_2 x_{2i} + \cdots + \beta_K x_{Ki} + \epsilon_i \qquad (13.7.1)$$

We begin by examining the rather pessimistic null hypothesis that *all* of the partial regression coefficients are 0, that is,

$$H_0: \quad \beta_1 = \beta_2 = \cdots = \beta_K = 0 \qquad (13.7.2)$$

If this hypothesis were true, the regression model 13.7.1 would reduce to

$$Y_i = \alpha + \varepsilon_i$$

implying that, *whatever the values of the independent variables,* the expected value of the dependent variable would be a constant number, α. This null hypothesis, then, implies that, taken as a group, the independent variables do not (linearly) influence the dependent variable. If this were the case, the attempt to explain the behavior of the dependent variable through the regression model would be regarded as a failure.

The basis of a test of this null hypothesis is provided by the sum of squares decomposition

$$SST = SSR + SSE \qquad (13.7.3)$$

introduced in Section 13.5. Recall that equation 13.7.3 expresses the total sample variability of the dependent variable as the sum of two parts. The first, SSR, is that part of total variability due to the regression on the independent variables, while the second, SSE, is that part of total variability that cannot be explained by the regression. Hence, if the null hypothesis that the independent variables do not (linearly) affect the dependent variable were true, we would expect SSR to be relatively small compared with SSE. Therefore, the larger the ratio of SSR to SSE, the less inclined would we be to believe the null hypothesis. In fact, in forming an appropriate test statistic, it is necessary to divide each of these sums of squares by their corresponding degrees of freedom. We have already seen that the degrees of freedom associated with SSE are $(n - K - 1)$. Also, since there are K partial regression coefficients, $\beta_1, \beta_2, \ldots, \beta_K$, there are K degrees of freedom associated with SSR.

It can be shown that, when the null hypothesis 13.7.2 is true, the random variable corresponding to

$$F = \frac{SSR/K}{SSE/(n - K - 1)} \qquad (13.7.4)$$

follows an[7] F distribution with numerator degrees of freedom K and denominator degrees of freedom $(n - K - 1)$. The verity of the null hypothesis would be in doubt if the regression sum of squares was large compared with the error sum of squares. Hence, the null hypothesis would be rejected for large values of the test statistic 13.7.4.

A Test on all the Parameters of a Regression Model

Consider the multiple regression model

$$Y_i = \alpha + \beta_1 x_{1i} + \beta_2 x_{2i} + \cdots + \beta_K x_{Ki} + \varepsilon_i$$

and the null hypothesis

$$H_0: \quad \beta_1 = \beta_2 = \cdots = \beta_K = 0$$

A test of H_0 against the alternative

$$H_1: \quad \text{At least one } \beta_i \neq 0$$

---

[7] The F distribution was introduced in Section 9.8. Cutoff points for this distribution are given in Table 7 of the Appendix.

at significance level α is based on the decision rule

$$\text{Reject } H_0 \text{ if } \quad \frac{\text{SSR}/K}{\text{SSE}/(n - K - 1)} > F_{K, n-K-1, \alpha}$$

where $F_{K, n-K-1, \alpha}$ is that number for which

$$P(F_{K, n-K-1} > F_{K, n-K-1, \alpha}) = \alpha$$

and the random variable $F_{K, n-K-1}$ follows an F distribution with numerator degrees of freedom K and denominator degrees of freedom $(n - K - 1)$.

The test can also be based directly on the coefficient of determination, R^2, since [8]

$$\frac{\text{SSR}/K}{\text{SSE}/(n - K - 1)} = \frac{(n - K - 1)}{K} \cdot \frac{R^2}{(1 - R^2)}$$

To illustrate this test, consider, for the savings and loan associations example, the null hypothesis

$$H_0: \quad \beta_1 = \beta_2 = 0$$

This is the hypothesis that, taken together, net revenues and the number of offices do not (linearly) affect profit margins. For these data, we have found

$$n = 25; \quad K = 2, \quad \text{SSE} = 0.0623, \quad \text{SST} = 0.4640$$

Hence, by subtraction,

$$\text{SSR} = \text{SST} - \text{SSE} = 0.4640 - 0.0623 = 0.4017$$

The test of the null hypothesis is then based on

$$\frac{\text{SSR}/K}{\text{SSE}/(n - K - 1)} = \frac{0.4017/2}{0.0623/22} = 70.92$$

From Table 7 of the Appendix, we find for a 1% level test,

$$F_{K, n-K-1, \alpha} = F_{2, 22, .01} = 5.72$$

Thus, the null hypothesis is very clearly rejected at the 1% significance level. The evidence against the null hypothesis that, taken together, net revenues per deposit dollar and the number of offices do not linearly influence percentage profit margins of savings and loan associations is overwhelming.

The calculations for this particular test are sometimes set out in a table called an

---

[8] This follows, as

$$R^2 = \frac{\text{SSR}}{\text{SST}} = 1 - \frac{\text{SSE}}{\text{SST}}$$

and hence,

$$\frac{R^2}{1 - R^2} = \frac{\text{SSR}/\text{SST}}{\text{SSE}/\text{SST}} = \frac{\text{SSR}}{\text{SSE}}$$

TABLE 13.3 Analysis of variance for savings and loan regression

SOURCES OF VARIATION	SUMS OF SQUARES	DEGREES OF FREEDOM	MEAN SQUARES	F RATIO
Regression	0.4017	2	0.20085	70.92
Error	0.0623	22	0.0028318	
Total	0.4640	24		

analysis of variance[9] table. Table 13.3 sets out the analysis of variance for the savings and loan data. The first column of numbers in the table lists the three sums of squares, while the second gives the corresponding degrees of freedom. The mean squares are the sums of squares divided by the associated degrees of freedom, while the ratio of these mean squares is given in the final column as the F ratio on which the test is based.

The logic behind the analysis of variance can be seen as follows. The error mean square is

$$s_e^2 = \frac{\text{SSE}}{(n - K - 1)}$$

As we saw in Section 13.5, this provides an unbiased estimate of σ_ε^2, the variance of the errors in the population regression model. It can further be shown that the regression mean square provides an unbiased estimate of the quantity

$$\sigma_\varepsilon^2 + \frac{\sum\limits_{i=1}^{n} [\beta_1(x_{1i} - \bar{x}_1) + \beta_2(x_{2i} - \bar{x}_2) + \cdots + \beta_K(x_{Ki} - \bar{x}_K)]^2}{K}$$

where $\bar{x}_j$ denotes the sample mean of the observations on the jth independent variable. Therefore, when the null hypothesis that all of the β_j are 0 is correct, the regression and error mean squares have the same expected value. However, when that hypothesis is not true, the expected value of the regression mean square exceeds that of the error mean square.

Example 13.3

For the data of Example 13.2 we will test the null hypothesis that the four independent variables, taken together, do not linearly affect the tender premium.

The null hypothesis, then, is

$$H_0: \quad \beta_1 = \beta_2 = \beta_3 = \beta_4 = 0$$

For these data, we have

$$n = 31; \quad K = 4; \quad R^2 = 0.72$$

The test of H_0 is based on

$$\frac{(n - K - 1)}{K} \cdot \frac{R^2}{(1 - R^2)} = \frac{26}{4} \cdot \frac{0.72}{0.28} = 16.7$$

_____

[9] We will discuss the analysis of variance in greater detail in Chapter 15.

From Table 7 of the Appendix, for a test at the 1% level, we have

$$F_{K, n-K-1, \alpha} = F_{4, 26, .01} = 4.14$$

The null hypothesis is therefore very clearly rejected at the 1% significance level. The evidence against the hypothesis that, taken together, the four independent variables do not linearly affect tender premiums is overwhelming.

So far we have developed tests of hypotheses for individual regression parameters and for all of the parameters taken together. As a final point, we note that it is also possible to test hypotheses about some subset of these parameters. Suppose that there are K independent variables and we are interested in the possibility that the coefficients on the first K_1 of them are 0. We denote the independent variables $X_1, X_2, \ldots, X_{K_1}, X_{K_1+1}, \ldots, X_K$, so that the regression model is

$$Y_i = \alpha + \beta_1 x_{1i} + \cdots + \beta_{K_1} x_{K_1 i} + \beta_{K_1+1} x_{K_1+1, i} \qquad (13.7.5)$$
$$+ \cdots + \beta_K x_{Ki} + \varepsilon_i$$

and the null hypothesis to be tested is

$$H_0: \quad \beta_1 = \beta_2 = \cdots = \beta_{K_1} = 0$$

If true, this hypothesis implies that, given that $X_{K_1+1}, \ldots, X_K$ are to be used, the addition of the independent variables $X_1, X_2, \ldots, X_{K_1}$ contributes nothing further toward explaining the behavior of the dependent variable. The test of this hypothesis is carried out by also estimating the regression of Y on $X_{K_1+1}, \ldots, X_K$ only, that is,

$$Y_i = \alpha^* + \beta^*_{K_1+1} x_{K_1+1, i} + \cdots + \beta^*_K x_{Ki} + \varepsilon^*_i \qquad (13.7.6)$$

where the purpose of the *s is to indicate that the quantities in 13.7.6 will differ from the corresponding values in 13.7.5 if the null hypothesis is false. Now, if H_0 is in fact true, we would not expect the sum of squared errors, SSE, from the full regression model 13.7.5 to differ much from the sum of squared errors, SSE*, from the regression model 13.7.6. The test is based on the difference between these quantities, and the decision rule is

$$\text{Reject } H_0 \text{ if} \quad \frac{(\text{SSE}^* - \text{SSE})/K_1}{\text{SSE}/(n - K - 1)} > F_{K_1, n-K-1, \alpha}$$

Example 13.4   A fast-food franchise operates in twenty regions of the country. It was believed that sales in these regions were likely to be influenced by promotional expenditures. Accordingly, a linear regression model was fitted to twenty pairs of observations. The dependent variable was the percentage increase in sales per outlet over the previous year and the independent variable was percentage change in promotional expenditures. The sum of squared errors from this fitted regression was found to be 78.21. An economist consulted by the franchise claimed that two other factors would also influence sales. Following the economist's advice, the franchise estimated a multiple regression equation. The dependent variable was defined as before. In addition to the independent variable previously used, two additional variables—the percentage change in regional unemployment rates and the percentage change in the numbers of outlets in the regions—were included. It was found that the sum of squared errors from this fitted multiple regression model was 66.73.

Test the null hypothesis that, given that promotional expenditures are used, unemployment and the number of outlets contribute nothing further toward explaining variability in sales.

The regression model here is

$$Y_i = \alpha + \beta_1 x_{1i} + \beta_2 x_{2i} + \beta_3 x_{3i} + \varepsilon_i$$

where

Y_i = Percentage increase in sales
x_{1i} = Percentage change in unemployment
x_{2i} = Percentage change in number of outlets
x_{3i} = Percentage change in promotional expenditures

The null hypothesis to be tested is

$$H_0: \quad \beta_1 = \beta_2 = 0$$

In the notation established previously, we have

$$n = 20; \quad K_1 = 2; \quad K = 3; \quad SSE^* = 78.21; \quad SSE = 66.73$$

Our test is based on

$$\frac{(SSE^* - SSE)/K_1}{SSE/(n - K - 1)} = \frac{(78.21 - 66.73)/2}{66.73/16} = 1.38$$

From Table 7 of the Appendix we find, for a 5% level test,

$$F_{K_1, n-K-1, \alpha} = F_{2, 16, .05} = 3.63$$

Thus, the null hypothesis that unemployment and the number of outlets contribute nothing further to explaining sales is not rejected by a 5% level test.

13.8 PREDICTION

As in the case of the linear regression model, an important use of the multiple regression model is in the prediction of values of the dependent variable for specified values of the set of independent variables. Suppose that the K independent variables are equal to the particular values $x_{1, n+1}, x_{2, n+1}, \ldots, x_{K, n+1}$. Assume further that the regression model continues to hold, so that the corresponding value for the dependent variable is

$$Y_{n+1} = \alpha + \beta_1 x_{1,n+1} + \beta_2 x_{2,n+1} + \cdots + \beta_K x_{K,n+1} + \varepsilon_{n+1} \qquad (13.8.1)$$

where ε_{n+1} is a random variable with mean 0. We want to predict the value Y_{n+1} will actually take. An obvious choice of predictor is obtained by replacing the unknown parameters, $\alpha, \beta_1, \beta_2, \ldots, \beta_K$, in 13.8.1 by their least squares estimates, $a, b_1, b_2, \ldots, b_K$, and the random variable ε_{n+1} by its expected value, 0.

Prediction from Multiple Regressions

Suppose that the population regression model

$$Y_i = \alpha + \beta_1 x_{1i} + \beta_2 x_{2i} + \cdots + \beta_K x_{Ki} + \varepsilon_i \qquad (i = 1, 2, \ldots, n + 1)$$

holds, and that the assumptions of Section 13.3 are valid. Let $a, b_1, b_2, \ldots, b_K$ be the least squares estimates of $\alpha, \beta_1, \beta_2, \ldots, \beta_K$, based on $(x_{11}, x_{21}, \ldots, x_{K1}, y_1), (x_{12}, x_{22}, \ldots, x_{K2}, y_2), \ldots, (x_{1n}, x_{2n}, \ldots, x_{Kn}, y_n)$. Then, given $x_{1,n+1}, x_{2,n+1}, \ldots, x_{K,n+1}$, the best linear unbiased predictor of Y_{n+1} is, by virtue of the Gauss-Markov theorem,

$$\hat{Y}_{n+1} = a + b_1 x_{1,n+1} + b_2 x_{2,n+1} + \cdots + b_K x_{K,n+1}$$

Returning once more to the savings and loan example, suppose we want to predict profit margins in a year in which net revenue is 4.50 and there are 9,000 offices. Then, in our notation,

$$x_{1,n+1} = 4.50 \qquad \text{and} \qquad x_{2,n+1} = 9,000$$

Our optimal point predictor of profit margins is, therefore,

$$\hat{Y}_{n+1} = a + b_1 x_{1,n+1} + b_2 x_{2,n+1}$$
$$= 1.565 + (0.237)(4.50) - (0.000249)(9000) = 0.39$$

Thus, in a year in which net revenues per deposit dollar are 4.50 and the number of offices is 9,000, we predict a value 0.39 for percentage profit margin of savings and loan associations.

Given the further assumption that the error terms, ε_i, are normally distributed, formulas can be developed for calculating confidence intervals for predictions from regression models. These formulas are, in general, rather complicated. However, many multiple regression computer packages have the facility for computing interval forecasts for predictions.

EXERCISES

8. In the study of Exercise 2, where the sample regression was based on 20 observations, the estimated standard errors were

$$s_{b_1} = 1.6; \qquad s_{b_2} = 2.2; \qquad s_{b_3} = 0.2; \qquad s_{b_4} = 1.1$$

(a) Find 90% and 95% confidence intervals for β_1.

(b) Find 95% and 99% confidence intervals for β_2.

(c) Test against a two-sided alternative the null hypothesis that, all else equal, the ratio of productivity of foreign to United States car workers does not influence the sales of imported cars as a percentage of total car sales in the United States.

(d) Test the null hypothesis that, all else equal, the difference between average gas mileage ratings of United States and imported cars does not influence the sales of imported cars as a percentage of total car sales in the United States, against the alternative that, the larger this difference, the higher the percentage of foreign car sales.

9. In the study of Exercise 3, where the sample regression was based on twenty-five observations, the estimated standard errors were

$$s_{b_1} = 0.012; \qquad s_{b_2} = 0.030; \qquad s_{b_3} = 0.259$$

(a) Test against the appropriate one-sided alternative the null hypothesis that, all else equal, supermarket profits are not affected by the value of food sales.

(b) Test against the appropriate one-sided alternative the null hypothesis that, all else equal, supermarket profits are not affected by the value of nonfood sales.

(c) Find 90%, 95%, and 99% confidence intervals for β_3.

10. Refer to the data of Example 13.1.

(a) Test the null hypothesis that, all else equal, the BTU per hour rating of air conditioners does not affect their price.

(b) Test the null hypothesis that, all else equal, the energy efficiency ratio of air conditioners does not affect their price.

11. Refer to the data of Example 13.2.

(a) Find 90% and 95% confidence intervals for the expected change in tender premium resulting from a 1 unit increase in soliciting fee per share, as a percentage of closing market price, when the values of all other independent variables remain unchanged.

(b) Find 95% and 99% confidence intervals for the expected change in tender premium resulting from a 1 unit increase in the percentage of shares sought, when the values of all other independent variables remain unchanged. Comment on these intervals.

12. In the study of Exercise 4, where the sample regression was based on thirty observations, the estimated standard errors were

$$s_{b_1} = 0.23; \qquad s_{b_2} = 5.98; \qquad s_{b_3} = 2.12$$

(a) Test against the appropriate one-sided alternative the null hypothesis that, all else equal, the number of passengers on a flight is not affected by the price of the ticket.

(b) Find 90% and 95% confidence intervals for β_2.

(c) Find 95% and 99% confidence intervals for β_3.

13. A student wanted to analyze the factors influencing the proportion of baseball games won per season, over a 21-year period, by the Cincinnati Reds. For these data, he estimated by least squares the following regression:

$$y = 1.781 - 0.686x_1 - 1.388x_2 + 1.165x_3 - 0.104x_4 + 0.045x_5; \qquad R^2 = 0.80$$
$$\quad (3.246) \quad (1.402) \quad (3.288) \quad (0.755) \quad (0.017) \quad (0.039)$$

where

y = Proportion of games won
x_1 = Team batting average
x_2 = Team fielding average

x_3 = Team slugging average
x_4 = Team earned run average
x_5 = Runs scored per game

The values in parentheses beneath the coefficient estimates are their estimated standard errors.

(a) Judged by the coefficient of determination, how well do the five independent variables explain the variability in the dependent variable?

(b) Estimate the effect on the proportion of games won of an increase of 1 in team earned run average, all else equal.

(c) Test against a one-sided alternative the null hypothesis that, all else equal, team earned run average does not affect the proportion of games won.

(d) Find a 90% confidence interval for the increase in proportion of games won resulting from an increase of 0.05 in team slugging average, holding all other independent variables constant.

14. In a study[10] aimed at explaining the demand for alcoholic beverages in the United Kingdom, the following regression equation was fitted to twenty annual observations:

$$y = -0.014 - 0.354x_1 + 0.0018x_2 + 0.657x_3 + 0.0059x_4; \qquad \bar{R}^2 = 0.689$$
$$\quad (0.012) \quad (0.268) \qquad (0.00053) \quad (0.266) \qquad (0.0034)$$

where

y = Annual change in pure alcohol consumption per adult
x_1 = Annual change in real price index of alcoholic drinks
x_2 = Annual change in real disposable income per person
x_3 = (Annual change in number of licensed premises) $\div$ (Adult population)
x_4 = Annual change in real advertising expenditure on alcoholic drinks per adult

The values in parentheses beneath the coefficient estimates are their estimated standard errors.

(a) Judged by the corrected coefficient of determination, how well do the four independent variables explain variability in the dependent variable?

(b) Interpret the point estimates of the four partial regression coefficients.

(c) Test, against two-sided alternatives, the null hypotheses $\beta_i = 0$, for each of the four partial regression coefficients.

15. For a random sample of forty territories, sales of a brand of pen were related to three explanatory variables—television advertising, the number of sales representatives, and a measure of wholesaler efficiency. The estimated regression equation obtained was

$$y = 31.150 + 12.968x_1 + 41.246x_2 + 11.524x_3; \qquad \bar{R}^2 = 0.881$$
$$\quad (34.175) \quad (2.737) \qquad (7.280) \qquad (7.691)$$

where

y = Sales in thousands
x_1 = Number of television advertising spots per month
x_2 = Number of sales representatives
x_3 = Wholesaler efficiency index (1 = poor, 2 = average, 3 = good, 4 = outstanding)

---

[10] T. McGuinness, "An econometric analysis of total demand for alcoholic beverages in the U.K.," *Journal of Industrial Economics, 29* (1980), 85–109.

Figures in parentheses beneath the coefficient estimates are their estimated standard errors.

(a) Test against the appropriate one-sided alternative the null hypothesis that, for a given number of sales representatives and a given wholesaler efficiency index, the number of sales is not affected by the number of television advertising spots.

(b) Find a 90% confidence interval for the increase in sales resulting from a 1 unit increase in wholsesaler efficiency for fixed levels of television advertising spots and number of sales representatives.

(c) Find a 99% confidence interval for the increase in sales from one additional sales representative for fixed levels of television advertising and wholesaler efficiency.

16. In a study[11] of the demand for imports in Jamaica, the following model was fitted to 19 years of data:

$$y = -58.9 + 0.20x_1 - 0.10x_2; \qquad \bar{R}^2 = 0.96$$
$$(0.0092) \quad (0.084)$$

where

y = Quantity of imports
x_1 = Personal consumption expenditures
x_2 = (Price of imports) ÷ (Domestic prices)

Figures in parentheses beneath the coefficient estimates are their estimated standard errors.

(a) Find a 95% confidence interval for β_1.

(b) Test, against the appropriate one-sided alternative, the null hypothesis that $\beta_2 = 0$.

17. For a sample of twenty-five unskilled workers, a labor economist obtained the estimated multiple regression

$$y = 2.46 + 0.43x_1 + 0.029x_2 + 0.052x_3; \qquad R^2 = 0.76$$
$$(0.10) \quad (0.011) \quad (0.047)$$

where

y = Worker's hourly wage rate (in dollars)
x_1 = Number of years employed by present firm
x_2 = Percentage of employees unionized in the worker's industry
x_3 = Size of the worker's firm (in thousands of employees)

Figures in parentheses beneath the coefficient estimates are their estimated standard errors.

(a) Find a 99% confidence interval for β_1, and interpret your result.

(b) Test the null hypothesis that $\beta_2 = 0$, and provide an interpretation of your test results.

(c) Test the null hypothesis that $\beta_3 = 0$, and provide an interpretation of your test results.

18. The manager of a student physical education facility wants to develop a model to predict the number of injuries in a week. For a period of 32 weeks he estimated the regression equation

---

[11] See J. Gafar, "Devaluation and the balance of payments adjustment in a developing economy: An analysis relating to Jamaica," *Applied Economics, 13* (1981), 151–65.

$$y = 3.12 + 0.013x_1 + 0.0058x_2 + 0.0093x_3; \qquad R^2 = 0.67$$
$$(0.004) \quad (0.0041) \quad (0.0047)$$

where

y = Number of injuries reported in the week
x_1 = Number of basketball players using the facility in the week
x_2 = Number of swimmers using the facility in the week
x_3 = Number of racquetball players using the facility in the week

Figures in parentheses beneath the coefficient estimates are their estimated standard errors.

(a) Find and interpret a 95% confidence interval for β_1.

(b) Test the null hypothesis that $\beta_2 = 0$ and provide an interpretation of your test results.

(c) Test the null hypothesis that $\beta_3 = 0$ and provide an interpretation of your test results.

19. Suppose that a dependent variable is related to K independent variables through a multiple regression model. Let R^2 denote the coefficient of determination, and $\bar{R}^2$ the corrected coefficient. Suppose that n sets of observations are used to fit the regression.

(a) Show that
$$\bar{R}^2 = \frac{(n-1)R^2 - K}{n - K - 1}$$

(b) Show that
$$R^2 = \frac{(n - K - 1)\bar{R}^2 + K}{n - 1}$$

(c) Show that the statistic for testing the null hypothesis that all the partial regression coefficients are 0 can be written
$$\frac{\text{SSR}/K}{\text{SSE}/(n - K - 1)} = \frac{(n - K - 1)}{K} \cdot \frac{\bar{R}^2 + A}{1 - \bar{R}^2}$$

where
$$A = \frac{K}{(n - K - 1)}$$

20. For the study on the influence of financial institutions on share prices of Exercise 1, 48 quarterly observations were used, and the corrected coefficient of determination was found to be $\bar{R}^2 = 0.463$. Test the null hypothesis
$$H_0: \quad \beta_1 = \beta_2 = 0$$

21. Refer to the study on the sales of imported cars in the United States, described in Exercises 2 and 5.

(a) Test the null hypothesis
$$H_0: \quad \beta_1 = \beta_2 = \beta_3 = \beta_4 = 0$$

(b) Construct the analysis of variance table for this regression.

22. Refer to the study on profits of supermarkets, described in Exercises 3 and 6.

(a) Test the null hypothesis
$$H_0: \quad \beta_1 = \beta_2 = \beta_3 = 0$$

(b) Set out the analysis of variance table for the fitted regression.

23. Refer to the study on the number of passengers on flights of the commercial airline, described in Exercises 4 and 7.

(a) Test the null hypothesis

$$H_0: \quad \beta_1 = \beta_2 = \beta_3 = 0$$

(b) Set out the analysis of variance table.

24. Refer to Exercise 14. Test the null hypothesis that, taken together, price, disposable income, number of licensed premises, and advertising expenditure do not linearly affect the demand for alcoholic beverages in the United Kingdom.

25. Refer to the study of Exercise 16. Test the null hypothesis that, taken together, consumption expenditures and the relative price of imports do not linearly affect the demand for imports in Jamaica.

26. Refer to the data of Example 13.1. Test the null hypothesis that, taken together, BTU per hour rating, energy efficiency ratio, and the number of settings do not linearly influence the price of air conditioners.

27. Using the data of Exercise 17, test the null hypothesis that, taken together, number of years employed, industry unionization, and firm size do not linearly affect a worker's hourly wage rate.

28. A dependent variable is regressed on K independent variables, using n sets of sample observations. We denote by SSE the error sum of squares and by R^2 the coefficient of determination for this estimated regression. We want to test the null hypothesis that K_1 of these independent variables, taken together, do not linearly affect the dependent variable, given that the other $(K - K_1)$ independent variables are also to be used. Suppose the regression is reestimated, with the K_1 independent variables of interest excluded. Let SSE* denote the error sum of squares, and R_*^2 the coefficient of determination for this regression. Show that the statistic for testing our null hypothesis, introduced in Section 13.7, can be expressed as

$$\frac{(\text{SSE}^* - \text{SSE})/K_1}{\text{SSE}/(n - K - 1)} = \frac{R^2 - R_*^2}{1 - R^2} \cdot \frac{n - K - 1}{K_1}$$

29. In Exercise 17 we present results for the regression of wage rates on years employed, unionization, and firm size. When the simple linear regression of wage rates on number of years employed alone was estimated, the coefficient of determination was found to be 0.54. Test the null hypothesis that, given that years employed are also used, unionization and firm size, taken together, do not linearly affect workers' hourly wage rates. [*Hint:* Use the result of Exercise 28.]

30. Using the information in Exercise 13, predict the proportion of games the Cincinnati Reds would win in a year in which team batting average was 0.25, team fielding average was 0.98, team slugging average was 0.4, team earned run average was 3.75, and 4 runs per game were scored.

31. Based on the results of Exercise 15, predict pen sales for a territory in which there were ten television advertising spots per month, six sales representatives, and wholesaler efficiency was "outstanding."

32. Using the results of Exercise 17, predict the hourly wage rate of an unskilled worker who has been employed for 5 years by his present firm, which has 5,000 employees, in an industry which is 60% unionized.

33. Using the results of Exercise 18, predict the number of injuries in a week in which the facility is used by 500 basketball players, 1,000 swimmers, and 300 racquetball players.

13.9 COMPUTER PACKAGES FOR REGRESSION CALCULATIONS

As we have already noted, multiple regression analysis calculations are almost invariably carried out on an electronic computer. Indeed, without this aid, the arithmetic involved would be formidably difficult if the model contained more than a very small number of independent variables.

Most modern computing facilities have available prewritten program packages for carrying out regression analyses. The user need only supply the data and specify the model that is to be fitted. All of these packages produce the basic information we have discussed in this chapter, and generally contain options to allow the production of much more numerical and graphical output.

Table 13.4 lists part of the output from a standard regression package, the Statistical Analysis System (SAS) program, used in the analysis of our savings and loan data. The upper half of the table shows the analysis of variance for the fitted regression. This reproduces the information in Table 13.3, but with the "Regression" source of variation referred to as MODEL. The final row of this table contains the sum of the degrees of freedom and sums of squares for the model and error. The coefficient of determination, R-SQUARE, is also shown in the upper portion of the output.

The lower half of the table shows the least squares parameter estimates and their estimated standard errors. The ratios of these estimates to their standard errors are labelled T FOR H0: PARAMETER = 0, and are the statistics discussed in Section 13.6 for testing the null hypotheses that the population regression parameters are 0.

Table 13.5 shows part of the output from a second package—the MINITAB program—used to perform the savings and loan regression. The output of the MINITAB program begins with a summary of the least squares estimated regression model. Next, the coefficient estimates, together with their associated standard errors and t ratios, are shown. The ST. DEV. OF Y ABOUT REGRESSION LINE is an estimate of the standard deviation of the error terms, ε_i, in the population regression. Thus, it provides an estimate of the standard deviation of the unexplained part of the dependent variable. Here, we have

TABLE 13.4 Part of output of SAS program for savings and loan regression

DEPENDENT VARIABLE: Y PROFIT MARGIN

SOURCE	DF	SUM OF SQUARES	MEAN SQUARE	F VALUE	R-SQUARE
MODEL	2	0.4017	0.20085	70.92	0.866
ERROR	22	0.0623	0.0028318		
CORRECTED TOTAL	24	0.4640			

PARAMETER	ESTIMATE	T FOR H0: PARAMETER = 0	STD. ERROR OF ESTIMATE
INTERCEPT	1.565		
X1	0.237	4.27	0.0555
X2	-0.000249	-7.78	0.0000320

STD. DEV. OF Y ABOUT REGRESSION LINE

$$= s_e$$

$$= \sqrt{\frac{SSE}{(n - K - 1)}}$$

$$= \sqrt{\frac{0.0623}{22}} = 0.0532$$

The coefficient of determination, R^2, is shown next, followed by the adjusted coefficient, $\bar{R}^2$. Finally, the analysis of variance table is displayed.

As a final example, Table 13.6 shows part of the output from the Statistical Package for the Social Sciences (SPSS) program for our savings and loan data. Many of the quantities reported by the SPSS program correspond in an obvious way to those reported by the other programs. The STANDARD ERROR is s_e, the estimated standard deviation of the error terms of the population regression model. Instead of reporting the t ratios, the SPSS program gives F values, which are in fact the squares of the t ratios.

It is obviously the case that the development of multiple regression computer packages has greatly eased the arithmetic burden involved in regression computations. Largely as a result of this factor, the technique is now extremely widely used. This is not entirely an unmixed blessing. Perhaps because the procedure is so easy to carry out, regression analyses based on an inadequate amount of forethought are frequently reported. In fact, there are a number of pitfalls involved in the analysis of data through this technique. Some of these will be discussed in the following chapter.

TABLE 13.5 Part of output of MINITAB program for savings and loan regression

```
THE REGRESSION EQUATION IS
Y = 1.565 + 0.237 X 1 - 0.000249 X 2

                                    ST. DEV.              T-RATIO =
                 COEFFICIENT        OF COEF.              COEF/S.D.
                   1.565
X1                 0.237            0.0555                  4.27
X2                -0.000249         0.00032                 -7.78

THE ST. DEV. OF Y ABOUT REGRESSION LINE IS S = 0.0532
WITH (25-3) = 22 DEGREES OF FREEDOM

R-SQUARED = 86.6 PERCENT
R-SQUARED = 85.4 PERCENT, ADJUSTED FOR D.F.

ANALYSIS OF VARIANCE

DUE TO              DF              SS              MS = SS/DF
REGRESSION          2              0.4017          0.20085
RESIDUAL            22             0.0623          0.0028318
TOTAL               24             0.4640
```

TABLE 13.6 Part of output of SPSS program for savings and loan regression

```
DEPENDENT VARIABLE . . Y PROFIT MARGIN

MULTIPLE R          0.930
R SQUARE            0.866
ADJUSTED R SQUARE   0.854
STANDARD ERROR      0.0532

ANALYSIS OF VARIANCE
                         SUM OF
                 DF      SQUARES      MEAN SQUARE      F
REGRESSION        2      0.4017       0.20085          70.92
RESIDUAL         22      0.0623       0.0028318

VARIABLE        B          STD. ERROR B         F
X1              0.237      0.0555               18.23
X2             -0.000249   0.0000320            60.53
(CONSTANT)      1.565
```

EXERCISES

34. In order to assess the impact in a state of banks' economic power on their political power, the following model was hypothesized and fitted to data from fifty states.[12]

$$Y = \alpha + \beta_1 x_1 + \beta_2 x_2 + \beta_3 x_3 + \beta_4 x_4 + \beta_5 x_5 + \varepsilon$$

where

Y = Ratio of bank provisions for state and local taxes (in thousands of dollars) to total state and local tax revenues (in millions of dollars)

x_1 = 3-Bank organization state concentration ratio (a measure of the concentration of banking resources)

x_2 = Per capita income in the state (in thousands of dollars)

x_3 = Ratio of nonfarm income to the sum of farm and nonfarm income

x_4 = Ratio of banks' net income after tax to banks' assets (multiplied by 1,000)

x_5 = Average of banks' assets (divided by 10,000)

Part of the computer output from the estimated regression is shown here. Write a report summarizing the findings of this study.

		R-SQUARE 0.515	
PARAMETER	ESTIMATE	T FOR H0: PARAMETER = 0	STD. ERROR OF ESTIMATE
INTERCEPT	10.60	2.41	4.40
X1	-0.90	-0.69	1.31
X2	0.14	0.50	0.28
X3	-11.85	-2.83	4.18
X4	0.080	0.50	0.160
X5	0.100	5.00	0.020

35. A random sample of 93 freshmen at the University of Illinois were asked to rate on a scale from 1 (low) to 10 (high) their overall opinion of residence hall life. They were also asked to rate, on similar scales, their levels of satisfaction with roommates, with the floor, with the hall, and with the resident adviser. (Information on satisfaction with the room itself was obtained, but this was later discarded as it provided no useful additional power in explaining overall opinion.) The following model was estimated:

$$Y = \alpha + \beta_1 x_1 + \beta_2 x_2 + \beta_3 x_3 + \beta_4 x_4 + \varepsilon$$

where

Y = Overall opinion of residence hall

x_1 = Satisfaction with roommates

x_2 = Satisfaction with floor

x_3 = Satisfaction with hall

x_4 = Satisfaction with resident adviser

Use the accompanying portion of the computer output from the estimated regression to write a report summarizing the findings of this study.

---

[12] Results quoted are adapted from C. A. Glassman, "The impact of banks' statewide economic power on their political power: An empirical analysis," *Atlantic Economic Journal, 19* (1981), No. 2, 53–56.

```
DEPENDENT VARIABLE: Y OVERALL OPINION

                         SUM OF    MEAN
SOURCE            DF     SQUARES   SQUARE    F VALUE    R-SQUARE

MODEL             4      37.016    9.2540    9.958      0.312
ERROR             88     81.780    0.9293
CORRECTED TOTAL   92     118.796

                                  T FOR HO:           STD. ERROR OF
PARAMETER      ESTIMATE        PARAMETER = 0            ESTIMATE

INTERCEPT      3.950               5.84                 0.676
X1             0.106               1.69                 0.063
X2             0.122               1.70                 0.072
X3             0.092               1.75                 0.053
X4             0.169               2.64                 0.064
```

36. The following model[13] was fitted to 47 monthly observations, in an attempt to explain the difference between secondary market certificate of deposit rates and Treasury bill rates:

$$Y = \alpha + \beta_1 x_1 + \beta_2 x_2 + \varepsilon$$

where

Y = Secondary market certificate of deposit rate less Treasury bill rate
x_1 = Treasury bill rate
x_2 = Ratio of loans and investments to capital

Use the part of the computer output from the estimated regression shown here to write a report summarizing the findings of this analysis.

```
                           R-SQUARE
                            0.730

                                  T FOR HO:           STD. ERROR OF
PARAMETER      ESTIMATE        PARAMETER = 0            ESTIMATE

INTERCEPT      -5.559             -4.14                 1.343
X1             0.186               5.64                 0.033
X2             0.450               2.08                 0.216
```

37. [*This problem requires either the material in Appendix A13.1 or a computer program to carry out the multiple regression computations.*] The accompanying table gives 10 years of annual data on traffic death rates per 100 million vehicle miles (y), total travel in billion vehicle miles (x_1), and the average speed in miles per hour of all vehicles (x_2). Compute the multiple regression of y on x_1 and x_2, and write a report discussing your findings.

---

[13] The results quoted here are adapted from I. H. Giddy, "Moral hazard and Central Bank rescues in an international context," *The Financial Review, 15* (1980), No. 2, 50–56.

y	x_1	x_2
5.2	738	52.6
5.3	767	53.8
5.4	805	55.8
5.6	847	55.9
5.5	889	56.4
5.7	930	57.3
5.5	962	58.0
5.4	1,016	59.0
5.2	1,071	60.0
4.9	1,121	59.2

38. [*This problem requires either the material in Appendix A13.1 or a computer program to carry out the multiple regression computations.*] The table shows, for twelve territories, product sales (y), expenditure on advertising (x_1), and expenditure on salesmen (x_2), all in thousands of dollars. Compute the multiple regression of y on x_1 and x_2, and write a report summarizing your findings.

y	x_1	x_2	y	x_1	x_2
127	18	10	161	25	14
149	25	11	128	16	12
106	19	6	139	17	12
163	24	16	144	23	12
102	15	7	159	22	14
180	26	17	138	15	15

39. [*This problem requires a computer program to carry out the multiple regression computations.*] For a random sample of twenty in-flight aircraft accidents, in which not all passengers were killed, the accompanying table shows the percentage of fatalities (y), the altitude (in hundreds of feet) of the lowest layer of clouds at the site of the crash (x_1), the total number of hours (in hundreds) logged by the pilot (x_2), the velocity (in knots) of surface winds at the accident site (x_3), and the visibility (in nautical miles) at the site of the crash (x_4). Compute the multiple regression of y on x_1, x_2, x_3, and x_4, and write a report summarizing your findings.

y	x_1	x_2	x_3	x_4
50.0	12	11.0	6	5
60.0	28	24.7	15	5
66.7	100	2.1	30	5
33.3	100	1.0	10	5
12.5	8	57.0	5	4
33.3	2	1.0	5	4
50.0	100	2.5	15	5
33.3	100	100.5	10	5
50.0	5	151.7	8	2
40.0	10	45.0	12	5

y	x_1	x_2	x_3	x_4
75.0	70	10.8	13	5
25.0	15	3.5	30	5
33.3	2	81.0	5	2
75.0	12	1.2	8	3
75.0	50	3.7	7	5
50.0	100	2.2	10	5
70.0	3	20.6	6	1
33.3	8	131.6	5	1
50.0	100	118.4	20	5
33.3	18	5.5	10	3

40. [*This problem requires a computer program to carry out the multiple regression computations.*] The table gives data, for a random sample of thirty-four countries, on gross national product per capita in U.S. dollars (y), population in millions (x_1), land in thousands of square kilometers (x_2), percentage of literate adults in the population (x_3), and the percentage of the economy attributed to agriculture (x_4). Compute the multiple regression of y on x_1, x_2, x_3, and x_4, and write a report summarizing your findings.

y	x_1	x_2	x_3	x_4
90	81.2	144	22	55
130	4.4	26	23	25
160	4.9	1,267	8	47
190	16.4	945	66	45
240	1.3	30	40	30
270	1.5	1,031	17	26
300	133.5	2,027	62	31
140	5.6	118	25	47
240	8.1	587	58	40
320	7.9	1,001	44	32
420	43.8	541	82	27
450	5.1	753	39	14
550	18.3	447	28	21
730	2.8	407	80	35
840	5.0	49	60	20
1,110	17.0	2,382	35	8
1,160	10.6	757	88	10
1,360	116.1	8,512	76	12
1,960	21.7	256	85	16
2,850	3.6	21	88	7
1,240	2.1	51	88	21
860	5.9	164	38	17
690	7.5	322	20	25
430	5.2	196	10	28
2,880	3.2	70	98	22
6,130	7.5	84	99	5
7,590	9.8	31	99	2
8,550	4.0	324	99	6
8,520	220.0	9,363	99	3
7,340	14.1	7,687	100	5
6,680	2.8	1,760	45	3
910	9.6	115	96	21
1,580	21.6	238	98	31
3,150	34.7	313	98	16

41. After estimating a multiple regression model, we can compute the decomposition of the total sum of squares into the sum of regression sum of squares and error sum of squares. Explain what can be learned from such a decomposition.

42. It is believed that the values of two independent variables, X_1 and X_2, influence the behavior of a dependent variable, Y. We can estimate a multiple regression of Y on X_1 and X_2. An alternative possibility might be to fit simple linear regressions of Y on X_1 and X_2 separately. Explain why the former course of action is preferred.

43. Explain the popularity of the method of least squares in estimating the parameters of multiple regression models.

44. Given a sample of n sets of observations $(x_{11}, x_{21}, \ldots, x_{K1}, y_1), \ldots, (x_{1n}, x_{2n}, \ldots, x_{Kn}, y_n)$, it is required to estimate the expected value of the dependent variable Y_{n+1}, when the independent variables take the specific values $x_{1,n+1}, x_{2,n+1}, \ldots, x_{K,n+1}$. What is the relevance of the Gauss-Markov theorem to this problem?

45. Suppose that, for n sets of observations, the models

$$Y_i = \alpha + \beta_1 x_{1i} + \cdots + \beta_K x_{Ki} + \varepsilon_i$$

and

$$Y_i = \alpha + \beta_1 x_{1i} + \cdots + \beta_K x_{Ki} + \beta_{K+1} x_{K+1,i} + \varepsilon_i$$

are estimated by least squares. Show that the error sum of squares must be smaller for the second model than for the first. What is the relevance of this result to the interpretation of the coefficient of determination?

46. [*This exercise requires the material in Appendix A13.1.*] Suppose that the regression model

$$Y_i = \alpha + \beta_1 x_{1i} + \beta_2 x_{2i} + \varepsilon_i$$

is estimated by least squares. Show that the residuals, e_i, from the fitted model sum to 0.

47. In a study of the factors influencing the use of a county park on weekends in the summer months, the following model was fitted to fifteen sample observations:

$$y = -1,483 + 37.1x_1 - 178.2x_2; \qquad R^2 = 0.623$$
$$ (14.6) \qquad (63.7)$$

where

y = Number of people using the park
x_1 = Temperature maximum (in degrees Fahrenheit)
x_2 = Rainfall (in inches)

Figures in parentheses beneath the coefficient estimates are their estimated standard errors.

(a) Interpret the estimated partial regression coefficients.

(b) Interpret the coefficient of determination.

(c) Test the null hypothesis that, all else equal, use is not influenced by rainfall. Use a 5% significance level, and test against the appropriate one-sided alternative.

(d) Find a 95% confidence interval for the increased use resulting from a 1° increase in maximum temperature, all else equal.

(e) Estimate the level of usage on a weekend in which the maximum temperature is 75° Fahrenheit and there is 0.2 inch of rainfall.

(f) Test the null hypothesis that, taken together, temperature and rainfall do not influence use of the park.

48. A study[14] was aimed at assessing the influence of various factors on the start of new firms in the electronics components industry in urban areas. For a sample of seventy standard metropolitan statistical areas, the following model was estimated:

$$y = -59.31 + 4.983x_1 + 2.198x_2 + 3.816x_3 - 0.310x_4$$
$$(1.156)\quad (0.210)\quad (2.063)\quad (0.330)$$

$$-0.886x_5 + 3.215x_6 + 0.085x_7; \qquad R^2 = 0.766$$
$$(3.055)\quad (1.568)\quad (0.354)$$

where

y = New business starts in the industry
x_1 = Population (in millions)
x_2 = Industry size
x_3 = Measure of economic quality of life
x_4 = Measure of political quality of life
x_5 = Measure of environmental quality of life
x_6 = Measure of health–educational quality of life
x_7 = Measure of social quality of life

Figures in parentheses beneath the coefficient estimates are their estimated standard errors.

(a) Interpret the estimated partial regression coefficients.

(b) Interpret the coefficient of determination.

(c) Find a 90% confidence interval for the increase in new business starts resulting from a 1 unit increase in the economic quality of life, with all other variables unchanged.

(d) Test, against a two-sided alternative at the 5% significance level, the null hypothesis that, all else equal, the environmental quality of life does not influence new business starts.

(e) Test, against a two-sided alternative at the 5% significance level, the null hypothesis that, all else equal, the health-educational quality of life does not influence new business starts.

(f) Test the null hypothesis that, taken together, these seven independent variables do not influence new business starts.

49. A survey research group conducts regular studies of households through mail questionnaires and is concerned about the factors influencing the response rate. In an experiment, thirty sets of questionnaires were mailed to potential respondents. The regression model fitted to the resulting data set was

$$Y = \alpha + \beta_1 x_1 + \beta_2 x_2 + \varepsilon$$

where

Y = Percentage of responses received
x_1 = Number of questions asked
x_2 = Length of questionnaire, in terms of number of words

---

[14] J. M. Pennings, "The urban quality of life and entrepreneurship," *Academy of Management Journal*, 25 (1982), 63–79.

Part of the SAS computer output from the estimated regression is shown here.

		R-SQUARE 0.637	
PARAMETER	ESTIMATE	T FOR HO: PARAMETER = 0	STD. ERROR OF ESTIMATE
INTERCEPT	74.3652		
X1	-1.8345	-2.89	0.6349
X2	-0.0162	-1.78	0.0091

(a) Interpret the estimated partial regression coefficients.
(b) Interpret the coefficient of determination.
(c) Test at the 1% significance level the null hypothesis that, taken together, the two independent variables do not linearly influence the response rate.
(d) Find and interpret a 99% confidence interval for β_1.
(e) Test the null hypothesis

$$H_0: \quad \beta_2 = 0$$

against the alternative

$$H_1: \quad \beta_2 < 0$$

and interpret your findings.

50. A consulting group offers courses in financial management for executives. At the end of these courses, participants are asked to provide overall ratings of the value of the course. To assess the impact of various factors on ratings, the model

$$Y = \alpha + \beta_1 x_1 + \beta_2 x_2 + \beta_3 x_3 + \varepsilon$$

was fitted for twenty-five such courses, where

Y = Average rating by participants of the course
x_1 = Percentage of course time spent in group discussion sessions
x_2 = Amount of money (in dollars) per course member spent on the preparation of subject matter material
x_3 = Amount of money per course member spent on the provision of non–course-related material (food, drinks, etc.)

Part of the SAS computer output for the fitted regression is shown here.

		R-SQUARE 0.579	
PARAMETER	ESTIMATE	T FOR HO: PARAMETER = 0	STD. ERROR OF ESTIMATE
INTERCEPT	42.9712		
X1	0.3817	1.89	0.2018
X2	0.5172	2.64	0.1957
X3	0.0753	1.09	0.0693

(a) Interpret the estimated partial regression coefficients.

(b) Interpret the coefficient of determination.

(c) Test at the 5% level the null hypothesis that, taken together, the three independent variables do not linearly influence the course rating.

(d) Find and interpret a 90% confidence interval for β_1.

(e) Test the null hypothesis

$$H_0: \quad \beta_2 = 0$$

against the alternative

$$H_1: \quad \beta_2 > 0$$

and interpret your result.

(f) Test at the 10% significance level the null hypothesis

$$H_0: \quad \beta_3 = 0$$

against the alternative

$$H_1: \quad \beta_3 \neq 0$$

and interpret your result.

51. [*This exercise requires a computer program to carry out the multiple regression calculations.*] The accompanying table shows, for a particular town, the selling prices of a random sample of twenty houses, together with the size of the house (in square feet of floor space), the size of the lot (in square feet), the number of bedrooms, and the number of bathrooms. Compute the multiple regression of price on house size, lot size, number of bedrooms, and number of bathrooms, and write a report on your findings.

PRICE	HOUSE SIZE	LOT SIZE	BEDROOMS	BATHROOMS
61,500	1,250	4,700	3	1.5
72,500	1,570	5,300	3	1.5
78,000	1,680	5,800	4	1.5
89,000	1,830	7,100	4	2.5
86,500	1,870	6,900	4	2.5
89,500	1,930	6,800	4	2.5
78,500	1,750	6,500	3	2.5
95,500	2,300	7,900	4	2.5
96,500	2,300	7,700	4	3.5
79,500	1,950	6,400	3	2.5
63,000	1,400	4,500	3	1.5
76,500	1,750	6,400	3	2.5
110,000	2,400	8,300	4	3.5
107,000	2,370	7,800	5	3.5
132,000	3,100	8,200	5	3.5
92,500	2,100	5,900	4	2.5
102,500	2,350	7,300	4	3.5
149,000	4,100	9,800	5	3.5
61,000	1,400	3,900	3	1.5
74,000	1,640	5,300	3	1.5

52. [*This exercise requires a computer program to carry out the multiple regression calculations.*] At the end of classes, professors are rated by their students on a scale from 1 (poor) to 5 (excellent). Students are also asked what course grades they expect, and these are coded as A = 4, B = 3, and so on. In the table we show, for a random sample of twenty classes, ratings of professors, average expected grades, and the numbers of students in the classes. Compute the multiple regression of rating on expected grade and number of students, and write a report on your findings.

RATING	EXPECTED GRADE	NUMBER OF STUDENTS	RATING	EXPECTED GRADE	NUMBER OF STUDENTS
4.1	3.4	45	3.5	3.0	40
3.4	3.1	52	4.0	3.5	29
3.3	3.0	47	3.6	3.3	38
3.0	2.8	63	3.1	3.1	67
4.7	3.7	20	3.3	3.3	61
4.6	3.5	32	4.5	3.9	50
3.0	2.9	51	2.8	2.9	63
4.6	3.7	32	3.7	3.2	47
4.6	3.5	21	3.8	3.4	51
3.6	3.2	33	3.9	3.4	31

APPENDIX A13.1

In Section 13.2 we defined the least squares estimates of the parameters of the multiple regression model as those values $a, b_1, b_2, \ldots, b_K$ for which the sum of squared discrepancies

$$SS = \sum_{i=1}^{n} (y_i - a - b_1 x_{1i} - b_2 x_{2i} - \cdots - b_K x_{Ki})^2$$

is a minimum. It can be shown[15] that this minimization problem leads to the values $b_1, b_2, \ldots, b_K$ which are the solutions of the K equations

$$\sum (x_{1i} - \bar{x}_1)(y_i - \bar{y}) = b_1 \sum (x_{1i} - \bar{x}_1)^2 + b_2 \sum (x_{1i} - \bar{x}_1)(x_{2i} - \bar{x}_2)$$

$$+ \cdots + b_K \sum (x_{1i} - \bar{x}_1)(x_{Ki} - \bar{x}_K)$$

$$\sum (x_{2i} - \bar{x}_2)(y_i - \bar{y}) = b_1 \sum (x_{2i} - \bar{x}_2)(x_{1i} - \bar{x}_1) + b_2 \sum (x_{2i} - \bar{x}_2)^2$$

$$+ \cdots + b_K \sum (x_{2i} - \bar{x}_2)(x_{Ki} - \bar{x}_K)$$

$$\vdots \qquad\qquad \vdots$$

$$\sum (x_{Ki} - \bar{x}_K)(y_i - \bar{y}) = b_1 \sum (x_{Ki} - \bar{x}_K)(x_{1i} - \bar{x}_1) + b_2 \sum (x_{Ki} - \bar{x}_K)(x_{2i} - \bar{x}_2)$$

$$+ \cdots + b_K \sum (x_{Ki} - \bar{x}_K)^2$$

and that the intercept estimate a is given by

$$a = \bar{y} - b_1 \bar{x}_1 - b_2 \bar{x}_2 - \cdots - b_K \bar{x}_K$$

where $\bar{x}_1, \bar{x}_2, \ldots, \bar{x}_K$ are the sample means of the independent variables.

The preceding equations are called the **normal equations.** The arithmetic involved in their solution is quite tedious. However, if there are only two independent variables, hand calculation is not too burdensome. In that case we have

$$\sum (x_{1i} - \bar{x}_1)(y_i - \bar{y}) = b_1 \sum (x_{1i} - \bar{x}_1)^2 + b_2 \sum (x_{1i} - \bar{x}_1)(x_{2i} - \bar{x}_2)$$

$$\sum (x_{2i} - \bar{x}_2)(y_i - \bar{y}) = b_1 \sum (x_{2i} - \bar{x}_2)(x_{1i} - \bar{x}_1) + b_2 \sum (x_{2i} - \bar{x}_2)^2$$

---

[15] The result follows by setting to 0 the partial derivatives of SS with respect to $a, b_1, b_2, \ldots, b_K$.

The solution of this pair of equations is

$$b_1 = \frac{\Sigma(x_{2i} - \bar{x}_2)^2 \, \Sigma(x_{1i} - \bar{x}_1)(y_i - \bar{y}) - \Sigma(x_{1i} - \bar{x}_1)(x_{2i} - \bar{x}_2) \, \Sigma(x_{2i} - \bar{x}_2)(y_i - \bar{y})}{\Sigma(x_{1i} - \bar{x}_1)^2 \, \Sigma(x_{2i} - \bar{x}_2)^2 - [\Sigma(x_{1i} - \bar{x}_1)(x_{2i} - \bar{x}_2)]^2}$$

and

$$b_2 = \frac{\Sigma(x_{1i} - \bar{x}_1)^2 \, \Sigma(x_{2i} - \bar{x}_2)(y_i - \bar{y}) - \Sigma(x_{1i} - \bar{x}_1)(x_{2i} - \bar{x}_2) \, \Sigma(x_{1i} - \bar{x}_1)(y_i - \bar{y})}{\Sigma(x_{1i} - \bar{x}_1)^2 \, \Sigma(x_{2i} - \bar{x}_2)^2 - [\Sigma(x_{1i} - \bar{x}_1)(x_{2i} - \bar{x}_2)]^2}$$

Finally, the estimate of the intercept is

$$a = \bar{y} - b_1\bar{x}_1 - b_2\bar{x}_2$$

We will use these formulas to calculate least squares estimates for the parameters of the savings and loan model. Using the data of Table 13.1, we calculate

$$\sum x_{1i} = 96.34; \qquad \sum x_{2i} = 181{,}083; \qquad \sum y_i = 16.86$$

$$\sum x_{1i}^2 = 379.2928; \qquad \sum x_{2i}^2 = 1{,}335{,}795{,}900; \qquad \sum x_{1i}x_{2i} = 710{,}932.32$$

$$\sum x_{1i}y_i = 63.6124; \qquad \sum x_{2i}y_i = 119{,}215.94$$

The sample means are, therefore,

$$\bar{x}_1 = \frac{\sum x_{1i}}{n} = \frac{96.34}{25} = 3.8536$$

$$\bar{x}_2 = \frac{\sum x_{2i}}{n} = \frac{181{,}083}{25} = 7{,}243.32$$

$$\bar{y} = \frac{\sum y_i}{n} = \frac{16.86}{25} = 0.6744$$

The quantities required in the calculation of the least squares estimates are, then,

$$\sum(x_{1i} - \bar{x}_1)^2 = \sum x_{1i}^2 - n\bar{x}_1^2 = 379.2928 - (25)(3.8536)^2 = 8.03698$$

$$\sum(x_{2i} - \bar{x}_2)^2 = \sum x_{2i}^2 - n\bar{x}_2^2 = 1{,}335{,}795{,}900 - (25)(7{,}243.32)^2 = 24{,}153{,}800$$

$$\sum(x_{1i} - \bar{x}_1)(x_{2i} - \bar{x}_2) = \sum x_{1i}x_{2i} - n\bar{x}_1\bar{x}_2 = 710{,}932.32 - (25)(3.8536)(7{,}243.32)$$

$$= 13{,}110.88$$

$$\sum (x_{1i} - \bar{x}_1)(y_i - \bar{y}) = \sum x_{1i}y_i - n\bar{x}_1\bar{y} = 63.6124 - (25)(3.8536)(0.6744)$$

$$= -1.359296$$

$$\sum (x_{2i} - \bar{x}_2)(y_i - \bar{y}) = \sum x_{2i}y_i - n\bar{x}_2\bar{y} = 119{,}215.94 - (25)(7{,}243.32)(0.6744)$$

$$= -2{,}906.43$$

We can now use the formulas to compute the least squares estimates:

$$b_1 = \frac{(24{,}153{,}800)(-1.359296) - (13{,}110.88)(-2{,}906.43)}{(8.03698)(24{,}153{,}800) - (13{,}110.88)^2} = 0.237$$

and

$$b_2 = \frac{(8.03698)(-2{,}906.43) - (13{,}110.88)(-1.359296)}{(8.03698)(24{,}153{,}800) - (13{,}110.88)^2} = -0.000249$$

Finally, the intercept term is estimated by

$$a = \bar{y} - b_1\bar{x}_1 - b_2\bar{x}_2 = 0.6744 - (0.237)(3.8536) - (-0.000249)(7{,}243.32)$$

$$= 1.565$$

When the regression model contains only two independent variables, manageable expressions can also be obtained for the estimated standard deviations of the least squares estimators, discussed in Section 13.6. For the estimated variances, it can be shown that

$$s_{b_1}^2 = \frac{s_e^2 \sum (x_{2i} - \bar{x}_2)^2}{\sum (x_{1i} - \bar{x}_1)^2 \sum (x_{2i} - \bar{x}_2)^2 - [\sum (x_{1i} - \bar{x}_1)(x_{2i} - \bar{x}_2)]^2}$$

and

$$s_{b_2}^2 = \frac{s_e^2 \sum (x_{1i} - \bar{x}_1)^2}{\sum (x_{1i} - \bar{x}_1)^2 \sum (x_{2i} - \bar{x}_2)^2 - [\sum (x_{1i} - \bar{x}_1)(x_{2i} - \bar{x}_2)]^2}$$

where

$$s_e^2 = \frac{SSE}{(n - K - 1)}$$

is the estimate of the variance of the error terms ε_i, in the population regression.

For the savings and loan association data, we have

$$s_e^2 = \frac{0.0623}{22} = 0.0028318$$

Hence,

$$s_{b_1}^2 = \frac{(0.0028318)(24{,}153{,}800)}{(8.03698)(24{,}153{,}800) - (13{,}110.88)^2} = 0.003077$$

so that, by taking square roots, we obtain

$$s_{b_i} = 0.0555$$

Also,

$$s_{b_2}^2 = \frac{(0.0028318)(8.03698)}{(8.03698)(24,153,800) - (13,110.88)^2} = 0.0000000010238$$

and so

$$s_{b_2} = 0.0000320$$

additional topics in regression analysis

CHAPTER FOURTEEN

14.1 MODEL BUILDING METHODOLOGY

In the previous two chapters, the technique of regression analysis has been discussed. The objective of fitting a regression equation is to use information on the independent variables to explain the behavior of the dependent variable. Subsequently, the estimated regression model might be employed to derive predictions of the dependent variable. In the development of a regression equation, a rich variety of alternative specifications is possible and a number of problems may arise. In this chapter, we will consider some of the issues that occur in practical model building.

As a prelude, a general strategy for the construction of statistical models, such as a regression equation, is outlined. We live in an extremely complex world. No one really believes that the behavior of business and economic variables can be *precisely* described by a simple equation or system of equations. Nevertheless, it may be possible to find a relatively simple formal model that provides a sufficiently close representation of the complex reality to provide useful insights. The art of model building is to recognize the impossibility of accounting for the myriad of individual influences on a variable of interest and to try, rather, to pick out the most influential factors. Next, it is necessary to formulate a model to depict the interaction of these factors. The goal is to achieve a model that is sufficiently simple to allow convenient interpretation, but not so oversimplified that important influences are ignored.

The process of statistical model building is problem-specific. Just what is, and can be, done will depend on what is known about the behavior of the quantities under study and on what data are available. It is, however, possible to make some generalizations about the various stages of a model building exercise.

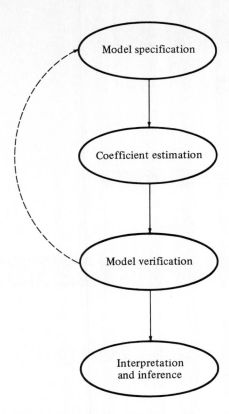

FIGURE 14.1 The stages of statistical model building

Figure 14.1 shows schematically the steps involved in statistical modelling. We will consider these in turn.

(i) MODEL SPECIFICATION

The initial problem is to specify a formal algebraic model that might provide a convenient, useful, and reasonably accurate description of the system under study. The regression equations of Chapters 12 and 13 have postulated a linear relationship between the expected value of a dependent variable and the values taken by independent variables. Often this formulation will provide an adequate description of the underlying reality. However, this is by no means invariably the case. In Sections 14.2–14.4 we will discuss certain elaborations and modifications of the multiple regression model.

It is important to keep in mind that any assumptions about the properties of the random variables involved in a statistical model constitute part of the postulated model specification. For example, in working with regression models to this point, we have assumed that the error terms all have the same variance and are uncorrelated with one another. Any inference derived from the estimated regression equation depends, for its validity, on these assumptions being appropriate.

(ii) COEFFICIENT ESTIMATION

A statistical model, once specified, typically involves a number of unknown coefficients, or **parameters.** The next stage of the model building exercise is to employ available data in the estimation of these coefficients. Both point estimates and interval estimates may be obtained. Thus, in the multiple regression model

$$Y_i = \alpha + \beta_1 x_{1i} + \beta_2 x_{2i} + \cdots + \beta_K x_{Ki} + \varepsilon_i \qquad (14.1.1)$$

observations on the dependent and independent variables are used to estimate the coefficients $\alpha, \beta_1, \beta_2, \ldots, \beta_K$.

The appropriate method of estimation depends on the model specification. We have seen that, given certain assumptions about the statistical properties of the error terms ε_i of 14.1.1, the method of least squares is appropriately used to estimate the partial regression coefficients. However, as we will see, if these assumptions fail to hold, least squares estimators can be quite inefficient. In general, the Gauss-Markov theorem, which provides a strong justification for the use of least squares, holds only when these assumptions are true.

(iii) MODEL VERIFICATION

In postulating a model, an investigator incorporates insights into the behavior of the underlying system. However, in translating these into algebraic form, certain simplifications and assumptions, which in the event might prove untenable, will have been made. Accordingly, it is important to check the adequacy of the model.

Having estimated a regression equation, an analyst may find that the estimates achieved simply do not make sense, given what is known or strongly suspected about the system under study. For instance, suppose that an estimated model suggests that, holding all other relevant factors fixed, the demand for imported cars *increases* as their price increases. Certainly such a conclusion runs counter to elementary economic theory—and to common sense! Generally, when a least squares point estimate has the "wrong sign," the problem is caused by insufficient data to allow much precision in the estimator. In our illustration, it is likely that, all else equal, at least a modest negative relationship exists between the demand for imported automobiles and their price. However, in that event, this can be masked by sampling error in the coefficient estimates. More seriously, it sometimes occurs that coefficient estimates with the "wrong sign" appear to differ significantly from 0 at the usual levels. It is likely, in such circumstances, that the cause of this phenomenon is a faulty model specification. The investigator should then give further thought to the original specification. Perhaps an important explanatory variable has been overlooked, or possibly the assumed functional form of the relationship is inappropriate. Certainly, checking the plausibility of the estimated model is an important ingredient of model verification.

It is also important to check the assumptions made about the statistical properties of the random variables in the model. In the regression equation 14.1.1 it is often assumed initially that the error terms ε_i all have the same variance and are uncorrelated with one another. In Sections 14.7 and 14.8, we will see how these assumptions can be checked, on the basis of evidence provided by the available data.

Any checks on the adequacy of a model may lead the analyst to conjecture an alternative specification. For that reason, we have connected the verification stage back to the specification stage with a dashed arrow in Figure 14.1. Viewed in this way, model building can be thought of as an iterative process of specification, estimation, and verification, the process being continued until an apparently satisfactory model is achieved.

(iv) INTERPRETATION AND INFERENCE

Once a model has been constructed, it can be used to learn something about the system under study. In regression analysis, this may involve finding confidence intervals for the model parameters, testing hypotheses of interest, or forecasting future values of the dependent variable, given assumed values of the independent variables. It is important to recognize that inference of this sort is based on the assumption of an appropriate model specification. The more severe are any specification errors, the less valuable, in general, is any inference derived from the fitted model.

14.2 DUMMY VARIABLES

It often happens that factors which are not directly quantifiable exert an important influence on the behavior of a dependent variable of interest. For example, the sales of many products will fluctuate according to the time of year, exports will be affected by dock strikes, and the rate of wage inflation will be influenced by government-mandated controls. Often such factors can be taken into account in a regression model through the introduction of *dummy variables*.

To illustrate the ideas involved, we will consider a study[1] on the taxable capacity of less developed countries. It was believed that tax revenues (Y) as a percentage of gross national product would depend on exports (x_1) as a percentage of gross national product and on per capita income (x_2). However, it was further suspected that tax revenues would be lower in those countries that participated in some form of economic integration (such as a free trade area, customs union, or common market), than in those that did not. In order to allow for the influence of this last factor, another independent variable (x_3) is created. This variable is constructed so that it takes the value 1 for those countries that take part in some form of economic integration and

---

[1] T. V. Truong and D. N. Gash, "Less developed countries' taxable capacity and economic integration: A cross-sectional analysis," *Review of Economics and Statistics, 61* (1979), 312–16.

0 for those countries that do not. The postulated multiple regresion model is then

$$Y_i = \alpha + \beta_1 x_{1i} + \beta_2 x_{2i} + \beta_3 x_{3i} + \varepsilon_i \qquad (14.2.1)$$

where

Y_i = Tax revenue as a percentage of gross national product in country i
x_{1i} = Exports as a percentage of gross national product in country i
x_{2i} = Income per capita (in U.S. dollars) in country i
$x_{3i} = \begin{cases} 1 & \text{if country } i \text{ participates in some form of economic integration} \\ 0 & \text{otherwise} \end{cases}$

Variables such as x_{3i} in equation 14.2.1 are called **dummy variables.** The interpretation of β_1 and β_2 in 14.2.1 is precisely the same as in regression models that do not contain dummy variables. Thus, β_2 is the expected increase in tax revenues (as a percentage of gross national product) resulting from a 1 unit increase in income per capita, with the values of the other independent variables held constant. To interpret β_3, notice that there are just two possibilities for the dummy variable. For countries participating in a form of economic integration, x_{3i} takes the value 1, so that equation 14.2.1 becomes

$$Y_i = \alpha + \beta_1 x_{1i} + \beta_2 x_{2i} + \beta_3 + \varepsilon_i$$

Similarly, for the other countries, since x_{3i} takes the value 0, equation 14.2.1 may be written

$$Y_i = \alpha + \beta_1 x_{1i} + \beta_2 x_{2i} + \varepsilon_i$$

Hence, we see that for fixed levels of exports and income per capita, β_3 is the difference between expected tax revenues (as a percentage of gross national product) in those countries that participate in some form of economic integration and those that do not.

Dummy Variables

Suppose it is believed that a dependent variable can be influenced by some factor which might be in one of two states, A or B. This factor is taken into account by adding to the regression a **dummy variable**—that is, an independent variable defined to take the value 1 when this factor is in state A, and 0 otherwise.

The partial regression coefficient on the dummy variable is the difference between the expected values of the dependent variable when the factor is in state A or B, and all other variables are held constant.

Regression equations involving dummy variables can be estimated by least squares, exactly as in Chapter 13. When the model 14.2.1 was fitted to data on

TABLE 14.1 Part of output from SAS program for tax revenues example

R-SQUARE

0.503

PARAMETER	ESTIMATE	T FOR H0: PARAMETER = 0	STD. ERROR OF ESTIMATE
INTERCEPT	9.188	5.58	1.646
X1	0.315	5.16	0.061
X2	0.00286	1.05	0.00272
X3	-3.664	-2.60	1.412

forty-three less developed countries, the estimated equation resulting was

$$y = 9.188 + 0.315x_1 + 0.00286x_2 - 3.664x_3; \qquad R^2 = 0.503$$
$$(1.646) \quad (0.061) \quad (0.00272) \quad (1.412)$$

Thus, taken together, the three independent variables explain 50.3% of the variability in the dependent variable for this sample of countries. Table 14.1 shows part of the computer output for the fitted regression relationship.

Based on the estimated equation, our point estimate is that, for given levels of exports and income per capita, expected tax revenues as a percentage of gross national product are lower by 3.664 in those countries that participate in some form of economic integration than in those that do not. This result is illustrated in Figure 14.2, which shows the estimated relation between tax revenues and exports, for a given level of income per capita. There are two possible relations—one for those countries participating in some form of integration and one for those that do not. These relationships are shown as two parallel lines, indicating that, for each level of exports, the expected tax revenues are higher by 3.664 in the latter group than in the former.

Confidence intervals for, and tests of hypotheses about, the model parameters can be obtained in the usual way when dummy variables are included in a regression model. Referring to the model 14.2.1, we will test the null hypothesis

$$H_0: \quad \beta_3 = 0$$

against the alternative

$$H_1: \quad \beta_3 < 0$$

using the data on forty-three less developed countries. The null hypothesis is that, all else equal, expected tax revenues in countries that participate in some form of economic integration are the same as in those that do not. The alternative hypothesis is that expected tax revenues are lower in those countries participating in some form of economic integration.

The test is based on the ratio of the estimated coefficient to its estimated standard error, that is,

$$\frac{b_3}{s_{b_3}} = \frac{-3.664}{1.412} = -2.595$$

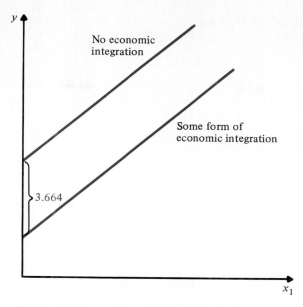

FIGURE 14.2 Estimated relationships between tax revenues as percentage of gross national product (y) and exports as percentage of gross national product (x_1), for a fixed level of income per capita

as can also be read directly from Table 14.1. Since there are 43 observations and 3 independent variables, we have

$$n = 43 \qquad \text{and} \qquad K = 3$$

so that the appropriate comparison with the tabulated Student's t distribution is $-t_{n-K-1,\alpha}$. From Table 6 of the Appendix, we have

$$-t_{39,\,.01} \approx -2.423 \qquad \text{and} \qquad -t_{39,\,.005} \approx -2.704$$

Hence the null hypothesis can be rejected at the 1% level, but not at the 0.5% level of significance. The evidence in favor of the alternative, that tax revenues are higher in the absence of participation in a form of economic integration, is very strong.

14.3 LAGGED DEPENDENT VARIABLES

The subject matter of this section is relevant when time series data are analyzed—that is, when measurements on the quantities of interest are taken through time. We might, for instance, have annual, quarterly, or monthly observations. To make the point notationally transparent, we will index the observations by the subscript t, denoting time, rather than the subscript i that we have previously used. Thus, a regression model in which the value taken by the dependent variable is influenced only by the values taken by K independent variables in the same time period will be written

$$Y_t = \alpha + \beta_1 x_{1t} + \beta_2 x_{2t} + \cdots + \beta_K x_{Kt} + \varepsilon_t$$

In many applications involving time series data, an elaboration of this model is desirable. The value Y_t taken by the dependent variable in time period t is often related also to the value taken by this variable in the previous time period—that is, to Y_{t-1}. The value of a dependent variable in an earlier time period is called a **lagged dependent variable.**

Regressions Involving Lagged Dependent Variables

Consider the following regression model linking a dependent variable, Y, and K independent variables, $x_1, x_2, \ldots, x_K$:

$$Y_t = \alpha + \beta_1 x_{1t} + \beta_2 x_{2t} + \cdots + \beta_K x_{Kt} + \gamma Y_{t-1} + \varepsilon_t$$

where $\alpha, \beta_1, \beta_2, \ldots, \beta_K, \gamma$ are fixed coefficients.

(i) An increase of 1 unit in the independent variable x_j in time period t, with all other independent variables held fixed, leads to an expected increase in the dependent variable of β_j in period t, $\beta_j \gamma$ in period $(t + 1)$, $\beta_j \gamma^2$ in period $(t + 2)$, $\beta_j \gamma^3$ in period $(t + 3)$, and so on. The total expected increase over all future time periods is $\beta_j/(1 - \gamma)$.

(ii) The coefficients $\alpha, \beta_1, \beta_2, \ldots, \beta_K, \gamma$ can be estimated by least squares in the usual manner.

(iii) Confidence intervals and hypothesis tests for the regression coefficients are obtained precisely as for the ordinary multiple regression model. (Strictly speaking, when the regression equation contains lagged dependent variables these procedures are only approximately valid. The quality of the approximation improves, all other things equal, as the number of sample observations increases.)

To illustrate the calculation of regression estimates and inference based on the fitted regression equation when the model includes lagged dependent variables, we consider a model aimed at explaining local advertising expenditure per household.[2] Table 14.2 gives 23 annual observations on this quantity.

It was believed that local advertising per household would depend on retail sales per household. (Data for this latter variable are given in Table 12.5.) Also, since advertisers may be unwilling, or unable, to adjust their plans to sudden changes in the level of retail sales, the value of local advertising expenditures per household in the previous year was also included in the model. Thus, advertising expenditures

---

[2] The model discussed here is fitted by N. K. Dhalla, "Short-term forecasts of advertising expenditures," *Journal of Advertising Research, 19* (1979), No. 1, 7–14.

TABLE 14.2 Data on local advertising expenditures per household in the United States (1972 dollars)

YEAR	ADVERTISING	YEAR	ADVERTISING
0	115.80	12	132.27
1	117.66	13	134.69
2	115.62	14	138.62
3	110.79	15	136.15
4	119.22	16	144.17
5	120.78	17	154.03
6	110.20	18	161.39
7	110.86	19	157.72
8	114.06	20	145.37
9	120.87	21	152.73
10	127.03	22	155.70
11	132.08		

in the current year are related to retail sales (x_t) in the current year and advertising expenditures (Y_{t-1}) in the previous year. The model to be fitted is, then,

$$Y_t = \alpha + \beta x_t + \gamma Y_{t-1} + \varepsilon_t$$

where

Y_t = Local advertising per houshold in year t

x_t = Retail sales per household in year t

In Table 14.3, the data are set out in the appropriate form for least squares estimation. The y_t are the observations on the dependent variable. For the purposes of carrying through the regression calculations, the lagged dependent variable, y_{t-1}, is treated simply as a second independent variable. Thus, the regression is computed as the regression of the dependent variable y_t on the pair of "independent variables," y_{t-1} and x_t.

TABLE 14.3 Data on local advertising expenditures per household (y_t) and retail sales per household (x_t)

YEAR	y_t	y_{t-1}	x_t	YEAR	y_t	y_{t-1}	x_t
1	117.66	115.80	5,492	12	132.27	132.08	5,907
2	115.62	117.66	5,540	13	134.69	132.27	6,124
3	110.79	115.62	5,305	14	138.62	134.69	6,186
4	119.22	110.79	5,507	15	136.15	138.62	6,224
5	120.78	119.22	5,418	16	144.17	136.15	6,496
6	110.20	120.78	5,320	17	154.03	144.17	6,718
7	110.86	110.20	5,538	18	161.39	154.03	6,921
8	114.06	110.86	5,692	19	157.72	161.39	6,471
9	120.87	114.06	5,871	20	145.37	157.72	6,394
10	127.03	120.87	6,157	21	152.73	145.37	6,555
11	132.08	127.03	6,342	22	155.70	152.73	6,755

TABLE 14.4 Part of output from SAS program for local advertising expenditures example

PARAMETER	ESTIMATE	T FOR HO: PARAMETER = 0	STD. ERROR OF ESTIMATE
INTERCEPT	-41.87		
X1	0.0185	6.61	0.0028
YT-1	0.480	5.58	0.086

In this event, the estimated regression obtained is

$$y_t = -41.87 + 0.0185x_t + 0.480y_{t-1}$$
$$(0.0028) \quad (0.086)$$

$$(14.3.1)$$

where, as usual, the figures in parentheses beneath the coefficient estimates are the corresponding estimated standard errors.

Part of the SAS computer output for the estimated regression is shown in Table 14.4. Notice that, in presenting the regression results, we have not given the value of the coefficient of determination, R^2. Although this coefficient is frequently reported in practice, its interpretation is problematic and can lead to misleading conclusions. For example, a high value for R^2 in the present context would *not* necessarily indicate a strong relationship between local advertising on the one hand and retail sales on the other. Rather, it is a well-known empirical fact that the time plots of many business and economic time series exhibit a rather smooth evolutionary pattern through time. This fact alone is enough to ensure a high value for the coefficient of determination when a lagged dependent variable is included in the regression model. As a practical matter, the reader is advised to pay relatively little attention to the value of R^2 for such models.

The estimated regression equation 14.3.1 can be interpreted as follows. Suppose that retail sales per household increase by \$1 in the current year. The expected impact on local advertising per household is an increase of \$0.0185 in the current year, a further increase of

$$(0.480)(0.0185) = \$0.0089$$

next year, a further increase of

$$(0.480)^2(0.0185) = \$0.0043$$

in 2 years, and so on. The total effect on all future advertising expenditure per household is an expected increase of

$$\frac{0.0185}{1 - 0.480} = \$0.0356$$

Thus, we see that the expected effect of an increase in sales is an immediate increase in advertising expenditure, a further smaller increase in the following year, a yet smaller increase 2 years ahead, and so on. This is illustrated in Figure 14.3, from which we see that, as time goes on, the effect of an increase in sales in the current year on advertising in distant future years becomes negligible.

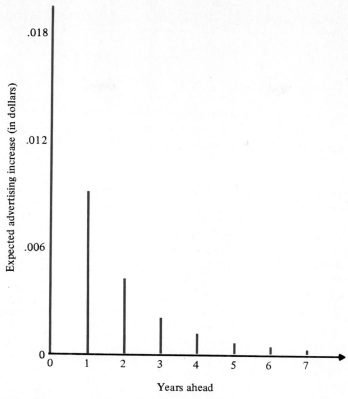

FIGURE 14.3 Expected future increases in local advertising per household

As we have already mentioned, no new principles are involved in finding confidence intervals for, or testing hypotheses about, the partial regression coefficients. To illustrate, we will find a 95% confidence interval for the coefficient β on x_t—that is, the expected increase in the same year in local advertising expenditure per household resulting from a 1 unit increase in retail sales per household. From equation 14.3.1, we have

$$b = 0.0185 \quad \text{and} \quad s_b = 0.0028$$

Since there are $n = 22$ observations and $K = 2$ "independent variables" (including the lagged dependent variable), we have from the tabulated Student's t distribution, for a 95% interval,

$$t_{n-K-1,\alpha/2} = t_{19,.025} = 2.093$$

Therefore, the 95% confidence interval for β is

$$b - t_{n-K-1,\alpha/2}s_b < \beta < b + t_{n-K-1,\alpha/2}s_b$$

that is,

$$0.0185 - (2.093)(0.0028) < \beta < 0.0185 + (2.093)(0.0028)$$

or

$$0.0126 < \beta < 0.0244$$

Thus, if retail sales per household increase by \$1 in the current year, the 95% confidence interval for the expected increase in local advertising expenditures per household this year runs from \$0.0126 to \$0.0244.

Finally, we test the null hypothesis that the coefficient, γ, on the lagged dependent variable is 0. The null hypothesis

$$H_0: \quad \gamma = 0$$

implies that any increase in sales would impact on advertising expenditures in the current year only—not also in future years. We will test against the alternative

$$H_1: \quad \gamma > 0$$

The test is based on the least squares estimate, c, of γ, and the corresponding estimated standard error s_c. Here we have, from the estimated regression 14.3.1,

$$c = 0.480 \qquad \text{and} \qquad s_c = 0.086$$

As usual, the test statistic is the ratio of the coefficient estimate to its standard error; that is,

$$\frac{c}{s_c} = \frac{0.480}{0.086} = 5.581$$

as can be read directly from the printout of Table 14.4. From the tabulated Student's t distribution, we find

$$t_{19, .005} = 2.861$$

Hence, the null hypothesis is very clearly rejected, even at the 0.5% significance level. This implies quite strongly the desirability of including the lagged dependent variable in this regression model if the only other independent variable to be used is the current level of retail sales per household.

EXERCISES

1. The following model was fitted[3] to eight annual sets of observations from 1972 to 1979, in an attempt to explain Organization of Petroleum Exporting Countries (OPEC) pricing behavior.

$$y = 0.37x_1 + 5.22x_2$$
$$(0.029) \quad (0.50)$$

where

---

[3] See W. D. Nordhaus, "Oil and economic performance in industrial countries," *Brookings Papers on Economic Activity* (1980), 341–88.

y = Difference between OPEC price in the current year and OPEC price in the previous year (in dollars per barrel)

x_1 = Difference between spot price in the current year and OPEC price in the previous year

x_2 = Dummy variable taking the value 1 in 1974 and 0 otherwise, to represent the specific effect of the oil embargo of that year

Interpret verbally and graphically the estimated coefficient on the dummy variable.

2. The following model was fitted[4] to data on forty-eight firms.

$$y = 1.399 + 0.246x_1 - 9.507x_2 - 0.016x_3 + 1.490x_4; \qquad \overline{R}^2 = 0.26$$
$$(0.056) \quad (4.244) \quad (0.017) \quad (1.380)$$

where

y = Rate of return on equity for the firm
x_1 = Market share
x_2 = Measure of firm size
x_3 = Industry growth rate
x_4 = Dummy variable taking the value 1 for firms in industries with high or moderate product differentiation, and 0 otherwise

(a) Interpret the estimated coefficient on the dummy variable.
(b) Test the null hypothesis that the true coefficient on the dummy variable is 0, against the alternative that it is positive.
(c) Find a 99% confidence interval for the expected effect of a 1 unit increase in measure of firm size on rate of return on equity, all else equal.

3. In order to try to explain variability in the prices paid by students for apartments, a random sample of thirty apartments was selected, and the following regression equation was estimated:

$$y = 20.5 + 42.5x_1 - 5.2x_2 + 36.9x_3; \quad R^2 = 0.78$$
$$(10.4) \quad (4.1) \quad (5.3)$$

where

y = Rent per month (in dollars)
x_1 = Size of apartment (in hundreds of square feet)
x_2 = Distance from campus (in miles)
x_3 = Dummy variable taking the value 1 if the apartment is furnished, and 0 otherwise

Figures in parentheses beneath the coefficient estimates are their estimated standard errors.

(a) Interpret the estimated coefficient on the dummy variable.
(b) Find a 90% confidence interval for the expected difference in rent per month between a furnished and an unfurnished apartment, all else equal.
(c) Test against the appropriate one-sided alternative the null hypothesis that, all else equal, the rent of an apartment is not affected by its distance from campus.

---

[4] J. A. Dalton and S. L. Levin, "Market power: Concentration and market share," *Industrial Organization Review*, 5 (1977), 27–36.

4. A law school dean wanted to assess the importance of factors that might help in predicting success in law school. For a random sample of fifty students, data were obtained when they graduated from law school, and the following model was fitted.

$$Y_i = \alpha + \beta_1 x_{1i} + \beta_2 x_{2i} + \beta_3 x_{3i} + \varepsilon_i$$

where

Y_i = Score reflecting overall performance while in law school
x_{1i} = Undergraduate grade point average
x_{2i} = Score on Law School Admission Test
x_{3i} = Dummy variable taking the value 1 if the student's letters of recommendation are unusually strong, and 0 otherwise

Use the portion of the computer output from the estimated regression shown here to write a report summarizing the findings of this study.

SOURCE	DF	SUM OF SQUARES	MEAN SQUARE	F VALUE	R-SQUARE
MODEL	3	641.04	213.68	8.48	0.356
ERROR	46	1159.66	25.21		
CORRECTED TOTAL	49	1800.70			

PARAMETER	ESTIMATE	T FOR H0: PARAMETER = 0	STD. ERROR OF ESTIMATE
INTERCEPT	6.512		
X1	3.502	1.45	2.419
X2	0.491	4.59	0.107
X3	10.327	2.45	4.213

5. A manufacturer of canned soda takes a random sample of twenty supermarkets in different parts of the country. These supermarkets serve markets of roughly equal sizes. In order to see what factors influence the sales of this brand of soda, the following model was fitted:

$$Y_i = \alpha + \beta_1 x_{1i} + \beta_2 x_{2i} + \beta_3 x_{3i} + \varepsilon_i$$

where

Y_i = Number of cases sold in the week of the study
x_{1i} = Price of the soda relative to average price of competitors' brands
x_{2i} = Average daily high temperature for the week of the study
x_{3i} = Dummy variable taking the value 1 if a special advertising display for the soda was used in the store, and 0 otherwise

Part of the computer output from the estimated regression is shown here. Write a report summarizing the findings of this study.

SOURCE	DF	SUM OF SQUARES	MEAN SQUARE	F VALUE	R-SQUARE
MODEL	3	145.32	48.44	3.80	0.416
ERROR	16	204.00	12.75		
CORRECTED TOTAL	19	349.32			

PARAMETER	ESTIMATE	T FOR H0: PARAMETER = 0	STD. ERROR OF ESTIMATE
INTERCEPT	10.51		
X1	-8.07	-2.04	3.95
X2	1.49	1.47	1.03
X3	6.71	1.29	5.21

6. [*This problem requires a computer program to carry out the multiple regression computations.*] In a student survey of twenty-seven undergraduates at the University of Illinois, the accompanying results were obtained on grade point average (y), number of hours per week spent on studying (x_1), average number of hours spent preparing for tests (x_2), number of hours per week spent in bars (x_3), whether students take notes or mark highlights when reading texts ($x_4 = 1$ if yes, $x_4 = 0$ if no), and average number of credit hours taken per semester (x_5). Estimate the regression of grade point average on the five independent variables, and write a report on your findings.

y	x_1	x_2	x_3	x_4	x_5
4.8	25	5	6	1	16
4.3	22	2	1	1	15
3.8	9	3	4	0	15
3.8	15	2	8	1	17
4.2	15	3	4	0	15
4.3	30	5	3	1	13
3.8	20	6	7	1	17
4.3	10	3	5	1	19
4.0	30	2	6	1	19
3.8	20	6	6	0	15
3.1	10	4	7	0	16
3.9	18	3	5	1	15
3.2	10	3	4	0	16
4.9	18	3	4	0	17
4.4	12	2	4	1	16
4.5	12	1	4	0	17
4.6	10	10	3	1	17
4.0	28	1	4	0	15
3.7	15	4	6	1	14
3.5	12	2	7	0	17
2.8	28	4	6	1	15
4.3	10	8	5	1	15
5.0	25	4	5	1	19
3.0	25	1	4	1	16
4.1	30	3	6	1	18
4.1	25	3	7	1	17
4.6	25	4	7	1	15

7. [*This problem requires a computer program to carry out the multiple regression computations.*] For a random sample of twelve divorces, the data shown in the table were obtained on length of marriage, in years (y), age of wife at time of marriage (x_1), whether wife worked ($x_2 = 1$ if yes, $x_2 = 0$ if no), and whether the couple was strongly religious ($x_3 = 1$ if yes, $x_3 = 0$ if no). Estimate the regression of duration of marriage on these three independent variables, and write a report on your findings.

y	x_1	x_2	x_3	y	x_1	x_2	x_3
20	20	0	0	3	42	1	0
18	21	0	1	15	24	0	0
10	20	1	0	15	23	1	0
15	22	0	1	5	39	1	0
8	20	1	0	10	25	1	1
9	20	1	1	10	24	0	0

8. The accompanying table[5] shows annual percentage growth rates for employment (x_1) and productivity (y) for twelve countries.

	x_1	y		x_1	y
Japan	5.8	7.8	Netherlands	1.4	4.1
Italy	3.9	4.2	Belgium	1.2	3.9
West Germany	2.8	4.5	Norway	0.2	4.4
Austria	2.2	4.2	Canada	2.1	1.3
France	1.8	3.8	United Kingdom	0.4	2.8
Denmark	2.5	3.2	United States	0.0	2.6

(a) Estimate the linear regression model

$$Y_i = \alpha + \beta x_{1i} + \varepsilon_i$$

and calculate any statistics needed to make inference about the dependence of productivity growth on employment growth.

(b) A colleague now suggests that the experience of Japan has been rather special. To take account of this, you estimate the regression

$$Y_i = \alpha + \beta_1 x_{1i} + \beta_2 x_{2i} + \varepsilon_i$$

where x_{2i} is a dummy variable taking the value 1 for Japan and 0 for the other eleven countries. Carry out the regression computations.

(c) Compare and comment on the results in parts (a) and (b).

(d) Show that the results in part (b) could have been obtained by fitting the simple linear regression of productivity growth on employment growth for the eleven countries remaining when Japan is excluded. Explain why this is so.

9. The equation

$$Y_t = \alpha + \beta x_t + \gamma Y_{t-1} + \varepsilon_t$$

---

[5] These data are given in N. Kaldor, *Strategic Factors in Economic Development,* New York State School of Industrial and Labor Relations, Cornell University: Ithaca, New York (1967).

was fitted to 56 quarterly observations, where

Y_t = United States consumption
x_t = United States income

The least squares estimates of β and γ (with standard errors in parentheses) were

$$b = 0.31 \quad (0.09) \qquad \text{and} \qquad c = 0.66 \quad (0.11)$$

Write a brief report analyzing these results.

10. [*This problem requires either the material in Appendix A13.1 or a computer program to carry out the multiple regression computations.*] In Chapter 12, using the data of Table 12.4, we estimated the regression model

$$Y_t = \alpha + \beta x_t + \varepsilon_t$$

where

Y_t = Retail sales per household
x_t = Disposable income per household

Use these same data to estimate the regression model

$$Y_t = \alpha + \beta x_t + \gamma Y_{t-1} + \varepsilon_t$$

and test the null hypothesis that $\gamma = 0$.

11. [*This problem requires a computer program to carry out the multiple regression computations.*] The table[6] shows 28 quarterly observations from the United Kingdom on quantity of money, M3 in million pounds (y), income in million pounds (x_1), and the local authority interest rate (x_2). Estimate the model

$$Y_t = \alpha + \beta_1 x_{1t} + \beta_2 x_{2t} + \gamma Y_{t-1} + \varepsilon_t$$

and write a report on your findings.

y	x_1	x_2	y	x_1	x_2
17,602.5	14,744	0.0805	17,965.1	15,950	0.0582
17,746.9	14,516	0.0828	18,651.9	15,957	0.0482
17,769.0	14,815	0.0781	19,352.7	16,031	0.0480
17,909.1	14,900	0.0738	20,446.1	16,295	0.0513
17,855.0	14,829	0.0798	20,835.3	16,151	0.0762
17,470.8	14,900	0.0914	21,827.4	16,803	0.0791
17,352.0	14,980	0.0957	22,375.2	17,528	0.1009
17,481.6	15,085	0.0922	23,217.0	17,301	0.0919
17,240.2	14,973	0.0910	24,011.6	17,503	0.1173
17,467.2	15,359	0.0813	24,975.2	17,455	0.1411
17,619.7	15,362	0.0754	24,736.3	16,620	0.1566
17,683.8	15,540	0.0718	23,407.3	17,779	0.1333
17,954.1	15,404	0.0753	23,560.7	18,040	0.1313
17,734.9	15,649	0.0666	23,421.2	17,827	0.1263

[6] See T. C. Mills, "The functional form of the UK demand for money," *Applied Statistics, 27* (1978), 52–57.

12. [*This problem requires a computer program to carry out the multiple regression computations.*] In Chapter 13, using the savings and loan association data of Table 13.1, we estimated the regression model

$$Y_t = \alpha + \beta_1 x_{1t} + \beta_2 x_{2t} + \varepsilon_t$$

where

Y_t = Percentage profit margin
x_{1t} = Net revenues per deposit dollar
x_{2t} = Number of offices

Use the same data to estimate the regression model

$$Y_t = \alpha + \beta_1 x_{1t} + \beta_2 x_{2t} + \gamma Y_{t-1} + \varepsilon_t$$

and test the null hypothesis that $\gamma = 0$.

13. [*This problem requires a computer program to carry out the multiple regression computations.*] The accompanying table[7] shows 20 quarterly observations on income (y) and money supply (x) in Canada. Estimate the model

$$Y_t = \alpha + \beta x_t + \gamma Y_{t-1} + \varepsilon_t$$

and write a report on your findings.

y	x	y	x
116,652	42,011	168,732	63,738
120,392	43,313	173,980	66,338
124,572	44,808	182,744	68,694
132,624	47,324	190,172	72,238
139,656	50,094	191,592	74,544
145,320	52,117	195,600	77,300
150,164	54,253	201,204	80,021
153,560	56,512	204,160	83,482
157,328	59,243	210,780	85,868
161,740	60,783	214,712	87,911

14. [*This problem requires a computer program to carry out the multiple regression computations.*] The table[8] shows 20 annual observations on the first confinement resulting in a live birth of the current marriage (y) and the number of first marriages (for females) in the previous year (x) in Australia. Estimate the model

$$Y_t = \alpha + \beta x_t + \gamma Y_{t-1} + \varepsilon_t$$

and write a report on your findings.

---

[7] These data are taken from C. Hsiao, "Autoregressive modeling of Canadian money and income data," *Journal of American Statistical Association, 74* (1979), 553–60.

[8] These data are taken from J. McDonald, "Modeling demographic relationships: An analysis of forecast functions for Australian births," *Journal of American Statistical Association, 76* (1981), 782–92.

y	x	y	x
65,792	63,488	76,127	87,110
65,431	65,471	81,341	90,608
66,717	65,956	85,650	96,553
66,890	65,902	88,412	102,186
70,177	67,077	95,418	105,235
68,310	68,609	91,683	106,337
69,130	70,849	85,707	102,106
68,581	72,833	86,248	99,950
70,197	77,670	81,543	98,031
73,462	84,850	78,086	90,010

14.4 NONLINEAR MODELS

In our discussion of regression modelling so far, we have assumed that the relationship between a dependent variable and a set of independent variables is *linear*. The assumption of linearity frequently provides a useful and easily analyzed approximation to a far more complex reality, at least within relevant ranges of the independent variables. However, on many occasions, it is both necessary and desirable to abandon the linearity assumption. Often subject matter theory, or data, or both will suggest that the appropriate formulation is *nonlinear*. The difficulty lies in the fact that, once our horizons are expanded in this way, an infinity of nonlinear functional forms is possible. Moreover, in business and economic applications, only very rarely does subject matter theory postulate a specific credible functional form. Nevertheless, nonlinear models are built on occasion, and frequently this can be achieved through only minor modifications of the techniques we have discussed so far.

To illustrate, consider the case of a dependent variable, Y, and a single independent variable, X_1. The linear regression model of Chapter 12 is

$$Y_i = \alpha + \beta x_{1i} + \varepsilon_i$$

One possible alternative to this model is to postulate **quadratic** dependence—that is, the dependence of Y on both X and X^2. The model then is

$$Y_i = \alpha + \beta_1 x_{1i} + \beta_2 x_{1i}^2 + \varepsilon_i \qquad (14.4.1)$$

But this is simply a multiple regression equation involving three unknown coefficients, α, β_1, and β_2. These can be estimated by least squares, using precisely the procedures described in the previous chapter. The model is simply treated as one relating the dependent variable to a pair of "independent variables," x_{1i} and x_{1i}^2. To make the point transparent, we can write equation 14.4.1 as

$$Y_i = \alpha + \beta_1 x_{1i} + \beta_2 x_{2i} + \varepsilon_i$$

where

$$x_{2i} = x_{1i}^2$$

Confidence intervals and hypothesis tests for the parameters β_1 and β_2 of the model 14.4.1 are obtained in the usual way. In particular, the desirability of employing the quadratic rather than the linear form can be checked by testing the null hypothesis that β_2 is 0. If that hypothesis were indeed correct, then 14.4.1 would simplify to the linear model.

A more common alternative to the linear model in business and economic applications is the **log linear model.** In this formulation, a linear relationship is postulated, not between the variables themselves, but between their *logarithms*. With just a single independent variable, this model is written

$$\log Y_i = \alpha + \beta \log x_i + \varepsilon_i$$

Such models are easily estimated by least squares, and analyzed through the methods described in Chapter 12. All that is necessary is to take logarithms of the observations on the dependent and independent variables first and proceed with these quantities through the remainder of the analysis.

The log linear model depicts a curvilinear relationship between the values of the dependent and independent variables, as illustrated in Figure 14.4. Part (a) of this figure shows the case where an increase in the independent variable leads to an expected increase in the dependent variable, while part (b) shows an increase in the former leading to an expected decrease in the latter. One way to recognize the possibility that a log linear specification might be appropriate is to graph the data and check whether one of the curvilinear forms of Figure 14.4 seems, from a visual perspective, to provide a better representation of the relationship than a linear form.

The log linear model can be extended to the case of several independent variables. As we note in the box, interpretation of the model parameters is somewhat different in this model than in the multiple regression model of the previous chapter.

The Log Linear Model

Suppose that a dependent variable Y is related to K independent variables, X_1, X_2, . . . , X_K. If the independent variables take the specific values $x_{1i}, x_{2i}, . . . , x_{Ki}$, the population **log linear model** is of the form

$$\log Y_i = \alpha + \beta_1 \log x_{1i} + \beta_2 \log x_{2i} + \cdots + \beta_K \log x_{Ki} + \varepsilon_i$$

where $\alpha, \beta_1, \beta_2, . . . , \beta_K$ are constants, and ε_i is a random variable with mean 0.

The coefficient β_j is the expected *percentage* increase in the dependent variable, resulting from a one *percent* increase in the independent variable X_j, when the values of all the other independent variables are held fixed.

The interpretation of the coefficients of the log linear model points up a feature that makes this formulation attractive in some applications. If, whatever its value, a

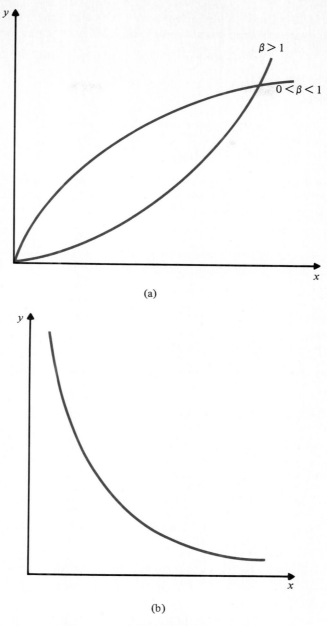

(b)

FIGURE 14.4 Log linear relationships between y and x

1% increase in the value of an independent variable is expected to lead to a fixed *percentage* increase in the dependent variable, then the log linear model is likely to be the preferred functional form.

Example 14.1

In a study[9] of the determinants of household expenditures on vacation travel, data were obtained from a sample of 2,246 households. The model estimated was

$$\log y = -4.0464 + 1.1556 \log x_1 - 0.4408 \log x_2; \quad R^2 = 0.168$$
$$ (0.0546) (0.0490)$$

where

y = Expenditure on vacation travel
x_1 = Total annual consumption expenditure
x_2 = Number of members in household

The figures in parentheses under the least squares parameter estimates are the corresponding estimated standard deviations. Part of the computer output from the fitted regression is shown here.

		R-SQUARE 0.168	
PARAMETER	ESTIMATE	T FOR H0: PARAMETER=0	STD. ERROR OF ESTIMATE
INTERCEPT	-4.0464		
LOG X1	1.1556	21.16	0.0546
LOG X2	-0.4408	-9.00	0.0490

The reported estimates have the following implications:

(a) For a fixed number of household members, a 1% increase in total annual consumption expenditure leads to an expected increase of 1.1556% in expenditure on vacation travel.

(b) For a fixed total annual consumption expenditure, a 1% increase in the number of household members leads to an expected *decrease* of 0.4408% in expenditure on vacation travel. More realistically, a 50% increase in the number of household members leads to an expected decrease of $(50)(0.4408) = 22.04\%$ in expenditure on vacation travel.

(c) For this sample, 16.8% of the variability in $\log y$ is explained by its linear dependence on $\log x_1$ and $\log x_2$.

Confidence intervals for the regression coefficients can be obtained in the usual way. To illustrate, we find a 95% confidence interval for the parameter β_1. This interval is of the form

$$b_1 - t_{n-K-1,.025} s_{b_1} < \beta_1 < b_1 + t_{n-K-1,.025} s_{b_1}$$

where

$$n = 2,246; \quad K = 2; \quad b_1 = 1.1556; \quad s_{b_1} = 0.0546$$

---

[9]See R. P. Hagermann, "The determinants of household vacation travel: some empirical evidence," *Applied Economics*, 13 (1981), 225–34.

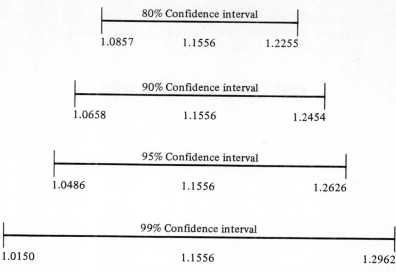

FIGURE 14.5 80%, 90%, 95%, and 99% confidence intervals for expected percentage increase in expenditure on vacation travel resulting from a 1% increase in total annual consumption expenditure, when the number of household members remains fixed

Table 6 of the Appendix does not give cutoff points for the Student's t distribution with 2,243 degrees of freedom. However, for such a large number of degrees of freedom, a very close approximation is provided by the values in the last row of that table, where the degrees of freedom are infinite. Thus, we have

$$t_{2,243,.025} \approx t_{\infty,.025} = 1.96$$

Our confidence interval is then

$$1.1556 - (1.96)(0.0546) < \beta_1 < 1.1556 + (1.96)(0.0546)$$

or

$$1.0486 < \beta_1 < 1.2626$$

Thus, the 95% confidence interval for the expected percentage increase in expenditure on vacation travel resulting from a 1% increase in total annual consumption expenditure, when the number of household members remains fixed, ranges from 1.0486% to 1.2626%. Figure 14.5 shows also 80%, 90%, and 99% confidence levels.

14.5 SPECIFICATION BIAS

The specification of a statistical model that adequately depicts real-world behavior is a delicate and difficult task. Certainly no simple model can describe perfectly the precise nature of the actual determinants of a quantity of interest. The objective of model building, rather, is to attempt to discover a straightforward formulation that is not too drastically at variance with the complex underlying reality. While simplicity of model form is certainly an advantage, substantial divergence of the model from actuality can result in seriously erroneous conclusions about the behavior of the system under study.

One important aspect of model formulation, briefly discussed in Section 14.4, is the appropriate specification of the functional form linking the dependent and independent variables. If the assumed form differs substantially from the true form, then any conclusions drawn from the estimated model will be of dubious value. Another important part of model specification concerns the assumptions made about the statistical properties of the error terms in a regression equation. In our analyses so far, we have assumed that these errors all have the same variance and are uncorrelated with one another. If these assumptions are in fact true, then we have seen that the least squares method and the inferential procedures flowing from it provide a convenient procedure for learning about the process under study. However, if these assumptions are seriously violated, this is no longer necessarily the case. This aspect of model specification will be considered in greater detail in Sections 14.7 and 14.8.

In the present section, we will discuss one specific form of model misspecification and highlight its potential consequences. In formulating a regression model, an investigator attempts to relate the dependent variable of interest to all its *important* determinants. Thus, if the linear model is adopted as the appropriate form, one wants to include among the independent variables all those quantities that it is believed might markedly influence the dependent variable of interest. In formulating the regression model

$$Y_i = \alpha + \beta_1 x_{1i} + \beta_2 x_{2i} + \cdots + \beta_K x_{Ki} + \varepsilon_i$$

it is implicitly assumed that the set of independent variables, $X_1, X_2, \cdots, X_K$, contains *all* those quantities that impact significantly on the behavior of the dependent variable, Y. Certainly, in any practical problem, there will be other factors which, in some minor way, also affect the dependent variable. The joint influence of these factors is absorbed within the error term, ε_i. It is, however, potentially of very great importance that no important explanatory variable be omitted from the list of independent variables.

Except in the very special (and rare) case where omitted variables are uncorrelated with those independent variables included in the regression model, very serious consequences can follow from this type of misspecification. In particular, the least squares estimates will generally be biased and the usual inferential statements derived from confidence intervals or tests of hypotheses can be seriously misleading.

To illustrate this particular type of specification bias, we will consider further an example used in Chapter 13. Table 13.1 gives 25 years of annual data on the percentage profit margin (y) of savings and loan associations, their percentage net revenues per deposit dollar (x_1), and the number of offices (x_2). In Sections 13.2 and 13.6, we found the estimated regression equation linking profits to revenues and number of offices to be

$$y = 1.565 + 0.237x_1 - 0.000249x_2; \qquad R^2 = 0.87$$
$$(0.079) \quad (0.0555) \quad (0.0000320)$$

(14.5.1)

One conclusion that follows from this analysis is that, for a fixed number of offices, a 1 unit increase in net revenues per deposit dollar leads to an expected *increase* of 0.237 unit in profit margin.

Now, suppose that our only interest is in the effect of net revenues on profit margins. One approach to this problem might be to estimate the regression of profit margin on net revenue, using these 25 pairs of observations. Such an analysis, in fact, yields the fitted model

$$y = 1.326 - 0.169x_1; \qquad R^2 = 0.50$$
$$(0.139) \quad (0.036)$$

(14.5.2)

Comparing the fitted models 14.5.1 and 14.5.2, we immediately notice that one consequence of ignoring the number of offices in the analysis is a substantial reduction in R^2, the proportion of the variability in the dependent variable explained by the regression.

There is, however, a more serious consequence. The fitted model 14.5.2 implies that an increase of 1 unit in percentage net revenues per deposit dollar leads to an expected *decrease* of 0.169 unit in the percentage profit margin. Moreover, comparing the coefficient estimate with its estimated standard error, we see that the null hypothesis of no linear relation between these variables is comfortably rejected against the alternative that an increase in net revenues leads to an expected *decrease* in profit margins. But such a conclusion is surely counterintuitive! While it is not completely beyond the realm of possibility, we would certainly expect, all else equal, to find high net revenues associated with high profit margins. But, over the 25-year period for which the model 14.5.2 was estimated, all else was *not* equal. In particular, another potentially important variable—the number of savings and loan offices—changed markedly over this period. When this relevant variable is incorporated into the regression analysis, in 14.5.1, an opposite conclusion is reached. It now emerges, as we might have predicted, that the association between profits and net revenues is positive, once the influence of the number of offices is taken into account.

This example rather nicely illustrates the point. If an important explanatory variable is not included in the regression model, then any conclusions drawn about the impacts of other independent variables can be seriously misleading. In this particular case, we have seen that adding a relevant variable could well alter the impression of a significant negative association to the conclusion of significant positive association. Further insight can be gained from casual inspection of the data in Table 13.1. Over the latter part of the period, at least, profit margins fell and net revenues rose, suggesting a negative association between these variables. However, a further look at the data reveals an increase in the number of offices over this same period, suggesting the possibility that this factor could be the root cause of the declining profit margins. The only legitimate way to disentangle the separate effects of the two independent variables on the dependent variable is to model them jointly in a regression equation. This example illustrates the importance of using the multiple regression model rather than simple linear regression equations when there is more than one relevant independent variable.

14.6 MULTICOLLINEARITY

If a regression model is correctly specified, then the least squares estimates are, in the sense of the Gauss-Markov theorem, the best that can be achieved. Nevertheless, in some circumstances, they may not be very good!

To illustrate, consider again the savings and loan association example of Chapter 13. In this particular study, we observed twenty-five pairs of annual values of net revenues per deposit dollar and the number of savings and loan offices, together with the corresponding profit margins. These data were then used to estimate, through a multiple regression model, the separate effects of the two independent variables on profit margins. Imagine, now, that you wanted to study this problem, but were in the fortunate position of the laboratory scientist, able to *design the experiment*. That is, rather than taking what nature happened to give, suppose that you were free to choose twenty-five pairs of values for the independent variables and to observe the impacts of your choices on profit margins. The best possible choice depends somewhat on the objectives of the analysis and involves the interesting statistical topic of **design of experiments.** Since in business and economic applications we are rarely in the position of being able to select a design, we will not pursue this question further. It is, however, instructive to ask what are the *worst* choices you could possibly make.

If you could choose pairs of values for net revenues (x_1) and number of offices (x_2), what would be the silliest choices you could make? One obvious answer is to select the same pair of values each time for the independent variables. Clearly, if all twenty-five experiments were run with net revenues set at 3.0 and the number of offices at 7,000, we could not learn anything about the influence of these factors on profit margins. Only by varying the independent variables is it possible to learn about their impact on the dependent variable.

The following choice is almost equally absurd:

x_{1i}	3.0	3.1	3.2	3.3	3.4	3.5
x_{2i}	8,000	7,900	7,800	7,700	7,600	7,500

and so on. The values of the independent variables are now changed from one

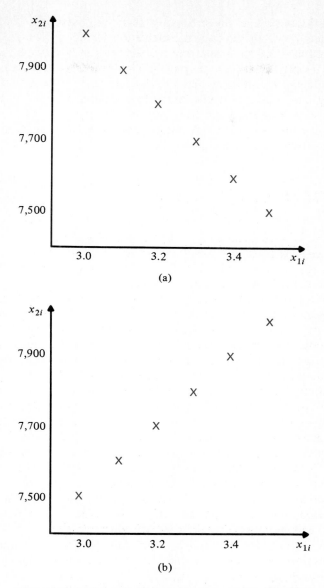

FIGURE 14.6 Two designs with perfect multicollinearity

experimental run to the next, but they are changed in unison. Indeed, these values can be represented as points on a (downward-sloping) straight line, as in Figure 14.6(a). [An essentially similar situation arises if the data points lie on an upward-sloping straight line, as in part (b) of the figure.] Thus, in the particular design given in the table, whenever x_1 is increased, x_2 is decreased by a proportionate amount. To see the futility of such a choice, suppose that you observe that an increase in net revenues, together with the associated decrease in number of offices, leads to an increase in profit margins. To what would you attribute this rise in profit margin? Was it in fact caused by revenue growth, by the decrease in the number of offices, or by some combination of the two? The short answer is that it is impossible to tell! If we want to assess the separate effects of the independent variables, it is essential that they not move in unison through the experiment.

In fact, in the extreme case of the previous example, if we attempted to estimate the coefficients β_1 and β_2 of the regression

$$Y_i = \alpha + \beta_1 x_{1i} + \beta_2 x_{2i} + \varepsilon_i \qquad (14.6.1)$$

by least squares, we would find it impossible to do so. This is because β_1 and β_2 are measures of the *separate* effects of the independent variables, and the data cannot possibly contain any useful information about these quantities. The fifth of the standard assumptions for a multiple regression analysis, set out in Section 13.3, is designed to exclude cases of this sort.

This, then, is clearly a very bad choice of design. A slightly less extreme case is illustrated in Figures 14.7(a) and (b). Here, the design points do not all lie on single straight lines, but are very close to doing so. In this situation, the experimental results are able to provide *some* information about the separate influences of the independent variables, *but not very much*. It will now be possible to calculate least squares point estimates of β_1 and β_2 in 14.6.1, but these will be very imprecise. This imprecision will be reflected in very large standard errors for the estimated coefficients, suggesting statistical insignificance even when, in fact, the relationships involved are quite strong. This phenomenon is referred to as **multicollinearity.**

In the vast majority of practical cases involving business and economic applications, we are not able to design the experiments. Rather, we are constrained to work with the particular experimental design that fate has given us. In this context, then, multicollinearity is a problem arising, not from our own bad choice of design, but from the one with which we must work. More generally, in regression equations involving several independent variables, the multicollinearity problem arises from patterns of strong intercorrelations among the independent variables—the case, in fact, where assumption 5 of Section 13.3 is close to being violated. Perhaps the most frustrating aspect of the problem, which can be summarized as having data that are not very informative about the parameters of interest, is that typically little can be done about it. It is, however, still important to be aware of the problem and to watch for its occurrence. A clear indication of the likely presence of multicollinearity occurs when, taken as a group, a set of independent variables appears to exert considerable influence on the dependent variable, but when looked at individually, through tests of hypotheses, all appear individually to be insignificant. It would certainly be unwise in these

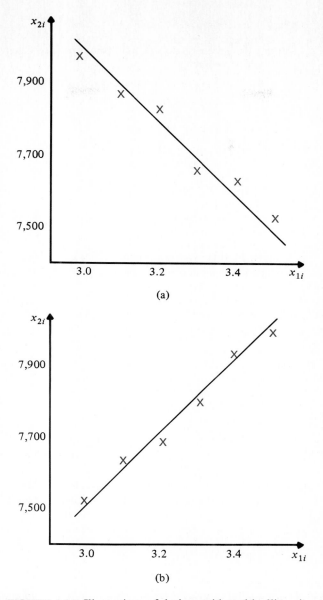

FIGURE 14.7 Illustrations of designs with multicollinearity

circumstances to jump to the conclusion that a particular independent variable did not affect the dependent variable. Rather, it is preferable to acknowledge that, while the group as a whole is clearly influential, the data are not sufficiently informative to allow the disentangling, with any precision, of its members' separate effects.

15. In a study[10] of credit unions, the following regression equation was fitted to eighty-seven quarterly observations:

$$\log y = 0.47 + 0.31 \log x_1 + 0.47 \log x_2; \qquad \bar{R}^2 = 0.86$$
$$(0.13) (0.21)$$

where

y = Output of loans
x_1 = Expenditure on capital equipment
x_2 = Salaries and wages paid to employees

and the figures in parentheses beneath the coefficient estimates are their estimated standard errors.

(a) Interpret the estimated partial regression coefficients.

(b) Test the null hypothesis that, all else equal, an increase in expenditure on capital equipment does not affect the output of loans.

(c) All else equal, find 90%, 95%, and 99% confidence intervals for the percentage increase in output of loans resulting from a 1% increase in salaries and wages paid to employees.

16. From a sample of thirty observations, the following regression model was estimated in an attempt to explain variability in the demand for butter:

$$\log y = -0.05 - 0.29 \log x_1 + 0.23 \log x_2 + 0.09 \log x_3$$
$$(0.12) (0.10) (0.08)$$

where

y = Quantity of butter purchased
x_1 = Price of butter
x_2 = Price of margarine
x_3 = Income of consumers

and the figures in parentheses below the coefficient estimates are their estimated standard errors.

(a) Interpret the estimated partial regression coefficients.

(b) Test the null hypothesis that, all else equal, the price of margarine does not affect the demand for butter.

(c) Test the null hypothesis that, all else equal, the income of consumers does not affect the demand for butter.

(d) Find 80%, 90%, and 95% confidence intervals for the percentage decrease in quantity of butter purchased resulting from a 1% increase in the price of butter, all else equal.

17. [*This problem requires a computer program to carry out the multiple regression calculations.*] Use the data of Exercise 11 to fit the regression model

$$\log Y_t = \alpha + \beta_1 \log x_{1t} + \beta_2 \log x_{2t} + \gamma \log Y_{t-1} + \varepsilon_t$$

and write a report summarizing your findings.

---

[10] These results are given by Y. L. Mahajan, "A macroeconometric model of the credit unions," *Atlantic Economic Journal*, 9 (1981), No. 2, 40–48.

18. [*This problem requires a computer program to carry out the regression calculations.*] The accompanying table shows twenty-five annual observations[11] on sales (y) and advertising expenditure (x), both in thousands of dollars, of Lydia E. Pinkham. Estimate the model

$$\log Y_t = \alpha + \beta \log x_t + \gamma \log Y_{t-1} + \varepsilon_t$$

and write a report on your findings.

y	x	y	x
1,103	339	1,984	981
1,266	562	1,787	974
1,473	745	1,689	766
1,423	749	1,866	920
1,767	862	1,896	964
2,161	1,034	1,684	811
2,336	1,054	1,633	789
2,602	1,164	1,657	802
2,518	1,102	1,569	770
2,637	1,145	1,390	639
2,177	1,012	1,387	644
1,920	836	1,289	564
1,910	941		

19. For a random sample of fifty observations, an economist estimated the regression model

$$\log Y_i = \alpha + \beta_1 \log x_{1i} + \beta_2 \log x_{2i} + \beta_3 \log x_{3i} + \beta_4 \log x_{4i} + \varepsilon_i$$

where

Y_i = Gross revenue from a medical practice
x_{1i} = Average number of hours worked by physicians in the practice
x_{2i} = Number of physicians in the practice
x_{3i} = Number of allied health personnel (such as nurses) employed in the practice
x_{4i} = Number of rooms used in the practice

Use the portion of the computer output shown here to write a report on these results.

```
                              R-SQUARE
                               0.927

                            T FOR HO:          STD. ERROR OF
PARAMETER      ESTIMATE      PARAMETER=0        ESTIMATE

INTERCEPT      2.347
LOG X1         0.239         3.27               0.073
LOG X2         0.673         8.31               0.081
LOG X3         0.279         6.64               0.042
LOG X4         0.082         1.61               0.051
```

---

[11] Data given in G. M. Erickson, "Using ridge regression to estimate directly lagged effects in marketing," *Journal of American Statistical Association, 76* (1981), 766–73, and K. S. Palda, *The Measurement of Cumulative Advertising Effects,* Prentice-Hall (1964).

20. For a sample of thirty-five industries, an economist estimated the regression model

$$\log Y_i = \alpha + \beta_1 \log x_{1i} + \beta_2 \log x_{2i} + \varepsilon_i$$

where

Y_i = Industry concentration ratio, measured by the percentage of output accounted for by the largest four firms in the industry

x_{1i} = Amount of capital required for entry into the industry, as measured by the size of a single efficient plant

x_{2i} = Measure of economies of scale in the industry, measured by the ratio of average plant size for the largest firms (those that account for 50% of industry output) to the total output in the industry

Part of the computer output is shown here. Write a report on these results.

```
                             R-SQUARE
                              0.825


                             T FOR HO:        STD. ERROR OF
PARAMETER      ESTIMATE      PARAMETER=0         ESTIMATE

INTERCEPT       3.621
LOG X1          0.234          3.55               0.066
LOG X2          0.267          3.18               0.084
```

21. The accompanying table shows total state government revenues per $1,000 of personal income (x) and state appropriations for higher education per $1,000 of personal income (y), for forty-nine states in 1980. (Alaska is excluded.)

	x	y		x	y
Alabama	138.48	16.02	Nebraska	103.86	12.72
Arizona	124.35	13.41	Nevada	112.25	9.13
Arkansas	138.85	13.00	New Hampshire	90.75	4.65
California	129.83	14.14	New Jersey	100.02	6.23
Colorado	110.42	11.41	New Mexico	211.45	15.78
Connecticut	98.65	7.68	New York	137.25	10.57
Delaware	164.44	10.71	North Carolina	129.71	15.82
Florida	96.46	9.37	North Dakota	167.74	16.18
Georgia	117.39	11.30	Ohio	88.01	7.93
Hawaii	195.89	15.95	Oklahoma	127.41	11.13
Idaho	133.88	13.58	Oregon	136.33	12.62
Illinois	100.38	8.76	Pennsylvania	112.33	8.12
Indiana	93.41	9.93	Rhode Island	150.09	10.23
Iowa	116.19	13.10	South Carolina	140.44	16.31
Kansas	98.83	12.91	South Dakota	136.90	10.54
Kentucky	143.62	12.98	Tennessee	111.06	11.15
Louisiana	157.30	12.39	Texas	101.43	13.08
Maine	152.65	8.34	Utah	158.34	16.93
Maryland	124.87	9.34	Vermont	174.41	8.46
Massachusetts	131.55	6.88	Virginia	112.81	11.24

	x	y		x	y
Michigan	118.71	10.37	Washington	128.64	14.59
Minnesota	145.95	14.53	West Virginia	157.32	12.88
Mississippi	165.42	17.59	Wisconsin	139.75	13.30
Missouri	91.40	8.81	Wyoming	179.54	14.12
Montana	156.57	11.42			

(a) Estimate the regression model

$$\log Y_i = \alpha + \beta \log x_i + \varepsilon_i$$

and comment on the results.

(b) In fact, for Alaska, state government revenues were \$661.13 and state appropriations for higher education were \$16.42 per thousand dollars of personal income. Why do you think the observation was omitted?

22. In the problem of Exercise 3, suppose that a regression of apartment rent on apartment size and distance from campus had been run, ignoring the dummy variable. Under what circumstances would you expect to find serious bias in the estimators of the model parameters?

23. Suppose that a regression relationship is given by

$$Y_i = \alpha + \beta_1 x_{1i} + \beta_2 x_{2i} + \varepsilon_i$$

If the simple linear regression of Y_i on x_{1i} is estimated from a sample of n observations, the resulting slope estimate will generally be biased for β_1. However, in the special case where the sample correlation between the x_{1i} and x_{2i} is 0, this will not be so. In fact, in that case, the same estimate results whether or not x_{2i} is included in the regression equation.

(a) Explain verbally why this statement is true.

(b) Show algebraically, using the material in Appendix A13.1, that this statement is true.

24. An economist wants to estimate a regression equation relating demand for a product (y) to its price (x_1) and income (x_2). It is to be based on 12 years of quarterly data. However, it is known that, for this product, demand is seasonal—that is, it is higher at certain times of the year than at others.

(a) One possibility for accounting for seasonality is to estimate the model

$$Y_t = \alpha + \beta_1 x_{1t} + \beta_2 x_{2t} + \beta_3 x_{3t} + \beta_4 x_{4t} + \beta_5 x_{5t} + \beta_6 x_{6t} + \varepsilon_t$$

where $x_{3t}, x_{4t}, x_{5t}, x_{6t}$ are dummy variables, with

$x_{3t} = 1$ in first quarter of each year, and 0 otherwise
$x_{4t} = 1$ in second quarter of each year, and 0 otherwise
$x_{5t} = 1$ in third quarter of each year, and 0 otherwise
$x_{6t} = 1$ in fourth quarter of each year, and 0 otherwise

Explain why this model cannot be estimated by least squares.

(b) A model that can be estimated is

$$Y_t = \alpha + \beta_1 x_{1t} + \beta_2 x_{2t} + \beta_3 x_{3t} + \beta_4 x_{4t} + \beta_5 x_{5t} + \varepsilon_t$$

Interpret the coefficients on the dummy variables in this model.

25. In the regression model

$$Y_i = \alpha + \beta_1 x_{1i} + \beta_2 x_{2i} + \varepsilon_i$$

the extent of any multicollinearity can be evaluated by finding the correlation between x_{1i} and x_{2i} in the sample. Explain why this is so.

26. An economist estimates the regression model

$$Y_i = \alpha + \beta_1 x_{1i} + \beta_2 x_{2i} + \varepsilon_i$$

The estimates of the parameters β_1 and β_2 are not very large, compared with their respective standard errors. On the other hand, the size of the coefficient of determination indicates quite a strong relationship between the dependent variable and the pair of independent variables. Having obtained these results, the economist strongly suspects the presence of multicollinearity. Since his chief interest is in the influence of x_1 on the dependent variable, he decides that he will avoid the problem of multicollinearity by regressing Y on x_1 alone. Comment on this strategy.

14.7 HETEROSCEDASTICITY

The least squares estimation procedure and the inferential procedures based on it are predicated on certain assumptions, discussed in Section 13.3. When these assumptions do, in fact, hold, the methods discussed in Chapter 13 provide a powerful set of tools for practical analysis of data in many fields. However, it can be the case that when one or more of the assumptions break down, least squares estimation of the coefficients of a regression model is inefficient, and the inferences drawn could be badly misleading.

In Section 14.5, one possible difficulty—specification bias—was considered. We saw there that, if important explanatory variables are omitted from a regression model, any inference drawn about the influences of the remaining variables is suspect. In this and the next section we will consider the possibility of violation of two of the standard assumptions about the error terms, ε_i, of the model

$$Y_i = \alpha + \beta_1 x_{1i} + \beta_2 x_{2i} + \cdots + \beta_K x_{Ki} + \varepsilon_i$$

Specifically, we have assumed that these error terms all have the same variance and are uncorrelated with one another. In the following section we will examine the possibility of correlated errors. For the moment, we consider the assumption of fixed variance.

That this assumption may not be plausible can be illustrated by a simple example. Suppose we are interested in the factors affecting output in a particular industry. Data are collected from several firms on the volume of output and such determinants as the quantities of labor and capital employed. Now, these firms will not generally be of the same size, and it might be expected that, for those firms with the highest outputs, the error terms from the postulated model will, on average, be larger in magnitude than those for the smallest firms. Thus, the error variances will not be the same for all firms, but rather an increasing function of firm size.

Models in which the error terms do not all have the same variance are said to exhibit **heteroscedasticity**. When this phenomenon is present, least squares is not the most efficient procedure for estimating the coefficients of the regression model. Moreover, the usual procedures for deriving confidence intervals and tests of hypotheses for these coefficients are no longer valid. It is, therefore, extremely important to have techniques for detecting the presence of heteroscedasticity. Most of the approaches in common use aim to check the assumption of constant error variance against some plausible alternative. It may be, for instance, that the size of the error variance is an increasing function of the size of one or another of the independent variables. Another possibility is that the error variance depends on the expected value of the dependent variable—that is, on $(\alpha + \beta_1 x_{1i} + \cdots + \beta_K x_{Ki})$.

Now, let $a, b_1, b_2, \ldots, b_K$ be the usual least squares estimates of the coefficients $\alpha, \beta_1, \beta_2, \ldots, \beta_K$. Then, natural estimates of the expected values of the dependent variables, given the independent variables, are provided by

$$\hat{y}_i = a + b_1 x_{1i} + b_2 x_{2i} + \cdots + b_K x_{Ki}$$

Similarly, the error terms ε_i are estimated by the **residuals**

$$e_i = y_i - \hat{y}_i$$

Graphical techniques for detecting heteroscedasticity are frequently of value. In practice, a series of graphs, in which the residuals e_i are plotted against the individual independent variables or against the expected values $\hat{y}_i$ can be examined. To illustrate, consider Figures 14.8 (a) and (b), which show possible plots of e_i against the independent variable x_{1i}. In part (a) of the figure we see that the magnitudes of the errors tend to increase with increasing x_i. This suggests that the error variances are not constant.

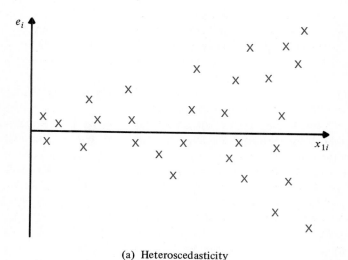

(a) Heteroscedasticity

FIGURE 14.8 Plots of residuals against an independent variable

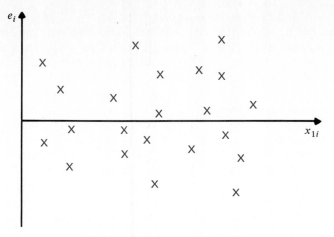

(b) No apparent heteroscedasticity

FIGURE 14.8 Plots of residuals against an independent variable (cont.)

By contrast, in part (b) of the figure, there appears to be no systematic relationship. In this latter case, then, there is no evidence to suggest the absence of constant error variance.

In Chapter 13, we related savings and loan association profit margins to net revenues per deposit dollar (x_1) and the number of offices (x_2) through the model

$$Y_i = \alpha + \beta_1 x_{1i} + \beta_2 x_{2i} + \varepsilon_i$$

The coefficients of this model were estimated by least squares, with the implicit assumption that the error terms all have the same variance. We will now check this assumption. Figures 14.9 (a) and (b) show the plots of the residuals from the least squares regression against the two independent variables. (The independent variables are tabulated in Table 13.1 and the residuals in Table 13.2.) From these two graphs there appears to be no systematic relationship between the magnitude of the residuals and the values of either independent variable.

In Figure 14.10, the residuals from the least squares regression are plotted against the predicted values, $\hat{y}_i$, of the dependent variable. (The values $\hat{y}_i$ for the fitted regression are given in Table 13.2.) Once again, there appears to be no strong relationship between the magnitudes of the residuals and the sizes of the predicted values of the dependent variables. Taken together, then, for these particular data, Figures 14.9 and 14.10 do not suggest the presence of heteroscedasticity.

In the remainder of this section, we will consider more formal procedures for detecting heteroscedasticity and for estimating the coefficients of regression models when it is strongly suspected that the assumption of constant error variances is untenable. In fact, a multitude of such procedures exists, tailored to meet the myriad of ways in which departures from this assumption might occur. We will deal here with a single possibility, which commonly arises in practice. Specifically, we will entertain

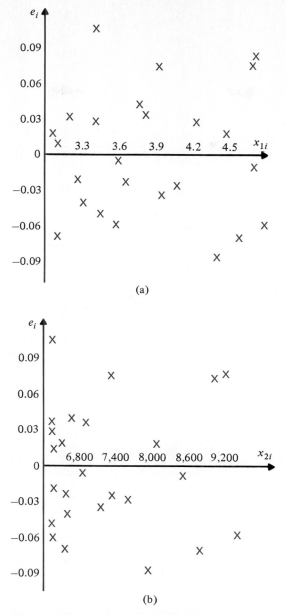

FIGURE 14.9 Plots of residuals against the independent variables for savings and loan data

the alternative hypothesis that the variance of the regression error terms, ε_i, depends on the expected value of the dependent variable, given the values of the independent variables. The test procedure is specified in the box.

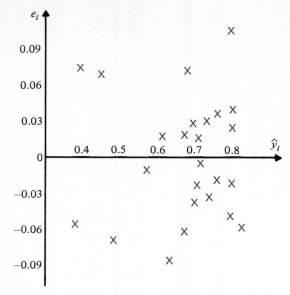

FIGURE 14.10 Plot of e_i against $\hat{y}_i$ for savings and loan data

---

A Test for Heteroscedasticity

Consider a regression model

$$Y_i = \alpha + \beta_1 x_{1i} + \beta_2 x_{2i} + \cdots + \beta_K x_{Ki} + \varepsilon_i$$

linking a dependent variable to K independent variables, and based on n sets of observations. Let $a, b_1, b_2, \ldots, b_K$ be the usual least squares estimates of the coefficients of this model, so that the predicted values of the dependent variable are

$$\hat{y}_i = a + b_1 x_{1i} + b_2 x_{2i} + \cdots + b_K x_{Ki}$$

and the residuals from the fitted model are

$$e_i = y_i - \hat{y}_i$$

In order to test the null hypothesis that the error terms, ε_i, all have the same variance, against the alternative that their variances depend on the expected values

$$\alpha + \beta_1 x_{1i} + \beta_2 x_{2i} + \cdots + \beta_K x_{Ki}$$

we estimate a simple linear regression. In this regression, the dependent variable is the square of the residual—that is, e_i^2—and the independent variable is the predicted value, $\hat{y}_i$.

Let R^2 be the coefficient of determination in this auxiliary regression. Then, for a test of significance level α, the null hypothesis is rejected if nR^2 is bigger than $\chi_{1,\alpha}^2$, where $\chi_{1,\alpha}^2$ is that number exceeded with probability α by a chi-square random variable with 1 degree of freedom.

TABLE 14.5 Quantities required for a test of heteroscedasticity in the savings and loan regression

e_i^2	$\hat{y}_i$	e_i^2	$\hat{y}_i$
0.005329	0.677	0.001600	0.800
0.000016	0.714	0.001225	0.755
0.001521	0.699	0.001156	0.734
0.003969	0.673	0.000625	0.705
0.000256	0.684	0.000729	0.693
0.000144	0.708	0.007225	0.635
0.000900	0.740	0.000289	0.613
0.000361	0.759	0.000100	0.570
0.011025	0.795	0.004900	0.480
0.000676	0.794	0.005184	0.438
0.002401	0.799	0.005476	0.396
0.003364	0.828	0.003364	0.378
0.000484	0.802		

We will carry out this test for the savings and loan example. The quantities employed in the auxiliary regression are tabulated in Table 14.5. The predicted values $\hat{y}_i$ are taken directly from Table 13.2, while the e_i^2 are the squares of the residuals e_i, also given in Table 13.2.

When the regression of these squared residuals on the predicted values is estimated by least squares, we obtain

$$e^2 = 0.00470 - 0.00328\hat{y}; \qquad R^2 = 0.027$$
$$(0.00408)$$

Part of the SAS computer output for this auxiliary regression is shown in Table 14.6.

Since there are $n = 25$ sets of observations, the test is based on

$$nR^2 = (25)(0.027) = 0.675$$

From Table 5 of the Appendix, we find, for a 10% level test,

$$\chi^2_{1,.10} = 2.71$$

Therefore, the null hypothesis that the error terms in the regression of profit margins on net revenues and number of offices all have the same variance cannot be rejected

TABLE 14.6 Part of output from SAS program for auxiliary regression for testing for heteroscedasticity in the savings and loan model

		R-SQUARE 0.027	
PARAMETER	ESTIMATE	T FOR H0: PARAMETER = 0	STD. ERROR OF ESTIMATE
INTERCEPT	0.00470		
YHAT	-0.00328	-0.80	0.00408

at this level. This formal test confirms our impression from Figures 14.9 and 14.10 that there is little evidence in these data to suggest that heteroscedasticity is a serious problem here.

When it is strongly suspected that the regression errors do not all have the same variance, least squares estimation may not be the most appropriate procedure. Several alternatives can be used, the choice depending on how the error variances are thought to behave. Here, we briefly mention one possibility. Suppose that the variance of the error terms is, in fact, directly proportional to the square of the expected value of the dependent variable, given the independent variables, that is,

$$\text{Var}(\varepsilon_i) \propto (\alpha + \beta_1 x_{1i} + \beta_2 x_{2i} + \cdots + \beta_K x_{Ki})^2$$

When this particular form is assumed, a simple two-stage procedure is used to estimate the parameters of the regression model. At the first stage, the model is estimated by least squares in the usual way and the predicted values, $\hat{y}_i$, of the dependent variable are recorded. At the second stage, we estimate the regression equation

$$\frac{Y_i}{\hat{y}_i} = \alpha \frac{1}{\hat{y}_i} + \beta_1 \frac{x_{1i}}{\hat{y}_i} + \beta_2 \frac{x_{2i}}{\hat{y}_i} + \cdots + \beta_K \frac{x_{Ki}}{\hat{y}_i} + u_i \qquad (14.7.1)$$

where u_i is an error term that will have approximately constant variance. Thus, equation 14.7.1 represents a regression model in which the dependent variable is $Y_i/\hat{y}_i$ and the independent variables are $1/\hat{y}_i$, $x_{1i}/\hat{y}_i$, $x_{2i}/\hat{y}_i$, . . . , $x_{Ki}/\hat{y}_i$. (Notice that there is no intercept term.) The coefficients of 14.7.1 are then estimated by least squares, the resulting estimates of α, β_1, β_2, . . . , β_K being retained.

14.8   AUTOCORRELATED ERRORS

We will now examine the possibility of the failure of the assumption that the error terms in a regression model are uncorrelated with one another. In situations where the data analyzed consist of a random sample from some population, this assumption is quite justifiable. However, for another extremely important class of regression problems, it must be regarded as somewhat suspect. When time series data are analyzed, the error term represents the amalgam of all those factors (apart from the independent variables) that influence the behavior of the dependent variable, and hence some correlation between errors might be anticipated. The behavior of these factors in the current time period might be quite similar to their behavior in the previous time period, suggesting the possibility of a positive correlation between those errors close together in time.

To emphasize that the subject matter of this section is restricted to time series data, we will, as in Section 14.3, subscript observations by t and write the regression model as

$$Y_t = \alpha + \beta_1 x_{1t} + \beta_2 x_{2t} + \cdots + \beta_K x_{Kt} + \varepsilon_t \qquad (14.8.1)$$

We have assumed that the error terms, ε_t, of such models are not correlated with one another. The consequences of proceeding with the usual least squares analysis when this assumption does not hold can be very serious indeed. In particular, the usual inferential statements based on confidence intervals or hypothesis tests for the model parameters might be very badly misleading.

It is, therefore, critically important in regressions involving time series data to test the hypothesis that the error terms are not correlated with one another. The possibility that they are correlated through time is referred to as the problem of **autocorrelated errors.** When approaching this problem, it can be useful to have in mind some specific plausible autocorrelation structure. One appealing possibility is that the error ε_t in time period t is quite strongly correlated with the error ε_{t-1} in the previous time period, rather less strongly correlated with the error ε_{t-2} two time periods earlier, and so on, so that the correlation between error terms separated by a considerable period of time is relatively weak. Let us denote by ρ the correlation between error terms in adjacent time periods, so that

$$\text{Corr}(\varepsilon_t, \varepsilon_{t-1}) = \rho$$

where, since ρ is a correlation coefficient, it will be less than 1 in absolute value. (In most applications it will be reasonable to take this correlation to be either 0 or positive.) A particularly simple formulation, in which the correlation between error terms decreases the further apart they are in time, is to have

$$\text{Corr}(\varepsilon_t, \varepsilon_{t-2}) = \rho^2$$
$$\text{Corr}(\varepsilon_t, \varepsilon_{t-3}) = \rho^3$$

and so on. Thus, for errors separated by j units of time, we have

$$\text{Corr}(\varepsilon_t, \varepsilon_{t-j}) = \rho^j \qquad (j = 1, 2, 3, \ldots) \qquad (14.8.2)$$

Suppose, for example, that the correlation between adjacent errors is 0.5, and that the autocorrelation structure 14.8.2 holds. We then have

$$\text{Corr}(\varepsilon_t, \varepsilon_{t-1}) = 0.5$$
$$\text{Corr}(\varepsilon_t, \varepsilon_{t-2}) = (0.5)^2 = 0.25$$
$$\text{Corr}(\varepsilon_t, \varepsilon_{t-3}) = (0.5)^3 = 0.125$$
$$\text{Corr}(\varepsilon_t, \varepsilon_{t-4}) = (0.5)^4 = 0.0625$$
$$\vdots$$

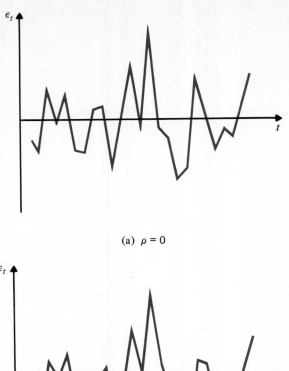

(a) $\rho = 0$

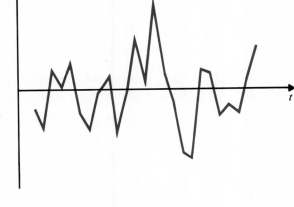

(b) $\rho = 0.3$

FIGURE 14.11 Time plots of residuals from regressions whose error terms follow a first-order autoregressive process

We see then that the correlation between errors far apart in time is relatively weak, while that between errors closer to one another in time is possibly quite strong.

Now, if we assume that the errors ε_t all have the same variance, it is possible to show that the autocorrelation structure 14.8.2 corresponds to the model

$$\varepsilon_t = \rho \varepsilon_{t-1} + u_t \qquad (14.8.3)$$

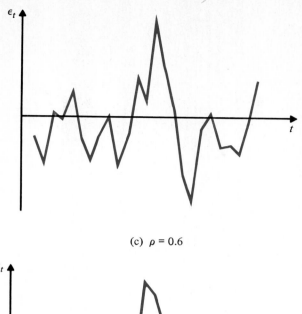

(c) $\rho = 0.6$

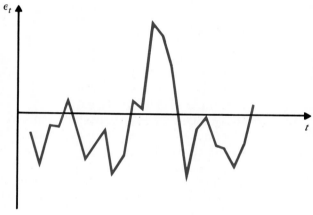

(d) $\rho = 0.9$

FIGURE 14.11 Time plots of residuals from regressions
whose error terms follow a first-order autoregressive
process (cont.)

where the random variable u_t has mean 0, constant variance, and is not autocorrelated.
Equation 14.8.3 depicts the **first-order autoregressive model** of autocorrelated behavior. Looking at this equation, we see that the value taken by the error at time t,
ε_t, depends on its value in the previous time period (the strength of that dependence
being determined by the correlation coefficient ρ) and on a second random term u_t.
This model is illustrated in Figure 14.11, which shows time plots of errors generated
by the model 14.8.3 for values of $\rho = 0, 0.3, 0.6, 0.9$. The case $\rho = 0$ corresponds
to no autocorrelation in the errors. In part (a) of the figure it can be seen that there is
no apparent pattern in the progression through time of the errors. The value taken by
one does not influence the values of others. As we move from relatively weak

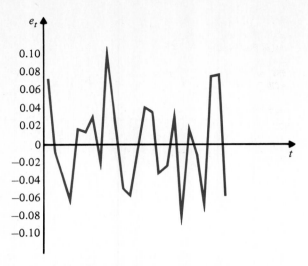

FIGURE 14.12 Time plot of residuals from savings and loan regression

autocorrelation ($\rho = 0.3$) to quite strong autocorrelation ($\rho = 0.9$), in parts (b)–(d), it emerges that the pattern through time of the errors becomes increasingly less jagged, so that in part (d) of the figure it is quite clear that an error is likely to be relatively close in value to its immediate neighbor.

Examination of Figure 14.11 suggests that graphical methods might be useful in detecting the presence of autocorrelated errors. Ideally, we would like to plot the errors ε_t of model 14.8.1 against time. However, since the actual errors will be unknown, estimates of them are obtained as the residuals, e_t, when the model is estimated by least squares. Thus, for practical purposes, it is the time plot of these residuals that is examined. Figure 14.12 shows the time plot of the residuals from our savings and loan regression. (These residuals are set out in Table 13.2.)

Looking at the time plot in Figure 14.12, one does not get the impression of any strong autocorrelation in the residuals. They do not follow the kind of smooth path through time depicted in Figure 14.11(d). On this evidence alone, then, our suspicions of autocorrelated errors would not be terribly strong. However, since the problem is so important, it is desirable to have a more formal test of the hypothesis of no autocorrelation in the errors of a regression model.

The test that is most often used is called the **Durbin-Watson test** and is based on the residuals, e_t, from the least squares estimated regression. The test statistic that is calculated is

$$d = \frac{\sum\limits_{t=2}^{n} (e_t - e_{t-1})^2}{\sum\limits_{t=1}^{n} e_t^2}$$

The computations needed to find this test statistic for the savings and loan regression are set out in Table 14.7. From this table, we find

$$\sum_{t=2}^{25} (e_t - e_{t-1})^2 = 0.121615 \qquad \text{and} \qquad \sum_{t=1}^{25} e_t^2 = 0.062319$$

Hence, the test statistic for this regression is

$$d = \frac{0.121615}{0.062319} = 1.95$$

When regressions involving time series data are reported, it is common practice to include among the summary information the calculated Durbin-Watson statistic. Our savings and loan regression would therefore be reported as

$$y = 1.565 + 0.237x_1 - 0.000249x_2; \qquad R^2 = 0.87; \qquad d = 1.95$$
$$\quad (0.079) \quad (0.0555) \quad (0.0000320)$$

Now, the Durbin-Watson statistic can be written approximately as

$$d = 2(1 - r) \tag{14.8.4}$$

where r is the sample estimate of the correlation, ρ, between adjacent errors.[12] Therefore, if the errors are not autocorrelated, we would expect the value of d to be

TABLE 14.7 Computations for Durbin-Watson statistic for savings and loan regression

t	e_t	e_t^2	$(e_t - e_{t-1})$	$(e_t - e_{t-1})^2$
1	0.073	0.005329		
2	−0.004	0.000016	−0.077	0.005929
3	−0.039	0.001521	−0.035	0.001225
4	−0.063	0.003969	−0.024	0.000576
5	0.016	0.000256	0.079	0.006241
6	0.012	0.000144	−0.004	0.000016
7	0.030	0.000900	0.018	0.000324
8	−0.019	0.000361	−0.049	0.002401
9	0.105	0.011025	0.124	0.015376
10	0.026	0.000676	−0.079	0.006241
11	−0.049	0.002401	−0.075	0.005625
12	−0.058	0.003364	−0.009	0.000081
13	−0.022	0.000484	0.036	0.001296
14	0.040	0.001600	0.062	0.003844
15	0.035	0.001225	−0.005	0.000025
16	−0.034	0.001156	−0.069	0.004761
17	−0.025	0.000625	0.009	0.000081
18	0.027	0.000729	0.052	0.002704
19	−0.085	0.007225	−0.112	0.012544
20	0.017	0.000289	0.102	0.010404
21	−0.010	0.000100	−0.027	0.000729
22	−0.070	0.004900	−0.060	0.003600
23	0.072	0.005184	0.142	0.020164
24	0.074	0.005476	0.002	0.000004
25	−0.058	0.003364	−0.132	0.017424
	Sums	0.062319		0.121615

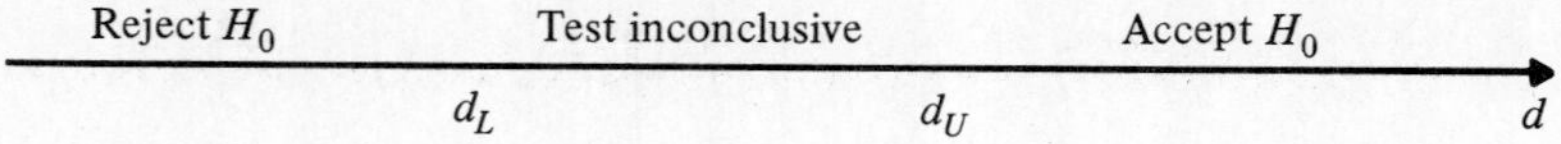

$$d_L \qquad\qquad\qquad\qquad d_U \qquad\qquad\qquad\qquad d$$

FIGURE 14.13 Decision rule for carrying out Durbin-Watson d test of null hypothesis, H_0, of no autocorrelation in regression errors

quite close to 2. On the other hand, positive autocorrelation in the errors would tend to produce a positive value for the sample correlation, r, and hence a value for d that is lower than 2. There is a theoretical difficulty involved in basing tests for auto-correlated errors on the Durbin-Watson statistic. The problem is that the actual sampling distribution of d, even when the hypothesis of no autocorrelation is true, depends on the particular values of the independent variables. It is obviously not possible to tabulate the distribution for every possible set of values of the independent variables. Fortunately, it is known that, whatever the independent variables, the distribution of d lies between that of two other random variables, whose percentage points can be tabulated. For tests of significance levels 1% and 5%, cutoff points for these random variables are tabulated in Table 10 of the Appendix. For various combinations of n, the number of sets of observations, and K, the number of independent variables, these tables give values d_L and d_U. The null hypothesis of no auto-correlation is rejected against the alternative of positive autocorrelation in the errors if the calculated d is less than d_L. The null hypothesis is accepted if d is bigger than d_U, while, if d lies between d_L and d_U, the test is inconclusive. This is illustrated in Figure 14.13.

To illustrate, we will test the null hypothesis of no autocorrelation in the errors of the savings and loan model, against the alternative that these errors are positively autocorrelated, at the 5% level of significance. In this model, we have

$$n = 25 \qquad \text{and} \qquad K = 2$$

so that, from Table 10,

$$d_L = 1.21 \qquad \text{and} \qquad d_U = 1.55$$

Since the calculated d, 1.95, is bigger than d_U, the null hypothesis of no auto-correlation in the errors is accepted, confirming our visual impression from Figure 14.12. This test is shown in Figure 14.14.

In fact, the detection of autocorrelated errors in regression analysis is so important that most modern regression computer packages have the calculation of the Durbin-Watson test statistic as an option. Thus, if such a package is available, it is not necessary to carry out by hand the computations of Table 14.7.

---

[12] This follows since

$$d = \frac{\sum_{t=2}^{n} (e_t - e_{t-1})^2}{\sum_{t=1}^{n} e_t^2} = \frac{\sum_{t=2}^{n} e_t^2}{\sum_{t=1}^{n} e_t^2} + \frac{\sum_{t=2}^{n} e_{t-1}^2}{\sum_{t=1}^{n} e_t^2} - 2\frac{\sum_{t=2}^{n} e_t e_{t-1}}{\sum_{t=1}^{n} e_t^2}$$

The first two terms on the right-hand side of this expression are very close to 1, while the ratio of sums in the final term provides an estimate, r, of the correlation between ε_t and ε_{t-1}.

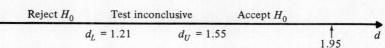

FIGURE 14.14 Test of null hypothesis, H_0, of no autocorrelation in the errors of the savings and loan regression model

---

The Durbin-Watson Test

Consider the regression model

$$Y_t = \alpha + \beta_1 x_{1t} + \beta_2 x_{2t} + \cdots + \beta_K x_{Kt} + \varepsilon_t$$

based on n sets of observations. We contemplate the possibility that the error terms ε_t are autocorrelated according to

$$\text{Corr}(\varepsilon_t, \varepsilon_{t-j}) = \rho^j$$

or, equivalently, that the ε_t can be represented by the process

$$\varepsilon_t = \rho \varepsilon_{t-1} + u_t$$

where u_t is not autocorrelated.

The test of the null hypothesis of no autocorrelation

$$H_0: \quad \rho = 0$$

is based on the Durbin-Watson statistic

$$d = \frac{\sum\limits_{t=2}^{n} (e_t - e_{t-1})^2}{\sum\limits_{t=1}^{n} e_t^2}$$

where the e_t are the residuals when the regression equation is estimated by least squares.

When the alternative hypothesis is of positive autocorrelation in the errors, that is,

$$H_1: \quad \rho > 0$$

the decision rule is:

Reject H_0 if $d < d_L$

Accept H_0 if $d > d_U$

Test inconclusive if $d_L < d < d_U$

where d_L and d_U are tabulated for values of n and K, and for significance levels of 1% and 5% in Table 10 of the Appendix.

Occasionally, one wants to test against the alternative of negative autocorrelation; that is,

$$H_1: \quad \rho < 0$$

The appropriate test is precisely the same as for the alternative of positive autocorrelation, except that it is based on the statistic $(4 - d)$ rather than d. This quantity is then compared with tabulated d_L and d_U.

Example
14.2

A student wanted to run a regression model to explain variability in the birth rate in the United States. She believed that relevant explanatory variables might be the percentage of women in the labor market, the unemployment rate, and the divorce rate. An attempt was also made, through the use of dummy variables, to take into account the effects of wars. The model was fitted using the 30 years of annual data (from 1941 to 1970) shown in Table 14.8. The equation obtained by least squares estimation was

$$y = 4.30 - 0.0477x_1 - 0.0389x_2 - 0.0094x_3 - 0.321x_4 + 0.0205x_5$$
$$\quad\quad (0.0155) \quad (0.0233) \quad (0.0166) \quad (0.077) \quad (0.0905)$$

$$R^2 = 0.73; \quad\quad d = 0.55$$

where

$y =$ Birth rate
$x_1 =$ Percentage of women in the labor force
$x_2 =$ Percentage of labor force unemployed

TABLE 14.8 Data for birth rate regression of Example 14.2

y_t	x_{1t}	x_{2t}	x_{3t}	x_{4t}	x_{5t}
2.03	25.4	9.9	17	1	0
2.22	26.7	4.7	18	1	0
2.27	29.1	1.9	23	1	0
2.12	29.2	1.2	28	1	0
2.04	29.2	1.9	30	1	0
2.41	27.8	3.9	27	0	0
2.66	27.4	3.9	24	0	1
2.49	28.0	3.8	23	0	1
2.45	28.3	5.9	25	0	1
2.41	28.8	5.3	23	1	0
2.49	29.3	3.3	24	1	0
2.51	29.4	3.0	25	1	0
2.50	29.2	2.9	25	1	0
2.53	29.4	5.5	25	0	0
2.50	30.2	4.4	25	0	1
2.52	31.0	4.1	24	0	1
2.53	31.2	4.3	25	0	1
2.45	31.5	6.8	25	0	0
2.40	31.7	5.5	26	0	0
2.37	32.3	5.5	27	0	0
2.33	32.6	6.7	26	0	0
2.24	32.7	5.5	26	0	0
2.17	33.2	5.7	26	0	0
2.10	33.6	5.2	26	0	0
1.94	34.0	4.5	27	1	0
1.84	34.6	3.8	27	1	0
1.78	35.1	3.8	27	1	0
1.75	35.5	3.6	28	1	0
1.78	36.3	3.5	30	1	0
1.84	36.7	4.9	33	1	0

x_3 = Divorce rate
x_4 = Dummy variable taking value 1 during war years, and 0 otherwise
x_5 = Dummy variable taking value 1 in the three years following the first year of peace after a war, and 0 otherwise

We want to test the null hypothesis of no autocorrelation in the errors of the regression equation, against the alternative of positive autocorrelation. Here we have,

$$n = 30 \quad \text{and} \quad K = 5$$

Thus, from Table 10 of the Appendix, for a 1% level test, the points for comparison with the calculated Durbin-Watson statistic are

$$d_L = 0.88 \quad \text{and} \quad d_U = 1.61$$

Since the calculated test statistic, $d = 0.55$, is below the lower bound, d_L, the null hypothesis of no autocorrelation in the errors can be rejected at the 1% significance level. These data, then, cast considerable doubt on the validity of the assumption that the error terms are uncorrelated with one another.

ESTIMATION OF REGRESSIONS WITH AUTOCORRELATED ERRORS

When, as in Example 14.2, it appears likely that the regression errors are auto-correlated, least squares estimates and the inference based on them can be very unreliable. In these circumstances, it is preferable to use an alternative estimator. To motivate the estimation procedure to be used, we write the regression

$$Y_t = \alpha + \beta_1 x_{1t} + \beta_2 x_{2t} + \cdots + \beta_K x_{Kt} + \varepsilon_t \qquad (14.8.5)$$

so that at time $(t - 1)$ we have

$$Y_{t-1} = \alpha + \beta_1 x_{1,t-1} + \beta_2 x_{2,t-1} + \cdots + \beta_K x_{K,t-1} + \varepsilon_{t-1} \quad (14.8.6)$$

Multiplying through equation 14.8.6 by ρ, the correlation between adjacent errors, gives

$$\rho Y_{t-1} = \alpha\rho + \beta_1 \rho x_{1,t-1} + \beta_2 \rho x_{2,t-1} + \cdots$$
$$+ \beta_K \rho x_{K,t-1} + \rho \varepsilon_{t-1} \qquad (14.8.7)$$

Subtracting equation 14.8.7 from 14.8.5 then gives

$$Y_t - \rho Y_{t-1} = \alpha(1 - \rho) + \beta_1(x_{1t} - \rho x_{1,t-1}) + \beta_2(x_{2t} - \rho x_{2,t-1}) + \cdots$$
$$+ \beta_K(x_{Kt} - \rho x_{K,t-1}) + u_t \qquad (14.8.8)$$

where, as in 14.8.3,

$$u_t = \varepsilon_t - \rho\varepsilon_{t-1}$$

and the random variable u_t has constant variance, and is not autocorrelated. Equation 14.8.8 can therefore be regarded as a regression model linking the dependent variable $(Y_t - \rho Y_{t-1})$ to the K independent variables $(x_{1t} - \rho x_{1,t-1})$, . . . , $(x_{Kt} - \rho x_{K,t-1})$. The parameters of this model are precisely the same as those of the original model 14.8.5, except that the intercept term in 14.8.8 is $\alpha(1 - \rho)$, rather than α. There is, however, one very important difference between the models 14.8.5 and 14.8.8. In the former, the error terms ε_t are autocorrelated, so that least squares estimation of the model parameters is inappropriate. On the other hand, since the errors, u_t, of 14.8.8 are not autocorrelated, the usual least squares inferential procedures when applied to this model are perfectly valid.

The foregoing discussion suggests that the problem of autocorrelated errors can be circumvented by estimating by least squares a regression equation with dependent variable $(Y_t - \rho Y_{t-1})$ and independent variables $(x_{1t} - \rho x_{1,t-1})$, . . . , $(x_{Kt} - \rho x_{K, t-1})$. Unfortunately, this is not possible in practice, since the true value of ρ will be unknown. However, we can replace ρ by its sample estimate r, which, by virtue of 14.8.4, can be found directly from the Durbin-Watson statistic as

$$r = 1 - \frac{d}{2}$$

The procedure is described in the box.

Estimation of Regression Models with Autocorrelated Errors

Suppose we want to estimate the parameters of the regression model

$$Y_t = \alpha + \beta_1 x_{1t} + \beta_2 x_{2t} + \cdots + \beta_K x_{Kt} + \varepsilon_t$$

when the error term ε_t is autocorrelated.

This can be accomplished in two stages, as follows:

(i) Estimate the model by least squares, obtaining the Durbin-Watson statistic, d, and hence the estimate

$$r = 1 - \frac{d}{2}$$

of the autocorrelation parameter.

(ii) Estimate a second regression, by least squares, in which the dependent variable

---

is $(Y_t - rY_{t-1})$ and the independent variables are $(x_{1t} - rx_{1, t-1})$, ..., $(x_{Kt} - rx_{K, t-1})$. The parameters $\beta_1, \beta_2, \ldots, \beta_K$ are estimated by the estimated partial regression coefficients from this second model. An estimate of α is obtained by dividing by $(1 - r)$ the estimated intercept for the second model. Hypothesis tests and confidence intervals for the β_i can be carried out using the standard methods on the output of the second regression.

Example 14.3

In Example 14.2 we estimated by least squares a regression model seeking to explain variability in birth rates in the United States. The possibility of autocorrelation in the errors of that model was strongly indicated by the Durbin-Watson statistic

$$d = 0.55$$

We now want to reestimate the model parameters, allowing for the presence of autocorrelated errors. First, the sample autocorrelation is

$$r = 1 - \frac{d}{2} = 1 - \frac{0.55}{2} = 0.725$$

Next, we must construct new variables for a second regression. The dependent variable for this regression is $(Y_t - rY_{t-1})$. Referring to Table 14.8, we find

$$y_2 - ry_1 = 2.22 - (0.725)(2.03) = 0.748$$

$$y_3 - ry_2 = 2.27 - (0.725)(2.22) = 0.661$$

and so on. The new independent variables are found in the same way, so that, for example,

$$x_{1, 2} - rx_{1, 1} = 26.7 - (0.725)(25.4) = 8.285$$

$$x_{1, 3} - rx_{1, 2} = 29.1 - (0.725)(26.7) = 9.743$$

In this way, the new variables can be computed, as summarized in Table 14.9. The resulting estimated regression was

$$(y_t - ry_{t-1}) = 1.383 - 0.0667\,(x_{1, t} - rx_{1, t-1}) - 0.0083\,(x_{2, t} - rx_{2, t-1})$$
$$\phantom{(y_t - ry_{t-1}) = 1.383 - }(0.0203)\phantom{\,(x_{1, t} - rx_{1, t-1})} (0.0133)$$

$$- 0.0200(x_{3, t} - rx_{3, t-1}) - 0.106(x_{4, t} - rx_{4, t-1}) + 0.0322\,(x_{5, t} - rx_{5, t-1})$$
$$(0.0109)\phantom{(x_{3, t} - rx} (0.0617)\phantom{(x_{4, t} - rx_{4, t-1})} (0.0526)$$

Now, the intercept term in the original regression is estimated by

$$\frac{1.383}{1 - r} = \frac{1.383}{1 - 0.725} = 5.029$$

The estimated model is, therefore,

$$y = 5.029 - 0.0667x_1 - 0.0083x_2 - 0.0200x_3 - 0.106x_4 + 0.0322x_5$$
$$(0.0203) (0.0133) (0.0109) (0.0617) (0.0526)$$

TABLE 14.9 Variables used for estimating birth rate regression, allowing for autocorrelated errors

$y_t - ry_{t-1}$	$x_{1t} - rx_{1,t-1}$	$x_{2t} - rx_{2,t-1}$	$x_{3t} - rx_{3,t-1}$	$x_{4t} - rx_{4,t-1}$	$x_{5t} - rx_{5,t-1}$
0.748	8.285	−2.478	5.675	0.275	0
0.661	9.743	−1.508	9.950	0.275	0
0.474	8.103	−0.178	11.325	0.275	0
0.503	8.030	1.030	9.700	0.275	0
0.931	6.630	2.523	5.250	−0.725	0
0.913	7.245	1.073	4.425	0	1
0.562	8.135	0.973	5.600	0	0.275
0.645	8.000	3.145	8.325	0	0.275
0.634	8.283	1.023	4.875	1	−0.725
0.743	8.420	−0.543	7.325	0.275	0
0.705	8.158	0.608	7.600	0.275	0
0.680	7.885	0.725	6.875	0.275	0
0.718	8.230	3.398	6.875	−0.725	0
0.666	8.885	0.413	6.875	0	1
0.708	9.105	0.910	5.875	0	0.275
0.703	8.725	1.328	7.600	0	0.275
0.616	8.880	3.683	6.875	0	−0.725
0.624	8.863	0.570	7.875	0	0
0.630	9.318	1.513	8.150	0	0
0.612	9.183	2.713	6.425	0	0
0.551	9.065	0.643	7.150	0	0
0.546	9.493	1.713	7.150	0	0
0.527	9.530	1.068	7.150	0	0
0.418	9.640	0.730	8.150	1	0
0.434	9.950	0.538	7.425	0.275	0
0.446	10.015	1.045	7.425	0.275	0
0.460	10.053	0.845	8.425	0.275	0
0.511	10.563	0.890	9.700	0.275	0
0.550	10.383	2.363	11.250	0.275	0

Given the estimated model, confidence intervals for, and tests of hypotheses about, the individual regression parameters can be carried out in the usual way. To illustrate, we will test the null hypothesis that, all else equal, the percentage of women in the labor force (x_1) does not influence the birth rate. That is the hypothesis

$$H_0: \quad \beta_1 = 0$$

This will be tested against the alternative

$$H_1: \quad \beta_1 < 0$$

The test is based on

$$\frac{b_1}{s_{b_1}} = \frac{-0.0667}{0.0203} = -3.286$$

In the second regression, we have $n = 29$ observations and $K = 5$ independent variables. Thus, for a 0.5% level test,

$$-t_{n-K-1, \alpha} = -t_{23, .005} = -2.807$$

The null hypothesis can, therefore, be rejected at this level.

CONSEQUENCES OF AUTOCORRELATED ERRORS

The consequences of proceeding with the ordinary least squares analysis of a regression model that has autocorrelated errors can be extremely severe. If the presence of autocorrelated errors is ignored, then:

(i) The least squares estimators of the parameters of the regression model are likely to be inefficient. Generally, superior estimators will be obtained if autocorrelated errors are taken into account.

(ii) Forecasts of future values of the dependent variable, obtained through the standard analysis, will be unnecessarily inefficient.

(iii) The usual inferential statements based on confidence intervals or tests of hypotheses can be very badly misleading.

We conclude this section by noting two special aspects of the problem of autocorrelated errors in models such as those discussed in Section 14.3, which contain lagged dependent variables. First, in such models, there is a sense in which ignoring the problem of autocorrelated errors causes even more severe difficulties than in the ordinary regression model. In addition to the points mentioned previously, least squares estimators of the regression parameters will not be consistent[13] in models with lagged dependent variables when autocorrelated errors are ignored. Second, the Durbin-Watson test is not appropriate for this model. However, it is possible to develop tests, whose details we will not discuss here, of autocorrelated errors that are approximately valid in large samples.

14.9 SUMMARY

In this chapter we have attempted to show that regression modelling involves considerably more than the routine exercise described in the previous two chapters. In practice, a considerable amount of art is involved. Care must be taken in the specification of a model. In particular, important explanatory variables should not be ignored. Some important explicands may require the use of dummy variables. Further possibilities that may need to be taken into account in model specification include the use of lagged dependent variables and the consideration of nonlinear models, such as the log linear form.

It is also extremely important to check any assumptions made about the behavior of the error terms in a regression model. Tests for heteroscedasticity and auto-correlated errors can be carried out, and, if either of these problems appears to be present, the regression model should be reestimated.

These considerations by no means exhaust the possibilities for departure from the standard regression treatment. They are, however, among the most important in business and economic applications, and a great many well-conceived practical studies involve at least one of them.

---

[13] The property of consistency of an estimator was discussed in Section 7.3.

27. In Chapter 12, we estimated the regression of retail sales per household on disposable income per household. The data are given in Table 12.5, and the residuals from the fitted regression are tabulated in Table 12.7.

    (a) Graphically check for heteroscedasticity in the regression errors.

    (b) Check for heteroscedasticity by using a formal test.

28. In Exercise 14 of Chapter 12, ten pairs of observations were used to estimate a regression of grade point average on hours per week of study time.

    (a) Graphically check for heteroscedasticity in the regression errors.

    (b) Check for heteroscedasticity by using a formal test.

29. In Exercise 33 of Chapter 12, for a sample of twelve families, a linear regression of savings on income was estimated.

    (a) Graphically check for heteroscedasticity in the regression errors.

    (b) Check for heteroscedasticity by using a formal test.

30. In Exercise 38 of Chapter 13, for a sample of twelve territories, a regression of product sales on advertising expenditure and expenditure on salesmen was estimated.

    (a) Graphically check for heteroscedasticity in the regression errors.

    (b) Use a formal test to check for heteroscedasticity.

31. In Exercise 40 of Chapter 13, for a sample of thirty-four countries, a regression of gross national product on population, area, percentage of literate adults, and percentage of economy devoted to agriculture was estimated.

    (a) Graphically check for heteroscedasticity in the regression errors.

    (b) Use a formal test to check for heteroscedasticity.

32. Consider the regression model

$$Y_i = \alpha + \beta_1 x_{1i} + \beta_2 x_{2i} + \cdots + \beta_K x_{Ki} + \varepsilon_i$$

Show that, if

$$\text{Var}\,(\varepsilon_i) = Kx_i^2 \qquad (K > 0)$$

then

$$\text{Var}\left(\frac{\varepsilon_i}{x_i}\right) = K$$

Discuss the possible relevance of this result in treating a form of heteroscedasticity.

33. [*This problem requires a computer program to carry out the regression calculations.*] The accompanying table gives, for eighteen countries, data on total expenditures on education as a percentage of gross national product (y), per capita income (x_1), median educational attainment (in years) of the population over 25 years of age (x_2), and the ratio of the population aged 0–14 to the total population (x_3). Estimate the multiple regression

$$Y_i = \alpha + \beta_1 x_{1i} + \beta_2 x_{2i} + \beta_3 x_{3i} + \varepsilon_i$$

What can you learn from an examination of the residuals from the fitted model?

	y	x_1	x_2	x_3
Argentina	2.0	1,060	5.72	0.30
Australia	4.3	2,300	8.05	0.29

	y	x_1	x_2	x_3
Austria	4.7	1,470	3.16	0.24
Canada	8.6	2,650	9.10	0.30
Czechoslovakia	4.4	1,370	5.72	0.24
Finland	6.3	1,980	4.36	0.26
France	3.5	2,460	4.36	0.25
Hungary	4.5	1,100	7.20	0.21
Ireland	4.8	1,110	5.38	0.31
Israel	5.4	1,570	7.56	0.33
Italy	4.3	1,400	3.60	0.24
Japan	4.1	1,430	5.86	0.24
Netherlands	7.3	1,760	4.42	0.27
Norway	5.9	2,160	5.20	0.25
Switzerland	4.2	2,700	5.38	0.23
United States	6.3	4,240	12.20	0.27
USSR	6.8	1,200	5.97	0.28
Venezuela	4.5	1,000	1.32	0.47

34. [*This problem requires a computer program to carry out the regression calculations.*] The table shows actual percentage income growth (y) for twenty corporations, together with predictions (x_1, x_2) made by two financial analysts. Estimate the multiple regression

$$Y_i = \alpha + \beta_1 x_{1i} + \beta_2 x_{2i} + \varepsilon_i$$

and check for heteroscedasticity in the errors.

y	x_1	x_2	y	x_1	x_2
8.2	7.2	8.3	9.8	10.4	9.6
7.5	7.6	8.4	9.5	10.6	8.5
8.7	8.0	9.7	8.3	10.6	8.4
9.1	8.7	11.0	10.0	11.4	9.6
9.0	9.2	11.2	12.0	11.4	15.4
9.8	9.4	8.7	9.8	11.7	13.1
8.7	9.5	10.1	11.3	12.4	12.6
11.2	9.8	10.2	15.1	12.7	11.0
7.6	10.1	6.6	12.0	12.8	12.0
9.8	10.1	11.4	15.0	14.2	12.9

35. Refer to the regression, in Chapter 12, of retail sales per household on disposable income per household.
   (a) Calculate the Durbin-Watson d statistic.
   (b) Test the null hypothesis of no autocorrelation in the regression errors.
   (c) If necessary, reestimate the model allowing for autocorrelated errors.

36. For a series of thirty-one annual observations, a student estimated the following regression model:

$$y_t = -509 + 0.93x_{1t} + 18.07x_{2t} - 22.30x_{3t} - 0.422x_{4t} + 8.36x_{5t}$$
$$ (23.4) \quad (3.24) \quad (53.46) \quad (0.175) \quad (5.69)$$

$$R^2 = 0.899; \qquad d = 1.055$$

where

y_t = Dow-Jones industrials yearly average
x_{1t} = Ratio of annual corporate profit to annual corporate sales
x_{2t} = Index of industrial production
x_{3t} = Corporate bond yield
x_{4t} = Disposable income per capita
x_{5t} = Consumer price index

The figures in parentheses below the estimated coefficients are their estimated standard errors. What can you conclude from these results?

37. [*This problem requires either the material in Appendix A13.1 or a computer program to carry out the regression calculations.*] The accompanying table shows 36 monthly observations on housing starts (in thousands) in the North Central region of the United States (y), the interest rate charged to buyers of new homes (x_1), and a dummy variable (x_2) defined to take the value 1 in the three winter months and 0 otherwise.

y_t	x_{1t}	x_{2t}	y_t	x_{1t}	x_{2t}
434	8.82	0	308	10.02	0
476	8.84	0	407	10.06	0
490	8.85	0	486	10.20	0
451	8.87	1	527	10.39	0
406	8.93	1	439	10.49	0
437	8.96	1	457	10.73	0
440	9.03	0	454	10.72	0
469	9.07	0	427	10.91	0
433	9.14	0	239	11.04	0
472	9.23	0	179	11.30	1
469	9.34	0	67	11.48	1
417	9.45	0	92	11.60	1
421	9.50	0	142	12.25	0
431	9.60	0	180	12.64	0
543	9.63	0	178	13.26	0
436	9.76	1	250	12.24	0
70	9.92	1	245	12.08	0
88	9.94	1	232	11.84	0

(a) Estimate the regression of housing starts on interest rates and the dummy variable.
(b) Calculate the Durbin-Watson d statistic.
(c) Test the null hypothesis of no autocorrelation in the regression errors.
(d) If necessary, reestimate the regression, allowing for autocorrelated errors.

38. A student wanted to explore possible causes of motor vehicle deaths in the United States. In order to do so, he estimated the following model for 26 annual observations:

$$y_t = -51,514 + 1.52x_{1t} + 1,681x_{2t} - 0.0171x_{3t}; \quad R^2 = 0.948; \quad d = 0.952$$
$$\quad\quad\quad\quad (0.45) \quad\quad (695) \quad\quad (0.0142)$$

where

y_t = Total motor vehicle deaths in the United States
x_{1t} = Personal consumption of alcoholic beverages (in millions of dollars) in the United States

x_{2t} = Average highway speed (in miles per hour) of all motor vehicles in the United States

x_{3t} = Miles of travel (in millions) by motor vehicles in the United States

The figures in parentheses beneath coefficient estimates are their estimated standard errors. What can you conclude from these results?

39. The accompanying table shows, for a consumer goods corporation, 20 consecutive years of data on sales (y) and advertising (x).

y	x	y	x
102	61	133	75
92	45	198	61
93	53	222	86
98	54	220	86
93	53	251	102
105	55	273	136
118	52	318	148
109	58	335	161
109	61	344	180
115	54	292	185

(a) Estimate the regression

$$Y_t = \alpha + \beta x_t + \varepsilon_t$$

(b) Check for autocorrelated errors in this model.

(c) If necessary, reestimate the model, allowing for autocorrelated errors.

40. The omission of an important independent variable from a time series regression model can result in the appearance of autocorrelated errors. In Section 14.5, we estimated the model

$$Y_t = \alpha + \beta_1 x_{1t} + \varepsilon_t$$

relating profit margins to net revenues for our savings and loan data. Carry out a Durbin-Watson test on the residuals from this model. What can you infer from the results?

REVIEW EXERCISES

41. Write a brief report outlining the use of dummy variables in regression analysis. Provide examples to illustrate how these quantities can be of practical use.

42. In Section 14.2, we discussed the fitting of the model

$$Y = \alpha + \beta_1 x_1 + \beta_2 x_2 + \beta_3 x_3 + \varepsilon$$

where

Y = Tax revenue as a percentage of gross national product in a country
x_1 = Exports as a percentage of gross national product in the country
x_2 = Income per capita in the country
x_3 = Dummy variable taking the value 1 if country participates in some form of economic integration, and 0 otherwise

This provides a means of allowing for the effects on tax revenue of participation in some form of economic integration. Another possibility would be to estimate the regression

$$Y = \alpha + \beta_1 x_1 + \beta_2 x_2 + \varepsilon$$

separately for those countries which did and did not participate in some form of economic integration. Explain how these approaches to the problem differ.

43. The model

$$Y_t = \alpha + \beta x_t + \gamma Y_{t-1} + \varepsilon_t$$

was fitted to data on income (x) and consumption (Y). The least squares estimates of β and γ were

$$b = 0.38 \quad \text{and} \quad c = 0.54$$

Interpret these estimates.

44. Discuss the following statement: "In many practical regression problems, multi-collinearity is so severe that it would be best to run separate simple linear regressions of the dependent variable on each independent variable."

45. Explain the nature of, and the difficulties caused by, each of the following
   (a) Heteroscedasticity
   (b) Autocorrelated errors

46. In Exercise 61 of Chapter 12, a regression of the number of promotions per year on the number of years with the firm was estimated. Investigate the possibility of hetero-scedasticity in the errors of the linear regression model.

47. In Exercise 63 of Chapter 12, a regression of percentage change in sales volume on percentage change in price was fitted. Check for the possibility of heteroscedastic errors in the population regression model.

48. In Exercise 51 of Chapter 13, we considered the regression of house price on house size, lot size, number of bedrooms, and number of bathrooms. Check for the possibility of heteroscedasticity in the error terms of the population regression model.

49. In a study[14] of the effects of competition, a regression equation was estimated to explain variability in the average costs of municipally-owned electric utility corporations. The independent variables included sales, capacity utilization, various factor costs, con-sumption rates, and dummy variables to allow for behavior in different states. Also included was a dummy variable taking the value 1 for companies facing competition and 0 for others. The estimated partial regression coefficient associated with this dummy variable was -0.7954, and the null hypothesis that the population parameter was 0 was rejected at the 1% significance level against the alternative that the true value was negative. What is the implication of this finding?

50. The accompanying table shows data[15] collected over a period of 25 years, on the market return (x) of the Standard and Poor's 500 stocks and the percentage (y) of portfolios in

---

[14] W. J. Primeaux, "An assessment of X-efficiency gained through competition," *Review of Economics and Statistics*, 59 (1977), 105–108.

[15] See W. S. Bauman and C. M. McLaren, "An asset allocation model for active portfolios," *Journal of Portfolio Management*, 8 (1982), No. 2, 76–86.

common stocks at market value at the end of the year for noninsurance private pension funds. Estimate the regression model

$$Y_t = \alpha + \beta x_t + \gamma Y_{t-1} + \varepsilon_t$$

and write a report on your findings.

	y	x		y	x
1955	30	31.6	1968	63	11.1
1956	32	6.6	1969	63	−8.5
1957	30	−10.8	1970	62	4.0
1958	38	43.4	1971	68	14.3
1959	43	12.0	1972	73	19.0
1960	43	0.5	1973	68	−14.7
1961	49	26.9	1974	56	−26.5
1962	45	−8.7	1975	60	37.2
1963	49	22.8	1976	62	23.8
1964	52	16.5	1977	56	−7.2
1965	55	12.5	1978	53	6.6
1966	53	−10.1	1979	55	18.7
1967	59	24.0			

51. A study[16] was conducted on the man-hour costs of Federal Deposit Insurance Corporation (FDIC) examinations of banks. Data were obtained on ninety-one such examinations. Some of these were conducted by FDIC alone, and some jointly with state examiners. Examiners rated banks' management as good, satisfactory, fair, or unsatisfactory. The model estimated was

$$\log y = 2.41 + 0.3674 \log x_1 + 0.2217 \log x_2 + 0.0803 \log x_3$$
$$(0.0477) \qquad (0.0628) \qquad (0.0287)$$

$$- 0.1755 x_4 + 0.2799 x_5 + 0.5634 x_6 - 0.2572 x_7; \qquad R^2 = 0.766$$
$$(0.2905) \quad (0.1044) \quad (0.1657) \quad (0.0787)$$

where

y = FDIC examiner man-hours
x_1 = Total assets of bank
x_2 = Total number of offices in bank
x_3 = Ratio of classified loans to total loans for bank
x_4 = 1 if management rating was "good," and 0 otherwise
x_5 = 1 if management rating was "fair," and 0 otherwise
x_6 = 1 if management rating was "unsatisfactory," and 0 otherwise
x_7 = 1 if examination was conducted jointly with the state, and 0 otherwise

The figures in parentheses beneath coefficient estimates are the associated estimated standard errors. Write a report on these results.

---

[16] R. J. Miller, "Examination man-hour cost for independent, joint, and divided examination programs," *Journal of Bank Research, 11* (1980), 28–35.

52. The accompanying table[17] shows, for a period of 20 years, data from Britain on days of incapacity through sickness per person at risk (y), the male unemployment rate (x_1), the ratio of sickness benefits to earnings (x_2), and the real wage rate (x_3). Estimate the model

$$\log Y_t = \alpha + \beta_1 \log x_{1t} + \beta_2 \log x_{2t} + \beta_3 \log x_{3t} + \varepsilon_t$$

and write a report on your findings. Include in your analysis a check on the possibility of autocorrelated errors and, if necessary, a correction for this problem.

Year	y	x_1	x_2	x_3	Year	y	x_1	x_2	x_3
1	12.2	1.5	.367	1.190	11	13.5	2.0	.446	1.362
2	12.2	1.2	.394	1.216	12	13.9	1.8	.493	1.355
3	11.7	1.3	.371	1.250	13	14.1	2.0	.686	1.364
4	11.9	1.7	.355	1.267	14	15.1	3.2	.732	1.383
5	12.1	2.4	.440	1.273	15	15.4	3.4	.728	1.407
6	12.3	2.5	.419	1.299	16	15.8	3.3	.710	1.406
7	11.9	1.9	.395	1.318	17	15.4	3.6	.727	1.454
8	12.1	1.9	.443	1.329	18	15.0	4.7	.779	1.501
9	12.8	2.5	.430	1.321	19	15.7	5.1	.737	1.593
10	12.9	3.1	.474	1.342	20	16.2	3.6	.706	1.660

[17] Taken from R. B. Thomas, "Wages, sickness benefits, and absenteeism," *Journal of Economic Studies, 7* (1980), No. 1, 51–61, © M.C.B. Publications Limited, Bradford, United Kingdom.

analysis
of variance

CHAPTER FIFTEEN

15.1 COMPARISON OF SEVERAL POPULATION MEANS

In Section 9.6, we saw how to test the hypothesis of equality of two population means. In fact, two distinct tests were developed, the appropriate test depending on the **experimental design**—that is, the mechanism employed in the generation of sample observations. Specifically, our tests assumed either paired observations or independent random samples. This distinction is important, and to clarify it, we pause to consider a simple illustration. Suppose it is our objective to compare the miles per gallon achieved by two different makes of automobile, A-cars and B-cars. We could randomly select ten people to drive these cars over a specified distance, each driver being assigned to a car of each type, so that any particular driver will drive both an A-car and a B-car. The twenty resulting miles per gallon figures obtained will consist of ten pairs, each pair corresponding to a single driver. This is the **matched pairs** design, and its attraction lies in its ability to produce a comparison between the quantities of interest (in this case, miles per gallon achieved by the two types of car) while making allowance for the possible importance of an additional relevant factor (individual driver differences). Thus, if a significant difference between the performance of A-cars and B-cars is found, we have some assurance that this is not a result of differences in driver behavior. An alternative design would be to take twenty drivers and randomly assign ten of them to A-cars and ten to B-cars (though, in fact, there is no need to have equal numbers of trials for each type of car). The twenty resulting miles per gallon figures would then constitute a pair of **independent random samples** of ten observations each on A-cars and B-cars.

609

TABLE 15.1 Miles per gallon figures from three independent random samples

	A-CARS	B-CARS	C-CARS
	22.2	24.6	22.7
	19.9	23.1	21.9
	20.3	22.0	23.3
	21.4	23.5	24.1
	21.2	23.6	22.1
	21.0	22.1	23.4
	20.3	23.5	
Sums	146.3	162.4	137.4

For these two types of design, we discussed in Section 9.6 appropriate procedures for testing the null hypothesis of equality of a pair of population means. In this chapter, our aim is to extend these procedures to the development of tests for the equality of several population means. Suppose, for example, that our study was to include a third make of automobile, the C-car. The null hypothesis of interest would then be that the population mean miles per gallon is the same for all three makes of car. We will see how tests for such hypotheses can be constructed, beginning with the case where independent random samples are taken. In Section 15.4 the extension of the test based on matched pairs will be discussed.

Suppose that, of twenty drivers, seven are randomly assigned to A-cars, seven to B-cars, and six to C-cars. Table 15.1 shows the miles per gallon figures obtained at the completion of these trials.

Now, since our objective is the comparison of population means, an obvious starting place is the calculation of the sample means. From Table 15.1, we obtain for these data:

$$\text{Sample mean for A-cars} = \frac{146.3}{7} = 20.9$$

$$\text{Sample mean for B-cars} = \frac{162.4}{7} = 23.2$$

$$\text{Sample mean for C-cars} = \frac{137.4}{6} = 22.9$$

Not surprisingly, these sample means are not all the same. As always, however, when testing hypotheses, we are interested in the likelihood of such differences arising by chance if, in fact, the null hypothesis were true. If it is concluded that such discrepancies would be very unlikely to arise by chance, then considerable skepticism about the truth of the null hypothesis would arise.

To clarify the issues involved, consider Figure 15.1, which depicts two hypothetical sets of data. The sample means in part (a) of the figure are precisely the same as those in part (b). The crucial difference is that, in the former, the observations are tightly clustered about their respective sample means, while in the latter there is much greater dispersion. Visual inspection of part (a) suggests very strongly the conjecture that the data in fact arise from three populations with different means. On the other

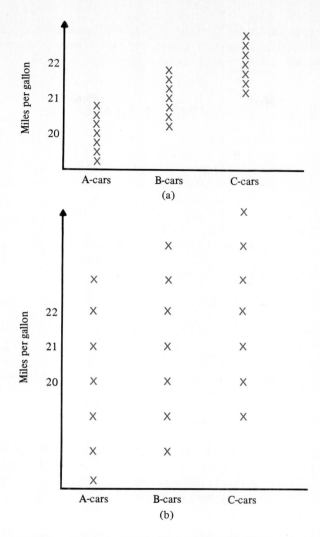

FIGURE 15.1 Two sets of sample miles per gallon data on three makes of automobile

hand, looking at part (b) of the figure, we would not be terribly surprised to learn that these data came from a common population.

This illustration serves to point out the very essence of the test for equality of population means. The critical factor is the *variability* involved in the data. If the variability *around* the sample means is small compared with the variability *among* the sample means, as in Figure 15.1(a), we would be inclined to doubt the null hypothesis that the population means are equal. On the other hand, if, as in Figure 15.1(b), the variability around the sample means is large compared with the variability among them, the evidence against this null hypothesis is rather flimsy. This being the case, it seems reasonable to expect that an appropriate test will be based on estimates of variance. This is indeed the case, and for this reason the general technique employed is referred to as the **analysis of variance.**

15.2 ONE-WAY ANALYSIS OF VARIANCE

The problem introduced in the previous section can be treated quite generally. Suppose that we want to compare the means of K populations, *each of which is assumed to have the same variance.* Independent random samples of $n_1, n_2, \ldots, n_K$ observations are taken from these populations. We will use the symbol x to denote the actual sample values, and will index this with a double subscript, so that x_{ij} denotes the jth observation on the ith population. Then, using the format of Table 15.1, we can display the sample data as in Table 15.2.

The procedure for testing the equality of population means in this setup is called **one-way analysis of variance,** a terminology that will become clearer when we discuss other analysis of variance models.

The Framework for One-Way Analysis of Variance

Suppose we have independent random samples of $n_1, n_2, \ldots, n_K$ observations from K populations. If the population means are denoted $\mu_1, \mu_2, \ldots, \mu_K$, **the one-way analysis of variance** framework is designed to test the null hypothesis

$$H_0: \quad \mu_1 = \mu_2 = \cdots = \mu_K$$

In this section we will develop a test of the null hypothesis that the K population means are equal, given independent random samples from those populations. The obvious first step is to calculate the sample means for the K groups of observations. These sample means will be denoted $\bar{x}_1, \bar{x}_2, \ldots, \bar{x}_K$. Formally, then,

$$\bar{x}_i = \frac{\sum_{j=1}^{n_i} x_{ij}}{n_i} \qquad (i = 1, 2, \ldots, K)$$

TABLE 15.2 Sample observations from independent random samples of K populations

POPULATION			
1	2	$\cdots$	K
x_{11}	x_{21}		x_{K1}
x_{12}	x_{22}		x_{K2}
.	.		.
.	.		.
.	.		.
x_{1n_1}	x_{2n_2}		x_{Kn_K}

where n_i denotes the number of sample observations in group i. In this notation, we have already found for the data of Table 15.1,

$$\bar{x}_1 = 20.9; \qquad \bar{x}_2 = 23.2; \qquad \bar{x}_3 = 22.9$$

Now, the null hypothesis of interest specifies that the K populations have a common mean. A logical step, then, is to form an estimate, from the sample data, of that common mean. A sensible choice for such an estimate is the *overall mean* of the sample observations. This is just the sum of all the sample values, divided by their total number. If we let n denote the total number of sample observations, then

$$n = \sum_{i=1}^{K} n_i$$

so that, in our example, $n = 20$. The overall mean of the sample observations can then be expressed as

$$\bar{x} = \frac{\displaystyle\sum_{i=1}^{K} \sum_{j=1}^{n_i} x_{ij}}{n}$$

where the double summation notation indicates that we sum over all groups and over all observations within each group—that is, we sum all of the available sample observations. An equivalent expression is

$$\bar{x} = \frac{\displaystyle\sum_{i=1}^{K} n_i \bar{x}_i}{n}$$

For the automobile mileage data of Table 15.1, the overall mean is

$$\bar{x} = \frac{(7)(20.9) + (7)(23.2) + (6)(22.9)}{20} = 22.305$$

Hence, if in fact the population mean miles per gallon is the same for A-cars, B-cars, and C-cars, we estimate that common mean to be 22.305 miles per gallon.

As indicated in the previous section, the test of equality of population means is based on a comparison of two types of variability exhibited by the sample members. The first is variability about the individual sample means within the K groups of observations. It is convenient to refer to this as **within groups variability**. Second, we are interested in the variability among the K group means. This is called **between groups variability**. We now seek measures, based on the sample data, of these two types of variability.

To begin, consider variability within the groups. To measure variability in the first group, we calculate the sum of squared deviations of the observations about their sample mean $\bar{x}_1$, that is,

$$SS_1 = \sum_{j=1}^{n_1} (x_{1j} - \bar{x}_1)^2$$

Similarly, for the second group, whose sample mean is $\bar{x}_2$, we calculate

$$SS_2 = \sum_{j=1}^{n_2} (x_{2j} - \bar{x}_2)^2$$

and so on. The total within groups variability, denoted SSW, is then the sum of these sums of squares over all K groups, that is,

$$SSW = SS_1 + SS_2 + \cdots + SS_K$$

or

$$SSW = \sum_{i=1}^{K} \sum_{j=1}^{n_i} (x_{ij} - \bar{x}_i)^2$$

For the data on miles per gallon, we have

$$SS_1 = (22.2 - 20.9)^2 + (19.9 - 20.9)^2 + \cdots + (20.3 - 20.9)^2 = 3.76$$
$$SS_2 = (24.6 - 23.2)^2 + (23.1 - 23.2)^2 + \cdots + (23.5 - 23.2)^2 = 4.96$$
$$SS_3 = (22.7 - 22.9)^2 + (21.9 - 22.9)^2 + \cdots + (23.4 - 22.9)^2 = 3.46$$

The within groups sum of squares is therefore

$$SSW = SS_1 + SS_2 + SS_3 = 3.76 + 4.96 + 3.46 = 12.18$$

Next, we need a measure of variability between groups. A natural measure is based on the discrepancies between the individual group means and the overall mean. In fact, as before, these discrepancies are squared, giving

$$(\bar{x}_1 - \bar{x})^2, \quad (\bar{x}_2 - \bar{x})^2, \quad \ldots, \quad (\bar{x}_K - \bar{x})^2$$

In computing the total between groups sum of squares, SSG, we weight each squared discrepancy by the number of sample observations in the corresponding group (so that most weight is given to the squared discrepancies in groups with most observations), giving

$$SSG = \sum_{i=1}^{K} n_i (\bar{x}_i - \bar{x})^2$$

Thus, for our miles per gallon data,

$$SSG = 7(20.9 - 22.305)^2 + 7(23.2 - 22.305)^2 + 6(22.9 - 22.305)^2$$
$$= 21.5495$$

Another sum of squares is often calculated. This is the sum of squared discrepancies of *all* the sample observations about their *overall* mean. This is called the **total sum of squares,** and is expressed as

$$SST = \sum_{i=1}^{K} \sum_{j=1}^{n_i} (x_{ij} - \bar{x})^2$$

In fact, it can be shown that the total sum of squares is the sum of the within groups and between groups sums of squares, that is,

$$SST = SSW + SSG$$

Hence, for the miles per gallon data, we have

$$SST = 12.18 + 21.5495 = 33.7295$$

In the accompanying box, we summarize the results obtained so far.

Sums of Squares Decomposition for One-Way Analysis of Variance

Suppose we have independent random samples of $n_1, n_2, \ldots, n_K$ observations from K populations. Denote by $\bar{x}_1, \bar{x}_2, \ldots, \bar{x}_K$ the K group sample means, and by $\bar{x}$ the overall sample mean.

We define the following **sums of squares.**

$$\text{WITHIN GROUPS:} \quad SSW = \sum_{i=1}^{K} \sum_{j=1}^{n_i} (x_{ij} - \bar{x}_i)^2$$

$$\text{BETWEEN GROUPS:} \quad SSG = \sum_{i=1}^{K} n_i (\bar{x}_i - \bar{x})^2$$

$$\text{TOTAL:} \quad SST = \sum_{i=1}^{K} \sum_{j=1}^{n_i} (x_{ij} - \bar{x})^2$$

where x_{ij} denotes the jth sample observation in the ith group.

Then

$$SST = SSW + SSG$$

The decomposition of the total sum of squares into the sum of two components—within groups and between groups sums of squares—provides the basis for the analysis of variance test of equality of group population means. We can view this decomposition as expressing the total variability of all the sample observations about their overall mean as the sum of variability within groups and variability between groups. Schematically, this is shown in Figure 15.2.

Our test of the equality of population means is based on the assumption that the K populations have a common variance. If the null hypothesis that the population means are all the same is true, then each of the sums of squares, SSW and SSG, can be used as the basis for an estimate of the common population variance. In order to obtain these estimates, the sums of squares must be divided by the appropriate number of degrees of freedom.

First, it can be shown that an unbiased estimate of the population variance results if SSW is divided by $(n - K)$. The resulting estimate is called the **within groups mean square,** denoted MSW, so that

$$MSW = \frac{SSW}{(n - K)}$$

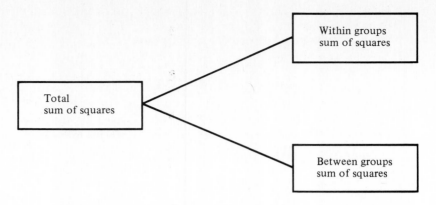

FIGURE 15.2 Sum of squares decomposition for one-way analysis of variance

For our data, we have

$$\text{MSW} = \frac{12.18}{(20 - 3)} = 0.7165$$

If the population means are equal, another unbiased estimate of the population variance is obtained by dividing SSG by $(K - 1)$. The resulting quantity is called the **between groups mean square,** denoted MSG, and hence

$$\text{MSG} = \frac{\text{SSG}}{(K - 1)}$$

For the miles per gallon data,

$$\text{MSG} = \frac{21.5495}{(3 - 1)} = 10.7748$$

When the population means are *not* equal, the between groups mean square does *not* provide an unbiased estimate of the common population variance. Rather, the expected value of the corresponding random variable exceeds the common population variance, as it also carries information about the squared differences of the true population means.

If the null hypothesis were true, we would now be in possession of two unbiased estimates of the same quantity, the common population variance. It would be reasonable to expect these estimates to be quite close to one another. The greater the discrepancy between these two estimates, all else equal, the stronger would be our suspicion that the null hypothesis is not true. The test of the null hypothesis is based on the ratio of mean squares

$$F = \frac{\text{MSG}}{\text{MSW}} \qquad\qquad (15.2.1)$$

If this ratio is quite close to 1, there would be little cause to doubt the null hypothesis of equality of population means. On the other hand, if the variability between groups is large compared to the variability within groups, then, as we have already noted, we would suspect the null hypothesis to be false. This is the case where a value considerably larger than 1 arises for the ratio 15.2.1. Thus, the null hypothesis is rejected for large values of this ratio.

A formal test follows from the fact that, if the null hypothesis of equality of population means is true, the random variable corresponding to 15.2.1 follows the F distribution (discussed in Section 9.8) with numerator degrees of freedom $(K - 1)$ and denominator degrees of freedom $(n - K)$, assuming the population distributions to be normal.

Hypothesis Test for One-Way Analysis of Variance

Suppose we have independent random samples of $n_1, n_2, \ldots, n_K$ observations from K populations. Denote by n the total sample size, so that

$$n = n_1 + n_2 + \cdots + n_K$$

We define the **mean squares** as follows.

$$\text{WITHIN GROUPS:} \quad \text{MSW} = \frac{\text{SSW}}{(n - K)}$$

$$\text{BETWEEN GROUPS:} \quad \text{MSG} = \frac{\text{SSG}}{(K - 1)}$$

The null hypothesis to be tested is that the K population means are equal, that is,

$$H_0: \quad \mu_1 = \mu_2 = \cdots = \mu_K$$

We make the following additional assumptions:

(i) The population variances are equal.
(ii) The population distributions are normal.

Then, a test of significance level α is provided by the decision rule

$$\text{Reject } H_0 \text{ if } \quad \frac{\text{MSG}}{\text{MSW}} > F_{K-1, n-K, \alpha}$$

where $F_{K-1, n-K, \alpha}$ is that number for which

$$P(F_{K-1, n-K} > F_{K-1, n-K, \alpha}) = \alpha$$

and the random variable $F_{K-1, n-K}$ follows an F distribution with numerator degrees of freedom $(K - 1)$ and denominator degrees of freedom $(n - K)$.

For the automobile data, we find

$$\frac{\text{MSG}}{\text{MSW}} = \frac{10.7748}{0.7165} = 15.04$$

TABLE 15.3 General format of one-way analysis of variance table

SOURCE OF VARIATION	SUMS OF SQUARES	DEGREES OF FREEDOM	MEAN SQUARES	F RATIO
Between groups	SSG	$K - 1$	$MSG = \dfrac{SSG}{(K - 1)}$	$\dfrac{MSG}{MSW}$
Within groups	SSW	$n - K$	$MSW = \dfrac{SSW}{(n - K)}$	
Total	SST	$n - 1$		

TABLE 15.4 One-way analysis of variance table for automobile miles per gallon data

SOURCE OF VARIATION	SUMS OF SQUARES	DEGREES OF FREEDOM	MEAN SQUARES	F RATIO
Between groups	21.5495	2	10.7748	15.04
Within groups	12.1800	17	0.7165	
Total	33.7295	19		

The numerator and denominator degrees of freedom are, respectively, $(K - 1) = 2$ and $(n - K) = 17$. Thus, for a 1% level test, from Table 7 of the Appendix, we have

$$F_{2, 17, .01} = 6.11$$

Hence, these data allow us to reject, at the 1% significance level, the null hypothesis that population mean miles per gallon is the same for all three types of automobile.

The computations involved in carrying out this test are very conveniently summarized in a **one-way analysis of variance table.** The general form of the table is set out in Table 15.3. For the automobile data, the analysis of variance is set out in Table 15.4. We note that, in some expositions, the within groups sum of squares is referred to as the **error sum of squares.**

Example 15.1

In Example 8.3, we discussed a study[1] aimed at assessing the readability of financial report messages. The technique for assessing the effectiveness of the written message is called the *cloze readability procedure*. Financial reports were given to independent random samples from three groups—certified public accountants, chartered financial analysts, and commercial bank loan officer trainees. The cloze procedure was then administered, and the scores for the sample members were recorded. The null hypothesis of interest is that the population mean scores for the three groups are identical. The sums of squares and degrees of freedom are given in the table. Complete the analysis of variance table and test the null hypothesis.

---

[1] A. H. Adelberg, "A methodology for measuring the understandability of financial report messages," *Journal of Accounting Research, 17* (1979), 565–92.

SOURCE OF VARIATION	SUMS OF SQUARES	DEGREES OF FREEDOM
Between groups	5,165	2
Within groups	120,802	1,005
Total	125,967	1,007

The mean squares are obtained by dividing the sums of squares by their associated degrees of freedom and the F ratio is obtained as the ratio of the mean squares. The analysis of variance table may then be completed as shown.

SOURCE OF VARIATION	SUMS OF SQUARES	DEGREES OF FREEDOM	MEAN SQUARES	F RATIO
Between groups	5,165	2	2,582.5	21.49
Within groups	120,802	1,005	120.2	
Total	125,967	1,007		

The numerator and denominator degrees of freedom are, respectively, 2 and 1,005, so that for a test of significance level α, we require $F_{2, 1,005, \alpha}$. Referring to Table 7 of the Appendix, we find that values are not tabulated for this large number of denominator degrees of freedom. However, we can approximate the required value by using an infinite number of denominator degrees of freedom. Thus, for a 1% level test,

$$F_{2, 1,005, .01} \approx F_{2, \infty, .01} = 4.61$$

The calculated F ratio considerably exceeds this tabulated value. Hence, the null hypothesis that the population mean cloze scores are the same for these three groups is very clearly rejected at the 1% level of significance.

POPULATION MODEL FOR ONE-WAY ANALYSIS OF VARIANCE

It is instructive to view the one-way analysis of variance model in a different light. Let the random variable X_{ij} denote the jth observation from the ith population, and μ_i the mean of this population. Then, X_{ij} can be viewed as the sum of two parts—its mean and a random variable ε_{ij} having mean 0. Therefore, we can write

$$X_{ij} = \mu_i + \varepsilon_{ij} \qquad (15.2.2)$$

Now, because independent random samples are taken, the random variables ε_{ij} will be uncorrelated with one another. Moreover, given our assumption that the population variances are all the same, it follows that the ε_{ij} all have the same variances. Hence, these random variables satisfy the standard assumptions (see Section 13.3) imposed on the error terms of a multiple regression model. Thus, equation 15.2.2 can be viewed as such a model, with unknown parameters $\mu_1, \mu_2, \ldots, \mu_K$. The null hypothesis of interest,

$$H_0: \quad \mu_1 = \mu_2 = \cdots = \mu_K$$

is a test on these parameters, facilitated by the further assumption of normality.

The model 15.2.2 can be written in a slightly different manner. Let μ denote the overall mean of the K combined populations and G_i the discrepancy between the population mean for the ith group and this overall mean, so that

$$G_i = \mu_i - \mu \qquad \text{or} \qquad \mu_i = \mu + G_i$$

Substituting into equation 15.2.2 then gives

$$X_{ij} = \mu + G_i + \varepsilon_{ij}$$

so that an observation is made up of the sum of an overall mean μ, a group-specific term G_i, and a random error ε_{ij}. Then our null hypothesis is that every population mean μ_i is the same as the overall mean, or

$$H_0: \quad G_1 = G_2 = \cdots = G_K = 0$$

This population model and some of the assumptions are illustrated in Figure 15.3. For each type of car, actual miles per gallon achieved in any trial can be represented by a normally distributed random variable. The population mean miles per gallon, μ_1, μ_2, and μ_3, for A-cars, B-cars, and C-cars, respectively, determine the centers of these distributions. According to our assumption, these population distributions must have the same variance. The figure also shows the mean μ of the three

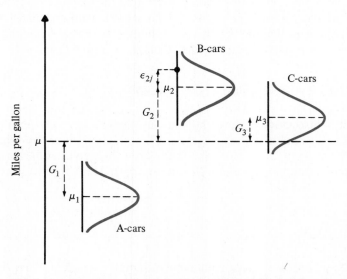

FIGURE 15.3 Illustration of the population model for the one-way analysis of variance

combined populations, and the differences G_i between the individual population means and the overall mean. Finally, for B-cars, we have marked by a dot the jth sample observation. The random variable ε_{2j} is then the difference between the observed value and the corresponding population mean. Thus, we can think of the observed value as the sum of three parts—the overall mean μ, the difference G_2 between the population mean miles per gallon for B-cars and the overall mean, and the discrepancy (due to sampling variability) of the observed value and the mean of the population from which it is drawn.

15.3 THE KRUSKAL-WALLIS TEST

As we have already noted, the one-way analysis of variance test of Section 15.2 generalizes to the multipopulation case the t test for comparing two population means when independent random samples are available. The test is based on an assumption that the underlying population distributions are normal. In Section 10.4 we introduced the Mann-Whitney test, a nonparametric test that is valid for the comparison of the central locations of two populations based on independent random samples, even when the population distributions are not normal. It is also possible to develop a nonparametric alternative to the one-way analysis of variance test. This is known as the **Kruskal-Wallis test,** and is employed when an investigator has strong grounds for suspecting that the parent population distributions may be markedly different from the normal.

   Like the majority of the nonparametric tests we have already encountered, the Kruskal-Wallis test is based on the *ranks* of the sample observations. We will illustrate the computation of the test statistic by reference to the automobile mileage data of Table 15.1 The sample values are all pooled together and ranked in ascending order, as in Table 15.5, using the average of adjacent ranks in the case of ties.

   The test is based on the sums of the ranks $R_1, R_2, \ldots, R_K$, for the K samples. In the automobile mileage example,

$$R_1 = 32; \qquad R_2 = 101.5; \qquad R_3 = 76.5$$

Now, the null hypothesis to be tested is that the three population means are the same. We would be suspicious of that hypothesis if there were substantial differences among the average ranks for the K samples. In fact, our test is based on the statistic

TABLE 15.5 Miles per gallon figures and ranks from three independent random samples

A-CARS	RANK	B-CARS	RANK	C-CARS	RANK
22.2	11	24.6	20	22.7	12
19.9	1	23.1	13	21.9	7
20.3	2.5	22.0	8	23.2	14
21.4	6	23.5	16.5	24.1	19
21.2	5	23.6	18	22.1	9.5
21.0	4	22.1	9.5	23.4	15
20.3	2.5	23.5	16.5		
Rank sums	32		101.5		76.5

$$W = \frac{12}{n(n+1)} \sum_{i=1}^{K} \frac{R_i^2}{n_i} - 3(n+1) \qquad (15.3.1)$$

where n_i are the sample sizes in the K populations and n is the total number of sample observations. The null hypothesis would be in doubt if a large value for 15.3.1 were observed. The basis for the test follows from the fact that, unless the sample sizes are very small, the random variable corresponding to the test statistic has, under the null hypothesis, a distribution that is well approximated by the χ^2 distribution with $(K-1)$ degrees of freedom.

The Kruskal-Wallis Test

Suppose we have independent random samples of $n_1, n_2, \ldots, n_K$ observations from K populations. Let

$$n = n_1 + n_2 + \cdots + n_K$$

denote the total number of sample observations. Denote by $R_1, R_2, \ldots, R_K$ the sums of ranks for the K samples when the sample observations are pooled together and ranked in ascending order. The test of the null hypothesis, H_0, of equality of the population means is based on the statistic

$$W = \frac{12}{n(n+1)} \sum_{i=1}^{K} \frac{R_i^2}{n_i} - 3(n+1)$$

A test of significance level α is given by the decision rule

$$\text{Reject } H_0 \text{ if } \quad W > \chi^2_{K-1, \alpha}$$

where $\chi^2_{K-1, \alpha}$ is the number that is exceeded with probability α by a χ^2 random variable with $(K-1)$ degrees of freedom.

This test procedure is approximately valid provided the sample contains at least five observations from each population.

For our automobile mileage data, we find

$$W = \frac{12}{(20)(21)} \left[\frac{(32)^2}{7} + \frac{(101.5)^2}{7} + \frac{(76.5)^2}{6} \right] - (3)(21) = 11.10$$

Here, we have $(K-1) = 2$ degrees of freedom so that, for a 0.5% significance level test, we find from Table 5 of the Appendix,

$$\chi^2_{2, .005} = 10.60$$

Hence, the null hypothesis that the population mean miles per gallon are the same for the three types of automobiles can be rejected even at the 0.5% significance level. Of course, we also rejected this hypothesis using the analysis of variance test of Section 15.2. However, here we have been able to do so without imposing the assumption of normality of the population distributions.

Example
15.2

Independent random samples of 101 lower class, 112 middle class, and 96 upper class women were asked to rate, on a scale from 1 to 7, the importance they attached to brand name when purchasing soft drinks. The value of the Kruskal-Wallis statistic for this study[2] was reported as 25.22. Test the null hypothesis that the population mean ratings are the same for these three populations.

The calculated test statistic is

$$W = 25.22$$

Since there are $K = 3$ groups, we have for a 0.5% level test,

$$\chi^2_{K-1, \alpha} = \chi^2_{2, .005} = 10.60$$

Thus, the null hypothesis that the three population mean ratings are the same is very clearly rejected on the evidence of this sample, even at the 0.5% level of significance.

EXERCISES

1. A manufacturer of diet soda is considering three alternative can colors—red, yellow, and blue. In order to check whether such considerations have any effect on sales, sixteen stores of approximately equal size are chosen. Red cans are sent to six of these stores, yellow cans are sent to five others, and blue cans to the remaining five. After a few days, a check is made on the number of sales in each store. The results (in tens of cans) shown in the table were obtained.

CAN COLOR		
RED	YELLOW	BLUE
46	52	61
52	37	29
59	38	38
76	64	53
61	74	79
84		

(a) Calculate the within groups, between groups, and total sums of squares.

(b) Complete the analysis of variance table and test the null hypothesis that the population mean sales levels are the same for all three can colors.

2. An instructor has a class of twenty-three students. At the beginning of the semester, each student is randomly assigned to one of four teaching assistants—Smiley, Haydon, Alleline, or Bland. The students are encouraged to meet with their assigned teaching assistant to discuss difficult course material. At the end of the semester, a common examination is administered. The scores obtained by students working with these teaching assistants are shown in the accompanying table.

---

[2] See P. E. Murphy, "The effect of social class on brand and price consciousness for supermarket products," *Journal of Retailing, 55* (1978), No. 2, 33–42.

TEACHING ASSISTANT			
SMILEY	HAYDON	ALLELINE	BLAND
72	81	77	79
69	93	68	70
84	79	59	61
76	97	75	74
64	88	82	85
	84	65	61

(a) Calculate the within groups, between groups, and total sums of squares.

(b) Complete the analysis of variance table and test the null hypothesis of equality of population mean scores for the four teaching assistants.

3. The lifetimes of three different types of lightbulb—Quick Glow, Watts New, and Slow Burn—were compared. Independent random samples of six bulbs of each type were tested. The resulting lifetimes (in hours) are listed in the table.

BULB BRAND		
QUICK GLOW	WATTS NEW	SLOW BURN
76	81	74
68	77	69
64	68	71
79	80	75
83	80	70
62	76	55

(a) Set out the analysis of variance table for these data.

(b) Test the null hypothesis that the population mean lifetimes are the same for all three types of bulb.

4. A corporation is trying to decide which of three makes of automobile to order for its fleet—domestic, Japanese, or European. Five cars of each type were ordered and, after 10,000 miles of driving, the operating cost per mile (in cents) of each was assessed. The accompanying results were obtained.

AUTOMOBILE		
DOMESTIC	JAPANESE	EUROPEAN
18.2	20.1	19.3
17.6	15.6	17.4
15.4	16.1	15.1
19.1	15.3	18.6
17.2	15.4	16.1

(a) Set out the analysis of variance table for these data.

(b) Test the null hypothesis that the population mean operating costs per mile are the same for these three types of car.

5. A manufacturer wants to compare the flammability of three fabric types that are being considered for use in the manufacture of children's clothing. Independent random samples of seven garments made from each fabric were tested. In each trial, the total time (in seconds) taken to burn was measured. The results are shown in the table.

FABRIC		
A	B	C
22.3	19.6	18.9
21.2	21.4	21.2
24.7	20.8	21.6
23.6	22.7	20.8
22.5	23.1	23.2
21.6	22.1	19.6
20.7	20.6	20.3

(a) Set out the analysis of variance table.

(b) Test the null hypothesis that the population mean burn times are the same for all three fabrics.

6. An executive of a preprepared frozen foods company is interested in the amounts of money spent on such products by families in different income ranges. Independent random samples of six families with incomes under $15,000 per year, five families with incomes between $15,000 and $30,000, and four families with incomes over $30,000 a year were taken. The estimates of monthly expenditures (in dollars) on preprepared frozen foods given by the sample members are shown in the table.

ANNUAL INCOME		
UNDER $15,000	$15,000–$30,000	OVER $30,000
45.2	48.2	50.7
63.7	51.6	71.6
52.8	63.7	61.3
31.7	46.8	49.8
33.6	49.2	
39.4		

(a) Set out the analysis of variance table.

(b) Test the null hypothesis that population mean expenditures on preprepared frozen foods are the same for all three income groups.

7. Sales of a particular fast-food chain have increased over last year. A marketing manager is interested in whether this improved performance is uniformly distributed across the country. Independent random samples of five outlets from each of four regions of the

country were taken, and the percentage increases in sales over the year were recorded. The results are given in the accompanying table.

REGION			
EAST	SOUTH	MIDWEST	WEST
10.4	12.8	11.2	13.9
12.8	14.2	9.8	14.2
15.6	16.3	10.7	12.7
9.2	10.1	6.3	15.0
8.7	12.0	12.4	13.7

(a) Set out the analysis of variance table.

(b) Test the null hypothesis that population mean sales growth rates were the same in all four regions.

8. Two tutoring services offer crash courses in preparation for the C.P.A. exam. To check on the effectiveness of these services, fifteen students were chosen. Five students were randomly assigned to service A, five to service B, while the remaining five took no crash course. Their scores on the examination, expressed as percentages, are given in the table.

TUTORING SERVICE		
A	B	NONE
83	74	72
74	69	71
92	87	81
67	81	61
85	64	59

(a) Set out the analysis of variance table.

(b) Test the null hypothesis that the three population mean scores are the same.

9. In an experiment aimed at assessing aids to the success of interviews of subordinates carried out by managers, interviewers were randomly assigned to one of three interview modes—"feedback," "feedback and goal setting," and "control." For the feedback mode, interviewers had the opportunity to examine and discuss their subordinates' reactions to previous interviews. In addition, in the feedback and goal setting mode, managers were encouraged to set goals for the forthcoming interview. For the control group, interviews were conducted in the usual manner, without feedback or goal setting. After the interviews were completed, the satisfaction levels of the subordinates with the interviews were assessed.[3] For the 45 people in the feedback group, the mean satisfaction level was 13.98. The 49 people in the feedback and goal setting group had mean satisfaction level 15.12, while the 41 control group members had mean satisfaction level 13.07. The F ratio calculated from these responses was 4.12.

---

[3] This study is due to W. F. Nemeroff and J. Cosentino, "Utilizing feedback and goal setting to increase performance appraisal interviewer skills of managers," *Academy of Management Journal, 22* (1979), 516–26.

(a) Set out the complete analysis of variance table.

(b) Test the null hypothesis that the population mean satisfaction levels are the same for all three types of interview.

10. A study[4] classified each of 134 managers into one of four groups, on the basis of an interview and observation. The 62 managers in group A were characterized as having high levels of stimulation and support, and average levels of public spirit. The 52 managers in group B had low stimulation, average support, and high public spirit. Group C contained 7 managers with average stimulation, low support, and low public spirit. The 13 managers in group D were rated low on all three criteria. Salary levels for these groups were compared. The sample means were 7.87 for group A, 7.47 for group B, 5.14 for group C, and 3.69 for group D. The calculated F ratio was 25.60.

(a) Set out the complete analysis of variance table.

(b) Test the null hypothesis of identical population mean salary levels for these four groups.

11. In the study of Exercise 10, the breadth of education of the managers was also assessed. The sample means were 2.63 for group A, 2.15 for group B, 1.29 for group C, and 3.31 for group D. The F ratio was found to be 3.61.

(a) Set out the complete analysis of variance table.

(b) Test the null hypothesis of identical population mean levels for breadth of education for these four groups.

12. For the data of Exercise 1, use the Kruskal-Wallis test to test the null hypothesis that the population mean sales levels are identical for the three can colors.

13. Using the data of Exercise 2, perform a Kruskal-Wallis test of the null hypothesis that the population mean test scores are the same for students assigned to the four teaching assistants.

14. Using the data of Exercise 3, carry out a test of the null hypothesis of equality of the three population mean bulb lifetimes, without assuming normality of population distributions.

15. For the data of Exercise 4, test the null hypothesis that the population mean operating costs per mile are the same for all three types of automobile, without assuming normal population distributions.

16. Using the data of Exercise 5, carry out a nonparametric test of the null hypothesis of equality of population mean burning times for the three fabric types.

17. Based on the data of Exercise 6, use the Kruskal-Wallis procedure to test the null hypothesis of equal population mean expenditures on preprepared frozen foods for the three income groups.

18. For the data of Exercise 7, test the null hypothesis of equal population mean sales growth rates in the four regions, without assuming a normal population distribution.

19. Based on the data of Exercise 8, perform the Kruskal-Wallis test of the null hypothesis of equal population mean scores on the C.P.A. exam for students using no tutoring service and those using services A and B.

20. Independent random samples of 101 lower class women, 112 middle class women, and 96 upper class women were asked to rate, on a scale from 1 to 7, the importance attached to brand name when purchasing paper towels. The value of the Kruskal-Wallis statistic obtained[5] was 0.17.

(a) What null hypothesis can be tested using this information?

(b) Carry out this test.

[4] See D. C. Pheysey, "Managers' occupational histories, organizational environments, and climates for management development," *Journal of Management Studies, 14* (1977), 58–79.

[5] See P. E. Murphy, "The effect of social class on brand and price consciousness for supermarket products," *Journal of Retailing, 55* (1978), No. 2, 33–42.

15.4 TWO-WAY ANALYSIS OF VARIANCE: ONE OBSERVATION PER CELL, RANDOMIZED BLOCKS

It is often the case that, while our prime interest lies in the analysis of one particular feature of an experiment, we suspect that a second factor may well exert an important influence on the outcome. In the earlier sections of this chapter, we discussed an experiment in which the objective was to compare the performances, in terms of miles per gallon, of three types of automobile. Data were collected from three independent random samples of trials, and analyzed through a one-way analysis of variance. In effect, it was assumed that the variability in the sample data was due to two causes—genuine differences between the performance characteristics of the three types of car and random variation. In fact, we might suspect that part of the observed random variability could be explained by differences in driver habits. Now, if in some way this last factor could be isolated, then the amount of random variability in the experiment would be reduced accordingly. This, in turn, might make it easier to detect differences in the performances of the automobiles. In other words, by designing an experiment to account for differences in driver characteristics we hope to achieve a more powerful test of the null hypothesis that population mean miles per gallon is the same for all types of automobiles.

In fact, it is quite straightforward to design an experiment in such a way that the influence of a second factor of this kind can be taken into account. Suppose, once again, that we have three makes of automobile (say, α-cars, β-cars, and γ-cars) whose fuel economies we wish to compare. We will consider an experiment in which six trials are to be run with each type of car. If these trials are conducted using six drivers, each of whom drives a car of all three types, then, since every car will have been tested by every driver, it will be possible to extract from the results information about driver variability as well as information about differences among the three types of car. The additional variable—in this case, drivers—is sometimes called a **blocking variable.** The experiment is said to be arranged in **blocks,** where, in our example, there would be six blocks (one for each driver).

This kind of blocked design can be used to obtain information about two factors simultaneously. For example, suppose that we want to compare mileages obtained, not only by different types of automobile, but also by different types of drivers. In particular, we may be interested in the effect of driver age on miles per gallon obtained. To do this, drivers can be subdivided into age categories. We might use the following six age classes (in years).

1. 25 and under
2. 26–35
3. 36–45
4. 46–55
5. 56–65
6. Over 65

Then, we can arrange our experiment so that an automobile from each group is driven by a driver from each age class. In this way, in addition to testing the hypothesis that

TABLE 15.6 Sample observations on miles per gallon achieved by three types of automobile driven by six drivers

DRIVER CLASSES	AUTOMOBILES			
	α-CARS	β-CARS	γ-CARS	SUMS
1	25.1	23.9	26.0	75.0
2	24.7	23.7	25.4	73.8
3	26.0	24.4	25.8	76.2
4	24.3	23.3	24.4	72.0
5	23.9	23.6	24.2	71.7
6	24.2	24.5	25.4	74.1
Sums	148.2	143.4	151.2	442.8

population mean miles per gallon are the same for each automobile type, we can also test the hypothesis that population mean miles per gallon are the same for each age class.

In fact, whether a car of each type is driven by each of six drivers, or a car of each type is driven by a driver from each of six age classes, the procedure for testing equality of population mean mileage for the automobile types is the same. In this section we will use the latter design for purposes of illustration.

Table 15.6 gives results for an experiment involving three automobile types and six driver age classes. Thus, the comparison of automobile types is the main focus of interest, and driver ages are used as a blocking variable.

This kind of design is called a **randomized blocks design.** The randomization arises because we randomly select one driver from the first age class to drive an α-car. This same procedure is repeated for each of the other driver classes. Also, if possible, the trials should be carried out in random order, rather than block by block.

In general, suppose that we have K groups and that there are H blocks. We will use x_{ij} to denote the sample observation corresponding to the ith group and the jth block. Thus, the sample data may be set out as in Table 15.7. Notice that the format here is simply an extension of the experimental form used for the paired observations test of Section 9.6, where we had only two groups. Thus, the development of this section extends the paired t test of Section 9.6 to allow us to test the equality of several population means.

TABLE 15.7 Sample observations on K groups and H blocks

BLOCK	GROUP			
	1	2	. . .	K
1	x_{11}	x_{21}	. . .	x_{K1}
2	x_{12}	x_{22}	. . .	x_{K2}
.	.	.		.
.	.	.		.
.	.	.		.
H	x_{1H}	x_{2H}	. . .	x_{KH}

In order to develop a test of the hypothesis that the population means are the same for all K groups, we require the sample means for these groups. For the mean of the ith group, we use the notation $\bar{x}_{i\cdot}$, so that

$$\bar{x}_{i\cdot} = \frac{\sum\limits_{j=1}^{H} x_{ij}}{H} \qquad (i = 1, 2, \ldots, K)$$

From Table 15.6, we obtain

$$\bar{x}_{1\cdot} = \frac{148.2}{6} = 24.7; \qquad \bar{x}_{2\cdot} = \frac{143.4}{6} = 23.9; \qquad \bar{x}_{3\cdot} = \frac{151.2}{6} = 25.2$$

We are also interested in the differences in the population block means. Hence, we require the sample means for the H blocks. We use $\bar{x}_{\cdot j}$ to denote the sample mean for the jth block, so that

$$\bar{x}_{\cdot j} = \frac{\sum\limits_{i=1}^{K} x_{ij}}{K} \qquad (j = 1, 2, \ldots, H)$$

For the mileage data of Table 15.6, we have

$$\bar{x}_{\cdot 1} = \frac{75.0}{3} = 25.0; \qquad \bar{x}_{\cdot 2} = \frac{73.8}{3} = 24.6; \qquad \bar{x}_{\cdot 3} = \frac{76.2}{3} = 25.4;$$

$$\bar{x}_{\cdot 4} = \frac{72.0}{3} = 24.0; \qquad \bar{x}_{\cdot 5} = \frac{71.7}{3} = 23.9; \qquad \bar{x}_{\cdot 6} = \frac{74.1}{3} = 24.7$$

Finally, we require the overall mean of the sample observations. If n denotes the total number of observations, then

$$n = HK$$

and the sample mean of all the observations is

$$\bar{x} = \frac{\sum\limits_{i=1}^{K} \sum\limits_{j=1}^{H} x_{ij}}{n} = \frac{\sum\limits_{i=1}^{K} \bar{x}_{i\cdot}}{K} = \frac{\sum\limits_{j=1}^{H} \bar{x}_{\cdot j}}{H}$$

For the data of Table 15.6,

$$\bar{x} = \frac{442.8}{18} = 24.6$$

Before proceeding to consider the form of an appropriate test for the hypothesis of interest, it is useful to examine the population model that is implicitly being assumed. Let the random variable X_{ij} correspond to the observation for the ith group and jth block. This value is then regarded as the sum of the following four components:

(i) An "overall" mean, μ
(ii) A parameter G_i, which is specific to the ith group and measures the discrepancy between the mean for that group and the overall mean

(iii) A parameter B_j, which is specific to the jth block, and measures the discrepancy between the mean for that block and the overall mean

(iv) A random variable ε_{ij}, which represents experimental error, or that part of the observation not explained by either the overall mean or the group or block membership

We can therefore write

$$X_{ij} = \mu + G_i + B_j + \varepsilon_{ij} \qquad (15.4.1)$$

The error term ε_{ij} is taken to obey the standard assumptions of the multiple regression model. In particular, then, we assume independence and equality of variances.

It is now convenient to write 15.4.1 as

$$X_{ij} - \mu = G_i + B_j + \varepsilon_{ij} \qquad (15.4.2)$$

Now, given sample data, the overall mean μ is estimated by the overall sample mean $\bar{x}$, so that an estimate of the left-hand side of 15.4.2 is provided by $(x_{ij} - \bar{x})$. The difference G_i between the population mean for the ith group and the overall population mean is estimated by the corresponding difference in sample means, that is, $(\bar{x}_{i\cdot} - \bar{x})$. Similarly, B_j is estimated by $(\bar{x}_{\cdot j} - \bar{x})$. Finally, by subtraction, we estimate the error term by

$$(x_{ij} - \bar{x}) - (\bar{x}_{i\cdot} - \bar{x}) - (\bar{x}_{\cdot j} - \bar{x}) = x_{ij} - \bar{x}_{i\cdot} - \bar{x}_{\cdot j} + \bar{x}$$

Thus, corresponding to equation 15.4.2, we have for the sample members

$$(x_{ij} - \bar{x}) = (\bar{x}_{i\cdot} - \bar{x}) + (\bar{x}_{\cdot j} - \bar{x}) + (x_{ij} - \bar{x}_{i\cdot} - \bar{x}_{\cdot j} + \bar{x}) \quad (15.4.3)$$

To illustrate, consider the miles per gallon obtained by a driver from the third class with an α-car. That is, from Table 15.6,

$$x_{13} = 26.0$$

The term on the left-hand side of equation 15.4.3 is

$$x_{13} - \bar{x} = 26.0 - 24.6 = 1.4$$

For the group (automobile) effect, we have

$$\bar{x}_{1\cdot} - \bar{x} = 24.7 - 24.6 = 0.1$$

(Notice that this term will result whenever the α-car is driven.) For the block (driver) effect, we have

$$\bar{x}_{\cdot 3} - \bar{x} = 25.4 - 24.6 = 0.8$$

Finally, the error term is

$$x_{13} - \bar{x}_1. - \bar{x}_{.3} + \bar{x} = 26.0 - 24.7 - 25.4 + 24.6 = 0.5$$

Thus, corresponding to equation 15.4.3, we have for this observation

$$1.4 = 0.1 + 0.8 + 0.5$$

We can interpret this equation as follows: When a driver from the third age class tested the α-car, she obtained 1.4 miles per gallon more than the average for all cars and drivers. Of this amount, it is estimated that 0.1 is due to the automobile, 0.8 to the driver age class, and the remaining 0.5 mile per gallon to other factors, which we put down to chance variability or experimental error.

Now, if both sides of equation 15.4.3 are squared and summed over all n sample observations, it can be shown that the result is

$$\sum_{i=1}^{K}\sum_{j=1}^{H}(x_{ij} - \bar{x})^2 = H\sum_{i=1}^{K}(\bar{x}_{i.} - \bar{x})^2 + K\sum_{j=1}^{H}(\bar{x}_{.j} - \bar{x})^2 + \sum_{i=1}^{K}\sum_{j=1}^{H}(x_{ij} - \bar{x}_{i.} - \bar{x}_{.j} + \bar{x})^2$$

This equation expresses the total sample variability of the observations about their overall mean as the sum of variabilities due to differences among groups, differences among blocks, and error, respectively. It is on this sums of squares decomposition that the analysis of experiments of this type is based. The analysis is called **two-way analysis of variance,** as the data are categorized in two ways, according to groups and blocks.

We illustrate this important sum of squares decomposition in Figure 15.4. Notice, by contrast with the decomposition for the one-way analysis of variance, that the total sum of squares of the sample observations about their overall mean is here broken down into *three* components. The extra component arises because of our ability to extract from the data information about differences among blocks.

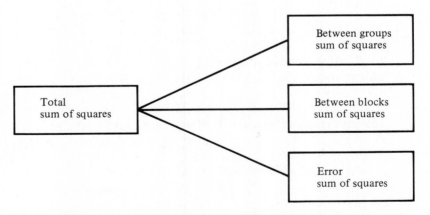

FIGURE 15.4 Sum of squares decomposition for two-way analysis of variance with one observation per cell

---

Sums of Squares Decomposition for Two-Way Analysis of Variance

Suppose that we have a sample of observations with x_{ij} denoting the observation in the ith group and jth block. Suppose also that there are K groups and H blocks, for a total of

$$n = KH$$

observations. Denote the group sample means by $\bar{x}_i.$ $(i = 1, 2, \ldots, K)$, the block sample means by $\bar{x}._j$ $(j = 1, 2, \ldots, H)$, and the overall sample mean by $\bar{x}$.

We define the following **sums of squares:**

TOTAL: $$\text{SST} = \sum_{i=1}^{K} \sum_{j=1}^{H} (x_{ij} - \bar{x})^2$$

BETWEEN GROUPS: $$\text{SSG} = H \sum_{i=1}^{K} (\bar{x}_i. - \bar{x})^2$$

BETWEEN BLOCKS: $$\text{SSB} = K \sum_{j=1}^{H} (\bar{x}._j - \bar{x})^2$$

ERROR: $$\text{SSE} = \sum_{i=1}^{K} \sum_{j=1}^{H} (x_{ij} - \bar{x}_i. - \bar{x}._j + \bar{x})^2$$

Then

$$\text{SST} = \text{SSG} + \text{SSB} + \text{SSE}$$

For the automobile mileage data of Table 15.6, we find

$$\text{SST} = (25.1 - 24.6)^2 + (24.7 - 24.6)^2 + \cdots + (25.4 - 24.6)^2 = 11.88$$

$$\text{SSG} = 6[(24.7 - 24.6)^2 + (23.9 - 24.6)^2 + (25.2 - 24.6)^2] = 5.16$$

$$\text{SSB} = 3[(25.0 - 24.6)^2 + (24.6 - 24.6)^2 + \cdots + (24.7 - 24.6)^2] = 4.98$$

and so, by subtraction,

$$\text{SSE} = \text{SST} - \text{SSG} - \text{SSB} = 11.88 - 5.16 - 4.98 = 1.74$$

From this point, the tests associated with the two-way analysis of variance proceed in a similar fashion to the one-way analysis of Section 15.2. First, the mean squares are obtained by dividing each sum of squares by the appropriate number of degrees of freedom. For the total sum of squares, the degrees of freedom are 1 less than the total number of observations, that is, $(n - 1)$. For the sum of squares between groups, the degrees of freedom are 1 less than the number of groups, or $(K - 1)$. Similarly, for the sum of squares between blocks, the number of degrees of freedom is $(H - 1)$. Hence, by subtraction, the degrees of freedom associated with the sum of squared errors are

$$(n-1) - (K-1) - (H-1) = n - K - H + 1$$
$$= KH - K - H + 1$$
$$= (K-1)(H-1)$$

The null hypothesis that the population group means are equal can then be tested through the ratio of the mean square for groups to the mean square error. Very often, a blocking variable is included in the analysis simply to reduce variability due to experimental error. However, it is sometimes the case that the hypothesis that the block population means are equal is also of interest. This can be tested through the ratio of the mean square for blocks to the mean square error. As in the case of the one-way analysis of variance, the relevant standard for comparison is obtained from a tail probability of the F distribution. The procedures are described in the box.

Hypothesis Tests for Two-Way Analysis of Variance

Suppose we have a sample observation for each group–block combination in a design containing K groups and H blocks. Define the following mean squares:

$$\text{BETWEEN GROUPS:} \quad \text{MSG} = \frac{\text{SSG}}{K-1}$$

$$\text{BETWEEN BLOCKS:} \quad \text{MSB} = \frac{\text{SSB}}{H-1}$$

$$\text{ERROR:} \quad \text{MSE} = \frac{\text{SSE}}{(K-1)(H-1)}$$

We assume that the error terms ε_{ij} in the model 15.4.1 are independent of one another and have the same variance. It is further assumed that these errors are normally distributed. Then:

(i) A test of significance level α of the null hypothesis H_0 that the K population group means are all the same is provided by the decision rule

$$\text{Reject } H_0 \text{ if } \quad \frac{\text{MSG}}{\text{MSE}} > F_{K-1,\,(K-1)(H-1),\,\alpha}$$

(ii) A test of significance level α of the null hypothesis H_0 that the H population block means are all the same is provided by the decision rule

$$\text{Reject } H_0 \text{ if } \quad \frac{\text{MSB}}{\text{MSE}} > F_{H-1,\,(K-1)(H-1),\,\alpha}$$

Here, $F_{\nu_1,\,\nu_2,\,\alpha}$ is that number exceeded with probability α by a random variable following an F distribution with numerator degrees of freedom ν_1 and denominator degrees of freedom ν_2.

For the automobile mileage data, the mean squares are

$$MSG = \frac{SSG}{K-1} = \frac{5.16}{2} = 2.58$$

$$MSB = \frac{SSB}{H-1} = \frac{4.98}{5} = 0.996$$

$$MSE = \frac{SSE}{(K-1)(H-1)} = \frac{1.74}{10} = 0.174$$

To test the null hypothesis that the population mean miles per gallon are the same for all three types of automobile, we require

$$\frac{MSG}{MSE} = \frac{2.58}{0.174} = 14.83$$

For a 1% level test, we have for comparison, from Table 7 of the Appendix,

$$F_{K-1,(K-1)(H-1),\alpha} = F_{2,10,.01} = 7.56$$

Therefore, on the evidence of these data, the hypothesis of equal mean population performances for the three types of automobile is clearly rejected at the 1% significance level.

In this particular example, the null hypothesis of equality of the population block means is the hypothesis that population values of mean miles per gallon are the same for each driver age class. The test is based on

$$\frac{MSB}{MSE} = \frac{0.996}{0.174} = 5.72$$

For a 1% level test, we have from Table 7 of the Appendix,

$$F_{H-1,(K-1)(H-1),\alpha} = F_{5,10,.01} = 5.64$$

Hence, the null hypothesis of equal population means for the six driver age classes is also rejected at the 1% significance level.

Once again, it is very convenient to summarize the computations in tabular form. The general setup for the **two-way analysis of variance table** is shown in Table 15.8.

TABLE 15.8 General format of two-way analysis of variance table

SOURCE OF VARIATION	SUMS OF SQUARES	DEGREES OF FREEDOM	MEAN SQUARES	F RATIOS
Between groups	SSG	$K-1$	$MSG = \dfrac{SSG}{K-1}$	$\dfrac{MSG}{MSE}$
Between blocks	SSB	$H-1$	$MSB = \dfrac{SSB}{H-1}$	$\dfrac{MSB}{MSE}$
Error	SSE	$(K-1)(H-1)$	$MSE = \dfrac{SSE}{(K-1)(H-1)}$	
Total	SST	$n-1$		

TABLE 15.9 Two-way analysis of variance table for automobile miles per gallon data

SOURCE OF VARIATION	SUMS OF SQUARES	DEGREES OF FREEDOM	MEAN SQUARES	F RATIOS
Automobiles	5.16	2	2.580	14.83
Drivers	4.98	5	0.996	5.72
Error	1.74	10	0.174	
Total	11.88	17		

For the automobile gas mileage data, this analysis of variance is set out in Table 15.9. The numbers of degrees of freedom are determined by the numbers of groups and blocks. The mean squares are obtained by dividing the sums of squares by their associated degrees of freedom. The mean square error is then the denominator in the calculation of the two F ratios on which our tests are based.

EXERCISES

21. Four financial analysts were asked to predict earnings growth over the coming year for five oil companies. Their forecasts, as projected percentage increases in earnings, are given in the accompanying table.

OIL COMPANY	ANALYST			
	A	B	C	D
1	10	12	7	13
2	9	9	8	12
3	12	10	9	10
4	11	10	10	12
5	9	8	10	14

(a) Set out the two-way analysis of variance table.
(b) Test the null hypothesis that the population mean growth forecasts are the same for all oil companies.

22. An agricultural experiment aimed at assessing differences in yields of corn for four different varieties, using three different fertilizers, produced the results (in bushels per acre) shown in the table.

FERTILIZER	VARIETY			
	A	B	C	D
1	86	88	77	84
2	92	91	81	93
3	75	80	83	79

(a) Set out the two-way analysis of variance table.

(b) Test the null hypothesis that the population mean yields are identical for all four varieties of corn.

(c) Test the null hypothesis that population mean yields are the same for all three brands of fertilizer.

23. A company test-markets three new brands of cereal in selected retail outlets over a period of 4 months. Sales achieved (in thousands of dollars) are given in the table.

	BRAND		
	A	B	C
MARCH	12	14	15
APRIL	16	17	19
MAY	18	16	13
JUNE	13	14	12

(a) Set out the two-way analysis of variance table.

(b) Test the null hypothesis that population mean sales are the same for all three brands of cereal.

24. A diet soda manufacturer wants to compare the effects on sales of three can colors—red, yellow, and blue. Four regions are selected for the test, and three stores are randomly chosen from each region, each to display one of these types of cans. The accompanying table shows sales (in tens of cans) at the end of the experimental period.

	CAN COLOR		
	RED	YELLOW	BLUE
EAST	47	52	60
SOUTH	56	54	52
MIDWEST	49	63	55
WEST	41	44	48

(a) Set out the appropriate analysis of variance table.

(b) Test the null hypothesis that population mean sales are the same for each can color.

25. An instructor in a business statistics course is undecided among three texts. Also she is not sure whether students prefer regular weekly quizzes. During the year she teaches six sections of this course, and randomly allocates a quiz–text strategy to each section. At the end of the course she obtains overall approval ratings from the student evaluations for each section. These ratings are given in the table.

	TEXT		
	A	B	C
QUIZ	4.8	5.4	4.7
NO QUIZ	4.5	5.0	4.4

(a) Set out the appropriate analysis of variance table.

(b) Test the null hypothesis of equality of population mean ratings for the three texts.

(c) Test the null hypothesis that population mean ratings are the same for the quiz and no quiz sections.

26. In equation 15.4.2, we considered, for the two-way analysis of variance, the population model

$$X_{ij} - \mu = G_i + B_j + \varepsilon_{ij}$$

For the data of Exercise 22, obtain sample estimates for each term on the right-hand side of this equation for the fertilizer 1–variety A combination.

27. For the data of Exercise 25, obtain sample estimates for each term on the right-hand side of equation 15.4.2 for the no quiz–text B combination.

28. Four real estate agents were asked to appraise the values of ten houses in a particular neighborhood. The appraisals were expressed in thousands of dollars, with the results shown in the table.

SOURCE OF VARIATION	SUMS OF SQUARES
Between agents	228
Between houses	1,152
Error	2,434

(a) Complete the analysis of variance table.

(b) Test the null hypothesis that mean assessments are the same for these four real estate agents.

29. Four television commercials were aired during six shows—a football game, a soap opera, a situation comedy, a drama series, a movie, and a current affairs program. The advertising agency then set out to measure recall—that is, the percentage of the audience who could correctly identify the product—one day after the show was seen. The results obtained are summarized in the accompanying table.

SOURCE OF VARIATION	SUMS OF SQUARES
Between commercials	123.2
Between shows	85.4
Error	111.9

(a) Complete the analysis of variance table.

(b) Test the null hypothesis that mean recall levels are the same for each commercial.

(c) Test the null hypothesis that mean recall levels are the same for viewers of each show.

30. Three television pilot shows for potential situation comedy series were shown to audiences in four regions of the country—East, South, Midwest, and West Coast. Based on audience reactions, a score (on a scale from 0 to 100) was obtained for each show. The

sums of squares between groups (shows) and between blocks (regions) were found to be

$$SSG = 95.2 \qquad and \qquad SSB = 75.6$$

and the error sum of squares was

$$SSE = 79.3$$

Set out the analysis of variance table and test the null hypothesis that the population mean scores for audience reactions are the same for all three shows.

31. Suppose that, in the two-way analysis of variance setup with one observation per cell, there are just two groups. Show that, in this case, the F ratio for testing the equality of the group population means is precisely the square of the test statistic discussed in Section 9.6 for testing equality of population means, given a sample of matched pairs. Hence, deduce that the two tests are equivalent in this particular case.

15.5 TWO-WAY ANALYSIS OF VARIANCE: MORE THAN ONE OBSERVATION PER CELL

In the two-way analysis of variance layout of Section 15.4, we can view the tabulated raw data (as in Tables 15.6 and 15.7) as being broken down into cells, where each cell refers to a particular group–block combination. Thus, for example, the results obtained when a driver from the fourth age class drives a β-car constitute a single cell. A feature of the design analyzed in Section 15.4 is that each cell contains just a single sample observation. Thus, a driver from the fourth age class tests a β-car only once.

In this section we consider the possibility of **replicating** the experiment, so that, for example, β-cars would be driven by more than one driver from the fourth age class. The data resulting from such a design would then involve more than just a single observation per cell. There are two major advantages in extending the sample in this way. First, the more sample data that are available, the more precise will be the resulting estimates and the more surely will we be able to distinguish differences among the population means. Second, a design with more than one observation per cell allows the isolation of a further source of variability—the **interaction** between groups and blocks. Such interactions occur when differences in group effects are not distributed uniformly across blocks. For example, drivers who achieve higher than average miles per gallon figures may be considerably more successful in getting better fuel economy than other drivers when driving an α-car than when driving a β-car. Thus, this better than average performance is not uniformly spread over all types of cars, but rather is more manifest in some types than others. This possibility of driver–car interaction can be taken into account in an analysis based on more than one observation per cell.

To illustrate the kind of data that can be analyzed, Table 15.10 contains results on miles per gallon achieved by drivers from five age classes with three types of automobile—X-cars, Y-cars, and Z-cars. The three observations in each cell refer to independent trials by drivers from a given age class with automobiles of a particular type.

TABLE 15.10 Sample observations on miles per gallon achieved by three types of automobile driven by five drivers; three observations per cell

DRIVER CLASSES	AUTOMOBILES								
	X-CARS			Y-CARS			Z-CARS		
1	25.0	25.4	25.2	24.0	24.4	23.9	25.9	25.8	25.4
2	24.8	24.8	24.5	23.5	23.8	23.8	25.2	25.0	25.4
3	26.1	26.3	26.2	24.6	24.9	24.9	25.7	25.9	25.5
4	24.1	24.4	24.4	23.9	24.0	23.8	24.0	23.6	23.5
5	24.0	23.6	24.1	24.4	24.4	24.1	25.1	25.2	25.3

TABLE 15.11 Sample observations on K groups and H blocks; L observations per cell

BLOCK	GROUP												
	1				2				...	K			
1	x_{111}	x_{112}	...	x_{11L}	x_{211}	x_{212}	...	x_{21L}	...	x_{K11}	x_{K12}	...	x_{K1L}
2	x_{121}	x_{122}	...	x_{12L}	x_{221}	x_{222}	...	x_{22L}	...	x_{K21}	x_{K22}	...	x_{K2L}
.			.				.					.	
.			.				.					.	
H	x_{1H1}	x_{1H2}	...	x_{1HL}	x_{2H1}	x_{2H2}	...	x_{2HL}	...	x_{KH1}	x_{KH2}	...	x_{KHL}

To denote the individual sample observations, we require a triple subscript, so that x_{ijl} will denote the lth observation in the (i, j)th cell, that is, the lth observation in the cell corresponding to the ith group and the jth block. As before, we will let K denote the number of groups and H the number of blocks. We denote by L the number of observations per cell. Hence, in the example of Table 15.10, $K = 3$, $H = 5$, and $L = 3$. This notation is illustrated in Table 15.11.

Based on the results of an experiment of this type, there are three null hypotheses that can be tested—no difference between group means, no difference between block means, and no group–block interaction. In order to carry out these tests, we will again calculate various sample means. These are defined and calculated as follows:

(i) GROUP MEANS

The mean of *all* the sample observations in the ith group is denoted $\bar{x}_{i\cdot\cdot}$, so that

$$\bar{x}_{i\cdot\cdot} = \frac{\sum\limits_{j=1}^{H} \sum\limits_{l=1}^{L} x_{ijl}}{HL}$$

From Table 15.10, we find

$$\bar{x}_{1\cdot\cdot} = \frac{25.0 + 25.4 + \cdots + 23.6 + 24.1}{15} = 24.86$$

$$\bar{x}_{2\cdot\cdot} = \frac{24.0 + 24.4 + \cdots + 24.4 + 24.1}{15} = 24.16$$

$$\bar{x}_{3\cdot\cdot} = \frac{25.9 + 25.8 + \cdots + 25.2 + 25.3}{15} = 25.10$$

(ii) BLOCK MEANS

The mean of all the sample observations in the jth block is denoted $\bar{x}_{\cdot j\cdot}$, so that

$$\bar{x}_{\cdot j\cdot} = \frac{\sum\limits_{i=1}^{K}\sum\limits_{l=1}^{L} x_{ijl}}{KL}$$

For the data of Table 15.10, we have

$$\bar{x}_{\cdot 1\cdot} = \frac{25.0 + 25.4 + \cdots + 25.8 + 25.4}{9} = 25.00$$

$$\bar{x}_{\cdot 2\cdot} = \frac{24.8 + 24.8 + \cdots + 25.0 + 25.4}{9} = 24.53$$

$$\bar{x}_{\cdot 3\cdot} = \frac{26.1 + 26.3 + \cdots + 25.9 + 25.5}{9} = 25.57$$

$$\bar{x}_{\cdot 4\cdot} = \frac{24.1 + 24.4 + \cdots + 23.6 + 23.5}{9} = 23.97$$

$$\bar{x}_{\cdot 5\cdot} = \frac{24.0 + 23.6 + \cdots + 25.2 + 25.3}{9} = 24.47$$

(iii) CELL MEANS

In order to check the possibility of group–block interactions, it is necessary to calculate the sample mean for each cell. Let $\bar{x}_{ij\cdot}$ denote the sample mean for the (i, j)th cell. Then

$$\bar{x}_{ij\cdot} = \frac{\sum\limits_{l=1}^{L} x_{ijl}}{L}$$

Hence we find, for the data of Table 15.10,

$$\bar{x}_{11\cdot} = \frac{25.0 + 25.4 + 25.2}{3} = 25.2$$

$$\bar{x}_{12\cdot} = \frac{24.8 + 24.8 + 24.5}{3} = 24.7$$

and similarly

$$\bar{x}_{13\bullet} = 26.2; \quad \bar{x}_{14\bullet} = 24.3; \quad \bar{x}_{15\bullet} = 23.9;$$

$$\bar{x}_{21\bullet} = 24.1; \quad \bar{x}_{22\bullet} = 23.7; \quad \bar{x}_{23\bullet} = 24.8; \quad \bar{x}_{24\bullet} = 23.9; \quad \bar{x}_{25\bullet} = 24.3;$$

$$\bar{x}_{31\bullet} = 25.7; \quad \bar{x}_{32\bullet} = 25.2; \quad \bar{x}_{33\bullet} = 25.7; \quad \bar{x}_{34\bullet} = 23.7, \quad \bar{x}_{35\bullet} = 25.2$$

(iv) OVERALL MEAN

We denote the mean of all the sample observations by $\bar{x}$, so that

$$\bar{x} = \frac{\sum\limits_{i=1}^{K}\sum\limits_{j=1}^{H}\sum\limits_{l=1}^{L} x_{ijl}}{KHL}$$

For our data, this quantity is simplest to calculate as the average of the three group sample means, giving

$$\bar{x} = \frac{24.86 + 24.16 + 25.10}{3} = 24.71$$

Now, to get a feeling for the analysis, it is convenient to think in terms of the assumed population model. Let X_{ijl} denote the random variable corresponding to the lth observation in the (i, j)th cell. Then, the model assumed in our analysis is

$$X_{ijl} = \mu + G_i + B_j + I_{ij} + \varepsilon_{ijl} \qquad (15.5.1)$$

The first three terms on the right-hand side of equation 15.5.1 are precisely the same as those in equation 15.4.1. As before, they represent an overall mean, a group-specific factor, and a block-specific factor. The next term I_{ij} represents the effect of being in the (i, j)th cell, given that the overall, group, and block effects are already accounted for. If there were no group–block interaction, this term would be 0. Its presence in the model allows us to check for interaction. Finally, the error term ε_{ijl} is a random variable representing experimental error.

We will rewrite equation 15.5.1 as

$$X_{ijl} - \mu = G_i + B_j + I_{ij} + \varepsilon_{ijl} \qquad (15.5.2)$$

From equation 15.5.2, it is seen that the total sum of squares can be decomposed as the sum of four terms, representing variability due to groups, blocks, interaction between groups and blocks, and error.

Without providing detailed derivations, we state in the box the decomposition on which the tests are based.

Two-Way Analysis of Variance: Several Observations per Cell

Suppose that we have a sample of observations on K groups and H blocks, with L observations per cell. Let x_{ijl} denote the lth observation in the cell for the ith group and jth block. Let $\bar{x}$ denote the overall sample mean, $\bar{x}_{i..}$ the group sample means, $\bar{x}_{.j.}$ the block sample means, and $\bar{x}_{ij.}$ the cell sample means.

Then, we define the following sums of squares and associated degrees of freedom:

	SUMS OF SQUARES	DEGREES OF FREEDOM
TOTAL:	$SST = \sum_i \sum_j \sum_l (x_{ijl} - \bar{x})^2$	$KHL - 1$
BETWEEN GROUPS:	$SSG = HL \sum_{i=1}^{K} (\bar{x}_{i..} - \bar{x})^2$	$K - 1$
BETWEEN BLOCKS:	$SSB = KL \sum_{j=1}^{H} (\bar{x}_{.j.} - \bar{x})^2$	$H - 1$
INTERACTION:	$SSI = L \sum_{i=1}^{K} \sum_{j=1}^{H} (\bar{x}_{ij.} - \bar{x}_{i..} - \bar{x}_{.j.} + \bar{x})^2$	$(K - 1)(H - 1)$
ERROR:	$SSE = \sum_i \sum_j \sum_l (x_{ijl} - \bar{x}_{ij.})^2$	$KH(L - 1)$

Then

$$SST = SSG + SSB + SSI + SSE$$

Division of the component sums of squares by their corresponding degrees of freedom yields the mean squares MSG, MSB, MSI, and MSE.

Tests of the hypotheses of no effects for groups, blocks, and interaction are based on the respective F ratios

$$\frac{MSG}{MSE}; \quad \frac{MSB}{MSE}; \quad \frac{MSI}{MSE}$$

The tests are carried out with reference to the F distributions with the corresponding numerator and denominator degrees of freedom. Their validity rests on the assumption that the ε_{ijl} in model 15.5.1 behave as a random sample from a normal distribution.

Figure 15.5 depicts the decomposition of the total sum of squares of the sample observations about their overall mean as the sum of four components. It differs from Figure 15.4 in that, as the experiment is replicated, we are now able to isolate an interaction sum of squares.

As before, the calculations involved can be conveniently summarized in an analysis of variance table. The general form of the table when there are L observations per cell in a two-way analysis of variance is shown in Table 15.12.

In fact, formulas that are computationally simpler exist for the calculation of the various sums of squares. Nevertheless, the arithmetic involved is still rather tedious. We will not go into further detail here, but will simply quote in Table 15.13 the results of the calculations for our data. In practice, analysis of variance computations are

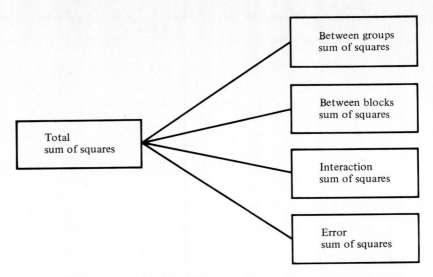

FIGURE 15.5 Sum of squares decomposition for a two-way analysis of variance with more than one observation per cell

typically carried out through a prewritten program package on an electronic computer. Thus, considerations of arithmetic complexity rarely impose any constraint on practical analyses.

The degrees of freedom in Table 15.13 follow from the fact that, for these data, we have

$$K = 3; \qquad H = 5; \qquad L = 3$$

The mean squares are obtained by dividing the sums of squares by their associated degrees of freedom. Finally, the F ratios follow from dividing, in turn, each of the first

TABLE 15.12 General format of the two-way analysis of variance table with L observations per cell

SOURCE OF VARIATION	SUMS OF SQUARES	DEGREES OF FREEDOM	MEAN SQUARES	F RATIOS
Between groups	SSG	$K - 1$	$\text{MSG} = \dfrac{\text{SSG}}{K - 1}$	$\dfrac{\text{MSG}}{\text{MSE}}$
Between blocks	SSB	$H - 1$	$\text{MSB} = \dfrac{\text{SSB}}{H - 1}$	$\dfrac{\text{MSB}}{\text{MSE}}$
Interaction	SSI	$(K - 1)(H - 1)$	$\text{MSI} = \dfrac{\text{SSI}}{(K - 1)(H - 1)}$	$\dfrac{\text{MSI}}{\text{MSE}}$
Error	SSE	$KH(L - 1)$	$\text{MSE} = \dfrac{\text{SSE}}{KH(L - 1)}$	
Total	SST	$KHL - 1$		

TABLE 15.13 Two-way analysis of variance table for automobile miles per gallon data of Table 15.10

SOURCE OF VARIATION	SUMS OF SQUARES	DEGREES OF FREEDOM	MEAN SQUARES	F RATIOS
Automobiles	7.1565	2	3.5783	92.46
Drivers	13.1517	4	3.2879	84.96
Interaction	6.6045	8	0.8256	21.33
Error	1.1600	30	0.0387	
Total	28.0727	44		

three mean squares by the error mean square.

Using the material in Table 15.13, we can test the three null hypotheses of interest. To begin, we test the null hypothesis that the population mean miles per gallon are the same for X-cars, Y-cars, and Z-cars. The test is based on the calculated F ratio 92.46. From Table 7, we find, for a 1% level test with numerator and denominator degrees of freedom 2 and 30, respectively,

$$F_{2, 30, .01} = 5.39$$

Hence, the null hypothesis of equality of the population means for automobile types is overwhelmingly rejected at the 1% significance level.

Next, we test the null hypothesis that the population mean miles per gallon is the same for all five driver age classes. From Table 15.13, the test is based on the calculated F ratio 84.96. Here, the numerator and denominator degrees of freedom are 4 and 30, so that, for a 1% level test,

$$F_{4, 30, .01} = 4.02$$

The null hypothesis of equality of population means for these driver age classes is, therefore, very clearly rejected at the 1% significance level.

Finally, we test the null hypothesis of no interaction between drivers and automobile type. This test is based on the calculated F ratio 21.33. Since the numerator and denominator degrees of freedom are 8 and 30, respectively, we have, from Table 7 of the Appendix,

$$F_{8, 30, .01} = 3.17$$

The null hypothesis of no interaction between car and driver type is, then, very clearly rejected at the 1% level of significance.

The evidence of our data points very firmly to the following three conclusions:

(i) The average performance levels, in terms of miles per gallon, are not the same for X-cars, Y-cars, and Z-cars.

(ii) The average performance levels are not the same for all driver classes.

(iii) The differences in driver performance are not spread evenly over all three types of automobile. Rather, compared with other drivers, a driver from a particular age class is likely to do relatively better in one automobile type than in another.

TABLE 15.14 Part of SPSS analysis of variance program output for automobile miles per gallon data

SOURCE OF VARIATION	SUM OF SQUARES	DF	MEAN SQUARE	F
MAIN EFFECTS	20.3082	6	3.3847	87.46
AUTOMOBILES	7.1565	2	3.5783	92.46
DRIVERS	13.1517	4	3.2879	84.96
2-WAY INTERACTIONS	6.6045	8	0.8256	21.33
AUTOMOBILES DRIVERS	6.6045	8	0.8256	21.33
EXPLAINED	26.9127	14	1.9223	49.67
RESIDUAL	1.1600	30	0.0387	
TOTAL	28.0727	44	0.6380	

Computer programs for carrying out analysis of variance calculations are now widely available, so that the computational burden involved in analyzing the models of this chapter, and more complex models, need not be severe. Table 15.14 shows part of the output from one prewritten package, the Statistical Package for the Social Sciences (SPSS) program, for our miles per gallon data. The output contains all of the information in our Table 15.13, with the error source labelled RESIDUAL. In addition, the total of the two main sums of squares is shown as MAIN EFFECTS, together with the associated degrees of freedom, mean square, and F ratio. Similarly, the sum of the MAIN EFFECTS and 2-WAY INTERACTIONS sums of squares is designated EXPLAINED sum of squares. This, too, is displayed, together with its degrees of freedom and the associated mean square and F ratio.

Example 15.3

In the development of this section, we have assumed that the number of observations is the same in each cell. However, this restriction is not necessary, and on occasion may be inconvenient for an investigator. In fact, the formulas for the computation of sums of squares can be modified to allow for unequal cell numbers. We will not be concerned here with the technical details of the calculation of appropriate sums of squares. Generally, an investigator will have available a computer package for this purpose. Rather, our interest lies in the analysis of the results.

A study[6] was designed to compare the satisfaction levels of introverted and extroverted workers performing stimulating and nonstimulating tasks. Thus, for the purpose of this study, there are two worker types and two task types, giving four combinations. The sample mean satisfaction levels reported by workers in these four combinations were as follows:

Introverted worker,   Nonstimulating task   (16 observations):   2.78

Extroverted worker,   Nonstimulating task   (15 observations):   1.85

---

[6] See J. S. Kim, "Relationships of personality to perceptual and behavioral responses in stimulating and nonstimulating tasks," *Academy of Management Journal, 23* (1980), 307–19.

| Introverted worker, | Stimulating task | (17 observations): | 3.87 |
| Extroverted worker, | Stimulating task | (19 observations): | 4.12 |

The table shows the calculated sums of squares and associated degrees of freedom. Complete the analysis of variance table and analyze the results of this experiment.

SOURCE OF VARIATION	SUMS OF SQUARES	DEGREES OF FREEDOM
Task	62.04	1
Worker type	0.06	1
Interaction	1.85	1
Error	23.31	63
Total	87.26	66

Once again, the mean squares are obtained from division of the sums of squares by their associated degrees of freedom. The F ratios then follow from division of the task, worker type, and interaction mean squares by the error mean square. The analysis of variance table may now be completed as shown.

SOURCE OF VARIATION	SUMS OF SQUARES	DEGREES OF FREEDOM	MEAN SQUARES	F RATIOS
Task	62.04	1	62.04	167.68
Worker type	0.06	1	0.06	0.16
Interaction	1.85	1	1.85	5.00
Error	23.31	63	0.37	
Total	87.26	66		

The analysis of variance table can be used as the basis for testing three null hypotheses. For the null hypothesis of equal mean population satisfaction levels with the two types of task, the calculated F ratio is 167.68. We have numerator degrees of freedom 1 and denominator degrees of freedom 63, so that, from Table 7 of the Appendix, for a 1% level test

$$F_{1,63,.01} = 7.07$$

Hence, the null hypothesis of equal population mean satisfaction levels for stimulating and nonstimulating tasks is very clearly rejected. Certainly, this result is not surprising! We would naturally expect workers to be more satisfied when performing stimulating rather than nonstimulating tasks.

Next, we test the null hypothesis that the population mean satisfaction levels are the same for introverted and extroverted workers. Here the calculated F ratio is 0.16. Again, the degrees of freedom are 1 and 63, so that for a 5% level test,

$$F_{1,63,.05} = 4.00$$

Therefore, the null hypothesis of equal mean levels of satisfaction for introverted and extroverted workers cannot be rejected at the 5% level of significance.

In many studies, the interaction term is not, in itself, of any great importance. The main reason for including it in the analysis is to "soak up" some of the variability in the data, rendering any differences between population means easier to detect. However, in this particular study, the interaction is of major interest. The null hypothesis of no interaction between task and worker type in determining worker satisfaction levels is tested through the calculated F ratio 5.00. Once again, the numerator and denominator degrees of freedom are 1 and 63, respectively. Hence, comparison with the tabulated values of the F distribution reveals that the null hypothesis of no interaction can be rejected at the 5%, but not at the 1% level of significance.

EXERCISES

32. A recent article[7] analyzed the scores given by judges to competitors in the figure skating events of the 1980 Winter Olympics. For the ladies' figure skating competition, there were 22 contestants and 9 judges. Each contestant was assessed by each judge in 7 subevents. The scores given can then be treated in the framework of a two-way analysis of variance with 198 contestant–judge cells and 7 observations per cell. The sums of squares are given in the table.

SOURCE OF VARIATION	SUMS OF SQUARES
Between contestants	364.50
Between judges	0.81
Interaction	4.94
Error	1,069.94

   (a) Complete the analysis of variance table.
   (b) Carry out the associated F tests, and interpret your findings.

33. Refer to Exercise 32. Twelve pairs were entered in the ice dancing competition. Once again, there were 9 judges, and contestants were assessed in 7 subevents. The sums of squares between groups (pairs of contestants) and between blocks (judges) were found to be

$$SSG = 60.10 \quad \text{and} \quad SSB = 1.65$$

while the interaction and error sums of squares were

$$SSI = 3.35 \quad \text{and} \quad SSE = 31.61$$

Analyze these results and verbally interpret the conclusions.

---

[7] G. Fenwick and S. Chatterjee, "Perception, preference, and patriotism: An exploratory analysis of the 1980 Winter Olympics," *The American Statistician*, 35 (1981), 170–73.

34. A psychologist is working with three types of aptitude tests that may be given to prospective management trainees. In deciding how to structure the testing process, an important issue is the possibility of interaction between test takers and test type. If there were no interaction, then at most one type of test would be needed. Three tests of each type are given to members of each of four groups of subject type. These were distinguished by ratings of poor, fair, good, and excellent in preliminary interviews. The scores obtained are listed in the table.

SUBJECT TYPE	TEST TYPE								
	PROFILE FIT			MINDBENDER			PSYCHE OUT		
Poor	65	68	62	69	71	67	75	75	78
Fair	74	79	76	72	69	69	70	69	65
Good	64	72	65	68	73	75	78	82	80
Excellent	83	82	84	78	78	75	76	77	75

(a) Set up the analysis of variance table.

(b) Test the null hypothesis of no interaction between subject type and test type.

35. A company employs four machine operators, each of whom works with a single machine. In an experiment aimed at assessing differences between machine and worker performances, each operator was assigned to each machine for two 1-hour periods. The numbers of finished parts produced during these trials were recorded, and are listed in the accompanying table.

OPERATOR	MACHINE							
	1		2		3		4	
A	14	12	15	15	18	14	20	22
B	16	16	17	15	17	17	19	21
C	15	11	14	12	16	14	17	17
D	10	10	12	14	13	13	14	16

(a) Set up the analysis of variance table.

(b) Test the null hypothesis that the mean number of finished parts is the same for all machines.

(c) Test the null hypothesis that the mean number of finished parts is the same for all workers.

(d) Test the null hypothesis of no interaction between workers and machines.

36. In some experiments with several observations per cell, the analyst is prepared to assume that there is no interaction between groups and blocks. Any apparent interaction found is then attributed to random error. When such an assumption is made, the analysis is carried out in the usual way, except that what were previously the interaction and error sums of squares are added together to form a new error sum of squares. Similarly, the corresponding degrees of freedom are also added. If the assumption of no interaction is correct, this approach has the advantage of providing more error degrees of freedom, and hence more powerful tests of the equality of group and block means. For the study of Exercise 35, suppose we now make the assumption of no interaction between operators and machines.

(a) State, in words, what is implied by this assumption.

(b) Given this assumption, set up the new analysis of variance table.

(c) Test the null hypothesis that the mean number of finished parts is the same for all four machines.

(d) Test the null hypothesis that the mean number of finished parts is the same for all four operators.

37. Refer to Exercise 22. Having carried out the experiment to compare mean yields per acre of four varieties of corn and three brands of fertilizer, an agricultural researcher suggested that there might be some interaction between variety and fertilizer. In order to check this possibility, another set of trials was carried out, producing the accompanying yields.

FERTILIZER	VARIETY			
	A	B	C	D
1	82	88	73	88
2	94	91	79	93
3	81	78	83	81

(a) What would be implied by an interaction between variety and fertilizer?

(b) Combine the data from the two sets of trials and set up an analysis of variance table.

(c) Test the null hypothesis that the mean yield is the same for all four varieties of corn.

(d) Test the null hypothesis that the mean yield is the same for all three brands of fertilizer.

(e) Test the null hypothesis of no interaction between variety of corn and brand of fertilizer.

38. Explain the uses of a two-way analysis of variance. Explain, as part of your answer, the notion of interaction, and provide an illustration of a situation where this factor might be expected to be important.

39. Having carried out the study of Exercise 25, the instructor decides to conduct another set of checks in the following year. The results obtained are shown in the table. Combining these results with those of Exercise 25, carry out an analysis of variance and discuss your findings.

	TEXT		
	A	B	C
QUIZ	4.6	5.3	4.8
NO QUIZ	4.8	5.2	4.6

40. Carefully distinguish between the one-way analysis of variance framework and the two-way analysis of variance framework. Provide illustrations, other than those discussed in the text and previous exercises, of business problems for which each might be appropriate.

41. Carefully explain what is meant by the interaction effect in the two-way analysis of variance with more than one observation per cell. Give illustrations of this effect in business-related problems.

42. The *fog index* is used to measure the reading difficulty of a written text; the higher the value of the index, the more difficult is the reading level. Independent random samples[8] of six advertisements were taken from *Scientific American, Fortune,* and *New Yorker,* and the fog indices for these 18 advertisements were measured and listed in the accompanying table. Test the null hypothesis that the population mean fog indices are the same for advertisements in these three magazines.

SCIENTIFIC AMERICAN	FORTUNE	NEW YORKER
15.75	12.63	9.27
11.55	11.46	8.28
11.16	10.77	8.15
9.92	9.93	6.37
9.23	9.87	6.37
8.20	9.42	5.66

43. In the study of Exercise 42, independent random samples of six advertisements from *True Confessions, People Weekly,* and *Newsweek* were taken. The fog indices for these advertisements are given in the table. Test the null hypothesis that the population mean fog indices are the same for advertisements in these three magazines.

TRUE CONFESSIONS	PEOPLE WEEKLY	NEWSWEEK
12.89	9.50	10.21
12.69	8.60	9.66
11.15	8.59	7.67
9.52	6.50	5.12
9.12	4.79	4.88
7.04	4.29	3.12

44. Independent random samples of the selling prices of houses in three districts were taken. The selling prices (in thousands of dollars) are shown in the accompanying table. Test the null hypothesis that the population mean selling prices are the same in all three districts.

---

[8] Data taken from F. K. Shuptrine and D. D. McVicker, "Readability levels of magazine advertisements," *Journal of Advertising Research, 21* (1981), No. 5, 45–50. The details of the construction of the fog index are set out in Exercise 73 of Chapter 9.

DISTRICT A	DISTRICT B	DISTRICT C
72.9	84.3	76.3
63.4	69.7	61.3
89.3	80.2	66.7
74.8	97.3	83.5
69.7	86.4	67.2
85.3	75.9	69.3
58.7	78.2	

45. For the one-way analysis of variance model, we write the jth observation from the ith group as

$$X_{ij} = \mu + G_i + \varepsilon_{ij}$$

where μ is the overall mean, G_i is the effect specific to the ith group, and ε_{ij} is a random error for the jth observation from the ith group. Consider the data of Exercise 42.

(a) Estimate μ.

(b) Estimate G_i for each of the three magazines.

(c) Estimate ε_{11}, the error term corresponding to the first observation (15.75) for *Scientific American*.

46. Use the model of the one-way analysis of variance for the data of Exercise 43.

(a) Estimate μ.

(b) Estimate G_i for each of the three magazines.

(c) Estimate ε_{21}, the error term corresponding to the first observation (9.50) for *People Weekly*.

47. For the data of Exercise 42, use the Kruskal-Wallis test to test the null hypothesis that the population mean fog indices are the same for advertisements in these three magazines.

48. For the data of Exercise 43, use the Kruskal-Wallis test to test the null hypothesis of equality of the three population means.

49. For the data of Exercise 44, use the Kruskal-Wallis test to test the null hypothesis that the population mean selling prices of houses are the same in the three districts.

50. Consider the one-way analysis of variance setup.

(a) Show that the within groups sum of squares can be written

$$\text{SSW} = \sum_{i=1}^{K} \sum_{j=1}^{n_i} x_{ij}^2 - \sum_{i=1}^{K} n_i \bar{x}_i^2$$

(b) Show that the between groups sum of squares can be written

$$\text{SSG} = \sum_{i=1}^{K} n_i \bar{x}_i^2 - n\bar{x}^2$$

(c) Show that the total sum of squares can be written

$$\text{SST} = \sum_{i=1}^{K} \sum_{j=1}^{n_i} x_{ij}^2 - n\bar{x}^2$$

51. Consider the two-way analysis of variance setup, with one observation per cell.

(a) Show that the between groups sum of squares can be written

$$SSG = H \sum_{i=1}^{K} \bar{x}_{i\bullet}^2 - n\bar{x}^2$$

(b) Show that the between blocks sum of squares can be written

$$SSB = K \sum_{j=1}^{H} \bar{x}_{\bullet j}^2 - n\bar{x}^2$$

(c) Show that the total sum of squares can be written

$$SST = \sum_{i=1}^{K} \sum_{j=1}^{H} x_{ij}^2 - n\bar{x}^2$$

(d) Show that the error sum of squares can be written

$$SSE = \sum_{i=1}^{K} \sum_{j=1}^{H} x_{ij}^2 - H \sum_{i=1}^{K} \bar{x}_{i\bullet}^2 - K \sum_{J=1}^{H} \bar{x}_{\bullet j}^2 + n\bar{x}^2$$

52. Information on consumer perceptions of three brand types—national, private, and generic—were obtained from a random sample[9] of 125 consumers. The sums of squares for these perception measures are given in the accompanying table. Complete the analysis of variance table and test the null hypothesis that the population mean perception levels are the same for all three brand types.

SOURCE OF VARIATION	SUMS OF SQUARES
Between consumers	37,571.5
Between brands	32,987.3
Error	55,710.7

53. A random sample of five supermarkets was taken, and the percentage increases in sales of three brands of cereals over the past year were measured for each. The results are shown in the table. Set out the analysis of variance table and test the null hypothesis that the population mean percentage sales increases are the same for all three brands.

SUPERMARKET	BRAND		
	A	B	C
1	10.2	13.4	8.3
2	3.7	5.7	4.0
3	6.4	6.1	5.1
4	8.3	8.2	6.2
5	5.2	7.9	6.0

---

[9] Data taken from J. A. Bellizzi, H. F. Krueckeberg, J. R. Hamilton, and W. S. Martin, "Consumer perceptions of national, private, and generic brands," *Journal of Retailing, 57* (1981), No. 4, 56–70.

54. Students were classified according to three parental income groups and also according to three possible score ranges in the SAT examination. One student was chosen randomly from each of the nine cross-classifications, and the grade point average of each sample member at the end of the sophomore year was recorded. The results are shown in the accompanying table.

SAT SCORE	INCOME GROUP		
	HIGH	MODERATE	LOW
VERY HIGH	3.7	3.6	3.6
HIGH	3.4	3.5	3.2
MODERATE	2.9	2.8	3.0

(a) Set out the analysis of variance table.

(b) Test the null hypothesis that the population mean grade point averages are the same for all three income groups.

(c) Test the null hypothesis that the population mean grade point averages are the same for all three SAT score groups.

55. For the two-way analysis of variance model with one observation per cell, we write the observation from the ith group and jth block as

$$X_{ij} = \mu + G_i + B_j + \varepsilon_{ij}$$

Refer to Exercise 53 and consider the observation on brand B and supermarket 1 ($x_{21} = 13.4$).

(a) Estimate μ.

(b) Estimate G_2.

(c) Estimate B_1

(d) Estimate ε_{21}.

56. Refer to Exercise 54 and consider the observation on low income group and very high SAT score ($x_{31} = 3.6$).

(a) Estimate μ.

(b) Estimate G_3.

(c) Estimate B_1.

(d) Estimate ε_{31}.

57. Consider the two-way analysis of variance setup, with L observations per cell.

(a) Show that the between groups sum of squares can be written

$$SSG = HL \sum_{i=1}^{K} \bar{x}_{i..}^2 - HKL\bar{x}^2$$

(b) Show that the between blocks sum of squares can be written

$$SSB = KL \sum_{j=1}^{H} \bar{x}_{.j.}^2 - HKL\bar{x}^2$$

(c) Show that the error sum of squares can be written

$$SSE = \sum_{i=1}^{K} \sum_{j=1}^{H} \sum_{l=1}^{L} x_{ijl}^2 - L \sum_{i=1}^{K} \sum_{j=1}^{H} \bar{x}_{ij}^2.$$

(d) Show that the total sum of squares can be written

$$SST = \sum_{i=1}^{K} \sum_{j=1}^{H} \sum_{l=1}^{L} x_{ijl}^2 - HKL\bar{x}^2$$

(e) Show that the interaction sum of squares can be written

$$SSI = L \sum_{i=1}^{K} \sum_{j=1}^{H} \bar{x}_{ij}^2 - HL \sum_{i=1}^{K} \bar{x}_{i..}^2 - KL \sum_{j=1}^{H} \bar{x}_{.j.}^2 + HKL\bar{x}^2$$

58. Purchasing agents were given information about a dictation system and asked to assess its quality.[10] The information given was identical except for two factors—price and country of origin. For price, there were three possibilities: $605, $495, or no price given. For country of origin, there were also three possibilities: United States, Brazil, or no country given. Part of the analysis of variance table for the quality assessments of the purchasing agents is shown here. Complete the analysis of variance table and provide a full analysis of these data.

SOURCE OF VARIATION	SUMS OF SQUARES	DEGREES OF FREEDOM
Between prices	0.178	2
Between countries	4.365	2
Interaction	1.262	4
Error	93.330	99

59. In the study of Exercise 58, information on the dictation system was also shown to M.B.A. students. Part of the analysis of variance table for their quality assessments is shown here. Complete the analysis of variance table and provide a full analysis of these data.

SOURCE OF VARIATION	SUMS OF SQUARES	DEGREES OF FREEDOM
Between prices	0.042	2
Between countries	15.319	2
Interaction	2.235	4
Error	70.414	50

60. Having carried out the study of Exercise 54, the investigator decided to take an independent random sample of one student from each of the nine income–SAT score categories. The grade point averages found are given in the accompanying table.

---

[10] This study is reported by D. R. Lambert, "Price as a quality cue in industrial buying," *Journal of Academy of Marketing Science, 9* (1981), 227–38; © 1981, reprinted by permission of the Academy of Marketing Science.

SAT SCORE	INCOME GROUP		
	HIGH	MODERATE	LOW
VERY HIGH	3.9	3.7	3.8
HIGH	3.2	3.6	3.4
MODERATE	2.7	3.0	2.8

(a) Set out the analysis of variance table.

(b) Test the null hypothesis that the population mean grade point averages are the same for all three income groups.

(c) Test the null hypothesis that the population mean grade point averages are the same for all three SAT score groups.

(d) Test the null hypothesis of no interaction between income group and SAT score.

61. [*This exercise requires a computer program to carry out the analysis of variance computations.*] An experiment was carried out to test the effects on yields of five varieties of corn and five types of fertilizer. For each variety–fertilizer combination, six plots were used and the yields recorded, with the results shown in the table.

FERTILIZER	VARIETY									
	A		B		C		D		E	
1	82	77	74	67	93	90	79	83	72	77
	79	83	73	65	87	82	87	88	79	83
	85	78	79	80	86	88	86	90	78	86
2	80	72	71	69	84	88	77	82	70	75
	76	73	75	62	90	79	84	87	80	80
	70	74	77	83	83	80	82	83	74	81
3	85	87	76	73	88	94	81	86	77	83
	80	79	77	70	89	86	90	90	87	79
	87	80	83	80	89	93	87	88	86	88
4	80	79	74	77	86	87	80	77	79	85
	82	77	69	78	90	85	90	84	88	80
	85	80	74	76	83	88	80	88	87	82
5	75	79	75	80	92	88	82	78	80	87
	86	82	84	80	89	94	85	86	90	83
	79	83	72	77	86	90	82	89	86	92

(a) Test the null hypothesis that the population mean yields are the same for all five varieties of corn.

(b) Test the null hypothesis that the population mean yields are the same for all five brands of fertilizer.

(c) Test the null hypothesis of no interaction between variety and fertilizer.

index numbers

CHAPTER SIXTEEN

16.1   THE INDEX NUMBER PROBLEM

What changes have occurred in the price of automobiles built in the United States in the last 10 years? It almost goes without saying that their prices have risen, but how can this price rise be described *quantitatively*? On the surface, this question may not seem very difficult to answer. One could collect price information about these automobiles in each of the last 10 years, and tabulate the data or graph them in a time plot.

However, when one thinks a little more deeply about the problem, a number of difficulties emerge. The first, and perhaps must crucial, is that automobiles are not homogeneous. It is meaningless to ask, "What is the price of an automobile?" It is necessary to be more specific. The price of a large luxury car is very much higher than that of a small subcompact. Perhaps, then, one could proceed by comparing the *average* prices of all automobiles built in the United States in each of the last 10 years. Unfortunately, such a comparison might prove misleading. Imagine a year in which relatively many luxury cars and very few subcompacts were sold, compared with a second year in which there was a preponderance of small compact sales with very few luxury cars sold. In the second year, the average price of cars sold would be low compared with that of the previous year, simply because of a change in the market mix. Table 16.1 gives a hypothetical example of a market in which there are just two types of automobile. We see from the table that the price of each type is *higher* in year 2 than in year 1. However, because of the difference in the product mixes sold in the two years, we see from the final column of the table that the average price of all cars sold in year 2 is *lower* than the average price of all cars sold in year 1. Obviously, we would not want to conclude from these data that automobile prices were lower in year 2 than in year 1! Hence, this kind of average is of no value for our purposes.

TABLE 16.1 Hypothetical data on automobile prices and sales

	SUBCOMPACT CARS		LUXURY CARS		ALL CARS
	PRICE (thousand dollars)	NUMBER SOLD (thousands)	PRICE (thousand dollars)	NUMBER SOLD (thousands)	AVERAGE PRICE (thousand dollars)
YEAR 1	5	5	12	15	10.25
YEAR 2	6	15	14	5	8

Another possible solution is to compute the average price of a *single* car of each type. That this procedure is not free from difficulty of interpretation is clear from a second hypothetical example, shown in Table 16.2. Here, we have a market in which subcompact cars are considerably more popular than luxury cars. The price of subcompacts is the same in the two years, while that of luxury cars doubles from year 1 to year 2. Hence, as we see in the final column of the table, the average price of a single subcompact car and a single luxury car is considerably higher in the second year than in the first. But this procedure does not present an accurate picture of the trend in the automobile market, since it gives equal weights to the prices of the two types of car, even though relatively few luxury cars were sold.

These examples demonstrate that, in order to form a reliable picture of the overall price pattern through time, it is necessary to take carefully into account the quantities purchased in each time period. In Section 16.4, we will see how this can be done.

A similar problem arises because cars are sold with optional "extras," which naturally affect the price. Thus, the average price of subcompacts sold in a year in which consumers opt for many of these extras will be higher than in a year in which they do not, even if the base price is the same in each year. One way out of this difficulty would be to look only at the price of the "stripped down" version in each year, in order to obtain a valid comparison.

Another difficulty arises because of technological improvements. Perhaps, because of technological advance, a subcompact car of the current model year is a superior product to that sold 10 years ago. It can be argued that some of the increase

TABLE 16.2 Hypothetical data on automobile prices and sales

	SUBCOMPACT CARS		LUXURY CARS		AVERAGE PRICE OF A SINGLE CAR OF EACH TYPE (thousand dollars)
	PRICE (thousand dollars)	NUMBER SOLD (thousands)	PRICE (thousand dollars)	NUMBER SOLD (thousands)	
YEAR 1	6	100	12	1	9
YEAR 2	6	100	24	1	15

in price over these 10 years can be ascribed to the purchase of additional quality. If this factor is ignored, then a simple comparison of prices over this period would overstate the extent of the inflation. We will not discuss further the problem of accounting for quality changes when comparing prices over time. It is, nevertheless, a factor that should not be ignored in assessing such comparisons.

As a final point, we note that the trend in dollar prices of automobiles may not be the most interesting or useful thing to examine. After all, the dollar price of virtually everything has risen steeply in the last 10 years. Perhaps it would be more relevant to compare the price increases of automobiles with those of other products.

The *index number problem* illustrated in this section arises through our desire to say something about the movement of prices for a *group* of commodities. For example, the price of common stock in each company whose shares are traded on the New York Stock Exchange will change over a 1-month period. We would like to produce a measure of the *aggregate* change in prices. **Index numbers** are designed to attack such problems.

16.2 PRICE INDEX FOR A SINGLE ITEM

We begin our discussion of index numbers with the simple case in which we trace the price movements of a single item. The second column of Table 16.3 gives the price of Ford Motor Company stock for each of the first 12 complete weeks of 1981. A time plot of these data, presented in Figure 16.1, allows a visual picture to be formed of the price movement over this period. As they stand, however, the actual price figures themselves are not easy to interpret without a little mental arithmetic. This task can be made more simple by expressing each price as a percentage of a single price.

This is done in the third column of Table 16.3. We have chosen the first week as a **base,** and expressed each price as a percentage of the price (20$\frac{2}{8}$) in the base

TABLE 16.3 Prices and price index numbers for Ford Motor Company stock in the first 12 weeks of 1981

WEEK	PRICE	PRICE INDEX
1	20$\frac{2}{8}$	100.0
2	19$\frac{7}{8}$	98.1
3	19	93.8
4	19$\frac{6}{8}$	97.5
5	20$\frac{2}{8}$	100.0
6	19$\frac{7}{8}$	98.1
7	19$\frac{3}{8}$	95.7
8	19$\frac{5}{8}$	96.9
9	21$\frac{1}{8}$	104.3
10	22$\frac{3}{8}$	110.5
11	25	123.5
12	23	113.6

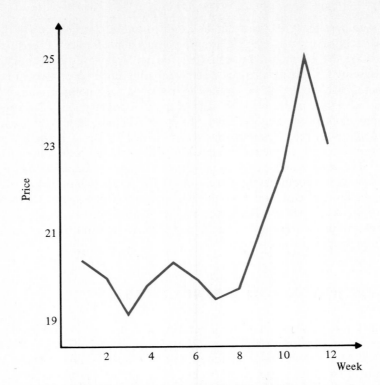

FIGURE 16.1 Time plot of prices of Ford Motor Company stock in the first 12 weeks of 1981

week. For example, the price in the second week is 19⅞, which is a percentage

$$100\left(\frac{19⅞}{20⅝}\right) = 98.1$$

of the base week price. The percentages calculated in this fashion are called **index numbers** of price. The choice of base period is arbitrary. We could have chosen any other week as our base and expressed all prices as a percentage of the price for that week.

Figure 16.2 shows a time plot of the index numbers taken from Table 16.3. It is identical to Figure 16.1, except that, in the second figure, the vertical axis is graduated in terms of percentages of the base period price, rather than the actual price. The advantage of using index numbers lies in the greater ease of interpretation of the figures themselves. We see immediately from Table 16.3, for instance, that the price rose by 13.6% in the first 12 weeks of 1981.

Calculating Price Indices for a Single Item

Suppose that we have a series of observations through time on the price of a single item. To form a **price index,** one time period is chosen as a **base** and the price in every

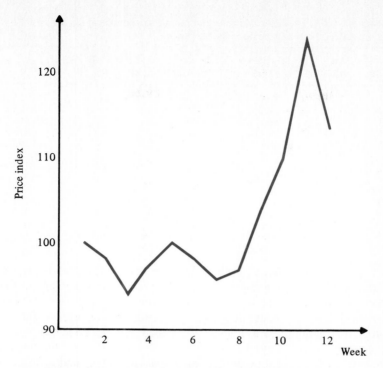

FIGURE 16.2 Time plot of price index of Ford Motor Company stock in the first 12 weeks of 1981 (week 1 = 100)

period is expressed as a percentage of the base period price. Thus, if p_0 denotes the price in the base period and p_1 the price in a second period, then the price index for this second period is

$$100\left(\frac{p_1}{p_0}\right)$$

16.3 UNWEIGHTED AGGREGATE PRICE INDICES

We now move on to consider our major problem—how to represent, in the aggregate, price movements for a group of items. Table 16.4 shows the prices of common stock of the four major United States automobile manufacturers in the first 12 weeks of 1981. The table also shows one way of achieving an aggregate price index for automobile stock. The final two columns in the table give the average price of a share in these four companies for each week, and an index of these averages, taking the first week as base.

TABLE 16.4 Prices of stock in four automobile corporations for the first 12 weeks of 1981, with the unweighted aggregate index of prices

WEEK	FORD	AMERICAN MOTORS	CHRYSLER	GENERAL MOTORS	AVERAGE	INDEX OF AVERAGE
1	$20\frac{2}{8}$	$4\frac{1}{8}$	$5\frac{2}{8}$	$46\frac{1}{8}$	18.93750	100.0
2	$19\frac{7}{8}$	$4\frac{1}{8}$	6	$45\frac{2}{8}$	18.81250	99.3
3	19	$4\frac{1}{8}$	$5\frac{4}{8}$	$45\frac{2}{8}$	18.46875	97.5
4	$19\frac{6}{8}$	$4\frac{1}{8}$	$5\frac{5}{8}$	46	18.87500	99.7
5	$20\frac{2}{8}$	$3\frac{7}{8}$	6	$48\frac{2}{8}$	19.59375	103.5
6	$19\frac{7}{8}$	$3\frac{7}{8}$	$5\frac{3}{8}$	$48\frac{5}{8}$	19.43750	102.6
7	$19\frac{3}{8}$	4	$5\frac{3}{8}$	$47\frac{6}{8}$	19.12500	101.0
8	$19\frac{5}{8}$	4	$5\frac{3}{8}$	$50\frac{1}{8}$	19.78125	104.5
9	$21\frac{1}{8}$	$4\frac{1}{8}$	$5\frac{6}{8}$	$51\frac{4}{8}$	20.62500	108.9
10	$22\frac{3}{8}$	$4\frac{3}{8}$	$5\frac{3}{8}$	51	20.78125	109.7
11	25	$4\frac{6}{8}$	$7\frac{2}{8}$	54	22.75000	120.1
12	23	$4\frac{3}{8}$	$6\frac{5}{8}$	$52\frac{6}{8}$	21.68750	114.5

The resulting **unweighted aggregate index of prices** is easy to calculate. It simply expresses the average price for each week as a percentage of the average price in the base week. It does, however, suffer from a serious flaw—the index is dependent on units of measurement. Suppose, for example, that prior to this period, General Motors had announced a stock split, which had no effect on subsequent relative movements in its share prices. General Motors stock would then have had less influence on the overall average, calculated as shown, and a different set of index numbers for average price would have resulted.

One way around this particular difficulty would be to form index numbers for the price of stock in each of the four companies, and to use the average of these indices as an aggregate index for industry stock. This is done in Table 16.5. The index numbers for Ford are taken from Table 16.3, while those for the other companies are found in a similar fashion.

TABLE 16.5 Price indices of stock in four automobile corporations for the first 12 weeks of 1981, with the average of price relatives

WEEK	FORD	AMERICAN MOTORS	CHRYSLER	GENERAL MOTORS	AVERAGE OF INDICES
1	100.0	100.0	100.0	100.0	100.0
2	98.1	100.0	114.3	98.1	102.6
3	93.8	100.0	104.8	98.1	99.2
4	97.5	100.0	107.1	99.7	101.1
5	100.0	93.9	114.3	104.6	103.2
6	98.1	93.9	102.4	105.4	100.0
7	95.7	97.0	102.4	103.5	99.7
8	96.9	97.0	102.4	108.7	101.3
9	104.3	100.0	109.5	111.7	106.4
10	110.5	106.1	102.4	110.6	107.4
11	123.5	115.2	138.1	117.1	123.5
12	113.6	106.1	126.2	114.4	115.1

These **averages of price relatives** are again quite straightforward to calculate, but once again, however, there is a serious objection to their use as an overall indicator of price trends. This index gives equal weights to relative price changes in all the individual items. For instance, in the calculation of Table 16.5, relative price movements in the stock of American Motors or Chrysler are given the same weight as relative changes in the price of General Motors stock, even though vastly more money is spent on the latter. For most applications, this does not seem reasonable, and in the next section we will discuss the construction of price indices designed to circumvent this problem.

Two Unweighted Price Indices

Suppose that we have a series of observations through time on the prices of a group of K items. As before, one time period is chosen as a base.

The **unweighted aggregate index of prices** is obtained by calculating the average price of these items in each time period, and calculating an index for these average prices. That is, the average price in every period is expressed as a percentage of the average price in the base period. Let p_{0i} denote the price of the ith item in the base period and p_{1i} the price of this item in a second period. The unweighted aggregate index of prices for this second period is

$$100 \left(\frac{\sum\limits_{i=1}^{K} p_{1i}}{\sum\limits_{i=1}^{K} p_{0i}} \right)$$

The **average of price relatives** is obtained by calculating individual price indices for each of the K items, and forming the average of these individual indices. Thus, in the notation of the previous paragraph, the average of price relatives for the second period is

$$100 \frac{\sum\limits_{i=1}^{K} \left(\frac{p_{1i}}{p_{0i}} \right)}{K}$$

16.4 WEIGHTED AGGREGATE PRICE INDICES

We now consider the possibility of using information on the quantities purchased in calculating aggregate price indices. Table 16.6 gives this quantity information for the stocks of the four United States automobile corporations. In this section, we will discuss three ways of taking into account quantities traded in forming an aggregate price index.

LASPEYRES PRICE INDEX

The **Laspeyres price index** uses the quantities traded in the base period only, and compares with the total cost of purchasing these quantities in the base period, what

TABLE 16.6 Volume of transactions (in hundreds of thousands) in shares of four automobile corporations for the first 12 weeks of 1981

WEEK	FORD	AMERICAN MOTORS	CHRYSLER	GENERAL MOTORS
1	8.2	4.3	14.4	27.1
2	6.3	1.5	16.0	12.9
3	6.7	1.3	6.9	12.1
4	4.5	1.9	4.4	13.6
5	4.3	2.7	5.0	21.9
6	5.4	1.5	3.8	17.3
7	3.8	1.7	3.1	11.7
8	4.3	1.5	3.8	23.8
9	5.4	1.8	4.9	17.0
10	9.5	3.5	4.1	21.4
11	13.7	4.4	18.1	25.0
12	8.3	2.6	11.3	20.5

would have been the total cost of purchasing *these same quantities* in each other period. An index of such total costs is then constructed.

Using the data of Tables 16.4 and 16.6, we see that the total cost of purchasing these quantities (in hundred thousand dollars) in the first week, which we will use as base period, was

$$(8.2)(20\tfrac{2}{8}) + (4.3)(4\tfrac{1}{8}) + (14.4)(5\tfrac{2}{8}) + (27.1)(46\tfrac{1}{8}) = 1,509$$

Now, in week 2, at the prices then prevailing, the total cost of purchasing the base period quantities would have been

$$(8.2)(19\tfrac{7}{8}) + (4.3)(4\tfrac{1}{8}) + (14.4)(6) + (27.1)(45\tfrac{2}{8}) = 1,493$$

The Laspeyres index for the second week is, therefore,

$$100\left(\frac{1,493}{1,509}\right) = 98.9$$

Table 16.7 shows the complete index, calculated in this way, for these data. A time plot is given in Figure 16.3.

TABLE 16.7 Laspeyres price index for common stocks of United States automobile manufacturers for first 12 weeks of 1981

WEEK	LASPEYRES INDEX	WEEK	LASPEYRES INDEX
1	100.0	7	102.6
2	98.9	8	107.0
3	98.0	9	110.6
4	99.9	10	110.1
5	104.5	11	118.8
6	104.4	12	114.8

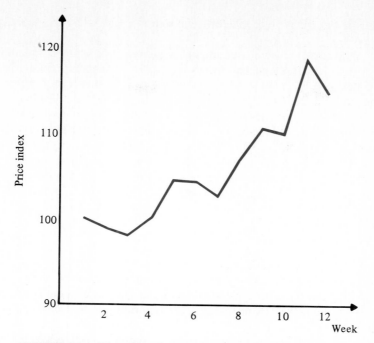

FIGURE 16.3 Time plot of Laspeyres price index for common stocks of automobile manufacturers for the first 12 weeks of 1981 (week 1 = 100)

The Laspeyres Price Index

Suppose we have a group of K commodities for which price information is available over a period of time. One period is selected as the base for an index. The **Laspeyres price index** in any period is the total cost of purchasing the quantities traded in the base period at the prices in the period of interest, expressed as a percentage of the total cost of purchasing these same quantities in the base period.

Let p_{0i} denote the price and q_{0i} the quantity purchased of the ith item in the base period. If p_{1i} is the price of the ith item in a second period, then the Laspeyres price index for this period is

$$100\left(\frac{\sum_{i=1}^{K} q_{0i}p_{1i}}{\sum_{i=1}^{K} q_{0i}p_{0i}}\right)$$

Comparison of the formula for the Laspeyres price index with that for the unweighted aggregate index of prices is instructive. The difference is that, in forming the Laspeyres index, the price of each item is *weighted* by the quantity traded in the base period.

A feature of the construction of Laspeyres price index numbers is that quantity information from only the base period is employed. This is particularly valuable when the collection of quantity information in every time period is either impossible or prohibitively expensive. On the other hand, this could be a disadvantage if, for some reason, the quantities traded in the period chosen as a base happened to be un-representative. This could occur, for example, if the Laspeyres index is computed over a long period of time. The allocation of quantities among the various items may undergo substantial changes over time, so that the original base quantities become outdated. One way around this problem is to construct a **moving Laspeyres price index,** in which the base period is changed from time to time, through the acquisition of quantity information for new base periods.

PAASCHE PRICE INDEX

By contrast with the Laspeyres index, the **Paasche price index** utilizes quantity information from each time period. The Paasche index compares, for a particular period, the total cost of purchasing the quantities *traded in that period* with what would have been the total cost of purchasing these same quantities in the base period.

To illustrate, we return again to the automobile stock price data, using as before week 1 as base. To find the index for week 2, we see from Tables 16.4 and 16.6 that the total cost of purchases in week 2 was

$$(6.3)(19\tfrac{7}{8}) + (1.5)(4\tfrac{1}{8}) + (16.0)(6) + (12.9)(45\tfrac{2}{8}) = 811.125$$

The total cost of these same quantities at the base period prices would have been

$$(6.3)(20\tfrac{2}{8}) + (1.5)(4\tfrac{1}{8}) + (16.0)(5\tfrac{2}{8}) + (12.9)(46\tfrac{1}{8}) = 812.775$$

The Paasche index for week 2 is, therefore,

$$100\left(\frac{811.125}{812.775}\right) = 99.8$$

Continuing in this manner, we see that the total cost of week 3 purchases was

$$(6.7)(19) + (1.3)(4\tfrac{1}{8}) + (6.9)(5\tfrac{1}{8}) + (12.1)(45\tfrac{2}{8}) = 718.1375$$

TABLE 16.8 Paasche price index for common stocks of United States automobile manufacturers for first 12 weeks of 1981

WEEK	PAASCHE INDEX	WEEK	PAASCHE INDEX
1	100.0	7	102.5
2	99.8	8	107.7
3	97.7	9	110.6
4	99.7	10	110.4
5	104.4	11	119.5
6	104.4	12	114.8

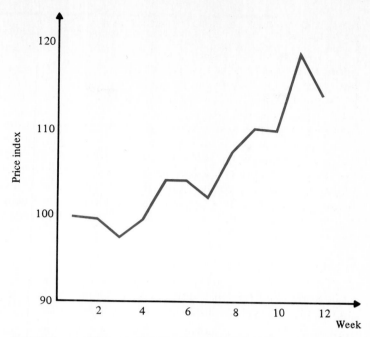

FIGURE 16.4 Time plot of Paasche price index for common stocks of automobile manufacturers for the first 12 weeks of 1981 (week 1 = 100)

The total cost of the week 3 quantities, had they been purchased at base period prices, would have been

$$(6.7)(20\tfrac{2}{8}) + (1.3)(4\tfrac{1}{8}) + (6.9)(5\tfrac{2}{8}) + (12.1)(46\tfrac{1}{8}) = 735.375$$

Thus, the Paasche price index for week 3 is

$$100\left(\frac{718.1375}{735.375}\right) = 97.7$$

In this way, the complete index can be calculated. The Paasche price index for our automobile stock data is set out in Table 16.8 and plotted in Figure 16.4.

The Paasche Price Index

The **Paasche price index** in any period is the total cost of purchases in that period, expressed as a percentage of what the total cost of these same quantities would have been, had they been purchased in the base period.

Let p_{0i} denote the price of the ith item in the base period, with p_{1i} the price and

---

q_{1i} the quantity purchased of that item in a second period. The Paasche price index for the second period is then

$$100\left(\frac{\sum_{i=1}^{K} q_{1i}p_{1i}}{\sum_{i=1}^{K} q_{1i}p_{0i}}\right)$$

The Paasche price index requires knowledge of the quantities traded in every time period. This presents a drawback to its use in circumstances where the acquisition of this information is costly. Moreover, its interpretation is less clear-cut than that of the Laspeyres index. In a Laspeyres index, the total costs being compared from year to year are those of the same quantities of goods. However, this is not true of the Paasche index.

FISHER'S IDEAL PRICE INDEX

For our automobile stock price data, we see from Tables 16.7 and 16.8, and from Figures 16.3 and 16.4, that the Laspeyres and Paasche price indices are very close to one another. However, this need not necessarily be the case. Sometimes, a compromise between the two, known as **Fisher's ideal price index,** is then calculated. The Fisher index is simply the square root of the product of the Laspeyres and Paasche indices. Thus, referring to Tables 16.7 and 16.8, we can compute the Fisher index for week 2 as

$$\sqrt{(98.9)(99.8)} = 99.3$$

Fisher's Ideal Price Index

Fisher's ideal price index provides a compromise between the Laspeyres and Paasche indices, and is defined as

$$\text{Fisher's ideal index} = \sqrt{(\text{Laspeyres index})(\text{Paasche index})}$$

TABLE 16.9 Fisher's ideal price index for price of common stocks of United States automobile manufacturers for first 12 weeks of 1981

WEEK	FISHER'S IDEAL INDEX	WEEK	FISHER'S IDEAL INDEX
1	100.0	7	102.5
2	99.3	8	107.3
3	97.8	9	110.6
4	99.8	10	110.2
5	104.4	11	119.1
6	104.4	12	114.8

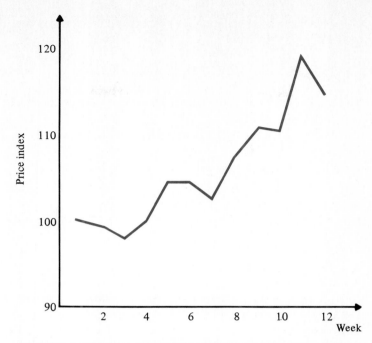

FIGURE 16.5 Time plot of Fisher ideal price index for common stocks of automobile manufacturers for the first 12 weeks of 1981 (week 1 = 100)

The complete index calculated in this manner is shown in Table 16.9 and graphed in Figure 16.5.

It follows, from its definition, that Fisher's ideal price index will be between the Laspeyres price index and the Paasche price index.

Example 16.1

The accompanying table shows United States farm prices (in dollars per bushel) and production (in millions of bushels) of wheat, corn, and soybeans in 10 consecutive crop years. Using crop year 1970–71 as base, construct Laspeyres, Paasche, and Fisher ideal price indices for this group of commodities.

YEAR	WHEAT PRICE	WHEAT PRODUCTION	CORN PRICE	CORN PRODUCTION	SOYBEANS PRICE	SOYBEANS PRODUCTION
1970–71	1.33	1,352	1.33	4,152	2.85	1,127
1971–72	1.34	1,618	1.08	5,641	3.03	1,176
1972–73	1.76	1,545	1.57	5,573	4.37	1,271
1973–74	3.95	1,705	2.55	5,647	5.68	1,547
1974–75	4.09	2,122	3.03	5,829	6.64	1,547
1975–76	3.56	2,142	2.54	6,266	4.92	1,288
1976–77	2.73	2,026	2.15	6,357	6.81	1,716
1977–78	2.33	1,799	2.02	7,082	6.42	1,843
1978–79	2.97	2,134	2.25	7,939	6.12	2,268
1979–80	3.78	2,370	2.52	6,648	6.28	1,817

We begin by calculating the Laspeyres index for crop year 1971–72. For the base year 1970–71, we have

$$\sum q_{0i}p_{0i} = (1{,}352)(1.33) + (4{,}152)(1.33) + (1{,}127)(2.85) = 10{,}532.27$$

The total cost of the base year output at 1971–72 prices would have been

$$\sum q_{0i}p_{1i} = (1{,}352)(1.34) + (4{,}152)(1.08) + (1{,}127)(3.03) = 9{,}710.65$$

Therefore, the Laspeyres index for 1971–72 is

$$100 \frac{\sum\limits_{i=1}^{K} q_{0i}p_{1i}}{\sum\limits_{i=1}^{K} q_{0i}p_{0i}} = 100\left(\frac{9{,}710.65}{10{,}532.27}\right) = 92.2$$

We now find the Paasche index for this same crop year. For 1971–72 we have

$$\sum q_{1i}p_{1i} = (1{,}618)(1.34) + (5{,}641)(1.08) + (1{,}176)(3.03) = 11{,}823.68$$

The total cost of the 1971–72 output in the base year would have been

$$\sum q_{1i}p_{0i} = (1{,}618)(1.33) + (5{,}641)(1.33) + (1{,}176)(2.85) = 13{,}006.07$$

Hence, the Paasche index for 1971–72 is

$$100 \frac{\sum\limits_{i=1}^{K} q_{1i}p_{1i}}{\sum\limits_{i=1}^{K} q_{1i}p_{0i}} = 100\left(\frac{11{,}823.68}{13{,}006.07}\right) = 90.9$$

It then follows that, for this year,

$$\text{Fisher's ideal index} = \sqrt{(\text{Laspeyres index})(\text{Paasche index})}$$
$$= \sqrt{(92.2)(90.9)} = 91.5$$

The three complete indices, calculated in the above manner, are shown in the table.

YEAR	LASPEYRES	PAASCHE	FISHER IDEAL
1970–71	100.0	100.0	100.0
1971–72	92.2	90.9	91.5
1972–73	131.2	130.1	130.6
1973–74	212.0	210.9	211.4
1974–75	243.0	244.4	243.7
1975–76	198.5	201.2	199.8
1976–77	192.7	192.5	192.6

YEAR	LASPEYRES	PAASCHE	FISHER IDEAL
1977–78	178.2	177.7	177.9
1978–79	192.3	191.7	192.0
1979–80	215.1	216.2	215.6

16.5 WEIGHTED AGGREGATE QUANTITY INDICES

Price indices are designed to provide a representation of the changes through time in overall prices of a group of commodities. We might also want a picture of the evolution of overall quantity traded. Once again, any reasonable approach to this problem is likely to result in the calculation of a weighted quantity index, since we would presumably want to give more weight to an increase in the quantity purchased of a very expensive item than to the same increase in purchases of an inexpensive item. We will see in this section how Laspeyres and Paasche indices, similar to the corresponding price indices of the previous section, can be calculated for quantities.

LASPEYRES QUANTITY INDEX

The **Laspeyres quantity index** weights the individual quantities by the prices prevailing in the base period. To illustrate, consider again the automobile stock data of Tables 16.4 and 16.6. We saw in the previous section that the total cost of the base period (week 1) purchases at base period prices was 1,509. To obtain a quantity index for week 2, we compare with the total cost of all week 2 purchases, had week 1 prices prevailed; that is,

$$(6.3)(20\tfrac{2}{8}) + (1.5)(4\tfrac{1}{8}) + (16.0)(5\tfrac{2}{8}) + (12.9)(46\tfrac{1}{8}) = 813$$

The Laspeyres quantity index for any period expresses the total cost of the quantities traded in that period, had they been purchased at base period prices, as a percentage of the total cost of quantities traded in the base period. This index for week 2 is then

$$100\left(\frac{813}{1,509}\right) = 53.9$$

Proceeding in this way, we can calculate the index numbers for the other ten weeks. The results are set out in Table 16.10.

TABLE 16.10 Laspeyres quantity index for common stocks of United States automobile manufacturers for first 12 weeks of 1981

WEEK	LASPEYRES INDEX	WEEK	LASPEYRES INDEX
1	100.0	7	42.4
2	53.9	8	80.3
3	48.7	9	61.4
4	49.7	10	80.5
5	75.2	11	102.3
6	61.9	12	78.4

Suppose we have a group of K items for which information on quantity (produced or traded) is available over a period of time. One period is selected as a base for the index. The **Laspeyres quantity index** in any period is then the total cost of the quantities traded in that period, had the base period prices prevailed, expressed as a percentage of the total cost of the base period quantities.

Let q_{0i} and p_{0i} denote the quantity and price of the ith item in the base period, and q_{1i} the quantity of that item in the period of interest. The Laspeyres quantity index for that period is then

$$100\left(\frac{\sum_{i=1}^{K} q_{1i}p_{0i}}{\sum_{i=1}^{K} q_{0i}p_{0i}}\right)$$

PAASCHE QUANTITY INDEX

The Laspeyres quantity index provides a reasonable picture provided the base period prices remain representative. By contrast, the Paasche quantity index uses price information in every time period. Thus, the **Paasche quantity index** for a particular period expresses the total cost of the quantities purchased in that period as a percentage of what the base period quantities would have cost, had they been purchased at the prices of the period of interest.

To illustrate, we will find the Paasche quantity index for week 2 of our automobile stock data. In the previous section we saw that the total cost of week 2 quantities at week 2 prices was 811.125. Based on the data of Tables 16.4 and 16.6, it follows that the total cost of week 1 (base period) quantities at week 2 prices would have been

$$(8.2)(19\tfrac{7}{8}) + (4.3)(4\tfrac{1}{8}) + (14.4)(6) + (27.1)(45\tfrac{2}{8}) = 1{,}493.3875$$

The Paasche quantity index for week 2 is then

$$100\left(\frac{811.125}{1{,}493.3875}\right) = 54.3$$

The complete index is given in Table 16.11.

TABLE 16.11 Paasche quantity index for common stocks of United States automobile manufacturers for first 12 weeks of 1981

WEEK	PAASCHE INDEX	WEEK	PAASCHE INDEX
1	100.0	7	42.4
2	54.3	8	80.8
3	48.6	9	61.4
4	49.6	10	80.7
5	75.1	11	102.9
6	61.9	12	78.4

TABLE 16.12 Fisher's ideal quantity index for common stocks of United States automobile manufacturers for first 12 weeks of 1981

WEEK	FISHER'S IDEAL INDEX	WEEK	FISHER'S IDEAL INDEX
1	100.0	7	42.4
2	54.1	8	80.5
3	48.6	9	61.4
4	49.6	10	80.6
5	75.1	11	102.6
6	61.9	12	78.4

The Paasche Quantity Index

The **Paasche quantity index** in any period is the total cost of purchases in that period, as a percentage of what the base period quantities would have cost, had they been purchased in the period of interest.

Let q_{0i} denote the quantity of the ith item in the base period, with q_{1i} the quantity purchased, and p_{1i} the price, of that item in a second period. The Paasche quantity index for the second period is then

$$100\left(\frac{\sum_{i=1}^{K} q_{1i}p_{1i}}{\sum_{i=1}^{K} q_{0i}p_{1i}}\right)$$

FISHER'S IDEAL QUANTITY INDEX

Once again, a compromise index can be obtained. **Fisher's ideal quantity index** is the square root of the product of the Laspeyres and Paasche quantity indices. Thus, using the results of Tables 16.10 and 16.11 for our automobile stock data, we find that the Fisher index for week 2 is

$$\sqrt{(53.9)(54.3)} = 54.1$$

The complete index, found in this way, is given in Table 16.12.

Fisher's Ideal Quantity Index

Fisher's ideal quantity index provides a compromise between the Laspeyres and Paasche quantity indices, and is defined as

$$\text{Fisher's ideal index} = \sqrt{(\text{Laspeyres index})(\text{Paasche index})}$$

16.6 A VALUE INDEX

Another type of index is sometimes calculated. In a **value index,** the total cost of items purchased in a particular period (at the prices of that period) is compared with the total cost of items purchased in the base period (at base period prices). The value index therefore reflects movements in both quantities and prices through time.

Value Index

Suppose we have a group of K items for which price and quantity information is available over a period of time. One period is selected as a base for the index. The **value index** then expresses the total cost of the quantities in a particular period as a percentage of the total cost of the quantities in the base period.

Let p_{0i} and q_{0i} be the price and quantity of the ith item in the base period, and p_{1i} and q_{1i} the price and quantity of this item in a second period. The value index for this second period is then

$$100\left(\frac{\sum_{i=1}^{K} q_{1i}p_{1i}}{\sum_{i=1}^{K} q_{0i}p_{0i}}\right)$$

Example 16.2

Using the data of Example 16.1, derive a value index for wheat, corn, and soybeans, using crop year 1970–71 as base.

In Example 16.1, we determined that the total cost in the base year was

$$\sum q_{0i}p_{0i} = 10{,}532.27$$

For crop year 1971–72, the total cost was found to be

$$\sum q_{1i}p_{1i} = 11{,}823.68$$

Hence, the value index for this crop year is

$$100\left(\frac{\sum q_{1i}p_{1i}}{\sum q_{0i}p_{0i}}\right) = 100\left(\frac{11{,}823.68}{10{,}532.27}\right) = 112.3$$

Thus, the total value of the 1971–72 crop was 12.3% higher than that of the 1970–71 crop.

The complete index is shown in the accompanying table.

YEAR	VALUE INDEX	YEAR	VALUE INDEX
1970–71	100.0	1975–76	283.7
1971–72	112.3	1976–77	293.2
1972–73	161.6	1977–78	288.0
1973–74	284.1	1978–79	361.6
1974–75	347.6	1979–80	352.5

EXERCISES

1. An agricultural economist has available data on the price per pound of beef, lamb, pork, and chicken over the last 15 years. She wants to obtain a measure of aggregate price movements of meat over this period, and asks your advice. What factors do you believe should be taken into consideration? Write a brief memorandum for this economist, outlining the problems involved and indicating what further information might be useful.
2. The accompanying table shows the price per share of stock in Bank of New York, Inc. for the first 12 weeks of 1981.

WEEK	PRICE	WEEK	PRICE
1	35	7	35
2	$35\frac{7}{8}$	8	$34\frac{6}{8}$
3	$34\frac{6}{8}$	9	$34\frac{6}{8}$
4	$34\frac{3}{8}$	10	$35\frac{2}{8}$
5	35	11	$38\frac{6}{8}$
6	$34\frac{7}{8}$	12	$37\frac{1}{8}$

   (a) Form a price index with week 1 as base.
   (b) Form a price index with week 4 as base.
3. The table shows the price per share of stock in Mobil Corp. for the first 12 weeks of 1981.

WEEK	PRICE	WEEK	PRICE
1	$78\frac{3}{8}$	7	$69\frac{4}{8}$
2	$81\frac{7}{8}$	8	$68\frac{2}{8}$
3	$77\frac{7}{8}$	9	$63\frac{5}{8}$
4	$73\frac{1}{8}$	10	$65\frac{5}{8}$
5	$72\frac{2}{8}$	11	$64\frac{4}{8}$
6	70	12	$66\frac{2}{8}$

   (a) Form a price index with week 1 as base.
   (b) Form a price index with week 4 as base.
4. The accompanying table shows total regulated margin credit at brokers and dealers for the first 7 months of 1981.

MONTH	CREDIT (millions of dollars)	MONTH	CREDIT (millions of dollars)
January	14,242	May	14,951
February	14,171	June	15,126
March	14,243	July	15,134
April	14,869		

(a) Form an index for the amount of credit with January as base.
(b) Form an index with April as base.

5. The table shown here gives United States farm prices of rye for 10 consecutive crop years.

CROP YEAR	PRICE (dollars per bushel)	CROP YEAR	PRICE (dollars per bushel)
1970–71	0.99	1975–76	2.36
1971–72	0.90	1976–77	2.47
1972–73	0.96	1977–78	1.20
1973–74	1.91	1978–79	1.99
1974–75	2.51	1979–80	2.06

(a) Form a price index with crop year 1970–71 as base.
(b) Form a price index with crop year 1975–76 as base.

6. The table shows the values (in billions of dollars) of corporate stock and homes in the United States from 1975 through 1980.

YEAR	CORPORATE STOCK	HOMES
1975	660	1,380
1976	772	1,572
1977	729	1,828
1978	797	2,130
1979	885	2,425
1980	1,037	2,612

(a) Form separate indices, each based on 1975, for values of corporate stock and homes.
(b) On the same graph, draw time plots of these two indices and comment on the comparison between their movements over this period.

7. The accompanying table shows the price per share of common stock in two major steel corporations for the first 12 weeks of 1981.

WEEK	BETHLEHEM STEEL	U.S. STEEL	WEEK	BETHLEHEM STEEL	U.S. STEEL
1	$25\frac{1}{8}$	$24\frac{7}{8}$	7	24	$27\frac{5}{8}$
2	$24\frac{7}{8}$	$24\frac{5}{8}$	8	26	28
3	23	24	9	$28\frac{4}{8}$	31
4	$25\frac{7}{8}$	$27\frac{2}{8}$	10	$28\frac{6}{8}$	31
5	$25\frac{1}{8}$	$27\frac{4}{8}$	11	$30\frac{3}{8}$	33
6	$24\frac{2}{8}$	$27\frac{4}{8}$	12	$30\frac{7}{8}$	$34\frac{4}{8}$

(a) Using week 1 as base, find the unweighted aggregate index of prices.
(b) With the same base, find the averages of price relatives.

8. The table given here shows the United States farm prices (in dollars per bushel) for oats and barley for crop years 1974–75 through 1979–80.

CROP YEAR	OATS	BARLEY
1974–75	1.53	2.80
1975–76	1.46	2.42
1976–77	1.56	2.25
1977–78	1.14	1.80
1978–79	1.20	1.92
1979–80	1.37	2.29

(a) Using crop year 1974–75 as a base, find the unweighted aggregate index of prices.

(b) With the same year as base, find the averages of price relatives for these commodities.

9. A restaurant offers three "specials"—steak, seafood, and chicken. Their average prices (in dollars) for the 12 months of last year are shown in the table.

MONTH	STEAK	SEAFOOD	CHICKEN
January	7.25	6.42	5.37
February	7.41	6.40	5.21
March	7.45	6.25	5.25
April	7.70	6.60	5.40
May	7.72	6.70	5.45
June	7.75	6.85	5.60
July	8.10	6.90	5.54
August	8.15	6.84	5.70
September	8.20	6.96	5.72
October	8.30	7.10	5.69
November	8.45	7.10	5.85
December	8.71	7.23	6.10

(a) Using January as a base, find the unweighted aggregate index of prices.

(b) With the same month as base, find the averages of price relatives for these three specials.

10. The accompanying table shows the price per share of common stock in four electronic office equipment manufacturers for the first 12 weeks of 1981.

WEEK	TEXAS INSTRUMENTS	HEWLETT PACKARD	IBM	XEROX
1	$116\frac{5}{8}$	$86\frac{2}{8}$	$67\frac{3}{8}$	$61\frac{7}{8}$
2	$117\frac{6}{8}$	$81\frac{5}{8}$	$66\frac{7}{8}$	$58\frac{2}{8}$
3	$111\frac{6}{8}$	81	65	57
4	$110\frac{2}{8}$	$82\frac{2}{8}$	$64\frac{3}{8}$	56
5	$115\frac{4}{8}$	88	$64\frac{2}{8}$	58
6	116	$84\frac{7}{8}$	$61\frac{6}{8}$	56

WEEK	TEXAS INSTRUMENTS	HEWLETT PACKARD	IBM	XEROX
7	109 2/8	85 4/8	60 4/8	56
8	111 7/8	87 4/8	64 2/8	58 2/8
9	107 2/8	83 4/8	61 7/8	54 7/8
10	112 5/8	83 6/8	63 2/8	56 5/8
11	115 6/8	87 3/8	63 1/8	59
12	117 6/8	90 3/8	61 6/8	56 1/8

(a) Using week 1 as base, find the unweighted aggregate index of prices.

(b) With the same week as base, find the averages of price relatives for these four corporations.

11. The table shows the volume of shares (in hundreds of thousands) traded in Bethlehem Steel and U.S. Steel corporations in the first 12 weeks of 1981. Use the price data of Exercise 7 and take week 1 as base.

WEEK	BETHLEHEM STEEL	U.S. STEEL	WEEK	BETHLEHEM STEEL	U.S. STEEL
1	3.5	6.4	7	1.1	5.6
2	2.0	3.5	8	4.5	11.7
3	2.0	2.2	9	11.9	19.3
4	3.2	16.8	10	9.3	10.9
5	3.4	7.0	11	7.8	17.7
6	1.5	5.6	12	6.7	13.3

(a) Compute the Laspeyres aggregate price index.

(b) Compute the Paasche aggregate price index.

(c) Find Fisher's ideal price index.

12. The accompanying table shows total production (in millions of bushels) of oats and barley for crop years 1974–75 through 1979–80. Use the price data of Exercise 8 and take crop year 1974–75 as base.

CROP YEAR	OATS	BARLEY
1974–75	642	374
1975–76	546	372
1976–77	748	416
1977–78	601	447
1978–79	527	383
1979–80	458	359

(a) Compute the Laspeyres aggregate price index.

(b) Compute the Paasche aggregate price index.

(c) Find Fisher's ideal price index.

13. In Exercise 9, we analyzed monthly data on the prices of three specials in a restaurant. The numbers of orders for these specials in each month are given in the table. Use the prices in Exercise 9 and take January as base.

MONTH	STEAK	SEAFOOD	CHICKEN
January	123	169	243
February	110	160	251
March	115	181	265
April	101	152	231
May	118	140	263
June	100	128	237
July	92	129	221
August	87	130	204
September	123	164	293
October	131	169	301
November	136	176	327
December	149	193	351

(a) Compute the Laspeyres aggregate price index.
(b) Compute the Paasche aggregate price index.
(c) Find Fisher's ideal price index.

14. The accompanying table shows the volume of shares (in hundreds of thousands) traded in the stock of four electronic office equipment manufacturers in the first 12 weeks of 1981. Use the price data of Exercise 10 and take week 1 as base.

WEEK	TEXAS INSTRUMENTS	HEWLETT PACKARD	IBM	XEROX
1	3.3	4.2	39.6	12.0
2	2.3	4.4	30.2	8.1
3	2.4	5.2	22.0	8.8
4	1.9	3.6	16.3	7.0
5	2.5	3.9	19.1	9.4
6	2.3	5.4	17.3	7.2
7	5.4	4.4	15.7	6.7
8	4.1	4.0	27.2	12.2
9	3.5	3.9	15.7	7.1
10	3.2	2.9	21.0	11.3
11	2.5	5.2	28.3	13.3
12	2.7	5.0	34.2	11.7

(a) Compute the Laspeyres aggregate price index.
(b) Compute the Paasche aggregate price index.
(c) Find Fisher's ideal price index.

15. In what circumstances would you expect substantial differences between the Laspeyres and Paasche aggregate price indices? Develop an illustrative example, complete with hypothetical data.

16. Using the data in Exercises 7 and 11, and taking week 1 as base, develop each of the following indices for the volume of shares traded in Bethlehem Steel and U.S. Steel:
    (a) Laspeyres aggregate quantity index
    (b) Paasche aggregate quantity index
    (c) Fisher's ideal quantity index

17. Using the data in Exercises 8 and 12, and taking crop year 1974–75 as base, develop each of the following indices for the total production of oats and barley:
    (a) Laspeyres aggregate quantity index
    (b) Paasche aggregate quantity index
    (c) Fisher's ideal quantity index

18. Using the data in Exercises 9 and 13, and taking January as base, develop each of the following indices for orders of steak, seafood, and chicken specials:
    (a) Laspeyres aggregate quantity index
    (b) Paasche aggregate quantity index
    (c) Fisher's ideal quantity index

19. Using the data in Exercises 10 and 14, and taking week 1 as base, develop each of the following indices for the volume of shares traded in Texas Instruments, Hewlett Packard, IBM, and Xerox:
    (a) Laspeyres aggregate quantity index
    (b) Paasche aggregate quantity index
    (c) Fisher's ideal quantity index

20. In what circumstances would you expect substantial differences between the Laspeyres and Paasche aggregate quantity indices? Develop an illustrative example, complete with hypothetical data.

21. Show that

$$\frac{\text{Laspeyres price index}}{\text{Laspeyres quantity index}} = \frac{\text{Paasche price index}}{\text{Paasche quantity index}}$$

22. Using the data of Exercises 7 and 11, and taking week 1 as base, develop an aggregate value index for shares traded in Bethlehem Steel and U.S. Steel.

23. Using the data of Exercises 8 and 12, and taking crop year 1974–75 as base, develop an aggregate value index for production of oats and barley.

24. Using the data of Exercises 9 and 13, and taking January as base, develop an aggregate value index for orders of steak, seafood, and chicken specials.

25. Using the data of Exercises 10 and 14, and taking week 1 as base, develop an aggregate value index for shares traded in Texas Instruments, Hewlett Packard, IBM, and Xerox.

26. Explain verbally the distinctions between aggregate indices of price, quantity, and value.

27. Show that

$$\text{Value index} = \frac{\text{Laspeyres price index} \times \text{Paasche quantity index}}{100}$$

$$= \frac{\text{Paasche price index} \times \text{Laspeyres quantity index}}{100}$$

28. Show that

$$\text{Value index} = \frac{\text{Fisher's ideal price index} \times \text{Fisher's ideal quantity index}}{100}$$

16.7 CHANGE IN BASE PERIOD

From time to time, officially published series of index numbers are updated by bringing forward the base period. In these circumstances, the value of the original index at the new base point is typically given. As an illustration, columns 2 and 3 of Table 16.13 give Laspeyres price indices for wheat, corn, and soybeans. These indices have been calculated from the data of Example 16.1. The second column shows the price index for crop years 1970–71 through 1975–76, using 1970–71 as base. The third column gives the Laspeyres price index for crop years 1975–76 through 1979–80, with 1975–76 as base. These indices are plotted in Figure 16.6, where the discontinuity in 1975–76 is obvious.

Given only the information in columns 2 and 3 of Table 16.13, it is difficult to obtain a clear picture of the overall progression of price through this entire period. In order to do so, we would like to put these two indices on the same footing, obtaining a **spliced index.** In our new index, crop year 1975–76 will be fixed as base, the index

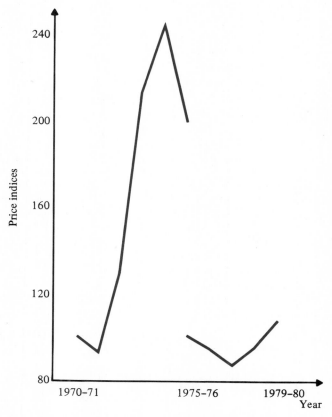

FIGURE 16.6 Time plots of Laspeyres aggregate price indices for wheat, corn, and soybeans for 1970–71 through 1975–6 (1970–71 = 100), and 1975–6 through 1979–80 (1975–6 = 100)

TABLE 16.13 Aggregate Laspeyres price indices for wheat, corn, and soybeans (with base periods 1970–71 and 1975–76) and a spliced index (based on 1975–76)

	BASE YEAR		SPLICED INDEX
YEAR	1970–71	1975–76	(BASE 1975–76)
1970–71	100.0		50.4
1971–72	92.2		46.4
1972–73	131.2		66.1
1973–74	212.0		106.8
1974–75	243.0		122.4
1975–76	198.5	100.0	100.0
1976–77		94.0	94.0
1977–78		86.7	86.7
1978–79		94.9	94.9
1979–80		107.0	107.0

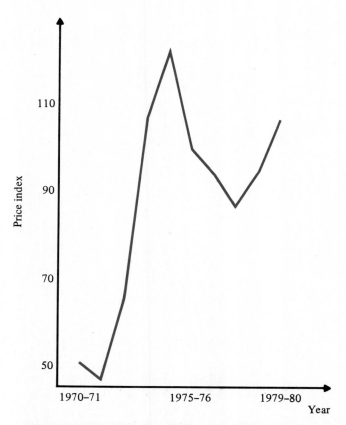

FIGURE 16.7 Spliced aggregate Laspeyres price index for wheat, corn, and soybeans (1975–76 = 100)

for that year being set at 100. In the original index, based on 1970–71, the value for 1975–76 was 198.5. In order to transform this to 100, we divide by 198.5 and multiply by 100. Similarly, the other index numbers based on 1970–71 can be converted to base 1975–76 by dividing each by 198.5 and multiplying the result by 100. For example, the new figure for 1974–75 is

$$243.0\left(\frac{100}{198.5}\right) = 122.4$$

The spliced index obtained in this manner is set out in the last column of Table 16.13, and graphed in Figure 16.7, which now presents a continuous picture of price evolution for these commodities over the entire period.

EXERCISES

29. The table given here shows the Consumer Price Index for entertainment from 1967 through 1980. Obtain a spliced index, with base year 1976.

	BASE YEAR	
YEAR	1967	1976
1967	100.0	
1968	105.7	
1969	111.0	
1970	116.7	
1971	122.9	
1972	126.5	
1973	130.0	
1974	139.8	
1975	152.2	
1976	159.8	100.0
1977		104.9
1978		110.3
1979		117.4
1980		127.5

30. The accompanying table shows an index of output per hour of all persons in the nonfarm business sector of the economy, for the period 1973 to 1980.

	BASE YEAR	
YEAR	1973	1977
1973	100.0	
1974	97.7	
1975	99.7	
1976	102.9	
1977	104.9	100.0

	BASE YEAR	
YEAR	1973	1977
1978		99.8
1979		99.1
1980		98.8

(a) Obtain a spliced index with base year 1977.

(b) Obtain a spliced index with base year 1973.

REVIEW EXERCISES

31. Universities incur many costs in their operation, including the costs of energy, books, laboratory and other equipment, stationery, and labor. Suppose you are asked to show how price levels faced by your university have changed over the last 10 years. What difficulties would you expect to encounter and how would you attempt to proceed?

32. The table shows the 3-months Treasury bill yield in Canada over a period of 20 years:

YEAR	YIELD	YEAR	YIELD	YEAR	YIELD
1961	2.83	1968	6.24	1975	8.37
1962	4.01	1969	7.14	1976	8.89
1963	3.57	1970	6.12	1977	7.35
1964	3.74	1971	3.62	1978	8.58
1965	3.97	1972	3.55	1979	11.57
1966	5.00	1973	5.39	1980	12.70
1967	4.59	1974	7.78		

(a) Form an index of yields with 1965 as base.

(b) Form an index with 1975 as base.

33. A chemical corporation purchases quantities of three raw materials. The table given here shows, for a period of 12 years, the quantities purchased and the prices of these materials. (Quantities are given in thousands of tons, and prices in dollars per ton.) Use 1971 as the base period.

	MATERIAL A		MATERIAL B		MATERIAL C	
YEAR	PRICE	QUANTITY	PRICE	QUANTITY	PRICE	QUANTITY
1971	127	31	182	21	199	42
1972	131	33	183	24	198	40
1973	134	30	187	25	200	38

	MATERIAL A		MATERIAL B		MATERIAL C	
YEAR	PRICE	QUANTITY	PRICE	QUANTITY	PRICE	QUANTITY
1974	132	28	192	26	206	37
1975	138	26	193	27	212	39
1976	143	28	199	28	220	40
1977	145	29	207	30	222	41
1978	144	35	215	31	231	43
1979	149	36	218	33	241	43
1980	160	34	229	31	250	45
1981	173	32	241	32	272	46
1982	189	37	262	34	298	48

(a) Find the unweighted aggregate index of prices.
(b) Find the averages of price relatives.
(c) Compute the Laspeyres aggregate price index.
(d) Compute the Paasche aggregate price index.
(e) Find Fisher's ideal price index.
(f) Compute the Laspeyres aggregate quantity index.
(g) Compute the Paasche aggregate quantity index.
(h) Find Fisher's ideal quantity index.
(i) Compute an aggregate value index.

34. A library purchases both books and journals. The accompanying table shows the average prices (in dollars) paid for each and the quantities purchased over a period of 6 years. Use 1977 as the base period.

	BOOKS		JOURNALS	
YEAR	PRICE	QUANTITY	PRICE	QUANTITY
1977	20.2	694	30.1	153
1978	22.3	723	33.4	159
1979	23.3	687	36.0	160
1980	24.6	731	39.8	163
1981	27.0	742	45.7	160
1982	29.2	739	51.2	155

(a) Find the unweighted aggregate index of prices.
(b) Find the averages of price relatives.
(c) Compute the Laspeyres aggregate price index.
(d) Compute the Paasche aggregate price index.
(e) Find Fisher's ideal price index.
(f) Compute the Laspeyres aggregate quantity index.
(g) Compute the Paasche aggregate quantity index.
(h) Find Fisher's ideal quantity index.
(i) Compute an aggregate value index.

35. A food processing company produces three products. The table shows the quarterly price per unit (in dollars) of each over a period of 3 years, together with the quantities sold (in thousands). Use March 1980 as the base period.

	PRODUCT A		PRODUCT B		PRODUCT C	
	PRICE	QUANTITY	PRICE	QUANTITY	PRICE	QUANTITY
March 1980	1.25	43	1.53	62	1.83	73
June 1980	1.27	47	1.54	65	1.85	70
Sept. 1980	1.26	41	1.52	63	1.88	71
Dec. 1980	1.30	40	1.56	60	1.87	75
March 1981	1.32	45	1.57	58	1.90	74
June 1981	1.35	46	1.59	61	1.91	72
Sept. 1981	1.35	49	1.60	62	1.93	76
Dec. 1981	1.40	47	1.58	62	1.96	77
March 1982	1.44	50	1.63	64	1.97	79
June 1982	1.46	51	1.64	67	1.99	81
Sept. 1982	1.45	53	1.68	68	2.00	82
Dec. 1982	1.48	54	1.69	70	2.04	85

(a) Find the unweighted aggregate index of prices.

(b) Find the averages of price relatives.

(c) Compute the Laspeyres aggregate price index.

(d) Compute the Paasche aggregate price index.

(e) Find Fisher's ideal price index.

(f) Compute the Laspeyres aggregate quantity index.

(g) Compute the Paasche aggregate quantity index.

(h) Find Fisher's ideal quantity index.

(i) Compute an aggregate value index.

36. The table given here shows an index of the unemployment rate in Canada for the years 1966 through 1981. Obtain a spliced index, with base year 1973.

	BASE YEAR	
YEAR	1966	1973
1966	100.0	
1967	111.8	
1968	132.4	
1969	129.4	
1970	167.6	
1971	182.4	
1972	182.4	
1973	161.8	100.0
1974		96.4
1975		125.5
1976		129.1

	BASE YEAR	
YEAR	1966	1973
1977		147.3
1978		152.7
1979		136.4
1980		136.4
1981		138.2

time series analysis and forecasting

CHAPTER SEVENTEEN

17.1 TIME SERIES DATA: PROBLEMS AND OPPORTUNITIES

In this chapter we will deal with some of the issues involved in analyzing a special type of data set. Specifically, we will be interested in measurements *through time* on a particular variable. Examples include monthly product sales, quarterly corporate earnings, and daily closing prices for shares of common stock.

Definition

A **time series** is a set of measurements, ordered through time, on a particular quantity of interest.

Time series data typically possess special characteristics that necessitate the development of new statistical methods for their analysis. Virtually all of the techniques of data analysis that we have developed so far are based on an assumption of *sample randomness*—that is, on the assumption that the available data consist of *independent* observations on the phenomenon of interest. Only very rarely will this assumption of independence be tenable for time series data. For example, consider a series of monthly sales of an engineering product. Lack of independence might be suspected for at least two reasons. First, if sales were relatively buoyant last month, it is reasonable to suspect that they are more likely than not to remain so in the current month. The general economic condition that led to high sales volume last month are not likely to undergo an abrupt change in 1 month, so that the market features faced by the engineering firm in the current month will, in all probability, be broadly similar to those of the previous month. Thus, we can expect some similarity between sales in adjacent months. Another feature of sales data for many products is *seasonality*. Sales tend to peak at about the same time every year. Hence, if June has historically been a good month for sales of the engineering product in past years, it is more likely than not that June sales in the current year will be relatively high. This, too, implies a lack of independence in the monthly sales figures.

We have seen that time series are likely to be characterized by certain types of dependence. Thus, an important assumption underlying the great majority of the statistical procedures discussed so far in this text, will very probably not hold for time series data. Furthermore, it is the case that this assumption is typically rather crucial, so that the analysis of a time series as if it consisted of independent measurements can produce seriously misleading conclusions. This is the fundamental problem of time series analysis, and the reason we must devote a separate chapter to this special type of data. We have, in fact, already met the problem in the context of regression analysis in Section 14.8. There we dealt with the problem of estimating regression models when the error terms were autocorrelated—that is, correlated through time. In this chapter we will restrict ourselves to the analysis of single time series.

In this section we have discussed a negative aspect of the typical kinds of dependency patterns likely to be present in time series data. Certainly these patterns do create problems, necessitating the development of special techniques of data analysis. However, inherent in this same phenomenon lies an opportunity. It is often possible to exploit any dependencies revealed in the past to produce *forecasts* of future values of a time series. For example, if sales in the current month are rather similar to sales in the immediately preceding months, and to sales in the same month in previous years, this information can be used to predict sales in future months.

17.2 A NONPARAMETRIC TEST FOR RANDOMNESS

Before discussing techniques for dealing with time series data exhibiting typical patterns of nonrandomness, we will consider a test for randomness in a time series. Among several such tests, the **runs test** is particularly easy to carry out. It is nonparametric—that is to say, no assumption is made about the distribution from which the observations were drawn.

To illustrate the test, we will analyze a series of 14 monthly yields on public utilities bonds. The data are given in Table 17.1 and plotted in Figure 17.1. We have

TABLE 17.1 Yields on public utilities bonds

1980	June	July	Aug.	Sept.	Oct.	Nov.	Dec.
	11.87	12.12	12.82	13.29	13.53	14.07	14.48
1981	Jan.	Feb.	March	April	May	June	July
	14.22	14.84	14.86	15.32	15.84	15.27	15.87

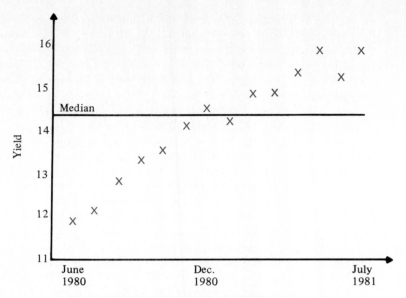

FIGURE 17.1 Yields on public utilities bonds

also drawn in this figure a line at the median yield—that is, the average of the middle pair when the data are arranged in ascending order. For these data, we have

$$\text{Median} = \frac{14.22 + 14.48}{2} = 14.35$$

Now, if this series of observations were random, the yield in one month would be independent of that in others, so that a high yield would be no more likely to be followed by another high yield than by a low yield. Even a very cursory inspection of Figure 17.1 suggests that this is not in fact so. The first six yields are all below the median, while the last six are all above.

The previous argument suggests that, in assessing the likelihood that our series is random, we might look at a sequence that depicts whether observations are above or below the median. If we let "+" denote an observation above the median, and " − " an observation below, then the sequence for these particular data is

$$- - - - - - + - + + + + + +$$

This sequence consists of a **run** of six −, followed by a run of one +, a run of one −, and finally, a run of six +. In total, there are therefore

$$R = 4$$

runs.

Next we ask how likely it would be to find this few or fewer runs, if the series were truly random. In order to answer this question, it is necessary to know the distribution of the number of runs when the null hypothesis of randomness is, in fact, correct. This cumulative distribution function is tabulated in Table 11 of the Appendix. We read directly from this table that, for a series with $n = 14$ observations, the probability, under the null hypothesis, of finding 4 or fewer runs is 0.025. If there is positive association between adjacent observations in a time series, so that high values are likely to be followed by high, and low by low, we would expect our sequence to contain relatively few runs. This is the usual hypothesis against which we would like to test. It follows from our results that, for these data, the null hypothesis of randomness can be rejected against the alternative of positive association at significance levels of 2.5% or higher. These data, then, present fairly strong evidence of positive association.

The Runs Test

Suppose that we have a time series of n observations. (For simplicity, it will be assumed that n is even.) A sequence of signs, with "+" denoting a value above the median and "−" a value below, is formed from these data. Let R denote the number of runs in the sequence. The null hypothesis to be tested is of randomness in the time series. Table 11 of the Appendix gives the smallest significance level against which this null hypothesis can be rejected against the alternative of positive association between observations that are adjacent in time.

If the alternative is the two-sided hypothesis of nonrandomness, then the significance level must be doubled if it is less than 0.5. Alternatively, if the significance level, α, read from the table is bigger than 0.5, the appropriate significance level for the test against the two-sided alternative is $2(1 - \alpha)$.

For time series of more than twenty observations, the distribution of the number of runs under the null hypothesis can be well approximated by the normal distribution. In fact, it can be shown that, under the null hypothesis, the random variable corresponding to

$$ Z = \frac{R - \dfrac{n}{2} - 1}{\sqrt{\dfrac{n^2 - 2n}{4(n - 1)}}} $$

has a standard normal distribution. This result can then be used as the basis of a test for randomness when a large number of observations is available. The procedure is described in the box.

The Runs Test: Large Samples

Suppose that we have a time series of n observations, where n is even and moderately large (greater than 20). Define the number of runs, R, as previously. We want to test the null hypothesis

$$H_0: \quad \text{The series is random}$$

The following tests have significance level α:

(i) If the alternative hypothesis is of positive association between adjacent observations, the decision rule is

$$\text{Reject } H_0 \text{ if } \quad \frac{R - \dfrac{n}{2} - 1}{\sqrt{\dfrac{n^2 - 2n}{4(n-1)}}} < -z_\alpha$$

(ii) If the alternative is two-sided, of nonrandomness, the decision rule is

$$\text{Reject } H_0 \text{ if } \quad \frac{R - \dfrac{n}{2} - 1}{\sqrt{\dfrac{n^2 - 2n}{4(n-1)}}} < -z_{\alpha/2} \quad \text{or} \quad \frac{R - \dfrac{n}{2} - 1}{\sqrt{\dfrac{n^2 - 2n}{4(n-1)}}} > z_{\alpha/2}$$

Example 17.1

In this example we consider a classic, frequently analyzed, set of data[1] consisting of 30 annual observations on sales (in thousands of dollars) of Lydia E. Pinkham.

The data, together with the corresponding sequence of $+$ and $-$ are shown in the table. Their median is

$$\frac{1{,}767 + 1{,}770}{2} = 1{,}768.5$$

YEAR	SALES		YEAR	SALES	
1931	1,806	+	1946	2,177	+
1932	1,644	−	1947	1,920	+
1933	1,814	+	1948	1,910	+
1934	1,770	+	1949	1,984	+
1935	1,518	−	1950	1,787	+
1936	1,103	−	1951	1,689	−
1937	1,266	−	1952	1,866	+
1938	1,473	−	1953	1,896	+
1939	1,423	−	1954	1,684	−
1940	1,767	−	1955	1,633	−
1941	2,161	+	1956	1,657	−
1942	2,336	+	1957	1,569	−
1943	2,602	+	1958	1,390	−
1944	2,518	+	1959	1,387	−
1945	2,637	+	1960	1,289	−

[1] These data are given in G. M. Erickson, "Using ridge regression to estimate directly lagged effects in marketing," *Journal of American Statistical Association, 76* (1981), 766–73, and K. S. Palda, *The Measurement of Cumulative Advertising Effects,* Prentice-Hall (1964).

We see from the table that the number of runs is $R = 8$. Since we have $n = 30$ observations, the value of the test statistic is

$$\frac{R - \dfrac{n}{2} - 1}{\sqrt{\dfrac{n^2 - 2n}{4(n - 1)}}} = \frac{8 - 15 - 1}{\sqrt{\dfrac{900 - 60}{116}}} = -2.97$$

From Table 3 of the Appendix, we see that the α value corresponding to $z_\alpha = 2.97$ is 0.0015. Hence, the null hypothesis of randomness can be rejected against the alternative of positive association between adjacent observations at any significance level above 0.15%. The evidence in favor of this alternative is quite overwhelming. It is highly improbable that this time series is random.

17.3 COMPONENTS OF A TIME SERIES

In this and the following two sections we will consider some descriptive ways of measuring the progression through time of a quantity of interest. We will let the series of interest be denoted $X_1, X_2, \ldots, X_n$, so that at time t, the observed value of a series is represented by X_t.

One way of thinking about the behavior of an actual observed series is to regard it as being made up of various **components.** Traditionally, four possible components have been considered, with the notion that any or all might be present in any particular series. These components are as follows:

(i) Trend component
(ii) Seasonality component
(iii) Cyclical component
(iv) Irregular component

Many time series met in practice exhibit a tendency either to grow or to decrease fairly steadily over time, and this pattern is identified as **trend.** For example, despite short-run deviations from the trend, measures of national wealth, such as gross domestic product, have moved steadily upward over the years. An example of a series exhibiting a steady upward trend is shown in the data of Table 17.2, graphed in Figure 17.2, on car ownership in Great Britain[2] over a period of 24 years.

It is clear from the table, and even more obvious from the graph, that the behavior of this time series is characterized by a fairly steady upward trend. There is a suggestion that the rate of growth of car ownership is less rapid in the latter part of the period than in the middle, raising the possibility of a slowly evolving trend pattern. Certainly the progress through time is not completely smooth, but the behavior of this time series does appear to be dominated by its (possibly slowly changing) trend.

---

[2] Data are taken from R. J. Brooks, et al., "A note on forecasting car ownership," *Journal of Royal Statistical Society A, 141* (1978), 64–68.

TABLE 17.2 Car ownership in Great Britain

YEAR	NUMBER OF CARS PER CAPITA	YEAR	NUMBER OF CARS PER CAPITA
1951	0.049	1963	0.142
1952	0.051	1964	0.157
1953	0.056	1965	0.169
1954	0.063	1966	0.180
1955	0.071	1967	0.194
1956	0.078	1968	0.202
1957	0.084	1969	0.209
1958	0.091	1970	0.214
1959	0.098	1971	0.223
1960	0.108	1972	0.234
1961	0.116	1973	0.248
1962	0.127	1974	0.251

Indeed, this picture is so clear-cut that it is very doubtful that any more sophisticated analysis of these data alone would reveal, with any worthwhile level of certainty, much more of value about car ownership in Great Britain.

Although a simple graphical presentation is usually not sufficient in the analysis of time series data, it does constitute an invaluable first step.

Many business and economic time series met in practice consist of quarterly or monthly observations. It is often the case that such series exhibit the phenomenon of **seasonality**, such that patterns are repeated from year to year. As a rather obvious example, retail sales of many products tend to be relatively high in December, because of Christmas shopping. Also, construction activity in the Midwest is typically low in

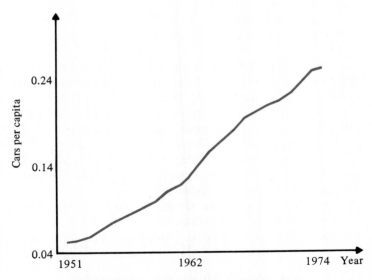

FIGURE 17.2 Car ownership in Great Britain

TABLE 17.3 Earnings per share of Burroughs Corporation

YEAR	QUARTER			
	1	2	3	4
1971	0.300	0.460	0.345	0.910
1972	0.330	0.545	0.440	1.040
1973	0.495	0.680	0.545	1.285
1974	0.550	0.870	0.660	1.580
1975	0.590	0.990	0.830	1.730
1976	0.610	1.050	0.920	2.040
1977	0.700	1.230	1.060	2.320
1978	0.820	1.410	1.250	2.730

the winter quarter, due to inclement weather. This seasonal behavior is generally easily spotted when a times series is graphed.

Table 17.3 and Figure 17.3 show earnings per share of Burroughs Corporation over a period of 8 years. These earnings figures are available quarterly, and from the table one can see evidence of seasonal behavior. The fourth quarter figures tend to be relatively high, while those in the first quarter are quite low.

This seasonal behavior is quite clear from Figure 17.3, where there is an obvious pattern repeating each year. The earnings in the second quarter are somewhat higher

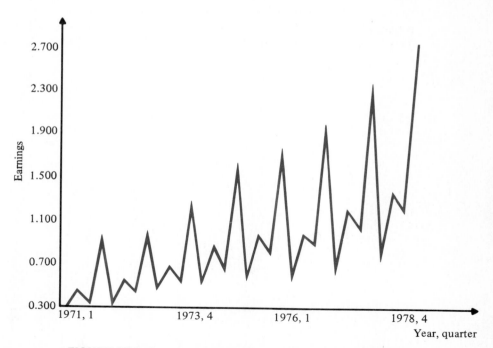

FIGURE 17.3 Quarterly earnings per share of Burroughs Corporation

Components of a Time Series **695**

than those of the immediately preceding or succeeding quarter, while those of the fourth quarter are much higher yet. This figure also makes clear that there is another component of the time series. Apart from the obvious seasonality, there is a noticeable upward trend in earnings per share over the period covered by our data. How one approaches the phenomenon of seasonality depends on the objectives. In some applications, such as routine sales forecasting for purposes of inventory control, it is important to obtain as good an assessment as possible of the likely outcome in each future month. In that case, it is clear that any pronounced seasonal pattern, which might reasonably be expected to recur in the future, will provide an important constituent in forecast derivation.

On the other hand, for some purposes, seasonality is rather a nuisance. In many applications, the analyst requires an assessment of overall movements in a time series, uncontaminated by the influence of seasonal factors. For instance, suppose that we have just received the most recent fourth quarter earnings figures from Burroughs Corporation. We already know that these will very likely be a good deal higher than those of the previous quarter. What we would like to do is assess how much of this increase in earnings is due to purely seasonal factors, and how much represents real underlying growth. In other words, we would like to produce a time series free from seasonal influence. Such a series is said to be **seasonally adjusted.** We will say a little more about seasonal adjustment in Section 17.5.

Seasonal patterns in a time series constitute one form of regular, oscillatory behavior. In addition, many business and economic time series met in practice appear to exhibit oscillatory, or **cyclical,** patterns unconnected with seasonal behavior. These patterns might, for example, mirror business cycles in the economy at large. They are not necessarily regular, but follow rather smooth patterns of upswings and downswings. To illustrate, in Figure 17.4 we graph the Lydia E. Pinkham sales data,

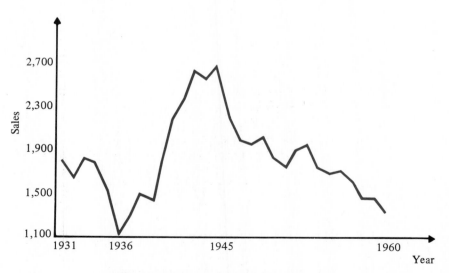

FIGURE 17.4 Sales of Lydia E. Pinkham

tabulated in Example 17.1. We see from the figure a decrease in sales to a trough in 1936, followed by an upswing to a peak in the mid-1940s, and thereafter a fairly steady decline. This kind of cyclical pattern is fairly common in business time series, and it is certainly convenient to describe historical behavior in terms of such cyclic movements. However, we are not suggesting that there is sufficient regularity in such historical patterns to allow the reliable prediction of future peaks and troughs. Indeed, the available evidence suggests that this is not the case.

So far we have discussed three sources of variability in a time series. If the only components of a series were trend, seasonality, and cycle, we would expect the time plot of that series to be very smooth and rather easily projected forward to produce forecasts. However, as the examples in this section indicate, actual data do not behave in this way. In addition to the components already considered, there will be an **irregular** element, induced by the multitude of factors influencing the behavior of any actual series and whose pattern looks rather unpredictable on the basis of past experience. We might think of this component in much the same way as the error term in a regression model. Its presence is clear in the sales data of Figure 17.4. We might easily draw a smooth curve, with a minimum in 1936 and a maximum in the mid-1940s quite close to these data points. However, the actual values will deviate to some extent from this curve in no apparently regular fashion.

The conceptual breakdown of a time series into trend, seasonal, cyclical, and irregular components provides us with a very useful vocabulary for describing its behavior. It is often convenient to go beyond verbal description and to think in terms of a more formal model. Let X_t denote the value of a series at time t. Then we might think of this series as the sum of its components, through the **additive model**

$$X_t = T_t + S_t + C_t + I_t \qquad (17.3.1)$$

where

T_t = Trend component
S_t = Seasonal component
C_t = Cyclical component
I_t = Irregular component

Alternatively, in some circumstances it might be more appropriate to view a series as the *product* of its constituent components, through the **multiplicative model**

$$X_t = T_t S_t C_t I_t \qquad (17.3.2)$$

In fact it is not necessary to restrict attention to just these two models. In some circumstances it may be convenient to treat some factors as additive and others as multiplicative.

Much of the early work in time series analysis concentrated on the isolation of the individual components from a series so that, at any point in time, an observation was expressed as a compound of trend, seasonality, cycle, and a residual irregular element. Often this breakdown was achieved through the use of *moving averages,* which will be discussed in the following two sections. This approach has recently been superceded by more modern approaches. An exception, however, is the problem of seasonal adjustment, which requires the extraction from a series of its seasonal component. In Section 17.7 we will discuss one procedure for estimating the components of a time series and show how it can be used in forecasting.

The more modern approach to time series analysis involves the construction of a formal model, in which the various components are either explicitly or implicitly present, to describe the behavior of a data series. In model building there are two possible treatments of series components. One possibility is to regard them as being *fixed* through time, so that, for example, trend might be represented by a straight line or some other convenient algebraic function. This approach is often valuable in the analysis of physical data, but is far less often appropriate in business and economic applications, where experience suggests that any apparently fixed regularities are all too often illusory on closer examination. To illustrate this point, suppose that we considered the Lydia E. Pinkham data for the years 1936 through 1943 only. We see from Figure 17.4 that, over this period, there appears to be a steady, fixed upward trend. However, had this "trend" been projected forward a few years from 1943, the resulting forecasts of future sales would have been horribly inaccurate. It is only when we look at the picture in future years that we see just how inappropriate a fixed trend model would have been.

For business and economic data there is a preferable treatment of the regular components of a time series. Rather than regarding them as being fixed for all time, it is generally more sensible to think of them as steadily *evolving* through time. Thus, we need not be committed to fixed trend or seasonal patterns, but can allow the possibility that these components change through time. Models of this sort will be considered after we have looked at moving averages.

17.4 MOVING AVERAGES

The irregular component in some time series may be so large that it obscures any underlying regularities, thus rendering difficult any visual interpretation of the time plot. In these circumstances, the actual plot will appear rather jagged, and we may want to *smooth* it to achieve a clearer picture.

This smoothing can be achieved through the method of **moving averages,** which is based on the idea that any large irregular component at any point in time will exert a smaller effect if the observation at that point is averaged with its immediate neighbors. The simplest technique of this kind is called a **simple centered $(2m + 1)$-point moving average.** The idea here is to replace each actual observation X_t by the average of itself and its m neighbors on either side, that is, replace X_t by

$$X_t^* = \frac{1}{2m+1} \sum_{j=-m}^{m} X_{t+j}$$

$$= \frac{X_{t-m} + X_{t-m+1} + \cdots + X_t + \cdots + X_{t+m-1} + X_{t+m}}{2m+1} \tag{17.4.1}$$

The moving average, X_t^*, is said to be *centered* because X_t is the central value of the sum in the numerator of equation 17.4.1.

For example, suppose we set m at 2, so that a 5-point moving average is formed. We then have

$$X_t^* = \frac{X_{t-2} + X_{t-1} + X_t + X_{t+1} + X_{t+2}}{5} \tag{17.4.2}$$

Since the first available observation is X_1, the first moving average of this sort that can be found is

$$X_3^* = \frac{X_1 + X_2 + X_3 + X_4 + X_5}{5}$$

This is, of course, just the average of the first five observations. Hence, for the Lydia E. Pinkham data of Example 17.1, we have for the year 1933,

$$X_3^* = \frac{1,806 + 1,644 + 1,814 + 1,770 + 1,518}{5} = 1,710.4$$

Similarly, X_4^* is the average of the second through sixth observations, and so on. Table 17.4 gives the original and the smoothed series. Notice that, when centered moving averages are computed in this way, we necessarily "lose" m observations at each end of the series. Thus, while the original series runs from 1931 through 1960, the smoothed series is obtained for 1933 through 1958.

Simple Centered ($2m + 1$)-Point Moving Averages

Let $X_1, X_2, \ldots, X_n$ be n observations on a time series of interest. A smoothed series can be obtained through use of a simple centered ($2m + 1$)-point moving average, yielding

$$X_t^* = \frac{1}{2m+1} \sum_{j=-m}^{m} X_{t+j} \quad (t = m + 1, m + 2, \ldots, n - m)$$

TABLE 17.4 Annual sales (X_t) of Lydia E. Pinkham, and simple centered 5-point moving average (X_t^*)

t	X_t	X_t^*	t	X_t	X_t^*
1	1,806		16	2,177	2,232.4
2	1,644		17	1,920	2,125.6
3	1,814	1,710.4	18	1,910	1,955.6
4	1,770	1,569.8	19	1,984	1,858.0
5	1,518	1,494.2	20	1,787	1,847.2
6	1,103	1,426.0	21	1,689	1,844.4
7	1,266	1,356.6	22	1,866	1,784.4
8	1,473	1,406.4	23	1,896	1,753.6
9	1,423	1,618.0	24	1,684	1,747.2
10	1,767	1,832.0	25	1,633	1,687.8
11	2,161	2,057.8	26	1,657	1,586.6
12	2,336	2,276.8	27	1,569	1,527.2
13	2,602	2,450.8	28	1,390	1,458.4
14	2,518	2,454.0	29	1,387	
15	2,637	2,370.8	30	1,289	

The smoothed series is graphed in Figure 17.5. It can be seen, by comparison with Figure 17.4, that the series of moving averages is indeed rather smoother than the original data series. This allows us to see even more clearly the underlying oscillatory behavior in these sales figures.

The kind of moving average discussed in this section is just one of many that might have been used. It is often deemed desirable to use a **weighted average,** in which most weight is given to the central observation, with weights for other values decreasing as their distance from that central observation increases. For example, if

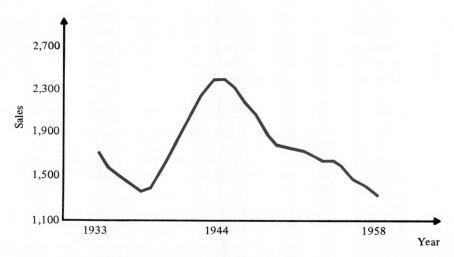

FIGURE 17.5 Simple centered 5-point moving average of sales of Lydia E. Pinkham

five points are to be used, we might employ, instead of the simple average 17.4.2, the weighted average

$$X_t^* = \frac{X_{t-2} + 2X_{t-1} + 4X_t + 2X_{t+1} + X_{t+2}}{10}$$

In any event, the objective in using moving averages remains the smoothing out of the irregular component, in order to allow us to form a clearer picture of the underlying regularities in a time series. The technique is perhaps of most value for descriptive purposes, in the production of graphs such as Figure 17.5.

17.5 EXTRACTION OF THE SEASONAL COMPONENT THROUGH MOVING AVERAGES

We now turn to an important application of moving averages. In Section 17.3 we noted that many business and economic time series contain a strong seasonal component. From some points of view, this component is rather a nuisance, and the analyst often wants to remove it from the series, to obtain a keener appreciation of the behavior through time of other components.

To be specific, suppose that we have a quarterly time series with a seasonal component. Now, suppose we produce a series of moving averages, whose first term is the average of the first four values of the original series, whose second term is the average of the second through fifth values of the original series, and so on. Then each member of the series of moving averages will be constituted from a single observation from each of the four quarters. The series formed in this way should, therefore, be free from strong seasonal patterns.

For the earnings per share data of Burroughs Corporation, given in Table 17.3, the first member of the series of 4-point moving averages is

$$\frac{0.300 + 0.460 + 0.345 + 0.910}{4} = 0.50375$$

while the second member is

$$\frac{0.460 + 0.345 + 0.910 + 0.330}{4} = 0.51125$$

The complete series is set out in the third column of Table 17.5.

Although this series of moving averages is free from seasonality, there remains one difficulty. The location in time of the members of the series of moving averages does not correspond precisely with that of the members of the original series. For instance, the first term is the average of the first four observations in the original series. We might therefore regard it as being centered between the second and third observations, and write it as

$$X_{2.5}^* = \frac{X_1 + X_2 + X_3 + X_4}{4}$$

TABLE 17.5 Actual (X_t) and centered 4-point moving average (X_t^*) earnings per share of Burroughs Corporation

t	EARNINGS (X_t)	4-POINT MOVING AVERAGES	CENTERED 4-POINT MOVING AVERAGES (X_t^*)
1	0.300		
2	0.460		
		0.50375	
3	0.345		0.5075
		0.51125	
4	0.910		0.5219
		0.53250	
5	0.330		0.5444
		0.55625	
6	0.545		0.5725
		0.58875	
7	0.440		0.6094
		0.63000	
8	1.040		0.6469
		0.66375	
9	0.495		0.6769
		0.69000	
10	0.680		0.7206
		0.75125	
11	0.545		0.7581
		0.76500	
12	1.285		0.7888
		0.81250	
13	0.550		0.8269
		0.84125	
14	0.870		0.8781
		0.91500	
15	0.660		0.9200
		0.92500	
16	1.580		0.9400
		0.95500	
17	0.590		0.9763
		0.99750	
18	0.990		1.0163
		1.03500	
19	0.830		1.0375
		1.04000	
20	1.730		1.0475
		1.05500	
21	0.610		1.0663
		1.07750	
22	1.050		1.1163
		1.15500	
23	0.920		1.1663
		1.17750	
24	2.040		1.2000
		1.22250	
25	0.700		1.2400
		1.25750	
26	1.230		1.2925
		1.32750	
27	1.060		1.3425
		1.35750	
28	2.320		1.3800
		1.40250	
29	0.820		1.4263
		1.45000	
30	1.410		1.5013
		1.55250	
31	1.250		
32	2.730		

Similarly, the second term could be written

$$X^*_{3.5} = \frac{X_2 + X_3 + X_4 + X_5}{4}$$

This difficulty is easily overcome. We can center our series of 4-point moving averages by calculating the averages of adjacent pairs. This yields a series whose first value is

$$X^*_3 = \frac{X^*_{2.5} + X^*_{3.5}}{2} = \frac{0.50375 + 0.51125}{2} = 0.5075$$

and which constitutes the centered moving average corresponding to the third observation of the original series. The remainder of the series of centered moving averages is set out in the final column of Table 17.5. Notice that, when moving averages are calculated in this way, we "lose" two observations from each end of the series.

The series of centered moving averages is plotted in Figure 17.6. For purposes of comparison with the original data series, the scale here is the same as in Figure 17.3. Obviously, most of the seasonality has been removed. Moreover, a by-product of using moving averages to eliminate seasonality is that the irregular component is also smoothed. The resulting picture thus allows us to judge the nonseasonal regularities in the data. Figure 17.6 is, of course, dominated by upward trend. Closer examination reveals steady earnings growth in the early part of the series, a central portion of rather slower growth, followed in the last part of the period by a resumption of a pattern similar to the early one.

Moving averages can be used as an aid to the seasonal adjustment of series of any period, as described in the box.

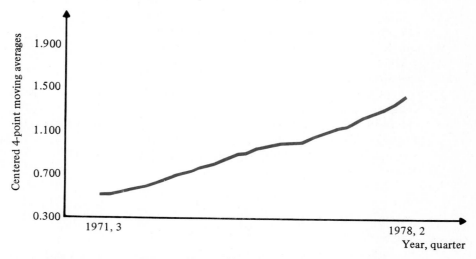

FIGURE 17.6 Centered 4-point moving averages for earnings per share of Burroughs Corporation

---

A Simple Moving Average Procedure for Seasonal Adjustment

Let X_t ($t = 1, 2, \ldots , n$) be a seasonal time series of period s (so that $s = 4$ for quarterly data and $s = 12$ for monthly data). A centered s-point moving average series, X_t^*, is obtained through the following steps, where it is assumed (as is usually the case) that s is even:

(i) Form the s-point moving averages

$$X_{t+0.5}^* = \frac{\sum_{j=-s/2+1}^{s/2} X_{t+j}}{s} \qquad \left(t = \frac{s}{2}, \frac{s}{2} + 1, \ldots , n - \frac{s}{2} \right)$$

(ii) Form the centered s-point moving averages

$$X_t^* = \frac{X_{t-0.5}^* + X_{t+0.5}^*}{2} \qquad \left(t = \frac{s}{2} + 1, \frac{s}{2} + 2, \ldots , n - \frac{s}{2} \right)$$

We have seen that the series of centered s-point moving averages can be a useful tool for gaining descriptive insight into the structure of a time series. Since it is largely free from seasonality and embodies a smoothing of the irregular component, it is well suited for the identification of trend and cycle. This series of moving averages has an additional value. It forms the basis for many practical seasonal adjustment procedures. How the adjustment is carried out depends on a number of factors, including the amount of stability one assumes in the seasonal pattern and whether seasonality is viewed as additive or multiplicative. One approach to a multiplicative model is to take logarithms of the data.

We will discuss here a seasonal adjustment approach that is based on an implicit assumption of a very stable seasonal pattern through time. This is known as the **seasonal index method.** Essentially, the assumption is that for any given month or quarter, in each year, the effect of seasonality is to raise or lower the observation by a constant proportionate amount, compared with what it would have been in the absence of seasonal influences.

To illustrate the seasonal index method, we return to our analysis of the Burroughs earnings data. The seasonally adjusted series is computed in Table 17.6. First, we show the original series, X_t, and the series, X_t^*, of centered 4-point moving averages. These are taken from Table 17.5. As a means of assessing the influence of seasonality, the next step is to express X_t as a percentage of X_t^*. Thus, for example, for the third quarter of 1971, we have

$$100\left(\frac{X_3}{X_3^*}\right) = 100\left(\frac{0.345}{0.5075}\right) = 67.98$$

These percentages are also entered in Table 17.7, where the calculation of the seasonal indices is shown. To assess the effect of seasonality in the first quarter, we find the median of the seven percentages for that quarter. This is the fourth value when they are arranged in ascending order—that is, 60.43. In a similar way, we find the median of X_t as a percentage of X_t^* for each of the other quarters.

TABLE 17.6 Seasonal adjustment of earnings per share of Burroughs Corporation by the seasonal index method

DATE	X_t	X_t^*	$100\left(\dfrac{X_t}{X_t^*}\right)$	SEASONAL INDEX	ADJUSTED SERIES
1971,1	0.300			61.06	0.4913
2	0.460			96.15	0.4784
3	0.345	0.5075	67.98	72.95	0.4729
4	0.910	0.5219	174.36	169.84	0.5358
1972,1	0.330	0.5444	60.62	61.06	0.5405
2	0.545	0.5725	95.20	96.15	0.5668
3	0.440	0.6094	72.20	72.95	0.6032
4	1.040	0.6469	160.77	169.84	0.6123
1973,1	0.495	0.6769	73.13	61.06	0.8107
2	0.680	0.7206	94.37	96.15	0.7072
3	0.545	0.7581	71.89	72.95	0.7471
4	1.285	0.7888	162.91	169.84	0.7566
1974,1	0.550	0.8269	66.51	61.06	0.9008
2	0.870	0.8781	99.08	96.15	0.9048
3	0.660	0.9200	71.74	72.95	0.9047
4	1.580	0.9400	168.09	169.84	0.9303
1975,1	0.590	0.9763	60.43	61.06	0.9663
2	0.990	1.0163	97.41	96.15	1.0296
3	0.830	1.0375	80.00	72.95	1.1378
4	1.730	1.0475	165.16	169.84	1.0186
1976,1	0.610	1.0663	57.21	61.06	0.9990
2	1.050	1.1163	94.06	96.15	1.0920
3	0.920	1.1663	78.88	72.95	1.2611
4	2.040	1.2000	170.00	169.84	1.2011
1977,1	0.700	1.2400	56.45	61.06	1.1464
2	1.230	1.2925	95.16	96.15	1.2793
3	1.060	1.3425	78.96	72.95	1.4531
4	2.320	1.3800	168.12	169.84	1.3660
1978,1	0.820	1.4263	57.49	61.06	1.3429
2	1.410	1.5013	93.92	96.15	1.4665
3	1.250			72.95	1.7135
4	2.730			169.84	1.6074

To obtain seasonal indices, one further minor adjustment is needed. We would like the average of the four seasonal indices to be 100%. However, we see in Table 17.7 that the four medians sum to 395.88. The desired result can be achieved by multiplying each median by (400/395.88). Thus, for the first quarter, we have

$$\text{Seasonal index} = 60.43 \cdot \frac{400}{395.88} = 61.06$$

We interpret this figure as estimating that the effect of seasonality is to lower first quarter earnings to 61.06% of what they would have been in the absence of seasonal factors.

The seasonal indices, taken from the last row of Table 17.7, are entered in the fifth column of Table 17.6 Notice that the same index is used for any particular quarter in every year. Finally, we obtain our seasonally adjusted series as

TABLE 17.7 Calculation of seasonal indices for earnings per share data of Burroughs Corporation

| | QUARTER | | | | |
	1	2	3	4	SUMS
1971			67.98	174.36	
1972	60.62	95.20	72.20	160.77	
1973	73.13	94.37	71.89	162.91	
1974	66.51	99.08	71.74	168.09	
1975	60.43	97.41	80.00	165.16	
1976	57.21	94.06	78.88	170.00	
1977	56.45	95.16	78.96	168.12	
1978	57.49	93.92			
Median	60.43	95.16	72.20	168.09	395.88
Seasonal Index	61.06	96.15	72.95	169.84	400

$$\text{Adjusted value} = \text{Original value} \cdot \frac{100}{\text{Seasonal index}}$$

For example, for the third quarter of 1971, the seasonally adjusted value is

$$0.345 \cdot \frac{100}{72.95} = 0.4729$$

The complete seasonally adjusted series obtained in this way is given in the final column of Table 17.6.

The seasonal index method of seasonal adjustment presented here gives one possible simple attack on the problem. Many important economic time series, such as gross domestic product and its components, employment and unemployment, prices and wages, have strong seasonal components. Generally, data on such quantities are published by government agencies in both unadjusted and adjusted form. Although they are more complex than the method described here, official adjustment procedures are typically based on moving averages. The seasonal adjustment procedure most commonly employed in official United States government publications is the Census X-11 method. It differs from the seasonal index method in allowing for a steadily evolving seasonal pattern through time. It can be shown that, in its additive version, X-11 estimates the seasonal component of a monthly time series, to a close approximation, by

$$S_t = \frac{Z_{t-36} + 2Z_{t-24} + 3Z_{t-12} + 3Z_t + 3Z_{t+12} + 2Z_{t+24} + Z_{t+36}}{15}$$

where

$$Z_t = X_t - X_t^*$$

with X_t the original value of the series at time t, and X_t^* the corresponding centered 12-point moving average. Of course, if such a procedure is used, some special treatment is needed for values toward the end of the time series, as the expression for the seasonal factor will involve values of the time series that have not yet occurred.

A possible way of accomplishing this is to replace unknown future values of a series, in the moving average, by *forecasts* based on the available data. Figure 17.7 shows

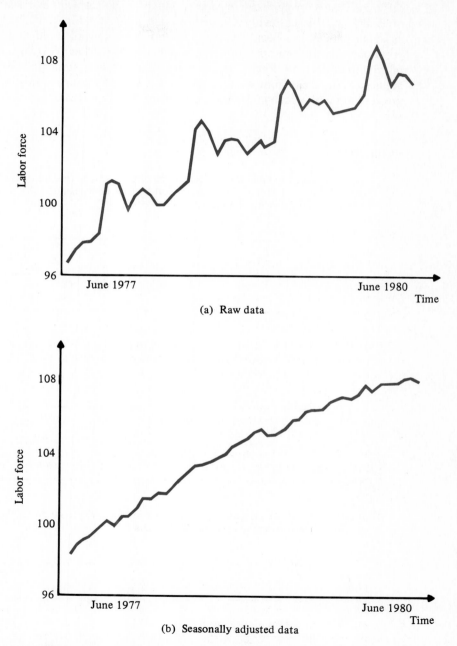

(a) Raw data

(b) Seasonally adjusted data

FIGURE 17.7 Total monthly United States labor force (in millions), 1977 to 1980

the official published monthly figures for the total United States labor force for the 4-year period from 1977 through 1980. In part (a) of the figure the raw data are plotted, revealing a general upward trend throughout the period. However, superimposed on this trend is a recurring seasonal pattern, with pronounced peaks in the summer months of each year. The seasonally adjusted data are graphed in Figure 17.7(b). The marked seasonal pattern has been removed, revealing an upward trend that is rather less pronounced in the latter part of the period than in the earlier. This decrease in the rate of growth of the labor force emerges more clearly from examination of the seasonally adjusted series than from the raw data. The value of seasonal adjustment lies in the greater ease with which underlying movements in a time series can be detected in this way.

EXERCISES

1. The accompanying table shows the volume of transactions (in hundreds of thousands) in shares of General Motors Corporation for the first 12 weeks of 1981. Use the runs test to test this series for randomness.

WEEK	1	2	3	4	5	6
VOLUME	27.1	12.9	12.1	13.6	21.9	17.3
WEEK	7	8	9	10	11	12
VOLUME	11.7	23.8	17.0	21.4	25.0	20.5

2. The table given here shows an index of real average hourly earnings of private nonfarm employees in the United States over the period from 1967 to 1980. Test this series for randomness, using the runs test.

YEAR	1967	1968	1969	1970	1971	1972	1973
EARNINGS	100.0	102.0	103.2	103.9	106.7	110.0	110.1
YEAR	1974	1975	1976	1977	1978	1979	1980
EARNINGS	107.4	107.0	107.3	108.4	108.9	105.5	101.4

3. The table shows the price per share of stock in Mobil Corporation for the first 12 weeks of 1981. Test for randomness, using the runs test.

WEEK	1	2	3	4	5	6
PRICE	$78\frac{3}{8}$	$81\frac{7}{8}$	$77\frac{7}{8}$	$73\frac{1}{8}$	$72\frac{2}{8}$	70
WEEK	7	8	9	10	11	12
PRICE	$69\frac{4}{8}$	$68\frac{2}{8}$	$63\frac{5}{8}$	$65\frac{5}{8}$	$64\frac{4}{8}$	$66\frac{2}{8}$

4. The accompanying table shows the volume of shares (in hundreds of thousands) traded in the stock of Xerox Corporation in the first 12 weeks of 1981. Use the runs test to test this series for randomness.

WEEK	1	2	3	4	5	6
VOLUME	12.0	8.1	8.8	7.0	9.4	7.2
WEEK	7	8	9	10	11	12
VOLUME	6.7	12.2	7.1	11.3	13.3	11.7

5. The table shown here lists sales of Kennecott Corporation over a period of 28 years. Use the large-sample variant of the runs test to test this series for randomness.

YEAR	SALES	YEAR	SALES	YEAR	SALES
1951	451	1961	501	1970	1,133
1952	472	1962	507	1971	1,053
1953	477	1963	505	1972	1,145
1954	424	1964	544	1973	1,395
1955	548	1965	666	1974	1,664
1956	567	1966	740	1975	769
1957	468	1967	489	1976	956
1958	395	1968	725	1977	925
1959	437	1969	1,050	1978	1,886
1960	497				

6. The table shows earnings per share of Cominco Ltd. over a period of 28 years. Use the large-sample variant of the runs test to test this series for randomness.

YEAR	EARNINGS	YEAR	EARNINGS	YEAR	EARNINGS
1951	50.2	1961	20.6	1970	22.4
1952	33.9	1962	21.6	1971	12.3
1953	20.6	1963	27.6	1972	20.9
1954	25.4	1964	36.5	1973	42.8
1955	32.9	1965	49.3	1974	86.3
1956	31.3	1966	45.4	1975	73.6
1957	18.8	1967	35.6	1976	47.7
1958	14.5	1968	30.1	1977	56.6
1959	17.5	1969	25.5	1978	53.1
1960	23.6				

7. Draw a time plot of the Kennecott Corporation sales data of Exercise 5, and comment on the components of the series revealed by this plot.

8. Draw a time plot of the Cominco Ltd. earnings data of Exercise 6, and comment on the components of the series shown by this plot.

9. The accompanying table shows retail sales (in thousands) of domestic new passenger cars in the United States over a period of 3 years. Draw a time plot of these data, and discuss the components of the time series that are revealed.

	1978	1979	1980
January	538	636	588
February	620	666	592
March	869	853	670
April	849	753	541
May	947	787	499
June	931	691	511
July	746	680	542
August	740	698	487
September	654	593	486
October	874	721	664
November	759	600	530
December	637	554	472

10. The table shows quarterly earnings per share of CBS, Incorporated over a period of 8 years. Draw a time plot of these data. What does your graph reveal about the components of this time series?

	QUARTER			
YEAR	1	2	3	4
1971	0.216	0.559	0.657	0.800
1972	0.420	0.710	0.720	1.030
1973	0.600	0.830	0.820	1.070
1974	0.730	1.030	0.900	1.140
1975	0.840	1.190	1.020	1.250
1976	0.970	1.670	1.430	1.680
1977	1.160	1.950	1.570	1.820
1978	1.220	2.140	1.750	2.040

11. Compute a simple centered 3-point moving average series for the hourly earnings index of Exercise 2. Plot the smoothed data and discuss the resulting graph.

12. Compute simple centered 5-point moving averages for the Kennecott sales data of Exercise 5. Draw a time plot of the smoothed series and comment on your results.

13. Compute simple 7-point moving averages of the Cominco earnings data of Exercise 6. Based on a time plot of the smoothed series, what can you say about its regular components?

14. The table shows the percentage of the United States female population, over 16 years of age, that is in the civilian labor force, for the period 1960 to 1980. Compute the series of 5-point moving averages. Draw a time plot of this smoothed series, and comment on its features.

YEAR	PERCENTAGE	YEAR	PERCENTAGE	YEAR	PERCENTAGE
1960	37.8	1967	41.2	1974	45.7
1961	38.1	1968	41.6	1975	46.4
1962	38.0	1969	42.7	1976	47.4
1963	38.3	1970	43.4	1977	48.5
1964	38.7	1971	43.4	1978	50.1
1965	39.3	1972	43.9	1979	51.1
1966	40.3	1973	44.7	1980	51.7

15. Let

$$X_t^* = \frac{1}{2m + 1} \sum_{j=-m}^{m} X_{t+j}$$

be a simple centered $(2m + 1)$-point moving average. Show that

$$X_{t+1}^* = X_t^* + \frac{X_{t+m+1} - X_{t-m}}{2m + 1}$$

How might this result be used in the efficient computation of series of centered moving averages?

16. Compute the series of centered 12-point moving averages for adjusted retail sales of domestic new passenger cars, based on the data of Exercise 9. Draw a time plot of this moving average series, and comment on the resulting picture.

17. Refer to Exercise 10. Use the seasonal index method to obtain a seasonally adjusted earnings per share series of CBS, Incorporated. Plot the seasonally adjusted series and discuss the patterns revealed.

18. The accompanying table shows earnings per share of Central Louisiana Energy Corporation over a period of 7 years.

	QUARTER			
YEAR	1	2	3	4
1972	0.362	0.370	0.621	0.384
1973	0.389	0.389	0.639	0.431
1974	0.411	0.448	0.712	0.584
1975	0.620	0.620	0.891	0.570
1976	0.540	0.690	0.870	0.680
1977	0.780	0.440	0.800	0.780
1978	0.690	0.400	1.030	0.940

(a) Draw a time plot of the data. Does your graph indicate the presence of a strong seasonal component in the series?

(b) Using the seasonal index method, obtain a seasonally adjusted series. Graph this series and comment on its behavior.

19. The table given here lists quarterly product sales over a period of 7 years.

	QUARTER			
YEAR	1	2	3	4
1976	890	475	372	507
1977	961	528	402	367
1978	876	401	307	336
1979	892	448	335	267
1980	738	257	289	311
1981	1,036	601	527	341
1982	1,050	625	429	410

(a) Draw a time plot of these sales data. Does this time series seem to have a strong seasonal component?

(b) Use the seasonal index method to obtain a seasonally adjusted series. Plot this series and discuss its behavior.

20. The table shows the United States Consumer Price Index over a period of 4 years.

	1977	1978	1979	1980
January	175.3	187.2	204.7	233.2
February	177.1	188.4	207.1	236.4
March	178.2	189.8	209.1	239.8
April	179.6	191.5	211.5	242.5
May	180.6	193.3	214.1	244.9
June	181.8	195.3	216.6	247.6
July	182.6	196.7	218.9	247.8
August	183.3	197.8	221.1	249.4
September	184.0	199.3	223.4	251.7
October	184.5	200.9	225.4	253.9
November	185.4	202.0	227.5	256.2
December	186.1	202.9	229.9	258.4

(a) Find the series of centered 12-point moving averages.

(b) Use the seasonal index method to obtain a seasonally adjusted series.

21. (a) Show that the centered s-point moving average series of Section 17.5 can be written

$$X_t^* = \frac{X_{t-(s/2)} + 2(X_{t-(s/2)+1} + \cdots + X_{t+(s/2)-1}) + X_{t+(s/2)}}{2s}$$

(b) Show that

$$X_{t+1}^* = X_t^* + \frac{X_{t+(s/2)+1} + X_{t+(s/2)} - X_{t-(s/2)+1} - X_{t-(s/2)}}{2s}$$

Discuss the computational advantages of this formula in the seasonal adjustment of monthly time series.

17.6 SIMPLE EXPONENTIAL SMOOTHING

In this and the remaining sections of this chapter we turn to the possibility of using current and past observations on a time series to obtain forecasts of future values. This problem, though easily stated, can be tricky to resolve satisfactorily. A vast array of forecasting methods are in common use and, to a great extent, the eventual choice will be problem-specific, depending on the resources and objectives of the analyst and the nature of the available data.

In any event, our aim is to use the available observations, $X_1, X_2, \ldots, X_n$ on a series to predict the unknown future values $X_{n+1}, X_{n+2}, \ldots$. Forecasting is clearly of crucial importance in the business environment so that a rational basis for decision-making is obtained. For example, monthly product sales are predicted as a basis for inventory control policy. Again, forecasts of future corporate earnings are employed in investment decision-making.

In this section we introduce a simple forecasting procedure which is in itself often valuable, and which forms the basis of some more elaborate methods, one of which we will consider in Section 17.7. This method, known as **simple exponential smoothing,** is appropriate when the series to be predicted is nonseasonal and has no consistent upward or downward trend.

In the absence of trend and seasonality, the objective is to estimate the current *level* of the time series. This estimate is then used as the forecast of all future values. Our position, then, is that we are standing at time n, looking back on the series of observations $X_n, X_{n-1}, X_{n-2}, \ldots$, and we want to form an assessment of the current level of the series. As a prelude, we will consider two extreme possibilities. First, we might simply use the most recent observation, X_n, so that the forecast of all future values would be the latest observation. For some important business series, particularly prices in speculative markets, this is about the best that can be done if forecasts are to be based exclusively on past history. However, in many applications where the series contains a substantial irregular component, it would be rash to restrict ourselves to only a single value of the series, which could be subject to severe random fluctuations. Rather, we would want to take into account earlier observations also.

At the opposite extreme, we might use as our estimate of current level the simple average of *all* the observations. A moment's reflection will suggest that often this would be absurd, for in forming the average each value is treated equally. Thus, for example, if future product sales were to be predicted in this way, the same weight would be given to sales many years earlier as to those in the current time period. However, distant experience is surely far less likely to be relevant to future patterns than are the most recent figures.

Simple exponential smoothing allows a compromise between these extremes, providing a forecast based on a *weighted average* of current and past values. In forming this average, most weight is given to the most recent observation, rather less weight to the immediately preceding value, less to the one before that, and so on. One way to accomplish this is to estimate the level at the current time n by $\bar{X}_n$, where

$$\overline{X}_n = (1 - \alpha)X_n + \alpha(1 - \alpha)X_{n-1} + \alpha^2(1 - \alpha)X_{n-2} + \cdots \quad (17.6.1)$$

where α is any number between 0 and 1. For example, if α is fixed at 0.5, the forecast of future observations of the series is

$$\overline{X}_n = 0.5X_n + 0.25X_{n-1} + 0.125X_{n-2} + \cdots$$

so that a weighted average, with declining weights, is applied to current and past observations in computing the forecasts.

By analogy with 17.6.1, the level of the series at any time t is estimated by

$$\overline{X}_t = (1 - \alpha)X_t + \alpha(1 - \alpha)X_{t-1} + \alpha^2(1 - \alpha)X_{t-2} + \cdots \quad (17.6.2)$$

and, similarly, the level at the previous time period, $(t - 1)$, would be estimated by

$$\overline{X}_{t-1} = (1 - \alpha)X_{t-1} + \alpha(1 - \alpha)X_{t-2} + \alpha^2(1 - \alpha)X_{t-3} + \cdots \quad (17.6.3)$$

Multiplying through equation 17.6.3 by α gives

$$\alpha\overline{X}_{t-1} = \alpha(1 - \alpha)X_{t-1} + \alpha^2(1 - \alpha)X_{t-2} + \cdots \quad (17.6.4)$$

Hence, on subtracting 17.6.4 from 17.6.2, we have

$$\overline{X}_t - \alpha\overline{X}_{t-1} = (1 - \alpha)X_t$$

or

$$\overline{X}_t = \alpha\overline{X}_{t-1} + (1 - \alpha)X_t \qquad (0 < \alpha < 1) \qquad (17.6.5)$$

Equation 17.6.5 provides a convenient recursive algorithm for calculating the level estimates. It expresses the level, $\overline{X}_t$, at time t as a weighted average of the previous estimate of level, $\overline{X}_{t-1}$, and the new observation X_t. The weights given to each depend on the choice of α, which is sometimes referred to as the **smoothing constant**.

In order to begin the calculations, we set

$$\overline{X}_1 = X_1$$

and then apply the formula 17.6.5 in turn for $t = 2, 3, \ldots, n$. To illustrate, we will apply this approach to the Lydia E. Pinkham sales data of Example 17.1. For con-

TABLE 17.8 Simple exponential smoothing ($\alpha = 0.4$) of Lydia E. Pinkham sales data

t	X_t	$\bar{X}_t$	t	X_t	$\bar{X}_t$
1	1,806	1,806.0	16	2,177	2,336.5
2	1,644	1,708.8	17	1,920	2,086.6
3	1,814	1,771.9	18	1,910	1,980.6
4	1,770	1,770.8	19	1,984	1,982.6
5	1,518	1,619.1	20	1,787	1,865.2
6	1,103	1,309.4	21	1,689	1,759.5
7	1,266	1,283.4	22	1,866	1,823.4
8	1,473	1,397.2	23	1,896	1,867.0
9	1,423	1,412.7	24	1,684	1,757.2
10	1,767	1,625.3	25	1,633	1,682.7
11	2,161	1,946.7	26	1,657	1,667.3
12	2,336	2,180.3	27	1,569	1,608.3
13	2,602	2,433.3	28	1,390	1,477.3
14	2,518	2,484.1	29	1,387	1,423.1
15	2,637	2,575.8	30	1,289	1,342.6

venience, the data are shown again in the X_t column of Table 17.8. We begin by setting

$$\bar{X}_1 = X_1 = 1,806$$

Our estimates of levels will be based on choosing the smoothing constant

$$\alpha = 0.4$$

so that equation 17.6.5 becomes

$$\bar{X}_t = 0.4\bar{X}_{t-1} + 0.6X_t$$

Hence, we have

$$\bar{X}_2 = 0.4\bar{X}_1 + 0.6X_2$$
$$= (0.4)(1,806) + (0.6)(1,644) = 1,708.8$$

Similarly,

$$\bar{X}_3 = 0.4\bar{X}_2 + 0.6X_3$$
$$= (0.4)(1,708.8) + (0.6)(1,814) = 1,771.9$$

Continuing in this way, we complete the $\bar{X}_t$ column of Table 17.8.

We see from the table that the most recent estimate of level is provided by

$$\bar{X}_n = \bar{X}_{30} = 1,342.6$$

This value is then used as the forecast of sales in all future years. The observed series and these forecasts are graphed in Figure 17.8.

Forecasting Through Simple Exponential Smoothing

Let $X_1, X_2, \ldots , X_n$ be a set of observations on a nonseasonal time series with

---

no consistent upward or downward trend. The **simple exponential smoothing** method of forecasting then proceeds as follows:

(i) Obtain the *smoothed* series, $\bar{X}_t$, as

$$\bar{X}_1 = X_1$$

$$\bar{X}_t = \alpha\bar{X}_{t-1} + (1 - \alpha)X_t \qquad (0 < \alpha < 1; t = 2, 3, \ldots, n)$$

where α is a **smoothing constant** whose value is fixed between 0 and 1.

(ii) Standing at time n, we obtain forecasts of future values, X_{n+h}, of the series by

$$\hat{X}_{n+h} = \bar{X}_n \qquad (h = 1, 2, 3, \ldots)$$

So far we have said little about the choice of the smoothing constant, α, in practical applications of simple exponential smoothing. In practice, this choice may be based on either subjective or objective grounds. One possibility is to rely on experience and judgment. For instance, an analyst who wants to predict product demand may have had considerable experience in working with data on similar product lines, and thus may know which values of the smoothing constant produce accurate forecasts in this area. Visual inspection of a graph of the available data can also be useful in suggesting an appropriate value for the smoothing constant. If the series appears to contain a substantial irregular element, we would not want to give too much weight to the most recent observation alone, as it may not be strongly indicative of what to expect in the future. In line with equation 17.6.5, this would suggest a relatively high value for the smoothing constant. On the other hand, if the series is rather smooth, we would use a lower value for α.

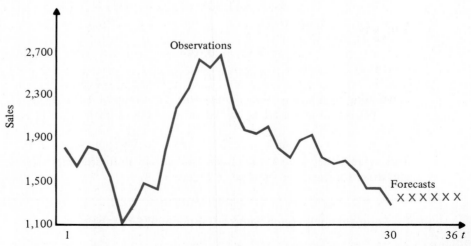

FIGURE 17.8 Sales of Lydia E. Pinkham, and forecasts based on simple exponential smoothing

A more objective approach is to try several different values and see which would have been most successful in predicting historical movements in the time series. We might, for example, compute the smoothed series for values of α of 0.2, 0.4, 0.6, and 0.8. If $\bar{X}_{t-1}$ is the forecast of X_t made at time $(t-1)$, then the error in this forecast will be

$$e_t = X_t - \bar{X}_{t-1}$$

One possibility is to compute, for each trial value of α, the sum of squared forecast errors

$$SS = \sum_{t=2}^{n} e_t^2 = \sum_{t=2}^{n} (X_t - \bar{X}_{t-1})^2$$

That value of α for which this sum of squared forecast errors is smallest will then be used in the prediction of future observations.

Whatever value of the smoothing constant is used, the basic equation 17.6.5 of simple exponential smoothing can be regarded as an **updating mechanism**. At time $(t-1)$, the level of the series is estimated by $\bar{X}_{t-1}$. Then, in the next time period, the new observation X_t is used to update this estimate, so that the new estimate of level is a weighted average of the previous estimate and the new observation.

17.7 THE HOLT-WINTERS EXPONENTIAL SMOOTHING FORECASTING MODEL

Many business forecasting procedures in common use are elaborations of the simple exponential smoothing approach. In this section we will describe one such method, known as the **Holt-Winters model**. Our objective is to allow for trend, and possibly also seasonality, in a time series.

We will begin with the problem of prediction for nonseasonal time series. Here, the objective is to estimate not only the current level of the series but also the trend, where, for this purpose, trend is regarded as the difference between the current and the preceding level.

The observed value of the series at time t will be denoted X_t, while $\bar{X}_t$ will again be used to represent the estimate of level. The trend estimate is represented as T_t. The principle behind the estimation of these two quantities is much the same as in the simple exponential smoothing algorithm. The two estimating equations are

$$\bar{X}_t = A(\bar{X}_{t-1} + T_{t-1}) + (1-A)X_t \qquad (0 < A < 1) \qquad (17.7.1)$$

and

$$T_t = BT_{t-1} + (1-B)(\bar{X}_t - \bar{X}_{t-1}) \qquad (0 < B < 1) \qquad (17.7.2)$$

where A and B are smoothing constants whose values are set between 0 and 1.

TABLE 17.9 Holt-Winters computations for car ownership in Great Britain ($A = 0.2$, $B = 0.2$)

t	X_t	$\bar{X}_t$	T_t	t	X_t	$\bar{X}_t$	T_t
1	0.049			13	0.142	0.1409	0.0135
2	0.051	0.0510	0.0020	14	0.157	0.1565	0.0151
3	0.056	0.0554	0.0039	15	0.169	0.1696	0.0135
4	0.063	0.0623	0.0063	16	0.180	0.1806	0.0115
5	0.071	0.0705	0.0079	17	0.194	0.1936	0.0127
6	0.078	0.0781	0.0076	18	0.202	0.2029	0.0099
7	0.084	0.0843	0.0065	19	0.209	0.2098	0.0075
8	0.091	0.0910	0.0066	20	0.214	0.2147	0.0054
9	0.098	0.0979	0.0069	21	0.223	0.2224	0.0073
10	0.108	0.1074	0.0089	22	0.234	0.2331	0.0100
11	0.116	0.1161	0.0087	23	0.248	0.2470	0.0131
12	0.127	0.1266	0.0102	24	0.251	0.2528	0.0073

As in the case of simple exponential smoothing, equations 17.7.1 and 17.7.2 can be viewed as updating formulas, in which previous estimates are modified in light of a new observation. The estimate of level, $\bar{X}_{t-1}$, made at time $(t - 1)$, taken in conjunction with the trend estimate, T_{t-1}, suggests for time t a level $(\bar{X}_{t-1} + T_{t-1})$. This estimate is modified, in light of the new observation, X_t, to obtain an updated estimate of level, $\bar{X}_t$, in equation 17.7.1.

In the same way, trend at time $(t - 1)$ is estimated as T_{t-1}. However, once the new observation X_t is available, an estimate of trend is suggested as the difference between the two most recent estimates of level. The trend estimate at time t is then the weighted average given by equation 17.7.2.

In order to begin the computations, we set

$$T_2 = X_2 - X_1 \quad \text{and} \quad \bar{X}_2 = X_2$$

Formulas 17.7.1 and 17.7.2 are then applied in turn for $t = 3, 4, \ldots, n$.

We will illustrate these calculations for the data on British car ownership of Table 17.2. These values are set out again in the X_t column of Table 17.9. Our initial estimates of trend and level, in period 2, are

$$T_2 = X_2 - X_1 = 0.051 - 0.049 = 0.002$$

and

$$\bar{X}_2 = X_2 = 0.051$$

We will use as smoothing constants

$$A = 0.2 \quad \text{and} \quad B = 0.2$$

so that equations 17.7.1 and 17.7.2 become

$$\bar{X}_t = 0.2(\bar{X}_{t-1} + T_{t-1}) + 0.8X_t \qquad (17.7.3)$$

and

$$T_t = 0.2T_{t-1} + 0.8(\bar{X}_t - \bar{X}_{t-1}) \qquad (17.7.4)$$

From equation 17.7.3, we have

$$\bar{X}_3 = 0.2(\bar{X}_2 + T_2) + 0.8X_3 = (0.2)(0.051 + 0.002) + (0.8)(0.056) = 0.0554$$

and from equation 17.7.4,

$$T_3 = 0.2T_2 + 0.8(\bar{X}_3 - \bar{X}_2) = (0.2)(0.002) + (0.8)(0.0554 - 0.051) = 0.00392$$

Continuing, we again use equation 17.7.3 to give

$$\bar{X}_4 = 0.2(\bar{X}_3 + T_3) + 0.8X_4$$
$$= (0.2)(0.0554 + 0.00392) + (0.8)(0.063) = 0.062264$$

and equation 17.7.4 to obtain

$$T_4 = 0.2T_3 + 0.8(\bar{X}_4 - \bar{X}_3)$$
$$= (0.2)(0.00392) + (0.8)(0.062264 - 0.0554) = 0.0062752$$

The remaining calculations are carried out in precisely the same way, and the results are given in Table 17.9.

We now consider the use of these level and trend estimates in forecasting future observations. Given a series $X_1, X_2, \ldots, X_n$, the most recent level and trend estimates are $\bar{X}_n$ and T_n, respectively. In the production of forecasts, it is assumed that this latest trend will continue from the most recent level. Thus, the next value of the series is predicted as

$$\hat{X}_{n+1} = \bar{X}_n + T_n$$

and the following one as

$$\hat{X}_{n+2} = \bar{X}_n + 2T_n$$

In general, standing at time n, and looking h time periods into the future, we predict the value of X_{n+h} to be

$$\hat{X}_{n+h} = \bar{X}_n + hT_n$$

From Table 17.9, the most recent level and trend estimates are

$$\bar{X}_{24} = 0.2528 \qquad \text{and} \qquad T_{24} = 0.0073$$

Thus, the forecast for the next year's car ownership is

$$\hat{X}_{25} = 0.2528 + 0.0073 = 0.2601$$

Similarly, the predictions 2 and 3 years ahead are

$$\hat{X}_{26} = 0.2528 + 2(0.0073) = 0.2674$$

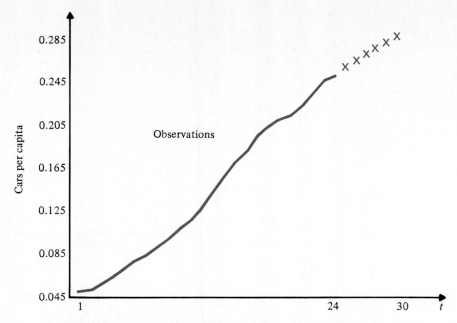

FIGURE 17.9 Car ownership in Great Britain, with forecasts based on the Holt-Winters method

and

$$\hat{X}_{27} = 0.2528 + 3(0.0073) = 0.2747$$

Figure 17.9 shows the data series and the first few forecasts.

Forecasting Through the Holt-Winters Method: Nonseasonal Series

Let $X_1, X_2, \ldots, X_n$ be a set of observations on a nonseasonal time series. The Holt-Winters method of forecasting proceeds as follows:

(i) Obtain estimates of level $\bar{X}_t$ and trend T_t as

$$\bar{X}_2 = X_2; \qquad T_2 = X_2 - X_1$$

$$\bar{X}_t = A(\bar{X}_{t-1} + T_{t-1}) + (1 - A)X_t \qquad (0 < A < 1; t = 3, 4, \ldots, n)$$

$$T_t = BT_{t-1} + (1 - B)(\bar{X}_t - \bar{X}_{t-1}) \qquad (0 < B < 1; t = 3, 4, \ldots, n)$$

where A and B are smoothing constants whose values are fixed between 0 and 1.

(ii) Standing at time n, we obtain forecasts of future values, X_{n+h}, of the series by

$$\hat{X}_{n+h} = \bar{X}_n + hT_n \qquad (h = 1, 2, 3, \ldots)$$

FORECASTING SEASONAL TIME SERIES

We now turn to an extension of the Holt-Winters method, allowing for seasonality. In most practical applications, the seasonal factor is taken to be multiplicative, so that, for example, in dealing with monthly sales figures, we might think of January sales in terms of a proportion of average monthly sales. As before, the trend component is assumed to be additive.

As for the nonseasonal case, we will use X_t, $\bar{X}_t$, and T_t to denote, respectively, the observed value and level and trend estimates at time t. The seasonal factor will be denoted F_t, so that, if the time series contains s periods per year, the seasonal factor for the corresponding period in the previous year will be F_{t-s}.

In the Holt-Winters model, the estimates of level, trend, and the seasonal factor are updated by the following three equations:

$$\bar{X}_t = A(\bar{X}_{t-1} + T_{t-1}) + (1 - A)\left(\frac{X_t}{F_{t-s}}\right) \qquad (0 < A < 1) \qquad (17.7.5)$$

$$T_t = BT_{t-1} + (1 - B)(\bar{X}_t - \bar{X}_{t-1}) \qquad (0 < B < 1) \qquad (17.7.6)$$

$$F_t = CF_{t-s} + (1 - C)\left(\frac{X_t}{\bar{X}_t}\right) \qquad (0 < C < 1) \qquad (17.7.7)$$

where A, B, and C are smoothing constants whose values are set between 0 and 1.

In equation 17.7.5, the term $(\bar{X}_{t-1} + T_{t-1})$ represents an estimate of the level at time t, formed one time period earlier. This estimate is then updated when the new observation X_t becomes available. However, here it is necessary to remove the influence of seasonality from that observation by deflating it by the latest available estimate, F_{t-s}, of the seasonal factor for that period. The updating equation for trend, 17.7.6, is identical to that used previously, 17.7.2.

Finally, the seasonal factor is estimated by equation 17.7.7. The most recent estimate of the factor, available from the previous year, is F_{t-s}. However, dividing the new observation X_t by the level estimate $\bar{X}_t$ suggests a seasonal factor $X_t/\bar{X}_t$. The new estimate of the seasonal factor is then a weighted average of these two quantities.

Again, we require some preliminary values to begin the computations. As a starting point, we use the first 3 years of data to obtain centered s-point moving averages, as described in Section 17.5.

To illustrate the computations, we will employ the series on earnings per share of Burroughs Corporations, given in Table 17.5. These data are partially reproduced in the second column of Table 17.10. The final column of that table shows the centered 4-point moving averages, taken from Table 17.5. These are our preliminary estimates of level.

In particular, we will require from this table the level estimate for period 10:

$$\bar{X}_{10} = 0.7206$$

TABLE 17.10 Initial values for Holt-Winters forecasts of earnings per share of Burroughs Corporation

t	X_t	$\bar{X}_t$
1	0.300	
2	0.460	
3	0.345	0.5075
4	0.910	0.5219
5	0.330	0.5444
6	0.545	0.5725
7	0.440	0.6094
8	1.040	0.6469
9	0.495	0.6769
10	0.680	0.7206
11	0.545	
12	1.285	

The trend in period 10 can be estimated as the difference in levels between periods 10 and 9; that is,

$$T_{10} = \bar{X}_{10} - \bar{X}_9 = 0.7206 - 0.6769 = 0.0437$$

Now, Table 17.10 also provides us with two estimates of the seasonal factor in each of the four quarters. For example, for the third quarter of the year, we have from period 3, the factor $0.345/0.5075$, and from period 7, the factor $0.440/0.6094$. We take, as our initial estimate of the seasonal factor in the third quarter, the average of these two estimates; that is,

$$F_7 = \frac{1}{2}\left(\frac{0.345}{0.5075} + \frac{0.440}{0.6094}\right) = 0.701$$

Similarly, for the other three quarters, we have

$$F_8 = \frac{1}{2}\left(\frac{0.910}{0.5219} + \frac{1.040}{0.6469}\right) = 1.676$$

$$F_9 = \frac{1}{2}\left(\frac{0.330}{0.5444} + \frac{0.495}{0.6769}\right) = 0.669$$

$$F_{10} = \frac{1}{2}\left(\frac{0.545}{0.5725} + \frac{0.680}{0.7206}\right) = 0.948$$

Given these initial estimates, the remaining values can be computed. We will use the smoothing constants

$$A = 0.5; \quad B = 0.5; \quad C = 0.3$$

Given these values, equations 17.7.5 to 17.7.7 become, for our quarterly model,

$$\bar{X}_t = 0.5(\bar{X}_{t-1} + T_{t-1}) + 0.5\left(\frac{X_t}{F_{t-4}}\right) \qquad (17.7.8)$$

$$T_t = 0.5T_{t-1} + 0.5(\bar{X}_t - \bar{X}_{t-1}) \qquad (17.7.9)$$

$$F_t = 0.3F_{t-4} + 0.7\left(\frac{X_t}{\bar{X}_t}\right) \qquad (17.7.10)$$

The earnings series, X_t, is set out in the second column of Table 17.11.

Formulas 17.7.8 to 17.7.10 are now applied, in turn, for $t = 11, 12, \ldots$. Thus, for period 11, we have from 17.7.8,

$$\bar{X}_{11} = 0.5(\bar{X}_{10} + T_{10}) + 0.5\left(\frac{X_{11}}{F_7}\right)$$

$$= 0.5(0.7206 + 0.0437) + 0.5\left(\frac{0.545}{0.701}\right) = 0.7709$$

TABLE 17.11 Holt-Winters computations for earnings of Burroughs Corporation ($A = 0.5$, $B = 0.5$, $C = 0.3$)

t	X_t	$\bar{X}_t$	T_t	F_t
1	0.300			
2	0.460			
3	0.345			
4	0.910			
5	0.330			
6	0.545			
7	0.440			0.701
8	1.040			1.676
9	0.495			0.669
10	0.680	0.7206	0.0437	0.948
11	0.545	0.7709	0.0470	0.705
12	1.285	0.7923	0.0342	1.638
13	0.550	0.8243	0.0331	0.668
14	0.870	0.8876	0.0482	0.971
15	0.660	0.9359	0.0483	0.705
16	1.580	0.9743	0.0433	1.627
17	0.590	0.9506	0.0098	0.635
18	0.990	0.9902	0.0247	0.991
19	0.830	1.0960	0.0652	0.742
20	1.730	1.1124	0.0408	1.577
21	0.610	1.0571	−0.0072	0.594
22	1.050	1.0547	−0.0048	0.994
23	0.920	1.1451	0.0428	0.785
24	2.040	1.2409	0.0693	1.624
25	0.700	1.2440	0.0362	0.572
26	1.230	1.2587	0.0254	0.982
27	1.060	1.3173	0.0420	0.799
28	2.320	1.3941	0.0594	1.652
29	0.820	1.4433	0.0543	0.569
30	1.410	1.4665	0.0388	0.968
31	1.250	1.5351	0.0537	0.810
32	2.730	1.6206	0.0696	1.675

Next, using 17.7.9, we have for the trend factor

$$T_{11} = 0.5T_{10} + 0.5(\bar{X}_{11} - \bar{X}_{10})$$
$$= (0.5)(0.0437) + 0.5(0.7709 - 0.7206) = 0.0470$$

Finally, from equation 17.7.10, the seasonal factor is estimated as

$$F_{11} = 0.3F_7 + 0.7\left(\frac{X_{11}}{\bar{X}_{11}}\right)$$
$$= (0.3)(0.701) + 0.7\left(\frac{0.545}{0.7709}\right) = 0.705$$

Continuing in this way, we can obtain the corresponding estimates for period 12. Thus, from 17.7.8, we have the level estimate

$$\bar{X}_{12} = 0.5(\bar{X}_{11} + T_{11}) + 0.5\left(\frac{X_{12}}{F_8}\right)$$
$$= 0.5(0.7709 + 0.0470) + 0.5\left(\frac{1.285}{1.676}\right) = 0.7923$$

From 17.7.9, the trend estimate is

$$T_{12} = 0.5T_{11} + 0.5(\bar{X}_{12} - \bar{X}_{11})$$
$$= (0.5)(0.0470) + 0.5(0.7923 - 0.7709) = 0.0342$$

and from 17.7.10, the seasonal factor is estimated by

$$F_{12} = 0.3F_8 + 0.7\left(\frac{X_{12}}{\bar{X}_{12}}\right)$$
$$= (0.3)(1.676) + 0.7\left(\frac{1.285}{0.7923}\right) = 1.638$$

The remaining estimates of level, trend, and the seasonal factors are obtained in precisely the same fashion, the results being set out in the final three columns of Table 17.11.

The reader will have formed the impression that the arithmetic burden of carrying out these computations is rather tedious. However, the recursive nature of the updating equations 17.7.5 to 17.7.7, in which the same three simple algebraic manipulations are undertaken at each time period, renders them particularly efficient as the basis of an algorithm to be programmed for an electronic computer. With this aid, the arithmetic can be performed very speedily. Moreover, a further advantage for this purpose is that very little information needs to be stored. Once the calculations have begun, it is not necessary to retain all past observations of the time series. All that is necessary is to retain the most recent estimates of level and trend, and of the seasonal factors for each period of the last year. Thus, for quarterly data, standing at time t, we need retain only the level and trend estimates $\bar{X}_t$ and T_t, and the estimated seasonal factors F_t, F_{t-1}, F_{t-2}, and F_{t-3}. These are then updated when the next observation, X_{t+1}, becomes available. The computational efficiency of this, and other exponential smoothing forecasting methods, makes it particularly useful in routine sales forecasting for inventory control, where predicted sales for a large number of product lines are required on a regular basis.

Having computed estimates of level, trend, and the seasonal factors, we can then exploit these in the production of forecasts of future values of the series. For our earnings example, we need for this purpose the latest estimates of level and trend

$$\bar{X}_{32} = 1.6206 \quad \text{and} \quad T_{32} = 0.0696$$

and the four most recent estimates of the seasonal factors

$$F_{29} = 0.569; \quad F_{30} = 0.968; \quad F_{31} = 0.810; \quad F_{32} = 1.675$$

We now consider the prediction of the next member of the series, X_{33}. To begin, we take our level estimate, and add it to the estimate of the trend. Next, we must take into account that observation 33 occurs in the first quarter of the year, so that our result must be multiplied by a seasonal factor for the first quarter, the latest available estimate of which is F_{29}. Hence, X_{33} is predicted by

$$\hat{X}_{33} = (\bar{X}_{32} + T_{32})F_{29} = (1.6206 + 0.0696)0.569 = 0.9617$$

Now we consider the prediction of the next value, X_{34}. Since this is two periods ahead of the base from which forecasts are calculated, we add twice the trend estimate to the level estimate. Finally, this must be multiplied by F_{30}, the latest estimate of the seasonal factor in the second quarter of the year. Our forecast then is

$$\hat{X}_{34} = (\bar{X}_{32} + 2T_{32})F_{30} = [1.6206 + 2(0.0696)]0.968 = 1.7035$$

We can calculate forecasts as far ahead as needed by continuing in this fashion. Figure 17.10 shows the actual data, together with predicted earnings for the next 2 years. These forecasts seem to reproduce rather well the recent trend and seasonal patterns in the series.

Forecasting Through the Holt-Winters Method: Seasonal Series

Let $X_1, X_2, \ldots, X_n$ be a set of observations on a seasonal time series of period s (so that $s = 4$ for quarterly data and $s = 12$ for monthly data). The **Holt-Winters** method of forecasting for such series proceeds as follows:

(i) First we require initial estimates of level, trend, and seasonality. These can be obtained through the method of moving averages. Set

$$\bar{X}_t = \frac{X_{t-(s/2)} + 2(X_{t-(s/2)+1} + \cdots + X_{t+(s/2)-1}) + X_{t+(s/2)}}{2s}$$

for $t = (s/2) + 1, (s/2) + 2, \ldots, (5s/2)$. The estimate $\bar{X}_{5s/2}$ provides the first needed estimate of level. Trend in this period is estimated by

$$T_{5s/2} = \bar{X}_{5s/2} - \bar{X}_{(5s/2)-1}$$

Initial estimates of the s seasonal factors are then provided by

$$F_{(5s/2)-j} = \frac{1}{2}\left(\frac{X_{(5s/2)-j}}{\bar{X}_{(5s/2)-j}} + \frac{X_{(3s/2)-j}}{\bar{X}_{(3s/2)-j}}\right) \qquad (j = 0, 1, \ldots, s - 1)$$

(ii) Beginning at period $[(5s/2) + 1]$ we then apply in turn the updating equations

$$\bar{X}_t = A(\bar{X}_{t-1} + T_{t-1}) + (1 - A)\left(\frac{X_t}{F_{t-s}}\right) \qquad (0 < A < 1)$$

$$T_t = BT_{t-1} + (1 - B)(\bar{X}_t - \bar{X}_{t-1}) \qquad (0 < B < 1)$$

$$F_t = CF_{t-s} + (1 - C)\left(\frac{X_t}{\bar{X}_t}\right) \qquad (0 < C < 1)$$

where A, B, and C are smoothing constants, for $t = (5s/2) + 1, \ldots, n$.

(iii) Standing at time n, we compute forecasts of future values, X_{n+h}, of the series by

$$\hat{X}_{n+h} = (\bar{X}_n + hT_n)F_{n+h-s} \qquad (h = 1, 2, \ldots, s)$$

$$= (\bar{X}_n + hT_n)F_{n+h-2s} \qquad (h = s + 1, s + 2, \ldots, 2s)$$

and so on.

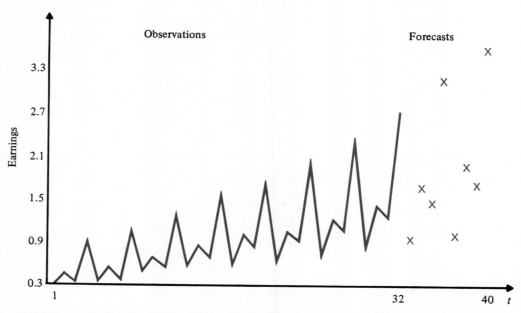

FIGURE 17.10 Earnings of Burroughs Corporation, with forecasts based on the Holt-Winters method for seasonal series

The actual forecasts obtained through the Holt-Winters approach will depend on the specific values chosen for the smoothing constants. As in our discussion of exponential smoothing in Section 17.6, this choice could be based on either subjective or objective criteria. The analyst's experience with similar data sets might suggest suitable values of the smoothing constants. Alternatively, several sets of possible values can be tried on the available data, and the set that would have yielded the best forecasts can then be retained for future use. The latter approach involves a heavier computational burden, but provides the safest course of action when attacking an unfamiliar problem.

EXERCISES

22. Based on the data of Exercise 2, use the method of simple exponential smoothing to derive predictions of the index of real average hourly earnings of private nonfarm employees in the United States over the next 4 years. Employ a smoothing constant of $\alpha = 0.4$. Plot the observations and forecasts.

23. Using the data of Exercise 3 on the price per share of stock in Mobil Corporation, use the method of simple exponential smoothing, with a smoothing constant of $\alpha = 0.2$, to predict price per share over the next 3 weeks.

24. Use the method of simple exponential smoothing, with a smoothing constant of $\alpha = 0.3$, to obtain forecasts of earnings per share of Cominco Ltd. for the next 5 years, based on the data of Exercise 6.

25. The accompanying table shows earnings per share of Olympia Brewing over a period of 20 years.

YEAR	EARNINGS	YEAR	EARNINGS	YEAR	EARNINGS
1959	3.11	1966	6.42	1973	3.54
1960	3.15	1967	7.01	1974	1.65
1961	3.63	1968	6.37	1975	2.15
1962	3.62	1969	5.82	1976	6.09
1963	3.66	1970	4.98	1977	5.95
1964	5.31	1971	3.43	1978	6.26
1965	6.14	1972	3.40		

(a) Using smoothing constants of $\alpha = 0.2, 0.4, 0.6,$ and 0.8, find forecasts based on simple exponential smoothing.

(b) Which of these forecasts would you choose to use?

26. (a) If forecasts are based on simple exponential smoothing, with $\bar{X}_t$ denoting the smoothed value of the series at time t, show that the error made in forecasting X_t, standing at time $(t - 1)$, can be written

$$e_t = X_t - \bar{X}_{t-1}$$

(b) Hence, show that we can write

$$\bar{X}_t = X_t - \alpha e_t$$

from which we see that the most recent observation and the most recent forecast error are used to compute the next forecast.

27. Suppose that, in the simple exponential smoothing method, the smoothing constant α is set equal to 1. What forecasts will result?

28. Comment on the following statement: "We know that all business and economic time series exhibit variability through time. Yet, if simple exponential smoothing is used, the same forecast results for all future values of the time series. Since we know that all future values will not be the same, this is absurd."

29. Use the Holt-Winters procedure, with smoothing constants $A = 0.2$ and $B = 0.6$, to predict the next 5 years of sales of Kennecott Corporation, based on the data of Exercise 5. Plot the original series and your forecasts.

30. The table shows capital expenditures of Scott Paper Company over a period of 20 years. Obtain forecasts of expenditures over the next 6 years, using the Holt-Winters method with smoothing constant $A = 0.3$ and $B = 0.7$. Graph the data and your forecasts.

YEAR	EXPENDITURES	YEAR	EXPENDITURES	YEAR	EXPENDITURES
1959	17	1966	68	1973	80
1960	37	1967	75	1974	147
1961	26	1968	50	1975	210
1962	19	1969	68	1976	196
1963	22	1970	72	1977	149
1964	66	1971	48	1978	134
1965	61	1972	47		

31. The table given here shows the seasonally adjusted United States Consumer Price Index over a period of 24 months. Obtain forecasts for the next 6 months, based on the Holt-Winters procedure with smoothing constants $A = 0.2$ and $B = 0.2$. Plot the data and the forecasts.

	1979	1980
January	205.4	233.8
February	207.7	237.0
March	209.8	240.4
April	211.8	242.6
May	214.0	244.7
June	216.2	247.1
July	218.5	247.1
August	220.7	248.8
September	223.3	251.3
October	225.5	253.8
November	227.8	256.3
December	230.6	259.0

32. The accompanying table shows the percentage of the total United States labor force that was female over the period 1960 to 1980. Obtain forecasts for the next 4 years, using the Holt-Winters method with smoothing constants $A = 0.4$ and $B = 0.4$. Graph the data and your forecasts.

YEAR	PERCENTAGE	YEAR	PERCENTAGE	YEAR	PERCENTAGE
1960	32.25	1967	35.14	1974	38.49
1961	32.64	1968	35.54	1975	39.12
1962	32.74	1969	36.26	1976	39.74
1963	33.17	1970	36.73	1977	40.25
1964	33.59	1971	36.96	1978	40.96
1965	33.98	1972	37.44	1979	41.45
1966	34.64	1973	37.96	1980	41.87

33. The table shows dividends per share for Cheesebrough-Pond's, Inc., over a period of 10 years. Obtain forecasts for the next 2 years, using the Holt-Winters method with $A = 0.6$ and $B = 0.6$.

YEAR	DIVIDEND	YEAR	DIVIDEND
1972	0.54	1977	0.84
1973	0.55	1978	0.94
1974	0.62	1979	1.08
1975	0.68	1980	1.20
1976	0.76	1981	1.52

34. The table given here shows percentage profit margins of Alabama-Tennessee Natural Gas over a period of 11 years. Obtain forecasts for the next 2 years, using the Holt-Winters method with smoothing constants $A = 0.6$ and $B = 0.6$.

YEAR	PROFIT MARGIN	YEAR	PROFIT MARGIN
1970	8.4	1976	7.7
1971	7.4	1977	7.1
1972	7.4	1978	8.5
1973	7.2	1979	7.0
1974	6.3	1980	5.7
1975	7.9		

35. Use the Holt-Winters seasonal method to obtain forecasts of earnings per share up to 8 quarters ahead for CBS, Incorporated, based on the data of Exercise 10. Employ the smoothing constants $A = 0.4$, $B = 0.4$, and $C = 0.4$. Plot the data and the forecasts.

36. Using the Holt-Winters seasonal method, with smoothing constants $A = 0.5$, $B = 0.4$, and $C = 0.3$, obtain forecasts up to 8 quarters ahead of earnings per share of Central Louisiana Energy Corporation, based on the data of Exercise 18. Draw a time plot showing the data and these forecasts.

37. Based on the Holt-Winters seasonal approach, with smoothing constants $A = 0.6$, $B = 0.5$, and $C = 0.4$, obtain sales forecasts up to 8 quarters ahead from the data of Exercise 19. Plot the data and forecasts.

17.8 AUTOREGRESSIVE MODELS

A rather different approach to time series forecasting involves using the available data to construct a **model** that might have generated the series of interest. In this section we will consider a very useful class of such models, while an important broader class will be briefly discussed in the following section.

We have already introduced, in the context of regression models in Section 14.8, the simplest of the models that will be the focus of our attention in this section. Essentially, the idea is to regard a time series as a series of random variables. For practical purposes, we might often be prepared to assume that these random variables all have the same means and variances. However, it would be rash, to say the least, to assume that they were independent of one another. Consider, for example, a series of annual values of product sales. We might suspect that the level of sales in the current period would be related to the levels in the immediately preceding years. Thus, we might expect to find a pattern of correlation through time in our series. Correlation patterns of this kind are sometimes referred to as **autocorrelation.**

Now, in principle, any number of autocorrelation patterns are possible. However, some are considerably more likely to arise in practice than others. A particularly attractive possibility arises when we think of the case of a fairly strong correlation between adjacent observations in time, a less strong correlation between observations two time periods apart, a weaker correlation yet between values three time periods apart, and so on. A very simple autocorrelation pattern of this sort arises when the correlation between adjacent values is some number, say ϕ_1, that between values two time periods apart is ϕ_1^2, that between values three time periods apart is ϕ_1^3, and so on. Thus, if we let X_t denote the value of the series at time t, we have under this model of autocorrelation

$$\text{Corr}(X_t, X_{t-j}) = \phi_1^j \qquad (j = 1, 2, 3, \ldots) \qquad (17.8.1)$$

It can be shown that a model of the time series giving rise to the autocorrelation structure 17.8.1 is

$$X_t = \gamma + \phi_1 X_{t-1} + a_t \qquad (17.8.2)$$

where γ and ϕ_1 are fixed parameters, and the random variables a_t have means 0 and fixed variances for all t, and are not correlated with one another. In 17.8.2 the purpose of the parameter γ is to allow for the possibility that the series X_t has some mean other than 0. Otherwise, this model is the one we used in Section 14.8 to represent autocorrelation in the error terms of a regression equation. It is called a *first-order autoregressive model.*

The first-order autoregressive model 17.8.2 expresses the current value, X_t, of a series in terms of the previous value, X_{t-1}, and a nonautocorrelated random variable a_t. Since the random variable a_t is not autocorrelated, it is unpredictable. It therefore follows that, for series generated by the first-order autoregressive model, forecasts of future values depend only on the most recent value of the series. However, in many applications, we would want to use more than this one observation as a basis for forecasting. An obvious extension of the model 17.8.2 would be to make the current value of the series dependent on the *two* most recent observations. Thus, we could use a model

$$X_t = \gamma + \phi_1 X_{t-1} + \phi_2 X_{t-2} + a_t$$

where γ, ϕ_1, and ϕ_2 are fixed parameters. This is called a **second-order autoregressive model.**

More generally, for any positive integer p, the current value of the series can be made (linearly) dependent on the p previous values, through the **autoregressive model of order p:**

$$X_t = \gamma + \phi_1 X_{t-1} + \phi_2 X_{t-2} + \cdots + \phi_p X_{t-p} + a_t \qquad (17.8.3)$$

where γ, ϕ_1, ϕ_2, . . . , ϕ_p are fixed parameters. Equation 17.8.3 depicts the general autoregressive model. In the remainder of this section we will consider the fitting of such models, and their exploitation in forecasting future values.

Suppose that we have a series of observations, $X_1, X_2, \ldots, X_n$. We want to use these to estimate the unknown parameters γ, ϕ_1, ϕ_2, . . . , ϕ_p of the autoregressive model 17.8.3. This can be done through the method of *least squares*. The parameter estimates are taken as those values of γ, ϕ_1, ϕ_2, . . . , ϕ_p for which the sum of squared discrepancies

$$SS = \sum_{t=p+1}^{n} (X_t - \gamma - \phi_1 X_{t-1} - \cdots - \phi_p X_{t-p})^2$$

is smallest. Hence, estimation can be carried out using a multiple regression program, through the techniques discussed in Chapter 13.

For the data on sales of Lydia E. Pinkham, given in Example 17.1, autoregressive models of orders up to four were estimated by least squares. The fitted first-order model was

$$X_t = 193.27 + 0.883 X_{t-1} + a_t$$
$$(188.92) \quad (0.097) \qquad\qquad\qquad (17.8.4)$$

For the second-order model we obtained

$$X_t = 313.68 + 1.180X_{t-1} - 0.358X_{t-2} + a_t \qquad (17.8.5)$$
$$(192.56) \quad (0.187) \qquad (0.191)$$

The estimated third- and fourth-order models were, respectively,

$$X_t = 322.29 + 1.188X_{t-1} - 0.317X_{t-2} - 0.057X_{t-3} + a_t \qquad (17.8.6)$$
$$(215.72) \quad (0.206) \qquad (0.308) \qquad (0.209)$$

and

$$X_t = 446.22 + 1.194X_{t-1} - 0.439X_{t-2} + 0.286X_{t-3} - 0.291X_{t-4} \qquad (17.8.7)$$
$$(232.77) \quad (0.211) \qquad (0.324) \qquad (0.317) \qquad (0.210)$$

In equations 17.8.4 to 17.8.7, the figures in parentheses beneath the coefficient estimates are the corresponding estimated standard errors.

Given access to a multiple regression computer program, these autoregressive models can be fitted quickly and inexpensively. Table 17.12 shows part of the output for the second-order autoregressive model 17.8.5.

If an autoregressive model is to be used to generate forecasts of future values of a time series, it is necessary to fix a value for p, the order of the autoregression. In making this choice, two considerations must be balanced. We want to choose the order sufficiently large to account for all the important autocorrelation behavior of the series. On the other hand, too large a value of p will lead to a model with irrelevant parameters, and consequently inefficient estimation of those parameters that are important.

One possibility is to fix the value of p arbitrarily, perhaps on the basis of past experience with similar data sets. An alternative approach is to set some maximal order, K, of the autoregression, and fit in turn models of order $p = K$, $(K - 1)$, $(K - 2)$, ... For each value of p, the null hypothesis that the final autoregressive parameter, ϕ_p, of the model is 0 is tested against a two-sided alternative. The pro-

TABLE 17.12 Part of SAS program output for second-order autoregressive model fitted to Lydia E. Pinkham data

PARAMETER	ESTIMATE	T FOR H0: PARAMETER = 0	STD. ERROR OF ESTIMATE
INTERCEPT	313.68		
XT − 1	1.180	6.31	0.187
XT − 2	−0.358	−1.87	0.191

cedure terminates when we find a value p for which this null hypothesis is not rejected. Our aim then is to test the null hypothesis

$$H_0: \quad \phi_p = 0$$

against the alternative

$$H_1: \quad \phi_p \neq 0$$

The test is based on the fact that, to a good approximation, the parameter estimator divided by its estimated standard error follows a standard normal distribution when the null hypothesis is true. The decision rule, then, is

$$\text{Reject } H_0 \text{ if } \quad \frac{\hat{\phi}_p}{s_p} < -z_{\alpha/2} \text{ or } \frac{\hat{\phi}_p}{s_p} > z_{\alpha/2}$$

where α is the significance level of the test, $\hat{\phi}_p$ and s_p are the parameter estimate and its standard error, and $z_{\alpha/2}$ is that number for which

$$P(Z > z_{\alpha/2}) = \frac{\alpha}{2}$$

where Z is a standard normal random variable.

We will apply this approach to the Lydia E. Pinkham data, using a 10% significance level for our tests, so that, from Table 3 of the Appendix,

$$z_{a/2} = z_{.05} = 1.645$$

We will fix at 4 the maximum order of autoregression contemplated. Beginning with the fourth-order fitted model 17.8.7 we find

$$\frac{\hat{\phi}_4}{s_4} = \frac{-0.291}{0.210} = -1.386$$

so that the null hypothesis that $\phi_4 = 0$ is not rejected. Turning to the estimated third-order model 17.8.6, we have

$$\frac{\hat{\phi}_3}{s_3} = \frac{-0.057}{0.209} = -0.273$$

Hence, the null hypothesis that $\phi_3 = 0$ is not rejected in this model. Next, consider the estimated second-order autoregressive model 17.8.5. There we find

$$\frac{\hat{\phi}_2}{s_2} = \frac{-0.358}{0.191} = -1.874$$

For this model, then, the null hypothesis that $\phi_2 = 0$ is rejected by a test of significance level 10%. Accordingly, we will proceed with the second-order model.

Having obtained an appropriate estimated autoregressive model, it is a relatively straightforward matter to compute forecasts of future values of a time series. We will illustrate the procedure by forecasting future sales from the Lydia E. Pinkham data, using the estimated second-order autoregressive model 17.8.5. The last two values in this series were

$$X_{29} = 1,387 \quad \text{and} \quad X_{30} = 1,289$$

We now want to predict the next value, X_{31}. Setting $t = 31$ in 17.8.5 gives

$$X_{31} = 313.68 + 1.180X_{30} - 0.358X_{29} + a_{31}$$

Now, a_{31} is simply a random variable with mean 0, uncorrelated with anything that is known at the time the forecast is made. Our best prediction of this term is therefore 0. Thus, our forecast of X_{31} is

$$\hat{X}_{31} = 313.68 + 1.180X_{30} - 0.358X_{29}$$
$$= 313.68 + 1.180(1,289) - 0.358(1,387) = 1,338.15$$

Setting $t = 32$ in 17.8.5, we have

$$X_{32} = 313.68 + 1.180X_{31} - 0.358X_{30} + a_{32}$$

Once again, our best prediction of a_{32} is simply 0. Moreover, we do not know X_{31}, but we do have its forecast, $\hat{X}_{31}$, so that a natural prediction for sales 2 years ahead is

$$\hat{X}_{32} = 313.68 + 1.180\hat{X}_{31} - 0.358X_{30}$$
$$= 313.68 + (1.180)(1,338.15) - (0.358)(1,289) = 1,431.24$$

Continuing in exactly the same way, we obtain

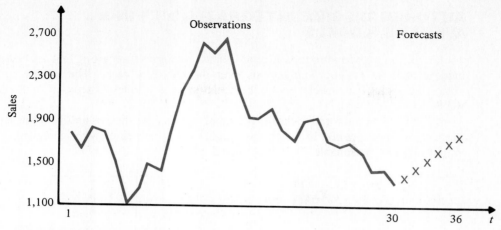

FIGURE 17.11 Sales of Lydia E. Pinkham, and forecasts based on a fitted second-order autoregressive model

$$X_{33} = 313.68 + 1.180X_{32} - 0.358X_{31} + a_{33}$$

Replacing X_{32}, X_{31}, and a_{33} by their forecasts, $\hat{X}_{32}$, $\hat{X}_{31}$, and 0, then yields the forecast for X_{33} of

$$
\begin{aligned}
\hat{X}_{33} &= 313.68 + 1.180\hat{X}_{32} - 0.358\hat{X}_{31} \\
&= 313.68 + (1.180)(1,431.24) - (0.358)(1,338.15) = 1,523.49
\end{aligned}
$$

Proceeding in this fashion, we can compute forecasts as far ahead as required. The data series and the first six forecasts are graphed in Figure 17.11.

The general procedure for computing forecasts based on an estimated autoregressive model is outlined in the box.

Forecasting from Estimated Autoregressive Models

Suppose that we have observations $X_1, X_2, \ldots, X_n$ from a time series, and that an autoregressive model of order p has been fitted to these data. Write the estimated model as

$$X_t = \hat{\gamma} + \hat{\phi}_1 X_{t-1} + \hat{\phi}_2 X_{t-2} + \cdots + \hat{\phi}_p X_{t-p} + a_t$$

Standing at time n, we obtain forecasts of future values of the series from

$$\hat{X}_{n+h} = \hat{\gamma} + \hat{\phi}_1 \hat{X}_{n+h-1} + \hat{\phi}_2 \hat{X}_{n+h-2} + \cdots + \hat{\phi}_p \hat{X}_{n+h-p} \quad (h = 1, 2, 3, \ldots)$$

where, for $j > 0$, $\hat{X}_{n+j}$ is the forecast of X_{n+j} standing at time n and, for $j \leq 0$, $\hat{X}_{n+j}$ is simply the observed value X_{n+j}.

17.9 AUTOREGRESSIVE INTEGRATED MOVING AVERAGE MODELS

In this section we will introduce an approach to time series forecasting that has become widely used in business applications over the past few years. The models to be discussed include as special cases the autoregressive models discussed in Section 17.8.

To introduce the rationale for the models, consider the possibility of an autoregressive representation in which the weight on X_{t-j} decreases with increasing j, as might often be reasonable. One such model is

$$X_t = \gamma^* + (\phi - \theta)X_{t-1} + \theta(\phi - \theta)X_{t-2} \\ + \theta^2(\phi - \theta)X_{t-3} + \cdots + a_t \qquad (17.9.1)$$

where γ^* is a fixed parameter and a_t represents a series of random variables with mean 0 and fixed variance that are not correlated with one another. We can view 17.9.1 as an autoregressive model, in which we assume that θ is between 0 and 1; thus, the autoregressive parameters $(\phi - \theta)$, $\theta(\phi - \theta)$, $\theta^2(\phi - \theta)$, . . . form a decreasing sequence. A model of this kind would imply that, in forecasting the next observation, most weight is given to the most recent value of the series, less weight to the preceding observation, and so on.

At time $(t - 1)$ the model 17.9.1 implies

$$X_{t-1} = \gamma^* + (\phi - \theta)X_{t-2} + \theta(\phi - \theta)X_{t-3} \\ + \theta^2(\phi - \theta)X_{t-4} + \cdots + a_{t-1} \qquad (17.9.2)$$

Multiplying through equation 17.9.2 by θ gives

$$\theta X_{t-1} = \theta\gamma^* + \theta(\phi - \theta)X_{t-2} + \theta^2(\phi - \theta)X_{t-3} + \cdots + \theta a_{t-1} \qquad (17.9.3)$$

Now, subtracting 17.9.3 from 17.9.1, we have

$$X_t - \theta X_{t-1} = \gamma^*(1 - \theta) + (\phi - \theta)X_{t-1} + a_t - \theta a_{t-1}$$

or

$$X_t = \gamma + \phi X_{t-1} + a_t - \theta a_{t-1} \qquad (17.9.4)$$

where

$$\gamma = \gamma^*(1 - \theta)$$

Now, in the model 17.9.4, the current value, X_t, of the series is related not only to the previous value, X_{t-1}, but also to the previous error term, a_{t-1}. This model is a simple example of a general class of time series models called **autoregressive moving average models.** More generally, for any positive integers p and q, X_t might be linearly related to $X_{t-1}, X_{t-2}, \ldots, X_{t-p}$, and $a_{t-1}, a_{t-2}, \ldots, a_{t-q}$, through the model

$$X_t = \gamma + \phi_1 X_{t-1} + \cdots + \phi_p X_{t-p} + a_t - \theta_1 a_{t-1} - \cdots - \theta_q a_{t-q} \qquad (17.9.5)$$

where $\gamma, \phi_1, \ldots, \phi_p, \theta_1, \ldots, \theta_q$ are fixed parameters. The model 17.9.5 is known as an **autoregressive moving average model of order** (p, q).

Autoregressive moving average models are frequently used to represent actual time series data. However, for business and economic series in which there is inertia in the system generating the data, a further refinement is often desirable. For instance, we would expect sales volume in the current week to be somewhat similar to that of the previous week, so that what we would really like to predict is the *change* in sales from one week to the next. This suggests that, rather than examining the original series, X_t, it might be sensible to look at the series of changes, or **first differences,**

$$W_t = X_t - X_{t-1}$$

We might then estimate an autoregressive moving average model for the first differences—that is, a model of the form 17.9.5, but with $W_t, W_{t-1}, W_{t-2}, \ldots, W_{t-p}$ replacing $X_t, X_{t-1}, X_{t-2}, \ldots, X_{t-p}$. Just occasionally it is desirable also to consider the possibility of fitting an autoregressive moving average model to the series of **second differences**

$$Z_t = W_t - W_{t-1}$$

When the possibility of differencing the data before fitting an autoregressive moving average model is considered, the resulting class of models is called **auto-regressive integrated moving average** or ARIMA. These models are now often used in business forecasting. Their popularity was stimulated, to a very large extent, by the publication of a book[3] by G. E. P. Box and G. M. Jenkins, who set out a practical methodology for building such forecasting models. For that reason, this approach to forecasting is sometimes referred to as the **Box-Jenkins approach.**

Here, we will be able to present only a brief outline of ARIMA model building methods.[4] Box and Jenkins describe an iterative three-part model building strategy for

[3] G. E. P. Box and G. M. Jenkins, *Time Series Analysis, Forecasting, and Control,* San Francisco: Holden-Day (1970).

[4] An excellent introduction is given by C. R. Nelson, *Applied Time Series for Managerial Forecasting,* San Francisco: Holden-Day (1973).

fitting an appropriate ARIMA model to a particular set of time series data. The three stages are as follows:

(i) Model selection
(ii) Parameter estimation
(iii) Model checking

At the initial stage, a specific model from the general ARIMA class is chosen, based on statistics calculated from the data. That is, a decision is made as to whether to work with the original series or one of its differences, and specific values are chosen for p and q, the orders of the autoregressive moving average model to be fitted. Although the data should contain useful information as to what might be an appropriate model, we will not be able to fix with certainty the right choice. The procedures used require a good deal of judgment, and thus are inexact. For that reason, the analyst is not irrevocably committed to the original model chosen. At a later stage in the analysis this model might be abandoned in favor of some alternative if the available evidence suggests the desirability of such a course.

Having chosen a specific model, we must then obtain estimates of its unknown parameters $\gamma, \phi_1, \ldots, \phi_p, \theta_1, \ldots, \theta_q$. The procedures used are statistically efficient, and are related to the least squares method described in the previous section. At the final stage of the model building cycle, checks are made on the adequacy of representation of the fitted model to the given data. Any inadequacies revealed at this stage might suggest an alternative model specification, in which case the cycle is repeated until a satisfactory model is achieved. That model can then be projected forward to obtain forecasts of future values of the time series.

This approach to time series, which has become popular in business forecasting, has several advantages:

(a) The models implied by many popular forecasting methods are, in fact, special cases of the ARIMA class. This is the case for the autoregressive models of the previous section, which are obtained by setting $q = 0$ in 17.9.5. It can also be shown that the simple exponential smoothing forecasts of Section 17.6 and the nonseasonal Holt-Winters forecasts of Section 17.7 can be generated by specific members of the ARIMA class.

(b) The approach to forecasting introduced here is extremely flexible. The analyst is not committed to a single specific forecast-generating mechanism, but rather has the ability to choose an appropriate model from a very broad general class. This choice is based on evidence contained in the historical record of the time series. Thus, for different kinds of data, different forecasting models can be used. For example, we have already seen in Figures 17.8 and 17.11 two sets of forecasts of the same set of sales data, based respectively on simple exponential smoothing and a second-order autoregressive model. The advantage of the flexibility given by ARIMA models is that we are provided with a sound basis for choosing among these and many other possible forecasting models, using our knowledge of past patterns.

(c) The models we have discussed in this section are designed for predicting nonseasonal time series. However, they can easily be extended to provide a class of models suitable for forecasting seasonal series.

(d) We have discussed methods for forecasting future values of a time series using just its own past values. However, when making predictions, the sensible analyst will want to use *all* available relevant information. The procedures of this section can be extended to the development of methods in which forecasts are based on past values of other related series, as well as on the past of the series itself.

(e) When this approach to forecasting has been compared with other methods,[5] using actual economic and business time series, it has usually been found to yield predictions that are more accurate, on the average, than those of its competitors. Thus, the procedure can be said to have survived the most important test of all: *In practice, it works!*

In concluding this brief discussion, we note that computer programs for carrying out an analysis through ARIMA models are now widely available. However, the method does have a drawback compared with some of the simpler procedures described in earlier sections. Because flexibility is allowed in choosing an appropriate model, the approach is more costly in terms of skilled manpower time than methods that force a single model structure onto every series.

EXERCISES

38. Using the data of Exercise 1, estimate a first-order autoregressive model for the volume of transactions in the shares of General Motors Corporation. Use the fitted model to obtain forecasts of volume for the next 2 weeks.

39. Using the data of Exercise 4, estimate a first-order autoregressive model for the volume of shares traded in the stock of Xerox Corporation. From this fitted model, obtain forecasts of volume traded for the next 4 weeks. Draw a time plot showing the original data and your forecasts.

40. [*This exercise requires access to a computer program to carry out the multiple regression calculations.*] Using the data of Exercise 5, estimate autoregressive models of orders one through four for the sales of Kennecott Corporation. Using the procedure of Section 17.8 for testing that the order of the autoregression is $(p - 1)$, against the alternative that the order is p, with a significance level of 10%, choose one of these models and compute sales forecasts for the next 5 years. Graph the original observations, together with your forecasts.

41. [*This exercise requires access to a computer program to carry out the multiple regression calculations.*] From the data of Exercise 6, fit autoregressive models of orders one through four to the earnings per share of Cominco Ltd. Using the method of Section 17.8 for testing that the order of the autoregression is $(p - 1)$, against the alternative that the order is p, with a significance level of 10%, select one of these models and calculate earnings forecasts for the next 5 years. Draw a time plot showing the original observations along with your forecasts. Would different forecasts result if a significance level of 5% was used for your tests?

42. [*This exercise requires access to a computer program to carry out the multiple regression calculations.*] Based on the data of Exercise 25, estimate autoregressive models of orders one through four for the earnings per share of Olympia Brewing. Through the method of Section 17.8 for testing that the autoregressive order is $(p - 1)$, against the alternative of order p, using a 10% significance level, choose one of these models and compute earnings forecasts for the next 6 years. Draw a graph showing the original observations

---

[5] See, for example, P. Newbold and C. W. J. Granger, "Experience with forecasting univariate time series and the combination of forecasts," *Journal of Royal Statistical Society A, 137,* 131–46.

together with these forecasts. Would the results differ if 5% significance levels had been used for the tests?

43. [*This exercise requires access to a computer program to carry out the multiple regression calculations.*] From the data of Exercise 30, estimate autoregressive models of orders one through five for capital expenditures of Scott Paper Company. Using the procedure of Section 17.8 for testing that the order of the autoregression is $(p - 1)$, against the alternative of order p, with a 10% significance level, select one of these models and calculate forecasts of capital expenditures for the next 4 years. Draw a graph showing the observations and forecasts. Would the same forecasts have resulted if 5% significance levels had been used for the tests?

44. A second-order autoregressive model was fitted to a time series of annual product sales data, yielding

$$X_t = 27 + 1.25X_{t-1} - 0.30X_{t-2} + a_t$$

In 1981 and 1982, sales levels were, respectively, 532 and 567. Obtain forecasts for the years 1983 through 1987.

45. It was found that, for a particular product, annual sales volume could be well represented by a third-order autoregressive model. The estimated model obtained was

$$X_t = 23 + 1.10X_{t-1} - 0.40X_{t-2} + 0.25X_{t-3} + a_t$$

For 1980, 1981, and 1982, sales were 469, 523, and 546, respectively. Calculate sales forecasts for the years 1983 through 1987.

46. For many time series, particularly prices in speculative markets, the *random walk* model has been found to give a good representation of actual data. This model is written

$$X_t = X_{t-1} + a_t$$

Show that, if this model is appropriate, then forecasts of X_{n+h}, standing at time n, are given by

$$\hat{X}_{n+h} = X_n \qquad (h = 1, 2, 3, \ldots)$$

47. Sometimes it is desirable to extend the model of Exercise 46 to allow for the possibility that the expected change from period to period is not 0. This augmented model, known as the *random walk with drift,* is written

$$X_t = \gamma + X_{t-1} + a_t$$

Show that, if this model is appropriate, then forecasts of X_{n+h}, standing at time n, are given by

$$\hat{X}_{n+h} = X_n + h\gamma \qquad (h = 1, 2, 3, \ldots)$$

48. [*This exercise requires access to a computer program to carry out the multiple regression calculations.*] Refer to the data in Exercise 32, showing the percentages of the total United States labor force that was female, in the period 1960 to 1980. Denote these observations $X_t (t = 1, 2, \ldots, 21)$. Now, form the series of first differences

$$Z_t = X_t - X_{t-1} \qquad (t = 2, 3, \ldots, 21)$$

Estimate autoregressive models of orders one through four for the series Z_t. Using the method of Section 17.8 for testing that the autoregressive order is $(p - 1)$, against the alternative of order p, using a 10% significance level, choose one of these models, and compute forecasts for Z_t, $t = 22, 23, 24$. Hence, obtain forecasts for the percentage of the total United States labor force that is female for the years 1981, 1982, 1983.

49. Explain the statement that a time series can be viewed as being made up of a number of components. Provide examples of business and economic time series for which you would expect particular components to be important.

50. In many business applications, forecasts of future values of time series, such as sales and earnings, are made exclusively on the basis of past information on the time series in question. What features of time series behavior are exploited in the production of such forecasts?

51. A manager in charge of inventory control requires sales forecasts for several products, on a monthly basis, over the next 6 months. This manager has available monthly sales records over the past 4 years for each of these products. He decides to use, as forecasts for each of the next 6 months, the average monthly sales over the previous 4 years. Do you think this is a good strategy? Provide reasons.

52. What is meant by the seasonal adjustment of a time series? Explain why government agencies expend a large amount of effort on the seasonal adjustment of economic time series.

53. The accompanying table shows the index of real output per hour of all persons in the nonfarm business sector in the United States over the period 1968 to 1981.

YEAR	OUTPUT PER HOUR	YEAR	OUTPUT PER HOUR
1968	86.7	1975	95.0
1969	86.4	1976	98.1
1970	86.7	1977	100.0
1971	89.6	1978	99.8
1972	93.0	1979	99.1
1973	95.3	1980	98.8
1974	93.1	1981	99.7

(a) Test this series for randomness, using the runs test.

(b) Draw a time plot of these data and discuss the features revealed by your graph.

54. The table given here shows the value of the U.S. dollar in Canadian dollars over the period 1970 to 1981.

YEAR	1970	1971	1972	1973	1974	1975
VALUE	1.044	1.010	0.991	1.000	0.978	1.017

YEAR	1976	1977	1978	1979	1980	1981
VALUE	0.986	1.064	1.140	1.172	1.169	1.199

(a) Test this series for randomness, using the runs test.

(b) Draw a time plot of these data, and discuss the resulting picture.

55. The table shows the market return on the Standard and Poor's 500 stocks for the period 1956 to 1979. Test this series for randomness, using the runs test.

YEAR	RETURN	YEAR	RETURN	YEAR	RETURN
1956	6.6	1964	16.5	1972	19.0
1957	−10.8	1965	12.5	1973	−14.7
1958	43.4	1966	−10.1	1974	−26.5
1959	12.0	1967	24.0	1975	37.2
1960	0.5	1968	11.1	1976	23.8
1961	26.9	1969	−8.5	1977	−7.2
1962	−8.7	1970	4.0	1978	6.6
1963	22.8	1971	14.3	1979	18.7

56. The accompanying table shows the 3-month Treasury bill yield in Canada over the years 1961 to 1980. Compute a simple centered 5-point moving average of these data. Draw a time plot of the smoothed series and comment on its features.

YEAR	YIELD	YEAR	YIELD	YEAR	YIELD
1961	2.83	1968	6.24	1975	8.37
1962	4.01	1969	7.14	1976	8.89
1963	3.57	1970	6.12	1977	7.35
1964	3.74	1971	3.62	1978	8.58
1965	3.97	1972	3.55	1979	11.57
1966	5.00	1973	5.39	1980	12.70
1967	4.59	1974	7.78		

57. The data given in the table are monthly values, over a period of 3 years, of the United States index of twelve leading indicators of economic activity. The data have been seasonally adjusted. Compute a simple centered 7-point moving average of these data. Draw a time plot of the smoothed series and comment on its features.

	1978	1979	1980
January	139.1	142.6	134.7
February	140.3	142.3	134.2
March	140.3	143.2	131.4
April	141.5	140.3	125.8
May	141.8	141.4	122.8
June	142.5	141.6	123.8
July	141.2	141.2	128.1
August	142.0	140.1	130.6
September	142.9	140.1	134.9
October	143.6	137.8	135.7
November	142.8	135.6	137.7
December	143.0	135.2	136.6

58. Listed in the accompanying table are data on earnings per share, on a quarterly basis, of Commonwealth Edison over a period of 7 years.

		QUARTER		
YEAR	1	2	3	4
1972	0.786	0.668	0.863	0.807
1973	0.802	0.670	0.885	0.805
1974	0.579	0.423	0.904	0.851
1975	0.430	0.409	1.120	0.958
1976	0.680	0.460	1.190	0.830
1977	0.760	0.440	1.020	0.630
1978	0.690	0.600	1.130	0.680

(a) Draw a time plot of these data. Does this plot suggest strong seasonality?

(b) Obtain a seasonally adjusted series, using the seasonal index method.

59. Based on the data of Exercise 54, use the simple exponential smoothing method, with smoothing constant $\alpha = 0.2$, to obtain forecasts of the value of the U.S. dollar in Canadian dollars for 1982 and 1983.

60. Based on the data of Exercise 57, use the simple exponential smoothing method, with smoothing constant $\alpha = 0.4$, to obtain forecasts of the index of leading indicators for the first 3 months of 1981.

61. The table shows annual percentage base wage increases in the private sector in Canada for the period 1967 to 1980. Use the simple exponential smoothing method, with smoothing constant $\alpha = 0.3$, to obtain forecasts for 1981 and 1982.

YEAR	WAGE INCREASE	YEAR	WAGE INCREASE
1967	7.8	1974	14.4
1968	8.1	1975	14.4
1969	8.6	1976	9.3
1970	8.6	1977	7.3
1971	8.0	1978	6.9
1972	9.2	1979	8.9
1973	10.1	1980	11.0

62. Refer to Exercise 53. Use the Holt-Winters procedure, with smoothing constants $A = 0.4$ and $B = 0.4$, to predict output per hour for the years 1982 to 1984.

63. Shown in the accompanying table is the unemployment rate in Canada for the years 1966 to 1981. Use the Holt-Winters procedure, with smoothing constants $A = 0.3$ and $B = 0.5$, to predict the unemployment rate for the years 1982 to 1985.

YEAR	UNEMPLOYMENT RATE	YEAR	UNEMPLOYMENT RATE
1966	3.4	1974	5.3
1967	3.8	1975	6.9
1968	4.5	1976	7.1
1969	4.4	1977	8.1
1970	5.7	1978	8.4

YEAR	UNEMPLOYMENT RATE	YEAR	UNEMPLOYMENT RATE
1971	6.2	1979	7.5
1972	6.2	1980	7.5
1973	5.5	1981	7.6

64. Quarterly product sales over a period of 5 years are given in the table. Use the Holt-Winters seasonal procedure, with smoothing constants $A = 0.5$, $B = 0.4$, and $C = 0.2$, to obtain forecasts for the four quarters of 1983.

	QUARTER			
YEAR	1	2	3	4
1978	70	60	97	109
1979	63	63	55	74
1980	65	77	94	122
1981	78	114	117	140
1982	91	137	173	221

65. A second-order autoregressive model[6] was fitted to monthly data on official gold and foreign exchange holdings in Sweden. The estimated model was

$$X_t = 1.38X_{t-1} - 0.38X_{t-2} + a_t$$

The last two observations in this time series were

$$X_{n-1} = 725 \quad \text{and} \quad X_n = 702$$

Obtain forecasts of X_{n+1}, X_{n+2}, and X_{n+3}.

66. [*This exercise requires access to a computer program to carry out the multiple regression calculations.*] Using the data of Exercise 63, estimate a third-order autoregressive model for the Canadian unemployment data and, on the basis of the fitted model, obtain forecasts for 1982 to 1984.

---

[6] Reported in C. W. J. Granger and P. Newbold, *Forecasting Economic Time Series,* New York: Academic Press (1977).

survey sampling methods

CHAPTER EIGHTEEN

18.1 INTRODUCTION

Much of statistical inference is concerned with problems of making statements about a population on the basis of information from a sample. In our discussions to this point, two important topics have been treated rather cursorily. First, very little has been said about how one would actually go about selecting the sample members. Second, it has generally been assumed that the number of population members is very large compared with the number of sample members. In this chapter we concentrate on the problem of a researcher who wants to discover something about a population that is not necessarily enormously large. The investigator intends to collect information on only a subset of the population members, and requires guidance as to how to proceed.

The general problem just discussed occurs frequently in business applications. Market researchers often survey human populations to elicit information about product preferences. Auditors typically select a sample of a corporation's accounts receivable, on the basis of which inference is made about the corresponding population. Personnel directors often require information on employees' attitudes to proposed new production methods, and find it convenient to sample the labor force. Of course, the use of sampling methods is extremely widespread, extending well beyond the field of business. Perhaps the best known example is the regular survey of voter preferences prior to a presidential election. The information gathered is of interest, not only to the general public, but also to those advisers of the candidates trying to determine where their efforts should be most heavily concentrated. Such surveys of voters have escalated to the point where voter opinions are sought on all aspects of public policy, and the professional pollster is becoming an increasingly important figure in the politician's entourage.

Before asking how a sample should be taken from a population, we pause to ask: "Why sample at all?" The alternative is to attempt to obtain information from every population member. This would be referred to as a **census,** rather than a sample. There are three good reasons why a sample is often preferred to a census. First, in many applications, taking a complete census would be enormously expensive, frequently prohibitively so. Second, information is often required fairly quickly, so that a full census, even when financially feasible, may take so long to complete that the value of the results is seriously diminished. Finally, with modern statistical methods, it is generally possible to obtain results of the desired level of precision through sampling. Any time and money expended on producing numbers whose apparent precision exceeds the investigator's needs could be better spent elsewhere. Moreover, if a relatively small sample is taken, the gains derived through extra effort in securing accurate information from the sample members could well outweigh the benefits of having information from a larger group which, because of time and cost constraints, may be less reliable. Taken together, these three factors—cost, time, and precision—far more often than not dictate in practice a preference for a sample rather than a census.

Suppose, now, that information about a population is required, and that the decision has been made to take a sample. It is convenient to think of a sampling study as involving six steps, each aimed at producing the answer to a single question, as set out schematically in Figure 18.1.

We will discuss each of these steps in turn, with reference to a market research problem. Suppose that a publisher intends to bring out a new business statistics text, and wants information on the current state of the market. For example, valuable information would include the number of students enrolled in business statistics courses, the levels of market penetration of existing texts, and instructors' views as to which topics are the most important for their courses. We will assume that the publisher decides to gather data from a sample of college campuses.

1. WHAT INFORMATION IS REQUIRED?

The answer to this question provides both the motivation and the starting point for the study. If the necessary information is either already available or impossible to obtain, there is no point in carrying through the survey. However, straightforward though the question seems, a rather delicate balance is often needed at this stage. The investigator may have in mind just a single requirement, or there may be several topics of interest. But, given that a survey is to be carried out, with all the attendant costs, it is typically worthwhile asking whether further potentially useful information can be gleaned from the population at minimal additional expense. For the publisher of the business statistics text, we have already noted that the most useful questions concern the topics that instructors view as most important, the size of the market, and the standing of competitors. Given that sample members are to be contacted to elicit this information, it may well be worthwhile to ask some additional questions. These may include whether the course is one or two semesters long, optional or compulsory, the de-

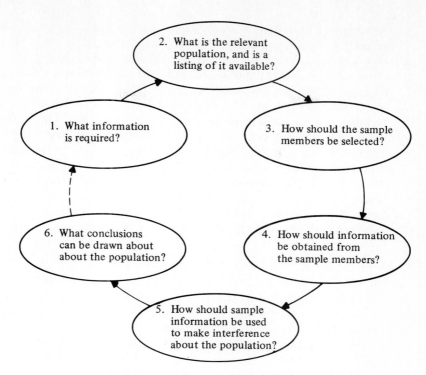

FIGURE 18.1 Steps in a sampling study

partment of the instructor, the mechanism for textbook adoption, the length of time the current text has been in use, and so on. Once having started along this road, we may be tempted to allow the list of questions to grow dramatically, because this would not generally increase greatly the cost of carrying out the study. However, there *is* a potential penalty. Respondents are more likely to cooperate in a study that asks relatively few questions, and consequently takes up little of their time. Thus, it is important for the investigator to strike a balance, in which questions on central issues are asked (for if an important omission is discovered it may prove too costly to repeat the whole exercise) and the number of questions asked remains tolerable to potential respondents.

2. WHAT IS THE RELEVANT POPULATION, AND IS A LISTING OF IT AVAILABLE?

It seems rather trite to point out that, if inference is to be made about a particular population, then that is the population that must be sampled. Nevertheless, dubious conclusions have often been reached, following an otherwise perfectly respectable analysis of survey data, precisely because this elementary point has been ignored. Many magazines invite the opinions of their readers on particular questions. It would, however, be dangerous to generalize their responses to a wider population. The population studied here is simply the readership of the magazine, and this readership is likely to be unrepresentative of the public at large. In many practical studies, the *real* population of interest may be impossible to define. For example, an organization attempting to predict the result of a presidential election is really interested in only those people who will, in fact, vote. However, although this is the relevant population, its members are not easy to distinguish. One possibility, of course, is to ask a sample member if he or she intends to vote. However, it is well known that the proportion answering such a question in the affirmative is higher than the proportion who do eventually vote. Another possiblilty is to ask whether the respondent voted in the previous election, but this too is far from completely satisfactory.

The textbook publisher is likely to regard the relevant population as all instructors (or, perhaps, all colleges) teaching business statistics courses. This population is quite easy to identify, and, as a result of earlier marketing activities, the publisher will almost certainly have a fairly accurate listing of its members.

3. HOW SHOULD THE SAMPLE MEMBERS BE SELECTED?

Much of the remainder of this chapter will be devoted to answering this question. At this point it should be said that there is no unique way to go about providing the "best" sampling scheme. The appropriate choice will generally depend on the problem at hand and on the resources of the investigator. In Chapter 6, we introduced the notion of **simple random sampling,** in which each potential sample of n members has an equal chance of being chosen. Indeed, all of the data-analytic tools introduced to this point have been based on an assumption that the sample was chosen in such a fashion. There are, however, many circumstances in which alternative sampling schemes might be preferred. As one example, suppose that our publisher is concerned about differences in the treatment of business statistics courses between 2-year and 4-year colleges. In that case it is important that the sample contain enough colleges of each type to allow reliable conclusions about both to be drawn. However, simple random sampling by no means guarantees to attain this objective. It is entirely possible, for example, that the sample chosen will contain an overwhelming preponderance of 4-year colleges. To guard against this possibility, one can draw separate simple random samples of 2-year and 4-year colleges from their respective populations. This is an example of **stratified sampling,** which will be discussed in more detail in Section 18.4. Another matter to be decided at this stage is the number of sample members. Essentially, the choice here depends on the degree of accuracy required and also on the costs involved. We will return to this question in Section 18.5.

4. HOW SHOULD INFORMATION BE OBTAINED FROM THE SAMPLE MEMBERS?

In many instances where human populations are surveyed, this is an extremely important question, which is the subject of much ongoing research. Broadly speaking, there are two important issues involved. First, the investigator will want to obtain answers from as high a proportion as possible of the sample members. If the number not responding is high, it will be difficult to be sure that those who do respond are representative of the population at large. For instance, professors failing to supply information to the textbook publisher may be those most heavily involved in research, consulting, or other activities, and their preferences about texts could well differ from those of their colleagues. We have already noted that the number of questions asked in a survey could affect the response rate. The manner in which sample members are contacted is also influential. Frequently, questionnaires are mailed to those selected for the sample, and it often happens that the proportion responding is disappointingly low. Many researchers attempt to improve the response rate by including a covering letter, explaining the purposes of the study and politely soliciting help. An assurance of anonymity may also be valuable. The inclusion of a stamped preaddressed envelope for returning the questionnaire is generally worthwhile, and some modest monetary inducement or gift might be promised. Nevertheless, there will almost inevitably be a fair proportion of nonrespondents, and it is good practice to institute a follow-up inquiry to try to learn something about them. More expensive contact methods, such as telephone calls or home visits by interviewers, are likely to achieve a higher level of response. However, such methods can be costly in terms of time as well as money, and the decision on how to collect information must depend on the investigator's resources and the extent to which nonresponse is thought to be a serious potential problem.

The textbook publisher may decide to mail questionnaires to sample members. Certainly this would be quite inexpensive, so that a relatively large initial sample could be drawn. In that case, the hope must be that the proportion of nonrespondents is not too high and that the responses obtained are reasonably representative. If, on the other hand, it is feared that nonresponse will induce substantial bias should a mail question-naire be used, then a smaller initial sample might be drawn and a greater effort made to contact individual sample members. In this particular instance, a feasible strategy might be to ask the company's representatives, who regularly visit campuses, to carry out interviews with the sample members on their next visits. Such a procedure should ensure quite a high response rate. Its major difficulty would be in the time taken before all the interviews were completed, rather than the additional costs, which would be quite low.

The second point is the obvious desideratum of obtaining answers that are as accurate and honest as possible. Nothing is to be gained from a highly sophisticated statistical analysis of basically unreliable information. There is an art in designing questions, whether asked through mail survey or by interviewer, in such a way as to extract honest and accurate replies. It is important that the questions be phrased as clearly and unambiguously as possible, so that subjects understand what is being asked. Also, it is well known that respondents can be biased toward providing partic-ular answers by the wording of the question, or even by the tone of an interviewer. Interviewers should in no way convey the impression of having strong views on the subject at hand, or of "wanting" a particular answer. It is also important not to "lead the witness"—questions should be phrased as neutrally as possible. As an extreme example, consider the following two methods of asking essentially the same question:

Q. Which three topics do you regard as most important in your business statistics course?

or

Q. Do you agree that modern methods in time series analysis and forecasting, because of their overwhelming importance in the business world, must now be considered one of the most important topics in any business statistics course?

Of course, no one interested in an accurate picture of instructors' opinions would ask the second question. However, much less clearly biased wording than this has been found to make an appreciable difference to subjects' replies.

As an illustration of this point, Opinion Research Corporation of Princeton, New Jersey, annually surveys public attitudes toward government. Concerned about the influence of question wording, the organization asked the same question in two ways in the 1981 survey. Respondents were asked what programs might best be sacrificed in the event of a severe budget squeeze. Among the list of candidates was "Aid to the needy." This item was selected for cutbacks by only 7% of all respondents. Two months later, when the same question was asked, the description "Aid to the needy" was replaced by "Public welfare." This time the item was chosen for cuts by 39% of respondents!

5. HOW SHOULD SAMPLE INFORMATION BE USED TO MAKE INFERENCE ABOUT THE POPULATION?

The greater part of this book has been devoted to providing answers to just this question. In subsequent sections of this chapter we will discuss methods of inference for particular sampling designs. The chief purpose of the present section is to note the importance of other aspects of a statistical inquiry.

6. WHAT CONCLUSIONS CAN BE DRAWN ABOUT THE POPULATION?

Finally, we come full circle and ask what can be said about the population under study as a result of a statistical investigation. Has the study produced clear answers to the questions that motivated it? Have additional important questions emerged in the course of the study? The investigator at this stage has the task of summarizing and presenting the information gathered. This may involve point or interval estimates, tabular summaries, or graphical presentation of results. What is our best estimate of the number of students enrolled in business statistics courses, and can confidence bands be put around this estimate? Which are the most popular texts at present? What topics do instructors consider most important? Are there significant differences between the 2-year and 4-year college markets? At this stage the task is to report the findings of the study and to decide how to proceed. It may happen that the analysis will suggest the desirability of gathering further information.

It often happens that important unanticipated issues arise during the course of a survey, and the investigator is stimulated to further study of the population. It is for this reason that we have joined the final box to the first, in Figure 18.1, with a dashed arrow. Suppose our publisher asks an open-ended question, such as

> **Q.** Our company is planning to bring to market a new business statistics text. Are there any features that you would particularly welcome in such a book?

Assume further that, when the questionnaires are returned, an appreciable number mention the possibility of simultaneously marketing, on cards or tape, a large data bank containing data on real business problems. Students could then spend a large portion of the course time in analyzing these data. Before going to the extra production expense, the publisher may well find it worthwhile to take another sample in order to assess the likely success of this venture.

18.2 SAMPLING AND NONSAMPLING ERRORS

When a sample is taken from a population, we will not be able to *know precisely* the value of any population parameter, such as the mean or proportion. Any point estimate that is produced will inevitably be in error. We have already discussed one source of error, that resulting from the fact that information is available on only a subset of all the population members. We call this **sampling error.** Given certain assumptions, statistical theory allows us to characterize the nature of the sampling error and to make well-defined probabilistic statements about population parameters. Confidence intervals are examples of such statements. In subsequent sections of this chapter we will discuss methods of statistical inference for various important sampling schemes. However, it is important to recognize first the potential for another source of error, which cannot be analyzed in such an elegant or clear-cut fashion.

In practical analyses, there is the possibility of an error which is not connected with the kind of sampling procedure used. Indeed, such errors could just as well arise if a complete census of the population were taken. These are referred to as **nonsampling errors.** In any particular survey, the potential for nonsampling error exists at a number of places. Examples include the following:

(a) **The population actually sampled is not, in fact, the relevant one.** A celebrated instance of this sort occurred in 1936, when the now defunct *Literary Digest* magazine confidently predicted that Alfred Landon would win the presidential election over Franklin Roosevelt. In the event, Roosevelt won by a very comfortable margin. This erroneous forecast resulted from the fact that the members of the *Digest*'s sample had been taken from telephone directories and other listings, such as magazine subscription lists and automobile registrations. These sources considerably under-represented the poor, who were predominantly Democratic supporters. The moral of the story is that if one wants to make inference about a population (in this case, the United States electorate), it is important to sample that population and not some subgroup of it, however convenient the latter course might appear.

(b) **A survey may produce inaccurate or dishonest answers from subjects.** This could arise because questions are phrased in a manner that is difficult to understand, or in a fashion that appears to make a particular answer seem more palatable. In addition, many questions which one might want to ask are so sensitive that it would be foolhardy to expect uniformly honest responses. Suppose, for example, that a plant manager wants to assess the annual losses to the company caused by employee thefts. In principle, a random sample of employees could be selected and sample members asked "What have you stolen from this plant in the past 12 months?" This is clearly not the most reliable means of obtaining the required information! In fact, we have already seen one promising possibility for getting around this kind of problem. For a description and illustration of this procedure—the *randomized response approach*—the interested reader can refer back to Examples 3.7 and 3.19.

(c) **Another possibility arises through nonresponse.** If this is substantial, it can induce additional sampling and nonsampling errors. The sampling error arises because the achieved sample size will be smaller than that intended. Nonsampling error possibly occurs because, in effect, the population being sampled is not that of interest. The results obtained can be regarded as a random sample *from the population of those willing to respond*. These people may differ in important ways from the larger population. If this is so, a bias will be induced in the resulting estimates.

If it is suspected that nonresponse bias is likely to be troublesome, three possibilities are open. First, the investigator can solicit information through one of those mechanisms known to achieve a relatively high response rate. Second, as far as possible, characteristics of respondents and nonrespondents can be compared, in such matters as age, sex, and race, to see if there are obvious differences between the two groups. Finally, an attempt can be made to contact nonrespondents, some of whom may well be prepared to provide answers to a few key questions. If these differ significantly from the answers of the original respondents, a correction for nonresponse bias can then be made.

There is no general procedure available for identifying and analyzing non-sampling errors. Nevertheless, they could be important. The main prescription is that the investigator take care in such matters as identifying the relevant population, designing the questionnaire, and dealing with nonresponse, in order to minimize their significance. In the remainder of this chapter it will be assumed that such care has been taken, and our discussion will center on the treatment of sampling errors.

EXERCISES

1. Suppose that you want to conduct a study to determine the views of business majors on your campus as to whether statistics should be a required course. Discuss the steps that you would take in setting up this study, the problems you might expect to encounter, and techniques you might employ to overcome them.

2. The manager of a campus bookstore is contemplating stocking *Playtime* magazine, but is unsure of prospective student demand. You are commissioned to carry out a survey on campus. Suggest how you would proceed, following along the lines of the six steps discussed in Section 18.1.

3. A supermarket manager wants to learn what types of customers purchase generic foods, and to ascertain the reasons for sales resistance to this type of product in some segments of the population. Discuss, in detail, how you might go about setting up a study to provide the required information.

4. A real estate executive wants to know which features of a house have been most influential in determining eventual purchase decisions in your city. Describe how you could set up a study aimed at obtaining this information. Follow the steps outlined in Section 18.1. What difficulties might you expect to meet and how would you try to overcome them?

5. A life insurance agent in your city wants to estimate the proportion of citizens in various demographic groupings who are considering the possibility of taking out further coverage

in the next 12 months. You are commissioned to carry out a study, with a total budget of $3,000. How would you proceed?

6. Refer to the study of Exercise 1.

    (a) Within the sampling framework you have designed, do you see any potential for nonsampling errors? If so, what steps would you take to minimize their magnitude?

    (b) Would you expect nonresponse to be a serious problem in carrying out the survey?

7. Refer to the study of Exercise 4.

    (a) Discuss likely sources of nonsampling errors and suggest how these could be kept to a minimum.

    (b) Is nonresponse likely to be a serious issue in this study? If so, what might be done about it?

8. For the study of Exercise 5, discuss the potential for nonsampling errors and nonresponse. Indicate how you would go about minimizing these problems.

9. One approach to nonresponse of a particular kind is the *recall method*. A survey of households is carried out by having interviewers call on a Thursday evening. Those households where no one was home are called again the following Thursday evening. This process can be continued, so that households which could not be reached at the first two attempts are recontacted on the next Thursday evening. What might be the value of information obtained in this fashion?

18.3 SIMPLE RANDOM SAMPLING

In the rest of this chapter, we will be interested in problems where a sample of n individuals or objects is to be drawn from a population containing a total of N individuals. In practical applications, many different schemes have been employed for the selection of such samples. The bulk of our discussion will be concentrated on **probability sampling** methods—that is, procedures where some mechanism involving *chance* is used to determine the sample members and the probability of any particular sample being drawn is known.

The simplest scheme of this sort is known as **simple random sampling,** which was introduced in Section 6.1. For convenience, we repeat the definition in the box.

Definition

Suppose that it is required to select a sample of n objects from a population of N objects. A **simple random sampling** procedure is one in which every possible sample of n objects is equally likely to be chosen.

If a listing of the population members is available, then the taking of a simple random sample is straightforward. Suppose we number the population members from 1 through N. A simple random sample could be achieved by placing N numbered balls in a box, mixing thoroughly, and drawing out n of them. As we noted in Section 6.1, a table of **random numbers** can be used to attain the same objective more efficiently. Essentially, such tables reproduce the properties of drawing balls from boxes, though with the proviso that a ball, once drawn, is replaced and could be redrawn. Some random numbers are listed in Table 4 of the Appendix.

Suppose that our population contains 1,000 individuals, numbered 1 through 1,000, and that a random sample of 100 population members is required. The procedure is to start at some arbitrary point in the table, say the top of the fifth column, and read off the last three digits of the random numbers. Thus, the first five are

319	499	166	082	420

The population members with these numbers are then the first five sample members. Continuing in this way, we obtain the rest of our sample, with 000 corresponding to population member 1,000. Any number drawn that has already been obtained is ignored, and the process continues until 100 different numbers have been obtained. This is known as **sampling without replacement.** (An alternative, **sampling with replacement,** which allows the possibility of an individual being included in the sample more than once, will not be discussed here.)

The reader may have guessed that the choice of a population of exactly 1,000 individuals helped make the exposition in the previous paragraph relatively straightforward. Unfortunately, in the real world, populations rarely come in powers of ten! For instance, suppose we want to sample 100 schools from a list of 1,395 colleges in the United States. Obviously, now, it is necessary to go to four-digit random numbers. Beginning again at the top of the fifth column in the table, we obtain the first five entries:

1319	4499	3166	5082	5420

Notice that only one of these numbers lies between 1 and 1,395. In principle, we could continue drawing until 100 different numbers in this range had been found, discarding all the others. However, this would be a rather time-consuming chore. It can be speeded up by noticing that only numbers up to 13 in the first two digits are of any use to us. Suppose, then, that we adopt the following strategy for the first two digits:

$$\text{Replace} \quad 00, \quad 14, \quad 28, \quad \ldots, \quad 84 \quad \text{by} \quad 00$$

$$\text{Replace} \quad 01, \quad 15, \quad 29, \quad \ldots, \quad 85 \quad \text{by} \quad 01$$

$$\vdots \qquad\qquad\qquad\qquad \vdots$$

$$\text{Replace} \quad 13, \quad 27, \quad 41, \quad \ldots, \quad 97 \quad \text{by} \quad 13$$

Numbers beginning 98 or 99 will still be ignored. Thus, the five entries given previously become

1319	0299	0366	0882	1220

Continuing in this way, we can complete our sample.

An even quicker procedure is possible if the population list is arranged in some fashion unconnected with the subject of interest. Then, if we want our sample to include, say, one-tenth of the population members, we simply take every tenth name on the list. This is known as **systematic sampling,** and is quite commonly used in practice. A systematic sample is analyzed in the same fashion as a simple random sample on the grounds that, relative to the subject of inquiry, the population listing is already in random order. The danger is that there could be some subtle, unsuspected link between the ordering of the population and the subject under study. If this were so, a bias would be induced if systematic sampling were employed.

ANALYSIS OF RESULTS FROM SIMPLE RANDOM SAMPLING

We turn now to the analysis of the sample results. Specifically, we will concentrate on estimating the population mean, total, and proportion. It will be assumed that the sample is sufficiently large that recourse to the Central Limit Theorem is appropriate.

Suppose that the population contains N members and that a simple random sample of size n is taken. We denote by μ the unknown population mean, and by x_1, x_2, . . . , x_n the actual observed sample values. Results for the estimation of the population mean are given in the box.

Estimation of the Population Mean

Let $x_1, x_2, \ldots, x_n$ denote the values observed from a simple random sample of size n, taken from a population of N members with mean μ. Then:

(i) The sample mean is an unbiased estimator of the population mean, μ. The point estimate is

$$\bar{x} = \frac{1}{n} \sum_{i=1}^{n} x_i$$

(ii) An unbiased estimation procedure for the variance of the sample mean yields the point estimate

$$\hat{\sigma}_{\bar{x}}^2 = \frac{s^2}{n} \cdot \frac{N-n}{N}$$

where

$$s^2 = \frac{1}{n-1} \sum_{i=1}^{n} (x_i - \bar{x})^2$$

is the sample variance.

(iii) Provided the sample size is large, $100(1 - \alpha)\%$ confidence intervals for the population mean are given by

$$\bar{x} - z_{\alpha/2}\hat{\sigma}_{\bar{x}} < \mu < \bar{x} + z_{\alpha/2}\hat{\sigma}_{\bar{x}}$$

where $z_{\alpha/2}$ is that number for which

$$P(Z > z_{\alpha/2}) = \frac{\alpha}{2}$$

and the random variable Z follows a standard normal distribution.

Notice that we use the **finite population correction factor** $(N - n)/N$. This factor is similar to one introduced in Section 6.2, and allows us to deal with cases where the number of sample members is not a negligible proportion of the number of population members.[1]

Example 18.1

An auditor, faced with a total of 950 accounts receivable, takes a random sample of 50, finding a sample mean of \$83.65 and sample standard deviation \$29.82. Estimate the mean value of accounts receivable for this population.

Denote the population mean by μ. We have

$$N = 950; \qquad n = 50; \qquad \bar{x} = 83.65; \qquad s = 29.82$$

The usual point estimate for the population mean is

$$\bar{x} = 83.65$$

To obtain interval estimates, we first find

$$\hat{\sigma}_{\bar{x}}^2 = \frac{s^2}{n} \cdot \frac{N - n}{N} = \frac{(29.82)^2}{50} \cdot \frac{900}{950} = 16.849$$

so that by taking square roots, we obtain

$$\hat{\sigma}_{\bar{x}} = 4.105$$

For a 95% confidence interval, from Table 3 of the Appendix,

$$z_{\alpha/2} = z_{.025} = 1.96$$

Hence, the 95% confidence interval for the mean value of accounts receivable in this population is

$$83.65 - (1.96)(4.105) < \mu < 83.65 + (1.96)(4.105)$$

or

---

[1] We saw in Section 6.2 that, when sampling from a finite population, the variance of the sample mean is $(\sigma^2/n)[(N - n)/(N - 1)]$, where σ^2 is the population variance. It can further be shown, for a finite population, that the expected value of the sample variance is $N\sigma^2/(N - 1)$. Hence, when the sample variance is used to estimate the population variance, the appropriate finite population correction factor is $(N - n)/N$.

$$75.60 < \mu < 91.70$$

That is, the interval runs from \$75.60 to \$91.70.

Frequently, interest centers on the population total rather than the mean. For example, the publisher of a business statistics text will want an estimate of the total number of students taking business statistics courses in all United States colleges. Inference about the population total is straightforward, the relevant results following from the fact that, in our notation,

$$\text{Population total} = N\mu$$

The details are provided in the box.

Estimation of the Population Total

Suppose that we have a simple random sample of size n from a population of size N, and that the quantity to be estimated is the population total, $N\mu$. Then:

(i) An unbiased estimation procedure for the population total, $N\mu$, yields the point estimate $N\bar{x}$.

(ii) An unbiased estimation procedure for the variance of our estimator of the population total yields the point estimate

$$N^2 \hat{\sigma}_{\bar{x}}^2 = \frac{s^2}{n} N(N - n)$$

(iii) Provided the sample size is large, $100(1 - \alpha)\%$ confidence intervals for the population total are obtained from

$$N\bar{x} - z_{\alpha/2} N\hat{\sigma}_{\bar{x}} < N\mu < N\bar{x} + z_{\alpha/2} N\hat{\sigma}_{\bar{x}}$$

Example 18.2

From a simple random sample of 400 of the 1,395 United States colleges, it was found that the sample mean enrollment during the past year in business statistics courses was 320.8 students, and the sample standard deviation was found to be 149.7 students. Estimate the total number of students enrolled in business statistics courses in the year.

If the population mean is μ, we require to estimate $N\mu$, given

$$N = 1,395; \quad n = 400; \quad \bar{x} = 320.8; \quad s = 149.7$$

Our point estimate for the total is

$$N\bar{x} = (1,395)(320.8) = 447,516$$

That is, we estimate a total of 447,516 students in business statistics courses. To obtain interval estimates, we require

$$N^2 \hat{\sigma}_{\bar{x}}^2 = \frac{s^2}{n} N(N - n) = \frac{(149.7)^2}{400} (1,395)(995) = 77,764,413$$

so that taking square roots yields

$$N\hat{\sigma}_{\bar{x}} = 8,818.4$$

For a 99% confidence interval, from Table 3 of the Appendix,

$$z_{\alpha/2} = z_{.005} = 2.575$$

Hence, the 99% confidence interval for the population total is

$$447,516 - (2.575)(8,818.4) < N\mu < 447,516 + (2.575)(8,818.4)$$

or

$$424,809 < N\mu < 470,223$$

Thus, our interval runs from 424,809 to 470,223 students.

Finally, we consider the case where it is required to estimate the proportion, p, of individuals in the population possessing some specific characteristic. Inference about this proportion should be based on the hypergeometric distribution of Section 4.6 when the number of sample members is not very small compared to the number of population members. However, we will again assume that the sample size is large enough to allow the Central Limit Theorem to be invoked. The main results are given in the box.

Estimation of the Population Proportion

Let $\hat{p}$ be the proportion possessing a particular characteristic in a random sample of n observations from a population, a proportion p of whose members possess that characteristic. Then:

(i) The sample proportion is an unbiased estimator of the population proportion, p.

(ii) An unbiased estimation procedure for the variance of our estimator of the population proportion yields the point estimate

$$\hat{\sigma}_{\hat{p}}^2 = \frac{\hat{p}(1 - \hat{p})}{n - 1} \cdot \frac{N - n}{N}$$

(iii) Provided the sample size is large, $100(1 - \alpha)\%$ confidence intervals for the population proportion are given by

$$\hat{p} - z_{\alpha/2} \hat{\sigma}_{\hat{p}} < p < \hat{p} + z_{\alpha/2} \hat{\sigma}_{\hat{p}}$$

Example 18.3

From a simple random sample of 400 of the 1,395 colleges in the United States, it was found that business statistics was a two-semester course in 141 of the sampled colleges. Estimate the proportion of all colleges for which the course is two semesters long.

We want to estimate the population proportion, p, given

$$N = 1{,}395; \qquad n = 400; \qquad \hat{p} = \frac{141}{400} = 0.3525$$

Our point estimate of p is simply $\hat{p} = 0.3525$. That is, we estimate that the course is two semesters long in 35.25% of all colleges. In order to calculate interval estimates, we need

$$\hat{\sigma}_{\hat{p}}^2 = \frac{\hat{p}(1 - \hat{p})}{n - 1} \cdot \frac{N - n}{N} = \frac{(0.3525)(0.6475)}{399} \cdot \frac{995}{1{,}395} = 0.0004079$$

so that

$$\hat{\sigma}_{\hat{p}} = 0.0202$$

For a 90% confidence interval, from Table 3 of the Appendix,

$$z_{\alpha/2} = z_{.05} = 1.645$$

The 90% interval is, therefore,

$$0.3525 - (1.645)(0.0202) < p < 0.3525 + (1.645)(0.0202)$$

or

$$0.3193 < p < 0.3857$$

Thus, the 90% confidence interval for the percentage of all colleges in which business statistics is a two-semester course runs from 31.93% to 38.57%. Figure 18.2 shows

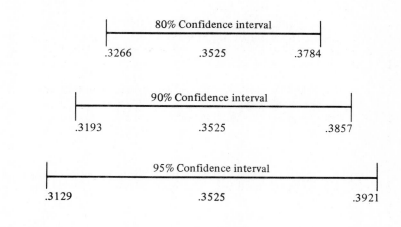

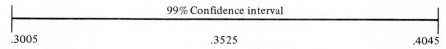

FIGURE 18.2 80%, 90%, 95%, and 99% confidence intervals for population proportion of all colleges in which business statistics is a two-semester course, based on the data of Example 18.3

also the 80%, 95%, and 99% confidence intervals for the population proportion, based on these data.

EXERCISES

10. Consult the day's *Wall Street Journal* to obtain a list of all stocks traded on the New York Stock Exchange. Using the random numbers of Table 4 of the Appendix, draw a simple random sample of twenty of these stocks. For your sample, find the mean percentage increase in price over the last week.

11. Obtain from your local newspaper a list of all houses advertised for sale in your city. Using the random numbers in Table 4 of the Appendix, draw a simple random sample of fifteen advertisements, and find the sample mean of the advertised prices.

12. A town contains 15,672 households. From a complete listing of these households, you want to draw a simple random sample of 100. Explain how you would use a table of random numbers to do this.

13. Using the listing of quoted mutual funds in the day's *Wall Street Journal,* take a simple random sample of thirty funds, and estimate the proportion of all funds that are no-load.

14. An auditor draws a simple random sample of 12, from a total of 184 accounts receivable. The amounts of these (in dollars) were

$$125.40; \quad 61.70; \quad 84.30; \quad 118.60; \quad 91.50; \quad 98.50;$$

$$61.70; \quad 66.00; \quad 87.90; \quad 131.30; \quad 84.50; \quad 91.60$$

(a) Find the sample mean.

(b) Using an unbiased estimation procedure, find an estimate of the variance of the sample mean.

15. In a particular city, a total of 1,118 mortgages were taken out last year. A random sample of 60 of these had mean amount $86,500 and standard deviation $18,900.

(a) Find an estimate of the variance of the sample mean, using an unbiased estimation procedure.

(b) Find a 95% confidence interval for the population mean.

16. A small corporation employs 526 manual workers. In a random sample of 80 of these, the mean length of time employed was 6.9 years and the sample standard deviation was 4.3 years. Find 90%, 95%, and 99% confidence intervals for the mean length of time employed for all these workers.

17. State whether each of the following statements is true or false:

(a) For a given number of population members and a given sample variance, the larger is the number of sample members, the wider is a 95% confidence interval for the population mean.

(b) For a given number of population members and a given number of sample members, the larger is the sample variance, the wider is a 95% confidence interval for the population mean.

(c) For a given number of sample members and a given sample variance, the larger is the number of population members, the wider is a 95% confidence interval for the population mean. Provide a *verbal* justification for your answer.

(d) For a given number of population members, a given number of sample members, and a given sample variance, a 95% confidence interval for the population mean is wider than a 90% confidence interval for the population mean.

18. Show that our estimate for the variance of the sample mean can be written

$$\hat{\sigma}_{\bar{x}}^2 = s^2 \left(\frac{1}{n} - \frac{1}{N} \right)$$

When $n = N$, it follows that $\hat{\sigma}_{\bar{x}}^2 = 0$. Why is this so?

19. For the data of Exercise 15, find a 90% confidence interval for the total value of all mortgages taken out in the city during last year.

20. Using the data of Exercise 16, find an 80% confidence interval for the total length of time employed by all 526 of this corporation's manual workers.

21. A simple random sample of 50 from a total of 430 accounts receivable found a mean value of $127.93 and a sample standard deviation of $42.27. Find 95% and 99% confidence intervals for the total value of all accounts receivable for this population.

22. The dean of a college wants to estimate the total amount of time spent by 120 faculty members in meetings during a given week. A random sample of 30 of the faculty were asked to keep diary records during the week. When the results were analyzed it was found that these faculty members spent a total of 126 hours in meetings. The sample standard deviation was 2.8 hours. Find a 90% confidence interval for the total number of hours spent in meetings during the week by these 120 faculty members.

23. A simple random sample of 400 from a total of 1,395 colleges in the United States contained 37 colleges that use the text *Statistics Made Difficult and Boring* by J. T. Ripper. Find a 95% confidence interval for the proportion of all colleges using Ripper's text.

24. A corporation is contemplating introducing a form of flexible-time working hours for 200 of its managers. A random sample of 40 of these managers found 23 in favor of this plan. Find 90% and 95% confidence intervals for the proportion of all managers favoring the plan.

25. A company receives a shipment of 150 parts. In a random sample of 45 of these parts, 11 failed to meet the required specifications. Find 95% and 99% confidence intervals for the proportion of all parts not meeting the specifications.

26. An auditor selects a random sample of 80 account balances from a population of 400. Of these sample balances, 12 were in error. Find a 95% confidence interval for the total number of balances in error in this population.

18.4 STRATIFIED SAMPLING

Suppose you decide to investigate the views of students on your campus on the issue of a university's holding shares in corporations that operate in South Africa. This is a sensitive topic, and the framing of appropriately worded questions could be difficult. It is likely that you would want to ask several questions of every sample member and so, given limited resources, would be able to take only a fairly small sample. You would presumably select a simple random sample of, say, 100 students from a list of all students on campus. Suppose, however, that on closer inspection of the records of the sample members, you find that only two of them are business majors, though the population proportion of business majors is far higher than this. Your problem at this stage is twofold. First, you may well be interested in comparing the views of business majors with those of the rest of the student population. This is hardly feasible given their minimal representation in your sample. Second, you may suspect that the views of business majors will differ on this question from those of their fellow students. If that is the case, you would worry about the reliability of inference based on a sample in which this group is seriously under-represented.

This second point is one on which, up to this stage, we have provided no guidance. You could perhaps console yourself with the thought that, since you have taken a random sample, any estimators derived in the usual way will be unbiased, and the resulting inference, in the statistical sense, will be strictly valid. However, a little reflection should convince you that this is scant consolation indeed! All that un-biasedness promises is that, if the sampling procedure is repeated a very large number of times, and the estimator calculated, its average will be equal to the corresponding population value. But, in fact, you are *not* going to repeat the sampling procedure a large number of times. You have to base your conclusions on *just a single sample,* and the fact that, in other samples you might have drawn, business majors could be over-represented, so that things "average out" in the long run, is not terribly useful.

A second tempting possibility, which is in many ways preferable to proceeding with the original sample, exists. You could simply curse your luck, discard the original sample, and take another. If the constitution of the sample achieved at the second attempt looks more representative of the population at large, you may well be better off to proceed with it. The difficulty, now, is that the sampling procedure you have adopted in effect—where the population is to be sampled until you achieve a sample you like the looks of—is very difficult to formalize, and consequently the sample results are very hard to analyze with any statistical validity. Certainly, this is no longer simple random sampling, and the procedures of the previous section are, therefore, strictly not valid.

Fortunately, a third alternative sampling scheme exists to afford protection against just this type of problem. If it is suspected at the outset that particular identifiable characteristics of population members are germane to the subject of inquiry, or if particular subgroups of the population are of special interest to the investigator, it is not necessary (and probably not desirable) to be content with simple random sampling as a means of selecting the sample members. Instead the population can be broken down into subgroups, or **strata,** and a simple random sample taken from each stratum. The only requirement is that each individual member of the population be identifiable as belonging to one, and only one, of the strata.

Stratified Random Sampling

Suppose that a population of N individuals can be subdivided into K mutually exclusive and collectively exhaustive groups, or **strata. Stratified random sampling** is the selection of independent simple random samples from each stratum of the population.

If the K strata in the population contain $N_1, N_2, \ldots, N_K$ members, then

$$N_1 + N_2 + \cdots + N_K = N$$

There is no need to take the same number of sample members from every stratum.

$\longrightarrow$

Denote the numbers in the sample by $n_1, n_2, \ldots, n_K$. Then the total number of sample members is

$$n = n_1 + n_2 + \cdots + n_K$$

The population of students, whose views are to be canvassed on the subject of investment in South Africa, could be divided into two strata—business majors and nonbusiness majors. Less straightforward stratification is also possible. Suppose, on some other topic, you believe that a student's sex and class year (senior, junior, sophomore, or freshman) are both potentially relevant. In that case, to satisfy the requirement that the strata be mutually exclusive and collectively exhaustive, eight strata—senior women, senior men, and so on—are needed.

ANALYSIS OF RESULTS FROM STRATIFIED RANDOM SAMPLING

The analysis of the results of a stratified random sample is relatively straightforward. We will denote by $\mu_1, \mu_2, \ldots, \mu_K$ the population means in the K strata, and by $\bar{x}_1, \bar{x}_2, \ldots, \bar{x}_K$ the corresponding sample means. Consider a particular stratum, say the jth. Then, since a simple random sample has been taken in this stratum, the stratum sample mean is an unbiased estimator of the population mean μ_j. Also, from an unbiased estimation procedure for the variance of the stratum sample mean, we have the point estimate

$$\hat{\sigma}_{\bar{x}_j}^2 = \frac{s_j^2}{n_j} \cdot \frac{N_j - n_j}{N_j}$$

where s_j^2 is the sample variance in the jth stratum. Inference about indivdual strata can, therefore, be made in the same way as in Section 18.3.

Generally, we also want to make inference about the overall population mean, μ, which is

$$\mu = \frac{N_1 \mu_1 + N_2 \mu_2 + \cdots + N_K \mu_K}{N} = \frac{1}{N} \sum_{j=1}^{K} N_j \mu_j$$

A natural point estimate is provided by

$$\bar{x}_{\text{st}} = \frac{1}{N} \sum_{j=1}^{K} N_j \bar{x}_j$$

An unbiased estimator of the variance of the estimator of μ follows from the fact that the samples in each stratum are independent of one another, and the point estimate is given by

$$\hat{\sigma}_{\bar{x}_{\text{st}}}^2 = \frac{1}{N^2} \sum_{j=1}^{K} N_j^2 \hat{\sigma}_{\bar{x}_j}^2$$

Inference about the overall population mean can be based on these results, as summarized in the box.

Estimation of the Population Mean from a Stratified Random Sample

Suppose that random samples of n_j individuals are taken from strata containing N_j individuals ($j = 1, 2, \ldots, K$). Let

$$\sum_{j=1}^{K} N_j = N \quad \text{and} \quad \sum_{j=1}^{K} n_j = n$$

Denote the sample means and variances in the strata by $\bar{x}_j$ and $s_j^2 (j = 1, 2, \ldots, K)$, and the overall population mean by μ. Then:

(i) An unbiased estimation procedure for the overall population mean μ yields the point estimate

$$\bar{x}_{st} = \frac{1}{N} \sum_{j=1}^{K} N_j \bar{x}_j$$

(ii) An unbiased estimation procedure for the variance of our estimator of the overall population mean yields the point estimate

$$\hat{\sigma}_{\bar{x}_{st}}^2 = \frac{1}{N^2} \sum_{j=1}^{K} N_j^2 \hat{\sigma}_{\bar{x}_j}^2$$

where

$$\hat{\sigma}_{\bar{x}_j}^2 = \frac{s_j^2}{n_j} \cdot \frac{N_j - n_j}{N_j}$$

(iii) Provided the sample size is large, $100(1 - \alpha)\%$ confidence intervals for the population mean are obtained from

$$\bar{x}_{st} - z_{\alpha/2} \hat{\sigma}_{\bar{x}_{st}} < \mu < \bar{x}_{st} + z_{\alpha/2} \hat{\sigma}_{\bar{x}_{st}}$$

Example 18.4

A small town contains a total of 4,200 households. The town is divided into three districts, containing respectively 1,150, 2,120, and 930 households. A stratified random sample of 450 households contains respectively 123, 227, and 100 households from these districts. Sample members were asked to estimate their annual incomes in the current year. The sample means and standard deviations (in dollars) were found to be

$$\bar{x}_1 = 16,228; \quad s_1 = 4,187$$

$$\bar{x}_2 = 21,593; \quad s_2 = 6,195$$

$$\bar{x}_3 = 32,711; \quad s_3 = 8,243$$

Estimate the mean income, μ, for this population.

In our notation, we have

$$N_1 = 1,150; \qquad N_2 = 2,120; \qquad N_3 = 930; \qquad N = 4,200$$

$$n_1 = 123; \qquad n_2 = 227; \qquad n_3 = 100; \qquad n = 450$$

Our estimate of the population mean is

$$\bar{x}_{st} = \frac{1}{N} \sum_{j=1}^{K} N_j \bar{x}_j = \frac{(1,150)(16,228) + (2,120)(21,593) + (930)(32,711)}{4,200}$$

$$= 22,585.85$$

Thus, the estimated mean family income for this town is \$22,585.85.

The next step is to calculate the quantities

$$\hat{\sigma}^2_{\bar{x}_1} = \frac{s_1^2}{n_1} \cdot \frac{N_1 - n_1}{N_1} = \frac{(4,187)^2}{123} \cdot \frac{1,027}{1,150} = 127,284$$

$$\hat{\sigma}^2_{\bar{x}_2} = \frac{s_2^2}{n_2} \cdot \frac{N_2 - n_2}{N_2} = \frac{(6,195)^2}{227} \cdot \frac{1,893}{2,120} = 150,963$$

$$\hat{\sigma}^2_{\bar{x}_3} = \frac{s_3^2}{n_3} \cdot \frac{N_3 - n_3}{N_3} = \frac{(8,243)^2}{100} \cdot \frac{830}{930} = 606,409$$

Together with the individual stratum sample means, these quantities can be used to compute confidence intervals for the population means of the three strata, precisely as in Example 18.1. Here, however, we concentrate on the overall population mean. To obtain confidence intervals for this quantity, we need

$$\hat{\sigma}^2_{\bar{x}_{st}} = \frac{1}{N^2} \sum_{j=1}^{K} N_j^2 \hat{\sigma}^2_{\bar{x}_j}$$

$$= \frac{(1,150)^2(127,284) + (2,120)^2(150,963) + (930)^2(606,409)}{(4,200)^2} = 77,738.3$$

and, on taking square roots,

$$\hat{\sigma}_{\bar{x}_{st}} = 278.82$$

For a 95% confidence interval for the population mean, from Table 3 of the Appendix,

$$z_{\alpha/2} = z_{.025} = 1.96$$

Thus, the 95% confidence interval for population mean household income is

$$22,585.85 - (1.96)(278.82) < \mu < 22,585.85 + (1.96)(278.82)$$

or

$$22,039.36 < \mu < 23,132.34$$

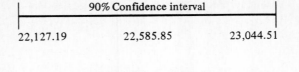

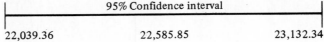

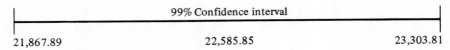

FIGURE 18.3 90%, 95%, and 99% confidence intervals for population mean household income, based on the data of Example 18.4

The 95% confidence interval runs from $22,039.36 to $23,132.34. Figure 18.3 shows also 90% and 99% confidence intervals for the mean of household incomes in this town.

Since the population total is the product of the population mean and the number of population members, these procedures can readily be modified to allow its estimation, as described in the box.

Estimation of the Population Total from a Stratified Random Sample

Suppose that we have random samples of n_j individuals from strata containing N_j individuals ($j = 1, 2, \ldots, K$), and that the quantity to be estimated is the population total, $N\mu$. Then:

(i) An unbiased estimation procedure for $N\mu$ leads to the point estimate

$$N\bar{x}_{st} = \sum_{j=1}^{K} N_j \bar{x}_j$$

(ii) An unbiased estimation procedure for the variance of our estimator of the population total yields the estimate

$$N^2 \hat{\sigma}_{\bar{x}_{st}}^2 = \sum_{j=1}^{K} N_j^2 \hat{\sigma}_{\bar{x}_j}^2$$

(iii) Provided the sample size is large, $100(1 - \alpha)\%$ confidence intervals for the population total are obtained from

$$N\bar{x}_{st} - z_{\alpha/2} N\hat{\sigma}_{\bar{x}_{st}} < N\mu < N\bar{x}_{st} + z_{\alpha/2} N\hat{\sigma}_{\bar{x}_{st}}$$

Example
18.5

Of the 1,395 colleges in the United States, 364 have 2-year programs and 1,031 are 4-year schools. A simple random sample of forty 2-year schools and an independent simple random sample of sixty 4-year schools were taken. The sample means and standard deviations of numbers of students enrolled in the past year in business statistics courses are given in the table. Estimate the total annual enrollment in business statistics courses.

	2-YEAR SCHOOLS	4-YEAR SCHOOLS
Mean	154.3	411.8
Standard deviation	87.3	219.9

In our notation, we have

$$N_1 = 364; \quad n_1 = 40; \quad \bar{x}_1 = 154.3; \quad s_1 = 87.3$$

$$N_2 = 1{,}031; \quad n_2 = 60; \quad \bar{x}_2 = 411.8; \quad s_2 = 219.9$$

Our estimate of the population total is

$$N\bar{x}_{st} = \sum_{j=1}^{K} N_j \bar{x}_j = (364)(154.3) + (1{,}031)(411.8) = 480{,}731$$

Thus, the estimated total number of students in business statistics courses is 480,731. Next, we require the quantities

$$\hat{\sigma}_{\bar{x}_1}^2 = \frac{s_1^2}{n_1} \cdot \frac{N_1 - n_1}{N_1} = \frac{(87.3)^2}{40} \cdot \frac{324}{364} = 169.59$$

$$\hat{\sigma}_{\bar{x}_2}^2 = \frac{s_2^2}{n_2} \cdot \frac{N_2 - n_2}{N_2} = \frac{(219.9)^2}{60} \cdot \frac{971}{1{,}031} = 759.03$$

Finally,

$$N^2 \hat{\sigma}_{\bar{x}_{st}}^2 = \sum_{j=1}^{K} N_j^2 \hat{\sigma}_{\bar{x}_j}^2 = (364)^2(169.59) + (1{,}031)^2(759.03) = 829{,}289{,}280$$

and, on taking square roots, we obtain

$$N\hat{\sigma}_{\bar{x}_{st}} = 28{,}797$$

For a 99% confidence interval, from Table 3 of the Appendix,

$$z_{\alpha/2} = z_{.005} = 2.575$$

The required 99% interval is, therefore,

$$480{,}731 - (2.575)(28{,}797) < N\mu < 480{,}731 + (2.575)(28{,}797)$$

or

$$406,579 < N\mu < 554,883$$

Thus, our 99% confidence interval runs from 406,579 to 554,883 students enrolled.

Next we consider the problem of estimating a population proportion, based on a stratified random sample. Let $p_1, p_2, \ldots, p_K$ be the population proportions in the K strata, and $\hat{p}_1, \hat{p}_2, \ldots, \hat{p}_K$ the corresponding sample proportions. Then, if p denotes the overall population proportion, its estimation is based on the fact that

$$p = \frac{N_1 p_1 + N_2 p_2 + \cdots + N_K p_K}{N} = \frac{1}{N} \sum_{j=1}^{K} N_j p_j$$

The procedures are described in the box.

Estimation of the Population Proportion from a Stratified Random Sample

Suppose that we have random samples of n_j individuals from strata containing N_j individuals ($j = 1, 2, \ldots, K$). Let p_j be the population proportion and $\hat{p}_j$ the sample proportion, in the jth stratum, of those possessing a particular characteristic. If p is the overall population proportion, then:

(i) An unbiased estimation procedure for p yields

$$\hat{p}_{st} = \frac{1}{N} \sum_{j=1}^{K} N_j \hat{p}_j$$

(ii) An unbiased estimation procedure for the variance of our estimator of the overall population proportion yields

$$\hat{\sigma}_{\hat{p}_{st}}^2 = \frac{1}{N^2} \sum_{j=1}^{K} N_j^2 \hat{\sigma}_{\hat{p}_j}^2$$

where

$$\hat{\sigma}_{\hat{p}_j}^2 = \frac{\hat{p}_j (1 - \hat{p}_j)}{n_j - 1} \cdot \frac{N_j - n_j}{N_j}$$

is the estimate of the variance of the sample proportion in the jth stratum.

(iii) Provided the sample size is large, $100(1 - \alpha)\%$ confidence intervals for the population proportion are obtained from

$$\hat{p}_{st} - z_{\alpha/2} \hat{\sigma}_{\hat{p}_{st}} < p < \hat{p}_{st} + z_{\alpha/2} \hat{\sigma}_{\hat{p}_{st}}$$

Example 18.6

In the study of Example 18.5, it was found that business statistics was taught by members of the economics department in 7 of the 2-year colleges and 13 of the 4-year colleges in the sample. Estimate the proportion of all colleges in which this course is taught in the economics department.

We have

$$N_1 = 364; \qquad n_1 = 40; \qquad \hat{p}_1 = \frac{7}{40} = 0.175$$

$$N_2 = 1{,}031; \qquad n_2 = 60; \qquad \hat{p}_2 = \frac{13}{60} = 0.217$$

Our estimate of the population proportion is

$$\hat{p}_{st} = \frac{1}{N} \sum_{j=1}^{K} N_j \hat{p}_j = \frac{(364)(0.175) + (1{,}031)(0.217)}{1{,}395} = 0.206$$

Thus, it is estimated that in 20.6% of all colleges, the course is taught by the economics department.

Next we need the quantities

$$\hat{\sigma}_{\hat{p}_1}^2 = \frac{\hat{p}_1(1 - \hat{p}_1)}{n_1 - 1} \cdot \frac{N_1 - n_1}{N_1} = \frac{(0.175)(0.825)}{39} \cdot \frac{324}{364} = 0.003295$$

$$\hat{\sigma}_{\hat{p}_2}^2 = \frac{\hat{p}_2(1 - \hat{p}_2)}{n_2 - 1} \cdot \frac{N_2 - n_2}{N_2} = \frac{(0.217)(0.783)}{59} \cdot \frac{971}{1{,}031} = 0.002712$$

Together with the individual stratum sample proportions, these values can be used to calculate confidence intervals for the two stratum population proportions, exactly as in Example 18.3. Here we will focus on interval estimation for the overall population proportion, for which we require

$$\hat{\sigma}_{\hat{p}_{st}}^2 = \frac{1}{N^2} \sum_{j=1}^{K} N_j^2 \hat{\sigma}_{\hat{p}_j}^2 = \frac{(364)^2(0.003295) + (1{,}031)^2(0.002712)}{(1{,}395)^2}$$

$$= 0.001706$$

so that, taking square roots yields

$$\hat{\sigma}_{\hat{p}_{st}} = 0.0413$$

For a 90% confidence interval, from Table 3 of the Appendix,

$$z_{a/2} = z_{.05} = 1.645$$

The 90% confidence interval for the population proportion is, then,

$$0.206 - (1.645)(0.0413) < p < 0.206 + (1.645)(0.0413)$$

or

$$0.138 < p < 0.274$$

Thus, our interval runs from 13.8% to 27.4% of all colleges.

ALLOCATION OF SAMPLE EFFORT AMONG STRATA

The question of the allocation of the sample effort among the various strata remains to be discussed. Assuming that a total of n sample members are to be selected, how many of these sample observations should be allocated to each stratum? In fact, the survey in question may have multiple objectives, so that no clear-cut answer is available. Nevertheless, it is possible to specify the following criteria for choice that the investigator might keep in mind:

(a) If little or nothing is known beforehand about the population, and if there are no strong requirements for the production of information about sparsely populated individual strata, then a natural choice is **proportional allocation.** Here, the proportion of sample members in any stratum is the same as the proportion of population members in that stratum. Thus, for the jth stratum, we have

$$\frac{n_j}{n} = \frac{N_j}{N}$$

so that

$$n_j = \frac{N_j}{N} \cdot n$$

This intuitively reasonable allocation mechanism is frequently employed in practice, and generally provides a satisfactory analysis.

(b) It is sometimes the case that strict adherence to proportional allocation will produce relatively few observations in strata in which the investigator is particularly interested. In that case, inference about the population parameters of these particular strata could be quite imprecise. In these circumstances, one might prefer to allocate more observations to such strata than is dictated by proportional allocation. In Examples 18.5 and 18.6, 364 of the 1,395 colleges are 2-year schools, and a sample of 100 observations is to be taken. If proportional allocation had been used, the number of 2-year schools in the sample would have been

$$n_1 = \frac{N_1}{N} \cdot n = \frac{364}{1,395} \cdot 100 = 26$$

Since the publisher was particularly interested in acquiring information about this market, it was thought that a sample of only 26 observations would be inadequate. For this reason, 40 of the 100 sample observations were allocated to this stratum.

(c) If the sole objective of a survey is to estimate as precisely as possible an overall population parameter, such as the mean, total, or proportion, and if enough is known about the population, then it is possible to derive an **optimal allocation.** If it is required to estimate an overall population mean or total, and if the population variances in the individual strata are denoted σ_j^2, then it can be shown that the most precise estimators are obtained when

$$n_j = \frac{N_j \sigma_j}{\sum\limits_{i=1}^{K} N_i \sigma_i} \cdot n \qquad (18.4.1)$$

This formula is intuitively plausible. Compared with proportional allocation, it allocates relatively more sample effort to those strata in which the population variance is highest. That is to say, a larger sample size is required where the greater population variability exists. Thus, in Example 18.4, where in fact proportional allocation was used, if the differences observed in the sample standard deviations correctly reflect differences in the population quantities, it would have been preferable to have taken more observations in the third stratum and less in the first.

An immediate objection arises to the use of formula 18.4.1. It requires knowledge of the population standard deviations, σ_j, whereas very often, one will not even have worthwhile estimates of these values before the sample is taken. We will return to this point in the final section of the chapter.

For estimating the overall population proportion, estimators with the smallest possible variance are obtained by choosing sample sizes

$$
n_j = \frac{N_j \sqrt{p_j(1 - p_j)}}{\sum\limits_{i=1}^{K} N_i \sqrt{p_i(1 - p_i)}} \cdot n
\qquad (18.4.2)
$$

Compared with proportional allocation, this formula allocates more sample observations to those strata in which the true population proportions are closest to 0.5, for if a proportion is close to 0 or 1, this can be learned with a fair amount of assurance from a relatively small sample. The difficulty in using formula 18.4.2 is that it involves the unknown proportions p_j—the very quantities that the survey is designed to estimate. Nevertheless, it is sometimes the case that prior knowledge about the population can provide at least a rough idea as to which strata have proportions closest to 0.5. In Example 18.6, the sample proportions suggest that the number of 2-year colleges in the sample should have been less than the number resulting from proportional allocation. The same conclusion also holds for this study when one compares the sample standard deviations of Example 18.5 with formula 18.4.1. In spite of this, it was decided that *more*, rather than fewer, 2-year colleges should be included in the sample. The reason for this decision was that, in this particular study, the publisher was anxious to obtain reliable information about both the 2-year and 4-year college markets.

This illustration serves as an example of an important point. Although the division of sample effort suggested by formulas 18.4.1 and 18.4.2 is often referred to as the *optimum allocation,* it is optimal only with regard to the narrow criterion of efficient estimation of overall population parameters. Frequently, in practice, surveys have broader objectives than this, in which case it may well be reasonable to depart from the optimum allocation.

EXERCISES

27. A restaurant chain has 50 restaurants in Illinois, 40 in Indiana, and 45 in Ohio. Management is considering adding to the menus a new delicacy item. In order to test the likely demand for this item, it was introduced on the menus of random samples of 12 restaurants in Illinois, 10 in Indiana, and 10 in Ohio, for a period of 1 week. The sample means and standard deviations for the numbers of orders per restaurant received are shown in the table.

	ILLINOIS	INDIANA	OHIO
Mean	21.2	13.3	26.1
Standard deviation	12.8	11.4	9.2

(a) Use an unbiased estimation procedure to find an estimate of the mean number of orders per restaurant that might be expected in a week for all 135 restaurants in the chain.

(b) Find an estimate of the variance of the estimator in part (a), using an unbiased estimation procedure.

(c) Find a 95% confidence interval for the population mean number of orders per restaurant.

28. A country club has two types of membership—regular and out-of-town. There are 350 regular members and 82 out-of-town members. The directors have formed an impression of an aging membership and are contemplating a campaign to recruit new younger members. In order to obtain a firmer assessment of the current picture, a random sample of 50 regular members is taken, revealing an average age of 52.4 years, and a sample standard deviation of 10.2 years. An independent random sample of 30 out-of-town members yielded a mean age of 47.6 years and a sample standard deviation of 8.5 years.

(a) Find a 90% confidence interval for the mean age of all regular members.

(b) Find an estimate for the mean age of all members, using an unbiased estimation procedure.

(c) Find 95% and 99% confidence intervals for the mean age of all members.

29. A company has four plants, employing a total of 2,150 blue-collar workers. Management has recently introduced a modified bonus scheme in each plant and wants to assess employee reaction. Random samples of workers from each plant were asked to rate, on a scale from 1 (very unfavorable) to 5 (very favorable), their attitudes to this new scheme. The results are summarized in the accompanying table.

	PLANT			
	1	2	3	4
N_i	720	550	490	390
n_i	50	40	30	20
$\bar{x}_i$	3.2	2.8	3.7	2.4
s_i	0.6	0.8	1.1	0.5

(a) Find a 99% confidence interval for the mean level of satisfaction for all employees in plant 1.

(b) Find an estimate of the mean level of satisfaction of all the company's employees, using an unbiased estimation procedure.

(c) Find 90% and 95% confidence intervals for the mean level of satisfaction of all the company's employees.

30. In a stratified random sample of United States colleges, the results given in the table were found for the number of years the current text had been employed in business statistics courses. (Adopting a new edition of the text currently used was not considered as switching texts.)

	COLLEGES	
	2-YEAR	4-YEAR
N_i	364	1,031
n_i	–40	60

	COLLEGES	
	2-YEAR	4-YEAR
$\bar{x}_i$	4.72	3.65
s_i	2.51	2.82

(a) Find a 95% confidence interval for the mean number of years the current text has been in use in 2-year colleges.

(b) Find a 95% confidence interval for the mean number of years the current text has been in use in 4-year colleges.

(c) Find a 95% confidence interval for the mean number of years the current text has been in use in all schools.

31. Using the information in Exercise 27, find a 99% confidence interval for the population total of all orders expected in a week for the delicacy item in the chain's 135 restaurants.

32. A corporation has 820 blue-collar workers. Of these, 430 have been with the company less than 5 years, 280 between 5 and 10 years, and 110 more than 10 years. The personnel director wants to estimate the total number of employee working days lost through sickness in the past year. Random samples of employees were taken and the results shown in the table were obtained.

	YEARS WITH COMPANY		
	LESS THAN 5	5–10	MORE THAN 10
n_i	20	25	30
$\bar{x}_i$	13.4	10.2	8.6
s_i	4.5	4.0	5.2

(a) Find an estimate of the total number of employee working days lost through sickness in a year by the corporation, using an unbiased estimation procedure.

(b) Find a 90% confidence interval for this total.

33. Of the 1,395 United States colleges, 364 have 2-year programs and 1,031 are 4-year schools. In a random sample of forty 2-year schools, it was found that the text *Statistics Can be Fun* by A. N. Optimist was used in ten of the schools. In an independent random sample of sixty 4-year schools, this text was used by eight of the sample members.

(a) Find an estimate of the proportion of all colleges using Optimist's text, using an unbiased estimation procedure.

(b) Find a 95% confidence interval for the proportion of all colleges using this text.

34. A consulting company has developed a short course on modern business forecasting methods for corporate executives. The first course was attended by 150 executives. From the information they supplied, it was concluded that the technical skills of 110 course members were more than adequate to follow the course material, while those of the remaining 40 were judged barely adequate. After the completion of the course, question-naires were sent to independent random samples of 25 people from each of these two groups, in order to obtain feedback that could lead to improved presentation in sub-sequent courses. Six of the more skilled and 13 of the less skilled group indicated that they believed the course had been too theoretical.

(a) Find an estimate of the proportion of all course members with this opinion, using an unbiased estimation procedure.

(b) Find 90% and 95% confidence intervals for this population proportion.

35. A village contains a total of 2,110 adult residents and is divided into three districts having, respectively, 820, 910, and 380 adult residents. The village council is considering converting a piece of wasteland into a public park. If undertaken, this project must be financed by an increase in property taxes. In order to assess local sentiment, independent random samples of 30 adults from each district were questioned. The numbers in favor of this scheme were, respectively, 10, 16, and 20.

   (a) Find an estimate of the proportion of all adults in the village favoring this scheme, using an unbiased estimation procedure.

   (b) Find 95% and 99% confidence intervals for this population proportion.

36. Refer to the data of Exercise 27. If a total sample of 32 restaurants is to be taken, determine how the sample would be allocated among the three states for each of the following schemes:

   (a) Proportional allocation

   (b) Optimum allocation, assuming the stratum population standard deviations are the same as the corresponding sample values

37. Refer to the data of Exercise 28. If a total sample of 80 club members is to be taken, determine how many out-of-town members would be included in the sample for each of the following schemes:

   (a) Proportional allocation

   (b) Optimum allocation, assuming the stratum population standard deviations are identical to the corresponding sample quantities

38. Refer to the data of Exercise 29. If a total of 140 workers are to be sampled, determine how many sample members would be selected from plant 4 under each of the following schemes:

   (a) Proportional allocation

   (b) Optimum allocation, assuming the stratum population standard deviations are the same as the corresponding sample quantities

39. Refer to the data of Exercise 30. If a total of 100 colleges are to be sampled, determine how many 2-year schools should be in the sample for each of the following schemes:

   (a) Proportional allocation

   (b) Optimum allocation, assuming the stratum population standard deviations are identical to the corresponding sample estimates

40. Refer to the data of Exercise 32. If a total of 75 employees are to be sampled, determine how many should be selected from those employees who have been with the company more than 10 years under each of the following schemes:

   (a) Proportional allocation

   (b) Optimum allocation, assuming the stratum population standard deviations are identical to the corresponding sample values

18.5 DETERMINING THE SAMPLE SIZE

An important aspect of the planning of any survey involves the determination of an appropriate number of sample members. Here, a number of factors may be relevant. If the procedure for contacting sample members is thought likely to lead to a high rate of nonresponse, this eventuality should be taken into account. In many instances the resources available to the investigator, in terms of time and money, will place constraints on what can be achieved. In this section, however, we abstract from such considerations and relate sample size to the variances of the estimators of population parameters, and consequently to the widths of resulting confidence intervals.

SAMPLE SIZES FOR SIMPLE RANDOM SAMPLING

We begin with the problem of estimating the population mean from a simple random sample of n observations. If the random variable $\bar{X}$ denotes the sample mean, we saw in Section 6.2 that the variance of this random variable is

$$\text{Var}(\bar{X}) = \sigma_{\bar{X}}^2 = \frac{\sigma^2}{n} \cdot \frac{N - n}{N - 1} \tag{18.5.1}$$

where σ^2 is the population variance. We now require to solve equation 18.5.1 for the sample size, n. Multiplying through by $(N - 1)n$ gives

$$(N - 1)\sigma_{\bar{X}}^2 n = N\sigma^2 - \sigma^2 n$$

so that

$$[(N - 1)\sigma_{\bar{X}}^2 + \sigma^2]n = N\sigma^2$$

and, finally,

$$n = \frac{N\sigma^2}{(N - 1)\sigma_{\bar{X}}^2 + \sigma^2} \tag{18.5.2}$$

If the population variance σ^2 is known, equation 18.5.2 allows the determination of the sample size n needed to achieve any specified value $\sigma_{\bar{X}}^2$ for the variance of the sample mean. Similar procedures are available if the quantity of interest is the population total.

Determination of Sample Size for Estimating Population Mean or Total Through Simple Random Sampling

Suppose that we want to estimate the mean of a population of N members, which has variance σ^2. If the desired variance, $\sigma_{\bar{X}}^2$, of the sample mean is specified, the required sample size is

$$n = \frac{N\sigma^2}{(N - 1)\sigma_{\bar{X}}^2 + \sigma^2}$$

Often it is more convenient to specify directly the width of confidence intervals for the population mean, rather than $\sigma_{\bar{X}}^2$. This is easily accomplished since, for example, a 95%

---

confidence interval for the population mean will extend an approximate amount $1.96\sigma_{\bar{x}}$ on each side of the sample mean.

   If the object of interest is the population total, we need only note that the variance of the sample estimator of this quantity is $N^2\sigma_{\bar{x}}^2$, and that confidence intervals for it extend an approximate amount $1.96N\sigma_{\bar{x}}$ on each side of $N\bar{x}$.

An obvious difficulty with the practical use of formula 18.5.2 is that it involves the population variance σ^2, which will typically be unknown. However, an investigator will often have some rough idea of the value of this quantity. In the next section we will see how it can sometimes be estimated from a preliminary sample of the population.

Example 18.7

As in Example 18.1, suppose that an auditor has a total of 950 accounts receivable, and intends to take a simple random sample to estimate their mean dollar value. From previous audits of this corporation's records, the auditor estimates that the population standard deviation is approximately \$32. The auditor wants to produce a 95% confidence interval for the population mean extending \$8 on each side of the sample mean. How many sample observations should be taken?

   We have

$$N = 950; \qquad \sigma = 32; \qquad 1.96\sigma_{\bar{x}} = 8 \quad (\text{so that } \sigma_{\bar{x}} = 4.08)$$

Then, the required sample size is

$$n = \frac{N\sigma^2}{(N-1)\sigma_{\bar{x}}^2 + \sigma^2} = \frac{(950)(32)^2}{(949)(4.08)^2 + (32)^2} = 57.8$$

Thus, a simple random sample of 58 observations will achieve the auditor's objective.

Next, we consider simple random sampling for the estimation of a population proportion, p. Let $\hat{p}_X$ be the random variable representing the sample proportion. Then, from the properties of the hypergeometric distribution discussed in Section 4.6, it follows that

$$\text{Var}(\hat{p}_X) = \sigma_{\hat{p}_X}^2 = \frac{p(1-p)}{n} \cdot \frac{N-n}{N-1}$$

where p is the population proportion. Solving this equation for the sample size gives

$$n = \frac{Np(1-p)}{(N-1)\sigma_{\hat{p}_X}^2 + p(1-p)} \tag{18.5.3}$$

Unfortunately, this expression involves the unknown population proportion p, whose estimation is the objective of the study. Two possibilities are open. We could either guess at the value of p, or follow the conservative option of replacing $p(1 - p)$ in 18.5.3 by its highest possible value, 0.25.

Determination of Sample Size for Estimating the Population Proportion Through Simple Random Sampling

Suppose that we want to estimate the proportion p of individuals in a population of size N who possess a certain attribute. If the desired variance, $\sigma_{\hat{p}_X}^2$, of the sample proportion is specified, the required sample size is

$$n = \frac{Np(1 - p)}{(N - 1)\sigma_{\hat{p}_X}^2 + p(1 - p)}$$

The largest possible value for this expression, whatever the value of p, is

$$n_{\max} = \frac{0.25N}{(N - 1)\sigma_{\hat{p}_X}^2 + 0.25}$$

A 95% confidence interval for the population proportion will extend an approximate amount $1.96\sigma_{\hat{p}_X}$ on each side of the sample proportion.

Example 18.8

As in Example 18.3, suppose we want to take a simple random sample of the 1,395 United States colleges to estimate the proportion in which the business statistics course is two semesters long. We want to ensure that, whatever the true proportion, a 95% confidence interval extends no further than 0.04 on each side of the sample proportion. How many sample observations should be taken?

We have

$$1.96\sigma_{\hat{p}_X} = 0.04$$

so that

$$\sigma_{\hat{p}_X} = 0.0204$$

The sample size needed is, then,

$$n_{\max} = \frac{0.25N}{(N - 1)\sigma_{\hat{p}_X}^2 + 0.25} = \frac{(0.25)(1,395)}{(1,394)(0.0204)^2 + 0.25} = 420.1$$

Hence, a simple random sample of 421 observations will suffice.

SAMPLE SIZES FOR STRATIFIED RANDOM SAMPLING

It is also possible to derive formulas for the sample size needed to yield a specified degree of precision when stratified random sampling is employed. Let the random variable $\bar{X}_{\text{st}}$ denote the estimator of the population mean from stratified sampling, and $\bar{X}_j (j = 1, 2, \ldots, K)$ the sample means for the individual strata. It then follows, since

$$\bar{X}_{\text{st}} = \frac{1}{N} \sum_{j=1}^{K} N_j \bar{X}_j$$

that the variance of $\bar{X}_{st}$ is

$$
\begin{aligned}
\operatorname{Var}(\bar{X}_{st}) = \sigma^2_{\bar{X}_{st}} &= \frac{1}{N^2} \sum_{j=1}^{K} N_j^2 \operatorname{Var}(\bar{X}_j) \\
&= \frac{1}{N^2} \sum_{j=1}^{K} N_j^2 \frac{\sigma_j^2}{n_j} \cdot \frac{N_j - n_j}{N_j - 1}
\end{aligned}
\tag{18.5.4}
$$

where the $\sigma_j^2 (j = 1, 2, \ldots, K)$ are the population variances for the K strata. Now, for any choice of $n_1, n_2, \ldots, n_K$, formula 18.5.4 can be used to derive the corresponding variance of the estimator of the population mean. However, the actual total sample size, n, required to achieve a particular value for this variance will depend on the manner in which the sample observations are allocated among the strata. In the previous section we discussed two frequently used procedures—proportional allocation and optimum allocation. In either case, we can substitute for n_j in 18.5.4, solve the resulting equation, and obtain the sample size n. The results are given in the box.[2]

Determination of the Sample Size for Stratified Random Sampling

Suppose that a population of N members is subdivided into K strata, containing $N_1, N_2, \ldots, N_K$ members. Let σ_j^2 denote the population variance in the jth stratum, and suppose we want to estimate the overall population mean. If the desired variance, $\sigma^2_{\bar{X}_{st}}$, of the sample estimator is specified, the required total sample size, n, is:

(i) Proportional Allocation

$$
n = \frac{\sum\limits_{j=1}^{K} N_j \sigma_j^2}{N \sigma^2_{\bar{X}_{st}} + \dfrac{1}{N} \sum\limits_{j=1}^{K} N_j \sigma_j^2}
$$

(ii) Optimal Allocation

$$
n = \frac{\dfrac{1}{N} \left(\sum\limits_{j=1}^{K} N_j \sigma_j \right)^2}{N \sigma^2_{\bar{X}_{st}} + \dfrac{1}{N} \sum\limits_{j=1}^{K} N_j \sigma_j^2}
$$

---

[2] In fact, in deriving these formulas, we used the approximation

$$
N_j / (N_j - 1) \approx 1
$$

which will cause no difficulty unless the population numbers in individual strata are very small.

If stratified random sampling is used to estimate a population proportion, the previous formulas are modified, $\sigma_{\bar{X}_{st}}^2$ being replaced by $\sigma_{\hat{p}_{st}}^2$, the variance of the relevant estimator, and σ_j^2 by $p_j(1 - p_j)$, where p_j is the population proportion in the jth stratum.

Example 18.9

As in Example 18.4, suppose we want to take a stratified random sample to estimate the mean household income in a town in which the numbers in the three strata (districts) are

$$N_1 = 1,150; \qquad N_2 = 2,120; \qquad N_3 = 930$$

Suppose, further, that the investigator's experience of similar districts in other towns suggests that the population standard deviations (in dollars) for incomes in these districts are likely to be approximately

$$\sigma_1 = 4,000; \qquad \sigma_2 = 6,000; \qquad \sigma_3 = 8,000$$

If we require a 95% confidence interval for the population mean extending $500 on each side of the sample estimate, how many sample observations, in total, are needed?
We have

$$1.96\sigma_{\bar{X}_{st}} = 500$$

so that

$$\sigma_{\bar{X}_{st}} = 255.1$$

We also require

$$\sum_{j=1}^{K} N_j\sigma_j^2 = (1,150)(4,000)^2 + (2,120)(6,000)^2 + (930)(8,000)^2$$

$$= 154,240,000,000$$

and

$$\frac{1}{N}\left(\sum_{j=1}^{K} N_j\sigma_j\right)^2 = \frac{[(1,150)(4,000) + (2,120)(6,000) + (930)(8,000)]^2}{4,200}$$

$$= 145,966,000,000$$

For proportional allocation, the sample size needed is

$$n = \frac{\sum N_j\sigma_j^2}{N\sigma_{\bar{X}_{st}}^2 + \dfrac{1}{N}\sum N_j\sigma_j^2}$$

$$= \frac{154,240,000,000}{(4,200)(255.1)^2 + 154,240,000,000/4,200} = 497.5$$

Thus, a sample of 498 observations will suffice to produce the desired level of precision.

If optimal allocation is to be used, we have

$$n = \frac{145{,}966{,}000{,}000}{(4{,}200)(255.1)^2 + 154{,}240{,}000{,}000/4{,}200} = 470.8$$

so that the same degree of reliability can be achieved with 471 observations if this method of allocation is used. Notice that the sample size necessary for optimum allocation is less than that for proportional allocation. The difference can be thought of as the payoff resulting from the use of the optimal allocation method.

18.6 OTHER SAMPLING METHODS

So far we have discussed in some detail simple and stratified random sampling. These are not the only procedures used in practice for choosing a sample. In this section we consider some alternative methods.

CLUSTER SAMPLING

Suppose that an investigator wants to survey a population spread over a wide geographical area, such as a large city or state. If either a simple random sample or stratified random sample is to be used, two immediate problems will arise. First, in order to draw the sample, the investigator will need a reasonably accurate listing of the population members. Such a list may not be available, or perhaps could be obtained only at prohibitively high cost. Second, even if the investigator does possess a list of the population, the resulting sample members will almost inevitably be thinly spread over a large area. In that case, contacting each individual sample member by interviewers will be quite costly. Of course, if a mail questionnaire is to be used, this latter problem does not arise. However, this means of contact may lead to an unacceptably high rate of nonresponse, inducing the investigator to prefer personal interviews.

Faced with the dilemma of either not having a reliable population listing, or of wanting to set up personal interviews with sample members when budget resources are tight, the investigator has an alternative sampling procedure available. This is known as **cluster sampling.** This approach is attractive when a population can conveniently be subdivided into relatively small, geographically compact units called **clusters.** For example, a city might be subdivided into political wards, or even residential blocks. This can generally be achieved even when a complete listing of residents or households is unavailable.

In cluster sampling, a simple random sample of *clusters* is selected from the population, and every individual in each of the sampled clusters is contacted; that is, a complete census is carried out in each of the chosen clusters. In the accompanying box, we list procedures for deriving valid inference about the population mean and proportion from the results of a cluster sample.

Inference from Cluster Sampling

Suppose that a population is subdivided into M clusters, that a simple random sample of m of these clusters is selected, and that information is obtained from every member of the sampled clusters. Let $n_1, n_2, \ldots, n_m$ denote the numbers of population members in the m sampled clusters. We denote by $\bar{x}_1, \bar{x}_2, \ldots, \bar{x}_m$ the means for these clusters, and by $p_1, p_2, \ldots, p_m$ the proportions of cluster members possessing an attribute of interest. Our objective is to estimate the overall population mean μ and proportion p. Then:

(i) Unbiased estimation procedures give

$$\bar{x}_c = \frac{\sum_{i=1}^{m} n_i \bar{x}_i}{\sum_{i=1}^{m} n_i}$$

and

$$\hat{p}_c = \frac{\sum_{i=1}^{m} n_i p_i}{\sum_{i=1}^{m} n_i}$$

(ii) Estimates of the variances of these estimators, following from unbiased estimation procedures, are

$$\hat{\sigma}_{\bar{x}_c}^2 = \frac{M-m}{Mm\bar{n}^2} \cdot \frac{\sum_{i=1}^{m} n_i^2 (\bar{x}_i - \bar{x}_c)^2}{m-1}$$

and

$$\hat{\sigma}_{\hat{p}_c}^2 = \frac{M-m}{Mm\bar{n}^2} \cdot \frac{\sum_{i=1}^{m} n_i^2 (p_i - \hat{p}_c)^2}{m-1}$$

where

$$\bar{n} = \frac{\sum_{i=1}^{m} n_i}{m}$$

is the average number of individuals in the sampled clusters.

(iii) Provided the sample size is large, $100(1-\alpha)\%$ confidence intervals are obtained from

$$\bar{x}_c - z_{\alpha/2}\hat{\sigma}_{\bar{x}_c} < \mu < \bar{x}_c + z_{\alpha/2}\hat{\sigma}_{\bar{x}_c}$$

and

$$\hat{p}_c - z_{\alpha/2}\hat{\sigma}_{\hat{p}_c} < p < \hat{p}_c + z_{\alpha/2}\hat{\sigma}_{\hat{p}_c}$$

Notice that inference can be carried out with relatively little prior information about the population. All that is required is a breakdown into identifiable clusters. We do not even need to know the total number of population members. It is sufficient to know the numbers in each of the *sampled* clusters, and these can be determined during the course of the survey, since a full census is taken in each cluster in the sample. In addition, since sample members will be geographically close to one another within clusters, their contact by interviewers is relatively inexpensive.

Example 18.10

A simple random sample of 20 blocks is taken from a residential area containing a total of 1,000 blocks. Each household in the sampled blocks is then contacted and information is obtained about family incomes. The accompanying table lists mean annual incomes, and the proportion of families with incomes below $15,000 per year, in the sampled blocks. For this residential area, estimate the mean family income and the proportion of families with incomes below $15,000 per year.

SAMPLED BLOCK	MEAN INCOME (IN DOLLARS)	PROPORTION BELOW $15,000	NUMBER OF HOUSEHOLDS
i	$\bar{x}_i$	p_i	n_i
1	26,283	.1304	23
2	19,197	.4516	31
3	37,911	.1250	24
4	14,527	.6585	41
5	16,753	.5143	35
6	28,312	.2692	26
7	21,646	.3548	31
8	29,312	.1563	32
9	31,829	.1333	30
10	18,412	.3846	39
11	33,893	.0769	26
12	38,409	.0476	21
13	43,911	0	20
14	14,699	.4375	32
15	24,921	.1111	36
16	31,827	.0909	33
17	34,436	.0833	24
18	37,647	.0400	25
19	30,026	.1081	37
20	16,493	.3659	41

In our notation, we have

$$m = 20 \quad \text{and} \quad M = 1,000$$

The total number of households in the sample is

$$\sum_{i=1}^{m} n_i = (23 + 31 + \cdots + 41) = 607$$

To obtain point estimates we need

$$\sum_{i=1}^{m} n_i \bar{x}_i = (23)(26,283) + (31)(19,197) + \cdots + (41)(16,493) = 15,848,158$$

and

$$\sum_{i=1}^{m} n_i p_i = (23)(.1304) + (31)(.4516) + \cdots + (41)(.3659) = 153$$

Our point estimates are therefore

$$\bar{x}_c = \frac{\sum n_i \bar{x}_i}{\sum n_i} = \frac{15,848,158}{607} = 26,109$$

and

$$\hat{p}_c = \frac{\sum n_i p_i}{\sum n_i} = \frac{153}{607} = .2521$$

Thus, on the basis of this sample evidence, we estimate that, for this residential area, mean annual household income is \$26,109 and 25.21% of households have income below \$15,000 per year.

In order to obtain interval estimates, we need the average cluster size

$$\bar{n} = \frac{\sum n_i}{m} = \frac{607}{20} = 30.35$$

Also

$$\frac{\sum_{i=1}^{m} n_i^2 (\bar{x}_i - \bar{x}_c)^2}{m-1} = \frac{(23)^2(26,283 - 26,109)^2 + \cdots + (41)^2(16,493 - 26,109)^2}{19}$$

$$= 69,270,551,000$$

so that

$$\hat{\sigma}_{\bar{x}_c}^2 = \frac{M - m}{Mm\bar{n}^2} \cdot \frac{\sum n_i^2 (\bar{x}_i - \bar{x}_c)^2}{m-1}$$

$$= \frac{(980)(69,270,551,000)}{(1,000)(20)(30.35)^2} = 3,684,914$$

and hence, taking square roots, we obtain

$$\hat{\sigma}_{\bar{x}_c} = 1,920$$

For a 95% confidence interval,

$$z_{\alpha/2} = z_{.025} = 1.96$$

Hence, a 95% confidence interval for the population mean is

$$26,109 - (1.96)(1,920) < \mu < 26,109 + (1.96)(1,920)$$

or

$$22{,}346 < \mu < 29{,}872$$

Our 95% confidence interval for the mean income of all families in this area therefore runs from \$22,346 to \$29,872.

To obtain interval estimates for the population proportion, we first require

$$\frac{\sum\limits_{i=1}^{m} n_i^2 (p_i - \hat{p}_c)^2}{m - 1} = \frac{(23)^2(.1304 - .2521)^2 + \cdots + (41)^2(.3659 - .2521)^2}{19}$$

$$= 38.1547$$

Then

$$\hat{\sigma}_{\hat{p}_c}^2 = \frac{M - m}{Mm\bar{n}^2} \cdot \frac{\sum n_i^2 (p_i - \hat{p}_c)^2}{m - 1}$$

$$= \frac{(980)(38.1547)}{(1{,}000)(20)(30.35)^2} = .0020297$$

and so, taking square roots, we obtain

$$\hat{\sigma}_{\hat{p}_c} = .0451$$

Thus, the 95% confidence interval for the population proportion is

$$.2521 - (1.96)(.0451) < p < .2521 + (1.96)(.0451)$$

or

$$.164 < p < .340$$

Our 95% confidence interval for the percentage of households with annual incomes below \$15,000 therefore runs from 16.4% to 34.0%. Figure 18.4 shows also 80%, 90%, and 99% confidence intervals for the proportion of all households in this residential area with annual incomes below \$15,000.

Cluster sampling has a superficial resemblance to stratified sampling. In both, the population is first divided into subgroups. However, the similarity is rather illusory. In stratified random sampling, a sample is taken from *every stratum* of the population, in an attempt to ensure that important segments of the population are given due weight. By contrast, in cluster sampling, a random sample of *clusters* is taken, so that some clusters will have no members in the sample. Since, within clusters, population members will probably be fairly homogeneous, the danger is that important subgroups of the population may be either not represented at all or grossly under-represented in the final sample. In consequence, while the great advantage of cluster sampling lies in its convenience, this convenience may well be at the cost of additional imprecision in the sample estimates. A further distinction between cluster sampling and stratified sampling is that, in the former, a complete census of cluster members is taken, while in the latter a random sample of stratum members is drawn. This

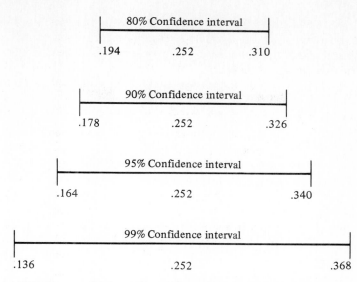

FIGURE 18.4 80%, 90%, 95%, and 99% confidence intervals for population proportion of households in a residential area with annual incomes below $15,000, based on the data of Example 18.10

difference, however, is not essential. Indeed, on occasions, an investigator may draw a random sample of cluster members, rather than take a full census.

TWO-PHASE SAMPLING

In many investigations, the population is not surveyed at a single step. Rather, it is often convenient to carry out an initial pilot study, in which a relatively small proportion of the sample members are contacted. The results obtained are then analyzed, prior to conducting the bulk of the survey. The chief disadvantage of such a procedure is that it can be quite time-consuming. However, this factor may be outweighed by several advantages. One important benefit is that the investigator is able, at modest cost, to try out the proposed questionnaire in order to ensure that the various questions can be thoroughly understood. The pilot study may also suggest additional questions, whose potential importance had previously been overlooked. Moreover, this study should also provide an estimate of the likely rate of nonresponse. Should this prove unacceptably high, some modification in the method of soliciting responses might appear desirable.

The carrying out of a survey in two stages, beginning with a pilot study, is called **two-phase sampling.** This approach has two further advantages. First, if stratified random sampling is employed, the pilot study can be used to provide estimates of the individual stratum variances. These, in turn, can be employed to estimate the optimum allocation of the sample among the various strata. Second, the results of the pilot study can be used to estimate the number of observations needed to obtain estimators of population parameters with a specified level of precision. The following examples serve to illustrate these points.

Example 18.11

We begin with a straightforward situation in which a simple random sample is to be used to estimate a population mean. At the outset, relatively little is known about this population, so that an initial pilot survey is to be carried out to get some idea of the sample size required.

An auditor wishes to estimate the mean value of accounts receivable in a total population of 1,120 accounts. He wants to produce a 95% confidence interval for the population mean extending approximately $4 on each side of the sample mean. To begin, he takes a simple random sample of 100 accounts, finding a sample standard deviation of $30.27. How many more accounts should be sampled?

Recall from Section 18.5 that the sample size needed can be expressed as

$$n = \frac{N\sigma^2}{(N-1)\sigma_{\bar{X}}^2 + \sigma^2}$$

where $N = 1,120$ is the number of population members in this particular case. In order for the 95% confidence interval to be of the required width, we need

$$1.96\sigma_{\bar{X}} = 4$$

so that $\sigma_{\bar{X}}$, the standard deviation of the sample mean, must be

$$\sigma_{\bar{X}} = \frac{4}{1.96} = 2.04$$

The population standard deviation, σ, is unknown. However, as a result of the initial study of 100 accounts receivable, we estimate it to be 30.27. The total number of sample observations needed is therefore

$$n = \frac{(1,120)(30.27)^2}{(1,119)(2.04)^2 + (30.27)^2} = 184.1$$

Since 100 observations have already been taken, an additional 85 will suffice to satisfy the auditor's objective.

Example 18.12

An investigator intends to take a stratified random sample to estimate mean family income in a town where the numbers in the three stratum districts are

$$N_1 = 1,150; \qquad N_2 = 2,120; \qquad N_3 = 930$$

To begin, he conducts a pilot sudy, sampling 30 households from each district, obtaining the sample standard deviations $3,657, $6,481, and $8,403, respectively. Suppose that his objective is to obtain, with as small a sample size as possible, a 95% confidence interval for the population mean extending $500 on each side of the sample estimate. How many additional observations should be taken in each district?

The requirement that a specified degree of precision be obtained with as few sample observations as possible implies that optimal allocation must be used. Recall from equation 18.4.1 that the numbers n_1, n_2, n_3 to be sampled in the three strata are as follows:

$$n_j = \frac{N_j\sigma_j}{\sum\limits_{i=1}^{3} N_i\sigma_i} \cdot n \qquad (j = 1, 2, 3)$$

where the σ_i are the stratum population standard deviations. Using our sample estimates in place of these quantities, we find

$$n_1 = \frac{(1,150)(3,657)}{(1,150)(3,657) + (2,120)(6,481) + (930)(8,403)} \cdot n = 0.163n$$

$$n_2 = \frac{(2,120)(6,481)}{(1,150)(3,657) + (2,120)(6,481) + (930)(8,403)} \cdot n = 0.533n$$

$$n_3 = \frac{(930)(8,403)}{(1,150)(3,657) + (2,120)(6,481) + (930)(8,403)} \cdot n = 0.303n$$

We have now specified the proportions of the total sample to be allocated to each stratum under the optimal scheme. It remains to determine the total number, n, of sample observations. From Section 18.5, we find

$$n = \frac{\frac{1}{N}\left(\sum_{j=1}^{3} N_j \sigma_j\right)^2}{N\sigma_{\bar{X}_{st}}^2 + \frac{1}{N}\sum_{j=1}^{K} N_j \sigma_j^2}$$

where $N = 4,200$ is the total number of population members, and $\sigma_{\bar{X}_{st}}^2$ is the variance of the estimator of the population mean. In order that the 95% confidence interval for the population mean extend \$500 on each side of the sample estimate, we need

$$1.96\sigma_{\bar{X}_{st}} = 500$$

so that

$$\sigma_{\bar{X}_{st}} = 255.1$$

As before, we use the estimated standard deviations from the pilot study in place of the unknown population quantities. Hence, we have

$$n = \frac{\frac{1}{4,200}[(1,150)(3,657) + (2,120)(6,481) + (930)(8,403)]^2}{(4,200)(255.1)^2 + \frac{1}{4,200}[(1,150)(3,657)^2 + (2,120)(6,481)^2 + (930)(8,403)^2]}$$

$$= 503.5$$

Rounding up, we conclude that the total number of sample observations should be 504. These are then allocated among the three strata as

$$n_1 = (0.163)(504) = 82$$
$$n_2 = (0.533)(504) = 269$$
$$n_3 = (0.303)(504) = 153$$

Since 30 households have already been sampled in each stratum, the numbers sampled in the second phase should be 52, 239, and 123.

NONPROBABILISTIC SAMPLING METHODS

We have considered various sampling schemes for which it is possible to specify the probability that any particular sample will be drawn from the population. It is because of this feature of the sampling methods that we are able to make valid statistical inference based on the sample results. Otherwise, the derivation of unbiased point estimates and confidence intervals with specified probability content could not be achieved with strict statistical validity.

Nevertheless, in many practical applications, **nonprobabilistic methods** are used for selecting sample members, primarily as a matter of convenience. For example, suppose you want to assess the reactions of students on your campus to some issue of topical interest. One possibility would be to ask all your friends how they feel about it. This group would not constitute a random sample from the population of all students. Accordingly, if you proceed to analyze the data as if it were obtained from a random sample, the resulting inference would lack proper statistical validity.

A more sophisticated version of the approach described in the previous paragraph, called **quota sampling,** is commonly used by polling organizations. Interviewers are assigned to a particular locale and instructed to contact specified numbers of people of certain age–race–sex characteristics. These assigned quotas represent what are thought to be appropriate proportions for the population at large. However, once the quotas are determined, interviewers are granted flexibility in the choice of sample members. Their choice is typically not random. It is certainly the case that quota sampling can, and often does, produce quite accurate estimates of population parameters. The drawback is that, since the sample is not chosen using probabilistic methods, there is no valid way of determining the reliability of the resulting estimates.

EXERCISES

41. It is required to estimate the mean amount of the 956 mortgages taken out in a city in the past year. Based on previous experience, a real estate broker estimates that the population standard deviation is likely to be about $20,000. If a 95% confidence interval for the population mean is to extend $2,000 on each side of the sample mean, how many sample observations are needed if a simple random sample is taken?

42. A department store has an inventory of 1,420 different products. In order to estimate the total dollar value of this inventory, an auditor intends to take a simple random sample of products. Based on last year's records, the population standard deviation is estimated to be $120. It is required that a 90% confidence interval for the population total extend $20,000 on each side of its sample estimate. How large a sample size is necessary to satisfy this requirement?

43. A corporation wants to poll a simple random sample of its 520 managers to estimate the percentage in favor of a new system of flexible-time working hours. The number of sample observations should be sufficiently large to ensure that a 99% confidence interval for the percentage of all managers in favor of this system extends no more than 5% on each side of the sample percentage in favor. How large a sample is needed?

44. A drug company employs 620 sales representatives. Its marketing department is contemplating introducing new promotion literature for one of its best selling lines, but wishes first to obtain reactions from representatives. A simple random sample is to be taken, and it is required that a 90% confidence interval for the proportion of all representatives enthusiastic about this promotion literature extend no more than 0.03 on each side of the sample proportion. How many sample observations must be obtained?

45. A fast-food chain has 420 outlets in Iowa, 380 in Nebraska, and 510 in Oklahoma. The chain is considering introducing passionfruit-flavored milk shakes, and wishes to estimate the mean number of orders per week per outlet that this product will command. When other exotic milk shakes were introduced in the past, company records showed that the standard deviations of weekly sales were 45 for the Iowa outlets, 30 for Nebraska outlets, and 50 for outlets in Oklahoma. It is required that a 99% confidence interval for the overall population mean weekly sales per outlet extend no more than 10 on each side of the sample estimate.

(a) How many total sample observations are needed if proportional allocation is used?

(b) How many total sample observations are required if optimal allocation is used?

46. An auditor wants to estimate the population mean value of a corporation's accounts receivable. The population is divided into four strata, for which the values given in the table have been established. If a 90% confidence interval for the overall population mean is to extend $20 on each side of the sample estimate, determine the total sample size required under each of the following schemes:

(a) Proportional allocation

(b) Optimal allocation

STRATUM	POPULATION SIZE	ESTIMATED STANDARD DEVIATION
1	800	$ 80
2	600	$120
3	500	$200
4	400	$400

47. A market research organization wishes to estimate the mean monthly expenditures on entertainment by households in a city that contains 65 precincts. A simple random sample of 10 precincts is selected, and every household in each sampled precinct is questioned. The accompanying results were obtained.

SAMPLED PRECINCT	NUMBER OF HOUSEHOLDS	MEAN EXPENDITURE (IN DOLLARS)
1	21	106
2	35	82
3	18	139
4	52	61
5	41	87
6	38	99
7	36	169
8	30	104
9	23	93
10	42	61

(a) Find a point estimate of the population mean expenditure per household on entertainment.

(b) Find a 90% confidence interval for the population mean.

48. A union executive wants to estimate the mean value of bonus payments made to a corporation's clerical employees in the first month of a new plan. This corporation has 52 subdivisions, and a simple random sample of 8 of these is taken. Information is then obtained from the payroll records of every clerical worker in each of the sampled subdivisions. The results obtained are shown in the table.

SAMPLED SUBDIVISION	NUMBER OF CLERICAL EMPLOYEES	MEAN BONUS (IN DOLLARS)
1	62	83
2	75	64
3	41	42
4	36	108
5	59	136
6	82	102
7	64	95
8	71	78

(a) Find a point estimate of the population mean bonus per clerical employee for this month.

(b) Find a 99% confidence interval for the population mean.

49. In the survey of Exercise 47, the households were asked if any of their members had been to the movies in the previous week. The numbers answering in the affirmative are given in the accompanying table.

PRECINCT NUMBER	1	2	3	4	5	6	7	8	9	10
	6	7	8	12	9	12	21	15	9	4

(a) Find a point estimate of the proportion of all households for which a member had been to the movies in the previous week.

(b) Find a 90% confidence interval for this population proportion.

50. In the survey of Exercise 48, the clerical employees in the 8 sampled subdivisions were asked if they were satisfied with the operation of the bonus plan. The results obtained are listed in the table.

SUBDIVISION	1	2	3	4	5	6	7	8
NUMBER SATISFIED	24	25	11	21	35	44	30	25

(a) Find a point estimate of the proportion of all clerical employees satisfied with the bonus plan.

(b) Find a 95% confidence interval for this population proportion.

51. A researcher wishes to estimate the proportion of all house sales in the past year in a city for which the buyer took up an assumable mortgage. A simple random sample of 12 of the city's 75 real estate agencies was taken, and the information in the accompanying table was obtained.

SAMPLED AGENCY	NUMBER OF SALES	NUMBER WITH ASSUMABLE MORTGAGE
1	43	14
2	36	9
3	59	20
4	74	21
5	82	26
6	61	15
7	30	6
8	35	12
9	40	18
10	66	37
11	26	11
12	31	9

(a) Find a point estimate of the proportion of all sales with an assumable mortgage.

(b) Find a 99% confidence interval for this population proportion.

52. A corporation employs 1,261 manual workers. A personnel manager wishes to compile a demographic picture of this work force. Initially, she takes a simple random sample of 40 of these employees and finds the sample standard deviation of their ages to be 14.8 years. If the personnel manager wants to obtain a 90% confidence interval for the population mean age extending 1.5 years on each side of the sample mean, how many additional sample observations must be taken?

53. A company leases both automobiles and trucks. An auditor needs to estimate the mean useful life (in months) for all these leases. In all, there are 10,000 automobile leases and 4,000 truck leases. To begin, simple random samples of 20 leases of each type are taken. The sample standard deviations for useful life were 9.2 months for automobile leases and 15.7 months for truck leases. It is required that a 99% confidence interval for the overall population mean extend 2 months on each side of the sample point estimate. Estimate the smallest total number of additional sample observations needed to achieve this objective.

54. A corporation has a fleet of 480 company cars—100 compact, 180 midsize, and 200 full size. In order to estimate the overall mean annual repair costs for these cars, a preliminary random sample of 10 cars of each type is selected. The sample standard deviations for repair costs were $105 for compacts, $142 for midsize, and $163 for full size cars. It is required that a 95% confidence interval for the overall population mean annual repair cost per car extend $20 on each side of the sample point estimate. Estimate the smallest total number of additional sample observations that must be taken.

REVIEW EXERCISES

55. (a) You have been asked to design and carry out a survey in your city on the effectiveness of a radio advertising campaign aimed at promoting a new movie. Outline how you would proceed.

(b) In the context of this survey, discuss the possibilities for nonsampling errors, and means for minimizing their importance.

(c) To what extent would you expect nonresponse to be a problem in this survey?

56. Based on a random sample of 10 members of your class, estimate the average amount of money per semester spent by class members on textbooks.

57. Carefully explain the distinction between stratified random sampling and cluster sampling. Provide illustrations of sampling problems where each of these techniques might be useful.

58. From a total of 150 accounts receivable, a random sample of 10 were drawn. Their amounts (in dollars) are listed in the accompanying table.

120.60	144.25	173.60	161.40	170.30
135.70	149.60	161.75	153.85	149.35

(a) Find a 90% confidence interval for the population mean value for these accounts receivable.

(b) Without doing the calculations, state whether a 95% confidence interval for the population mean would be wider or narrower than that found in part (a).

(c) Find a 99% confidence interval for the total value of the 150 accounts receivable in this population.

59. A department store has 230 delinquent accounts. A random sample of 50 of these accounts was taken. The sample mean amount of these accounts was $273.20, and the sample standard deviation was found to be $98.20.

(a) Find a 99% confidence interval for the population mean amount of these delinquent accounts.

(b) Find a 95% confidence interval for the total amount of the 230 delinquent accounts in this population.

(c) Without doing the calculations, state whether a 90% confidence interval for the population total would be wider or narrower than that found in part (b).

60. An economics class contains 95 students. Of a random sample of 40 of these students, 18 were found to be business majors.

(a) Find a 90% confidence interval for the proportion of all members of this class who are business majors.

(b) Without doing the calculations, state whether a 99% confidence interval for the population proportion would be wider or narrower than that found in part (a).

61. A restaurant chain has 150 outlets in the Midwest. A random sample of 50 of these was taken, and it was found that, for 32 of the sample members, profits were higher this year than in the previous year. Find a 95% confidence interval for the population proportion with higher profits.

62. A company has three subdivisions, employing a total of 970 managers. Independent random samples of managers were taken from each subdivision, and the number of years with the company was determined for each sample member. The results are summarized in the accompanying table.

	SUBDIVISION		
	1	2	3
N_i	352	287	331
n_i	30	20	20
$\bar{x}_i$	9.2	12.3	13.5
s_i	4.7	6.1	7.8

(a) Find a 99% confidence interval for the mean number of years with the company for managers in subdivision 1.

(b) Find a 99% confidence interval for the mean number of years with the company for all managers.

63. Of the 300 pages in a particular book, 180 are primarily nontechnical, while the remainder are technical. Independent random samples of technical and nontechnical pages were taken, and the numbers of errors per page were recorded. The results are summarized in the table.

	TECHNICAL	NONTECHNICAL
N_i	120	180
n_i	20	20
$\bar{x}_i$	1.65	0.75
s_i	0.98	0.55

(a) Find a 95% confidence interval for the mean number of errors per page in this book.

(b) Find a 99% confidence interval for the total number of errors in the book.

64. In the analysis of Exercise 63, it was found that 7 of the sampled technical pages and 15 of the sampled nontechnical pages contained no errors. Find a 90% confidence interval for the proportion of all pages in this book that have no errors.

65. Refer to the data of Exercise 62. If a total of 70 managers were sampled, determine how many sample members would be from subdivision 1 under each of the following schemes:

(a) Proportional allocation

(b) Optimum allocation, assuming the stratum population standard deviations are the same as the corresponding sample quantities

66. Refer to the data of Exercise 63. If a total of 40 pages are to be sampled, determine how many sampled pages would be technical under each of the following schemes:

(a) Proportional allocation

(b) Optimum allocation, assuming the stratum population standard deviations are the same as the corresponding sample quantities.

67. You intend to sample the students in your university to assess their views on the adequacy of space in the library. You decide to use a stratified sample by year—freshman, sophomore, and so forth. Discuss the factors you would take into account in deciding how many sample observations to take in each stratum.

68. It is required to estimate the mean selling price of the 725 homes sold in a city during the past year. Previous experience suggests that the standard deviation of selling prices is likely to be around $22,500. If a 90% confidence interval for the population mean is to extend $1,500 on each side of the sample mean, how large a sample is required if simple random sampling is employed?

69. A simple random sample is to be taken of the 930 adult residents in a small town in order to estimate the proportion in favor of a proposal to set aside a tract of land for a park. How many sample observations are needed to ensure that a 95% confidence interval for the population proportion in favor extends at most 0.04 on each side of the sample proportion?

70. An auditor wants to estimate the total value of a corporation's accounts receivable. The population is divided into three strata, for which the values in the accompanying table have been determined. If a 95% confidence interval for the population total is to extend

$30,000 on each side of the sample estimate, determine the total number of sample observations under each of the following schemes:

(a) Proportional allocation

(b) Optimal allocation

STRATUM	POPULATION SIZE	ESTIMATED STANDARD DEVIATION
1	800	$100
2	600	$150
3	400	$250

71. A market research group takes a random sample of 8 of a city's 50 precincts. Every household in each sampled precinct is questioned as to whether it has a central air conditioning system, and also on its summer monthly electricity costs. The results are shown in the table.

SAMPLED PRECINCT	NUMBER OF HOUSEHOLDS	MEAN MONTHLY ELECTRICITY COSTS (IN DOLLARS)	NUMBER WITH CENTRAL AIR CONDITIONING
1	32	75	28
2	43	64	22
3	24	49	8
4	37	68	25
5	30	53	13
6	25	77	21
7	47	42	14
8	36	55	17

(a) Find a 95% confidence interval for the population mean summer monthly electricity costs.

(b) Find a 95% confidence interval for the population proportion of households with central air conditioning.

statistical decision theory

CHAPTER NINETEEN

19.1 DECISION-MAKING UNDER UNCERTAINTY

The topic of this chapter could be characterized as capturing the essence of management problems in any commercial organization. Indeed, the applicability of the subject matter extends further, touching many aspects of our everyday lives. We will be concerned here with the situation in which an individual, a group, or a corporation has available several alternative feasible courses of action. The decision as to which course to follow must be made in a world in which there is uncertainty about the future behavior of those factors that will determine the consequences stemming from the action taken.

We are all constrained to operate in an environment about whose future direction we are uncertain. For example, you may consider attending a baseball game, but are doubtful because the possibility of rain is threatened. If you *knew* that it was not going to rain, you would go to the game. On the other hand, if you were *certain* that heavy rain was going to fall for several hours, you would not go. However, you are unable to predict the weather with complete assurance, and your decision must be made while contemplating an uncertain future. As another example, at some stage during your final year in college, you will have to decide what to do upon graduation. It is probable that you will have offers of employment from several sources. Graduate school, too, may be a possibility. The decision as to initial career direction is clearly an important one. Certainly, you will have acquired information about the alternatives. You will know what starting salaries are on offer, you will have learned something about the business operations of your future potential employers, and how you might fit into these operations. However, one really does not have a very clear picture of where one will be in one or two years if a particular offer is accepted. This important decision, then, is made in the face of uncertainty about the future.

In the business world, circumstances of this type often arise, as the following examples illustrate:

(i) In a recession, a company must decide whether to lay off employees. If the downturn in business activity is to be short-lived, it may be preferable to retain these workers, who might be difficult to replace when demand improves. On the other hand, if the recession is to be prolonged, their retention would be costly. Unfortunately, the art of economic forecasting has not reached the stage where it is possible to predict with great certainty the length or severity of a recession.

(ii) An investor may believe that, at the present time, interest rates are at a peak. In that case, long-term bonds would appear to be very attractive. However, it is impossible to be sure about the future direction of interest rates, and if they were to continue to rise, the decision to tie up funds in long-term bonds would have been suboptimal.

(iii) Contractors are often required to submit bids for a program of work. The decision to be made is the level at which the bid should be pitched. Two areas of uncertainty may be relevant here. First, the contractor will not know how low a bid will be necessary in order to secure the work. Second, he is not likely to be sure precisely how much it will cost to fulfill the contract. Again, in spite of this uncertainty, some decision must be made.

(iv) The cost of drilling exploratory offshore oil wells is enormous, and in spite of excellent geological advice, oil companies will not know, before a well is drilled, whether commercially viable quantities of oil will be discovered. The decision as to whether and where to drill in a particular field is one that must be made in an uncertain environment.

In this chapter, our objective is to study methods for attacking decision-making problems of the type just discussed. To make our analysis more concrete, consider the problem of a manufacturer planning to introduce a new candy bar. The manufacturer has available four alternative production processes, denoted A, B, C, and D, ranging in scope from a relatively minor modification of existing facilities to a quite major extension of the plant. The decision as to which course of action to follow must be made at a time when the eventual demand for the product will be unknown. For convenience, we will characterize this potential demand as either "low," "moderate," or "high." It will also be assumed that the manufacturer is able to calculate, for each production process, the profit over the lifetime of the investment for each of the three levels of demand. Table 19.1 shows these profit levels (in dollars) for each production process–level of demand combination.

TABLE 19.1 Estimated profits of candy bar manufacturer for different production process–level of demand combinations

PRODUCTION PROCESS	LEVEL OF DEMAND		
	LOW	MODERATE	HIGH
A	70,000	120,000	200,000
B	80,000	120,000	180,000
C	100,000	125,000	160,000
D	100,000	120,000	150,000

This particular problem serves to illustrate the general framework for our analysis. A decision-maker is faced with a finite number, K, of possible **actions,** which will be labelled $a_1, a_2, \ldots, a_K$. In our example these actions correspond to the four possible production process adoptions. At the time a particular action must be selected, the decision-maker is uncertain about the future of some factor that will determine the consequences of the chosen action. It is assumed that the possibilities for this factor can be characterized by a finite number, H, of **states of nature.** These will be denoted $s_1, s_2, \ldots, s_H$. In the candy bar example, there are three states of nature, corresponding to the three possible levels of demand for the product. Finally, it is assumed that the decision-maker is able to specify the monetary rewards, or **payoffs,** for each action–state of nature combination. We will let M_{ij} represent the payoff for action a_i in the event of the occurrence of state of nature s_j. Corresponding to Table 19.1 these payoffs can be displayed in a **payoff table,** whose general form is shown in Table 19.2.

TABLE 19.2 General form of payoff table for a decision problem with K possible actions and H states of nature; M_{ij} is the payoff corresponding to action a_i and state of nature s_j

ACTIONS	STATES OF NATURE			
	s_1	s_2	$\cdots$	s_H
a_1	M_{11}	M_{12}	$\cdots$	M_{1H}
a_2	M_{21}	M_{22}	$\cdots$	M_{2H}
$\vdots$	$\vdots$	$\vdots$		$\vdots$
a_K	M_{K1}	M_{K2}	$\cdots$	M_{KH}

Framework for a Decision Problem

(i) Decision-maker has available K courses of **action:**

$$a_1, a_2, \ldots, a_K$$

(ii) There are H possible uncertain **states of nature:**

$$s_1, s_2, \ldots, s_H$$

(iii) For each possible action–state of nature combination, there is an associated monetary **payoff,** M_{ij}, corresponding to action a_i and state of nature s_j.

The decision-making problem, as we have outlined it, is essentially **discrete** in character. That is to say, we have postulated a finite number of available alternative actions and a finite number of possible states of nature. However, many practical problems are **continuous.** The state of nature, for instance, may be more appropriately measured on a continuum rather than depicted by a number of discrete possibilities.

In our example of the candy bar manufacturer, it may be possible to anticipate a range of potential demand levels, rather than simply to specify three levels. Also, in some problems, the available actions are most appropriately represented by a continuum. This would be the case, for example, when a contractor must decide on the level at which to bid for a contract. In the remainder of this chapter, we will concentrate on the discrete case. The *principles* involved in the analysis of the continuous case are no different. However, the details of that analysis are based on calculus, and will not be considered further here.

When a decision-maker is faced with alternative courses of action, the appropriate choice will, to a considerable extent, depend on the objectives. It is possible to describe various lines of attack that have been employed in the solution of business decision-making problems. However, it must be kept in mind that each individual problem has its own special features, and that the objectives of decision-makers may vary considerably, and indeed be rather complex. A situation of this sort arises when one contemplates the position of a middle manager in a large corporation. In practice, this manager's objectives may differ somewhat from those of the corporation. In making decisions, the manager is very likely to be conscious of his or her own position as well as the overall good of the corporation.

In spite of the individual nature of decision-making problems, we can very simply specify one general rule. There may be some available actions that can be eliminated from further consideration under any circumstances. Referring to Table 19.1, consider the production process D. The payoff from this process will be precisely the same as that from process C if there is a low level of demand, and lower than that from process C if the level of demand were to be either moderate or high. It therefore makes no sense to choose option D, since there exists another available choice through which the payoffs can be no lower, and could be higher. Since action C is necessarily at least as rewarding as, and possibly more rewarding than, action D, we say that the action C **dominates** action D. If an action is dominated by another available alternative, it is said to be **inadmissible.** Such actions can then be removed from further consideration, as it would be suboptimal to adopt them.

Definitions

If the payoff for action a_j is at least as high as that for a_i, whatever the state of nature, and if the payoff for a_j is higher than that for a_i for at least one state of nature, then action a_j is said to **dominate** action a_i.

Any action that is dominated in this way is said to be **inadmissible.** Inadmissible actions are removed from the list of available possibilities prior to further analysis of a decision-making problem.

Any action that is not dominated by some other action, and is therefore not inadmissible, is said to be **admissible.**

In our analysis of the decision problem of the candy bar manufacturer, we have seen that the action of choosing production process D is inadmissible. Accordingly, this possibility will be dropped from further consideration and, in our subsequent analysis of this problem, we will entertain the possibility of adoption of processes A, B, and C only.

SOLUTIONS NOT INVOLVING SPECIFICATION OF PROBABILITIES

Before deciding which production process to employ, our manufacturer of candy bars is likely to ask the question, "What are the chances of each of these levels of demand actually materializing?" In the bulk of this chapter, we will discuss solutions to the decision-making problem that require the specification of outcome probabilities for the various states of nature. However, in this section we will briefly consider two choice criteria that are not based on such probabilities and in fact have no probabilistic content. Rather, these approaches (and others of the same type) depend only on the structure of the payoff table.

The two procedures considered in this section are called the *maximin criterion* and the *minimax regret criterion*. We will discuss each in relation to the payoff table for the candy bar manufacturer, with the inadmissible strategy of choosing production process D ignored. The manufacturer must therefore select from among three available actions, faced with three possible states of nature.

(i) MAXIMIN CRITERION

Here we consider the worst possible outcome for each action, whatever state of nature materializes. This *worst outcome* is simply the smallest payoff that could conceivably result. For the candy bar manufacturer's problem, the smallest payoff, whatever production process is used, in fact occurs at the low level of demand. The **maximin criterion** selects that action for which the minimum payoff is highest—that is, we *maximize* the *minimum* payoff. Clearly, as set out in Table 19.3, the maximum value of these minimum payoffs is $100,000, which will occur if production process C is used. The maximin criterion therefore selects this action.

TABLE 19.3 Choice of production process C by the maximin criterion

PRODUCTION PROCESS	LEVEL OF DEMAND			MINIMUM PAYOFF
	LOW	MODERATE	HIGH	
A	70,000	120,000	200,000	70,000
B	80,000	120,000	180,000	80,000
C	100,000	125,000	160,000	100,000 ← Maximin

Example 19.1   An investor wishes to choose between investing $10,000 for 1 year at an assured interest rate of 12%, and investing the same amount over that period in a portfolio of common stocks. If the fixed interest choice is made, she will be assured of a payoff

of $1,200. On the other hand, if the portfolio of stocks is chosen, the return will depend on the performance of the market over the year. If the market is buoyant, a profit of $2,500 is expected; if the market is steady the expected profit is $500, while for a depressed market a loss of $1,000 is expected. Set up the payoff table for this investor, and find the maximin choice of action.

The accompanying table shows the payoffs (in dollars), with a negative payoff indicating a loss. The minimum payoff for the fixed interest investment is $1,200, as this will occur whatever happens in the stock market. The minimum payoff from the stock portfolio is −$1,000, and occurs when the market is depressed. Hence, the largest minimum payoff arises from the fixed interest investment, which is, therefore, the action chosen by the maximin criterion.

INVESTMENT	STATE OF THE MARKET			MINIMUM PAYOFF
FIXED INTEREST	1,200	1,200	1,200	1,200 ← Maximin
STOCK PORTFOLIO	2,500	500	−1,000	−1,000

From these illustrations, the general form of the decision rule based on the maximin criterion is clear. This rule is set out in the box.

Decision Rule Based On Maximin Criterion

Suppose that a decision-maker has to choose from K admissible actions, a_1, a_2, . . . , a_K, given H possible states of nature. Let M_{ij} denote the payoff corresponding to the ith action and jth state.

For each action, we seek the smallest possible payoff. For the action a_1, for example, this is the smallest of $M_{11}, M_{12}, . . . , M_{1H}$. Suppose we denote this minimum M_1^*. Then

$$M_1^* = \text{Min } (M_{11}, M_{12}, . . . , M_{1H})$$

More generally, the smallest possible payoff for action a_i is given by

$$M_i^* = \text{Min } (M_{i1}, M_{i2}, . . . , M_{iH})$$

The **maximin criterion** then selects that action a_i for which the corresponding M_i^* is largest.

The positive feature of the maximin criterion for decision-making is that it produces the largest possible payoff that can be *guaranteed*. If production process C is used, the candy bar manufacturer is *assured* a payoff of at least $100,000, whatever the level of demand turns out to be. Similarly, for the investor of Example 19.1, the choice of fixed interest makes a *certain* profit of $1,200. In neither example can any available alternative action *guarantee* as much.

However, it is precisely within this guarantee that reservations about the maximin criterion arise, because one must often pay a price for such a guarantee. The price here lies in the foregoing of opportunities to receive a very much larger payoff, through the choice of some other action, *however unlikely* the worst-case situation seems to be. Thus, for example, the candy bar manufacturer may be virtually certain that a high level of demand will result, in which case production process C would be a poor choice, since it yields the lowest payoff at this demand level.

The maximin criterion, then, can be thought of as providing a very cautious strategy for choosing among alternative actions. Such a strategy may, in certain circumstances, be appropriate, but only an extreme pessimist would invariably use it.

(ii) MINIMAX REGRET CRITERION

The decision-maker wanting to use the minimax regret criterion must imagine himself in the position where a choice of action has been made, one of the states of nature has occurred, and he can look back on his choice either with satisfaction or with disappointment because, as things turned out, some alternative action would have been preferable. Consider, once again, our candy bar manufacturer. Suppose that the level of demand for the new product turns out to be low. In that case, the best choice of action would have been production process C, yielding a payoff of $100,000. Had this choice been made, the manufacturer would have had 0 **regret.** On the other hand, had process A been chosen, the resulting profit would have been only $70,000. The extent of the manufacturer's regret, in this eventuality, is the difference between the best payoff that could have been obtained ($100,000) and that resulting from, what turned out to be, an inferior choice of action. Thus, the regret would be $30,000. Similarly, given low demand, if process B had been chosen the regret would have been

$$\$100,000 - \$80,000 = \$20,000$$

Continuing in this way, we can calculate the regrets involved for moderate and high levels of demand. In each case the regret is $0 for what would have turned out to be the best choice of action (process C for moderate demand and process A for high demand).

In this way we can construct a **regret table,** with an entry for each action–state of nature combination. Table 19.4 gives regrets (in dollars) for the candy bar manufacturer's decision problem.

TABLE 19.4 Regret table for candy bar manufacturer

PRODUCTION PROCESS	LEVEL OF DEMAND		
	LOW	MODERATE	HIGH
A	30,000	5,000	0
B	20,000	5,000	20,000
C	0	0	40,000

Next we ask, for each possible course of action, what is the largest amount of regret that can result. From Table 19.4, these maxima are, respectively, $30,000, $20,000, and $40,000 for processes A, B, and C. The **minimax regret criterion** then selects that action for which the maximum regret is smallest. As set out in Table 19.5, the use of this criterion would dictate the choice of production process B.

TABLE 19.5

PRODUCTION PROCESS	LEVEL OF DEMAND			MAXIMUM REGRET	
	LOW	MODERATE	HIGH		
A	30,000	5,000	0	30,000	
B	20,000	5,000	20,000	20,000	← **Minimax regret**
C	0	0	40,000	40,000	

Example 19.2

Consider again the decision problem of the investor of Example 19.1. What action would be chosen if the minimax regret criterion were followed?

The calculations are set out in the accompanying table. Once again, 0 regret follows from the action that would, in the event, have proved the better alternative. We see, then, that the fixed interest investment is selected by the minimax regret criterion.

INVESTMENT	STATE OF THE MARKET			MAXIMUM REGRET	
	BUOYANT	STEADY	DEPRESSED		
FIXED INTEREST	1,300	0	0	1,300	← **Minimax regret**
STOCK PORTFOLIO	0	700	2,200	2,200	

The general decision rule based on the minimax regret criterion is stated in the box.

Decision Rule Based on Minimax Regret Criterion

Suppose that a payoff table is arranged as a rectangular array, with rows corresponding to actions, and columns to states of nature. Then, if each payoff in the table is subtracted from the largest payoff *in its column*, the resulting array is called a **regret table**.

Given the regret table, the action dictated by the **minimax regret criterion** is found as follows:

> **(i)** For each row (action), find the maximum regret.
>
> **(ii)** Choose that action corresponding to the *minimum* of these *maximum* regrets.

The minimax regret criterion for decision-making produces the smallest possible regret that can be *guaranteed*. It does, however, have the following serious drawbacks:

(a) The logic behind the criterion does not provide a compelling framework for analysis for a wide range of practical business decision-making problems. Certainly, there is something to be said for not having to shed too many tears over missed opportunities. Nevertheless, in a rational world, decisions ought to be made on rather more substantial grounds.

(b) Like the maximin criterion, the minimax regret criterion does not allow the decision-maker to inject his or her views as to the likelihood of occurrence of the states of nature into the decision-making process. Since, in most practical business problems, the decision-maker is presumably operating in an environment with which he or she is at least moderately familiar, this represents a waste of expertise.

EXERCISES

1. An investor wishes to choose among three alternatives—a savings account, a high-risk stock, and a low-risk stock—for a $10,000 investment. He considers the following three possible states of nature:

   s_1: Stock market rises
   s_2: Stock market remains flat
   s_3: Stock prices fall

   The payoff table (in dollars) is also shown here:

ACTIONS	STATES OF NATURE		
	s_1	s_2	s_3
SAVINGS	500	500	500
HIGH-RISK STOCK	1,500	100	−500
LOW-RISK STOCK	1,000	200	−100

   (a) Are any of these actions inadmissible?
   (b) Which action is chosen by the maximin criterion?
   (c) Which action is chosen by the minimax regret criterion?

2. A manufacturer of earth-moving equipment must decide if and when to market a new product line. The four possible actions are as follows:

a_1: Begin production immediately
a_2: Delay production 1 year
a_3: Delay production 2 years
a_4: Cancel the product line

The profitability of this new line will depend on the level of road construction activity over the next 5 years. The three states of nature envisaged are

s_1: Low activity level
s_2: Medium activity level
s_3: High activity level

The payoff table (in millions of dollars) is as shown here.

ACTIONS	STATES OF NATURE		
	s_1	s_2	s_3
a_1	-10	1	20
a_2	-7	4	8
a_3	-7	-6	6
a_4	0	0	0

(a) Are any of these actions inadmissible?

(b) Which action is chosen by the maximin criterion?

(c) Which action is chosen by the minimax regret criterion?

3. Consider a decision problem with two possible actions and two states of nature.

(a) Provide an example of a payoff table for which both actions are admissible, and the same action is chosen by both the maximin criterion and the minimax regret criterion.

(b) Provide an example of a payoff table for which different actions are chosen by the maximin criterion and the minimax regret criterion.

4. Consider a decision problem with two admissible actions and two possible states of nature. Formulate a description of the form that the payoff table must take in order that the same action be chosen by the maximin criterion and by the minimax regret criterion.

5. The prospective operator of a shoe store has the opportunity to locate in an established and successful shopping center. Alternatively, at lower cost, he can locate in a new center, whose development has recently been completed. If the new center turns out to be very successful it is expected that annual store profits from location in it would be $120,000. If the center is only moderately successful, annual profits would be $50,000. Finally, if the new center is unsuccessful, an annual loss of $10,000 would be expected. The profits to be expected from location in the established center will also depend to some extent on the degree of success of the new center, as potential customers may be drawn to it. If the new center were to be unsuccessful, annual profit for the shoe store located in the established center would be expected to be $90,000. However, if the new center was moderately successful, this expected profit would be $60,000, while it would be $30,000 if the new center turned out to be very successful.

(a) Set up the payoff table for the decision-making problem of this shoe store operator.

(b) Which action is chosen by the maximin criterion?

(c) Which action is chosen by the minimax regret criterion?

6. A company wishes to decide how to heat a new facility that is to be built. The following three alternative actions are being considered:

a_1: Complete solar system
a_2: Oil and natural gas
a_3: Combination of solar energy and oil and natural gas

The costs of heating the facility will depend on future energy prices. Three states of nature are contemplated, as follows:

s_1: Low increases in energy prices
s_2: Moderate increases in energy prices
s_3: Large increases in energy prices

The accompanying table shows heating *costs* (in tens of thousands of dollars), annualized over the expected life of the facility.

ACTIONS	STATES OF NATURE		
	s_1	s_2	s_3
a_1	8	8	8
a_2	4	9	13
a_3	6	8	10

(a) Set up the payoff table.

(b) Are any of the actions inadmissible?

(c) Which action is chosen by the maximin criterion?

(d) Which action is chosen by the minimax regret criterion?

19.3 EXPECTED MONETARY VALUE

We have already suggested rather strongly that an important ingredient in the analysis of a great many business decision-making problems is likely to be the decision-maker's assessment of the chances of occurrence of the various states of nature relevant in the determination of the eventual payoff. The criteria discussed in the previous section do not allow the incorporation of this kind of assessment into the decision-making process. However, a manager will almost invariably have a good feeling for the environment in which the decision is to be made, and will want this expertise to be taken into account before deciding on a course of action. The candy bar manufacturer will presumably have some experience of the market for his product and, on the basis of that experience, will be able to form a view as to the likelihood of occurrence of low, moderate, or high demand. In this section, we assume that a *probability* of occurrence can be attached to each state of nature, and we will see how these probabilities are employed in arriving at an eventual decision.

Suppose the candy bar manufacturer knows that, of all previous new introductions of this type of product, 10% have met low demand, 50% moderate demand, and 40% high demand. In the absence of any further information, it is then reasonable to postulate, for this particular market introduction, the following probabilities for the states of nature:

$$\text{Probability of low demand} = 0.1$$

$$\text{Probability of moderate demand} = 0.5$$

$$\text{Probability of high demand} = 0.4$$

Notice that since one, and only one, of the states of nature must occur, these probabilities necessarily sum to 1.

In solving the decision-making problem, the probabilities for the occurrences of the states of nature are to be employed, together with the payoffs corresponding to each action–state of nature combination. It is therefore convenient to add this probabilistic information to the payoff table, as in Table 19.6.

TABLE 19.6 Payoffs and states of nature probabilities for candy bar manufacturer

PRODUCTION PROCESS	LEVEL OF DEMAND		
	LOW	MODERATE	HIGH
PROBABILITIES	0.1	0.5	0.4
A	70,000	120,000	200,000
B	80,000	120,000	180,000
C	100,000	125,000	160,000

In general, when there are H possible states of nature, a probability must be attached to each. We will denote these probabilities by $p_1, p_2, \ldots, p_H$, so that probability p_j corresponds to state of nature s_j. Again, these probabilities must sum to 1, so that

$$\sum_{j=1}^{H} p_j = 1$$

Then the general set-up of our decision-making problem is shown in Table 19.7.

At the time when the decision-maker is to choose an action, he will see himself, for any particular choice, as having specific probabilities of receiving the various associated payoffs. He will therefore be able to calculate the **expected payoff** arising from each action. If the candy bar manufacturer adopts production process A, he will receive a payoff of $70,000 with probability 0.1, $120,000 with probability 0.5, and $200,000 with probability 0.4. The expected payoff for this action is then the sum of the individual payoffs, weighted by their associated probabilities. These expected

TABLE 19.7 Payoffs, M_{ij}, and states of nature probabilities, p_j, for a decision problem with K admissible actions and H possible states of nature

ACTIONS	STATES OF NATURE			
	s_1	s_2	$\cdots$	s_H
PROBABILITIES	p_1	p_2	$\cdots$	p_H
a_1	M_{11}	M_{12}	$\cdots$	M_{1H}
a_2	M_{21}	M_{22}	$\cdots$	M_{2H}
$\vdots$	$\vdots$	$\vdots$		$\vdots$
a_K	M_{K1}	M_{K2}	$\cdots$	M_{KH}

payoffs are often called the **expected monetary values** of the actions. For the candy bar manufacturer, the expected monetary values for the three admissible actions are as follows:

PROCESS A $(0.1)(70,000) + (0.5)(120,000) + (0.4)(200,000) = \$147,000$

PROCESS B $(0.1)(80,000) + (0.5)(120,000) + (0.4)(180,000) = \$140,000$

PROCESS C $(0.1)(100,000) + (0.5)(125,000) + (0.4)(160,000) = \$136,500$

      The general form of the definition of expected monetary value is stated in the box.

Expected Monetary Values

      Suppose that a decision-maker has K possible actions, $a_1, a_2, \ldots, a_K$, and is faced with H states of nature. Let M_{ij} denote the payoff corresponding to the ith action and jth state, and p_j the probability of occurrence of the jth state of nature, with

$$\sum_{j=1}^{H} p_j = 1$$

Then the **expected monetary value**, $EMV(a_i)$, of the action a_i is

$$EMV(a_i) = p_1 M_{i1} + p_2 M_{i2} + \cdots + p_H M_{iH}$$

$$= \sum_{j=1}^{H} p_j M_{ij}$$

      The expected monetary values associated with the alternative courses of action provide the decision-maker with a choice criterion that will be extremely attractive for a great many practical problems. By this criterion, the action with highest expected monetary value is adopted. Hence, following this rule, the candy bar manufacturer would choose production process A. It is interesting to note that neither the maximin criterion nor the minimax regret criterion of the previous section led to this particular choice. However, we have now added the additional information that a high level of demand appears much more likely than a low level. This renders process A a relatively attractive option.

Expected Monetary Value Criterion

Given a choice among alternative actions, the **expected monetary value criterion** dictates the choice of that action for which expected monetary value is highest.

The analysis of a decision problem, through the expected monetary value criterion, can be conveniently set out diagramatically through a mechanism called a **decision tree.** Such a diagram is shown, for the candy bar manufacturer, in Figure 19.1. Beginning at the left-hand side of the figure, branches emerge from the square junction there to represent the three possible actions. Junctions marked by squares are those at which decisions must be made. Next, we reach circular junctions, from which emerge branches, each representing a possible state of nature, to which we attach the associated probability. Finally, at the end of these last branches, the payoffs corresponding to each action–state of nature combination are inserted. The computations proceed from right to left, beginning with these payoffs. For each circular junction we

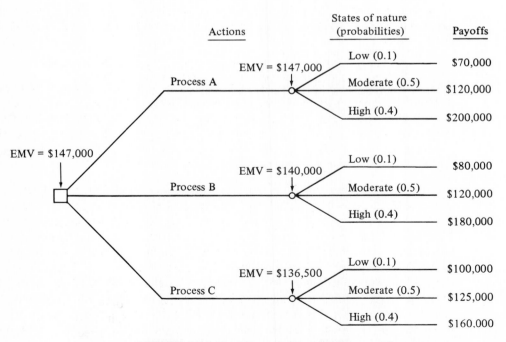

FIGURE 19.1 Decision tree for candy bar manufacturer

find the sum of probability times payoff for the emerging branches. This provides the EMV for each action. Finally, the highest of the EMVs is indicated at the square junction. We see that this results from process A, which is therefore chosen by the expected monetary value criterion. This choice of action results in an expected monetary value, or expected profit, of $147,000 for the candy bar manufacturer.

Example 19.3

Consider again the investor of Examples 19.1 and 19.2 who must decide between a fixed interest rate investment and a portfolio of stocks. Let us assume that, in fact, this investor is very optimistic about the future course of the stockmarket, believing the probability that it will be buoyant is 0.6, while the probability is 0.2 for each of the other two states. The payoffs and state of nature probabilities are, therefore, those given in the accompanying table. Which investment should be chosen according to the expected monetary value criterion?

INVESTMENT	STATE OF THE MARKET		
	BUOYANT	STEADY	DEPRESSED
PROBABILITIES	0.6	0.2	0.2
FIXED INTEREST	1,200	1,200	1,200
STOCK PORTFOLIO	2,500	500	−1,000

Since a payoff of $1,200 will result from the fixed interest investment, whatever happens in the stock market, the expected monetary value of this investment is $1,200. For the stock portfolio, we have

$$\text{EMV} = (0.6)(2,500) + (0.2)(500) + (0.2)(-1,000) = \$1,400$$

Since this is the higher expected monetary value, the investor would choose the portfolio of common stocks, according to this criterion.

The decision tree for this problem is shown in Figure 19.2. Notice that, for the fixed interest action, there are no sub-branches corresponding to states of nature, since the same payoff ($1,200) materializes whatever the state. This payoff, then, is the expected monetary value for that action.

Example 19.4

A manufacturer receives regular contracts for large consignments of parts for the automobile industry. This manufacturer's production process is such that, when it is functioning correctly, 10% of all parts produced do not meet industry specifications. However, the process is prone to a particular malfunction, whose presence can be checked at the beginning of a production run. When the process is operated with this malfunction, 30% of the parts produced fail to meet specifications. The manufacturer supplies parts under a certain contract, which will yield a profit of $2,000 if only 10% of the parts are defective, and a profit of $1,200 if 30% of the parts are defective.

The cost of checking for the malfunction is $100 and, if it turns out that a repair is needed, this costs an additional $200. If these costs are incurred, they must necessarily be subtracted from the profit of the contract. Historically, it has been found that

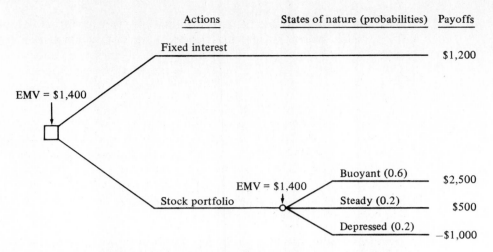

Actions   States of nature (probabilities)   Payoffs

Fixed interest $1,200

EMV = $1,400

Stock portfolio   EMV = $1,400

Buoyant (0.6) $2,500

Steady (0.2) $500

Depressed (0.2)   −$1,000

FIGURE 19.2 Decision tree for investor of Example 19.3

the process functions correctly 80% of the time. Based on the expected monetary value criterion, should the process be checked before production is begun?

As a first step, we reduce this problem to the framework of a payoff table. We can do this in the following steps:

(a) There are two possible actions—to check or not to check.
(b) There are two possible states of nature—repair is needed or repair is not needed.
(c) We now need the payoffs for each action–state of nature combination. If no check is made, the manufacturer's profit (payoff) will be $2,000 if repair was not needed and $1,200 if repair was needed. If a check is made and it emerges that no repair is necessary, this involves a subtraction of $100 from the $2,000 profit, leaving a payoff of $1,900. Finally, if the check reveals the necessity of repair work, a total cost of $300 must be subtracted from the $2,000 profit, leaving a payoff of $1,700.

We also know that the probability is 0.8 that repair will not be needed. Hence, the probability that repair will be needed is 0.2. We can now construct the accompanying payoff table, showing the states of nature probabilities.

ACTIONS	STATES OF NATURE	
	REPAIR NOT NEEDED	REPAIR NEEDED
PROBABILITIES	0.8	0.2
CHECK	1,900	1,700
NO CHECK	2,000	1,200

Now that we have put the problem into this standard format, the rest of the analysis can be carried out along familiar lines. The expected monetary values are as follows:

CHECK $\qquad (0.8)(1,900) + (0.2)(1,700) = \$1,860$

NO CHECK $\qquad (0.8)(2,000) + (0.2)(1,200) = \$1,840$

Hence, according to the expected monetary value criterion, the production process should be checked at the beginning of the production run. The decision tree is shown in Figure 19.3. This tree is constructed in exactly the same fashion as that of Figure 19.1 for the candy bar manufacturer. Referring to Figure 19.3, we see at the square junction on the left-hand side that the preferred course of action is to check the production process, for which the expected monetary value (that is, the expected profit to the manufacturer) is $1,860.

Example 19.5

This example illustrates a problem in which a *sequence* of decisions may be required. The use of decision trees is particularly helpful in solving such problems.

A drug manufacturer holds the patent rights to a new formula for arthritic pain relief. The manufacturer is able to sell the patent for $50,000 or to proceed with intensive tests of the drug's efficacy. The cost of carrying out these tests is $10,000. If the drug is found to be ineffective, it will not be marketed and the cost of the tests is written off as a loss. In the past, tests of drugs of this type have shown 60% to be effective and 40% ineffective.

If the tests should now reveal the drug to be effective, the manufacturer again has two options available. He can sell the patent rights and test results for $120,000, or market the drug himself. If the drug is marketed, it is estimated that profits on sales (exclusive of the cost of the tests) will amount to $180,000 if the sales campaign is

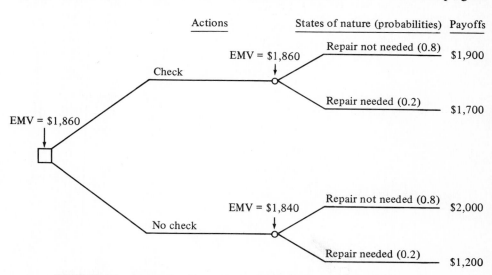

FIGURE 19.3 Decision tree for automobile parts supplier of Example 19.4

highly successful, but only $90,000 if it is just moderately successful. It is estimated that these two levels of market penetration are equally likely. According to the expected monetary value criterion, how should the drug manufacturer proceed?

It is best to attack this problem through the construction of a decision tree. The completed tree is shown in Figure 19.4. The branches are constructed, beginning on the left-hand side at the first decision point. The manufacturer may decide either to sell the patent, in which case there is nothing further to be done, or retain it and carry out tests on the drug's efficacy. Here there are two possible states of nature—the drug is either effective (with probability 0.6) or it is ineffective (with probability 0.4). In the latter case, the story ends. However, if the drug proves to be effective, a second decision must be made—whether to market it or to sell the patent rights and test result. If the former option is adopted, the eventual outcome is determined by the level of marketing success, which could be either moderate or high (each with probability 0.5).

Next, the payoffs resulting from all action–state of nature combinations are entered on the right-hand side of the diagram. We will begin at the bottom. If the manufacturer's original decision is to sell the patent, he receives $50,000. If the patent is kept, but the drug turns out to be ineffective, the manufacturer sustains a loss of $10,000, the cost of carrying out the tests. This is shown as a negative payoff in that amount. If the drug is found to be effective and the patent and test results are then sold, the manufacturer receives $120,000, from which must be subtracted the cost of the tests, leaving a payoff of $110,000. Finally, if the drug is marketed, the payoffs for moderate and high success are, respectively, $90,000 and $180,000, less the cost of the tests.

Having reached this point, we can solve the decision problem by working backward from right to left along the tree. This is necessary because the appropriate action at the first decision point cannot be determined until we have found the expected monetary value of the best available option at the second decision point.

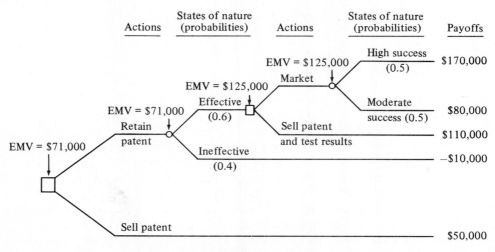

FIGURE 19.4 Decision tree for the drug manufacturer of Example 19.5

We therefore begin by supposing that, initially, the patent was retained and the tests proved the drug to be effective. If the patent and test results are sold, a profit of $110,000 will result. The expected monetary value from marketing the drug is

$$(0.5)(170,000) + (0.5)(80,000) = \$125,000$$

Since this exceeds $110,000, the better option at this stage, by the expected monetary value criterion, is to market the drug. This amount is, therefore, entered at the square junction of the second decision point, and is treated as the payoff that results if the manufacturer's initial decision is to retain the patent and the tests indicate that the drug is effective.

Hence, for the initial decision, the payoff table with states of nature probabilities is as shown here. The expected monetary value of selling the patent is the assured $50,000, while the expected monetary value of retaining it is

$$(0.6)(125,000) + (0.4)(-10,000) = \$71,000$$

Thus, by the expected monetary value criterion, the patent should be retained.

ACTIONS	STATES OF NATURE	
	EFFECTIVE	INEFFECTIVE
PROBABILITIES	0.6	0.4
RETAIN PATENT	125,000	−10,000
SELL PATENT	50,000	50,000

Thus, we have finally reached the conclusion that, if the objective is the maximization of expected monetary value (that is, expected profit) the manufacturer should retain the patent and, if tests prove the drug to be effective, market it himself. This strategy yields an expected profit of $71,000.

19.4 USE OF SAMPLE INFORMATION: BAYESIAN ANALYSIS

Decisions made in the business world can often involve considerable amounts of money, and the cost of making a suboptimal choice may turn out to be substantial. This being the case, it could well pay the decision-maker to make an effort to obtain as much relevant information as possible before the decision is made. In particular, he or she will want to become as thoroughly informed as is feasible about the chances of occurrence of the various states of nature that determine the eventual payoff.

This feature of any careful analysis of a decision problem has not been apparent in our discussion thus far. The candy bar manufacturer, in the previous section, assessed the probabilities of low, moderate, and high levels of demand for a new candy bar as 0.1, 0.5, and 0.4, respectively. However, this assessment reflected no more than the historical proportions achieved by previous products. In practice, he might well want to carry out some market research on the prospects for the new product. Given such research, these initial or **prior probabilities** may be modified, yielding new probabilities called **posterior probabilities** for the three demand levels. The information (in this case, the market research results) leading to the modification of probabilities for the states of nature will be referred to as **sample information.**

In fact, we saw in Section 3.8 the mechanism for modifying prior probabilities to produce posterior probabilities. This is accomplished through **Bayes' theorem,** which, for convenience, we restate in the box.[1]

Bayes' Theorem

Let $s_1, s_2, \ldots, s_H$ be H mutually exclusive and collectively exhaustive events, corresponding to the H states of nature of a decision problem. Let A be some other event. Denote the conditional probability that s_i will occur, given that A occurs, by $P(s_i|A)$, and the probability of A, given s_i, by $P(A|s_i)$.

Then the conditional probability of s_i, given A, can be expressed as

$$P(s_i|A) = \frac{P(A|s_i)P(s_i)}{P(A)}$$

$$= \frac{P(A|s_i)P(s_i)}{P(A|s_1)P(s_1) + P(A|s_2)P(s_2) + \cdots + P(A|s_H)P(s_H)}$$

In the terminology of this section, $P(s_i)$ is the **prior probability** of s_i, and is modified to the **posterior probability** $P(s_i|A)$, given the **sample information** that event A has occurred.

Now, suppose the candy bar manufacturer hires a market research organization to predict the level of demand for his new product.[2] The organization provides a rating of "poor," "fair," or "good," on the basis of its research. A review of the market research company's records reveals the quality of its past predictions in this field. Table 19.8 shows, for each level of demand outcome, the proportion of poor, fair, and

TABLE 19.8 Proportion of assessments of each type provided by market research organization for candy bars achieving given levels of demand

ASSESSMENT	LEVEL OF DEMAND		
	LOW (s_1)	MODERATE (s_2)	HIGH (s_3)
POOR	0.6	0.3	0.1
FAIR	0.2	0.4	0.2
GOOD	0.2	0.3	0.7

---

[1] Before proceeding further, the reader may wish to review Section 3.8.

[2] He will, of course, have to pay for this service. In Section 19.5 we discuss the question of whether the returns merit the cost involved.

good assessments. Thus, for example, on 10% of occasions that demand was high, the assessment was "poor." Thus, in the notation of conditional probability, denoting low, moderate, and high demand levels by s_1, s_2, and s_3, respectively, we have

$$P(\text{Poor}\,|\,s_1) = 0.6; \qquad P(\text{Poor}\,|\,s_2) = 0.3; \qquad P(\text{Poor}\,|\,s_3) = 0.1$$

and so on.

Suppose, now, that the market research firm is consulted, and produces an assessment of "poor" for the prospects of the candy bar. Given this new information, the prior probabilities

$$P(s_1) = 0.1; \qquad P(s_2) = 0.5; \qquad P(s_3) = 0.4$$

for the three demand levels can be modified using Bayes' theorem. For a low level of demand, the posterior probability is

$$P(s_1\,|\,\text{Poor}) = \frac{P(\text{Poor}\,|\,s_1)P(s_1)}{P(\text{Poor}\,|\,s_1)P(s_1) + P(\text{Poor}\,|\,s_2)P(s_2) + P(\text{Poor}\,|\,s_3)P(s_3)}$$

$$= \frac{(0.6)(0.1)}{(0.6)(0.1) + (0.3)(0.5) + (0.1)(0.4)} = \frac{0.06}{0.25} = 0.24$$

Similarly, for the other two demand levels, the posterior probabilities are

$$P(s_2\,|\,\text{Poor}) = \frac{(0.3)(0.5)}{0.25} = 0.60$$

$$P(s_3\,|\,\text{Poor}) = \frac{(0.1)(0.4)}{0.25} = 0.16$$

The posterior probabilities can then be employed to calculate the expected monetary values. Table 19.9 shows the payoffs, together with the posterior probabilities for the three demand levels. This is simply a modification of Table 19.6, with the posterior probabilities replacing the prior probabilities of that table.

TABLE 19.9 Payoffs for candy bar manufacturer, and posterior probabilities for states of nature, given an assessment of "poor" by market research organization

PRODUCTION PROCESS	LEVEL OF DEMAND		
	LOW	MODERATE	HIGH
POSTERIOR PROBABILITIES	0.24	0.60	0.16
A	70,000	120,000	200,000
B	80,000	120,000	180,000
C	100,000	125,000	160,000

The expected monetary values for the three production processes can be found in precisely the same manner as before. These are as follows:

PROCESS A $\quad$ $(0.24)(70,000) + (0.60)(120,000) + (0.16)(200,000) = \$120,800$

PROCESS B $\quad$ $(0.24)(80,000) + (0.60)(120,000) + (0.16)(180,000) = \$120,000$

PROCESS C $\quad$ $(0.24)(100,000) + (0.60)(125,000) + (0.16)(160,000) = \$124,600$

We see that if the assessment of market prospects is "poor," then according to the expected monetary value criterion, production process C should be used. The market research group's assessment has rendered low demand much more likely and high demand considerably less likely than was previously the case. This shift in the view of market prospects is sufficient to induce the candy bar manufacturer to switch his preference from process A (based on the prior probabilities) to process C.

Following the same line of argument, we can determine the decisions that would be made if the prospects for the candy bar's market success were rated either "fair" or "good." Again, the posterior probabilities for the three levels of demand can be obtained through Bayes' theorem. For a "fair" assessment, these are

$$P(s_1 \mid \text{Fair}) = \frac{1}{15}; \qquad P(s_2 \mid \text{Fair}) = \frac{2}{3}; \qquad P(s_3 \mid \text{Fair}) = \frac{4}{15}$$

and, for a "good" assessment,

$$P(s_1 \mid \text{Good}) = \frac{2}{45}; \qquad P(s_2 \mid \text{Good}) = \frac{1}{3}; \qquad P(s_3 \mid \text{Good}) = \frac{28}{45}$$

Using these posterior probabilities, we can now determine the expected monetary values of each of the production processes for each given assessment. Table 19.10, contains these quantities.

As we have already seen, if the assessment is "poor," process C is preferred by the expected monetary value criterion. On the other hand, if any other assessment is made, we would choose to use production process A, according to this criterion.

We recall that, for the candy bar manufacturer's problem, when the prior probabilities for levels of demand were used, the optimal decision according to the expected monetary value criterion was to use process A. It can be the case (if an assessment of "poor" is obtained) that a different decision will be made when these prior probabilities are modified by sample information. Hence, it turns out that

TABLE 19.10 Expected monetary values for candy bar manufacturer for each of three possible assessments by market research firm

PRODUCTION PROCESS	ASSESSMENT		
	POOR	FAIR	GOOD
A	120,800	138,000	167,556
B	120,000	133,333	155,556
C	124,600	132,667	145,667

consulting the market research organization could be valuable for the manufacturer. Of course, if the choice of process A had proved optimal, whatever the assessment, the sample information could not possibly be of value.

Example 19.6
In Example 19.4, a supplier of parts to the automobile industry had to decide whether to check the production process for a certain malfunction at the beginning of a production run. The two states of nature were

s_1: Repair not needed (10% of parts produced fail to meet specifications)
s_2: Repair needed (30% of parts produced fail to meet specifications)

The prior probabilities, obtained from the historical record for this production process, are

$$P(s_1) = 0.8 \quad \text{and} \quad P(s_2) = 0.2$$

Based on this information alone, the expected monetary value criterion, as we saw in Example 19.4, dictates that the production process should be checked at the outset. However, the manufacturer has an additional option. He can, before commencing the full production run, produce a single part and check whether it meets specifications. If this procedure is followed, what course should be adopted once this sample information is obtained?

The two possible sample outcomes are that the single part checked either does or does not meet specifications. If, in fact, repair is not needed, only 10% of all parts fail to meet specifications. Hence, the probability is 0.1 that the sampled part will not meet the specifications; that is

$$P(\text{Not meet specifications} \mid s_1) = 0.1$$

It necessarily follows that, if in fact repair is not needed, the probability that the sampled part will meet the specifications is 0.9; that is,

$$P(\text{Meets specifications} \mid s_1) = 0.9$$

Similarly, since 30% of all parts do not meet specifications when repair is needed, the conditional probabilities for the sampled part are

$$P(\text{Not meet specifications} \mid s_2) = 0.3$$

$$P(\text{Meets specifications} \mid s_2) = 0.7$$

We can now use Bayes' theorem to find the posterior probabilities. Consider first the case where the sampled part fails to meet specifications. The posterior probability for state s_1 (Repair not needed) is then

$$P(s_1 \mid \text{Not meet specs}) = \frac{P(\text{Not meet specs} \mid s_1)P(s_1)}{P(\text{Not meet specs} \mid s_1)P(s_1) + P(\text{Not meet specs} \mid s_2)P(s_2)}$$

$$= \frac{(0.1)(0.8)}{(0.1)(0.8) + (0.3)(0.2)} = 0.5714$$

Since the posterior probabilities for the two states must sum to 1, it follows that

$$P(s_2 \mid \text{Not meet specs}) = 0.4286$$

The payoff table, with the posterior states of nature probabilities, given that the sampled part did not meet the specifications, is shown here. The payoffs are the same

ACTIONS	STATES OF NATURE	
	REPAIR NOT NEEDED (s_1)	REPAIR NEEDED (s_2)
POSTERIOR PROBABILITIES	0.5714	0.4286
CHECK	1,900	1,700
NO CHECK	2,000	1,200

as those given in Example 19.4. The expected monetary values are then calculated in the usual manner. They are:

CHECK $\qquad$ $(0.5714)(1,900) + (0.4286)(1,700) = \$1,814.28$

NO CHECK $\qquad$ $(0.5714)(2,000) + (0.4286)(1,200) = \$1,657.12$

Hence, if the sampled part fails to meet the specifications, the production process should be checked, according to the expected monetary value criterion.

Next, we need to consider the case where the sampled part does satisfy the specifications. Once again, an appeal to Bayes' theorem provides the relevant posterior probabilities. For state s_1, we have

$$P(s_1 | \text{Meets specs}) = \frac{P(\text{Meets specs} | s_1)P(s_1)}{P(\text{Meets specs} | s_1)P(s_1) + P(\text{Meets specs} | s_2)P(s_2)}$$

$$= \frac{(0.9)(0.8)}{(0.9)(0.8) + (0.7)(0.2)} = 0.8372$$

Again, the posterior probabilities for the two states of nature must sum to 1, so that

$$P(s_2 | \text{Meets specs}) = 0.1628$$

The expected monetary values, given that this sample result has been observed, are

CHECK $\qquad$ $(0.8372)(1,900) + (0.1628)(1,700) = \$1,867.44$

NO CHECK $\qquad$ $(0.8372)(2,000) + (0.1628)(1,200) = \$1,869.76$

Therefore, if the sampled part does satisfy the specifications, the expected monetary value criterion reveals a (rather marginal) preference for proceeding with the production run without first checking for the malfunction.

Example 19.7

In Example 19.5, a drug manufacturer had to decide whether to sell the patent for a pain relief formula, before subjecting the drug to thorough testing. (Subsequently, if the patent was retained and the drug found to be effective, a second decision—to market the drug or sell the patent and test results—also had to be made.) For the initial decision, the two states of nature were

s_1:   Drug is effective
s_2:   Drug is ineffective

The associated prior probabilities, formed on the basis of previous experience, are

$$P(s_1) = 0.6 \quad \text{and} \quad P(s_2) = 0.4$$

The drug manufacturer has the option of carrying out, at modest cost, an initial test before the first decision is made. This test is not infallible. For drugs that have subsequently proved effective, the preliminary test result was "positive" on 60% of occasions, and "negative" on the remainder. For ineffective drugs, a "positive" preliminary test result was obtained 30% of the time, the other results being "negative." Given the results of the preliminary test, how should the drug manufacturer proceed? Assume that it is still possible to sell the patent for $50,000 if the preliminary test result is "negative."

First, we note that if the patent *is* retained and the exhaustive tests prove the drug to be effective, then, in the absence of any sample information on market conditions, the optimal decision at this stage, as in Example 19.5, is to market the drug. The information provided by the preliminary test is irrelevant in that particular decision. However, it could conceivably influence the initial decision as to whether to sell the patent. Accordingly, it is only on this decision that we need to concentrate.

The conditional probabilities of the sample outcomes, given the states of nature are

$$P(\text{Positive} \,|\, s_1) = 0.6; \qquad P(\text{Negative} \,|\, s_1) = 0.4$$

$$P(\text{Positive} \,|\, s_2) = 0.3; \qquad P(\text{Negative} \,|\, s_2) = 0.7$$

If the result of the preliminary test is "positive," the posterior probability for the state s_1 (Effective), given this information, is

$$P(s_1 \,|\, \text{Positive}) = \frac{P(\text{Positive} \,|\, s_1)P(s_1)}{P(\text{Positive} \,|\, s_1)P(s_1) + P(\text{Positive} \,|\, s_2)P(s_2)}$$

$$= \frac{(0.6)(0.6)}{(0.6)(0.6) + (0.3)(0.4)} = 0.75$$

Further, since the two posterior probabilities must sum to 1, we have

$$P(s_2 \,|\, \text{Positive}) = 0.25$$

The accompanying payoff table is the same as in Example 19.5, with these posterior probabilities added.

ACTIONS	STATES OF NATURE	
	EFFECTIVE	INEFFECTIVE
POSTERIOR PROBABILITIES	0.75	0.25
RETAIN PATENT	125,000	−10,000
SELL PATENT	50,000	50,000

The expected monetary value, if the patent is sold, is \$50,000, while, if the patent is retained, the expected monetary value is

$$(0.75)(125,000) + (0.25)(-10,000) = \$91,250$$

Therefore, if the initial test result is "positive," the patent should be retained, according to this criterion.

Next, we consider the case where the preliminary test result is "negative." The posterior probability for the state s_1 is, by Bayes' theorem,

$$P(s_1 \mid \text{Negative}) = \frac{P(\text{Negative} \mid s_1)P(s_1)}{P(\text{Negative} \mid s_1)P(s_1) + P(\text{Negative} \mid s_2)P(s_2)}$$

$$= \frac{(0.4)(0.6)}{(0.4)(0.6) + (0.7)(0.4)} = 0.4615$$

Hence, the posterior probability for the state s_2 is

$$P(s_2 \mid \text{Negative}) = 0.5385$$

Once more, if the patent is sold, the expected monetary value is the \$50,000 that will be received. If the patent is retained, the expected monetary value of this decision is

$$(0.4615)(125,000) + (0.5385)(-10,000) = \$52,302.50$$

Thus, even if the preliminary test result is "negative," the optimal decision, by the expected monetary value criterion, is to retain the patent.

In this particular example, then, whatever the sample information, the chosen action is the same. The manufacturer should retain the patent in the event of either result emerging from the preliminary test. Since the sample information cannot possibly affect the decision, there is, of course, no point in gathering it. In fact, since performing the preliminary test will not be costless, it will be suboptimal to do so. Thus, we conclude that, according to the expected monetary value criterion, the drug manufacturer should retain the patent and, if the thorough tests prove the drug to be effective, he should market it himself. The preliminary test should not be carried out.

19.5 THE VALUE OF SAMPLE INFORMATION

We have seen how sample information can be incorporated into the decision-making process. The potential value of such information lies, of course, in its provision of a better feel for the chances of occurrence of the relevant states of nature. This, in turn, can provide firmer ground on which to base a decision. In this section we will see how a *monetary* value can be attached to the sample information. This is important, since there will typically be some cost involved in obtaining the sample information, and the decision-maker will want to know whether the expected benefits exceed this cost.

In Example 19.7, we considered a situation where the same action was optimal, whatever the sample result. In that case, the sample information clearly has no value, since the same action would have been taken without it. This is a general rule: If the sample information cannot affect the choice of action, then it has value 0.

Accordingly, for the remainder of this section, we will concern ourselves only with those circumstances in which the sample result can affect the choice of action. Our example of the candy bar manufacturer planning to introduce a new product is such a case. This manufacturer has to choose from three production processes and is faced with three states of nature, representing different levels of demand for the product. In Section 19.3 we saw that, in the absence of sample information and using only the prior probabilities, we choose process A, with an expected monetary value of $147,000.

Now, in practice, having obtained sample information, the decision-maker will typically not *know* which state of nature will occur. He will, rather, have more firmly grounded probabilistic assessments for these states. However, before discussing the value of sample information in this general framework, it is useful to consider the extreme case where **perfect information** is obtainable—that is, the case where the decision-maker is able to gain information that will tell him *with certainty* which state will occur. We ask what is the value to the decision-maker of having such perfect information.

In the context of our candy bar manufacturer, perfect information corresponds to knowledge of which of the three possible demand levels will actually result. In the absence of any sample information and on the basis of the prior probabilities only, process A will be chosen. However, referring to Table 19.6, we see that if it were known that the level of demand would be low, the best choice would be process C. Since this has a payoff which exceeds by $30,000 that of process A, the value of knowing that demand would be low is $30,000. Similarly, if it were known that moderate demand would result, process C would again be chosen. Here, the payoff from the best available choice exceeds that of process A by $5,000, which is, accordingly, the value of knowing that demand will be moderate. On the other hand, if it were known that high demand would occur, process A would be chosen. Thus, this particular knowledge is of no value, since the same decision would have been made without it. We see, therefore, that the value of perfect information depends on what that information is. Using the prior probabilities of the various states of nature, we can find the **expected value of perfect information.**

For the candy bar manufacturer, the prior probabilities are 0.1 for low, 0.5 for moderate, and 0.4 for high demand. It therefore follows that, to this manufacturer, the value of perfect information is $30,000 with probability 0.1, $5,000 with probability 0.5, and $0 with probability 0.4. The expected value of perfect information is, accordingly,

$$(0.1)(30,000) + (0.5)(5,000) + (0.4)(0) = \$5,500$$

This dollar amount, then, represents the expected value to the candy bar manufacturer of knowing what level of demand will result.

We now show in the box the general procedure for computing the expected value of perfect information.

Expected Value of Perfect Information

Suppose a decision-maker has to choose from among K possible actions, in the face of H states of nature, $s_1, s_2, \ldots, s_H$. **Perfect information** corresponds to knowledge of which state of nature will arise. The expected value of perfect information is obtained as follows:

(i) Determine which action will be chosen if only the prior probabilities $P(s_1)$, $P(s_2), \ldots, P(s_H)$ are used.

(ii) For each possible state of nature, s_i, find the difference W_i, between the payoff for the best choice of action, if it were known that state would arise, and the payoff for the action chosen if only the prior probabilities are used. This is the value of perfect information, when it is known that s_i will occur.

(iii) The **expected value of perfect information** is then

$$P(s_1)W_1 + P(s_2)W_2 + \cdots + P(s_H)W_H$$

Now, although perfect information will typically not be available, the calculation of the expected value of perfect information can be useful. Since, of course, no sample information can be better than perfect, its expected value cannot be higher than that of the expected value of perfect information. Thus, the expected value of perfect information provides an *upper limit* for the expected value of any sample information. For example, if the candy bar manufacturer is offered information at a cost of $6,000, it is not necessary to inquire further about the quality of this information. It should not be purchased, however reliable, according to the expected monetary value criterion, since its expected value cannot be more than $5,500.

We now turn to the more general problem of assessing the value of sample information that is not necessarily perfect. Again, we will consider the decision-making problem of the candy bar manufacturer, who has the option of obtaining an assessment, from a market research organization, of the prospects for the new candy bar. These prospects will be rated either "poor," "fair," or "good." We saw in Section 19.4 that in either of these last two eventualities, process A will still be chosen. Thus, if a "fair" or "good" rating is obtained, the initial choice of action will remain unchanged, and, in the event, nothing will have been gained from consulting the market research company.

On the other hand, if the prospects are rated "poor," then, referring to Table 19.10 we see that the optimal choice is process C. This optimal choice would yield an expected monetary value of $124,600, whereas process A, which otherwise would have been used, gives an expected monetary value of $120,800. The difference in these amounts, $3,800, represents the gain from the sample information, *if the assessment is "poor."*

Thus, the gains from the sample information are $0 for ratings of "good" or "fair," and $3,800 for a rating of "poor."

We now need to know how likely these gains are to materialize, so that in our example we require to find the probability of a "poor" assessment. In general, if A denotes a piece of sample information, and $s_1, s_2, \ldots, s_H$ the H possible states of nature, then

$$P(A) = P(A|s_1)P(s_1) + P(A|s_2)P(s_2) + \cdots + P(A|s_H)P(s_H)$$

For the candy bar example, with s_1, s_2, s_3 denoting low, moderate, and high levels of demand, we have seen that

$$P(s_1) = 0.1; \qquad P(s_2) = 0.5; \qquad P(s_3) = 0.4$$

and

$$P(\text{Poor}|s_1) = 0.6; \qquad P(\text{Poor}|s_2) = 0.3; \qquad P(\text{Poor}|s_3) = 0.1$$

Therefore, the probability of a "poor" assessment is

$$P(\text{Poor}) = P(\text{Poor}|s_1)P(s_1) + P(\text{Poor}|s_2)P(s_2) + P(\text{Poor}|s_3)P(s_3)$$
$$= (0.6)(0.1) + (0.3)(0.5) + (0.1)(0.4) = 0.25$$

In the same way, using the conditional probabilities of Table 19.8, we obtain the following probabilities for the other two assessments:

$$P(\text{Fair}) = 0.30 \qquad \text{and} \qquad P(\text{Good}) = 0.45$$

We see then that the value of the sample information is \$3,800 with probability 0.25, \$0 with probability 0.30, and \$0 with probability 0.45. It therefore follows that the **expected value of the sample information** is

$$(0.25)(3,800) + (0.30)(0) + (0.45)(0) = \$950$$

This dollar amount, then, represents the expected value of the sample information to the decision-maker. In terms of the expected monetary value criterion, this sample information will be worth aquiring provided its cost is less than its expected value. We define the **expected net value of sample information** as the difference between its expected value and its cost.

Suppose that the market research group charges a fee of \$750 for its assessment. Then the expected net value of this assessment to the candy bar manufacturer is $(950 - 750) = \$200$. Thus, the manufacturer's expected payoff will be \$200 higher if the sample information is purchased than if it is not. This amount represents the expected worth of having that information, taking into account its cost. In this case, the manufacturer's optimal strategy is to purchase the market research report, and then use production process A if the assessment is either good or fair, and process C if the assessment is poor. The expected monetary value of this strategy is \$147,200—that is, \$147,000 that would result from no sample information plus the expected net value of the sample information.

The general framework for computing the value of sample information is set out in the box.

Expected Value of Sample Information

Suppose a decision-maker has to choose from K possible actions, in the face of H States of nature, $s_1, s_2, \ldots, s_H$. The decision-maker may obtain sample information. Let there be M possible sample results, $A_1, A_2, \ldots, A_M$.

The expected value of sample information is obtained as follows:

(i) Determine which action would be chosen if only the prior probabilities were used.

(ii) Determine the probabilities of obtaining each sample result:

$$P(A_i) = P(A_i|s_1)P(s_1) + P(A_i|s_2)P(s_2) + \cdots + P(A_i|s_H)P(s_H)$$

(iii) For each possible sample result, A_i, find the difference, V_i, between the expected monetary value for the optimal action and that for the action chosen if only prior probabilities are used. This is the value of the sample information, given that A_i was observed.

(iv) The **expected value of sample information** is then

$$P(A_1)V_1 + P(A_2)V_2 + \cdots + P(A_M)V_M$$

The **expected net value of sample information** is the difference between its expected value and its cost.

According to the expected monetary value criterion, the decision-maker should purchase the sample information if its expected net value is positive. Otherwise, the sample information should not be purchased.

Example
19.8

This is a continuation of Examples 19.4 and 19.6. A supplier of automobile parts must decide whether to check the production process for malfunction before a run is begun. The two states of nature are that repair is not needed (s_1) and repair is needed (s_2). The manufacturer has the alternative of producing and checking a single part before starting production. The sample information obtained will be that the part checked does or does not meet specifications.

In Example 19.4, we found that, given only the prior probabilities

$$P(s_1) = 0.8 \quad \text{and} \quad P(s_2) = 0.2$$

the optimal strategy (with an expected monetary value of $1,860) is to check the process.

In Example 19.6, it was found that if the sampled part does not meet specifications, the process should be checked, but otherwise, according to the expected monetary value criterion, it should not.

Find the expected value of this sample information.

If the sampled part fails to meet specifications, the action taken will be the same as when the prior probabilities were used. If the sampled part does satisfy the specifications, however, the process will not be checked. In that case, as we found in Example 19.6, this action has expected monetary value $1,869.76, while the expected monetary value from checking is $1,867.44. The difference, $2.32, represents the value of the sample information when the checked part meets specifications.

We now need the probability that the checked part will meet specifications—that is, the probability that this gain of $2.32 will indeed materialize. The following conditional probabilities were given in Example 19.6:

$$P(\text{Meets specs}\,|\,s_1) = 0.9 \qquad \text{and} \qquad P(\text{Meets specs}\,|\,s_2) = 0.7$$

Thus, we have

$$P(\text{Meets specs}) = P(\text{Meets specs}\,|\,s_1)P(s_1) + P(\text{Meets specs}\,|\,s_2)P(s_2)$$

$$= (0.9)(0.8) + (0.7)(0.2) = 0.86$$

Hence, the probability that the sampled part does not meet specifications is

$$P(\text{Not meet specs}) = 1 - 0.86 = 0.14$$

Thus, taking this sample information leads to an expected gain of $2.32 with probability 0.86, and no gain with probability 0.14. The expected value of sample information is, therefore,

$$(0.86)(2.32) + (0.14)(0) = \$1.99$$

According to the expected monetary value criterion, then, this sample information should be collected only if its cost is less than $1.99.

Example 19.8 is suggestive of a general class of problems that the decision-maker may have to consider. So far, we have considered using the expected monetary value criterion to determine whether sample information should be purchased. More generally, the relevant question is *how much*, if any, sample information should be obtained. This would be a more realistic way of viewing the problem of the supplier of automobile parts in Example 19.8. In that example we allowed only two possibilities—either no sample information is taken or just a single part is checked before beginning production. In practice, the supplier would also have the option of checking a sample of several parts before making a decision. It is possible, along the lines we have discussed here, to find the expected value of a sample of any specific number of parts. Given the costs of obtaining samples of each size, we would then obtain the expected net value of sample information for each possible number of sampled parts. The supplier would then choose that number of sample observations for which the expected net value of sample information is highest. All these calculations can be made before sampling begins, so that the initial decision to be made is the choice of sample size. Besides noting that this kind of problem can be incorporated into the decision-making framework so far introduced, we will not consider further the details of its analysis.

VALUE OF SAMPLE INFORMATION VIEWED THROUGH DECISION TREES

The expected value of sample information can be computed in an alternative (but equivalent) manner, which is arithmetically slightly more cumbersome. On the other hand, it does provide a convenient way of representing the problem in terms of a sequence of decisions, through the construction of a decision tree. The first decision to be made is whether to obtain the sample information. Next, it is necessary to decide which of the alternative actions should be followed.

The decision tree for the automobile parts supplier of Example 19.8 is set out in Figure 19.5. We will assume that the cost of obtaining the sample information is $5. From Example 19.8 we already know that, in fact, this should not be done, since its expected value is only $1.99. From Figure 19.5, we see that this same conclusion is reached through a slightly different route. The tree is constructed beginning at the

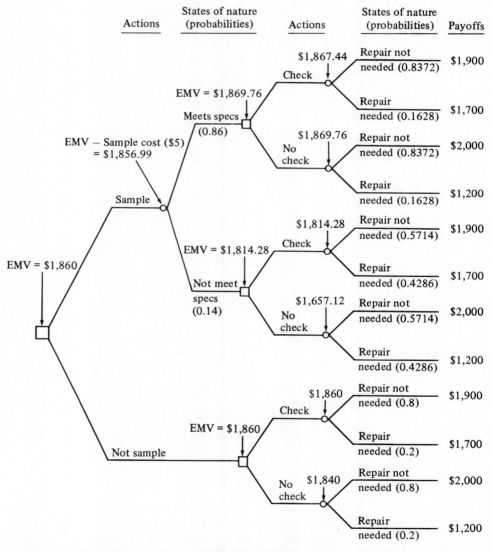

FIGURE 19.5 Decision tree for automobile parts supplier when sample information may be obtained

left-hand side with the decision as to whether to collect the sample information. If no sample is taken, then the supplier simply proceeds with the prior probabilities, and his subsequent decision problem—to check or not check the process for malfunction—is precisely that of Example 19.4. The lower part of the decision tree of Figure 19.5, stemming from the branch "Not sample," is, therefore, exactly the same as the decision tree of Figure 19.3. The optimal action is to check the process, the expected monetary value being $1,860.

If a sample observation is taken, there are two possible states of nature. The sampled part either will or will not meet specifications. In either case, a second decision—to check or not to check—must be made. Finally, there are two possible states of nature—repair is or is not needed. The *posterior probabilities* attached to these states of nature, given the sample result, were obtained in Example 19.6.

The computations are begun, as usual, at the right-hand side of the figure, by inserting the expected monetary values for the action–state of nature combinations. (These do not incorporate the cost of collecting the sample information, which will enter the analysis later.) As we have already noted, if the initial decision is not to sample, the calculations in that part of the tree are precisely those of Figure 19.3. We now need to compute expected monetary values when the sample information is taken.

Given the sample result, the expected monetary values for each of the actions "Check" and "No check" are obtained by weighting the payoffs by the posterior probabilities and summing. These values were derived in Example 19.6. We see that, if the sampled part meets the specifications, the optimal action is not to check, with an expected monetary value of $1,869.76. If the sampled part fails to meet specifications, the higher expected monetary value is $1,814.28, arising from checking the process for malfunction.

It emerges from our analysis that, if the sample information is obtained, an expected monetary value of $1,869.76 arises with probability 0.86 (the probability that the sampled part meets the specifications), and an expected monetary value of $1,814.28 with probability 0.14. Thus, the expected monetary value of this course of action is

$$(0.86)(1,869.76) + (0.14)(1,814.28) = \$1,861.99$$

From this we must subtract the $5 cost of sampling, leaving $1,856.99. This is compared with the expected monetary value of $1,860 from not sampling.

It can be seen from the decision tree that, according to the expected monetary value criterion, the supplier's best strategy is not to obtain the sample information, but to go ahead and check the process before the production run commences. The expected monetary value—that is, the supplier's expected profit—from following this strategy is $1,860.

EXERCISES

7. A manager has to choose between two possible actions, a_1 and a_2, in a situation in which two states of nature, s_1 or s_2, could arise. The payoffs are shown in the accompanying table. If the manager believes that each state of nature is equally likely to occur, which action should be taken under the expected monetary value criterion?

ACTIONS	STATES OF NATURE	
	s_1	s_2
a_1	10,000	5,000
a_2	8,000	6,000

8. During the last few months I have become increasingly skeptical about the existence of Santa Claus. I now have to decide whether to expend the effort (which I value at $1) in hanging up my stocking this year. I estimate that the value of my presents, if I receive any, will be $10. If there is no Santa, I will get these presents whatever I do. On the other hand, if Santa exists, I will get them only if I hang up my stocking.

(a) Set up the payoff table for my decision-making problem.

(b) Suppose I believe that the probability that Santa exists is 0.05. According to the expected monetary value criterion, should I hang up my stocking?

9. The investor of Exercise 1 believes that the probability is 0.5 that the stock market will rise, 0.3 that it will remain flat, and 0.2 that it will fall.

(a) According to the expected monetary value criterion, which investment should be chosen?

(b) Draw the decision tree.

10. Refer to Exercise 2. The manufacturer of earth-moving equipment believes that the probability of a low level of construction activity is 0.2, the probability of medium activity level is 0.4, and the probability of a high level of activity is 0.4.

(a) Which action should be taken, according to the expected monetary value criterion?

(b) Draw the decision tree for this problem.

11. Consider a decision problem with K possible actions and H possible states of nature. Show that, for any such problem, an inadmissible action cannot be chosen by the expected monetary value criterion.

12. Consider a decision problem with two admissible actions and two possible states of nature, each of which is equally likely to occur. Determine whether each of the following statements is true or false for such problems:

(a) The action chosen by the expected monetary value criterion will always be the same as that chosen by the maximin criterion.

(b) The action chosen by the expected monetary value criterion will always be the same as that chosen by the minimax regret criterion.

(c) The action chosen by the expected monetary value criterion will always be that for which the average of the possible payoffs is highest.

13. Refer to part (c) of Exercise 12. Would your answer be the same if the two states of nature were not equally likely to occur?

14. The shoe store operator of Exercise 5 believes that the probability is 0.5 that the new shopping center will be very successful, the probability is 0.3 that it will be moderately successful, and the probability that it will be unsuccessful is 0.2. According to the expected monetary value criterion, where should he locate the store?

15. The company of Exercise 6 believes that the probabilities for low, moderate, and large increases in future energy prices are, respectively, 0.1, 0.3, and 0.6.

(a) According to the expected monetary value criterion, which heating system should be chosen for this company's new facility?

(b) Draw the decision tree for the company's problem.

16. A contractor must decide whether to submit a bid for a building project. The cost of preparing this bid is $12,000, and this cost will be incurred whether or not the bid is accepted. The contractor intends to set the bid so that, if the bid is accepted, he will make a profit of $100,000 (less the cost of preparing the bid). On the basis of past records, he finds that 20% of bids prepared in this manner have been accepted.

(a) Set up the payoff table.

(b) According to the expected monetary value criterion, should a bid be prepared?

(c) Draw the contractor's decision tree.

17. A manufacturer must decide whether to institute, at a cost of $75,000, a new advertising campaign. It is estimated that, if the campaign is very successful, profits will increase by $300,000 (excluding the cost of the advertising campaign). If the campaign is moderately successful, profits will increase by $120,000, while, if the campaign is unsuccessful, profits will remain unchanged. It is estimated that the probabilities for very successful, moderately successful, and unsuccessful advertising campaigns are, respectively, 0.2, 0.4, and 0.4.

(a) Set up the payoff table.

(b) If the expected monetary value criterion is used, should the advertising campaign be instituted?

(c) Draw the manufacturer's decision tree.

18. On Thursday evening, the manager of a small branch of a car rental agency finds that he has available six cars for rental on the following day. However, he is able to request delivery of additional cars, at a cost of $20 each, from the regional depot. Each car that is rented produces an expected profit of $40. (The cost of delivery of the car must be subtracted from this profit.) On the other hand, each potential customer requesting a car when none is available is counted a $10 loss in goodwill. On reviewing his records for previous Fridays, the manager finds that the numbers of cars requested have ranged from six to ten, with the percentages shown in the accompanying table. The manager must decide how many cars, if any, to order from the regional depot.

NUMBER OF REQUESTS	6	7	8	9	10
PERCENT	10	20	40	20	10

(a) Set up the payoff table.

(b) If the expected monetary value criterion is used, how many cars should be ordered?

19. A contractor has decided to place a bid for a project. Bids are to be set in multiples of $20,000. It is estimated that the probability that a bid of $240,000 will secure the contract is 0.3, the probability that a bid of $220,000 will be successful is 0.6, while the probability that a bid of $200,000 will be accepted is 0.8. It is thought that any bid under $200,000 is certain to succeed, while any bid over $240,000 is certain to fail. If the manufacturer wins the contract, he must solve a design problem with two possible choices at this stage. He can hire outside consultants, who will guarantee a satisfactory solution, for a price of $80,000. Alternatively, he can invest $20,000 of his own resources in an attempt to solve the problem internally; if this effort fails, he must then engage the consultants. It is estimated that the probability of successfully solving the problem internally is 0.6. Once this problem has been solved, the additional cost of fulfilling the contract is $140,000.

(a) Potentially, this contractor has two decisions to make. What are they?

(b) Draw the decision tree.

(c) What is the optimal course of action, according to the expected monetary value criterion?

20. A publisher intends to sign a contract for an accounting text with one of three authors, Smith, Brown, or Jones. If the text turns out to be very successful, then profits (excluding any extraordinary advertising costs) will be $250,000, while if the book is only moderately successful, these profits will be $50,000. In the event that the text fails, a loss of $50,000 will result. The probabilities given in the first table have been conjectured for these states of nature for the three books.

	VERY SUCCESSFUL	MODERATELY SUCCESSFUL	FAILURE
SMITH	0.2	0.6	0.2
BROWN	0.1	0.8	0.1
JONES	0.3	0.2	0.5

The publisher also has the option of mounting, at a cost of $40,000, an extraordinary advertising campaign for the book, once it is published. It is estimated that, if this were carried out, the probabilities for the three states of nature would be as shown in the following table:

	VERY SUCCESSFUL	MODERATELY SUCCESSFUL	FAILURE
SMITH	0.4	0.4	0.2
BROWN	0.3	0.6	0.1
JONES	0.5	0.2	0.3

(a) Draw the publisher's decision tree.

(b) According to the expected monetary value criterion, which author should be signed, and should the extraordinary advertising campaign be mounted?

(c) Following the calculations in part (b), the publisher has signed a contract with the chosen author. At this point, it is discovered that a clerical error has been made in the marketing department, and that in fact the actual cost of the advertising campaign is $55,000. According to the EMV criterion, should the publisher offer to pay the chosen author to withdraw from the contract and, if so, what is the highest sum he should offer?

21. Consider a decision problem with two actions, a_1 and a_2, and two states of nature, s_1 and s_2. Let M_{ij} denote the payoff corresponding to action a_i and state of nature s_j. Assume that the probability of occurrence of state of nature s_1 is p, so that the probability of state s_2 is $(1 - p)$.

(a) Show that action a_1 is selected by the EMV criterion if

$$p(M_{11} - M_{21}) > (1 - p)(M_{22} - M_{12})$$

(b) Hence, show that if a_1 is an admissible action, there is some probability, p, for which it will be chosen. However, if a_1 is not admissible, it cannot be chosen whatever the value of p.

22. A consultant is considering preparing proposals for one of two contracts. He has the alternative of preparing proposals for neither, but time constraints prevent his preparing proposals for both. The costs of preparation of initial outline proposals are $500 for contract A and $750 for contract B. When the initial proposals have been submitted, responses are obtained from the potential clients. These responses can be categorized as either "positive," "noncommital," or "negative." The probabilities for the two contracts are shown in the accompanying table. If the response to the initial outline proposal is negative, no contract will be obtained. If the response is not negative, the consultant may provide a more detailed final proposal, at a cost of $1,000 for contract A and $1,500 for contract B. For contract A, the probabilities that this final proposal will be accepted are 0.9 if the initial response was positive and 0.4 if it was noncommital. For contract B, these probabilities are 0.8 and 0.2, respectively. If contract A is finally secured, the consultant's profit (from which costs of proposal preparations must be subtracted) is $5,000. For contract B the corresponding figure is $6,000. The consultant intends to follow the course of action dictated by the EMV criterion.

	POSITIVE	NONCOMMITAL	NEGATIVE
CONTRACT A	0.6	0.2	0.2
CONTRACT B	0.8	0.1	0.1

(a) Draw the decision tree for this consultant.

(b) Should the consultant submit an initial outline proposal and, if so, for which project?

(c) If the answer to part (b) is "yes," how should the consultant proceed if the response from the selected potential client to the initial proposal is "noncommital"?

(d) This consultant has recently hired a statistician who suggests that it would be better simply to submit the final proposals without going through the stage of preparing initial outline proposals. The costs of proposals prepared in this way would be $1,250 for contract A and $1,875 for contract B. Assuming that the probabilities of final acceptance remain unchanged, is the statistician correct?

23. A manufacturer makes precision instruments that have two potential faults. He has the ability to check for these faults before the instruments are shipped to customers. The cost of checking for the first fault is $20 and that of checking for the second is $30. If a fault is found, it can be fixed at negligible cost. Historically, it has been found that the first fault is present in 10% of all instruments and the second fault is also present in 10% of all instruments. Five percent of all instruments have both faults. This manufacturer estimates the damage done to his business, for each faulty instrument shipped out, as a cost of $300. He has the following possible strategies:

(i) All instruments could be shipped without checking for faults.

(ii) The check for the first fault could be made, but not for the second.

(iii) The check for the first fault could be made and, if an instrument passes this check, a check for the second fault could be made.

According to the expected monetary value criterion, which strategy should be followed?

24. A manufacturer must decide whether to mount, at a cost of $100,000, an advertising campaign for a product whose sales have been rather flat. It is estimated that a highly successful campaign would add $400,000 (from which the campaign's cost must be subtracted) to profits, a moderately successful campaign would similarly add $100,000, while an unsuccessful campaign would add nothing. Historically, 50% of all similar campaigns have been very successful, 30% moderately successful, and the remainder unsuccessful. This manufacturer consults a media consultant for a judgment on the potential effectiveness of the campaign. This consultant's record is such that she has reported favorably on 80% of campaigns that turned out to be highly successful, 40% of those that were moderately successful, and 10% of unsuccessful campaigns.

(a) Find the prior probabilities for the three states of nature.

(b) In the absence of any report from the media consultant, should this advertising campaign be mounted according to the EMV criterion?

(c) Find the posterior probabilities for the three states of nature, given that the media consultant reports favorably.

(d) Given a favorable report from the consultant, should the advertising campaign be mounted, according to the EMV criterion?

(e) Find the posterior probabilities for the three states of nature, given that the media consultant does not report favorably.

(f) If the consultant's report is not favorable, should the advertising campaign be mounted, according to the EMV criterion?

25. Refer to Exercise 2. The manufacturer of earth-moving equipment has available four courses of action. His decision will depend on the view he takes of the likely level of road construction activity in the next 5 years. Historically, activity has been low 20% of the time, medium 40% of the time, and high the remainder of the time. A forecasting group sells predictions of the future level of road construction activity. The accompanying table shows the proportions of "buoyant," "steady," and "weak" forecasts, given the particular level of activity actually resulting.

FORECAST	ACTIVITY LEVEL		
	LOW	MEDIUM	HIGH
WEAK	.6	.3	.2
STEADY	.3	.4	.3
BUOYANT	.1	.3	.5

(a) What are the prior probabilities for the three states of nature?

(b) If the manufacturer does not consult the forecasting group, what action should he take according to the EMV criterion?

(c) What are the posterior probabilities of the three states of nature, given a forecast of "weak"?

(d) According to the EMV criterion, given a forecast of "weak," what course of action should the manufacturer adopt?

(e) What are the posterior probabilities of the three states of nature, given a forecast of "steady"?

(f) According to the EMV criterion, given a forecast of "steady," which action should be chosen?

(g) What are the posterior probabilities of the three states of nature, given a forecast of "buoyant"?

(h) If the EMV criterion is followed, which action should be chosen given a forecast of "buoyant"?

26. The shoe store operator of Exercise 5 has available two courses of action. His decision is based on his view of the likely level of success of the new shopping center. Historically, 50% of new centers of this type have been very successful, 30% moderately successful, and 20% unsuccessful. A consulting group sells assessments of the prospects of this type of shopping center. The table given here shows the proportions of "good," "fair," and "poor" assessments, given the particular outcome actually resulting.

ASSESSMENT	LEVEL OF SUCCESS		
	VERY SUCCESSFUL	MODERATELY SUCCESSFUL	UNSUCCESSFUL
GOOD	.7	.3	.2
FAIR	.2	.4	.3
POOR	.1	.3	.5

(a) What are the prior probabilities for the three states of nature?

(b) If the shoe store operator does not seek advice from the consulting group, what action should he take according to the EMV criterion?

(c) What are the posterior probabilities of the three states of nature, given an assessment of "good"?

(d) According to the EMV criterion, given an assessment of "good," what course of action should the shoe store operator adopt?

(e) What are the posterior probabilities of the three states of nature, given an assessment of "fair"?

(f) According to the EMV criterion, given an assessment of "fair," which action should be chosen?

(g) What are the posterior probabilities of the three states of nature, given an assessment of "poor"?

(h) If the EMV criterion is followed, which action should be chosen given a forecast of "poor"?

27. The company of Exercise 6 has to choose from among three available heating systems. Its decision will be influenced by the view taken about the size of future increases in energy prices. The prior probabilities are 0.1 for low increases, 0.3 for moderate increases, and 0.6 for high increases. A forecasting group offers predictions of future energy price increases. The table shows the proportions of "low," "moderate," and "high" predictions, given the particular outcomes that actually materialized.

PREDICTIONS	OUTCOME		
	LOW	MODERATE	HIGH
LOW	0.6	0.2	0.1
MODERATE	0.3	0.5	0.2
HIGH	0.1	0.3	0.7

(a) If the company does not consult the forecasting group, what decision will be taken, according to the EMV criterion?

(b) What are the posterior probabilities of the three states of nature, given a prediction of low energy price increases?

(c) Given a prediction of low energy price increases, which action should be chosen, according to the EMV criterion?

(d) What are the posterior probabilities of the three states of nature, given a prediction of moderate energy price increases?

(e) If the prediction is for moderate energy price increases, what would be the optimal course of action, according to the EMV criterion?

(f) What are the posterior pobabilities of the three states of nature, given a prediction that energy price increases will be high?

(g) Given a prediction of high energy price increases, which action should be chosen according to the EMV criterion?

28. Consider the drug manufacturer of Example 19.7, who had to decide whether to sell the patent for a pain relief formula, before subjecting the drug to thorough testing. In the example we saw that, whatever the result of a certain preliminary test of the drug's efficacy, the optimal decision was to retain the patent. Subsequently, this manufacturer developed a superior preliminary test, which again could be carried out at modest cost. For drugs that subsequently proved effective, this new test gave a "positive" result 80% of the time, while a "positive" result was obtained for only 10% of those drugs that proved to be ineffective.

(a) Find the posterior probabilities of the two states of nature, given a "positive" result from this new preliminary test.

(b) According to the EMV criterion, should the patent be sold if the new test result is "positive"?

(c) Find the posterior probabilities of the two states of nature, given a "negative" result from the new preliminary test.

(d) According to the EMV criterion, should the patent be sold if the new test result is "negative"?

(e) Is the optimal choice of action affected by the result of the new preliminary test?

(f) Explain what feature of the preliminary test determines whether its outcome influences the optimal choice of action.

29. In Example 19.6 we considered the problem of a supplier of parts to the automobile industry, faced with the decision of whether to check the production process for a malfunction at the beginning of a production run. There are two possible states of nature, as follows:

s_1: Repair not needed (10% of parts produced fail to meet specifications)
s_2: Repair needed (30% of parts produced fail to meet specifications)

The prior probabilities are

$$P(s_1) = 0.8 \quad \text{and} \quad P(s_2) = 0.2$$

The payoffs are shown in the accompanying table. In example 19.6, we considered the case where a single part is made and checked prior to beginning the full production run. Depending on the outcome, a decision as to whether to check the production process for malfunction was made.

ACTIONS	STATES OF NATURE	
	s_1	s_2
CHECK	1,900	1,700
NO CHECK	2,000	1,200

Suppose now that, prior to full production, *two* parts are made and checked.

(a) If, in fact, repair is not needed, what are the probabilities that both parts, just one part, and neither part will fail to meet specifications?

(b) Compute the same probabilities as in part (a) given that repair is, in fact, needed.

(c) Using the expected monetary value criterion, determine the optimal action under each of the following circumstances:

   (i) Both parts fail to meet specifications

   (ii) Just one part fails to meet specifications

   (iii) Neither part fails to meet specifications

(d) Without doing the calculations, indicate how the decision problem could be attacked if, before beginning full production, ten parts were made and checked.

30. The Watts New Lightbulb Corporation ships large consignments of lightbulbs to big industrial users. When the production process is functioning correctly (which happens 90% of the time), 10% of all bulbs produced are defective. However, the process is susceptible to an occasional malfunction, leading to a defective rate of 20%. The Watts New Corporation counts the cost, in terms of goodwill, of a shipment with the higher defective rate to an industrial user as $5,000. If a consignment is suspected to contain this larger proportion of defectives, it can instead be sold to a chain of discount stores, though this involves a reduction of $600 in profits, whether or not the consignment does indeed contain a large proportion of defective bulbs. Decisions by this company are made through the EMV criterion.

(a) A consignment is produced. In the absence of any further information, should it be shipped to an industrial user or to the discount chain?

(b) Suppose that a single bulb from the consignment is checked. Determine where the consignment should be shipped under each of the following circumstances:

   (i) This bulb is defective

   (ii) This bulb is not defective

(c) Suppose that two bulbs from the consignment are checked. Determine where the consignment should be shipped for each of the following situations:

   (i) Both bulbs are defective

   (ii) Just one bulb is defective

   (iii) Neither bulb is defective

(d) Without doing the calculations, indicate how the decision problem could be attacked if 100 bulbs were checked prior to shipping the consignment.

31. Refer to the problem of the investor of Exercise 1.

(a) In the context of this problem, what would constitute perfect information?

(b) If the prior probabilities are 0.5 for a rise in the stock market, 0.3 for a flat market, and 0.2 for a fall in the market, what is the expected value of perfect information?

32. The manufacturer of earth-moving equipment of Exercise 2 is unsure about the future levels of road construction activity. The prior probabilities are 0.2 for low level, 0.4 for

medium level, and 0.4 for high level. What is the expected value of perfect information to this manufacturer?

33. For the shoe store operator of Exercise 5, the prior probabilities are 0.5 that the new shopping center will be very successful, 0.3 that it will be moderately successful, and 0.2 that it will be unsuccessful. What is the expected value of perfect information to this shoe store operator?

34. The company of Exercise 6 is trying to decide how to heat a new facility. The prior probabilities for the three possible states of nature are 0.1 for low energy price increases, 0.3 for moderate energy price increases, and 0.6 for large increases in energy prices. What is the most this company should pay for sample information, however good, according to the expected monetary value criterion?

35. In Section 19.5, before showing how to find the expected value of sample information, we discussed separately the determination of the expected value of perfect information. In fact, this was not necessary because perfect information is just a special kind of sample information. Given the general procedure for finding the expected value of sample information, show how to specialize this to the case of perfect information.

36. Refer to Exercise 24. A manufacturer is considering an advertising campaign, and first seeks the advice of a media consultant.
   (a) What is the expected value to the manufacturer of the media consultant's advice?
   (b) The media consultant charges a fee of $5,000. What is the expected net value of the consultant's advice?
   (c) This manufacturer faces a two-stage decision problem. First, he must decide whether to purchase advice from the media consultant. Next, he must decide whether to mount the advertising campaign. Draw the complete decision tree, and indicate how the manufacturer should proceed.

37. Refer to Exercise 25. The manufacturer of earth-moving equipment consults an economic forecasting group before deciding if, or when, to begin production of a new product line. Find the expected value to the manufacturer of the forecasting group's report.

38. For the shoe store operator of Exercise 26, find the expected value of an assessment of the shopping center's prospects provided by the consulting group.

39. Based on the information in Exercise 27, find the expected value to the company, in the context of its decision about which heating system to install, of a prediction about future energy price increases from the forecasting group.

40. Refer to Exercise 28. Before deciding whether to sell the patent of a new pain relief formula, the drug manufacturer carries out the new preliminary test. Find the expected value to the manufacturer of the test result.

41. Refer to Exercise 29. The automobile parts supplier examines two parts before deciding whether to check the process.
   (a) What is the expected value of this sample information?
   (b) If this initial examination costs $5 per part examined, what is the expected net value of the sample information?
   (c) Suppose this supplier is faced with a two-stage decision problem. At the first stage he can decide either to check no parts, or to check one or two parts. At the next stage he must decide whether to check the production process. Draw the complete decision tree and indicate how the supplier should proceed.
   (d) In fact, another strategy is possible. The supplier can check a single part and, *depending on the outcome of this check,* decide whether to check a second part. Thus, he is potentially faced with a three-stage decision problem. First, he must decide whether to check a part. Next, given the outcome of the first check (if it is carried out), he decides whether to check another part. Finally, the decision as to

whether to check the production process must be made. Draw the complete decision tree and indicate the optimal strategy, based on the expected monetary value criterion.

42. Consider the Watts New Corporation of Exercise 30. The corporation can check one or more lightbulbs before deciding whether to ship a consignment to an industrial user or to a discount chain.

   (a) What is the expected value to the corporation of checking a single lightbulb?
   (b) What is the expected value to the corporation of checking two lightbulbs?
   (c) What is the difference between expected values of checking two and one bulbs?
   (d) If the first bulb checked turns out to be defective, what is the expected value of checking the second?
   (e) If the first bulb checked turns out not be be defective, what is the expected value of checking the second?
   (f) Reconcile your answer to part (c) with those to parts (d) and (e).

19.6 ALLOWING FOR RISK: UTILITY ANALYSIS

The expected monetary value criterion provides a framework for decision-making that has wide practical applicability. That is to say, in many instances, an individual or corporation will believe that the action offering the highest expected monetary value is the preferred course. However, this is not invariably the case, as the following examples illustrate:

   (i) Many individuals purchase term life insurance through which, for a relatively modest outlay, the insured person's estate is generously compensated in the event of death during the term of the policy. Now, insurance companies are able to calculate the probability of the death of an individual of any given sex and age, during a specified period of time. Accordingly, their rates are set in such a way that the price of a policy exceeds the amount of money that is expected to be paid out. The amount of this excess covers the insurance company's costs and provides, on the average, a margin of profit. It then follows that, for the person insured, the expected payoff from the life insurance policy is less than its cost. Therefore, if everyone based decisions on the expected monetary value criterion, term life insurance would not be purchased. Nevertheless, many people do buy this form of insurance, demonstrating a willingness to sacrifice something in expected returns for the assurance that the heirs will be provided a financial cushion in the event of death.

   (ii) Suppose that an investor is considering purchasing shares in one or more of a group of corporations, whose prospects he regards as bright. In principle, it is possible to postulate the various states of nature that will influence the returns from investment in each of these corporations. In this way, the expected monetary value of an investment of a fixed amount in each corporation could be determined. According to the expected monetary value criterion, the investor should then put all of his available capital into that corporation for which the expected monetary value is highest. In fact, a great many investors in the stock market do not follow such a strategy. Rather, they spread their cash over a portfolio of stocks. The abandonment of the option of "putting all one's eggs in a single basket," while leading to a lower expected return, provides a hedge against the possibility of losing a good deal of money if the single stock with the highest expected return happens to perform badly. In opting for a portfolio of stocks, the investor is asserting a willingness to sacrifice something in expected monetary value for a smaller chance of a large financial loss.

In each of these examples, the decision-maker has exhibited a preference for a criterion of choice other than expected monetary value, and in each circumstance this preference seems to be extremely reasonable. The two examples involve a common ingredient, in addition to expected returns. In both cases, the decision-maker wants to take into account **risk.** The purchaser of term life insurance is prepared to accept a negative expected return as the price to be paid for the chance of a large positive return in the event of death. In doing so, he is expressing a *preference* for risk.[3] On the other hand, the investor who, in spreading his investment over a portfolio of stocks, accepts a lower expected return in order to reduce the chances of a large loss, is expressing an *aversion* to risk.

We have seen that the expected monetary value criterion is inappropriate for decision-makers who either prefer, or are averse to, risk. Fortunately, it is not too difficult to modify this criterion to handle situations in which risk is a relevant factor. Essentially, the idea is to replace the monetary payoffs by quantities that reflect, not only the dollar amounts to be received, but also the decision-maker's attitude to risk.

THE CONCEPT OF UTILITY

In Example 19.3 we considered the problem of an investor choosing between a guaranteed fixed interest investment and a portfolio of stocks. The former yielded a payoff of $1,200, while gains of $2,500 and $500 resulted for the latter if the stock market were to be buoyant or steady, but a loss of $1,000 resulted if it were to be depressed. This investor believed that the respective probabilities for these three states of nature were 0.6, 0.2, and 0.2. In that event, we saw that the expected monetary value from choosing the stock portfolio was $1,400, exceeding by $200 that of the fixed interest investment. At this juncture, we need to inquire whether this higher expected return merits the risk of losing $1,000, as would occur if the market were depressed. A very wealthy investor, who could quite comfortably sustain such a loss, would almost certainly decide that it did. However, the position of a relatively poor person, to whom a loss of $1,000 would be quite disastrous, may well be different. For such an investor, the payoffs must be replaced by some other quantities that more adequately reflect the calamitous nature of a loss of $1,000. These quantities must measure the value, or **utility,** to the investor of a loss of $1,000 as compared with, for example, gains of $500 or $2,500.

The concept of utility, which plays a central role in microeconomics, provides the basis for the solution of decision-making problems in the presence of risk preference or aversion. In order to employ it we need only fairly mild, and typically quite reasonable, assumptions. Suppose that an individual is faced with several possible payoffs, which may or may not be monetary in nature. It is assumed that the individual can rank in order (possibly with ties) the utility, or satisfaction that would be derived from each. Thus, if payoff A is preferred to B, and B is preferred to C, then A must be preferred to C.

---

[3] He is, of course, guarding *against* the risk that his family will be financially ill-prepared for his death.

We need also to assume that, if payoff A is preferred to B, and B is preferred to C, then there exists a gamble, which offers A with probability p and C with probability $(1 - p)$, such that the decision-maker will be indifferent between taking this gamble and receiving B with certainty. Given these and certain other, generally innocuous, assumptions whose details need not detain us, it is possible to show that the rational decision-maker will choose that action for which expected utility is highest. Consequently, we can analyze the decision problem precisely as in the previous three sections, *but with utilities instead of payoffs*. That is to say, we construct a utility table rather than a payoff table, and then employ states of nature probabilities to compare expected utilities.

We must now discuss how the utilities corresponding to the various payoffs are determined. The possible payoffs for our investor are $-\$1,000$, $\$500$, $\$1,200$, and $\$2,500$. The steps involved in finding the corresponding utilities are stated in the box.

Obtaining a Utility Function

Suppose that a decision-maker may receive several alternative payoffs. The transformation from payoffs to **utilities** is obtained as follows:

(i) The units in which utility is measured are arbitrary. Accordingly, we can fix a scale in any convenient fashion. Let L be the lowest and H the highest of all the payoffs. We will assign utility 0 to payoff L, and utility 100 to payoff H.
(ii) Let I be any payoff between L and H. We need to determine that probability p such that the decision-maker is indifferent between the following alternatives:
 (a) Receive payoff I with certainty.
 (b) Receive payoff H with probability p, and payoff L with probability $(1 - p)$.
(iii) The utility to the decision-maker of payoff I is then $100p$. The curve relating utility to payoff is called a **utility function**.

The first step is straightforward, and simply provides us with a convenient metric for measuring utility. The choices of the numbers 0 and 100 to represent the utilities of the lowest and highest payoffs is entirely arbitrary. Any other pair of numbers could equally well be used, provided the utility of the highest payoff was greater than that of the lowest, without affecting the remaining analysis.

As a practical matter, the second step is the most difficult, partly because it presupposes that the decision-maker can manipulate probabilities in a coherent way. In practice, the probability p must be determined by trial and error, through the presentation of questions such as the following:

Q. "Would you prefer to receive I with certainty, or a gamble in which you could obtain H with probability 0.9 and L with probability 0.1?"

Q. "Would you prefer to receive I with certainty, or a gamble in which you could obtain H with probability 0.8 and L with probability 0.2?"

This process is continued until the point of indifference is reached.

The logic of the final step is quite straightforward. Since H has utility 100 and L has utility 0, the **expected** utility if H is obtained with probability p and L with probability $(1 - p)$ is

$$100p + 0(1 - p) = 100p$$

Since the decision-maker is indifferent between this gamble and receiving I with certainty, the utility $100p$ is associated with the payoff I.

We now return to our investor. At the first step, we attach utility 0 to the lowest payoff, $-\$1,000$, and utility 100 to the highest, $\$2,500$.

It remains to determine the utilities for the intermediate payoffs, $\$500$ and $\$1,200$. This is achieved by posing to the decision-maker a series of questions such as

> **Q.** "Would you prefer to receive $\$500$ with certainty, or a gamble in which you could obtain a gain of $\$2,500$ with probability p, and a loss of $\$1,000$ with probability $(1 - p)$?"

Different values of the probability p are tried until that value at which the decision-maker is indifferent between the two alternatives is found. This process is repeated for the payoff of $\$1,200$.

Suppose that the investor is indifferent between a payoff of $\$500$ and this gamble with $p = 0.6$, and between a payoff of $\$1,200$ and the gamble with $p = 0.8$. The utilities for the intermediate payoffs are then

$$\text{PAYOFF} \quad \$500 \quad \text{Utility} = (100)(0.6) = 60$$

$$\text{PAYOFF} \quad \$1,200 \quad \text{Utility} = (100)(0.8) = 80$$

The four utilities for this investor are plotted against the corresponding payoffs as points in Figure 19.6. To indicate the general shape of this investor's utility function, we have drawn a curve through these points. The shape of this curve is interesting, since it characterizes the investor's attitude to risk. As must be the case, utility increases as the payoff increases. However, notice that the *rate of increase* of utility is highest at the lowest payoffs, and decreases as payoff increases. This implies a distaste for the lowest payoffs that is more than commensurate with their monetary amounts, indicating **aversion** to risk. This aversion can be seen from the investor's attitude to the gambles offered. For example, the investor is indifferent between a sure payoff of $\$500$, and a gamble in which $\$2,500$ might be won with probability 0.6 and $\$1,000$ lost with probability 0.4. The expected monetary value of this gamble is

$$(0.6)(2,500) + (0.4)(-1,000) = \$1,100$$

which considerably exceeds the equally preferred sure payoff of $\$500$. The amount of this difference provides a measure of the extent of the aversion to risk.

The shape of Figure 19.6 is typical of risk aversion. In Figure 19.7, we show three types of utility functions. The function in part (a) of the figure, where utility increases at a *decreasing* rate as payoff increases, has the same shape as Figure 19.6, once again reflecting an *aversion* to risk. In part (b) of the figure, utility increases at

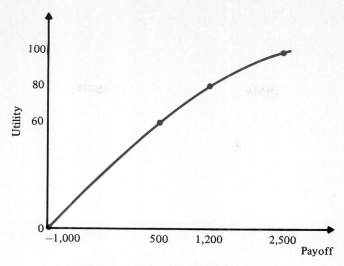

FIGURE 19.6 Utility function for an investor

an *increasing* rate as the payoffs become higher. This implies a taste for the highest payoffs that is more than commensurate with the monetary amounts involved, thus showing a *preference* for risk. Finally, part (c) of Figure 19.7 shows the intermediate case, with utility increasing at a *constant* rate for all payoffs. In this case, then, the monetary values of the payoffs provide a true measure of their utility to the decision-maker, who thus demonstrates *indifference* to risk.

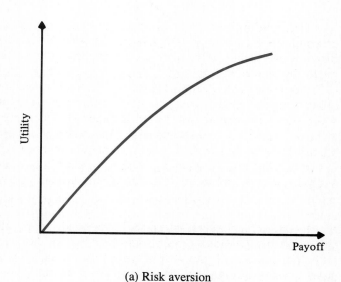

(a) Risk aversion

FIGURE 19.7 Utility functions

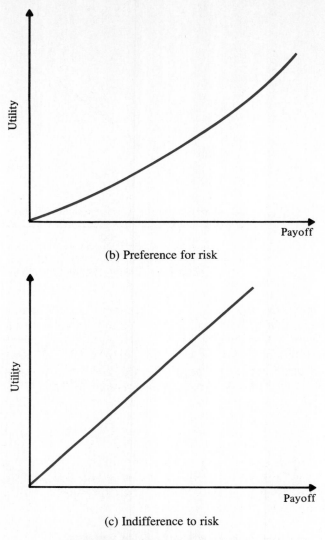

(b) Preference for risk

(c) Indifference to risk

FIGURE 19.7 Utility functions (cont.)

The three curves of Figure 19.7 characterize aversion for, preference for, and indifference to, risk. However, it is not necessarily the case that a decision-maker will exhibit just one of these attitudes over the whole range of possible payoffs. Figure 19.8 illustrates a more complex situation. Here, for payoffs in the range between M_1 and M_2, the utility function has the shape of Figure 19.7(a), indicating aversion to risk in this payoff range. However, for payoffs of monetary amounts between M_2 and M_3, this utility function has the shape of Figure 19.7(b). Hence, for this range of payoffs, the decision-maker exhibits a preference for risk. Finally, in the range of highest payoffs, between M_3 and M_4, the position is once again reversed, the decision-maker being

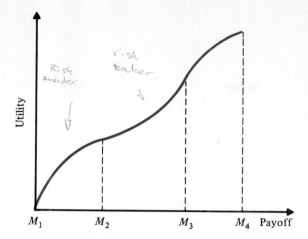

(handwritten annotations: "Risk Avoider", "risk taker")

FIGURE 19.8 Utility function showing aversion to risk for payoffs between M_1 and M_2, and between M_3 and M_4, and preference for risk between payoffs M_2 and M_3

averse to risk in this region. Such a utility function can arise in practical problems. For example, an investor may well be averse to sustaining a substantial loss, while being prepared to accept some risk to obtain a fairly high positive return rather than a modest one. However, if a satisfactorily high return can be achieved at modest risk, the investor may be reluctant to risk much more for the possibility of an even higher return.

EXPECTED UTILITY CRITERION FOR DECISION-MAKING

Having determined the appropriate utilities, it remains only to solve the decision-making problem by finding that course of action with the highest expected utility. These expected utilities are obtained in the usual manner, employing the probabilities of the states of nature, as described in the box.

The Expected Utility Criterion

Suppose that a decision-maker has K possible actions, $a_1, a_2, \ldots, a_K$, and is faced with H states of nature. Let U_{ij} denote the utility corresponding to the ith action and jth state, and p_j the probability of occurrence of the jth state of nature. Then the **expected utility**, $EU(a_i)$, of the action a_i is

$$EU(a_i) = p_1 U_{i1} + p_2 U_{i2} + \cdots + p_H U_{iH}$$

$$= \sum_{j=1}^{H} p_j U_{ij}$$

Given a choice between alternative actions, the **expected utility criterion** dictates the choice of that action for which expected utility is highest. Under generally reasonable assumptions, it can be shown that this criterion should be adopted by the rational decision-maker.

If the decision-maker is indifferent to risk, the expected utility criterion and the expected monetary value criterion are equivalent.

Table 19.11 shows the utilities and states of nature probabilities for our investor. If the fixed interest investment is chosen, a utility of 80 is assured, whichever state of nature prevails. For the portfolio of stocks, the expected utility is

$$(0.6)(100) + (0.2)(60) + (0.2)(0) = 72$$

Since this is less than 80, this investor should elect to make the fixed interest investment, according to the expected utility criterion.

TABLE 19.11 Utilities and states of nature probabilities for an investor

INVESTMENT	STATE OF THE MARKET		
	BUOYANT	STEADY	DEPRESSED
PROBABILITIES	0.6	0.2	0.2
FIXED INTEREST	80	80	80
STOCK PORTFOLIO	100	60	0

In Example 19.3, we found that investment in the portfolio of stocks was selected by the expected monetary value criterion. However, the incorporation into the analysis of another factor—the extent of this investor's aversion to risk—leads to the conclusion that the fixed interest option is the better choice. This example serves to illustrate that, on occasion, when risk is an important factor, the expected monetary value criterion is inadequate for solving decision-making problems.

The expected utility criterion is the most generally applicable and intellectually defensible of the criteria we have introduced for attacking decision-making problems. In practice, its chief drawback arises from the difficulty of eliciting information about which gambles are regarded as equally attractive as particular assured payoffs. As we have seen, this type of information is essential in the determination of utilities. For a wide range of problems, where indifference to risk can safely be assumed, the expected monetary value criterion remains applicable. This would typically be the case, for example, in decision-making in a mature corporation, when the payoffs involved represented only a small proportion of the corporation's total turnover. On the other hand, if (as may be the case in the development of a new commercial airliner, for example) possible losses from a project could threaten a corporation with insolvency, the utilities appropriate should reflect an aversion to risk. A company may attempt to spread this risk by forming partnerships with other firms in the industry, or with possible customers. Perhaps, also, government may be invited to share, or take on the bulk of, the risk.

EXERCISES

43. The investor of Exercise 1 has the following seven different possible payoffs (in dollars):

$$-500; \ -100; \ 100; \ 200; \ 500; \ 1,000; \ 1,500$$

We assign utility 0 to the payoff $-\$500$ and utility 100 to the payoff $\$1,500$. For each of the other five possible payoffs, the investor is asked the question:

Q. "Would you prefer to receive payoff I with certainty, or a gamble in which you could obtain a gain of $\$1,500$ with probability p, and a loss of $\$500$ with probability $(1 - p)$?"

The probability p, at which the investor is indifferent between these alternatives, is then recorded. The results obtained are shown in the accompanying table.

PAYOFF	-100	100	200	500	1,000
p	0.20	0.35	0.40	0.60	0.95

(a) What are the utilities for these five payoffs?

(b) Suppose the probabilities for the three states of nature are

$$P(s_1) = 0.5; \qquad P(s_2) = 0.3; \qquad P(s_3) = 0.2$$

Which investment should be chosen if expected utility is to be maximized?

44. A decision-maker is faced with a problem in which the possible payoffs (in dollars) are:

$$100; \quad 300; \quad 700; \quad 900; \quad 1,000; \quad 1,200$$

A utility of 0 is assigned to the payoff $\$100$, and a utility of 100 to payoff $\$1,200$. This decision-maker is indifferent to risk for payoffs in this region.

(a) What are the utilities for the five intermediate payoffs?

(b) For each individual intermediate payoff, I, find the probability p such that this decision-maker is indifferent between receiving I with certainty, and a gamble in which he would receive $\$1,200$ with probability p and $\$100$ with probability $(1 - p)$.

45. Refer to Exercise 2. After his inadmissible action is eliminated, the manufacturer of earth-moving equipment has seven different possible payoffs. These are (in millions of dollars):

$$-10; \quad -7; \quad 0; \quad 1; \quad 4; \quad 8; \quad 20$$

Assign utility 0 to a loss of $\$10$ million and utility 100 to a profit of $\$20$ million. For each intermediate payoff, I, the probabilities p such that the manufacturer is indifferent between receiving I with certainty, and a gamble in which he would gain $\$20$ million with probability p and lose $\$10$ million with probability $(1 - p)$ are shown in the table.

PAYOFF	-7	0	1	4	8
p		0.10	0.35	0.40	0.55	0.70

(a) What are the utilities for the intermediate payoffs?
(b) Suppose the probabilities for the three states of nature are

$$P(s_1) = 0.2; \qquad P(s_2) = 0.4; \qquad P(s_3) = 0.4$$

Which action should be followed, if expected utility is to be maximized?

46. The company of Exercise 6, faced with a decision as to how to heat a new facility, has the following six possible payoffs (in tens of thousands of dollars):

$$-13; \quad -10; \quad -9; \quad -8; \quad -6; \quad -4$$

Assign utility 0 to a cost of $130 thousand and utility 100 to a cost of $40 thousand. For each intermediate payoff, I, the probabilities p such that the company is indifferent between payoff I with certainty, and a cost of $40,000 with probability p and a cost of $130,000 with probability $(1 - p)$ are given in the table.

PAYOFF	-10	-9	-8	-6
p		0.30	0.35	0.40	0.65

(a) What are the utilities for the intermediate payoffs?
(b) Suppose the probabilities for the three states of nature are

$$P(s_1) = 0.1; \qquad P(s_2) = 0.3; \qquad P(s_3) = 0.6$$

What action should be taken if expected utility is to be maximized?

47. The contractor of Exercise 16 must decide whether to place a bid. State under what circumstances such a contractor would be indifferent as to whether to bid for the project.

48. The publisher of Exercise 20 faces a two-stage decision problem, for which the possible eventual payoffs (in dollars) are as follows:

$$-90,000; \quad -50,000; \quad 10,000; \quad 50,000; \quad 210,000; \quad 250,000$$

A utility of 0 is assigned to a loss of $90,000, and a utility of 100 to a profit of $250,000. For each intermediate payoff, I, the probabilities p such that the publisher is indifferent between payoff I with certainty and a gamble in which $250,000 is gained with probability p and $90,000 lost with probability $(1 - p)$ are given in the accompanying table. If this publisher wants to maximize expected utility, what strategy should he follow?

PAYOFF	-50,000	10,000	50,000	210,000
p		0.20	0.30	0.50	0.95

49. Consider the problem of the consultant of Exercise 22, and ignore the option in part (d). The possible payoffs (in dollars) are:

$$-2,250; \quad -1,500; \quad -750; \quad -500; \quad 3,500; \quad 3,750$$

For each intermediate payoff, I, the probabilities p such that the consultant is indifferent between receiving I with certainty and a gamble in which \$3,750 is gained with probability p and \$2,250 lost with probability $(1 - p)$ are given in the accompanying table. If the consultant wants to maximize expected utility, how should he proceed?

PAYOFF	$-1,500$	-750	-500	$3,500$
p	0.10	0.20	0.25	0.98

REVIEW EXERCISES

50. We have discussed the following four criteria for decision-making:
   (i) Maximin criterion
   (ii) Minimax regret criterion
   (iii) Expected monetary value criterion
   (iv) Expected utility criterion
   Briefly outline the philosophies behind these criteria, and discuss their advantages and disadvantages.

51. Of what potential value is sample information in the context of business decision-making? Provide examples of decision-making problems where it would be realistic to expect sample information to be useful.

52. Distinguish among aversion to risk, preference for risk, and indifference to risk. What is the relevance of these concepts to the analysis of business decision-making problems?

53. A consultant is considering submitting detailed bids for two possible contracts. The bid for the first contract costs \$100 to prepare, while that for the second contract costs \$150 to prepare. If the bid for the first contract is accepted and the work carried out, a profit of \$800 will result. If the bid for the second contract is accepted and the work is carried out, a profit of \$1,200 will result. Any costs of bid preparation must be subtracted from these profits. The consultant can, if he wishes, submit bids for both contracts. He does not, however, have the resources to carry out both pieces of work simultaneously. If a bid is submitted and accepted, and the consultant is then unable to carry out the work, he counts this as a cost of \$200 in lost goodwill. For this decision-making process there are four possible states of nature:

   s_1: Both bids rejected
   s_2: Bid for the first contract accepted, bid for the second contract rejected
   s_3: Bid for the second contract accepted, bid for the first contract rejected
   s_4: Both bids accepted

   (a) The consultant has four possible courses of action. What are they?
   (b) Set out the payoff table for this consultant's decision-making problem.
   (c) Which action is chosen by the maximin criterion?
   (d) Which action is chosen by the minimax regret criterion?

54. Refer to Exercise 53. The consultant believes that the probability is 0.8 that a bid for the first contract would be accepted, and 0.5 that a bid for the second contract would be accepted. He also believes that the acceptance of one bid is independent of acceptance of the other.

(a) What are the probabilities for the four states of nature?

(b) According to the expected monetary value criterion, which action should the consultant adopt, and what is the expected monetary value of this action?

(c) Draw the decision tree for the consultant's problem.

(d) What is the expected value of perfect information to this consultant?

(e) The consultant is offered "inside information" on the prospects of the bid for the first contract. This information is entirely reliable, in the sense that it would allow him to know for sure whether the bid would be accepted. However, no further information is available on the prospects of the bid for the second contract. What is the expected value of this "inside information"?

55. Refer to Exercises 53 and 54. There are nine possible payoffs for this consultant, as follows (in dollars):

$$-250; \quad -150; \quad -100; \quad 0; \quad 550; \quad 700; \quad 750; \quad 950; \quad 1{,}050$$

A utility of 0 is assigned to a loss of $250, and a utility of 100 to a profit of $1,050. For each intermediate payoff, I, the probabilities, p, such that the consultant is indifferent between payoff I with certainty, and a gamble in which $1,050 is gained with probability p and $250 lost with probability $(1 - p)$ are shown in the accompanying table. According to the expected utility criterion, which action should the consultant choose, and what is the expected utility of this action?

PAYOFF	-150	-100	0	550	700	750	950
p	0.05	0.10	0.20	0.60	0.70	0.75	0.85

tables

TABLE 1 Probability function of the binomial distribution

The table shows the probability of *x* successes in *n* independent trials, each with probability of success *p*. For example, the probability of 4 successes in 8 independent trials, each with probability of success 0.35, is 0.1875.

n	x	.05	.10	.15	.20	.25	.30	.35	.40	.45	.50
1	0	.9500	.9000	.8500	.8000	.7500	.7000	.6500	.6000	.5500	.5000
	1	.0500	.1000	.1500	.2000	.2500	.3000	.3500	.4000	.4500	.5000
2	0	.9025	.8100	.7225	.6400	.5625	.4900	.4225	.3600	.3025	.2500
	1	.0950	.1800	.2550	.3200	.3750	.4200	.4550	.4800	.4950	.5000
	2	.0025	.0100	.0225	.0400	.0625	.0900	.1225	.1600	.2025	.2500
3	0	.8574	.7290	.6141	.5120	.4219	.3430	.2746	.2160	.1664	.1250
	1	.1354	.2430	.3251	.3840	.4219	.4410	.4436	.4320	.4084	.3750
	2	.0071	.0270	.0574	.0960	.1406	.1890	.2389	.2880	.3341	.3750
	3	.0001	.0010	.0034	.0080	.0156	.0270	.0429	.0640	.0911	.1250
4	0	.8145	.6561	.5220	.4096	.3164	.2401	.1785	.1296	.0915	.0625
	1	.1715	.2916	.3685	.4096	.4219	.4116	.3845	.3456	.2995	.2500
	2	.0135	.0486	.0975	.1536	.2109	.2646	.3105	.3456	.3675	.3750
	3	.0005	.0036	.0115	.0256	.0469	.0756	.1115	.1536	.2005	.2500
	4	.0000	.0001	.0005	.0016	.0039	.0081	.0150	.0256	.0410	.0625
5	0	.7738	.5905	.4437	.3277	.2373	.1681	.1160	.0778	.0503	.0312
	1	.2036	.3280	.3915	.4096	.3955	.3602	.3124	.2592	.2059	.1562
	2	.0214	.0729	.1382	.2048	.2637	.3087	.3364	.3456	.3369	.3125

TABLE 1 Probability function of the binomial distribution (continued)

n	x	.05	.10	.15	.20	.25	.30	.35	.40	.45	.50
	3	.0011	.0081	.0244	.0512	.0879	.1323	.1811	.2304	.2757	.3125
	4	.0000	.0004	.0022	.0064	.0146	.0284	.0488	.0768	.1128	.1562
	5	.0000	.0000	.0001	.0003	.0010	.0024	.0053	.0102	.0185	.0312
6	0	.7351	.5314	.3771	.2621	.1780	.1176	.0754	.0467	.0277	.0156
	1	.2321	.3543	.3993	.3932	.3560	.3025	.2437	.1866	.1359	.0938
	2	.0305	.0984	.1762	.2458	.2966	.3241	.3280	.3110	.2780	.2344
	3	.0021	.0146	.0415	.0819	.1318	.1852	.2355	.2765	.3032	.3125
	4	.0001	.0012	.0055	.0154	.0330	.0595	.0951	.1382	.1861	.2344
	5	.0000	.0001	.0004	.0015	.0044	.0102	.0205	.0369	.0609	.0938
	6	.0000	.0000	.0000	.0001	.0002	.0007	.0018	.0041	.0083	.0156
7	0	.6983	.4783	.3206	.2097	.1335	.0824	.0490	.0280	.0152	.0078
	1	.2573	.3720	.3960	.3670	.3115	.2471	.1848	.1306	.0872	.0547
	2	.0406	.1240	.2097	.2753	.3115	.3177	.2985	.2613	.2140	.1641
	3	.0036	.0230	.0617	.1147	.1730	.2269	.2679	.2903	.2918	.2734
	4	.0002	.0026	.0109	.0287	.0577	.0972	.1442	.1935	.2388	.2734
	5	.0000	.0002	.0012	.0043	.0115	.0250	.0466	.0774	.1172	.1641
	6	.0000	.0000	.0001	.0004	.0013	.0036	.0084	.0172	.0320	.0547
	7	.0000	.0000	.0000	.0000	.0001	.0002	.0006	.0016	.0037	.0078
8	0	.6634	.4305	.2725	.1678	.1001	.0576	.0319	.0168	.0084	.0039
	1	.2793	.3826	.3847	.3355	.2670	.1977	.1373	.0896	.0548	.0312
	2	.0515	.1488	.2376	.2936	.3115	.2965	.2587	.2090	.1569	.1094
	3	.0054	.0331	.0839	.1468	.2076	.2541	.2786	.2787	.2568	.2188
	4	.0004	.0046	.0185	.0459	.0865	.1361	.1875	.2322	.2627	.2734
	5	.0000	.0004	.0026	.0092	.0231	.0467	.0808	.1239	.1719	.2188
	6	.0000	.0000	.0002	.0011	.0038	.0100	.0217	.0413	.0703	.1094
	7	.0000	.0000	.0000	.0001	.0004	.0012	.0033	.0079	.0164	.0312
	8	.0000	.0000	.0000	.0000	.0000	.0001	.0002	.0007	.0017	.0039
9	0	.6302	.3874	.2316	.1342	.0751	.0404	.0207	.0101	.0046	.0020
	1	.2985	.3874	.3679	.3020	.2253	.1556	.1004	.0605	.0339	.0176
	2	.0629	.1722	.2597	.3020	.3003	.2668	.2162	.1612	.1110	.0703
	3	.0077	.0446	.1069	.1762	.2336	.2668	.2716	.2508	.2119	.1641
	4	.0006	.0074	.0283	.0661	.1168	.1715	.2194	.2508	.2600	.2461
	5	.0000	.0008	.0050	.0165	.0389	.0735	.1181	.1672	.2128	.2461
	6	.0000	.0001	.0006	.0028	.0087	.0210	.0424	.0743	.1160	.1641
	7	.0000	.0000	.0000	.0003	.0012	.0039	.0098	.0212	.0407	.0703
	8	.0000	.0000	.0000	.0000	.0001	.0004	.0013	.0035	.0083	.0176
	9	.0000	.0000	.0000	.0000	.0000	.0000	.0001	.0003	.0008	.0020
10	0	.5987	.3487	.1969	.1074	.0563	.0282	.0135	.0060	.0025	.0010
	1	.3151	.3874	.3474	.2684	.1877	.1211	.0725	.0403	.0207	.0098
	2	.0746	.1937	.2759	.3020	.2816	.2335	.1757	.1209	.0763	.0439
	3	.0105	.0574	.1298	.2013	.2503	.2668	.2522	.2150	.1665	.1172
	4	.0010	.0112	.0401	.0881	.1460	.2001	.2377	.2508	.2384	.2051
	5	.0001	.0015	.0085	.0264	.0584	.1029	.1536	.2007	.2340	.2461
	6	.0000	.0001	.0012	.0055	.0162	.0368	.0689	.1115	.1596	.2051
	7	.0000	.0000	.0001	.0008	.0031	.0090	.0212	.0425	.0746	.1172
	8	.0000	.0000	.0000	.0001	.0004	.0014	.0043	.0106	.0229	.0439
	9	.0000	.0000	.0000	.0000	.0000	.0001	.0005	.0016	.0042	.0098
	10	.0000	.0000	.0000	.0000	.0000	.0000	.0000	.0001	.0003	.0010
11	0	.5688	.3138	.1673	.0859	.0422	.0198	.0088	.0036	.0014	.0005
	1	.3293	.3835	.3248	.2362	.1549	.0932	.0518	.0266	.0125	.0054

TABLE 1 Probability function of the binomial distribution (continued)

n	x	.05	.10	.15	.20	.25	.30	.35	.40	.45	.50
	2	.0867	.2131	.2866	.2953	.2581	.1998	.1395	.0887	.0513	.0269
	3	.0137	.0710	.1517	.2215	.2581	.2568	.2254	.1774	.1259	.0806
	4	.0014	.0158	.0536	.1107	.1721	.2201	.2428	.2365	.2060	.1611
	5	.0001	.0025	.0132	.0388	.0803	.1321	.1830	.2207	.2360	.2256
	6	.0000	.0003	.0023	.0097	.0268	.0566	.0985	.1471	.1931	.2256
	7	.0000	.0000	.0003	.0017	.0064	.0173	.0379	.0701	.1128	.1611
	8	.0000	.0000	.0000	.0002	.0011	.0037	.0102	.0234	.0462	.0806
	9	.0000	.0000	.0000	.0000	.0001	.0005	.0018	.0052	.0126	.0269
	10	.0000	.0000	.0000	.0000	.0000	.0000	.0002	.0007	.0021	.0054
	11	.0000	.0000	.0000	.0000	.0000	.0000	.0000	.0000	.0002	.0005
12	0	.5404	.2824	.1422	.0687	.0317	.0138	.0057	.0022	.0008	.0002
	1	.3413	.3766	.3012	.2062	.1267	.0712	.0368	.0174	.0075	.0029
	2	.0988	.2301	.2924	.2835	.2323	.1678	.1088	.0639	.0339	.0161
	3	.0173	.0852	.1720	.2362	.2581	.2397	.1954	.1419	.0923	.0537
	4	.0021	.0213	.0683	.1329	.1936	.2311	.2367	.2128	.1700	.1208
	5	.0002	.0038	.0193	.0532	.1032	.1585	.2039	.2270	.2225	.1934
	6	.0000	.0005	.0040	.0155	.0401	.0792	.1281	.1766	.2124	.2256
	7	.0000	.0000	.0006	.0033	.0115	.0291	.0591	.1009	.1489	.1934
	8	.0000	.0000	.0001	.0005	.0024	.0078	.0199	.0420	.0762	.1208
	9	.0000	.0000	.0000	.0001	.0004	.0015	.0048	.0125	.0277	.0537
	10	.0000	.0000	.0000	.0000	.0000	.0002	.0008	.0025	.0068	.0161
	11	.0000	.0000	.0000	.0000	.0000	.0000	.0001	.0003	.0010	.0029
	12	.0000	.0000	.0000	.0000	.0000	.0000	.0000	.0000	.0001	.0002
13	0	.5133	.2542	.1209	.0550	.0238	.0097	.0037	.0013	.0004	.0001
	1	.3512	.3672	.2774	.1787	.1029	.0540	.0259	.0113	.0045	.0016
	2	.1109	.2448	.2937	.2680	.2059	.1388	.0836	.0453	.0220	.0095
	3	.0214	.0997	.1900	.2457	.2517	.2181	.1651	.1107	.0660	.0349
	4	.0028	.0277	.0838	.1535	.2097	.2337	.2222	.1845	.1350	.0873
	5	.0003	.0055	.0266	.0691	.1258	.1803	.2154	.2214	.1989	.1571
	6	.0000	.0008	.0063	.0230	.0559	.1030	.1546	.1968	.2169	.2095
	7	.0000	.0001	.0011	.0058	.0186	.0442	.0833	.1312	.1775	.2095
	8	.0000	.0000	.0001	.0011	.0047	.0142	.0336	.0656	.1089	.1571
	9	.0000	.0000	.0000	.0001	.0009	.0034	.0101	.0243	.0495	.0873
	10	.0000	.0000	.0000	.0000	.0001	.0006	.0022	.0065	.0162	.0349
	11	.0000	.0000	.0000	.0000	.0000	.0001	.0003	.0012	.0036	.0095
	12	.0000	.0000	.0000	.0000	.0000	.0000	.0000	.0001	.0005	.0016
	13	.0000	.0000	.0000	.0000	.0000	.0000	.0000	.0000	.0000	.0001
14	0	.4877	.2288	.1028	.0440	.0178	.0068	.0024	.0008	.0002	.0001
	1	.3593	.3559	.2539	.1539	.0832	.0407	.0181	.0073	.0027	.0009
	2	.1229	.2570	.2912	.2501	.1802	.1134	.0634	.0317	.0141	.0056
	3	.0259	.1142	.2056	.2501	.2402	.1943	.1366	.0845	.0462	.0222
	4	.0037	.0348	.0998	.1720	.2202	.2290	.2022	.1549	.1040	.0611
	5	.0004	.0078	.0352	.0860	.1468	.1963	.2178	.2066	.1701	.1222
	6	.0000	.0013	.0093	.0322	.0734	.1262	.1759	.2066	.2088	.1833
	7	.0000	.0002	.0019	.0092	.0280	.0618	.1082	.1574	.1952	.2095
	8	.0000	.0000	.0003	.0020	.0082	.0232	.0510	.0918	.1398	.1833
	9	.0000	.0000	.0000	.0003	.0018	.0066	.0183	.0408	.0762	.1222
	10	.0000	.0000	.0000	.0000	.0003	.0014	.0049	.0136	.0312	.0611
	11	.0000	.0000	.0000	.0000	.0000	.0002	.0010	.0033	.0093	.0222
	12	.0000	.0000	.0000	.0000	.0000	.0000	.0001	.0005	.0019	.0056
	13	.0000	.0000	.0000	.0000	.0000	.0000	.0000	.0001	.0002	.0009
	14	.0000	.0000	.0000	.0000	.0000	.0000	.0000	.0000	.0000	.0001

TABLE 1 Probability function of the binomial distribution (continued)

n	x	.05	.10	.15	.20	.25	.30	.35	.40	.45	.50
15	0	.4633	.2059	.0874	.0352	.0134	.0047	.0016	.0005	.0001	.0000
	1	.3658	.3432	.2312	.1319	.0668	.0305	.0126	.0047	.0016	.0005
	2	.1348	.2669	.2856	.2309	.1559	.0916	.0476	.0219	.0090	.0032
	3	.0307	.1285	.2184	.2501	.2252	.1700	.1110	.0634	.0318	.0139
	4	.0049	.0428	.1156	.1876	.2252	.2186	.1792	.1268	.0780	.0417
	5	.0006	.0105	.0449	.1032	.1651	.2061	.2123	.1859	.1404	.0916
	6	.0000	.0019	.0132	.0430	.0917	.1472	.1906	.2066	.1914	.1527
	7	.0000	.0003	.0030	.0138	.0393	.0811	.1319	.1771	.2013	.1964
	8	.0000	.0000	.0005	.0035	.0131	.0348	.0710	.1181	.1647	.1964
	9	.0000	.0000	.0001	.0007	.0034	.0116	.0298	.0612	.1048	.1527
	10	.0000	.0000	.0000	.0001	.0007	.0030	.0096	.0245	.0515	.0916
	11	.0000	.0000	.0000	.0000	.0001	.0006	.0024	.0074	.0191	.0417
	12	.0000	.0000	.0000	.0000	.0000	.0001	.0004	.0016	.0052	.0139
	13	.0000	.0000	.0000	.0000	.0000	.0000	.0001	.0003	.0010	.0032
	14	.0000	.0000	.0000	.0000	.0000	.0000	.0000	.0000	.0001	.0005
	15	.0000	.0000	.0000	.0000	.0000	.0000	.0000	.0000	.0000	.0000
16	0	.4401	.1853	.0743	.0281	.0100	.0033	.0010	.0003	.0001	.0000
	1	.3706	.3294	.2097	.1126	.0535	.0228	.0087	.0030	.0009	.0002
	2	.1463	.2745	.2775	.2111	.1336	.0732	.0353	.0150	.0056	.0018
	3	.0359	.1423	.2285	.2463	.2079	.1465	.0888	.0468	.0215	.0085
	4	.0061	.0514	.1311	.2001	.2252	.2040	.1553	.1014	.0572	.0278
	5	.0008	.0137	.0555	.1201	.1802	.2099	.2008	.1623	.1123	.0667
	6	.0001	.0028	.0180	.0550	.1101	.1649	.1982	.1983	.1684	.1222
	7	.0000	.0004	.0045	.0197	.0524	.1010	.1524	.1889	.1969	.1746
	8	.0000	.0001	.0009	.0055	.0197	.0487	.0923	.1417	.1812	.1964
	9	.0000	.0000	.0001	.0012	.0058	.0185	.0442	.0840	.1318	.1746
	10	.0000	.0000	.0000	.0002	.0014	.0056	.0167	.0392	.0755	.1222
	11	.0000	.0000	.0000	.0000	.0002	.0013	.0049	.0142	.0337	.0667
	12	.0000	.0000	.0000	.0000	.0000	.0002	.0011	.0040	.0115	.0278
	13	.0000	.0000	.0000	.0000	.0000	.0000	.0002	.0008	.0029	.0085
	14	.0000	.0000	.0000	.0000	.0000	.0000	.0000	.0001	.0005	.0018
	15	.0000	.0000	.0000	.0000	.0000	.0000	.0000	.0000	.0001	.0002
	16	.0000	.0000	.0000	.0000	.0000	.0000	.0000	.0000	.0000	.0000
17	0	.4181	.1668	.0631	.0225	.0075	.0023	.0007	.0002	.0000	.0000
	1	.3741	.3150	.1893	.0957	.0426	.0169	.0060	.0019	.0005	.0001
	2	.1575	.2800	.2673	.1914	.1136	.0581	.0260	.0102	.0035	.0010
	3	.0415	.1556	.2359	.2393	.1893	.1245	.0701	.0341	.0144	.0052
	4	.0076	.0605	.1457	.2093	.2209	.1868	.1320	.0796	.0411	.0182
	5	.0010	.0175	.0668	.1361	.1914	.2081	.1849	.1379	.0875	.0472
	6	.0001	.0039	.0236	.0680	.1276	.1784	.1991	.1839	.1432	.0944
	7	.0000	.0007	.0065	.0267	.0668	.1201	.1685	.1927	.1841	.1484
	8	.0000	.0001	.0014	.0084	.0279	.0644	.1134	.1606	.1883	.1855
	9	.0000	.0000	.0003	.0021	.0093	.0276	.0611	.1070	.1540	.1855
	10	.0000	.0000	.0000	.0004	.0025	.0095	.0263	.0571	.1008	.1484
	11	.0000	.0000	.0000	.0001	.0005	.0026	.0090	.0242	.0525	.0944
	12	.0000	.0000	.0000	.0000	.0001	.0006	.0024	.0081	.0215	.0472
	13	.0000	.0000	.0000	.0000	.0000	.0001	.0005	.0021	.0068	.0182
	14	.0000	.0000	.0000	.0000	.0000	.0000	.0001	.0004	.0016	.0052
	15	.0000	.0000	.0000	.0000	.0000	.0000	.0000	.0001	.0003	.0010
	16	.0000	.0000	.0000	.0000	.0000	.0000	.0000	.0000	.0000	.0001
	17	.0000	.0000	.0000	.0000	.0000	.0000	.0000	.0000	.0000	.0000

TABLE 1 Probability function of the binomial distribution (continued)

n	x	.05	.10	.15	.20	.25	.30	.35	.40	.45	.50
18	0	.3972	.1501	.0536	.0180	.0056	.0016	.0004	.0001	.0000	.0000
	1	.3763	.3002	.1704	.0811	.0338	.0126	.0042	.0012	.0003	.0001
	2	.1683	.2835	.2556	.1723	.0958	.0458	.0190	.0069	.0022	.0006
	3	.0473	.1680	.2406	.2297	.1704	.1046	.0547	.0246	.0095	.0031
	4	.0093	.0700	.1592	.2153	.2130	.1681	.1104	.0614	.0291	.0117
	5	.0014	.0218	.0787	.1507	.1988	.2017	.1664	.1146	.0666	.0327
	6	.0002	.0052	.0301	.0816	.1436	.1873	.1941	.1655	.1181	.0708
	7	.0000	.0010	.0091	.0350	.0820	.1376	.1792	.1892	.1657	.1214
	8	.0000	.0002	.0022	.0120	.0376	.0811	.1327	.1734	.1864	.1669
	9	.0000	.0000	.0004	.0033	.0139	.0386	.0794	.1284	.1694	.1855
	10	.0000	.0000	.0001	.0008	.0042	.0149	.0385	.0771	.1248	.1669
	11	.0000	.0000	.0000	.0001	.0010	.0046	.0151	.0374	.0742	.1214
	12	.0000	.0000	.0000	.0000	.0002	.0012	.0047	.0145	.0354	.0708
	13	.0000	.0000	.0000	.0000	.0000	.0002	.0012	.0044	.0134	.0327
	14	.0000	.0000	.0000	.0000	.0000	.0000	.0002	.0011	.0039	.0117
	15	.0000	.0000	.0000	.0000	.0000	.0000	.0000	.0002	.0009	.0031
	16	.0000	.0000	.0000	.0000	.0000	.0000	.0000	.0000	.0001	.0006
	17	.0000	.0000	.0000	.0000	.0000	.0000	.0000	.0000	.0000	.0001
	18	.0000	.0000	.0000	.0000	.0000	.0000	.0000	.0000	.0000	.0000
19	0	.3774	.1351	.0456	.0144	.0042	.0011	.0003	.0001	.0000	.0000
	1	.3774	.2852	.1529	.0685	.0268	.0093	.0029	.0008	.0002	.0000
	2	.1787	.2852	.2428	.1540	.0803	.0358	.0138	.0046	.0013	.0003
	3	.0533	.1796	.2428	.2182	.1517	.0869	.0422	.0175	.0062	.0018
	4	.0112	.0798	.1714	.2182	.2023	.1491	.0909	.0467	.0203	.0074
	5	.0018	.0266	.0907	.1636	.2023	.1916	.1468	.0933	.0497	.0222
	6	.0002	.0069	.0374	.0955	.1574	.1916	.1844	.1451	.0949	.0518
	7	.0000	.0014	.0122	.0443	.0974	.1525	.1844	.1797	.1443	.0961
	8	.0000	.0002	.0032	.0166	.0487	.0981	.1489	.1797	.1771	.1442
	9	.0000	.0000	.0007	.0051	.0198	.0514	.0980	.1464	.1771	.1762
	10	.0000	.0000	.0001	.0013	.0066	.0220	.0528	.0976	.1449	.1762
	11	.0000	.0000	.0000	.0003	.0018	.0077	.0233	.0532	.0970	.1442
	12	.0000	.0000	.0000	.0000	.0004	.0022	.0083	.0237	.0529	.0961
	13	.0000	.0000	.0000	.0000	.0001	.0005	.0024	.0085	.0233	.0518
	14	.0000	.0000	.0000	.0000	.0000	.0001	.0006	.0024	.0082	.0222
	15	.0000	.0000	.0000	.0000	.0000	.0000	.0001	.0005	.0022	.0074
	16	.0000	.0000	.0000	.0000	.0000	.0000	.0000	.0001	.0005	.0018
	17	.0000	.0000	.0000	.0000	.0000	.0000	.0000	.0000	.0001	.0003
	18	.0000	.0000	.0000	.0000	.0000	.0000	.0000	.0000	.0000	.0000
	19	.0000	.0000	.0000	.0000	.0000	.0000	.0000.	.0000	.0000	.0000
20	0	.3585	.1216	.0388	.0115	.0032	.0008	.0002	.0000	.0000	.0000
	1	.3774	.2702	.1368	.0576	.0211	.0068	.0020	.0005	.0001	.0000
	2	.1887	.2852	.2293	.1369	.0669	.0278	.0100	.0031	.0008	.0002
	3	.0596	.1901	.2428	.2054	.1339	.0716	.0323	.0123	.0040	.0011
	4	.0133	.0898	.1821	.2182	.1897	.1304	.0738	.0350	.0139	.0046
	5	.0022	.0319	.1028	.1746	.2023	.1789	.1272	.0746	.0365	.0148
	6	.0003	.0089	.0454	.1091	.1686	.1916	.1712	.1244	.0746	.0370
	7	.0000	.0020	.0160	.0545	.1124	.1643	.1844	.1659	.1221	.0739
	8	.0000	.0004	.0046	.0222	.0609	.1144	.1614	.1797	.1623	.1201
	9	.0000	.0001	.0011	.0074	.0271	.0654	.1158	.1597	.1771	.1602
	10	.0000	.0000	.0002	.0020	.0099	.0308	.0686	.1171	.1593	.1762
	11	.0000	.0000	.0000	.0005	.0030	.0120	.0336	.0710	.1185	.1602

TABLE 1 Probability function of the binomial distribution (continued)

n	x	p									
		.05	.10	.15	.20	.25	.30	.35	.40	.45	.50
	12	.0000	.0000	.0000	.0001	.0008	.0039	.0136	.0355	.0727	.1201
	13	.0000	.0000	.0000	.0000	.0002	.0010	.0045	.0146	.0366	.0739
	14	.0000	.0000	.0000	.0000	.0000	.0002	.0012	.0049	.0150	.0370
	15	.0000	.0000	.0000	.0000	.0000	.0000	.0003	.0013	.0049	.0148
	16	.0000	.0000	.0000	.0000	.0000	.0000	.0000	.0003	.0013	.0046
	17	.0000	.0000	.0000	.0000	.0000	.0000	.0000	.0000	.0002	.0011
	18	.0000	.0000	.0000	.0000	.0000	.0000	.0000	.0000	.0000	.0002
	19	.0000	.0000	.0000	.0000	.0000	.0000	.0000	.0000	.0000	.0000
	20	.0000	.0000	.0000	.0000	.0000	.0000	.0000	.0000	.0000	.0000

Reproduced with permission from National Bureau of Standards, *Tables of the Binomial Probability Distribution*, United States Department of Commerce (1950).

TABLE 2 Values of $e^{-\lambda}$

λ	$e^{-\lambda}$	λ	$e^{-\lambda}$	λ	$e^{-\lambda}$	λ	$e^{-\lambda}$
0.00	1.000000	2.60	.074274	5.10	.006097	7.60	.000501
0.10	.904837	2.70	.067206	5.20	.005517	7.70	.000453
0.20	.818731	2.80	.060810	5.30	.004992	7.80	.000410
0.30	.740818	2.90	.055023	5.40	.004517	7.90	.000371
0.40	.670320	3.00	.049787	5.50	.004087	8.00	.000336
0.50	.606531	3.10	.045049	5.60	.003698	8.10	.000304
0.60	.548812	3.20	.040762	5.70	.003346	8.20	.000275
0.70	.496585	3.30	.036883	5.80	.003028	8.30	.000249
0.80	.449329	3.40	.033373	5.90	.002739	8.40	.000225
0.90	.406570	3.50	.030197	6.00	.002479	8.50	.000204
1.00	.367879	3.60	.027324	6.10	.002243	8.60	.000184
1.10	.332871	3.70	.024724	6.20	.002029	8.70	.000167
1.20	.301194	3.80	.022371	6.30	.001836	8.80	.000151
1.30	.272532	3.90	.020242	6.40	.001661	8.90	.000136
1.40	.246597	4.00	.018316	6.50	.001503	9.00	.000123
1.50	.223130	4.10	.016573	6.60	.001360	9.10	.000112
1.60	.201897	4.20	.014996	6.70	.001231	9.20	.000101
1.70	.182684	4.30	.013569	6.80	.001114	9.30	.000091
1.80	.165299	4.40	.012277	6.90	.001008	9.40	.000083
1.90	.149569	4.50	.011109	7.00	.000912	9.50	.000075
2.00	.135335	4.60	.010052	7.10	.000825	9.60	.000068
2.10	.122456	4.70	.009095	7.20	.000747	9.70	.000061
2.20	.110803	4.80	.008230	7.30	.000676	9.80	.000056
2.30	.100259	4.90	.007447	7.40	.000611	9.90	.000050
2.40	.090718	5.00	.006738	7.50	.000553	10.00	.000045
2.50	.082085						

TABLE 3 Cumulative distribution function of the standard normal distribution

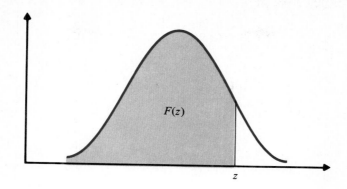

The table shows the probability, $F(z)$, that a standard normal random variable is less than the number z. For example, the probability is 0.9750 that a standard normal random variable is less than 1.96.

z	$F(z)$	z	$F(z)$	z	$F(z)$	z	$F(z)$
.00	.5000						
.01	.5040	.51	.6950	1.01	.8438	1.51	.9345
.02	.5080	.52	.6985	1.02	.8461	1.52	.9357
.03	.5120	.53	.7019	1.03	.8485	1.53	.9370
.04	.5160	.54	.7054	1.04	.8508	1.54	.9382
.05	.5199	.55	.7088	1.05	.8531	1.55	.9394
.06	.5239	.56	.7123	1.06	.8554	1.56	.9406
.07	.5279	.57	.7157	1.07	.8577	1.57	.9418
.08	.5319	.58	.7190	1.08	.8599	1.58	.9429
.09	.5359	.59	.7224	1.09	.8621	1.59	.9441
.10	.5398	.60	.7257	1.10	.8643	1.60	.9452
.11	.5438	.61	.7291	1.11	.8665	1.61	.9463
.12	.5478	.62	.7324	1.12	.8686	1.62	.9474
.13	.5517	.63	.7357	1.13	.8708	1.63	.9484
.14	.5557	.64	.7389	1.14	.8729	1.64	.9495
.15	.5596	.65	.7422	1.15	.8749	1.65	.9505
.16	.5636	.66	.7454	1.16	.8770	1.66	.9515
.17	.5675	.67	.7486	1.17	.8790	1.67	.9525
.18	.5714	.68	.7517	1.18	.8810	1.68	.9535
.19	.5753	.69	.7549	1.19	.8830	1.69	.9545
.20	.5793	.70	.7580	1.20	.8849	1.70	.9554
.21	.5832	.71	.7611	1.21	.8869	1.71	.9564
.22	.5871	.72	.7642	1.22	.8888	1.72	.9573
.23	.5910	.73	.7673	1.23	.8907	1.73	.9582
.24	.5948	.74	.7704	1.24	.8925	1.74	.9591
.25	.5987	.75	.7734	1.25	.8944	1.75	.9599
.26	.6026	.76	.7764	1.26	.8962	1.76	.9608

TABLE 3 Cumulative distribution function of the standard normal distribution (continued)

z	F(z)	z	F(z)	z	F(z)	z	F(z)
.27	.6064	.77	.7794	1.27	.8980	1.77	.9616
.28	.6103	.78	.7823	1.28	.8997	1.78	.9625
.29	.6141	.79	.7852	1.29	.9015	1.79	.9633
.30	.6179	.80	.7881	1.30	.9032	1.80	.9641
.31	.6217	.81	.7910	1.31	.9049	1.81	.9649
.32	.6255	.82	.7939	1.32	.9066	1.82	.9656
.33	.6293	.83	.7967	1.33	.9082	1.83	.9664
.34	.6331	.84	.7995	1.34	.9099	1.84	.9671
.35	.6368	.85	.8023	1.35	.9115	1.85	.9678
.36	.6406	.86	.8051	1.36	.9131	1.86	.9686
.37	.6443	.87	.8078	1.37	.9147	1.87	.9693
.38	.6480	.88	.8106	1.38	.9162	1.88	.9699
.39	.6517	.89	.8133	1.39	.9177	1.89	.9706
.40	.6554	.90	.8159	1.40	.9192	1.90	.9713
.41	.6591	.91	.8186	1.41	.9207	1.91	.9719
.42	.6628	.92	.8212	1.42	.9222	1.92	.9726
.43	.6664	.93	.8238	1.43	.9236	1.93	.9732
.44	.6700	.94	.8264	1.44	.9251	1.94	.9738
.45	.6736	.95	.8289	1.45	.9265	1.95	.9744
.46	.6772	.96	.8315	1.46	.9279	1.96	.9750
.47	.6803	.97	.8340	1.47	.9292	1.97	.9756
.48	.6844	.98	.8365	1.48	.9306	1.98	.9761
.49	.6879	.99	.8389	1.49	.9319	1.99	.9767
.50	.6915	1.00	.8413	1.50	.9332	2.00	.9772
2.01	.9778	2.51	.9940	3.01	.9987	3.51	.9998
2.02	.9783	2.52	.9941	3.02	.9987	3.52	.9998
2.03	.9788	2.53	.9943	3.03	.9988	3.53	.9998
2.04	.9793	2.54	.9945	3.04	.9988	3.54	.9998
2.05	.9798	2.55	.9946	3.05	.9989	3.55	.9998
2.06	.9803	2.56	.9948	3.06	.9989	3.56	.9998
2.07	.9808	2.57	.9949	3.07	.9989	3.57	.9998
2.08	.9812	2.58	.9951	3.08	.9990	3.58	.9998
2.09	.9817	2.59	.9952	3.09	.9990	3.59	.9998
2.10	.9821	2.60	.9953	3.10	.9990	3.60	.9998
2.11	.9826	2.61	.9955	3.11	.9991	3.61	.9998
2.12	.9830	2.62	.9956	3.12	.9991	3.62	.9999
2.13	.9834	2.63	.9957	3.13	.9991	3.63	.9999
2.14	.9838	2.64	.9959	3.14	.9992	3.64	.9999
2.15	.9842	2.65	.9960	3.15	.9992	3.65	.9999
2.16	.9846	2.66	.9961	3.16	.9992	3.66	.9999
2.17	.9850	2.67	.9962	3.17	.9992	3.67	.9999
2.18	.9854	2.68	.9963	3.18	.9993	3.68	.9999
2.19	.9857	2.69	.9964	3.19	.9993	3.69	.9999
2.20	.9861	2.70	.9965	3.20	.9993	3.70	.9999
2.21	.9864	2.71	.9966	3.21	.9993	3.71	.9999
2.22	.9868	2.72	.9967	3.22	.9994	3.72	.9999
2.23	.9871	2.73	.9968	3.23	.9994	3.73	.9999
2.24	.9875	2.74	.9969	3.24	.9994	3.74	.9999
2.25	.9878	2.75	.9970	3.25	.9994	3.75	.9999

TABLE 3 Cumulative distribution function of the standard normal distribution (continued)

z	F(z)	z	F(z)	z	F(z)	z	F(z)
2.26	.9881	2.76	.9971	3.26	.9994	3.76	.9999
2.27	.9884	2.77	.9972	3.27	.9995	3.77	.9999
2.28	.9887	2.78	.9973	3.28	.9995	3.78	.9999
2.29	.9890	2.79	.9974	3.29	.9995	3.79	.9999
2.30	.9893	2.80	.9974	3.30	.9995	3.80	.9999
2.31	.9896	2.81	.9975	3.31	.9995	3.81	.9999
2.32	.9898	2.82	.9976	3.32	.9996	3.82	.9999
2.33	.9901	2.83	.9977	3.33	.9996	3.83	.9999
2.34	.9904	2.84	.9977	3.34	.9996	3.84	.9999
2.35	.9906	2.85	.9978	3.35	.9996	3.85	.9999
2.36	.9909	2.86	.9979	3.36	.9996	3.86	.9999
2.37	.9911	2.87	.9979	3.37	.9996	3.87	.9999
2.38	.9913	2.88	.9980	3.38	.9996	3.88	.9999
2.39	.9916	2.89	.9981	3.39	.9997	3.89	1.0000
2.40	.9918	2.90	.9981	3.40	.9997	3.90	1.0000
2.41	.9920	2.91	.9982	3.41	.9997	3.91	1.0000
2.42	.9922	2.92	.9982	3.42	.9997	3.92	1.0000
2.43	.9925	2.93	.9983	3.43	.9997	3.93	1.0000
2.44	.9927	2.94	.9984	3.44	.9997	3.94	1.0000
2.45	.9929	2.95	.9984	3.45	.9997	3.95	1.0000
2.46	.9931	2.96	.9985	3.46	.9997	3.96	1.0000
2.47	.9932	2.97	.9985	3.47	.9997	3.97	1.0000
2.48	.9934	2.98	.9986	3.48	.9997	3.98	1.0000
2.49	.9936	2.99	.9986	3.49	.9998	3.99	1.0000
2.50	.9938	3.00	.9986	3.50	.9998		

TABLE 4 Some uniformly distributed random numbers

85387	51571	57714	00512	61319	69143	08881	01400	55061	82977
84176	03311	16955	59504	54499	32096	79485	98031	99485	16788
27258	51746	67223	98182	43166	54297	26830	29842	78016	73127
99398	46950	19399	65167	35082	30482	86323	41061	21717	48126
72752	89364	02150	85418	05420	84341	02395	27655	59457	55438
69090	93551	11649	54688	57061	77711	24201	16895	64936	62347
39620	54988	67846	71845	54000	26134	84526	16619	82573	01737
81725	49831	35595	29891	46812	57770	03326	31316	75412	80732
87968	85157	84752	93777	62772	78961	30750	76089	23340	64637
07730	01861	40610	73445	70321	26467	53533	20787	46971	29134
32825	82100	67406	44156	21531	67186	39945	04189	79798	41087
34453	05330	40224	04116	24597	93823	28171	47701	76201	68257
00830	34235	40671	66042	06341	54437	81649	70494	01883	18350
24580	05258	37329	59173	62660	72513	82232	49794	36913	05877
59578	08535	77107	19838	40651	01749	58893	99115	05212	92309
75387	24990	12748	71766	17471	15794	68622	59161	14476	75074
02465	34977	48319	53026	53691	80594	58805	76961	62665	82855

TABLE 4 Some uniformly distributed random numbers (continued)

49689	08342	81912	92735	30042	47623	60061	69427	21163	68543
60958	20236	79424	04055	54955	73342	14040	72431	99469	41044
79956	98409	79548	39569	83974	43707	77080	08645	20949	56932
04316	01206	08715	77713	20572	13912	94324	14656	11979	53258
78684	28546	06881	66097	53530	42509	54130	30878	77166	98075
69235	18535	61904	99246	84050	15270	07751	90410	96675	62870
81201	04314	92708	44984	83121	33767	56607	46371	20389	08809
80336	59638	44368	33433	97794	10343	19235	82633	17186	63902
65076	87960	92013	60169	49176	50140	39081	04638	96114	63463
90879	70970	50789	59973	47771	94567	35590	23462	33993	99899
50555	84355	97066	82748	98298	14385	82493	40182	20523	69182
48658	41921	86514	46786	74097	62825	46457	24428	09245	86069
26373	19166	88223	32371	11570	62078	92317	13378	05734	71778
20878	80883	26027	29101	58382	17109	53511	95536	21759	10630
20069	60582	55749	88068	48589	01874	42930	40310	34613	97359
46819	38577	20520	94145	99405	47064	25248	27289	41289	54972
83644	04459	73253	58414	94180	09321	59747	07379	56255	45615
08636	31363	56033	49076	88908	51318	39104	56556	23112	63317
92058	38678	12507	90343	17213	24545	66053	76412	29545	89932
05038	18443	87138	05076	25660	23414	84837	87132	84405	15346
41838	68590	93646	82113	25498	33110	15356	81070	84900	42660
15564	81618	99186	73113	99344	13213	07235	90064	89150	86359
74600	40206	15237	37378	96862	78638	14376	46607	55909	46398
78275	77017	60310	13499	35268	47790	77475	44345	14615	25231
30145	71205	10355	18404	85354	22199	90822	35204	47891	69860
46944	00097	39161	50139	60458	44649	85537	90017	18157	13856
85883	21272	89266	94887	00291	70963	28169	95130	27223	35387
83606	98192	82194	26719	24499	28102	97769	98769	30757	81593
66888	81818	52490	54272	70549	69235	74684	96412	65186	87974
63673	73966	34036	44298	60652	05947	05833	27914	57021	58566
37944	16094	39797	63253	64103	32222	65925	64693	34048	75394
93240	66855	29336	28345	71398	45118	01454	72128	09715	29454
40189	76776	70842	32675	81647	75868	21288	12849	94990	21513

Reprinted from page 259 of *A Million Random Digits With 100,000 Normal Deviates,* by the Rand Corporation. New York: The Free Press, 1955. Copyright 1955 by The Rand Corporation. Used by Permission.

TABLE 5 Cutoff points of the chi-square distribution function

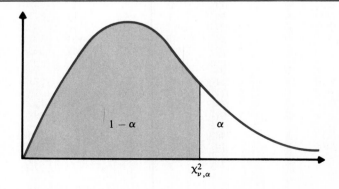

For selected probabilities α, the table shows the values $\chi^2_{\nu,\alpha}$ such that $\alpha = P(\chi^2_\nu > \chi^2_{\nu,\alpha})$, where χ^2_ν is a chi-square random variable with ν degrees of freedom. For example, the probability is 0.100 that a chi-square random variable with 10 degrees of freedom is greater than 15.99.

ν	α									
	.995	.990	.975	.950	.900	.100	.050	.025	.010	.005
1	0.0^4393	0.0^3157	0.0^3982	0.0^2393	0.0158	2.71	3.84	5.02	6.63	7.88
2	0.0100	0.0201	0.0506	0.103	0.211	4.61	5.99	7.38	9.21	10.60
3	0.072	0.115	0.216	0.352	0.584	6.25	7.81	9.35	11.34	12.84
4	0.207	0.297	0.484	0.711	1.064	7.78	9.49	11.14	13.28	14.86
5	0.412	0.554	0.831	1.145	1.61	9.24	11.07	12.83	15.09	16.75
6	0.676	0.872	1.24	1.64	2.20	10.64	12.59	14.45	16.81	18.55
7	0.989	1.24	1.69	2.17	2.83	12.02	14.07	16.01	18.48	20.28
8	1.34	1.65	2.18	2.73	3.49	13.36	15.51	17.53	20.09	21.96
9	1.73	2.09	2.70	3.33	4.17	14.68	16.92	19.02	21.67	23.59
10	2.16	2.56	3.25	3.94	4.87	15.99	18.31	20.48	23.21	25.19
11	2.60	3.05	3.82	4.57	5.58	17.28	19.68	21.92	24.73	26.76
12	3.07	3.57	4.40	5.23	6.30	18.55	21.03	23.34	26.22	28.30
13	3.57	4.11	5.01	5.89	7.04	19.81	22.36	24.74	27.69	29.82
14	4.07	4.66	5.63	6.57	7.79	21.06	23.68	26.12	29.14	31.32
15	4.60	5.23	6.26	7.26	8.55	22.31	25.00	27.49	30.58	32.80
16	5.14	5.81	6.91	7.96	9.31	23.54	26.30	28.85	32.00	34.27
17	5.70	6.41	7.56	8.67	10.09	24.77	27.59	30.19	33.41	35.72
18	6.26	7.01	8.23	9.39	10.86	25.99	28.87	31.53	34.81	37.16
19	6.84	7.63	8.91	10.12	11.65	27.20	30.14	32.85	36.19	38.58
20	7.43	8.26	9.59	10.85	12.44	28.41	31.41	34.17	37.57	40.00
21	8.03	8.90	10.28	11.59	13.24	29.62	32.67	35.48	38.93	41.40
22	8.64	9.54	10.98	12.34	14.04	30.81	33.92	36.78	40.29	42.80
23	9.26	10.20	11.69	13.09	14.85	32.01	35.17	38.08	41.64	44.18
24	9.89	10.86	12.40	13.85	15.66	33.20	36.42	39.36	42.98	45.56
25	10.52	11.52	13.12	14.61	16.47	34.38	37.65	40.65	44.31	46.93
26	11.16	12.20	13.84	15.38	17.29	35.56	38.89	41.92	45.64	48.29
27	11.81	12.88	14.57	16.15	18.11	36.74	40.11	43.19	46.96	49.64
28	12.46	13.56	15.31	16.93	18.94	37.92	41.34	44.46	48.28	50.99
29	13.12	14.26	16.05	17.71	19.77	39.09	42.56	45.72	49.59	52.34
30	13.79	14.95	16.79	18.49	20.60	40.26	43.77	46.98	50.89	53.67
40	20.71	22.16	24.43	26.51	29.05	51.81	55.76	59.34	63.69	66.77
50	27.99	29.71	32.36	34.76	37.69	63.17	67.50	71.42	76.15	79.49
60	35.53	37.48	40.48	43.19	46.46	74.40	79.08	83.30	88.38	91.95
70	43.28	45.44	48.76	51.74	55.33	85.53	90.53	95.02	100.4	104.2
80	51.17	53.54	57.15	60.39	64.28	96.58	101.9	106.6	112.3	116.3
90	59.20	61.75	65.65	69.13	73.29	107.6	113.1	118.1	124.1	128.3
100	67.33	70.06	74.22	77.93	82.36	118.5	124.3	129.6	135.8	140.2

Reproduced with permission from C. M. Thompson, "Tables of percentage points of the chi-square distribution," *Biometrika, 32* (1941).

TABLE 6 Cutoff points for the Student's *t* distribution

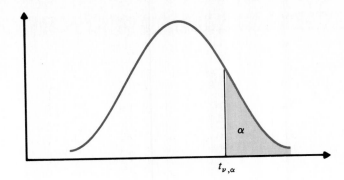

For selected probabilities, α, the table shows the values $t_{\nu,\alpha}$ such that $P(t_\nu > t_{\nu,\alpha}) = \alpha$, where t_ν is a Student's *t* random variable with ν degrees of freedom. For example, the probability is 0.10 that a Student's *t* random variable with 10 degrees of freedom exceeds 1.372.

ν	α				
	.100	.050	.025	.010	.005
1	3.078	6.314	12.706	31.821	63.657
2	1.886	2.920	4.303	6.965	9.925
3	1.638	2.353	3.182	4.541	5.841
4	1.533	2.132	2.776	3.747	4.604
5	1.476	2.015	2.571	3.365	4.032
6	1.440	1.943	2.447	3.143	3.707
7	1.415	1.895	2.365	2.998	3.499
8	1.397	1.860	2.306	2.896	3.355
9	1.383	1.833	2.262	2.821	3.250
10	1.372	1.812	2.228	2.764	3.169
11	1.363	1.796	2.201	2.718	3.106
12	1.356	1.782	2.179	2.681	3.055
13	1.350	1.771	2.160	2.650	3.012
14	1.345	1.761	2.145	2.624	2.977
15	1.341	1.753	2.131	2.602	2.947
16	1.337	1.746	2.120	2.583	2.921
17	1.333	1.740	2.110	2.567	2.898
18	1.330	1.734	2.101	2.552	2.878
19	1.328	1.729	2.093	2.539	2.861
20	1.325	1.725	2.086	2.528	2.845
21	1.323	1.721	2.080	2.518	2.831
22	1.321	1.717	2.074	2.508	2.819
23	1.319	1.714	2.069	2.500	2.807
24	1.318	1.711	2.064	2.492	2.797
25	1.316	1.708	2.060	2.485	2.787
26	1.315	1.706	2.056	2.479	2.779
27	1.314	1.703	2.052	2.473	2.771
28	1.313	1.701	2.048	2.467	2.763

TABLE 6 Cutoff points for the Student's *t* distribution (continued)

ν	α				
	.100	.050	.025	.010	.005
29	1.311	1.699	2.045	2.462	2.756
30	1.310	1.697	2.042	2.457	2.750
40	1.303	1.684	2.021	2.423	2.704
60	1.296	1.671	2.000	2.390	2.660
∞	1.282	1.645	1.960	2.326	2.576

Reproduced with permission of the trustees of Biometrika, from *Biometrika Tables for Statisticians*, Vol. 1 (1966).

TABLE 7 Cutoff points for the F distribution

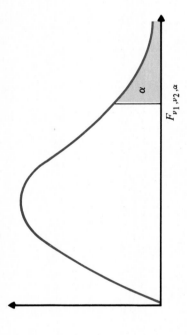

For probabilities $\alpha = 0.05$ and $\alpha = 0.01$, the tables show the values $F_{\nu_1, \nu_2, \alpha}$ such that $P(F_{\nu_1, \nu_2} > F_{\nu_1, \nu_2, \alpha}) = \alpha$, where F_{ν_1, ν_2} is an F random variable, with numerator degrees of freedom ν_1 and denominator degrees of freedom ν_2. For example, the probability is 0.05 that an $F_{3,7}$ random variable exceeds 4.35.

$\alpha = 0.05$

DENOMINATOR ν_2	NUMERATOR ν_1																		
	1	2	3	4	5	6	7	8	9	10	12	15	20	24	30	40	60	120	∞
1	161.4	199.5	215.7	224.6	230.2	234.0	236.8	238.9	240.5	241.9	243.9	245.9	248.0	249.1	250.1	251.1	252.2	253.3	254.3
2	18.51	19.00	19.16	19.25	19.30	19.33	19.35	19.37	19.38	19.40	19.41	19.43	19.45	19.45	19.46	19.47	19.48	19.49	19.50
3	10.13	9.55	9.28	9.12	9.01	8.94	8.89	8.85	8.81	8.79	8.74	8.70	8.66	8.64	8.62	8.59	8.57	8.55	8.53
4	7.71	6.94	6.59	6.39	6.26	6.16	6.09	6.04	6.00	5.96	5.91	5.86	5.80	5.77	5.75	5.72	5.69	5.66	5.63
5	6.61	5.79	5.41	5.19	5.05	4.95	4.88	4.82	4.77	4.74	4.68	4.62	4.56	4.53	4.50	4.46	4.43	4.40	4.36
6	5.99	5.14	4.76	4.53	4.39	4.28	4.21	4.15	4.10	4.06	4.00	3.94	3.87	3.84	3.81	3.77	3.74	3.70	3.67
7	5.59	4.74	4.35	4.12	3.97	3.87	3.79	3.73	3.68	3.64	3.57	3.51	3.44	3.41	3.38	3.34	3.30	3.27	3.23
8	5.32	4.46	4.07	3.84	3.69	3.58	3.50	3.44	3.39	3.35	3.28	3.22	3.15	3.12	3.08	3.04	3.01	2.97	2.93
9	5.12	4.26	3.86	3.63	3.48	3.37	3.29	3.23	3.18	3.14	3.07	3.01	2.94	2.90	2.86	2.83	2.79	2.75	2.71
10	4.96	4.10	3.71	3.48	3.33	3.22	3.14	3.07	3.02	2.98	2.91	2.85	2.77	2.74	2.70	2.66	2.62	2.58	2.54
11	4.84	3.98	3.59	3.36	3.20	3.09	3.01	2.95	2.90	2.85	2.79	2.72	2.65	2.61	2.57	2.53	2.49	2.45	2.40
12	4.75	3.89	3.49	3.26	3.11	3.00	2.91	2.85	2.80	2.75	2.69	2.62	2.54	2.51	2.47	2.43	2.38	2.34	2.30
13	4.67	3.81	3.41	3.18	3.03	2.92	2.83	2.77	2.71	2.67	2.60	2.53	2.46	2.42	2.38	2.34	2.30	2.25	2.21

DENOMINATOR v_2 / **NUMERATOR v_1**

v_2	1	2	3	4	5	6	7	8	9	10	12	15	20	24	30	40	60	120	∞
14	4.60	3.74	3.34	3.11	2.96	2.85	2.76	2.70	2.65	2.60	2.53	2.46	2.39	2.35	2.31	2.27	2.22	2.18	2.13
15	4.54	3.68	3.29	3.06	2.90	2.79	2.71	2.64	2.59	2.54	2.48	2.40	2.33	2.29	2.25	2.20	2.16	2.11	2.07
16	4.49	3.63	3.24	3.01	2.85	2.74	2.66	2.59	2.54	2.49	2.42	2.35	2.28	2.24	2.19	2.15	2.11	2.06	2.01
17	4.45	3.59	3.20	2.96	2.81	2.70	2.61	2.55	2.49	2.45	2.38	2.31	2.23	2.19	2.15	2.10	2.06	2.01	1.96
18	4.41	3.55	3.16	2.93	2.77	2.66	2.58	2.51	2.46	2.41	2.34	2.27	2.19	2.15	2.11	2.06	2.02	1.97	1.92
19	4.38	3.52	3.13	2.90	2.74	2.63	2.54	2.48	2.42	2.38	2.31	2.23	2.16	2.11	2.07	2.03	1.98	1.93	1.88
20	4.35	3.49	3.10	2.87	2.71	2.60	2.51	2.45	2.39	2.35	2.28	2.20	2.12	2.08	2.04	1.99	1.95	1.90	1.84
21	4.32	3.47	3.07	2.84	2.68	2.57	2.49	2.42	2.37	2.32	2.25	2.18	2.10	2.05	2.01	1.96	1.92	1.87	1.81
22	4.30	3.44	3.05	2.82	2.66	2.55	2.46	2.40	2.34	2.30	2.23	2.15	2.07	2.03	1.98	1.94	1.89	1.84	1.78
23	4.28	3.42	3.03	2.80	2.64	2.53	2.44	2.37	2.32	2.27	2.20	2.13	2.05	2.01	1.96	1.91	1.86	1.81	1.76
24	4.26	3.40	3.01	2.78	2.62	2.51	2.42	2.36	2.30	2.25	2.18	2.11	2.03	1.98	1.94	1.89	1.84	1.79	1.73
25	4.24	3.39	2.99	2.76	2.60	2.49	2.40	2.34	2.28	2.24	2.16	2.09	2.01	1.96	1.92	1.87	1.82	1.77	1.71
26	4.23	3.37	2.98	2.74	2.59	2.47	2.39	2.32	2.27	2.22	2.15	2.07	1.99	1.95	1.90	1.85	1.80	1.75	1.69
27	4.21	3.35	2.96	2.73	2.57	2.46	2.37	2.31	2.25	2.20	2.13	2.06	1.97	1.93	1.88	1.84	1.79	1.73	1.67
28	4.20	3.34	2.95	2.71	2.56	2.45	2.36	2.29	2.24	2.19	2.12	2.04	1.96	1.91	1.87	1.82	1.77	1.71	1.65
29	4.18	3.33	2.93	2.70	2.55	2.43	2.35	2.28	2.22	2.18	2.10	2.03	1.94	1.90	1.85	1.81	1.75	1.70	1.64
30	4.17	3.32	2.92	2.69	2.53	2.42	2.33	2.27	2.21	2.16	2.09	2.01	1.93	1.89	1.84	1.79	1.74	1.68	1.62
40	4.08	3.23	2.84	2.61	2.45	2.34	2.25	2.18	2.12	2.08	2.00	1.92	1.84	1.79	1.74	1.69	1.64	1.58	1.51
60	4.00	3.15	2.76	2.53	2.37	2.25	2.17	2.10	2.04	1.99	1.92	1.84	1.75	1.70	1.65	1.59	1.53	1.47	1.39
120	3.92	3.07	2.68	2.45	2.29	2.17	2.09	2.02	1.96	1.91	1.83	1.75	1.66	1.61	1.55	1.50	1.43	1.35	1.25
∞	3.84	3.00	2.60	2.37	2.21	2.10	2.01	1.94	1.88	1.83	1.75	1.67	1.57	1.52	1.46	1.39	1.32	1.22	1.00

TABLE 7 Cutoff points for the F distribution (continued)

$\alpha = 0.01$

NUMERATOR ν_1

DENOMINATOR ν_2	1	2	3	4	5	6	7	8	9	10	12	15	20	24	30	40	60	120	∞
1	4052	4999.5	5403	5625	5764	5859	5928	5982	6022	6056	6106	6157	6209	6235	6261	6287	6313	6339	6366
2	98.50	99.00	99.17	99.25	99.30	99.33	99.36	99.37	99.39	99.40	99.42	99.43	99.45	99.46	99.47	99.47	99.48	99.48	99.50
3	34.12	30.82	29.46	28.71	28.24	27.91	27.67	27.49	27.35	27.23	27.05	26.87	26.69	26.60	26.50	26.41	26.32	26.22	26.13
4	21.20	18.00	16.69	15.98	15.52	15.21	14.98	14.80	14.66	14.55	14.37	14.20	14.02	13.93	13.84	13.75	13.65	13.56	13.46
5	16.26	13.27	12.06	11.39	10.97	10.67	10.46	10.29	10.16	10.05	9.89	9.72	9.55	9.47	9.38	9.29	9.20	9.11	9.02
6	13.75	10.92	9.78	9.15	8.75	8.47	8.26	8.10	7.98	7.87	7.72	7.56	7.40	7.31	7.23	7.14	7.06	6.97	6.88
7	12.25	9.55	8.45	7.85	7.46	7.19	6.99	6.84	6.72	6.62	6.47	6.31	6.16	6.07	5.99	5.91	5.82	5.74	5.65
8	11.26	8.65	7.59	7.01	6.63	6.37	6.18	6.03	5.91	5.81	5.67	5.52	5.36	5.28	5.20	5.12	5.03	4.95	4.86
9	10.56	8.02	6.99	6.42	6.06	5.80	5.61	5.47	5.35	5.26	5.11	4.96	4.81	4.73	4.65	4.57	4.48	4.40	4.31
10	10.04	7.56	6.55	5.99	5.64	5.39	5.20	5.06	4.94	4.85	4.71	4.56	4.41	4.33	4.25	4.17	4.08	4.00	3.91
11	9.65	7.21	6.22	5.67	5.32	5.07	4.89	4.74	4.63	4.54	4.40	4.25	4.10	4.02	3.94	3.86	3.78	3.69	3.60
12	9.33	6.93	5.95	5.41	5.06	4.82	4.64	4.50	4.39	4.30	4.16	4.01	3.86	3.78	3.70	3.62	3.54	3.45	3.36
13	9.07	6.70	5.74	5.21	4.86	4.62	4.44	4.30	4.19	4.10	3.96	3.82	3.66	3.59	3.51	3.43	3.34	3.25	3.17
14	8.86	6.51	5.56	5.04	4.69	4.46	4.28	4.14	4.03	3.94	3.80	3.66	3.51	3.43	3.35	3.27	3.18	3.09	3.00
15	8.68	6.36	5.42	4.89	4.56	4.32	4.14	4.00	3.89	3.80	3.67	3.52	3.37	3.29	3.21	3.13	3.05	2.96	2.87
16	8.53	6.23	5.29	4.77	4.44	4.20	4.03	3.89	3.78	3.69	3.55	3.41	3.26	3.18	3.10	3.02	2.93	2.84	2.75
17	8.40	6.11	5.18	4.67	4.34	4.10	3.93	3.79	3.68	3.59	3.46	3.31	3.16	3.08	3.00	2.92	2.83	2.75	2.65
18	8.29	6.01	5.09	4.58	4.25	4.01	3.84	3.71	3.60	3.51	3.37	3.23	3.08	3.00	2.92	2.84	2.75	2.66	2.57
19	8.18	5.93	5.01	4.50	4.17	3.94	3.77	3.63	3.52	3.43	3.30	3.15	3.00	2.92	2.84	2.76	2.67	2.58	2.49
20	8.10	5.85	4.94	4.43	4.10	3.87	3.70	3.56	3.46	3.37	3.23	3.09	2.94	2.86	2.78	2.69	2.61	2.52	2.42
21	8.02	5.78	4.87	4.37	4.04	3.81	3.64	3.51	3.40	3.31	3.17	3.03	2.88	2.80	2.72	2.64	2.55	2.46	2.36
22	7.95	5.72	4.82	4.31	3.99	3.76	3.59	3.45	3.35	3.26	3.12	2.98	2.83	2.75	2.67	2.58	2.50	2.40	2.31
23	7.88	5.66	4.76	4.26	3.94	3.71	3.54	3.41	3.30	3.21	3.07	2.93	2.78	2.70	2.62	2.54	2.45	2.35	2.26
24	7.82	5.61	4.72	4.22	3.90	3.67	3.50	3.36	3.26	3.17	3.03	2.89	2.74	2.66	2.58	2.49	2.40	2.31	2.21
25	7.77	5.57	4.68	4.18	3.85	3.63	3.46	3.32	3.22	3.13	2.99	2.85	2.70	2.62	2.54	2.45	2.36	2.27	2.17
26	7.72	5.53	4.64	4.14	3.82	3.59	3.42	3.29	3.18	3.09	2.96	2.81	2.66	2.58	2.50	2.42	2.33	2.23	2.13
27	7.68	5.49	4.60	4.11	3.78	3.56	3.39	3.26	3.15	3.06	2.93	2.78	2.63	2.55	2.47	2.38	2.29	2.20	2.10
28	7.64	5.45	4.57	4.07	3.75	3.53	3.36	3.23	3.12	3.03	2.90	2.75	2.60	2.52	2.44	2.35	2.26	2.17	2.06
29	7.60	5.42	4.54	4.04	3.73	3.50	3.33	3.20	3.09	3.00	2.87	2.73	2.57	2.49	2.41	2.33	2.23	2.14	2.03
30	7.56	5.39	4.51	4.02	3.70	3.47	3.30	3.17	3.07	2.98	2.84	2.70	2.55	2.47	2.39	2.30	2.21	2.11	2.01
40	7.31	5.18	4.31	3.83	3.51	3.29	3.12	2.99	2.89	2.80	2.66	2.52	2.37	2.29	2.20	2.11	2.02	1.92	1.80
60	7.08	4.98	4.13	3.65	3.34	3.12	2.95	2.82	2.72	2.63	2.50	2.35	2.20	2.12	2.03	1.94	1.84	1.73	1.60
120	6.85	4.79	3.95	3.48	3.17	2.96	2.79	2.66	2.56	2.47	2.34	2.19	2.03	1.95	1.86	1.76	1.66	1.53	1.38
∞	6.63	4.61	3.78	3.32	3.02	2.80	2.64	2.51	2.41	2.32	2.18	2.04	1.88	1.79	1.70	1.59	1.47	1.32	1.00

TABLE 8 Cutoff points for the distribution of the Wilcoxon test statistic

For sample size n, the table shows, for selected probabilities, α, the numbers T_α such that $P(T \le T_\alpha) = \alpha$, where the distribution of the random variable T is that of the Wilcoxon test statistic under the null hypothesis.

n	α				
	.005	.010	.025	.050	.100
4	0	0	0	0	1
5	0	0	0	1	3
6	0	0	1	3	4
7	0	1	3	4	6
8	1	2	4	6	9
9	2	4	6	9	11
10	4	6	9	11	15
11	6	8	11	14	18
12	8	10	14	18	22
13	10	13	18	22	27
14	13	16	22	26	32
15	16	20	26	31	37
16	20	24	30	36	43
17	24	28	35	42	49
18	28	33	41	48	56
19	33	38	47	54	63
20	38	44	53	61	70

Reproduced with permission from R. L. McCornack, "Extended tables of the Wilcoxon matched pairs signed rank statistics," *Journal of American Statistical Association, 60* (1965).

TABLE 9 Cutoff points for the distribution of Spearman's rank correlation coefficient

For sample size n, the table shows, for selected probabilities, α, the numbers $r_{s,\alpha}$ such that $P(r_s > r_{s,\alpha}) = \alpha$, where the distribution of the random variable r_s is that of Spearman's rank correlation coefficient under the null hypothesis of no association.

n	α			
	.050	.025	.010	.005
5	0.900	—	—	—
6	0.829	0.886	0.943	—
7	0.714	0.786	0.893	—
8	0.643	0.738	0.833	0.881
9	0.600	0.683	0.783	0.833
10	0.564	0.648	0.745	0.794
11	0.523	0.623	0.736	0.818
12	0.497	0.591	0.703	0.780
13	0.475	0.566	0.673	0.745
14	0.457	0.545	0.646	0.716
15	0.441	0.525	0.623	0.689

TABLE 9 Cutoff points for the distribution of Spearman's rank correlation coefficient (continued)

n	α			
	.050	.025	.010	.005
16	0.425	0.507	0.601	0.666
17	0.412	0.490	0.582	0.645
18	0.399	0.476	0.564	0.625
19	0.388	0.462	0.549	0.608
20	0.377	0.450	0.534	0.591
21	0.368	0.438	0.521	0.576
22	0.359	0.428	0.508	0.562
23	0.351	0.418	0.496	0.549
24	0.343	0.409	0.485	0.537
25	0.336	0.400	0.475	0.526
26	0.329	0.392	0.465	0.515
27	0.323	0.385	0.456	0.505
28	0.317	0.377	0.448	0.496
29	0.311	0.370	0.440	0.487
30	0.305	0.364	0.432	0.478

Reproduced with permission from E. G. Olds, "Distribution of sums of squares of rank differences for small samples," *Annals of Mathematical Statistics, 9* (1938).

TABLE 10 Cutoff points for the distribution of the Durbin-Watson test statistic

Let d_α be that number such that $P(d < d_\alpha) = \alpha$, where the random variable d has the distribution of the Durbin-Watson statistic under the null hypothesis of no autocorrelation in the regression errors. For probabilities $\alpha = 0.05$ and $\alpha = 0.01$, the tables show, for numbers of independent variables, K, values d_L and d_U such that $d_L \leq d_\alpha \leq d_U$, for numbers n of observations.

$$\alpha = 0.05$$

n	\multicolumn{2}{c}{1}		\multicolumn{2}{c}{2}		\multicolumn{2}{c}{3}		\multicolumn{2}{c}{4}		\multicolumn{2}{c}{5}	
	d_L	d_U	d_L	d_U	d_L	d_U	d_L	d_U	d_L	d_U
15	1.08	1.36	0.95	1.54	0.82	1.75	0.69	1.97	0.56	2.21
16	1.10	1.37	0.98	1.54	0.86	1.73	0.74	1.93	0.62	2.15
17	1.13	1.38	1.02	1.54	0.90	1.71	0.78	1.90	0.67	2.10
18	1.16	1.39	1.05	1.53	0.93	1.69	0.82	1.87	0.71	2.06
19	1.18	1.40	1.08	1.53	0.97	1.68	0.86	1.85	0.75	2.02
20	1.20	1.41	1.10	1.54	1.00	1.68	0.90	1.83	0.79	1.99
21	1.22	1.42	1.13	1.54	1.03	1.67	0.93	1.81	0.83	1.96
22	1.24	1.43	1.15	1.54	1.05	1.66	0.96	1.80	0.86	1.94
23	1.26	1.44	1.17	1.54	1.08	1.66	0.99	1.79	0.90	1.92
24	1.27	1.45	1.19	1.55	1.10	1.66	1.01	1.78	0.93	1.90
25	1.29	1.45	1.21	1.55	1.12	1.66	1.04	1.77	0.95	1.89
26	1.30	1.46	1.22	1.55	1.14	1.65	1.06	1.76	0.98	1.88
27	1.32	1.47	1.24	1.56	1.16	1.65	1.08	1.76	1.01	1.86
28	1.33	1.48	1.26	1.56	1.18	1.65	1.10	1.75	1.03	1.85
29	1.34	1.48	1.27	1.56	1.20	1.65	1.12	1.74	1.05	1.84
30	1.35	1.49	1.28	1.57	1.21	1.65	1.14	1.74	1.07	1.83
31	1.36	1.50	1.30	1.57	1.23	1.65	1.16	1.74	1.09	1.83
32	1.37	1.50	1.31	1.57	1.24	1.65	1.18	1.73	1.11	1.82
33	1.38	1.51	1.32	1.58	1.26	1.65	1.19	1.73	1.13	1.81
34	1.39	1.51	1.33	1.58	1.27	1.65	1.21	1.73	1.15	1.81
35	1.40	1.52	1.34	1.58	1.28	1.65	1.22	1.73	1.16	1.80
36	1.41	1.52	1.35	1.59	1.29	1.65	1.24	1.73	1.18	1.80
37	1.42	1.53	1.36	1.59	1.31	1.66	1.25	1.72	1.19	1.80
38	1.43	1.54	1.37	1.59	1.32	1.66	1.26	1.72	1.21	1.79
39	1.43	1.54	1.38	1.60	1.33	1.66	1.27	1.72	1.22	1.79
40	1.44	1.54	1.39	1.60	1.34	1.66	1.29	1.72	1.23	1.79
45	1.48	1.57	1.43	1.62	1.38	1.67	1.34	1.72	1.29	1.78
50	1.50	1.59	1.46	1.63	1.42	1.67	1.38	1.72	1.34	1.77
55	1.53	1.60	1.49	1.64	1.45	1.68	1.41	1.72	1.38	1.77
60	1.55	1.62	1.51	1.65	1.48	1.69	1.44	1.73	1.41	1.77
65	1.57	1.63	1.54	1.66	1.50	1.70	1.47	1.73	1.44	1.77
70	1.58	1.64	1.55	1.67	1.52	1.70	1.49	1.74	1.46	1.77
75	1.60	1.65	1.57	1.68	1.54	1.71	1.51	1.74	1.49	1.77
80	1.61	1.66	1.59	1.69	1.56	1.72	1.53	1.74	1.51	1.77
85	1.62	1.67	1.60	1.70	1.57	1.72	1.55	1.75	1.52	1.77
90	1.63	1.68	1.61	1.70	1.59	1.73	1.57	1.75	1.54	1.78
95	1.64	1.69	1.62	1.71	1.60	1.73	1.58	1.75	1.56	1.78
100	1.65	1.69	1.63	1.72	1.61	1.74	1.59	1.76	1.57	1.78

$$\alpha = 0.01$$

n	K									
	1		2		3		4		5	
	d_L	d_U	d_L	d_U	d_L	d_U	d_L	d_U	d_L	d_U
15	0.81	1.07	0.70	1.25	0.59	1.46	0.49	1.70	0.39	1.96
16	0.84	1.09	0.74	1.25	0.63	1.44	0.53	1.66	0.44	1.90
17	0.87	1.10	0.77	1.25	0.67	1.43	0.57	1.63	0.48	1.85
18	0.90	1.12	0.80	1.26	0.71	1.42	0.61	1.60	0.52	1.80
19	0.93	1.13	0.83	1.26	0.74	1.41	0.65	1.58	0.56	1.77
20	0.95	1.15	0.86	1.27	0.77	1.41	0.68	1.57	0.60	1.74
21	0.97	1.16	0.89	1.27	0.80	1.41	0.72	1.55	0.63	1.71
22	1.00	1.17	0.91	1.28	0.83	1.40	0.75	1.54	0.66	1.69
23	1.02	1.19	0.94	1.29	0.86	1.40	0.77	1.53	0.70	1.67
24	1.04	1.20	0.96	1.30	0.88	1.41	0.80	1.53	0.72	1.66
25	1.05	1.21	0.98	1.30	0.90	1.41	0.83	1.52	0.75	1.65
26	1.07	1.22	1.00	1.31	0.93	1.41	0.85	1.52	0.78	1.64
27	1.09	1.23	1.02	1.32	0.95	1.41	0.88	1.51	0.81	1.63
28	1.10	1.24	1.04	1.32	0.97	1.41	0.90	1.51	0.83	1.62
29	1.12	1.25	1.05	1.33	0.99	1.42	0.92	1.51	0.85	1.61
30	1.13	1.26	1.07	1.34	1.01	1.42	0.94	1.51	0.88	1.61
31	1.15	1.27	1.08	1.34	1.02	1.42	0.96	1.51	0.90	1.60
32	1.16	1.28	1.10	1.35	1.04	1.43	0.98	1.51	0.92	1.60
33	1.17	1.29	1.11	1.36	1.05	1.43	1.00	1.51	0.94	1.59
34	1.18	1.30	1.13	1.36	1.07	1.43	1.01	1.51	0.95	1.59
35	1.19	1.31	1.14	1.37	1.08	1.44	1.03	1.51	0.97	1.59
36	1.21	1.32	1.15	1.38	1.10	1.44	1.04	1.51	0.99	1.59
37	1.22	1.32	1.16	1.38	1.11	1.45	1.06	1.51	1.00	1.59
38	1.23	1.33	1.18	1.39	1.12	1.45	1.07	1.52	1.02	1.58
39	1.24	1.34	1.19	1.39	1.14	1.45	1.09	1.52	1.03	1.58
40	1.25	1.34	1.20	1.40	1.15	1.46	1.10	1.52	1.05	1.58
45	1.29	1.38	1.24	1.42	1.20	1.48	1.16	1.53	1.11	1.58
50	1.32	1.40	1.28	1.45	1.24	1.49	1.20	1.54	1.16	1.59
55	1.36	1.43	1.32	1.47	1.28	1.51	1.25	1.55	1.21	1.59
60	1.38	1.45	1.35	1.48	1.32	1.52	1.28	1.56	1.25	1.60
65	1.41	1.47	1.38	1.50	1.35	1.53	1.31	1.57	1.28	1.61
70	1.43	1.49	1.40	1.52	1.37	1.55	1.34	1.58	1.31	1.61
75	1.45	1.50	1.42	1.53	1.39	1.56	1.37	1.59	1.34	1.62
80	1.47	1.52	1.44	1.54	1.42	1.57	1.39	1.60	1.36	1.62
85	1.48	1.53	1.46	1.55	1.43	1.58	1.41	1.60	1.39	1.63
90	1.50	1.54	1.47	1.56	1.45	1.59	1.43	1.61	1.41	1.64
95	1.51	1.55	1.49	1.57	1.47	1.60	1.45	1.62	1.42	1.64
100	1.52	1.56	1.50	1.58	1.48	1.60	1.46	1.63	1.44	1.65

Reproduced with permission from J. Durbin and G. S. Watson, "Testing for serial correlation in least squares regression, II," *Biometrika, 38* (1951).

TABLE 11 Cumulative distribution function of the runs test statistic

For a given number, n, of observations, the table shows the probability, for a random time series, that the number of runs will not exceed K.

n	2	3	4	5	6	7	8	9	10	11	12	13	14	15	16	17	18	19	20
6	.100	.300	.700	.900	1.000														
8	.029	.114	.371	.629	.886	.971	1.000												
10	.008	.040	.167	.357	.643	.833	.960	.992	1.000										
12	.002	.013	.067	.175	.392	.608	.825	.933	.987	.998	1.000								
14	.001	.004	.025	.078	.209	.383	.617	.791	.922	.975	.996	.999	1.000						
16	.000	.001	.009	.032	.100	.214	.405	.595	.786	.900	.968	.991	.999	1.000					
18	.000	.000	.003	.012	.044	.109	.238	.399	.601	.762	.891	.956	.988	.997	1.000	1.000			
20	.000	.000	.001	.004	.019	.051	.128	.242	.414	.586	.758	.872	.949	.981	.996	.999	1.000	1.000	1.000

Reproduced with permission from F. Swed and C. Eisenhart, "Tables for testing randomness of grouping in a sequence of alternatives," Annals of Mathematical Statistics, 14 (1943).

answers to selected even-numbered exercises

CHAPTER 2

2. (a) 29.25 (b) 29.0
4. (a) 4.818 (b) 3 (c) 2
6. (a) 15,660 (b) 15,800
8. (a) 7.81 (b) 6.3 (c) Calculate the weighted average as $(0.119)(6.0) + \cdots + (0.024)(12.5) = 6.2882$.
10. (a) $s^2 = 117.64$, $s = 10.85$ (b) 8.75 (c) 32 (d) First quartile = 20.25, third quartile = 38.75; interquartile range = 18.5
12. (a) $s^2 = 66.96$, $s = 8.18$ (b) 5.01 (c) First quartile = 2, third quartile = 5; inter-quartile range = 3
14. (a) $\sigma^2 = 4,894,400$ (b) 6,200
16. (a) 34.82 (b) 34.25 (c) $s^2 = 253.50$, $s = 15.92$ (d) First quartile = 19.275, third quartile = 50.100; interquartile range = 30.825
18. (a) At least 84% in range $82.50 to $287.50 (b) At least 50% in range $127.02 to $242.98
20. Safeco Growth Fund: $\sigma = 5.133$; General Motors: $\sigma = 5.205$
26. (a) 2.8 (b) 3 (c) 5 (d) $\sigma^2 = 3.6$, $\sigma = 1.897$
32. (b) 0.108, 0.267, 0.425, 0.200 (c) 0.108, 0.375, 0.800, 1 (d) 44.33 (e) $\sigma^2 = 327.89$, $\sigma = 18.11$ (f) 45.88 (g) First quartile = 30.47, Third quartile = 57.75, Interquartile range = 27.28 (h) 40–60
34. (b) 0.30, 0.20, 0.35, 0.10, 0.05 (c) 0.30, 0.50, 0.85, 0.95, 1 (d) 3.8 (e) $s^2 = 5.64$, $s = 2.38$ (f) 3.95 (g) First quartile = 1.58, Third quartile = 5.50, Interquartile range = 3.92 (h) 4–6
36. (b) 0.10, 0.17, 0.40, 0.20, 0.13 (c) 0.10, 0.27, 0.67, 0.87,

1   **(d)** $1,040   **(e)** $s = \$462.1$ **(f)** 1,033.3   **(g)** First quartile = 740.0,   Third quartile = 1,383.3, Interquartile range = 643.3   **(h)** 800–1,200

38.   **(b)** $27,050   **(c)** $\sigma = \$13,264$ **(d)** 23,571

70.   **(a)** 135.6   **(b)** 136   **(c)** $s^2 = 12.28$ **(d)** $s = 3.50$ **(e)** First quartile = 132.5, Third quartile = 138.0; Interquartile range = 5.5

CHAPTER 3

2.   **(a)** Yes   **(b)** Yes, provided we take the probability that the value remains unchanged to be zero.   **(c)** $P(A) = 0.8$, $P(B) = 0.2$ **(d)** 0.5

4.   **(a)** 0.12   **(b)** 0.88

6.   **(a)** 0.58   **(b)** 0.22

8.   **(a)** 24   **(b)** 0.25

10.   2,450

12.   90

14. $\frac{1}{15}$

16.   **(a)** 150   **(b)** $\frac{40}{150} = 0.27$ **(c)** $\frac{30}{150} = 0.2$

18.   0.45

20.   0.05

22. $(.98)(.99)(.95) = 0.922$

24.   **(a)** 0.625   **(b)** 0.357   **(c)** No

26.   0.02

28.   **(a)** 0.4725   **(b)** 0.1005

30.   0.1

32.   **(a)** 0.9   **(b)** 0.88   **(c)** 0.925

34.   **(a)** 0.10   **(b)** 0.67   **(c)** No   **(d)** 0.25   **(e)** No   **(f)** 0.80   **(g)** 0.30   **(h)** 0.90

36.   **(a)** 0.10   **(b)** 0.10   **(c)** No   **(d)** 0.20

38.   **(a)** 0.30   **(b)** 0.38   **(c)** 0.605   **(d)** 0.767   **(e)** 0.033

40.   **(a)** 0.30   **(b)** 0.319   **(c)** No

42.   **(a)** 0.02   **(b)** 0.56   **(c)** 0.536

44.   **(a)** 0.16   **(b)** 0.44   **(c)** 0.40   **(d)** 0.067

46.   **(a)** 0.889   **(b)** 0.6875   **(c)** 0.826   **(d)** 0.367

48.   **(a)** $(0.9057)^3 = 0.743$ **(b)** $1 - (0.0943)^3 = 0.9992$

52.   **(a)** True   **(b)** True   **(c)** True   **(d)** True   **(e)** False   **(f)** False   **(g)** False

54.   **(a)** False   **(b)** False   **(c)** True   **(d)** False   **(e)** False

56.   **(a)** 0.9   **(b)** 0.5   **(c)** $1 - (0.9)^3 = 0.271$

58.   0.833

60.   **(a)** $20!/12!\ 8! = 125,970$

      (b) $1 - \dfrac{(10!/8!\ 2!) + (10!/7!\ 3!)(10!/9!\ 1!)}{125,970} = 0.990$

62.   0.05

64.   **(a)** 0.40   **(b)** 0.31   **(c)** 0.20   **(d)** 0.258   **(e)** No

66.   **(a)** 0.5192   **(b)** 0.6482   **(c)** Negative

68.   0.483

70.   **(a)** 0.5   **(b)** 0.45

CHAPTER 4

2. **(c)** 0.83
4. **(c)** 0.78
6. $P_X(0) = {}^{28}\!/_{45}$, $P_X(1) = {}^{16}\!/_{45}$, $P_X(2) = {}^{1}\!/_{45}$; the outcome for the second lightbulb chosen will not be independent of that for the first.
8. $\mu = 1.25$, $\sigma^2 = 1.0075$
10. $\mu = \$70.7$, $\sigma = \$5.745$
12. $\mu = \$0.65$, $\sigma = \$63.28$
14. **(a)** $\mu = 16.3$, $\sigma = 0.9$ **(b)** Mean = 2.55 cents, standard deviation = 1.35 cents
16. **(a)** $\mu = 1.75$, $\sigma = 0.994$ **(b)** Mean = \$1,750, standard deviation = \$994
18. Strategy 1 has highest expected profit (\$650). However, investor should also consider risk in choosing strategy.
20. **(a)** 0.54 **(b)** $P_{X|Y}(0|3) = {}^{1}\!/_{6}$, $P_{X|Y}(1|3) = {}^{5}\!/_{12}$, $P_{X|Y}(2|3) = {}^{5}\!/_{12}$ **(c)** $P_{Y|X}(1|0) = {}^{10}\!/_{19}$; $P_{Y|X}(2|0) = {}^{5}\!/_{19}$, $P_{Y|X}(3|0) = {}^{4}\!/_{19}$ **(d)** 0.1091 **(e)** No
22. **(a)** $P_X(0) = 0.70$, $P_X(1) = 0.30$; $P_Y(0) = 0.52$, $P_Y(1) = 0.48$ **(b)** $P_{Y|X}(0|1) = 0.40$, $P_{Y|X}(1|1) = 0.60$ **(c)** No
24. Because of independence, can obtain joint probabilities by multiplying marginal probabilities, so that $P(0, 0) = 0.005$, and so on.
26. $\mu = 902.5$, $\sigma = 36.6$
28. **(a)** 0.00032 **(b)** 0.32768 **(c)** 0.26272
30. **(a)** 0.5904 **(b)** 0.0272
32. **(a)** 0.07776 **(b)** 0.31744
34. 0.032
36. **(a)** 0.1143 **(b)** Mean = \$810, standard deviation = \$270
38. Mean = 40, standard deviation = 4.899
40. **(a)** 0.8290 **(b)** 0.5490 **(c)** 0.1671
42. 0.875
44. **(a)** 0.272 **(b)** 0.550 **(c)** 0.250
46. 0.9762
48. 0.456
50. **(a)** 0.091 **(b)** 0.221
52. 0.433
54. 0.442
60. **(a)** 10 **(b)** 11.62
62. **(a)** $P(0) = 0.21$, $P(1) = 0.20$, $P(2) = 0.24$, $P(3) = 0.35$ **(b)** $P(0) = 0.20$, $P(1) = 0.17$, $P(2) = 0.19$, $P(3) = 0.20$, $P(4) = 0.24$ **(c)** 1.73 **(d)** 1.148 **(e)** 2.11 years **(f)** 1.455 years **(g)** 0.5597 **(h)** 0.737 **(i)** 3.46 **(ii)** 1.623 **(iii)** 0.9559
64. **(a)** 0.671 **(b)** 0.082 **(c)** 2.1 **(d)** Only that it is at least 2.1
66. **(a)** 0.323 **(b)** \$200 **(c)** \$134.16
68. The probability of doing as well as, or better than, the analyst by random selection is 0.167.
70. 0.433
72. 0.9596

CHAPTER 5

2. **(b)** $F_X(x) = 0.2x$, for $0 < x < 5$ **(c)** 0.2 **(d)** 0.4 **(e)** 0.2
4. 0.3
6. $\mu = \$180,000$, $\sigma = \$10,000$
8. $\mu = \$30,000$, $\sigma = \$7,500$
10. $\mu = \$1,000$, $\sigma = \$1,000$
12. $\mu = \$50,000$, $\sigma = \$6,481$
14. Mean $= -0.6$, standard deviation $= 1$
16. **(a)** 0.9082 **(b)** 0.1151 **(c)** 0.1587 **(d)** 0.9641 **(e)** 0.0233 **(f)** 0.8723
    (g) 0.1228
18. **(a)** 0.9082 **(b)** 0.2514 **(c)** 0.3707 **(d)** 0.8413 **(e)** 0.1596 **(f)** 0.7495
    (g) 0.2120
20. **(a)** 0.0228 **(b)** 0.0228 **(c)** 0.9544
22. **(a)** 0.7881 **(b)** 0.8849 **(c)** 0.5762 **(d)** 1,100–1,300 hours
24. **(a)** 0.9332 **(b)** 0.9332 **(d)** 0.1873
26. **(a)** 0.0062 **(b)** 0.0918 **(c)** 0.0974
28. The probability (0.9332) is higher for supplier A.
30. **(a)** 14.08 **(b)** 17.26
32. 72.08 or higher
34. 1.284
36. **(a)** 0.0475 **(b)** 0.31
38. **(a)** 0.97 **(b)** 0.0067
40. **(a)** 0.802 **(b)** 0.548 **(c)** 0.509
42. **(a)** 0.123 **(b)** 0.221 **(c)** 0.803
44. **(a)** Without correction, 0.683; with correction, 0.729 **(b)** Without correction 0.499; with correction, 0.539
50. **(b)** $F_X(x) = 0.5(x - 1)$, for $1 < x < 3$ **(c)** 0.5 **(d)** 0.875
54. **(a)** $662.50 **(b)** $86.25 **(c)** Mean $= \$165$, standard deviation $= \$34.50$
58. 0.86 hour
60. **(a)** 0.0764 **(b)** 24.896 **(c)** 0.522
62. Brand A (probability is 0.7881)
64. Virtually 1.0
66. 0.007 (using continuity correction)

CHAPTER 6

2. **(a)** 1,200 hours **(b)** 10,000 **(c)** 100 hours **(d)** 0.0668
4. **(a)** 0.6 mile **(b)** 0.2514 **(c)** 0.1587 **(d)** 0.9664 **(e)** Lower; lower; higher
6. **(a)** 0.8944 **(b)** 0.9938 **(c)** 0.9876 **(d)** 0.894
8. **(a)** $2 **(b)** 0.1587 **(c)** 0.0668 **(d)** 0.0456
10. **(a)** 0.20 gram **(b)** 0.15 gram **(c)** 0.17 gram
12. **(a)** 465 observations **(b)** Smaller **(c)** Larger
14. **(a)** Virtually 1.0 **(b)** 0.9616 **(c)** 0.8414 **(d)** 0.0046
16. **(a)** 0.2586 **(b)** 0.4972 **(c)** Higher

18. 0.6853
20. (a) 0.2 (b) 0.0013 (c) 0.0365 (d) 0.0853
22. (a) 0.015 (b) 0.0228 (c) 0.0918 (d) 0.7486 (e) Higher; higher; lower
24. (a) 0.4648 (b) 0.2354 (c) 0.4455
26. 0.05
28. (a) 0.047 (b) 0.060 (c) 0.042
30. 0.0934
32. (a) 0.03468 (b) 0.8749 (c) 0.616 (d) 0.594
34. 0.087
36. (a) No (b) No
38. (a) 1507.5 (b) 307.1
40. (a) $3,943 (b) $6,404
42. (a) 169% (b) 47%
44. (a) 171% (b) One possibility is $a = 24$ and $b = 229$.
46. The probability is bigger than 0.10.
52. The probability is higher for the sample of sixteen students.
54. (a) 0.030 (b) 0.056 (c) 0.134 (d) Higher; higher; higher
56. 0.95
58. (a) 0.916 (b) 0.755 (c) 0.549
60. (a) 0.574 (b) 0.640 (c) 0.032
62. (a) More than 0.10 (b) More than 0.10
64. Between 0.05 and 0.10

CHAPTER 7

2. (a) $\bar{x} = 4.83$, $s_x^2 = 26.57$, $s_x = 5.15$ (b) Mean and variance (c) 4.43
4. (b) $\bar{X}$ (c) 1.25
6. (a) $1,500 (b) $\sigma_1^2/15 + \sigma_2^2/12$ (c) 283,500
8. (a) 0.06 (b) $p_1(1 - p_1)/100 + p_2(1 - p_2)/200$
10. 0.456
12. 0.7
14. $\hat{p}(1 - \hat{p})/(n - 1)$, where $\hat{p} = X/n$
22. (a) 33.7 (b) 14.23 (c) 1.423 (d) 0.5 (e) 0.028
26. (a) 1.32 (b) 0.57
28. (b) $3(n + 1)/2(2n + 1)$

CHAPTER 8

2. (a) $4.162 < \mu < 4.318$ (b) Wider
4. (a) $19.800 < \mu < 20.020$ (b) Wider
6. $11.33 < \mu < 13.23$
8. $1.141 < \mu < 1.339$; $1.122 < \mu < 1.358$
10. $12.12 < \mu < 24.18$
12. $11.7 < \mu < 27.3$
14. $24.31 < \mu < 26.69$

16. $12.16 < \mu < 20.29$
18. $0.418 < p < 0.582$
20. $0.569 < p < 0.729$
22. 0.9676
24. $0.128 < p < 0.194; \quad 0.118 < p < 0.204$
28. $2,421,600 < \sigma^2 < 25,916,095$
30. (a) $2.99 < \sigma^2 < 13.85$ (b) Wider
32. $0.39 < \sigma^2 < 2.23$
34. $27.7 < \mu_X - \mu_Y < 48.3$
36. $1.06 < \mu_X - \mu_Y < 2.28$
38. $0.03 < \mu_X - \mu_Y < 2.37$
40. $-9.13 < \mu_X - \mu_Y < -1.51$
42. $1.33 < \mu_X - \mu_Y < 10.07$
44. $0.144 < p_X - p_Y < 0.376$
46. 0.4314
48. 246
50. 664
52. 1,037
54. No
58. (a) $8.87 < \mu < 9.87$ (b) Wider
60. $63.93 < \mu < 147.85$
62. (a) $0.441 < p < 0.517$ (b) 0.8472
64. (a) $0.43 < p < 0.56$ (b) Wider
66. $0.763 < \sigma < 1.584$, assuming normal population
68. $-0.04 < \mu_X - \mu_Y < 8.04$
70. $-1.86 < \mu_X - \mu_Y < 36.88; \quad -5.56 < \mu_X - \mu_Y < 40.58$
72. $-86.6 < \mu_X - \mu_Y < 5.5$, assuming normal populations with equal variances
74. $-0.10 < p_X - p_Y < 0.31$
76. 601

CHAPTER 9

2. 0.0668
4. Test statistic is -0.73; accept H_0 at 5% level; p-value is 0.4654
6. No
8. 0.0872
10. Test statistic is -1.907; can reject H_0 at 10%, but not at 5% level
12. (a) No (b) Yes
14. Test statistic is 1.93; can reject H_0 at 10%, but not at 5% level
16. Test statistic is 9.77; accept H_0
18. Test statistic is 13.8; can reject H_0 at 10%, but not at 5% level
20. Test statistic is 46.7; can reject H_0 at 5%, but not at 2% level
22. Test statistic is -1; lowest level for rejection is 0.3174 (31.74%)
24. Test statistic is -4.24; can reject H_0 at virtually any level
26. Test statistic is -2.78; lowest level for rejection is 0.0027 (0.27%)

28. Test statistic is 2.22; can reject H_0 at 5%, but not at 2.5% level
30. Test statistic is 1.68; can reject H_0 at 10%, but not at 5% level
32. Test statistic is -0.30; lowest level for rejection is 0.7642 (76.42%)
34. Test statistic is -1.299; cannot reject H_0 at 20% level
36. Test statistic is 1.282; cannot reject H_0 at 10% level
38. Test statistic is 1.41; lowest level for rejection is 0.1586 (15.86%)
40. Test statistic is 0.16; lowest level for rejection is 0.4364 (43.64%)
42. Test statistic is -4.55; can reject H_0 at virtually any level
44. Test statistic is 1.96; cannot reject H_0 at 10% level
46. Test statistic is 2.10; cannot reject H_0 at 10% level
48. 0.508
50. 0.095
52. 0.348
54. **(a)** 0.0047 **(b)** 0.0968 **(c)** 0.7257
60. Test statistic is -1.41; lowest level for rejection is 0.0793 (7.93%)
62. **(a)** Test statistic is -3.82; reject H_0 **(b)** Test statistic is 16.8; accept H_0
64. Test statistic is 1.48; reject H_0
68. Test statistic is 1.938; can reject H_0 at 10%, but not at 5% level
70. Test statistic is -5.16; can reject H_0 at virtually any level
72. Test statistic is 0.92; cannot reject H_0 at 20% level
74. Test statistic is 0.18; cannot reject H_0 at 20% level
76. 0.8572
78. No; the samples are not independent.

CHAPTER 10

2. Can reject H_0 against one-sided alternative at 5.5%, but not at 1.1% level
4. Test statistic is 0.82; lowest level for rejection is 0.2061 (20.61%)
6. Test statistic is -0.85; lowest level for rejection, against two-sided alternative, is 0.3954 (39.54%)
8. Test statistic is 11.5; cannot reject H_0 at 5% level
10. Test statistic is -0.15; lowest level for rejection , against two-sided alternative, is 0.8808 (88.08%)
12. Test statistic is -0.26; lowest level for rejection, against two-sided alternative, is 0.7948 (79.48%)
14. Test statistic is 1.81; lowest level for rejection, against two-sided alternative, is 0.0702 (7.02%)
16. Test statistic is 3.19; lowest level for rejection is 0.0007 (0.07%)
18. Test statistic is -4.85; can reject H_0 at virtually any level
22. Can reject, against two-sided alternative, null hypothesis that views are evenly balanced, at the 58.1% level, but not at the 26.7% level
24. Can reject null hypothesis, against this one-sided alternative, at significance levels above 0.0033 (0.33%)
26. Test statistic is 10.5; cannot reject H_0 against two-sided alternative at 10% level
28. Test statistic is -2.87; can reject H_0 against one-sided alternative at significance levels above 0.0021

CHAPTER 11

2. Test statistic is 24.24; can reject H_0 at 0.5% level
4. Test statistic is 30.72; can reject H_0 at 0.5% level
6. Test statistic is 0.48; H_0 is not rejected at the usual significance levels
8. With $\hat{p} = 0.29$, test statistic is 9.94; H_0 is rejected at 0.5% level
10. Test statistic is 6.25; can reject H_0 at 10% level
12. Test statistic is 17.1; can reject H_0 at 0.5% level
14. Test statistic is 18.5; can reject null hypothesis at 0.5% level
18. Test statistic is 1.67; cannot reject H_0 at 10% level
20. Test statistic is 5.92; can reject H_0 at 10%, but not at 5% level
22. Test statistic is 3.2; cannot reject H_0 at 10% level
24. Test statistic is 2.25; cannot reject H_0 at 10% level
26. Test statistic is 16.1; can reject H_0 at 0.5% level
28. Test statistic is 1.807; the data contain virtually no evidence supporting any association
30. Test statistic is 0.35; H_0 is not rejected at 10% level
32. Test statistic is 14.5; H_0 is rejected at 5% level
34. Test statistic is 31.8; H_0 is rejected at 1% level

CHAPTER 12

2. **(a)** 0.429 **(b)** Test statistic is 1.16; accept null hypothesis
4. Sample correlation is 0.057; test statistic is 0.22; H_0 is accepted at the usual significance levels
6. Test statistic is 0.80; H_0 is accepted at the usual levels
8. Rank correlation is 0.7857; can reject null hypothesis at 5% level, and very nearly at the 2.5% level
10. Rank correlation is 0.129 (using sample correlation of ranks, since there is a tie); null hypothesis of no association is not rejected at 10% level
12. Rank correlation is 0.916; null hypothesis is rejected at 0.5% level (However, the test is based on the assumption—which may well be false—that year-to-year observations are independent of one another.)
14. **(a)** $y = 3.26 + 0.038x$ **(b)** An increase of 0.038
16. **(a)** $y = 6,574 + 7,528x$ **(b)** An increase of 0.1 in grade point average leads to an expected increase of $752.8 in starting salary
18. **(a)** $y = -11.68 + 0.404x$ **(b)** Each additional 1 unit in aptitude test score leads to an expected increase of $404 in weekly sales
22. $R^2 = 0.889$; in this sample, 88.9% of the variability in sales is explained by their linear dependence on price.
24. $R^2 = 0.714$
26. $R^2 = 0.303$
28. **(a)** -0.906 **(b)** 0.820
30. **(a)** 0.135 **(b)** 0.00027 **(c)** $0.000 < \beta < 0.076$
32. **(a)** $y = 3.173 + 0.549x$ **(b)** $0.24 < \beta < 0.86$
34. Test statistic is 4.681; null hypothesis can be rejected at 0.5% level
36. Test statistic is 3.094; can reject H_0 at 2.5% level, but not at 1% level

40. Test statistic is -0.709; H_0 is not rejected at the 10% level
42. (a) 438.6 (b) (i) $405 < Y_{n+1} < 472$; (ii) $422 < Y_{n+1} < 455$
44. (a) $22.5 < Y_{n+1} < 31.8$ (b) $25.5 < Y_{n+1} < 28.8$
46. 90% interval is $11.7 < Y_{n+1} < 15.3$; 95% interval is $11.2 < Y_{n+1} < 15.8$
48. (a) 11.08 (b) (i) $7.7 < Y_{n+1} < 14.5$ (ii) $9.7 < Y_{n+1} < 12.5$
52. (a) 0.749 (b) Test statistic is 3.20; reject null hypothesis
54. Test statistic is 5.04; reject null hypothesis
56. Test statistic is 0.8286, H_0 cannot be rejected at 5% level
58. Rank correlation is 0.488 (using sample correlation of ranks, since there are ties); null hypothesis is not rejected at 10% level
62. (a) An increase of 1 unit in the predicted change is associated with an expected increase of 0.7916 unit in the actual change of the spot rate (b) 9.7% of the variability in actual change is explained, in this sample, by its linear relation with predicted change (c) Test statistic is 2.87; null hypothesis is rejected at 0.5% level (d) Test statistic is -0.76; null hypothesis is not rejected at 20% level
64. (a) A 1 unit increase in the compensation ratio change leads to an expected increase of 0.32 unit in the absence rate change. (b) 15% of the variability in absence rate changes, in this sample, is explained by their linear dependence on compensation ratio changes. (c) Test statistic is 3.2; H_0 is rejected at the 1% level

CHAPTER 13

2. All else equal, an increase of $1 in the price of gasoline leads to an expected increase of 10.9 in sales of imported cars as a percentage of total car sales, and so on.
4. All else equal, an increase of $1 in ticket price leads to an expected decrease of 0.92 in number of passengers, and so on.
6. (a) $R^2 = 0.929$; taken together, the three independent variables explain 92.9% of the variability in profits in this sample. (b) $\bar{R}^2 = 0.919$
8. (a) 90% interval is $8.1 < \beta_1 < 13.7$, 95% interval is $7.5 < \beta_1 < 14.3$ (b) 95% interval is $-1.1 < \beta_2 < 8.3$, 99% interval is $-2.9 < \beta_2 < 10.1$ (c) Test statistic is 1.00; null hypothesis is accepted at 20% level (d) Test statistic is 3.545; null hypothesis is rejected at 0.5% level
10. (a) Test statistic is 4.6; null hypothesis is rejected against two-sided alternative at the 1% level (b) Test statistic is 2.194; null hypothesis is rejected against two-sided alternative at the 5%, but not at the 2% level
12. (a) Test statistic is -4; null hypothesis is rejected at 0.5% level (b) 90% interval is $12.46 < \beta_2 < 32.86$, 95% interval is $10.37 < \beta_2 < 34.95$ (c) 95% interval is $1.44 < \beta_3 < 10.16$, 99% interval is $-0.09 < \beta_3 < 11.69$
14. (c) For β_1, test statistic is -1.321; H_0 is not rejected at 20% level. For β_2, test statistic is 3.396; H_0 is rejected at 1% level. For β_3, test statistic is 2.470; H_0 is rejected at 5%, but not at 2% level. For β_4, test statistic is 1.735; H_0 is rejected at 20%, but not at 10% level.
16. (a) $0.18 < \beta_1 < 0.22$ (b) Test statistic is -1.190; null hypothesis is not rejected at 10% level
18. (a) $0.005 < \beta_1 < 0.021$ (b) Test statistic is 1.415; null hypothesis is rejected against one-sided alternative at 10%, but not at 5% level (c) Test statistic is 1.979; null hypothesis is rejected against one-sided alternative at 5%, but not at 2.5% level
20. Test statistic is 21.3; H_0 is rejected at 1% level

22. **(a)** Test statistic is 91.9; H_0 is rejected at 1% level

(b)

SOURCE	SS	DF	MS	F RATIO
Regression	610.48	3	203.493	91.9
Error	46.50	21	2.214	
Total	656.98	24		

24. Test statistic is 11.5; null hypothesis is rejected at 1% level
26. Test statistic is 26.25; null hypothesis is rejected at 1% level
30. 0.505
32. $6.61
48. **(a)** An additional 1 million in population leads, all else equal, to an expected additional 4.983 new starts, and so on. **(b)** In this sample, 76.6% of the variability in new starts is explained by their linear relation with the seven independent variables. **(c)** $0.371 < \beta_3 < 7.261$ **(d)** Test statistic is -0.290; null hypothesis is not rejected **(e)** Test statistic is 2.050; null hypothesis is rejected **(f)** Test statistic is 29.4; null hypothesis is rejected at 1% level
50. **(a)** An increase of 1% in the time spent in group discussion leads, all else equal, to an expected increase of 0.3817 in average rating, and so on. **(b)** In this sample, 57.9% of the variability in ratings is explained by their linear relation with the three independent variables. **(c)** Test statistic is 9.63; null hypothesis is rejected at 1% level **(d)** $0.0344 < \beta_1 < 0.7290$ **(e)** Test statistic is 2.64; H_0 is rejected at 1%, but not at 0.5% level **(f)** Test statistic is 1.09; H_0 is not rejected at 20% level

CHAPTER 14

2. **(a)** All else equal, the expected rate of return on equity is higher by 1.490 for firms in industries with high or moderate product differentiation than for other firms. **(b)** Test statistic is 1.080; null hypothesis is not rejected at 10% level **(c)** $-20.94 < \beta_2 < 1.93$
16. **(a)** All else equal, a 1% increase in the price of butter leads to an expected decrease of 0.29% in the quantity of butter purchased, and so on. **(b)** Test statistic is 2.3; null hypothesis is rejected against two-sided alternative at the 5%, but not at the 2% level **(c)** Test statistic is 1.125; null hypothesis is not rejected against two-sided alternative at the 20% level **(d)** 80% interval is $0.13 < -\beta_1 < 0.45$, 90% interval is $0.09 < -\beta_1 < 0.49$, 95% interval is $0.04 < -\beta_1 < 0.54$
22. When the dummy variable is strongly correlated with apartment size and/or distance from campus
24. **(a)** For any observation, the values of the dummy variables sum to 1. Since the regression equation contains an intercept, there will then be perfect multicollinearity. **(b)** The estimate of β_3 provides an estimate of the difference between expected demand in the first and fourth quarters, for fixed levels of price and income, and so on.
26. It is likely to lead to serious specification bias if x_2 is an important determinant of Y.
36. The null hypothesis of no autocorrelation in the errors is rejected at the 5% level. It would be rash to make further inference until the model is reestimated, allowing for the possibility of autocorrelated errors.
38. Same as Exercise 36.
40. $d = 0.85$; null hypothesis of no autocorrelation in the errors is rejected at the 1% level

CHAPTER 15

2. **(a)** SSW = 1,303.33, SSG = 1,032.58, SST = 2,335.91

   (b)

SOURCE	SS	DF	MS	F RATIO
Between groups	1,032.58	3	344.193	5.02
Within groups	1,303.33	19	68.596	
Total	2,335.91	22		

Null hypothesis is rejected at 1% level

4. **(a)**

SOURCE	SS	DF	MS	F RATIO
Between groups	2.80	2	1.40	0.47
Within groups	36.12	12	3.01	
Total	38.92	14		

   (b) Cannot reject null hypothesis at 5% level

6. **(a)**

SOURCE	SS	DF	MS	F RATIO
Between groups	479.334	2	239.667	2.30
Within groups	1,248.830	12	104.069	
Total	1,728.164	14		

   (b) Cannot reject null hypothesis at 5% level

8. **(a)**

SOURCE	SS	DF	MS	F RATIO
Between groups	325.735	2	162.8675	1.88
Within groups	1,041.600	12	86.8000	
Total	1,367.335	14		

   (b) Cannot reject null hypothesis at 5% level

10. **(a)**

SOURCE	SS	DF	MS	F RATIO
Between groups	221.3401	3	73.7800	25.60
Within groups	374.6641	130	2.8820	
Total	596.0042	133		

   (b) Reject null hypothesis at 1% level

12. Test statistic is 1.181; null hypothesis is not rejected at 10% level

14. Test statistic is 4.012; null hypothesis is not rejected at 10% level
16. Test statistic is 3.614; null hypothesis is not rejected at 10% level
18. Test statistic is 6.86; null hypothesis is rejected at 10%, but not at 5% level
20. **(a)** The null hypothesis of equality of the centers of the population distributions **(b)** The null hypothesis is not rejected at the usual significance levels.

22. **(a)**

SOURCE	SS	DF	MS	F RATIO
Fertilizers	200.67	2	100.335	4.56
Varieties	62.25	3	20.750	0.94
Error	132.00	6	22.000	
Total	394.92	11		

(b) Cannot reject null hypothesis at 5% level **(c)** Cannot reject null hypothesis at 5% level

24. **(a)**

SOURCE	SS	DF	MS	F RATIO
Regions	230.92	3	76.973	3.22
Colors	74.00	2	37.000	1.55
Error	143.33	6	23.888	
Total	448.25	11		

(b) Null hypothesis is not rejected at 5% level

26. $G_1 = 0.25$, $B_1 = -0.33$, $\varepsilon_{11} = 2.00$

28. **(a)**

SOURCE	SS	DF	MS	F RATIO
Agents	228	3	76.000	0.84
Houses	1,152	9	128.000	1.42
Error	2,434	27	90.148	
Total	3,814	39		

(b) Null hypothesis is not rejected at 5% level

30.

SOURCE	SS	DF	MS	F RATIO
Shows	95.2	2	47.600	3.60
Regions	75.6	3	25.200	1.91
Error	79.3	6	13.217	
Total	250.1	11		

Null hypothesis is not rejected at 5% level

32. (a)

SOURCE	SS	DF	MS	F RATIO
Contestants	364.50	21	17.3571	19.27
Judges	0.81	8	0.1013	0.11
Interaction	4.94	168	0.0294	0.03
Error	1,069.94	1,188	0.9006	
Total	1,440.19	1,385		

(b) Null hypothesis of no difference between contestants is rejected at 1% level; the other two null hypotheses are not rejected at the 5% level.

34. (a)

SOURCE	SS	DF	MS	F RATIO
Subject types	389.00	3	129.667	5.31
Test types	57.56	2	28.780	1.18
Interaction	146.67	6	24.445	1.00
Error	586.00	24	24.417	
Total	1,179.23	35		

(b) Null hypothesis is not rejected at 5% level

36. (a) Any difference among machines is uniform among operators.

(b)

SOURCE	SS	DF	MS	F RATIO
Operators	94.375	3	31.458	15.69
Machines	120.375	3	40.125	20.01
Error	50.125	25	2.005	
Total	264.875	31		

(c) Reject null hypothesis at 1% level   (d) Reject null hypothesis at 1% level

42. Test statistic is 6.97; null hypothesis is rejected at 1% level

44. Test statistic is 2.47; null hypothesis is not rejected at 5% level

46. (a) 8.074 (b) *True Confessions*: $G_1 = 2.327$; *People Weekly*: $G_2 = -1.029$; *Newsweek*: $G_3 = -1.298$ (c) 2.455

48. Test statistic is 5.56, so null hypothesis is rejected at 10%, but not at 5% level.

52.

SOURCE	SS	DF	MS	F RATIO
Between consumers	37,571.5	124	303.00	1.35
Between brands	32,987.3	2	16,493.65	73.42
Error	55,710.7	248	224.64	
Total	126,269.5	374		

Null hypothesis is rejected at 1% level

54. **(a)**

SOURCE	SS	DF	MS	F RATIO
SAT scores	0.82667	2	0.41333	24.79
Incomes	0.00667	2	0.00333	0.20
Error	0.06667	4	0.01667	
Total	0.90000	8		

(b) Null hypothesis accepted at 5% level **(c)** Null hypothesis rejected at 1% level

56. **(a)** 3.3 **(b)** -0.033 **(c)** 0.333 **(d)** 0

58.

SOURCE	SS	DF	MS	F RATIO
Prices	0.178	2	0.0890	0.09
Countries	4.365	2	2.1825	2.32
Interaction	1.262	4	0.3155	0.33
Error	93.330	99	0.9427	
Total	99.135	107		

None of the three null hypotheses is rejected at 5% level

60. **(a)**

SOURCE	SS	DF	MS	F RATIO
SAT scores	2.20111	2	1.10056	66.02
Incomes	0.01778	2	0.00889	0.53
Interaction	0.10223	4	0.02556	1.53
Error	0.15000	9	0.01667	
Total	2.47112	17		

(b) Null hypothesis accepted at 5% level **(c)** Null hypothesis rejected at 1% level **(d)** Null hypothesis accepted at 5% level

CHAPTER 16

2. **(a)** 100, 102.5, 99.3, 98.2, 100.0, 99.6, 100.0, 99.3, 99.3, 100.7, 110.7, 106.1 **(b)** 101.8, 104.4, 101.1, 100, 101.8, 101.5, 101.8, 101.1, 101.1, 102.5, 112.7, 108.0

4. **(a)** 100, 99.5, 100.0, 104.4, 105.0, 106.2, 106.3 **(b)** 95.8, 95.3, 95.8, 100, 100.6, 101.7, 101.8

6. **(a)** Corporate stock: 100, 117.0, 110.5, 120.8, 134.1, 157.1; Homes: 100, 113.9, 132.5, 154.3, 175.7, 189.3

8. **(a)** 100, 89.6, 88.0, 67.9, 72.1, 84.5
(b) 100, 90.9, 91.2, 69.4, 73.5, 85.7

10. **(a)** 100, 97.7, 94.8, 94.2, 98.1, 95.9, 93.7, 96.9, 92.6, 95.2, 97.9, 98.2 **(b)** 100, 97.3, 94.6, 94.0, 97.5, 95.0, 93.3, 96.7, 92.3, 94.8, 97.4, 97.0

12. **(a)** 100, 90.8, 90.8, 69.2, 73.3, 85.5 **(b)** 100, 90.4, 91.1, 68.6, 72.8, 85.0 **(c)** 100, 90.6, 90.9, 68.9, 73.1, 85.3

14. **(a)** 100, 98.1, 95.4, 94.5, 96.0, 92.8, 91.1, 95.7, 91.7, 94.0, 95.2, 93.5 **(b)** 100, 98.0, 95.1, 94.4, 96.3, 93.7, 92.3, 95.8, 92.0, 93.9, 95.5,

93.7 **(c)** 100, 98.1, 95.3, 94.4, 96.2, 93.2, 91.7, 95.8, 91.9, 94.0, 95.4, 93.6

16. **(a)** 100, 55.6, 42.5, 201.6, 105.0, 71.6, 67.5, 163.5, 315.2, 204.3, 257.5, 202.0 **(b)** 100, 55.6, 42.2, 204.0, 105.3, 73.0, 69.4, 164.5, 314.4, 202.4, 258.6, 202.4 **(c)** 100, 55.6, 42.3, 202.8, 105.2, 72.3, 68.5, 164.0, 314.8, 203.3, 258.0, 202.2

18. **(a)** 100, 96.7, 104.2, 89.9, 96.5, 85.9, 81.7, 78.0, 107.2, 111.3, 118.0, 128.1, **(b)** 100, 96.6, 104.0, 89.7, 96.4, 85.8, 81.5, 77.9, 107.0, 111.0, 117.8, 128.0, **(c)** 100, 96.6, 104.1, 89.8, 96.4, 85.9, 81.6, 78.0, 107.1, 111.1, 117.9, 128.1

22. 100, 55.0, 40.0, 218.7, 112.5, 77.0, 73.3, 179.9, 379.3, 244.9, 332.2, 269.4

24. 100, 95.9, 103.0, 92.3, 100.0, 90.8, 87.1, 84.1, 116.6, 121.9, 131.4, 147.3

30. **(a)** 95.3, 93.1, 95.0, 98.1, 100, 99.8, 99.1, 98.8 **(b)** 100, 97.7, 99.7, 102.9, 104.9, 104.7, 104.0, 103.6

32. **(a)** 71.3, 101.0, 89.9, 94.2, 100, 125.9, 115.6, 157.2, 179.8, 154.2, 91.2, 89.4, 135.8, 196.0, 210.8, 223.9, 185.1, 216.1, 291.4, 319.9 **(b)** 33.8, 47.9, 42.7, 44.7, 47.4, 59.7, 54.8, 74.6, 85.3, 73.1, 43.2, 42.4, 64.4, 93.0, 100, 106.2, 87.8, 102.5, 138.2, 151.7

34. **(a)** 100, 110.7, 117.9, 128.0, 144.5, 159.8
    (b) 100, 110.7, 117.5, 127.0, 142.7, 157.3
    (c) 100, 110.5, 116.4, 124.4, 138.2, 150.9
    (d) 100, 110.5, 116.4, 124.4, 138.1, 150.6
    (e) 100, 110.5, 116.4, 124.4, 138.1, 150.8
    (f) 100, 104.1, 100.4, 105.6, 106.3, 105.2
    (g) 100, 104.1, 100.4, 105.6, 106.3, 105.0
    (h) 100, 104.1, 100.4, 105.6, 106.3, 105.1
    (i) 100, 115.1, 116.9, 131.4, 146.8, 158.5

36. 61.8, 69.1, 81.8, 80.0, 103.6, 112.7, 112.7, 100, 96.4, 125.5, 129.1, 147.3, 152.7, 136.4, 136.4, 138.2

CHAPTER 17

2. $R = 3$; null hypothesis is rejected against alternative of positive association at the 0.4% level, and against two-sided alternative at 0.8% level

4. $R = 7$; null hypothesis is rejected only against two-sided alternative at 78.4% level

6. $R = 7$, test statistic is -3.08; null hypothesis is rejected against alternative of positive association at 0.1% level, and against two-sided alternative at 0.2% level

12. Some values are 1953: 474.4; 1954: 497.6; 1975: 1,141.8; 1976: 1,240.0

14. Some values are 1962: 38.18; 1963: 38.48; 1977: 48.70; 1978: 49.76

16. Some values are July 1978: 767.75; August 1978: 773.75; May 1980: 558.25; June 1980: 551.92

18. Some values are 1972, Q1: 0.398; 1972, Q2: 0.430; 1978, Q3: 0.790; 1978, Q4: 1.015

22. Forecast is 103.5 for each year.

24. Forecast is 54.1 for each year.

30. 148, 148, 148, 147, 147, 146

32. 42.46, 42.98, 43.49, 44.00

34. 6.7, 6.5
36. 0.822, 0.620, 1.429, 1.224, 1.080, 0.801, 1.817, 1.535
38. $X_t = 18.515 - 0.32X_{t-1} + a_t$; 17.9, 17.9
44. 576, 577, 576, 573, 571
54. (a) $R = 3$; null hypothesis is rejected against alternative of positive association at 1.3% level, and against two-sided alternative at 2.6% level
56. Some values are 1963: 3.624; 1964: 4.058; 1977: 8.952; 1978: 9.818
58. (b) Some values are 1972, Q1: 0.880; 1972, Q2: 1.094; 1978, Q3: 0.804; 1978, Q4: 0.623
60. Forecast is 136.5 for each month.
62. 99.4, 99.3, 99.2

CHAPTER 18

14. (a) 91.92   (b) 42.88
16. 90% interval: $6.17 < \mu < 7.63$;   95% interval: $6.03 < \mu < 7.77$; 99% interval: $5.76 < \mu < 8.04$
20. $3,331.3 < N\mu < 3,927.5$
22. $416.6 < N\mu < 591.4$
24. 90% interval: $0.459 < p < 0.691$;   95% interval: $0.436 < p < 0.714$
26. $32 < Np < 88$
28. (a) $50.20 < \mu_1 < 54.60$ (b) 51.49   (c) 95% interval: $49.32 < \mu < 53.66$;   99% interval: $48.64 < \mu < 54.34$
30. (a) $3.99 < \mu_1 < 5.45$ (b) $2.96 < \mu_2 < 4.34$ (c) $3.32 < \mu < 4.41$
32. (a) 9,564   (b) $8,771 < N\mu < 10,357$
34. (a) 0.315   (b) 90% interval: $0.218 < p < 0.411$;   95% interval: $0.200 < p < 0.430$
36. (a) 12 from Illinois, 9 from Indiana, 11 from Ohio   (b) 13 from Illinois, 10 from Indiana, 9 from Ohio
38. (a) 25   (b) 17
40. (a) 10   (b) 12
42. 173
44. 341
46. (a) 256   (b) 179
48. (a) 88.90   (b) $67.33 < \mu < 110.47$
50. (a) 0.439   (b) $0.362 < p < 0.515$
52. 162
54. 340
58. (a) Based on the t distribution with 9 degrees of freedom, since the sample size is small, interval is $143.06 < \mu < 161.02$ (b) Wider   (c) $20,914 < N\mu < 24,698$
60. (a) $0.350 < p < 0.550$ (b) Wider
62. (a) $7.09 < \mu_1 < 11.31$ (b) $9.64 < \mu < 13.53$
64. $0.477 < p < 0.703$
66. (a) 16   (b) 22
68. 332
70. (a) 299   (b) 260

CHAPTER 19

2. **(a)** Action a_3 is inadmissible **(b)** Action a_4 **(c)** Action a.

6. **(a)** Same as the table given, except that payoff is minus cost **(b)** No **(c)** Action a_1
   (d) Action a_3

8. **(a)**

	SANTA EXISTS	DOES NOT EXIST
HANG UP	9	9
DO NOT HANG UP	0	10

   (b) No; not hanging up stocking has the higher EMV ($9.50)

10. **(a)** Action a_1, with EMV $6.4 million

14. Locate in the new center, with EMV $73,000

16. **(a)**

	BID ACCEPTED	NOT ACCEPTED
SUBMIT	88,000	−12,000
NOT SUBMIT	0	0

   (b) Should prepare and submit bid, with EMV $8,000

18. **(a)**

CARS ORDERED BY MANAGER	CARS REQUESTED BY CUSTOMERS				
	6	7	8	9	10
0	0	−10	−20	−30	−40
1	−20	20	10	0	−10
2	−40	0	40	30	20
3	−60	−20	20	60	50
4	−80	−40	0	40	80

   (b) Should order 2 cars, with EMV $20

20. **(b)** Should sign Jones and mount advertising campaign, with EMV $80,000 **(c)** Now the EMV for strategy in part (b) is $65,000, which is still higher than the EMV for Jones with no advertising campaign. However, the best strategy would now be to have the Smith book, without the advertising campaign. This has EMV $70,000. Hence, any payment up to $5,000 for Jones to withdraw from the contract is justified.

22. **(b)** An outline proposal for contract B should be submitted. This has EMV $1,890. **(c)** Detailed final proposal should not be submitted, since the expected return would be −$300. **(d)** Yes; the best strategy now is to submit a final proposal for contract B. This has EMV $2,085.

24. **(a)** 0.5 for highly successful, 0.3 for moderately successful, 0.2 for unsuccessful **(b)** Yes; the EMV is $130,000 **(c)** $20/27$ for highly successful, $6/27$ for moderately

successful, $\frac{1}{27}$ for unsuccessful   **(d)** Yes; the EMV is $218,519   **(e)** $\frac{5}{23}$ for highly successful; $\frac{9}{23}$ for moderately successful, $\frac{9}{23}$ for unsuccessful   **(f)** Yes; the EMV is $26,087

26. **(a)** 0.5 for very successful, 0.3 for moderately successful, 0.2 for unsuccessful **(b)** Locate in new center, with EMV $73,000   **(c)** $\frac{35}{48}$ for very successful, $\frac{9}{48}$ for moderately successful, $\frac{4}{48}$ for unsuccessful   **(d)** Locate in new center, with EMV $96,042   **(e)** $\frac{5}{14}$ for very successful, $\frac{6}{14}$ for moderately successful, $\frac{3}{14}$ for unsuccessful   **(f)** Locate in new center, with EMV $62,143   **(g)** $\frac{5}{24}$ for very successful, $\frac{9}{24}$ for moderately successful, $\frac{10}{24}$ for unsuccessful   **(h)** Locate in established center, with EMV $66,250

28. **(a)** $\frac{12}{13}$ for effective, $\frac{1}{13}$ for ineffective   **(b)** It should be retained; the EMV is $114,615. **(c)** 0.25 for effective, 0.75 for ineffective   **(d)** It should be sold; the EMV is $50,000.   **(e)** Yes

30. **(a)** To industrial user, with EMV $-$500   **(b) (i)** To discount chain, with EMV $-$600 **(ii)** To industrial user, with EMV $-$449   **(c) (i)** To discount chain   **(ii)** To discount chain   **(iii)** To industrial user, with EMV $-$404   **(d)** The binomial distribution can be used to compute the probabilities for different numbers of defectives in samples from either type of consignment.

32. $3.2 million

34. $4,000

36. **(a)** 0; the same action is taken whatever the advice   **(b)** $-$5,000

38. $6,400

40. $14,175

42. **(a)** $34   **(b)** $55.80   **(c)** $21.80   **(d)** None, since the result will not influence the optimal decision   **(e)** $24.49 [*Note:* Use the posterior probabilities from the check of the first bulb as prior probabilities for analyzing the results of the second check.] **(f)** Multiplying $24.49 by the probability (0.89) that the first bulb is not defective gives $21.80.

44. **(a)** 18.18 for $300, 54.55 for $700, 72.73 for $900, 81.82 for $1,000   **(b)** 0.1818, 0.5455, 0.7273, 0.8182

46. **(a)** 30, 35, 40, 65   **(b)** Action a_1, with expected utility 40

48. Should sign Smith, and not mount advertising campaign, with expected utility 54

54. **(a)** 0.1 for s_1, 0.4 for s_2, 0.1 for s_3, 0.4 for s_4 **(b)** Bid for both contracts, with EMV $590   **(d)** $215   **(e)** $60

index

Hypothesis tests (*cont.*)
   for correlation, 446–48
   for difference between population means, 361–69
   for difference between population proportions, 369–71
   for equality of population variances, 376–78
   for goodness of fit, 415–20
   nonparametric, 394–409
   in one-way analysis of variance, 617
   for partial regression coefficients, 515–16
   for population mean (large sample sizes), 346–47
   for population mean (variance known), 338–46
   for population mean (variance unknown), 348–51
   for population proportion, 357–59
   for population variance, 353–57
   for rank correlation, 450–51
   for regression slope, 477–78
   on sets of regression coefficients, 520–24
   in two-way analysis of variance:
      more than one observation per cell, 643
   in two-way analysis of variance:
      one observation per cell, 634

Inadmissible action, 798
Independence:
   of continuous random variables, 197
   and covariance, 158, 198
   of discrete random variables, 154
Independent attributes, 120
Independent events, 109
Independent variable, 458
Index number problem, 657–59
Index numbers, 657–87
Indifference to risk, 841
Inference about population, 228
Interaction, 639
International Telephone and Telegraph Corporation, 55–58
Interquartile range:
   for grouped data, 42–45
   of set of numbers, 25–26
Intersection of events:
   defined, 84–85
   probability of, 107
Interval estimate, 277
Interval estimator, 277
Irregular component, 697

Jaggi, B., 408
Jenkins, G. M., 737
Joint cumulative distribution function, 197
Joint cumulative probability function, 155
Joint distribution:
   of continuous random variables, 197–98
   of discrete random variables, 151–61
Joint probability function:
   defined, 151
   properties of, 152–53
Jones, W. H., 393

Kaldor, N., 504
Keyfitz, N., 229
Kim, J. S., 646
Kolrin, P., 320
Krueckeberg, H. F., 653
Kruskal-Wallis test, 621–23
Kummer, D. R., 347

LaBarbera, P. A., 311
Lachman, R., 439

LaForge, R. L., 314
Lagged dependent variables, 555–60
Lambert, D. R., 655
Lambert, Z. V., 321
Laspeyres price index, 663–66
Laspeyres quantity index, 671–72
Laumer, J. F., 299
Least squares estimation, 460–64, 492–93, 502–4
   sampling distribution of estimators, 474–75, 514
LeClaire, A., 375
Lee, T. A., 327
Lee, W. Y., 415
Leinhardt, S., 62
Lerner, D., 453
Level of confidence, 277
Level of test: *see* significance level
Levin, S. L., 561
Lightner, S. M., 329
Linear association:
   and correlation, 441–42
   and regression, 456
Loan status, 426
Log linear models, 568–71
Lying with statistics, 66–72
Lynn, J. R., 438

MacLachlan, J. M., 311, 374
Mahajan, V., 368
Mahajan, Y. L., 576
Maintained hypothesis, 332
Mandell, L., 439
Mann-Whitney test, 404–9
Marginal distribution function, 197
Marginal probability function, 152
Market failure, 300
Market for cigarettes, 242
Marshall, J., 326
Martin, P. Y., 449
Martin, W. S., 653
Matched pairs:
   and confidence intervals, 307–9
   and hypothesis tests, 361–63, 609
Maximin criterion, 799
May, F. E., 327
McDonald, J., 566
McElroy, B. F., 321, 361
McGuinness, T., 529
McLaren, C. M., 606
McVicker, D. D., 392, 651
Mean:
   and central limit theorem, 214
   of continuous random variable, 194
   of discrete random variable, 141
   for grouped data, 39–42
   for multiple-observation values, 37–39
   of set of numbers, 8–11
Mean absolute deviation, 22–24
Mean squares:
   for one-way analysis of variance, 615–17
   for two-way analysis of variance:
      more than one observation per cell, 643
   for two-way analysis of variance:
      one observation per cell, 634
Median:
   for grouped data, 42–45
   and relative efficiency, 266–67
   of set of numbers, 12–13
Melnik, A., 519